VII International Scientific Conference on Problems of Complex Development of Georesources (PCDG 2018)

E3S Web of Conferences Volume 56 (2018)

Khabarovsk, Russia
25 – 27 September 2018

Editors:

I. Rasskazov
S. Kornilkov
S. Tkach

ISBN: 978-1-5108-7254-7

Printed from e-media with permission by:

Curran Associates, Inc.
57 Morehouse Lane
Red Hook, NY 12571

Some format issues inherent in the e-media version may also appear in this print version.

Printed by Curran Associates, Inc. (2018)

For additional information, please contact EDP Sciences – Web of Conferences at the address below.

EDP Sciences – Web of Conferences
17, Avenue du Hoggar
Parc d'Activité de Courtabœuf
BP 112
F-91944 Les Ulis Cedex A
France

Phone: +33 (0) 1 69 18 75 75
Fax: +33 (0) 1 69 28 84 91

contact-edps@webofconferences.org

Additional copies of this publication are available from:

Curran Associates, Inc.
57 Morehouse Lane
Red Hook, NY 12571 USA
Phone: 845-758-0400
Fax: 845-758-2633
Email: curran@proceedings.com
Web: www.proceedings.com

TABLE OF CONTENTS

Dear colleagues!

We present you the proceedings VII of the International scientific conference "Problems of Complex Development of Georesources" which has taken place on September 25-27, 2018 in Khabarovsk city. They contain new results of basic and applied researches in area: geotechnological and geomechanical problems of development of a subsoil; problems of complex processing of mineral raw materials and new technologies of mineral processing; geoenvironmental and economic problems of subsurface use; strategic problems of mineral and raw and mining complexes.

A number of topics are discussed in the of the conference. Some of them are creation of new and increase in efficiency of the applied technologies of development, including nonconventional georesources; creation of new and improvement of the current methods of effective and complex extraction of useful components from refractory ores, alluvial materials and tailing dumps; reproduction, rational and complex use and preservation of a subsoil of Earth at development of active stocks and expected resources of natural and technogenic ore and alluvial deposits of precious metals, including georesources of cryogenic zony; estimation of influence of natural and technogenic factors on an intense and dynamic condition of burst hazard rock massif; estimation of regional geodynamic features of ore areas of the Far East of Russia and neighbouring territories; creations of methodical bases of complex monitoring of geo-media, including located in undermined and urbanized territories; development of technologies of rock destruction, including their explosive crushing; substatntiation of strategy of development for mineral different types of mineral resources according regional aspects and mining complexes and areas; increases in social and economic efficiency of environmental management with a growth of competitiveness of subsoil users in the world market of raw materials; creations of effective methods of the solution of environmental problems at development of mineral raw material resources.

The leading scientists and research specialists, representatives of universities of mining-and-geological profile, specialists of mining enterprises, representatives of state structure and other organizations from Russia, Germany, Australia, Japan, the USA, Kazakhstan, Kyrgystan and other countries have participated in the conference and prepared the reports.

Extensive geography of participants reflects both complexity, importance, relevance of the discussed scientific and technological, innovative mining problems and emphasizes the great interest of the mining community to the held conference which has become traditional.

Organizing committee of the conference hope that the materials will be useful to a wide range of scientists and practicians in the field of mining and will provide rational development and use of the Earth subsoil for the development and prosperity of the states and people of the world.

The conference is held with financial support of the Russian Fund of Fundamental Research (grant No. 18-05-20073 G).

Chairman of the Organizing Committee
VII International Scientific Conference
"Problems of Complex Development of Georesources"
Director of the Mining Institute FEB RAS
Prof. I.Yu. Rasskazov

edp sciences

Web of Conferences

Statement of Peer review

In submitting conference proceedings to *Web of Conferences*, the editors of the proceedings certify to the Publisher that

1. They adhere to its **Policy on Publishing Integrity** in order to safeguard good scientific practice in publishing.
2. All articles have been subjected to peer review administered by the proceedings editors.
3. Reviews have been conducted by expert referees, who have been requested to provide unbiased and constructive comments aimed, whenever possible, at improving the work.
4. Proceedings editors have taken all reasonable steps to ensure the quality of the materials they publish and their decision to accept or reject a paper for publication has been based only on the merits of the work and the relevance to the journal.

Title, date and place of the conference

VII International scientific conference "PROBLEMS OF COMPLEX DEVELOPMENT OF GEORESOURCES" (PCDG 2018), 25-27 September 2018, Mining Institute Far Eastern Branch Russian Academy of Sciences, 680000 Turgenev St., 51, Khabarovsk, Russian Federation

Proceedings editor(s):

Igor Rasskazov, Dr. of Engineering, Mining Institute Far Eastern Branch Russian Academy of Sciences
Sergei Kornilkov, Dr. of Engineering, Mining Institute of Ural branch Russian Academy of Sciences
Sergei Tkach, Dr. of Engineering, Mining Institute of the North, Siberian Branch, Russian Academy of Sciences

Date and editor's signature

11.09.2018

Influence of dynamics of development of open mining on technological features of blasting breaking

Sergey Zharikov[1*] and *Vyacheslav* Kutuev[1]

[1]Institute of Mining, Ural Branch of the Russian Academy of Sciences, Ekaterinburg, Russia.

Abstract. The paper desribes the results of the analysis of the drilling and blasting index at some large enterprises. The significant period of their work has been considered. The increase in the specific consumption of explosives by years has been shown and the reasons for this increase have been explained. The issue of the development prospects of the drilling and blasting complex in terms of reducing the cost of explosives, as well as some issues associated with the use of explosives, manufactured on site. It is suggested that in the near future the development of open mining will be associated with the acceleration of processes and a significant increase in their intensity, both in the development of new ones and in the refinement of existing large open-pit mines. In this regard, drilling and blasting operations are waiting for certain changes in both the mechanization of processes and the technological methods of production. An approach is proposed for improving energy efficiency and resource saving in the production of drilling and blasting operations, which consists in applying a horizontal cut along the bottom of the bench with blasting breaking of the rock mass. It has been previously shown that the use of this method of breaking can be quite effective in comparison with the traditional approach. **Key words:** drilling and blasting operations, explosive materials, horizontal cut, specific consumption of explosives, explosive breaking, horizontal drilling.

1 Introduction

Currently, in large open-pit mines, the DBO costs are 30% of total production costs and are expected to be increased as mining is reduced. From 1975 to 2015 the Institute of Mining of the Ural Branch of RAS had accumulated a significant amount of data on the BDO of iron ore and building material open-pit mines, which makes it possible to estimate the dynamics of changes in production indicators and to establish the trend of their progress in the future. It should be noted that the Uralasbest plant has data only since 1990. For the analysis, it was assumed that earlier indicators were at the level of the specified year. Therefore, a comparison of the DBO indicators and their averaging for the estimation of the dynamics of

* Corresponding author: 333vista@mail.ru

change has been carried out for 9 MPS and separately for 8 MPS of iron ore deposits without taking into account the data of the Uralasbest combine.

2 Theory

Figures 1 - 3 show that the most significant increase in the specific consumption of explosives is traced from 1988-1990. Approximately at this point, the widespread use of explosives manufactured on site, the specific consumption of which is higher than for standard explosives by 10 - 20% due to the density of matter. For each enterprise, this time was different, and also the increase in the specific consumption of explosives is different. However, taking into account the recalculation for standard explosive, the specific consumption of explosives has grown quite significantly, although in general the costs for explosive materials have decreased due to their cost. The introduction of explosives produced on site has significantly improved the economy of mining enterprises. At the same time, the way of breaking the excavating blocks remained practically unchanged. So the underestimation of pre-destructive impact in the rear of the rock mass resulted in some cases in a deterioration in the quality of crushing, which adversely affected the efficiency of drilling and blasting operations.

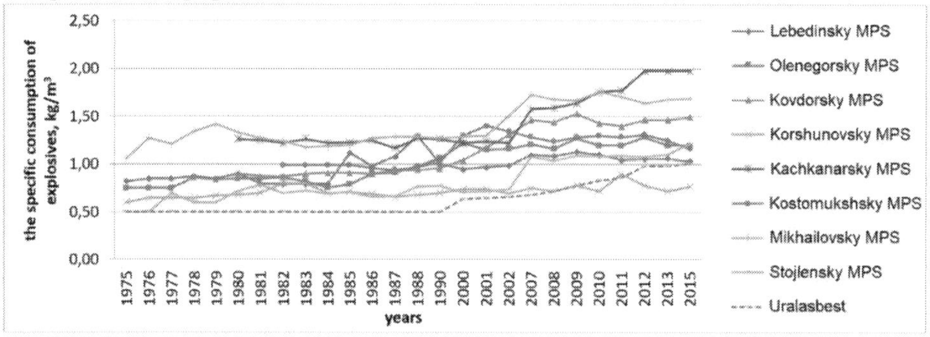

Fig. 1. Dynamics of change in specific consumption of explosives for 9 large MPS.

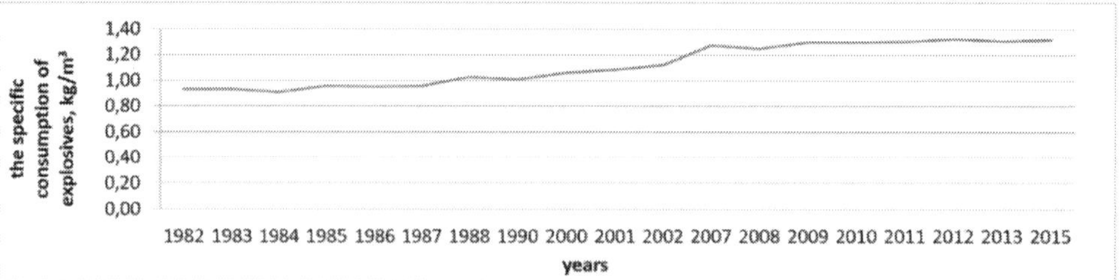

Fig. 2. Average dynamics of change in specific consumption of explosives for 8 large MPS in iron-mining industry.

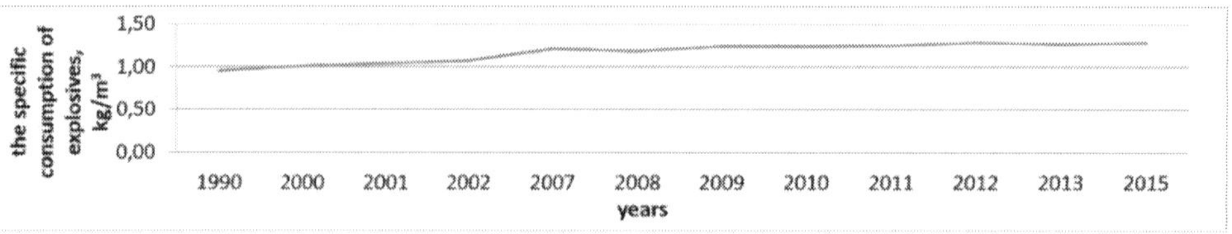

Fig. 3. Dynamics of change in specific consumption of explosives for 9 large MPS.

Change in the sphere of drilling operations is characterized by a steady increase in both drilling volumes and weighted average productivity. It is noted that the use of high-performance machines of foreign production does not always reflect the efficiency of technological solutions [2]. Therefore, an important technological issue of the near-term perspective at the final extraction of deposits is the utilization rate of the drilling rig.

Increasing the efficiency of drilling and blasting operations in general is represented in the reduction of costs. This approach is traditional, and as time shows, it results in the improvement not only of the production of technological explosions, but also of other mining operations. The analysis allows to draw a conclusion that if the goal is to reduce the costs of drilling and blasting operations, first of all it is necessary to reduce the main indicator - the specific consumption of explosives. The methods can be different. First of all, these are charge structures determined on the basis of the rapid study of the physical and mechanical properties of rock mass of excavating blocks prior to the charging process [3 - 5]; a variable array of boreholes installed directly in the process of drilling off the block [6]; as well as the technology for breaking off the excavation blocks, for example, using horizontal drilling (the question has not been practically investigated).

For a long time, it was believed that in order to improve the quality of crushing of very hard-shot viscous rocks, it is sufficient to increase the specific consumption of explosives. However, as practice shows, the narrowing of the borehole array and the increase in the specific consumption of explosives do not result in improvement of the crushing quality [7]. As a result, material resources are not spent efficiently. In support of this, an approximate calculation of the effect of the explosion at a distance from the charge has been made. For the calculation, the dependencies indicated in [1, 8 - 9] have been used .

Pressure on the wall of the borehole:

$$P = \frac{r_{chrg}^2}{2\, r_{bh}^2} \cdot \frac{\rho_{\exp} \cdot D^2}{4}, \tag{1}$$

where r_{chrg} - charge radius, m; r_{bh} - borehole radius, m; P_d - detonation wave pressure, MPa; ρ_{exp} - explosive density, kg/m^3; D - detonation velocity, m/s.

Pressure in the rock mass from the explosion of a cylindrical charge at a distance:

$$P_R = P\left(\frac{r_{bh}}{R}\right)^{1,5}, \tag{2}$$

where R - distance from charge explosion, m

In the calculation, the diameter of the borehole is 250 mm. The diameter of the borehole and the diameter of the charge are equal. The results are shown in Table 1. According to Table 1, after only 1 m, the pressure drops by 96% for all explosives. Let us consider an example for granulotol:

$$\frac{3125 - 138{,}11}{3125} 100\% = 96\%.$$

At a distance of 2 m from the charge, the pressure drops by 98%, and at a distance of 3 m - by 99%. The indicated distances correspond approximately to the zone of controlled crushing, which is expressed by 4-8 charge diameters [10]. With a charge of 250 mm, the zone of controlled crushing is 1 - 2 m. In this case, according to [10], the cracking zone amounts to 15 - 30 charge diameters, in our case it is 3.5 - 7.5 meters. If, as an example, we consider a rock with a limit strength in the sample: in compression being 200 MPa, in expansion being 20 MPa, and a coefficient of structural weakening in the rock mass being 0.1 (respectively, compression in rock mass being 20 MPa, expansion being 2 MPa), then the pressure from explosion in many explosives (see Table 1) overcomes the compressive stresses by 3 - 4 m from the charge, and expanding being up to 10 m.

Therefore, it is advisable to keep the distances between the boreholes in the block unchanged, and vice versa, to maximize the possible value (reduce the specific consumption of the explosive).

Table 1. Pressure at a distance from the explosion of certain explosives

Explosive	Detonation velocity, m/s	Load density of explosive, g/cm^3	Pressure on the walls of the borehole, MPa	Pressure from explosion, MPa									
				Distance from charge, m									
				1	2	3	4	5	6	7	8	9	10
granulotol	5000	1	3125.0	138.11	48.83	26.58	17.26	12.35	9.40	7.46	6.10	5.12	4.37
gramonite 79/21	3700	0.9	1540.1	68.06	24.06	13.10	8.51	6.09	4.63	3.68	3.01	2.52	2.15
poremite 1A	5000	1.2	3750.0	165.73	58.59	31.89	20.72	14.82	11.28	8.95	7.32	6.14	5.24
Nitronite E-70	4500	1.2	3037.5	134.24	47.46	25.83	16.78	12.01	9.13	7.25	5.93	4.97	4.25
siberite	4800	1.25	3600.0	159.10	56.25	30.62	19.89	14.23	10.83	8.59	7.03	5.89	5.03
granemite I-30	5000	1.35	4218.8	186.44	65.92	35.88	23.31	16.68	12.69	10.07	8.24	6.91	5.90
fortis edv. 80	5510	1.1	4174.5	184.49	65.23	35.50	23.06	16.50	12.55	9.96	8.15	6.83	5.83
emulsolite	5000	1.25	3906.3	172.63	61.04	33.22	21.58	15.44	11.75	9.32	7.63	6.39	5.46
slarrite	5000	1.25	3906.3	172.63	61.04	33.22	21.58	15.44	11.75	9.32	7.63	6.39	5.46
emulogram	4500	1.3	3290.6	145.43	51.42	27.99	18.18	13.01	9.90	7.85	6.43	5.39	4.60
emulite VET 500	4600	1.35	3570.8	157.81	55.79	30.37	19.73	14.11	10.74	8.52	6.97	5.84	4.99

As approximate calculations show, in principle it is possible to increase the distance between boreholes in strong rocks to 8 - 10 meters or more, if the issue is solved with the development of the resistance line on the bottom of the bench. At large distances between the boreholes, the issue of developing the bottom is decisive. As an alternative of the solution of this question, horizontal cutting can be used. Thus, the bottom can be developed without the use of vertical re-drills.

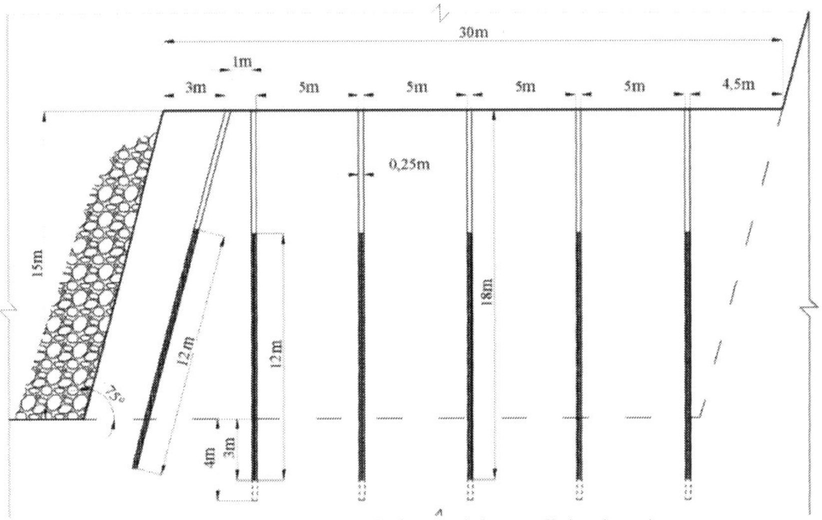

Fig. 4. The scheme of the location of charges with the retaining wall (option 1).

Below are the results of an approximate estimation of the efficiency of the concept of horizontal cutting (option 2) compared to conventional breaking technology (option 1). The conditions for estimating the options are as follows. The rock mass is composed of strong rocks with an average coefficient of strength 14. The ultimate strength in the sample for compression is 140 MPa, the tensile strength being 14 MPa, the density of rocks being 3.5 t/m^3, the coefficient of structural weakening in the rock mass being 0.1. Length excavating block 120 m. The explosive is Nitronite E-70 (charge density being 1.2 g/cm^3). Parameters (option 1) are shown in figure 4. In addition: the borehole diameter being 250 mm, the borehole array being 6 × 5 m, blasting on the retaining wall.

For option 2 of breaking parameters are defined in accordance with the value of the cracking radius using a specific explosive (expression (1) and (3)).

Cracking radius [5]:

$$R_{mp} = r_h \sqrt[3]{\left(P/\sigma_{\text{дуn}}\right)^2} , \qquad (3)$$

where r_h - borehole radius, m; P - pressure on the borehole walls, MPa; $\sigma_{дин}$ - dynamic strength of rock, MPa (adopted 0.13 σ_p = 1.82 MPa).

Table 2. The cracking radius at $\sigma_{дин}$ = 1.82 MPa

Explosive	Cracking radius, m									
	Parameters of charges $D_{скв}/d_{зар}$									
	90/90	110/90	120/90	160/90	110/110	120/110	160/110	120/120	160/120	160/160
granulotol	6.5	6.0	5.9	5.3	7.9	7.7	7.0	8.6	7.8	11.5
gramonite 79/21	4.0	3.8	3.7	3.3	4.9	4.8	4.3	5.4	4.9	7.2
poremite 1A	7.3	6.8	6.6	6.0	8.9	8.7	7.9	9.7	8.8	13.0
Nitronite E-70	6.3	5.9	5.8	5.2	7.7	7.5	6.8	8.4	7.7	11.3
siberite	7.1	6.6	6.4	5.9	8.7	8.4	7.6	9.5	8.6	12.6
granemite I-30	7.9	7.4	7.2	6.5	9.6	9.4	8.5	10.5	9.5	14.0
fortis edv. 80	7.8	7.3	7.1	6.5	9.6	9.3	8.4	10.4	9.5	13.9
emulsolite	7.5	7.0	6.8	6.2	9.2	8.9	8.1	10.0	9.1	13.3
slarrite	7.5	7.0	6.8	6.2	9.2	8.9	8.1	10.0	9.1	13.3
emulogram	6.7	6.2	6.1	5.5	8.2	7.9	7.2	8.9	8.1	11.9
emulite VET 500	7.1	6.6	6.4	5.8	8.6	8.4	7.6	9.4	8.5	12.5
granulite igdanit	2.5	2.4	2.3	2.1	3.1	3.0	2.7	3.4	3.1	4.5

The data in Table 2 allows you to select the explosive and the most rational design of the cuttings. The selection of explosives implies the provision of the smallest fracture zone, so in this case the cracking radius should be minimal. Otherwise, the effect of the cut-off slot on the lower horizon will be comparable with the effect of the overdrills. In this case, vertical boreholes can be underdrilled to the bottom since in this area cuttings work.

According to the data in Table 2, gramonite 79/21 and granulite igdanite have the smallest radii of cracking. With these explosives, the underdrill can average 2 - 4 m. The most acceptable is the diameter of the charge of 90 mm. This diameter ensures the smallest cracking and reduces the likelihood rambling. Thus, the ratios accepted for further calculation are 90/90, 110/90, 120/90. The results are shown in Table 3.

Table 3. Calculation of pressure from explosion of charge of cut slot at a distance

Charge parameters		Pressure from explosion of contour charge at a distance, MPa								
		Explosive being granulit Igdanite								
		Distance, m								
		1	2	3	4	5	6	7	8	9
D borehole, mm	**90**									
D charge, mm	90	7.39	2.61	1.42	0.92	0.66	0.50	0.40	0.33	0.27
D borehole, mm	110									
D charge, mm	90	6.69	2.36	1.29	0.84	0.60	0.46	0.36	0.30	0.25
D borehole, mm	**120**									
D charge, mm	90	6.40	2.26	1.23	0.80	0.57	0.44	0.35	0.28	0.24
		Explosive substance being grammonite 79/21								
D borehole, mm	**90**									
D charge, mm	90	14.70	5.20	2.83	1.84	1.31	1.00	0.79	0.65	0.54
D borehole, mm	110									
D charge, mm	90	13.30	4.70	2.56	1.66	1.19	0.90	0.72	0.59	0.49
D borehole, mm	**120**									
D charge, mm	90	12.73	4.50	2.45	1.59	1.14	0.87	0.69	0.56	0.47

The data in Table 3 show that in order to overcome the tensile stresses 1.4 MPa, the distances as follows. For grammonite 79/21 at a ratio $D_{скв}/d_{зар}$ 90/90 are equal 4 m with charging each borehole or 3 m with charging the next but one. The same is true for other borehole diameters. For grammonite igdanit at a ratio $D_{скв}/d_{зар}$ 90/90 are equal 3 m with charging each borehole or 2 m with charging the next but one. In other cases, is equal to 2 m with charging of each borehole. The distances between the boreholes for the maximum productivity of the machine should be as high as possible, therefore adopted the gramonite 79/21 at a ratio $D_{скв}/d_{зар}$ 90/90. According to Table 1, the pressure at the explosion of nitronite E-70 at a distance of 4 m from the charge exceeds the compressive strength in the rock mass, therefore the distances between between the rows boreholes can be up to 8 m. Is assumed to be 7.5 m, that between the boreholes in the row being 8 m. The rational size of the underdrill is determined by the length of the drill rods, the properties of rocks and the properties of explosives. In our case discussed a SBSH-250 machine. The length of the rod is usually 8 m (the first bar drills at 6 m). Two bars 14 m.

Parameters (option 1) are shown in figure 5. In addition, the distance between horizontal boreholes cut is 3 m with charging the next but one.

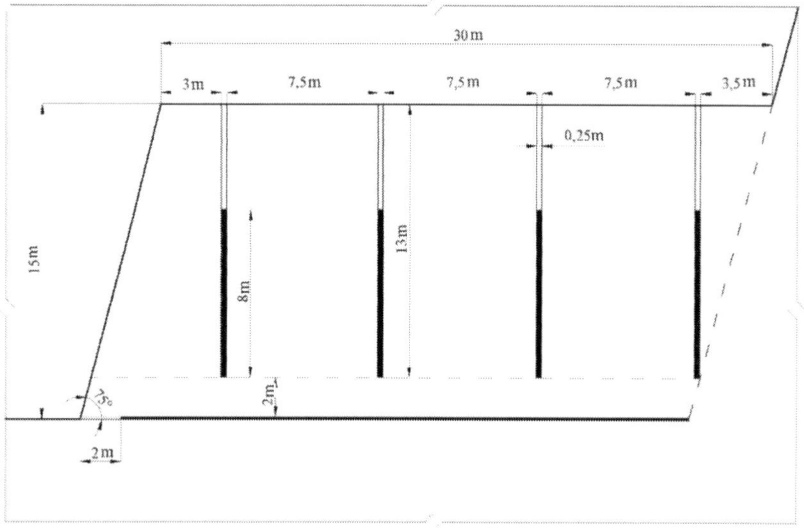

Fig. 5. The scheme of the location of charges for horizontal cut on the bottom of the bench.

3 Results

The approximate calculations presented in Table 4 show that the use of the method of breaking the extraction block with horizontal cut along the bottom of the bench can be quite efficient. Of course, the results are approximate and somewhat idealized, however, the obtained data indicate that the proposed method of breaking will significantly reduce the specific consumption of explosives in the explosive destruction of strong rock masses. Therefore, this direction, with the development of means of mechanization, providing drilling and the formation of long horizontal charges of explosives, can be quite relevant and have practical significance.

Table 4. The results of the calculation of the DBO parameters for two options of the cut-off block technology

Index	Value	
	Normal breaking (first option)	Breaking with the use of horizontal cut on the bottom of the bench (second option)
Number of boreholes in the block, pcs.	120	100
Drilling capacity, m	2280	1900
Volume of block, m^3	54000	54000
Number of explosives per block, kg	84780	32532
Specific consumption of explosive, kg/m^3	1.57	0.6

4 Conclusions

At present, in large open-pit mines, the distances between the boreholes are on the average 5-6 meters, and the size of the overdrills is from 2 to 3 m. The analysis has showed that it is possible to increase the distance between the boreholes in strong rocks to 8-10 meters and a corresponding decrease in the specific consumption of explosives. However, this is possible only if the issue is solved with the development of the resistance line on the bottom of the bench. At large distances between the boreholes, the issue of developing the bottom is decisive.

The approximate calculations has shown that the use of the method of breaking the excavating block with horizontal cut along the bottom of the bench can be quite efficient. Developing the bottom of the bench with a horizontal cut and increasing the distances between vertical boreholes makes it possible to reduce both the volume of drilling and the consumption of explosives, the latter being by more than 2 times. Therefore, this direction is topical and can have high practical significance.

The research has been carried out within the framework of the State Proposal 007-00293-18-00, themes No. 0405-2018-0015, No. 0405-2018-0001, project No. 18-5-5-10, as well as with additional attraction of contractual means.

References

1. I. F. Bondarenko, S. N. Zharikov, I. V. Zyryanov, V. G. Shemenyov, Blasting and Exploration Work at the Kimberlite Open-Pit Mines of Yakutia, (Moscow, 2017) 172 pp. [In Russian].
2. R. I. Sukhov, A. S. Regotunov, Bulletin of Scientific, Technical and Economic Information, Ferrous Metallurgy, **10**, 22 (2017).
3. I. A. Tangaev, Energy Intensity of Mining and Processing of Minerals, (Moscow, 1986) 231 pp. [In Russian].
4. I. V. Vorontsov, A. A. Kotyashev, A. S. Matorin, V. G. Shemenyov, Izvestiya Vuzov, Mining Magazine, **8** , 94 (2008).
5. S. N. Zharikov, V. G. Shemenyov, V. A. Kutuev, Sustainable Development of Mountain Territories **9**, 74 (2017).
6. S. N. Zharikov, Interrelation of Specific Energy Characteristics of the Roller Drilling Processes and Explosive Destruction of the Rock Mass, (Ekaterinburg, 2011) 139 pp. [In Russian].
7. M. B. Togunov, Yu. A. Shitov, G. V. Melik-Gaykazov, V. A. Fokine, N. A. Sverdlenko, Mountain Magazine, **9**, 40 (2007).
8. S. N. Zharikov, V. G. Shemenyov, Izvestiya Vuzov, Mining Magazine, **2**, 80 (2013).
9. S. N. Zharikov, V. A. Kutuev, Deep Structure, Geodynamics, Thermal Field of the Earth, Interpretation of Geophysical Fields: Conference Materials. 179 pp. (2017) [In Russian].
10. V. A. Kuznetsov, Justification of Technology of Drilling and Blasting Operations in Open-Pit Mines and Open Mining and Construction Workings on the Basis of Deformation Zoning of Blasting Benches. 225 pp. (2010) [In Russian].

Simulation and study of strength properties of rocks by measuring characteristics during the blast holes drilling

Andrey Regotunov[1*] and *Rudolf* Sukhov[1]

[1]Federal state budget-funded science establishment Institute of Mining of Ural branch of Russian Academy of Science, Ekaterinburg, Russia.

Abstract. The article shows the results of the development of a method for obtaining information on the strength properties of an massif in the process of drilling of blast holes. This method is developed for open-cast mining and serves to increase the effectiveness of mass explosions. The order of operations is shown and the equipment is described for obtaining the necessary measured characteristics for the subsequent simulation of the distribution of rocks in the volume of the local block of the massif. The paper presents an approach to simulating of a three-dimensional model of rock distribution from the difficulty of drilling in the volume of a local block that is subject to explosive destruction. The results of approbation of the equipment and simulation procedure at a large mining enterprise OJSC "Uralasbest" when blast holes are drilled with roller-type drill rig are presented. As a result of the approbation of the equipment, the values of the average power consumption for the drilling of one meter of the well were set by the machine tool, the fraction of power attributable to the auxiliary operations, and the average energy intensity of the fracture (by the rotator power, without taking into account the compressor's cost for blowing the well). Based on the test results, a digital model has been developed that has made it possible to improve the quality of studies on strength properties of rocks in an experimental local block prepared for explosive destruction. **Key words:** Strength properties, models, rocks, properties, drilling, blast holes, equipment.

1 Introduction

In Russia, during the development of solid minerals deposits in large quarries, up to 1 billion tons of rocks are mined, at least 80% of which are rocky and can be destroyed only with the use of drilling and blasting methods. The effectiveness of destruction of rocks by

*Corresponding author: pochta8400@inbox.ru

drilling and blasting is largely dependent on the correspondence of strength properties of rocks to the technical characteristics of drilling equipment and explosives. Existing methods for obtaining information on the strength of the rocks being developed are the operational additional exploration of mineral deposits, the method for determining the properties of the rocks of the massif in terms of the rock drillability index have certain certainty that should increase as the mineral deposit is mined.

The Institute of Mining of the Ural Branch of the Russian Academy of Sciences is carrying out research to improve the modeling and strength properties of rocks in the process of roller drilling blast holes in quarries. The method of obtaining information on the properties of the rock massif was first proposed by the Soviet scientist I.A. Tangayev [1] in the 70's of the XX century. Further research in this direction was developed by Russian [2-6] and foreign scientists [7-11].

2 Sequence of actions for obtaining the necessary data for modeling purposes

According to the studies [3], index of the difficulty of drilling based on the index of rock drillability, according to V.V. Rzhevsky, can be used as a criterion for evaluating rocks by strength.

To determine the difficulty of drilling index, the Institute of Mining of the Ural Branch of the Russian Academy of Sciences, in cooperation with JSC NPO Avtomatiki, developed a special set of equipment. The equipment is a system of sensors connected to the controller, which provides wireless data transfer to an external server (Fig. 1).

Fig. 1. The equipment installed in the cabin of the drilling rig SBSH-250MNA-32: 1- controller.

The sequence of actions performed to obtain the necessary information for the subsequent simulation of the distribution of rocks in terms of the difficulty of drilling in the volume of a local block is as follows.

Choosing a rational drilling technology and tools. Next, a roller-type drilling rig drills blast holes according to the project for drilling and blasting operations (Fig. 2). In the process of drilling blast holes, the chisel passes various strength rocks. At the same time, the sensor system measures the electric parameters of the drilling rig motors, and records the time and depth of drilling of each well.

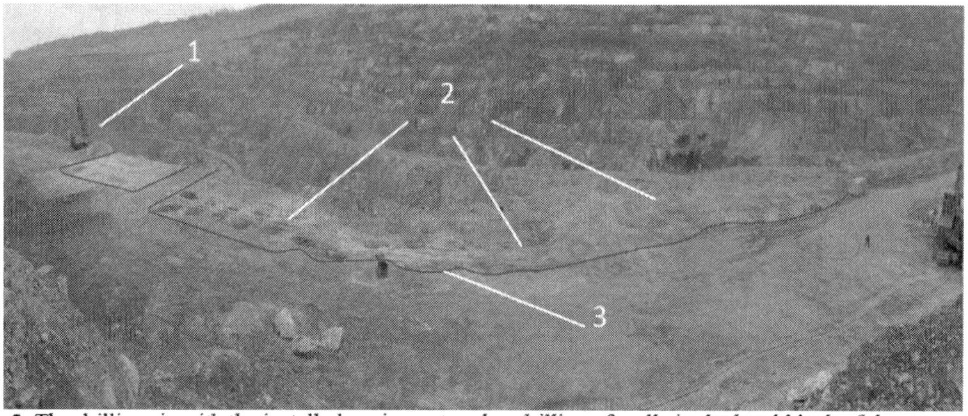

Fig. 2. The drilling rig with the installed equipment makes drilling of wells in the local block of the rock massif: 1 - drilling rig SBSH-250MNA-32; 2 - blasting wells; 3 - local block of the rock massif.

The obtained information on the parameters of the drilling process for blast holes is transferred via wireless communication to the server of the enterprise data center to calculate the drilling difficulty for each well. The calculation results are transferred to the database of the system and in the future they are used as basis for constructing of a digital three-dimensional model for the location of rocks and ores with different strengths within the boundaries of the local block of the massif.

3 Simulation of strength properties of rocks by drilling results

Based on the information received from the sensors, the drilling difficulty index, which objectively reflects the strength properties of the massif, is calculated. Thus, we obtain an array of data spatially distributed over each well in the local block. To build a volumetric model of rock distribution from the difficulty of drilling, it is necessary to determine the values of the index reflecting the strength properties of the rocks not only in each well, but also at points located between the wells. To solve the problem, the kriging geostatistical method [12], which is realized in the program environment R, is the best.

At the first stage of building a three-dimensional model, the grid of the block is calculated. To do this, the block is divided into the specified elements. The size of one element is: one meter in width, one meter in length and 0.5 meters in depth. You can specify other values for the elements. However, if you make a very small grid, the calculation of the model is much longer. At the final stage, a three-dimensional model of the rock distribution from the difficulty of drilling is calculated. The result of the model is shown in Fig. 3.

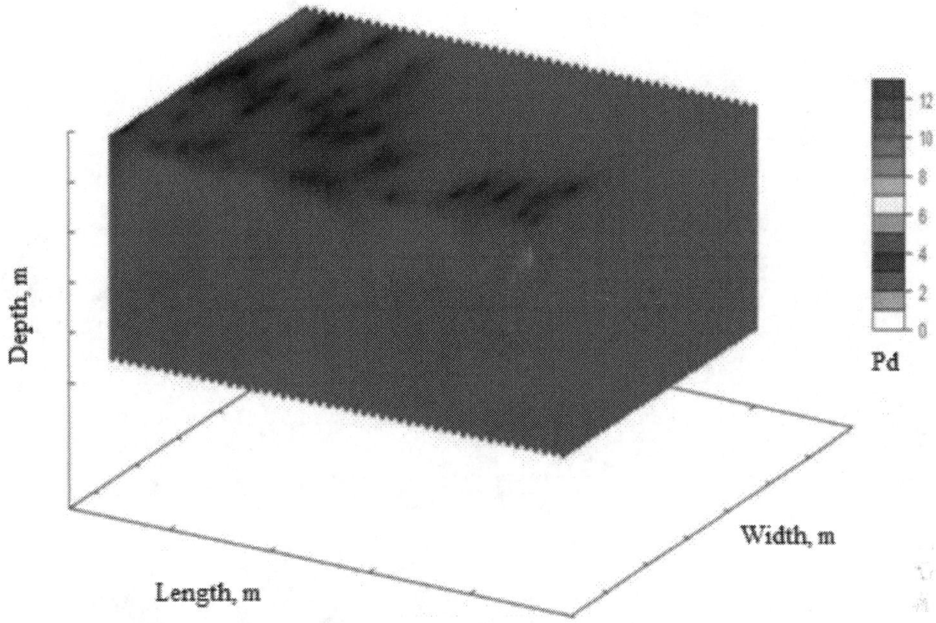

Fig. 3. Three-dimensional model of the distribution of rocks in the local block in terms of the difficulty of drilling index P_d.

To visualize the features of the distribution of rocks in terms of the difficulty of drilling index P_d into the depth of the massif prepared for explosive destruction, a number of sections through 0.35 m are carried out in a three-dimensional model. One of these sections of the model in the direction of changing the width of the block is shown at Fig. 4.

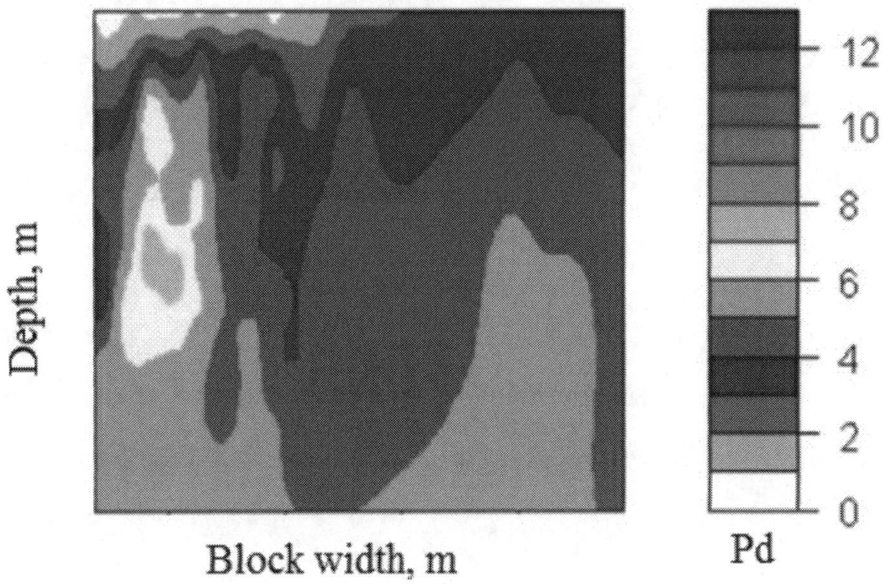

Fig. 4. Three-dimensional model direction by depth of the block.

4 Results and discussion

Approbation of the equipment and simulation procedure was carried out at quarries of an OJSC "Uralasbest" large mining enterprise when blasting wells were drilled on the test block with a SBSH-250MNA-32 roller-type drill rig.

In the process of measuring the parameters of drilling at the quarries of OJSC "Uralasbest" a large amount of observations were obtained in rocks of various petrographic composition represented by chrysotile serpentinites, amorphous and schistose crescents, gabbros and diorites.

Processing of data in a special software allowed the data obtained from the machine to be translated into the values of the difficulty of drilling index P_d. Further, using the kriging method, digital models were created.

As a result of distribution of rocks in the experimental block of the massif prepared for explosion destruction simulation, it was established that there is a section of very difficult rocks with high strength (Fig. 5).

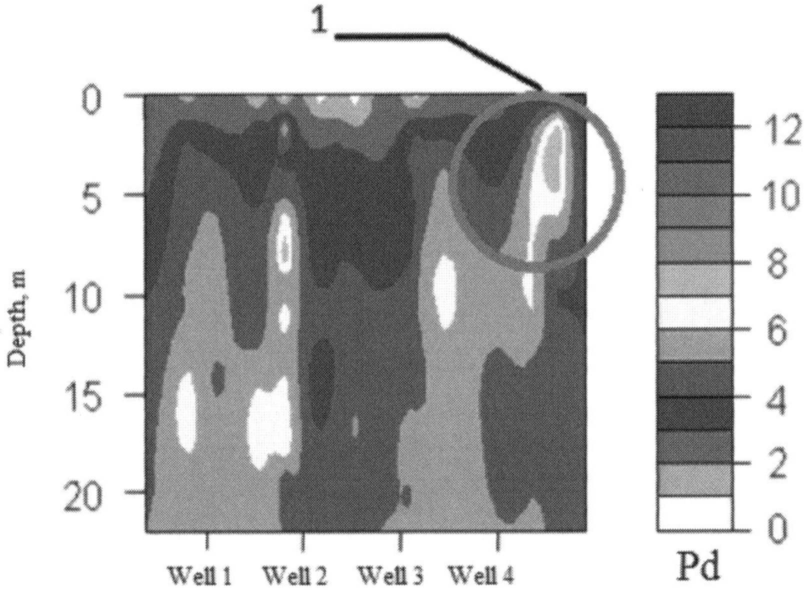

Fig. 5. Three-dimensional model direction on the difficulty of drilling index P_d: 1- section of very difficult rocks with high strength.

After carrying out of blasting operations in the field of prospective finding of a site of rocks with the high durability there was an undestroyed site of the diorite block (fig. 6). The Institute of Mining of the Ural Branch of the Russian Academy of Sciences has developed a procedure [13] for determining the specific explosive consumption, taking into consideration the data of the simulation of the rock distribution by the difficulty of drilling index. The technique allows to adjust the charge column and the mass of explosives located in the well, in accordance with the forecast. Thus, as a result of a mass explosion, it is possible to improve the quality of crushing of the rock mass.

Fig. 6. Experimental site after explosion: 1-unprocessed area of diorite; 2- destroyed area of the local massif

In general, studies have shown that in the rock massifs represented by the same type of rock, the difficulty of drilling index varied widely. The main factors determining a wide range of the indicator's change were heterogeneity, intermittency of the rock massif and the influence of the zones of technological disturbance. As a result of the analysis of the results of the experimental drilling in the quarries of OJSC "Uralasbest" it was also found that the SBS-250MNA-32 drilling rig for drilling one meter of a 18-meter-long well consumes an average of 134 kWh of electricity. Ancillary operations consume about 20 kWh of this amount. At the same time, the energy intensity of a blast hole drilling (according to the power of the rotor without taking into account the compressor's power consumption for blowing the well to take off the products of destruction) averages 1.4 kWh / m.

5 Conclusions

1. The technique of research of strength properties of local rock massifs in the process of drilling of blast holes is improved, based on measuring the parameters of the engine of the drilling rig. To implement the methodology, a special set of equipment was created.

2. Approbation of the device on a roller-type drilling rig was carried out at quarries of an OJSC "Uralasbest" large mining enterprise. Based on the test results, a digital model has been developed that makes it possible to accelerate the process and improve the quality of studies on the strength properties of rocks of a local block prepared for explosive destruction.

Studies carried out within the framework of the State order 007-00293-18-00, on subject No. 0405-2018-0015, No. 0405-2018-0001, project No.18-5-5-10, plus attraction of additional contractual means.

References

1. I.A. Tangayev, Drilling and Explosion of Rocks, 184 (1978).
2. B.S. Rogalsky, E.A. Shtogrin, I.S. Kushmir, Journal of mining, **7**, 41 (1985).
3. S.N. Zharikov, Journal of mining, **7**, 50 (2010).

4. S.N. Zharikov, Blasting work, **111-68**, 155 (2014).
5. R.I. Sukhov, Journal of mining, **3**, 26 (1991).
6. PI.Sukhov, A.S. Regotunov, Subsurface resources management issues. **4 (11)**, 121 (2016).
7. K. Thuro, Proceedings of the World Tunnel Congress '97 materials, **1**, 103 (1997).
8. N. Bilgin, S. Kahraman, 1a" International Mining Congress and Exhibition ot Turkey-IMCET, 177 (2003).
9. E. I.Luis, E. C.Luciano, Mining Technology (Trans. Inst. Min. Metall. A), **113**, A 225 (2004)
10. V.C. Kelessidis, The Journal of The Southern African Institute of Mining and Metallurgy, **111**, 429 (2011).
11. O.Yarali, E. Soyer, Tunnelling and underground space technology, **33**, 46 (2013)
12. A.A. Saveliev, S.S. Muharamova, A.G. Pilyugin, N.A. Chizhikova Geostatistical analysis of data in ecology and nature management (with the use of package R), 120(2012).
13. S.N. Zharikov, V.G. Shemenev, V.A. Kutuev, Sustainable Development of Mountain Territories, **9**, 74 (2017).

Estimation of Influence of Explosive Characteristics of Emulsion Explosives on Shotpile Width

Victor Sinitsyn[1*], *Pavel* Menshikov[1] and *Vyacheslav* Kutuev[1]

[1]Institute of Mining, Ural Branch of the Russian Academy of Sciences, Ekaterinburg, Russia.

Abstract. The article deals with the question of the effect of explosive characteristics of emulsion explosives on the shotpile width. Currently, there are two main points of view to select an efficient type of explosive, which contributes to the qualitative destruction (fragmentation) of coarse clastic rocks. The first is based on the assumption that the detonation velocity of explosives must correspond to the break-down point of the rock (dynamic compression). Another point of view is that the detonation pressure of explosives determines only the head part of the pulse, on which the rock fragmentation is dependent only near the charge, in the contact zone around the borehole. The fragmentation of the entire rock volume within a given borehole array depends on the total magnitude of the explosion pulse, determined not by the detonation velocity, but by the total energy reserve of the explosive charge. Experimental explosions with some of the most common industrial explosives have been carried out in the current conditions of blasting of borehole charges by various types of industrial explosives from the point of view to select the most important parameter, which determines its influence on the shotpile width The investigations have been carried out according to the data obtained to establish that the energy properties of explosives (heat of explosive transformation and density of explosives) determine the decisive influence on the shotpile width, and the operability, the volume of the released gases, the detonation velocity for the change in the shotpile width have very little effect and may not be taken into account in calculations for the prediction of the shotpile.

1 Introduction

Explosives used for the destruction of rocks are, for the most part, a mixture of two or more components. One of the directions is the influence of explosive characteristics on the shotpile width and the definition of the most important of them, which are necessary for an efficient explosion [1, 2].

Explosive destruction of rocks is a very complicated phenomenon. One of the main tasks of breaking rocks is the separation of the rock from the rock mass with its simultaneous crushing into pieces of a certain size and a compact arrangement of them on

* Corresponding author: slavik1988@mail.ru

the bottom of the ledge. The effect of the explosion in the medium and the intensity of crushing of rocks depend on many factors that characterize both the properties of the medium and the properties of explosives, as well as the conditions of explosion, in particular, the parameters of borehole breakdown. The selection of the type of explosives and the prediction of the optimal parameters of drilling and blasting operations on a particular physical model of the action of the explosion of a borehole charge. Such models are built on the basis of general laws of continuum mechanics taking into account experimental data. The medium is usually characterized as isotropic with physicomechanical properties averaged over the whole volume [3, 4].

2 Theory

Currently, there are two main points of view to select an efficient type of explosive, which contributes to the qualitative destruction (fragmentation) of coarse clastic rocks. The first is based on the assumption that the detonation velocity of explosives must correspond to the break-down point of the rock (dynamic compression). The criterion for this indicator is the ratio of the impedances of explosives and rocks, which should be as close as possible to unity. Another point of view is that the detonation pressure of explosives determines only the head part of the pulse, on which the rock fragmentation is dependent only near the charge, in the contact zone around the borehole. The fragmentation of the entire rock volume within a given borehole array depends on the total magnitude of the explosion pulse, determined not by the detonation velocity, but by the total energy reserve of the explosive charge. In the matter of the effect of explosive characteristics of explosives on the shotpile width there are many more different opinions [4, 5].

Experimental explosions with some of the most common industrial explosives have been carried out in the present conditions of blasting of borehole charges by various types of industrial explosives from the point of view to select the most important parameter, which determines its influence on the shotpile width (Table 1). Of course, all the above explosive characteristics of explosives are interrelated and a decrease or increase in one of them can lead to a change in the other. However, under the standard conditions of charge explosion, which are sought to be as close as possible in the production conditions of explosion, the characteristics of explosives [4] are practically constant values, usually determined by reference books or technical specifications [5-7].

The accuracy of the experimental determination of the heat of explosion depends on the conditions of the explosion (the diameter of the charge, the material and the thickness of the shell), the dispersity of the components and the uniformity of their mixing, the density of explosives and many other factors.

Table 1. Explosive characteristics of some of the most common industrial explosives

Name of explosive	Heat of explosion, J	Density of explosive, g/cm^3	Working capacity, cm^3	Detonation velocity, km/s	Volume of gases, l/kg	Volumetric energy of explosive, kJ
Granulotol	4100	0.95-1.0	290	5.5-6.5	1945	3895
Grammonit 79/21	4300	0.9-0.95	360	3.5-4.2	850	4085
Ammonite 6ZhV	4300	0.9-0.95	360-380	3.6-4.8	895	4085
Igdanit 94/6	3300	0.8-0.85	320-330	2.2-2.7	990	3230
Iphzanite T-20	3300	1.34-1.38	290	4.2-4.5	937	4488
Poramit	3200	1.25-1.35	300-310	4.5-5.0	890	4160

3 Results

Below are the results of an experimental determination of the heat of explosion (relative operability) of some industrial explosives (Table 2).

Table 1. Experimental and calculated values of the heat of explosion for various explosives

Explosive	Oxygen balance, %	Heat of explosion, kcal/kg			Density g/cm^3
		Q_{max}	Q_v	Q_{ex}	
TNT	-74	1300	985-1000	850	0.85
«-»	-74	1300	985-1000	1000-1160	1.5-1.6
Hexogen	-22	1510	1240-1330	1270-1325	1.5-1.6
Ten	-10	1550	1405	1400	1.5-1.6
TNT/hexogen (50/50)	-48	1405	1146	1140	1.5-1.6
Granulotol	-74	1300	975	870	1
Alumotol	-76.2	1831	1260	1130	1.1
Zernogranulite 30/70	-45.9	1200	950	870	1-1.1
Zernogranulite 50/50	-27.5	1120	985	910	0.9-1
Zernogranulite 79/21	0	1030	1030	967	0.9-1
Granulite AC-8	+0.3	1242	1242	1260	0.9-1
Granulite AC-4	+0.4	1080	1080	1128	0.9-1
Granulite M	+0.1	920	920	904	0.9-1
Aquatol 65/35	-12.5	837	767	707	1.45
Aquatol M-15	-21	1366	1167	1107	1.4

Thus, on the basis of the studies carried out, it is possible to draw conclusions that are very important for practical application:

- the energy properties of explosives (the heat of explosive transformation and the density of explosives) are critical for the shotpile width;
- the operability, the volume of released gases, the detonation velocity for the change in the shotpile width have a very small effect and can not be taken into account in calculations for the prediction of the shotpile.

The design characteristics of explosives in boreholes for carrying out experimental explosions at the open-pit mines of the Pervouralskoe Mine Administration are given in Table 3.

Table 3. Results of statistical comparison of measurement data for the shotpile width with basic characteristics of explosives

Comparable values	Number of measurements	Average value of compared values	Dispersion	Calculated value of coefficient $T_{0\,n}$	Tabular value $T_{0*0.5;\,n+m-2}$	Conclusion
Working capacity, cm^3	7	303	100.4	10.89	2.18	Significant difference
	7	311	53.2			
Detonation velocity, m/s	7	5132	988	10.13	2.18	Significant difference
	7	3663	4015			
Volume of gases, l/kg	7	1743	3726	10.37	2.18	Significant difference
	7	940	1868			
Heat of explosion, J	7	4136	851	10.92	2.18	Significant difference
	7	4284	435			
Load density of explosive, g/cm^3	7	0.966	0.00004	8.25	2.18	Significant difference
	7	0.935	0.00005			
Volumetric energy of explosive, J/kg	7	3998	72	0.76	2.18	Significant difference
	7	4002	1156			
Shotpile width, m	7	21.7	4.23	0.71	2.18	Significant difference
	7	22.6	4.49			

The values of the explosive characteristics of the emulsion explosives in the boreholes and the measurements of the shotpile width were compared among themselves in pairs by the Student-Gorset criterion. Calculation of the Student-Gorset design criteria was performed by comparing them with a tabular account of the significance level t(a) 0.05 and the degree of freedom $n_1+n_2 = 2$ [8, 9].

4 Conclusions

The obtained experimental parameters allow to establish the compliance of explosives with quality control standards, technical conditions and optimize rational parameters of drilling and blasting operations.

The research has been carried out within the framework of the State Proposal 007-00293-18-00, themes No. 0405-2018-0015, No. 0405-2018-0001, project No. 18-5-5-10, as well as with additional attraction of contractual means.

References

1. B. N. Kutuzov, Rock Disintegration by Explosion, (Moscow, 1992) 516 pp. [In Russian].
2. E. V. Bashkuev, V. F. Bogatsky et al., Engineering of Blasting Operations in Industry, (Moscow, 1983) 359 pp. [In Russian].
3. V. N. Rozhdestvenskii, Investigation of Ways to Control Shotpile during Blasting Rocks in Open-Pit Mines, (Yekaterinburg, 1997) 197 pp. [In Russian].
4. A. N. Khanukaev, Physical Processes during Rock Breaking by Explosion, (Moscow, 1974) [In Russian].
5. N. V. Melnikova, Theory and Practice of Open Development, (Moscow, 1974) 144 pp. [In Russian].
6. V. A. Sinitsyn, Increase in Efficiency of Preparation of Rock Mass in Open-Pit Mines with Use of Explosives Based on Inverse Emulsions, (Ekaterinburg, 2007) 143 pp. [In Russian].
7. A. A. Kushko, Research and Development of Methods to Control Parameters of Shotpile during Explosive Rock Breaking, (Krivoy Rog, 1976) 24 pp. [In Russian].
8. L. Z. Rumishsky, Mathematical Treatment of Experimental Results, (Moscow, 1974) 194 pp. [In Russian].
9. E. S. Wenzel, V. A. Ovcharov, Probability Theory and its Engineering Applications, (Moscow, 1982) 480 pp. [In Russian].

Main factors determining potential qualified appearance of pebble and effel dumps of depleted alluvial gold deposits

Gadzhi Mirzekhanov[1,2*], *Zoya* Mirzekhanova[3].

[1]Mining Institute of Far eastern branch of Russian Academy of Sciences, Khabarovsk, Russia.
[2]"SPC Kompas-Geoservis", LLC, Khabarovsk, Russia
[3]FSBIS "Institute of Water and Environmental Problems", Far East Division of Russian Academy of Sciences, Khabarovsk, Russia

Abstract: The influence of two most important factors determining gold losses during mining of alluvial deposits is considered. They include the granulometric composition of sands and the initial gold content in them. It is shown that the underestimation of these indicators in the assessment of anthropogenic deposits significantly reduces the resource attractiveness of depleted objects. Specific examples give calculations that allow for a more accurate predictive estimate of the resource potential of the dredging tailing (*effel*) dumps. Based on the consideration of natural and climatic factors in determining the conditions, the placer content is plotted for the gold content in the sands of various economic areas of Khabarovsk Region. It has been established that at the same values of the designed process loss and given the comparability of the main placer parameters, a high average gold content in the initial sands makes it possible to predict within the dump some objects with standard contents. **Keywords:** Alluvial deposits, factors, gold losses, dumps, conditions, effel, sand composition, initial contents

1. Introduction

Completeness of gold extraction from placer deposits is determined by the joint influence of various factors, which are manifested to some extent in the process of developing a site. There are more than 14 production and geological factors, which influence on gold losses can be taken into account to varying extent [1, 2, 3]. It is obvious that it is rather difficult to trace the influence of an individual factor *per se*, since they are interrelated, complementary or mutually exclusive. In general, such factors as gold granulometry and clayeyness, which determines the "rinsing" of sand, have been studied. The initial data of these parameters are used to calculate the process losses of metal during the mining of alluvial deposits. In most cases, the identified process losses are not so significant and amount to 5-10%. Accepting process losses as a constant value, the authors reasonably predict in pebble-effel dumps the gold content of 30-100 mg / m^3, which in most cases, using present process schemes, does

* Corresponding author: mgs_gold@mail.ru

not enable to include them in the list of objects suitable for profitable re-development [4]. The predicted estimation of the resource potential of the precious metal in anthropogenic dumps of alluvial gold deposits in the regions of Far Eastern Federal District is significant, based on the above [5, 6, 7, 8]. Meanwhile, the share of forecasted resources in highly promising deposits, which are suitable under modern conditions for industrial processing, is negligible. It does not exceed 10%. At the same time, studies and practical experience show that the influence on gold losses during the washing of productive sands of individual factors or their combination can rationalize the resource attractiveness of pebble-effel dumps, undoubtedly, suitable for repeated profitable mining. First of all, the granulometric composition of sands and the initial contents of gold in them can be classified as such.

2. Methodology and Objects

The work is based on the results of studies of pebble-effel dredged dumps of more than 50 of depleted and operating placer deposits of Far East. Wastes of alluvial gold mining are represented mainly by dumps of different genetic type, differing both by morphological, morphometric parameters, and by the composition of deposits and the resource possibilities of their subsequent processing.

The formation of the dump complex was studied by traditional methods: pebble-effel dumps of the spent placers were completely opened to the full depth by sweeps and ditches in which large-volume samples (0.5 – 3 m^3) were taken every 10 meters. The gold content and the total volume of the dump were identified, which were compared with the initial data of spent placers. In developing placers, daily quality control of washing was carried out. In the selected samples, the following were determined: the gold content, the nature of change in the process loss of gold, depending on the degree of the effel volume of the original sands.

3. Results and Discussion

3.1. Granulometric composition of sands

An important condition in the formation of the resource prospects of a deposit is the ratio in the original gold-bearing sands of the effel (less than 80 mm) and pebble (more than 80 mm) fractions. In most cases, the fraction of the effel fraction (Ke) in the sands varies in the range 60-80%. But the placers are not an exception, where the pebble component predominates over effel (Ke = 20-40%). It is extremely difficult to detect the influence of this factor on gold losses as a part of the combined effect of all factors during washing. The predominance of some or other fraction, of course, makes the process of sand disintegration difficult or, vice versa, easy. So increasing the share of a fraction less than 80 mm creates a viscous gravel-sand-water stream with a ratio of L:S (liquid to solid phase) of less than 10. The flow of such a consistency, falling on the sluice, does not give gold to completely precipitate, which leads to an increase in loss. The degree of influence of this factor on process loss is generally insignificant. And only in extremely abnormal cases, when the share of the effel component exceeds 90%, it is estimated at 20-30% of the calculated data [3]. Accordingly, such an influence, which is often compensated by the cumulative effect of other factors, is little manifested in the "industrial" form of the dumps.

It is another matter how the share of the effel component of the original sands is reflected in the gold content in the final products of their disintegration. When developing alluvial deposits, non-recoverable gold (process loss) is mainly concentrated in the effel dumps (90%), as specified in the principal process scheme for washing gold-bearing sands, which makes them divided into two main components (pebble and effel). Based on the sand's effel content, the same amount of lost metal will be in different volumes of the dump. For example, when washing 1,000 m^3 of sands containing 1 g / m^3 of gold with estimated process loss of 10%, 100 grams of gold will be found in dumps, of which 90 grams (90%) in effel. If in the sands 60% fall for effel size fraction, then 90 grams of metal are spread over the dump body of 600 m^3, and if the share of this fraction is only 10%, then it will be concentrated in a 100 m^3 dump. With a constant amount of lost gold (90 grams), its content in each case will vary: in the first case it is 150 mg / m^3, in the second - 900 mg / m^3, which is close to that in the original sands. Thus, the amount of technologically lost gold, expressed in terms of its content (mg / m^3) in the effel dump has an inverse dependence on the effel of the original sands (Fig. 1). This pattern gives grounds to predict the possibility of having in the waste space highly promising dumplings with gold content, allowing them to be estimated as anthropogenic placers with conditioning parameters. The identification of "high" contents of a useful component in the effel dumps is always an occasion to search for the reasons that caused these anomalies.

In practice, in most cases, high gold content in the tested dumps is classified as excess losses, which is fundamentally wrong. For example, with the above parameters of washed sands (1,000 mg / m^3), 20% of process loss and 10% effel, the gold content in the dump will be 1,800 mg / m^3, which is almost 2 times higher than its content in the original sands. At 30% process loss, this value will be 2,400 mg / m^3, etc.

In the field, it is not possible to identify the volumes of effel dumps of depleted placers. Anthropogenic formations are either reclaimed, or blurred by water flows. With a strict account of the volume of dumps and representative sampling with the identification of the gold content in them (C_e), using the coefficient of sand effel content (K_e) and the metal content in the initial sands (C_s), it is possible to calculate the actual loss percentage (L_{act}) (formula 1):

$$L_{act} = C_e \text{ x } K_e / C_s. \quad (1)$$

In other cases, on the contrary, knowing the average gold content in the original sands, the percentage of process loss and the sand effel factor, it is possible to predict possible gold contents in effel heaps using formula 2:

$$C_e = L_{act} \text{ x } C_s / K_e \quad (2)$$

On the one hand, this formula makes it possible to assess the resource potential of anthropogenic formations of the placer as a whole, on the other hand, to strictly control the annual data on gold loss at specific alluvial sites submitted by subsoil users.

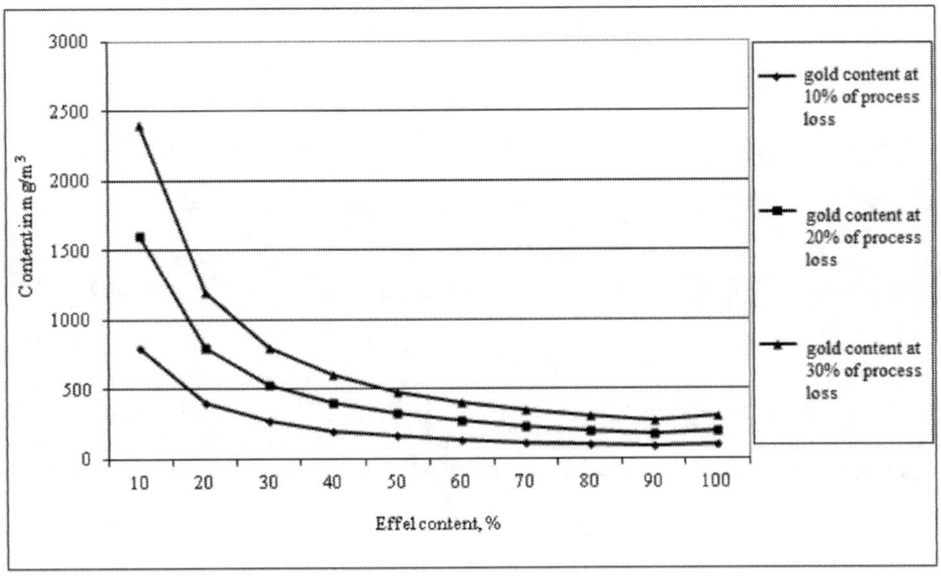

Fig. 1. Dependency of gold content in effel dumps on the effel content in sand at initial gold content in 1 g / m^3 of sand, daily washing of 1,000 m^3 and process loss of 10%, 20% and 30%.

So, according to the results of the study of the process sample (volume 2.5 m^3) from the effel dump placer of the Leviy Creek (the Limuri River basin), the gold content was 3 g / m^3, with 60% extracted by gravitational method. According to the operational data, the placer had a useful component content in the initial sands of 6 g / m^3, effel content about 60%, clayeyness of sands being insignificant 5%, washability of sands being good, process loss about 10%, heavy minerals were practically absent. Gold is small, often in the form of chats or with clay urges. Based on the available parameters (content, effel content, process loss) under normal conditions of washing, the gold content in the dumping site, according to the calculation by formula 2 should be close to 360 mg / m^3. However, in this case, the figure an order of magnitude higher than the control one. Factors that determined such losses could be: the presence of clay material, the incorrect identification of the proportion of the effel component or the content of gold in the sands, i.e., the subsoil user knowingly submitted distorted information.

Thus, the granulometric composition of sands, on the one hand, has a direct effect, especially in the case of increased effel content, when a viscous sand-water flow is created. At the same time, changes in estimated process loss are insignificant. On the other hand, the effect of indirect or consequent-false, as in the example considered above, forms the "standard" appearance of a possible anthropogenic placer.

3.2. Initial gold content in pillar sands

Regions of Far East are significantly different natural and climatic conditions, so for the reserves of placer deposits located in different climatic conditions, particular specifications have been developed. For example, only in Khabarovsk Region, this indicator is divided into three economic areas - southern, central, northern. Accordingly, the minimal commercial metal content, which makes it possible to profitably develop placers in these areas, is significantly different. Taking into account the current gradations of the conditioning requirements and the minimal commercial gold content, placers and exposures of placers of Khabarovsk Region are divided into 5 groups (Table 1).

Table. 1. Proportion of placers in terms of gold content in the sands of different economic areas of Khabarovsk Region

Economic area	No. of placers	Exposures of placers <0.2 g/m^3, %	Placer share in terms of gold content, %			
			0.2-0.8 g/m^3	0.9-2.0 g/m^3	2.1-5.0 g/m^3	> 5 g/m^3
Northern	530	4.9	34.3	41.1	16.0	3.8
Central	217	3.0	37.6	32	19.3	8.1
Southern	42	16.7	23.8	35.7	9.5	7.0

As can be seen from the table, the share of placers with concentrations of more than 2.1 g/m^3 in different economic areas varies from 16.5 to 27.4%. These placers account for up to 62% (from 25.7 to 79%) of reserves in Khabarovsk Region (Table 2). This indicator is highest in Ayano-Maysky, Okhotsky and Verkhnebureinsky districts of Khabarovsk Region.

Table 2. Distribution of gold reserves in the placers of Khabarovsk Region

Economic area	No. of placers	Share of reserves (%) in placers with different gold content in original sands, of the total in the area				
		<0.2 g/m^3	0.2-0.8 g/m^3	0.81-2.0 g/m^3	2.1-5 g/m^3	> 5 g/m^3
Northern	530	0.3	3.4	38.7	15.1	42.3
Central	217	0.08	9.6	10.4	8.5	71.5
Southern	42	2.3	40.6	4	24.8	0.9
TOTAL	789			54.8	13.8	48.5

The initial gold content in the sands, at first glance, does not have a direct effect on the volume of loss and is not used as an indicator in the calculation of process loss. But in practice it turns out that the higher the initial contents in the washed sands, the higher the share of actual metal loss and, accordingly, its absolute content in dumps. Perhaps this is due to a limited bed capacity of the log in the device. To avoid unnecessary losses in this case, it is possible to include in the process scheme a schedule of 2-3 times of washing-ups [9,10]. However, in reality, because of the additional loss of time and, accordingly, the reduction of the washing volumes, multiple washing-ups are practically not made. This problem can be solved by an extension of the logs, use of 2-3-band sluices, adding more sluices of small filling and other methods oriented to optimizing process schemes. Thus, high gold content in wash sands with non-compliance with process requirements lead to excess gold loss. Theoretically, it is assumed that the greatest correspondence between designed and real losses is achieved when washing the sands of those placers where the gold content is from 0.9 to 2.0 g / m^3. But there are observations when, against the background of more pronounced other factors that predetermine metal loss during the development of placers, the manifestation of this factor exacerbates their influence, causing high content of the useful component in anthropogenic objects. Experimental and methodological work on various alluvial objects confirmed that an increase in the initial contents in the sands to 5 g / m^3 and more determines a 2-fold increase in process loss [3].

It should be noted that the importance of the "primary content" factor can also be considered from other positions, in particular taking into account direct losses: the higher the gold content in the original sands, the greater its loss in quantitative terms, which is extremely important in assessing the resource potential of anthropogenic formations. Given the same values of the designed process loss and the comparability of the main placer's parameters, the high average gold content in the initial sands makes it possible to predict its increased contents in the dumping sites. Accordingly, one can expect in the northern areas of Khabarovsk Region, where placers with high minimum industrial contents are developed, higher gold content occurs in the dump sites. Theoretically, with the process losses identified at 10% with 80% of the sand's effel content, the gold content in dumps of placers can be as follows:

- 25-100 mg / m^3 - with initial metal content in the sands of 0.2-0.8 g / m^3;
- 113-250 mg / m^3 - with initial metal content in the sands of 0.9-2.0 g / m^3;
- up to 625 mg / m^3 - with initial metal content in the sands of 2.1-5.0 g / m^3;
- over 625 mg / m^3 - with initial metal content in the sands of 5.0 g / m^3.

In this case, a decrease in the share of the effel component can significantly increase the level of gold content in the dump sites and their resource attractiveness.

4 Conclusions

Thus, high gold content in the original sands combined with a low coefficient of efficiency (less than 50%) cause the formation of highly promising anthropogenic placers of the dumping site with metal contents in them of at least 500 mg / m^3. With the safety of the dump site and the presence of other positive parameters (volumes of rock mass, absence of reclamation) in the waste area of the alluvial deposit, it is possible to predict a new man-made placer meeting the standard requirements with a high degree of verification. When assessing the feasibility of identifying promising dump sites, analysis and consideration of these factors is mandatory, otherwise one will be limited to the forecast of man-made placers with 60-200 mg/m^3, which significantly reduces the present possibility of involving highly prospective objects in re-development.

Reference

1. O.V. Zamyatin, B.K. Kavchik, Gold mining. Information and advertising bulletin. **52**. 6-9 (2-3).
2. V.A. Makarov, Gold and technology. **3 (13)**. 42-49 (2011).
3. G.S. Mirzekhanov, Z.G. Mirzekhanova, Resource potential of anthropogenic formations of placer gold deposits (Moscow: MAKS Press 2013).
4. G.S. Mirzekhanov, Z.G. Mirzekhanova, Mine survey and subsoil use. **5(91).**14-20 (2017).
5. I.Iu. Rasskazov, V.S. Litvintsev, G.S. Mirzekhanov, T.S. Banschikova, Subsoil use, XXI century. **1 (58)**. 46-55 (2016).
6. B.I. Benevolsii, Gold of Russia: Problems of use and reproduction of mineral and raw materials base. 2nd ed., enlarged (Moscow: "Geoinformark", CJSC, 2002).
7. A.P. Sorokin, V.D. Glotov, Gold-bearing structural-material assemblages of Far East. Vladivostok (Dalnauka, 1997).
8. I.S. Litvinenko, I.S. Golubenko, Survey and protection of subsoil. **5**.17-24 (2015).
9. S.M. Shorokhov, Technology and complex mechanization of development of alluvial deposits. (Moscow: Nedra, 1973).
10. I.I. Kovlekov, Anthropogenic gold of Yakutia. (Moscow: MSGU, 2002).

The technology of development of residue objects of precious metals placer deposits

Viktor Litvintsev[1*], *Vladimir* Alekseyev[1] and *Irina* Kradenykh[1]

[1]Mining Institute of Far eastern branch of Russian Academy of Sciences, Khabarovsk, Russia

Abstract The domestic and foreign experience has been analyzed, showing that the problem of development of geographically close residue gold-ore deposits was practically not investigated. The necessity of finding new methodological and technological-economic approaches to effective and large-scale integrated management of territorially-contiguous residue placer deposits has been substantiated. A number of methodological provisions for the substantiation of this approach have been proposed, an algorithm and a computer program complex has been developed for calculating the technical and economic indicators for the simultaneous development of each of the geographically approximated residue objects of placer gold deposits. The practice of organization of merger processes in the mining sector of the economy has been analyzed, their features have been examined, the advantages and negative aspects of integration processes have been analyzed, which is an important factor in solving the economic problems of the development of these residue placers. **Keywords:** residue placer deposits, geographically close alluvial deposits, technologies for the integrated development of residue placers, software complex.

1 INTRODUCTION

The gold mining industry of Russia remains the most important branch of the economy, including for Far Eastern Federal District (FEFD). At the beginning of the 21st century, the gold mining industry underwent a structural transformation, the extraction of gold from ores (approximately 70%) began to predominate, whereas only 30% were extracted from placers, although it was the other way around in the previous century. This has led to a certain crisis in the economic and social spheres in the traditional gold mining regions, caused, basically, by a sharp decrease in the number of operating small and micro-enterprises. The following grouping of mining enterprises for gold production is presented in [1]: large (more than 2 tons), medium (0.5 to 2 tons), small (0.05 to 0.5 tons) and micro (less than 0.05 t). At present, the situation began to improve:according to [1] in 2015, compared to 2014, the number of gold mining enterprises in Russia increased by 5% (475 against 452 in 2014), mainly due to small and micro-enterprises, but the increase in gold production doubled in 10 years (from 2006 to 2015) occurred due to large organizations only. The extraction of gold by medium, small and micro businesses (especially small business) is practically not growing (table 1).

* Corresponding author: litvinzev@igd.khv.ru

Table 1. The extraction of gold from placers in 2006-2015, depending on the scale of business

Scale of business	The volume of gold mining from placers by years, tons									
	2006	2007	2007	2009	2010	2011	2012	2013	2014	2015
Large	85	97	107	125	130	138.6	146.1	158.7	175.4	173.7
Average	25	20	25	21.7	16.7	20	23.4	28.3	23	31.7
Small	30	28.3	21.7	23.3	23.3	20	26.7	31.7	30	26.7
Micro	In the range from 3 to 3.2									

This once again confirms the fact that in the areas of traditional placer gold mining in Russian Far East, rich and easily accessible placers have been worked out, and small in terms of stocks of metal (including those with sub-standard gold content) are of no interest to subsoil users.

Of course, it is necessary to continue the geological exploration of large placers, but there is already a significant stockpile of gold reserves and other valuable associated components in residue complexes of alluvial deposits. Modern approaches to the development of residue placers are not effective enough, requiring new methods and technological solutions. The direction of our research is aimed at solving this problem.

2 ANALYSIS OF PROBLEMS

After working off the geogenous placers, several billion cubic meters of pebble, dredging waste, peat dumps and reloaders from the dumps have accumulated [2]. The annual volume of gold mining from residue placers is about 10-15% of the extracted metal, which is completely inadequate in conditions of a significant decrease in the reserves and resources of geogenous placers. According to expert data [2], the total amount of gold in all residue placer formations is 5,022 tons or 57% of the total amount of gold mined in the history of placer exploitation. This convincingly confirms the thesis that residue placers are an important resource of the mineral and raw materials base of placer mining.

B.K. Kavchik and V.G. Pyatakov show that there was gold in each of the spent placers: in some placers - 10-15% of the mined one, in others - 50%. Despite the low average contents (up to 100-150 mg / m3), almost "... each residue placer has the objects with gold content, which is quite acceptable for profitable extraction" [3]. According to expert estimates, "... ancillary mining from residue and non-traditional sources can provide at least 7-10% of the total annual all-Russian production" [4].

Only in the Khabarovsk Krai during the period from 2000 to 2007, residue placer complexes of more than 400 million m3 were formed, characterized by a wide group of chemical elements with a toxic effect. Redisue ore and placer formations lead to a change in the natural landscape, geochemical and geomechanical fields in the rocks, being the bases of the residue massifs, as well as to the change in the hydrological regimes of these areas.

This is a huge environmental burden on the habitat of humans, flora and fauna; therefore the development of these facilities with the subsequent reclamation of disturbed lands is not only of economic but also of social nature.

According to enlarged analytical calculations, gold resources of only 149 residue alluvial deposits of the Khabarovsk Krai total 124.3 tons [5]. These residue placers are distinguished by ordinary and low metal content, calculated on the entire processed rock mass of the deposit. Residual gold is concentrated mainly in dredging waste; therefore, one should expect an actual increase in the metal content in them by 2-3 times (depending on the ratio of dredging waste and gravel fractions in the dump complex).

The results of many studies allow drawing the following conclusions:

a) Single processing of natural placer does not allow to extract gold completely, its losses were significant. The amount of abandoned metal (rafts, side and residual tiles) or sands lost during washing (dredging waste and pebble dumps, mud) reaches 10 to 150% or more of its volume in the primary placer [3, 6-8];

b) Repeated mining of residue placers exacerbates the impact on the environment, expanding the range and severity of environmental problems, so the full and integrated extraction of useful components will allow the territory to be cleaned of harmful substances and recultivated.

The results of the planned studies are a scientific and technological basis for large-scale involvement in the operation of complex resources of residue objects minerals various in terms of rock mass volumes. However, a special approach is required to solve this problem. We propose a new strategic approach to the integrated development of residue placers - the development of geographically close residue placers with the processing of rock mass on a stationary washing device (dressing plant).

All the factors listed above show that the creation of scientific and methodological foundations of technology for the development of geographically close residue placers is an actual fundamental and applied task aimed at addressing, among other things, the socio-economic and environmental problems of the region.

There is no exact analogue of the proposed line of research with respect to residue placers, but there are a number of related studies on the development of ore and placer deposits and methodological approaches to the organization of processing their rock mass.

On the example of sub-standard ores, which at present cannot be economically worked out as an independent raw material, Academician K.N. Trubetskoy reveals the possibilities of resource-reproducing functions of mining production by changing the conditions for the occurrence of the dredge in question: "With their indirect development by open method and separate stockpiling on the surface or in the formation of residue deposits with optimal parameters, they become a real resource, and their reserves can be developed with profit" [9].

This approach is certainly relevant for solving the problem of large-scale and complex development of residue placers.

Among the technical solutions close to the planned research, other technical solutions are known, both in Russia and abroad.

The thesis of V.R. Kabirov "Assessment of the economic efficiency of the development of a group of geographically-related ore (metal) deposits" lists the following stages of development of such deposits: 1 – Consideration of a group of geographically-adjacent deposits from the standpoint of technological possibilities for their development; 2 - Identify the possibilities of creating common links of infrastructure for a group of geographically adjacent deposits; 3 - Selection of a variant for the development of geographically adjacent deposits; 4 - Determination of the sequence of involvement of each field in the development [10]. The peculiarity of this research is that a single infrastructure base is formed taking into account the definition of the optimal location and the number of processing complexes.

In [11], D.S. Opryshko provides a rationale for a rational technology for the development of small continental placer gold deposits. The result of the study is the substantiation of the solution of the issue of effective technical and technological possibility of involving small gold-bearing placers, including residue ones that are currently unprofitable in the operation, by using the existing small-sized machinery and the improvement thereof, in relation to specific mining-geological and climatic conditions. This is an important aspect that must be taken into account when developing approximate placers.

The authors of [12] theoretically deeply substantiated the methods of estimating the increment and quality of reserves of approximate placers within a single river valley , where, along with available placers, exist the low-margin ones, which are often never involved in development, although, as practice shows, they are licensed by a single mining enterprise.

In some cases, the joint development of such placers may prove to be economically viable, and the active part of the mineral-raw-material base of placer gold can be substantially expanded. The resulting availability of the entire valley's reserves is determined by the availability of the base field, while the sequence of commissioning of other fields is consistent, determined by the subsoil user. This is one of the principal differences from the technology developed by us.

There is a well-known rationale for the expediency and order of commissioning of a group of kimberlite quarries that are part of the MIBA mining company (Democratic Republic of the Congo), taking into account the creation of a sustainable schedule of overburden for each of the quarry groups, as well as minimizing transportation costs for supplying several concentrator factories [13].

In the thesis, A.B. Anisimova implements the idea of the possibility of replenishing the mineral and raw materials base of solid minerals with the use of methods for the economic evaluation of the effective development of similar deposits of unallocated and unaccounted subsoil fund [14].

The analysis of domestic and foreign experience shows that the problem of developing geographically close residue placers has not been fully explored, therefore the planned study is important from both scientific positions, and from technological, economic and environmental ones.

Residue placers have features that differ from the pillar placers. These are: lower gold content (not always!); its granulometric features, expressed in the predominance of fine fractions; a significant dispersion of volumes of rock masses, depending on the methods of the primary placer mining; spatial disconnectedness of residue products promising for processing; good rinsing of residue formations, determined by the granulometric homogeneity of the material of the dumps; various degree of safety of residue formations in the modern relief, etc. In earlier studies [15, 16], the features of the resource potential of residue gold-ore deposits characterized by the effect of their renewability under the influence of natural processes have been noted, and new technological directions for the development of residue complexes of alluvial deposits have been proposed.

It is necessary to note especially the presence of other associated components of value in the sands, which, on the one hand, makes their exploitation difficult, and on the other hand, can increase their economic attractiveness [9]. Associated components of gold-bearing placers can be of considerable significance and used in various spheres of the national economy. The dumps of residue complexes of placers contain rare earth minerals suitable for the production of composites, high-magnetite sands - for further processing with the obtaining of products for the production of paints or iron ore, the rock mass of thin clay dredging waste heaps can be used as a material for road construction as gravel. Only in the Khabarovsk Krai, in addition to the balance gold-platinum complex placer deposits, dozens of reside alluvial gold deposits containing cassiterite, wolframite, magnetite, ilmenite, and rare earth elements are available.

3 METHODS AND DIRECTIONS OF RESEARCH

All of the above allows concluding that it is necessary to search for new methodological and technological and economic approaches to solving the problem of effective and large-scale integrated development of geographically close residue placers.

Currently, we have determined the number of methodological provisions for the substantiation of this approach, an algorithm has been created and a software package for a computer has been developed to calculate the technical and economic indicators for the simultaneous development of several selected placer-related residue objects of several placers, transporting it to a special site, processing the delivered rock mass on a stationary flushing device or dressing plant). The selection of the location of the stationary industrial device (dressing plant) is determined by the software complex, provided transportation and operating costs are minimized [17].

The sequence of acceptance of methodical, project-technological and financial-economic solutions for the development of a number of closely related residue placer sites includes the following stages:

1. The mining and geological parameters of geographically close residue placers of a particular region, the resources of the base metal and valuable passing components are determined by satellite imagery or, if available, based on the design and engineering documentation of the worked out geogenous placers.

2. If a license is available and the selected fields are included in the list of developed fields, a full range of geological exploration is carried out at all fields, or exploration work is carried out in the course of actual mining in order to clarify the complex resources of the fields.

3. The type of washing device (dressing plant) equipped with additional dressing equipment is chosen, its productivity is calculated. With the help of the developed software package, a sustainable (by economic, technological and ecological indicators) location of a stationary flushing device is determined. A procedure is being developed for the preparation of the stockpiled rock mass on a special site near the flushing device and its subsequent processing, provided that the reserves of all deposits are uniformly offset for the planning period set.

Let's consider in more detail the operation of the software.

Matlab R2013b has been selected to compile a mathematical model of the mining area. The selected mathematical package is well optimized for working with the matrix data type. The Matlab core allows you to easily and quickly work with matrices, which is an important factor for large amounts of data.

The main idea in creating a mathematical model is to break up the study area into elementary areas of a given size and to represent the resulting set of elementary sites and the entire territory as a whole into a matrix view.

To form a matrix representing the area under investigation, we use the developed application, implemented in C ++, using the Visual Studio 2013 software environment.

The investigated territory is divided into a matrix of a given size, where each element of the matrix is an elementary site of the terrain. An identifier is assigned to each site, defined by the program logic, depending on the type of site. When forming the matrix form of the investigated territory, the elementary parts of the terrain are divided into several types:

– accessible parts of the terrain - areas where the dressing plant can be located, and through which the route from the field to the plant can be laid, the identifier of section 1;

– inaccessible areas of the terrain - areas where it is impossible to locate the dressing plant and the route (elevations, ponds), site identifier 0;

– deposits - places where the rock is exported to the dressing plant, an identifier of the site 4;

– the proposed optimal location of the dressing plant, the site identifier 3;

– the route from field to plant, the site identifier 2.

The relevant areas are noted in the application that forms the matrix. The remaining sections are marked as available ones. The optimum location of the dressing plant and the routes from the fields to the plant are determined during the calculation of the mathematical model.

To determine the distance from the deposits to the prospective location of the dressing plant, an algorithm for finding the shortest path has been developed. The beginning of the algorithm operation is that starting from the selected elementary section, where the deposit is located, the distance from adjacent accessible elementary sites to the proposed location of the factory is calculated. Neighboring sites are classified in the sense of Moore's neighborhood, where all eight sites are considered to be neighboring. Unavailable sites, if any, are excluded from the calculation.

From the distance values obtained, the smallest is chosen; the segment with the smallest distance is defined as the route to be laid. The following calculation is made between the neighboring sections with the previous elementary sections obtained during the last calculation by the route laid. A site with the shortest distance to the proposed location of the plant becomes the next point on the route. The calculation is made before reaching the final section.

To determine the optimal location of the industrial device or dressing plant, a cell is selected that is accessible at the beginning of the matrix map and is marked as a cell on which the dressing plant is located. The program determines the shortest routes from the deposits to the prospective site of the dressing plant, then, based on the initial data on the fields and the transport and mining equipment used, calculates the profitability of the plant when it is placed in the specified cell. After that, the next available cell is selected, for which, on the same principle, the profitability of the plant is determined when it is placed in a new cell. Thus, all available cells are calculated. After calculation, the cell with the maximum profitability index is selected. The resulting cell on the plan is the most optimal location for the dressing plant.

Thus, a software method has been developed for determining the optimum location of the flushing device or dressing plant for effective development of the adjacent residue placers or adjacent small (possibly unprofitable) natural placers.

The proposed approach will require subsoil users to attract significant financial resources, use modern and efficient mining and processing equipment, therefore, mostly small and medium-sized enterprises in the extraction of metals may need a restructuring of their structure through, for example, horizontal integration. This is a real opportunity to carry out the large-scale development of residue placers with high efficiency even with minimal conditional contents of useful components.

4 CONCLUSIONS

An analysis of domestic and foreign experience in the study of the problems of the development of natural and residue placers shows that the strategic issues of developing geographically close residue gold-ore deposits have been poorly studied, but the importance of their solution, in conditions of a decrease in the level of placer gold mining, is increasing.

The search for new methodological and technological-economic approaches to solving the problem of effective and large-scale integrated development of geographically close residue placers is an important fundamental and applied task.

A number of methodological provisions for the substantiation of this approach as well as an algorithm and a software complex for computation of technical and economic indicators for the simultaneous development of each of the geographically approximated residue objects of gold-ore deposits have been developed.

The practice of organization of merger processes in the mining sector of the economy has been analyzed, their features have been examined, advantages and negative aspects of integration processes have been analyzed, which is an important factor in solving economic problems in the development of residue placers.

The most important result of the introduction of this technology will also be a significant increase in the social and environmental safety of the regions of traditional placer gold mining.

References

1. Goldmining Union Analytical review "Gold of Russia 2015", - access link: http://goldminingunion.ru/news/view/3667.htm
2. B.I. Benevolskii, T.P. Shevtsov, Concerning the potential of residue gold placers in the Russian Federation // Mineral Resources. **1**. P. 9-16. (2000)
3. B.K. Kavchik, V.G. Piatakov. Zolotodobycha. **135.** pp. 14-19. (2010).
4. S.V. Belov. Zolotodobyvayushchaya promyshlennost'. **3 (27)**. pp. 30-37. (2008).
5. A.P. Van-Van-E, Resource base of natural-residue gold-ore deposits "Gornaya Kniga" publishing house, Publishing house of Moscow State Mining University. pp 268. (2010).
6. Iu.A. Mamaev, A.P. Van-Van-Ye, A.P. Sorokin, V.S. Litvintsev, A.M. Puliaevskii, Problems of rational development of gold-ore deposits of Far East. (geology, mining, processing) / Vladivostok: Dalnauka. pp 200. (2002).
7. L.Z. Bykhovskii, L.V. Sporykhina, K.V. Tsvetkova. Placers and deposits of residue, current problems of study and development. Material of XIV International conference, Novosibirsk: Apelsin. pp. 141-148. (2010).
8. Z.G. Mirzekhanova, G.S. Mirzekhanov, I.D. Debelaia Residue formations of placer gold deposits: resource and technological aspects of mining. Khabarovsk, FED RAS. pp 297. (2014).
9. K.N. Trubetskoi. Mining magazine. **11**. pp. 18–23. (2009).
10. V.R. Kabirov, E.I. Reishakhrit. Proceedings of the 11th International Scientific and Practical Conference "Development of Mineral Resources of the North: Problems and Solutions". Vorkuta city: Vorkuta Mining Institute (division) FGBOU HPE "Gornyi" National Mineral and Raw University. pp. 545-548. (2013).
11. D.S. Opryshko. Collection: New ideas in sciences of earth. Moscow. **2**. pp: 81. (2005).
12. M. Kh. Peshkova, N. A. Matsko, M. iu. Kharitonova. Mining information-analytical bulletin. MGGU. **10.** pp. 29-36. (2007).
13. Bukasa Mukendi Pitchu, Abstract of a thesis for PhD in Technical Sciences degree majoring in 25.00.21 "Justification of the expediency and priority of the development of the Bakwang diamond core deposits by a group of quarries", link: http://www.dissercat.com/content/obosnovanie-tselesoobraznosti-i-ocherednosti-razrabotki-korennykh-mestorozhdenii-almazov-bak#ixzz4K61OONAL
14. A.B. Anisimova. The economic justification for involving the development of the adjacent iron ore deposits of the polar Urals // Abstract of a thesis for PhD in Economic Sciences degree, M.: FSBEI HPE " Russian State Geological Prospecting University". p. 24 (2011).
15. V.S. Litvintsev. Physical and technical problems of mineral development / Novosibirsk, Siberian Division of Russian Academy of Sciences. **1.** pp. 118 –126. (2013).
16. V.S. Litvintsev. "Physical and technical problems of mineral mining" magazine. **1.** pp. 97 – 104. (2015).
17. Certificate of state registration of the computer program No.2016619240 "Program for the search for the optimal location of the concentrator in the development of geographically-adjacent mineral deposits", the date of state registration in the Register of Computer Programs - August 16, 2016. Authors: V.S. Litvintsev, A.A. Sokol, V.S. Alekseyev, I.A. Kradenykh.

Improving processes and natural production minerals recycling

Natalia Khrunina[1*] and *Anton* Cheban[1]

[1]Mining Institute of Far eastern branch of Russian Academy of Sciences, Khabarovsk, Russia

Annotation. The article is devoted to improving the processes of extraction of natural carbon minerals through increased selectivity and reduce dilution, as well as processing of ash-slag wastes to reduce environmental burdens and increasing complexity of extracting minerals. Data are elemental and mineral composition, energy dispersive microanalysis by scanning electron microscope JEOL JCM-6000. As a result of natural objects analysis carbon minerals, with their typing on structural and mineralogical composition, marked by the characteristic difficult structural signs plastoobraznyh and lenticular deposits. A method of treatment of structured development of carbonaceous deposits of minerals based on the improved design of the working body of the harvester, that will increase the efficiency of minerals through mining selectivity. As a result, microanalysis samples ash wastes found the presence of native gold, platinum and silver. The application of ways of dressing of secondary raw materials using innovative technologies developed, based on the processes of gravitational and ultrasonic cavitation effects, provide comprehensiveness of allocating valuable components micron level and reduce the pollution of Wednesday.

1 The urgency of the problem

At present the most important problems of mineral development remain mineral loss, ensuring selectivity indentations, complexity of recovery of valuable components and reduce the adverse impact on the Wednesday [1, 2]. Gets the development direction research of processes of extraction of valuable components from waste carbon minerals of different genesis, it should be noted, however, that the bulk of the commodity mined open pit [3]. Prevalence is wide enough abundance of coal in the form of reservoirs, plastoobraznyh and lenticular deposits. The data also contain deposits of clay, sand, argillites, siltstones, sandstones. The content of toxic and reactive compounds pollution danger Wednesday [4]. One of the characteristic of spatial objects take major complexes in the far eastern region is the Luchegorskij Colliery, that includes sand-gravel and sandy-argillaceous rocks, fig. 1 [5-6].

To produce deeply selective dredging minerals from the treatment of structured arrays allow mountain combines. Application of combine harvesters allows to refuse from drilling and blasting, enhances the extraction ratio and to improve the quality of the extracted mineral resources [7-16]. Escalating of volumes of coal, provided for the energy strategy of Russia's development, and decrease its cost is impossible without technical re-equipment of

* Corresponding author: npetx@mail.ru

mining enterprises. As already mentioned, a number of coal and oil shale deposits are characterized by the complex structure, where the mineral layers interspersed with empty rocks, the power of layers and interlayers can range from a few metres to a few centimeters.

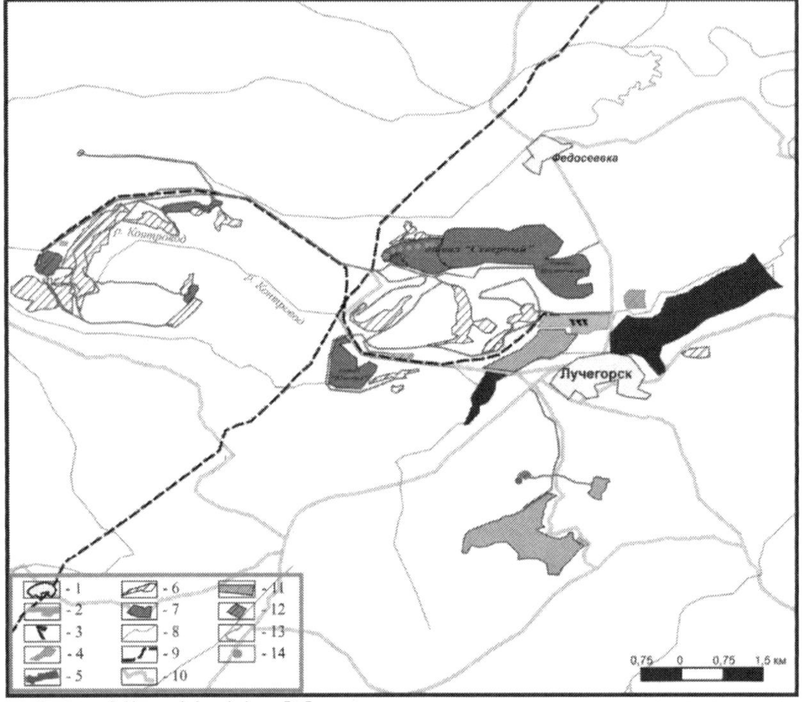

Fig. 1. Luchegorskij coal incision [5]: 1 – human settlements; 2 – technological complex of Seaside GRES; 3 – boilers, boiler emissions into the atmosphere; 4 – ash dumps; 5 – pond-cooler of Seaside GRES; 6 – coal cuts; 7 – dumps of rocks; 8 – river network; 9 – railways; 10 - road; 11 – auxiliary production; 12 – SLR «Jekomett»; 13 – coal road; 14 – the point of testing dumps

The development of such deposits using traditional technologies, in which loosening of the array is done by blasting method, and excavation of rock work is conducted by, not allows highly selective dredging, leads to mixing of mineral with an empty rock. And this in turn will affect the subsequent comprehensive recovery of precious components from waste coal (schist).

The aim of the study is a comprehensive solution to the problem, includes improving processes the natural production of carbon minerals through increased selectivity dredging and reduce the dilution, as well as the processing of ash-slag wastes to reduce environmental burdens and increasing complexity of extracting minerals.

2 Research methods

In the laboratory of mineral processing IGD FEB RAS studied material composition ash wastes Seaside GRES [17] and experimental study on intensification of extraction processes of gravity of noble metals. Using scanning electron microscope JEOL JCM-6000 PLUS NEOSCOPE (JEOL, Japan) carried out by energy dispersive microanalysis samples. Accelerating voltage 15kV, probe talk 7.475 nA. Systematic error of scanning electron microscope JEOL JCM-6000 PLUS NEOSCOPE When all requirements is within 0.01-0.001 [18-19]. When analyzing the current status of machinery and mineral extraction and processing technologies used phenomenological approach to creating effective technologies and means for their implementation.

3 Experimental results and discussion

Electron-microscopic examination of the original Ash on raster electronic microscope JEOL (Japan), with energy dispersing Analyzer "JCM-6000 PLUS" revealed the presence not only of gold, but Platinum and silver, fig. 2.

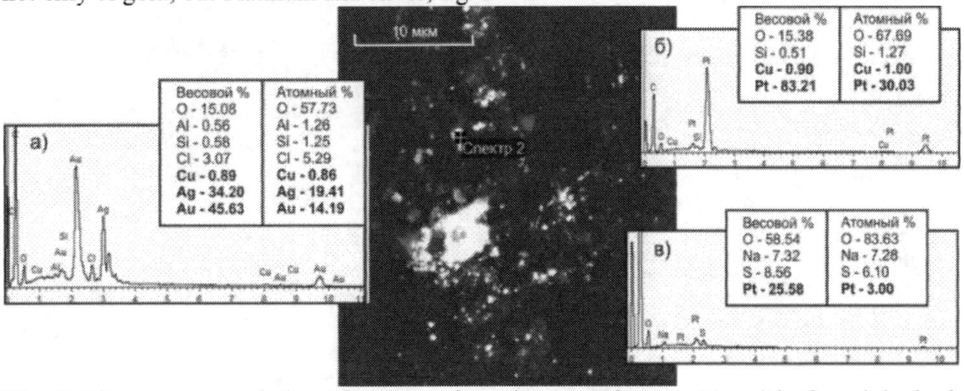

Fig. 2. Disperse accumulation of grains and precious metals (0.5 -10 μm) in the original ash (class size + 2.0): a) kjustelit Ag_3Au_2 (1 μm); b) samorodnaja Platinum (Pt) mixed with copper (0.5 μm); c) kuperit PtS (1 μm) in srastanii with tenarditom Na_2SO_4.

As a result of physical and chemical processing of ash formations before the gravitational enrichment process material reduced loss of fine gold.

Improvement of extracting carbon minerals affect the use of products of its enrichment and reprocessing of wastes are valuable components in a comprehensive range of source. With a view to intensify the destruction treatment of structured arrays of stacked rocks vary in strength properties, developed the executive bodies of the mountain combines with a combined cutting tools. To create executive bodies is to expand the technological capacity of mining equipment, reducing energy intensity and increasing the productivity of cutting process when working in periodically changing mining and geological conditions. The virtues of the techniques of dredging coal and shale oil an important drawback is the presence of a large number of small factions in the mined minerals, which reduces their quality and market value, as well as prejudice loss mountain mass loading, transportation, handling and storage.

Burning of coal containing significant amounts of fines, leads to a decrease in the efficiency of combustion devices, reduces efficiency and environmental safety of fuel used. To improve the quality of coal and oil shale producing their sorting with removing trivia, resulting in accumulating a huge amount of waste [20-24]. According to the domestic classification of coal size pieces divided into groups and classes, there are two groups of coals: varietal and combined. High-quality coal are in high demand and have a greater market value.

To comprehensively address the problem of extracting carbon raw materials and processing of secondary raw materials in the IGD FEB RAS method treatment of structured development of coal deposits using the career combine as well as gravity concentration technology secondary carbon minerals extraction-based small and fine particles of valuable components more environmentally and technologically effective means [25-26]. The authors developed and offered systems modelling processes of multistage Jet-acoustic disintegration mineral components of hydraulic fluid by using hydrodynamic kavitatorov [26] and ultrasonic processing complex [25]. Integrated coalfield development scheme (shale) and recycling (slags, slimes) are presented in figure 3.

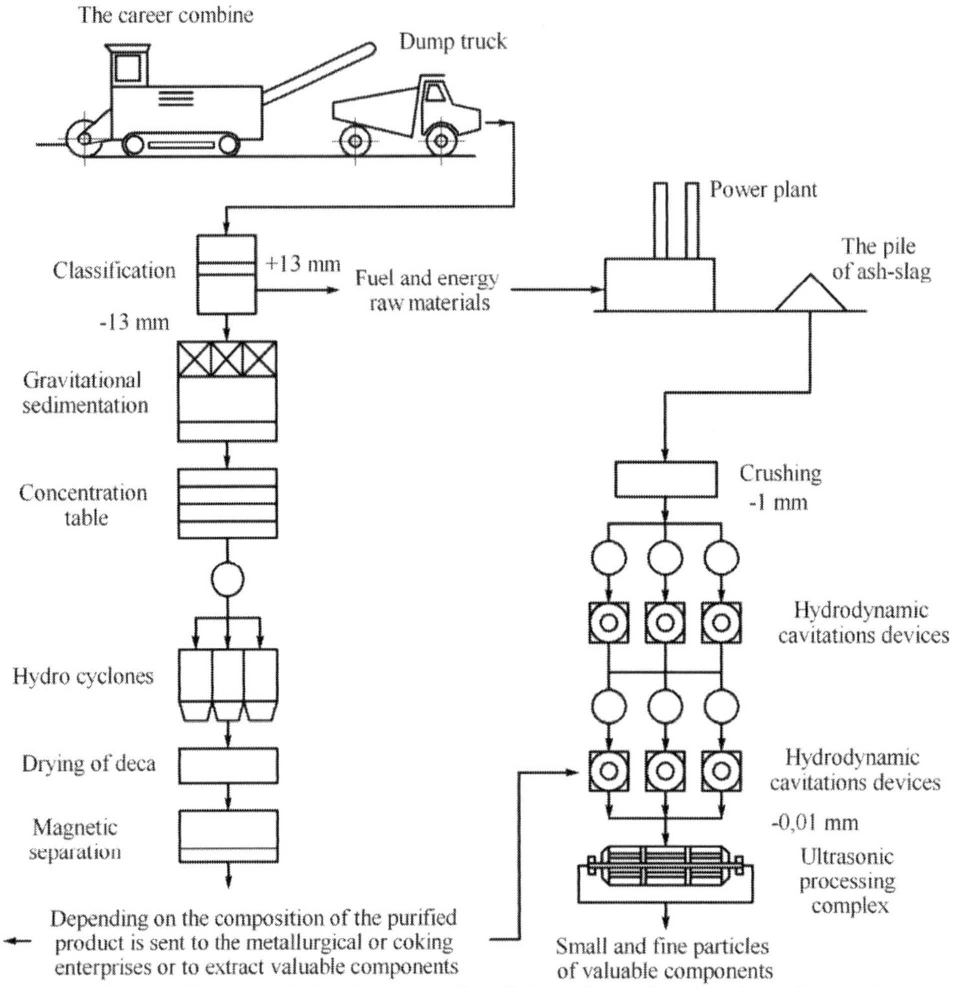

Fig. 3. Scheme of integrated development of coal deposits and processing of secondary raw materials

Method of treatment of structured development of coal deposits using the career combine and the Special Executive Body with variable placement scheme and long incisors includes the required stress concentrations by regulating cutting forcesin the area of processing and shaping in the surface layer of processed array destruction zones using cutting tools working body combine depending on strength characteristics of rock across the width of the workpiece when you change functional parameters-rotation speed shell working body and forward speed combine in their rational combination. Automatic control system of process includes fixation strength characteristics of the breed registration sensor strength. Selective development of ensuring reduction of coal grinding is performed by increasing the step placement coal cutting tools in relation to step placement rock cutting tools, when this fixation discs with cutting tools in working position is carried out automatically stops. The proposed body and how its application would increase selectivity when designing mountain combines treatment of structured deposits, reduce coal pereizmelchenie (Slate) and razubozhennost minerals.

To address the issue of mikrodezintegracii, with the purpose of extraction of fine particles of valuable components more technologically effective means can be used by system processes based on the cavitation acoustic impact on mineral content of hydraulic fluid [25-

28]. For the processing of ash-slag wastes, including fine component to retrieve the Micron particles of valuable components in IGD FEB RAS designed complex with an ultrasonic initiation based on the principle of cavitation effects on mineral gidrosmes in tonkoslojnom thread [25] and affusion hydrodynamic cavitations device [26].

Cavitation reactor systems provide tonkostrujnoe particle with increased acoustic cavitation effects on fine mineral content of hydraulic fluid. Ultrasonic processing complex [25] contains a multilevel partition Dec with riffles, loosening system and regulate the flow of pulp and flushing water. Surface emitting ultrasound transducers are aimed towards trap surfaces Dec taking into account the direction of flow of pulp. Intensifies the intensification of disaggregation and the destruction of relationships between particles of gangue and precious minerals. Processing complex is designed for high performance due to increased surface Dec installed a few levels. This will improve performance and removing particles of valuable minerals from the secondary minerals are smaller than 0.1 μm.

4 Conclusion

As a result of natural objects analysis carbon minerals, with their typing on structural and mineralogical composition, marked by characteristic signs of slozhnostrukturnye deposits. As a result, microanalysis samples ash wastes seaside gres found the presence of native gold, platinum and silver.

For a comprehensive solution to the problem of slozhnostrukturnogo and natural plastoobraznogo extraction and processing of secondary raw material carbon tehnological possibility and feasibility of introducing mountain combines improved designand the use of new gravitational ways of dressing of the secondary carbon minerals based on hydrodynamic cavitation and impact of ultrasound exposure at the micro level to retrieve valuable microparticles components.

Work is performed within the framework of the project № 15-1-2-060 Program FNI "Far East".

References

1. T. Moskalenko, E. V. Nap, Irkutsk State. Tech. Irkutsk, **10**, 98-102. (2014).
2. M. Y. Nazarenko, N. K. Kondrasheva, S. N. Saltykova, V.Y. Bazhin, News of Tomsk Polytechnic University. UN-Ty. Engineering for georesources (udsc), **327 (3)** 67-74. (2016).
3. Y. M. Linnik, Yuri Linnik, Mining equipment and electromechanics, **10**, 2-6. (2012).
4. N. P. Khrunina, A. Yu. Cheban, Vestnik Magnitogorskogo Gosudarstvennogo Tekhnicheskogo Universiteta im. G.I. Nosova, **4**, 50-55. (2015).
5. M. B. Bubnova, Problems of integrated development for georesources (udsc): proceedings of the VI Vserocsijskoj scientific conference with participation of foreign scientists (Khabarovsk, Russia, 05-07 October, 2016 г.)-Khabarovsk: IGD Feb RAS, 183-189 (2017)
6. I. Kalinin, Evaluation of structural-tectonic structure of landslide slopes coal incisions using electrical tomography: example Luchegorskogo lignite cut: thesis. candidate of technical sciences: 25.00.16, Vladivostok (2006).
7. A. Yu. Cheban, , G. V. Sekisov, N. P. Khrunina, S. A. Shemyakin, Eurasian mining, **1**, 22-24 (2014).
8. A. Grabski, Mining, **4**, 60-62. (2010).
9. A. Y. Cheban, N. P. Khrunina, S. A. Shemyakin, Proceedings of the Ural State. Mining University, **4(44).** 54-56. (2016).

10. Mohd Imran, International Journal of Technical Research and Applications, **01**, 224–226. (2016).
11. S.K. Palei, N.C. Karmakar, P. Paliwal, B. Schimm, International Journal of Research in Engineering and Technology, **02**, 393–396. (2013).
12. K. Dey, A. K. Ghose, Journal of Mines, Metals and Fuels, **56(5).** 85–92. (2008).
13. A. Y. Cheban, Magnitogorsk State. Tech. UN-Ty them. G.i. Nosova, **14(2).** 5-9. (2016).
14. A. K. Ghose, Journal of Mines Metals and Fuels, **56(3, 4)** 41–43. (2008).
15. K. Dey, K. Pathak, P. Sen, National Seminar on Mining in the New Millennium, 10–12 November 2000. – Hyderabad, 136–142. (2000).
16. M. Pichler, Y.B. Pankiewicz, in Nikolaev, mining, **3**, 20-22. (2006).
17. A.V. Rasskazova, N. A. Lavrik, N. M. Litvinova, R. V. Bogomjakov, Mountain information-analytical bulletin, **S21**, 282-296. (2016).
18. N. M. Litvinova, N.A. Lavryk, V. B. Bubnova, V. G. Kryukov, R.V. Bogomjakov, K.V. Prokhorov, The journal «Ecological chemistry», **27 (2),** 99-109. (2017).
19. R. V. Bogomjakov. N. M. Litvinov, A. V. Rasskazova, N. P. Chrunina, Inventions, utility models as a waitress. Newsletter, **7** (2018).
20. A. Y. Cheban, N. P. Khrunina, Science and education, **2(82)** 26-31 (2016).
21. I. Demchenko, A mountain of information and analitecheskij bulletin, **5**, 120-122. (1997).
22. Yuri Shuvalov, J. D. Tarasov, A. N. Nikulin, A mountain of information and analitecheskij bulletin, **8**, 243-247. (2011).
23. R. Leimbi-Merike, H. Tiina, L. Eneli, K. Rein, Oil shale, **31(2).** 147-160. (2014).
24. Q. Wang, J. Bai, J. Ge, Y.Z. Wie, S. Li, Oil shale, **31(3).** 266-277. (2014).
25. N. P. Khrunina, Inventions, utility models as a waitress. Newsletter, **18.** (2004).
26. N. P. Khrunina, Inventions, utility models as a waitress. Newsletter, **4.** (2014).
27. V. V .Elshin, S. A. Melnyk, Austrian Journal of Technical and Natural Sciences, **9–10.** 114–118. (2014).
28. Zhi-Jun Dai, Zhen-Bang Kuang, She-Xu Zhao, Transport in Porous Media, **65**, 237–264. (2006).

The technologies of application and design features of layer by layer milling machines

Anton Cheban[1*] and *Natalia* Khrunina[1]

[1]Mining Institute of Far eastern branch of Russian Academy of Sciences, Khabarovsk, Russia

Abstract. Significant quantity of surface miners, various designs and engineering factors are involved in open-cut mining. The article authors offer expanded classification of applied and perspective surface miners with regard to features of design of work tools, rock destruction tool, undercarriage and other elements. **Key words:** surface miner, massif destruction, picks, mills, undercarriage, layer by layer mining

1 Introduction

The technology of blastless mining of solid mineral deposits with the application of surface miners become more common in various world countries [1-3]. Surface miners mine the deposits of coal, bauxites, phosphorites, marls, gypsum, limestones and other minerals. The degree of the application of mineral deposits blastless mining technologies expand embracing new deposits built up with increasingly more coherent rocks. The transfer of enterprises to new technologies is conditioned by both economical factors and material enhancement of measures of ecological and social safety of construction and mining which are limited by drilling and blasting operations.

Loosening of massif using a blast has a row of disadvantages. In case if work front approaching close to inhabited localities or infrastructure facilities, drilling and blasting operations is completely excluded, at the same time, in a row of long operated deposits, basic and more quality reserves (for the reason of mining of other reserves) of minerals are found to be in the immediate proximity to residential space and other facilities.

Currently, the production of surface miners is performed by a great number of companies producing equipment for mining and construction, surface miners by Wirtgen, Man Takraf and Krupp companies got widespread. Active search for optimal engineering and layout designs in order to ensure high performance and reliability of machinery is conducted with the creation of everything new in mechanical engineering. As a result, by now, great number of miners significantly differing from one another by design, scope of application and operating principle has been created [2-3]. In order to ensure effective interaction of surface miners and other types of mining machinery operating together with them, it is necessary to optimize basic parameters by which the selection of equipment is made, to that end authors have made an attempt to classify known designs of surface miners as well as features of their work equipment and undercarriage.

* Corresponding author: chebanay@mail.ru

2 Scope of application and design features of surface miners

Layer by layer which got widespread, and benching technologies (fig. 1) may be distinguished by the technology of deposits mining with surface miners. In case of layer by layer mining technology, miner travels along pit face with rather high speed, surface miners designed for layer by layer mining of deposits are produced by such companies as Wirtgen, Man Takraf, Krupp and others. Such miners are simpler by design, more reliable, more compact and cheaper than machinery for mining of deposits by benching, and also guarantee more effective selective mining of mineral in case of mining of flat and shallow-dipping beds [3-5]. Miners performing mining of deposits by benching have gained lesser spread, they are produced by such companies as Alpine Westfalia, Eickhoff, "MGGU-Geofrez" and others. Work tool of such miners is located on a boom therefore these miners are called boom ones which allows mining a deposit by benches with the height of up to 10 m. In case of benching, surface miner travels across pit face manifold slower compared to machinery operating by layer by layer mining technology, due to this a row of advantages appears, namely: in order to ensure high-performance operation of a surface miner there is no need in creation of extraction front of significant extent; the integration of surface miners into mining process diagram becomes simpler which guaranties production flow; mineral mining selectivity quality improves with steeply dipping bed mining; miner connection to external power source becomes possible which reduces operational costs of a miner operation and increases ecological cleanliness of production.

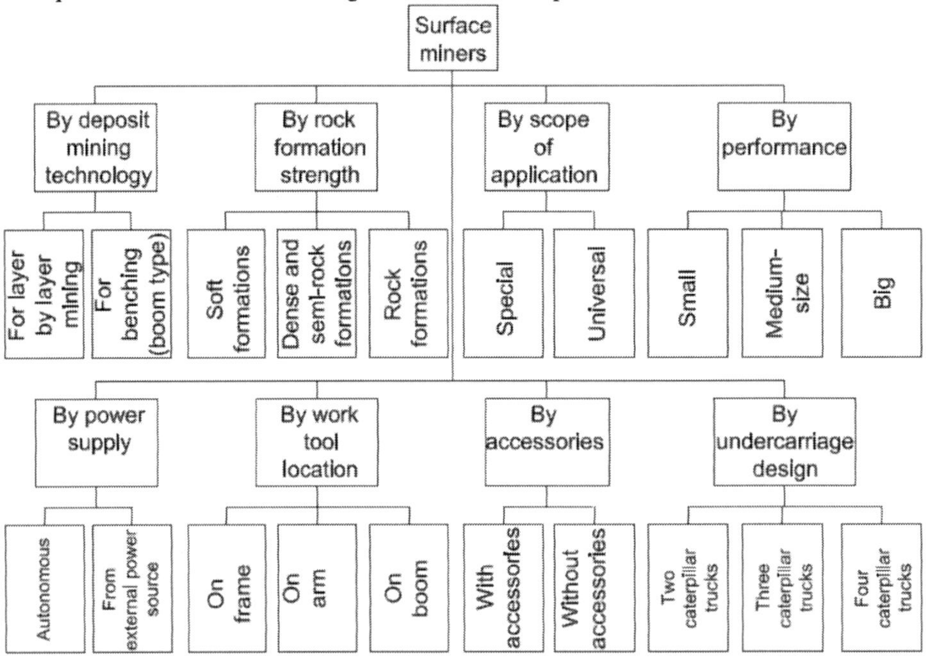

Fig. 1. Peculiarities of design and methods of surface miners operation

Surface miners are designed for the mining of a certain class of rock formations. First surface miners – excavating and milling machines appeared in the mid of the previous century were designed to mine soft formations: sand clays, loams, clays, etc., these machines turned out to be very effective at the construction of trunk lines, reservoirs and other large facilities. In 80's of the previous century, surface miners for blastless selective mining of dense (soft and medium-hard coals, marl, gypsum and others) and semi-rock (phosphorites, bauxites, limestones, shales, hard coals and others) formations [4, 6]. These machines successfully compete with bucket-wheel excavators, especially in case of the development of complex structure mineral deposits. In connection with the expansion of technologies of layer by layer mining in mining operations, by now, surface miners for effective extraction of easily mined rocks (with compressive strength of 50-80 MPa and more), for example, of limestones being row materials for cement production, have been created [4].

One of prototypes of modern surface miners along with excavating and milling machines was road milling machine designed for layer by layer destruction and scraping when repairing motor roads and aerodromes. Big road milling machines were advantageously operated in case of the development of complex structure deposits built-up with dense and semi-rock formations [6]. Consequently, surface miners may be subdivided by purpose into special, usually medium-size and large machines designed for overburden and mining operations in pits, and universal, as a rule small miners or big road milling machines which may be deployed both in mining operations and construction. It should be noted that big surface miners may be either used at especially large construction works [5].

Manufacturing companies produce various technical parameters of made miners (performance is given in m^3/h in dense mass, in m^3/h of excavated rock, t/h of excavated rock), and so far there is no shared vision of what is the main parameter of a miner, for example, if we judge by surface miner tagging, for machines by Wirtgen company – this will be a width of strip being milled in mm, for miners by Man Takraf company – rated capacity in solid in m^3/h, and for miners by Krupp – theoretical capacity by loosen rock in m^3/h [7-8]. With regard to national practice for the classification of bucket-wheel excavatorsby their rated technical performance for loosen mass, it is possible to accept given gradation as well as for surface miners within the limits as follows: small – to 630 m^3/h, medium-size – to 2500 m^3/h and big – over 2500 m^3/h [8].

Miners operating by layer by layer technology are actuated by diesel engine, this enables operation in autonomous mode irrespective of the availability of electrical networks in the area of mining operations which is especially important in areas where there are no centralized electrical networks and power is generated by means of diesel power plants. At that, a miner may easily move along an open-cut in case if it is necessary to operate at a new site. Engines of miners operating by benching, are basically actuated by external power sources.

Miners design maps provide for various variants of a work tool location. As much, surface miners by Man Takraf (model MTS) and Wirtgen (model SM) have a work tool mounted on a frame, accordingly, in a front and mid parts of the machine [5]. Miners by Krupp (model KSM) have a work tool mounted on an arm. All the afore-said models of miners are designed for layer by layer deposit mining technology. Surface miners developing a deposit by benching has a work tool mounted on a boom, such machines are produced by companies Alpine Westfalia (models VASM), "MGGU-Geofrez" (model KGF-80) and others.

Miners may be outfitted with accessories, for example, with bed boundary detection system which allows selective extraction of mineral and barren rock in automatic mode, some miners are equipped with a special shakers for separation of obtained mineral matter immediately in an open-cut [5].

Undercarriage of surface miners may consist of two, three or four caterpillar trucks. Two-caterpillar truck undercarriage ensures maximum mobility of a miner which is rather critical when operating at areas short in length. Design with three or four caterpillar trucks is used provided that a work tool is located in the central part of a miner, swing of such a miner is made by means of caterpillar trucks turning around hydraulic racks which these trucks are hung on [5].

3 Analysis of designs of surface miners work tools

By design, work tools of surface miners may be divided into three groups. Screw and milling work tools representing screws and drums, on which rock destruction tool is mounted, got widespread, they allow mining dense, semi-rock and rock formations [4]. Bucket work tools represented by bucket-wheels installed on machines type KSM by Krupp, ZFM type by Voronezh Plant and by some other companies may excavate soft, dense and semi-rock formations. Plough work tools gained limited application and mounted on machines designed for the development of soft and some dense formations (fig. 2).

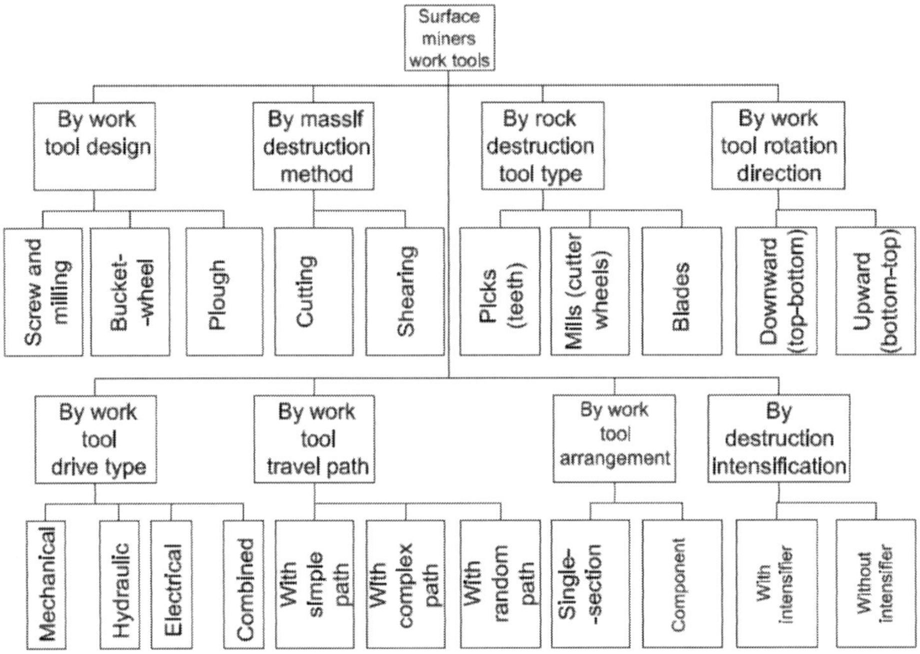

Fig. 2. Surface miners work tools design peculiarities

The destruction of formations with a surface miner is carried out via cutting or shearing massif using a rock destruction tools installed on a work tool. The destruction via cutting is applied in case of the development of soft and some dense formations and destruction by shear — dense, semi-rock and rock formations.

Picks (teeth), mills (cutter wheels) and blades may be used as rock destruction tool. Picks (teeth) gained utmost spread, they are installed on screw and milling and bucket-wheel work tools [5]. Mills (cutter wheels) and blades on surface miners gained limited application and are installed, correspondingly, on screw and milling and plough work tools.

Depending on the direction of work tool rotation, picks penetrate into formation from top to bottom (downward mining) or from bottom to top (upward mining). In case of downward mining, the contact between loose rock and a work tool drum is minimized resulting in the reduction of wear of picks and pick boxes, decrease in the intensity of adhesion of wet stock onto a work tool as well as the reduction of possibility of large rock blocks generation as a results of breaking of oversized flagstones in case of the mining of robust formations of stratified structure.

A surface miner work tool drive may be mechanical, electrical or combined (electrohydraulic) one. Torque of a diesel engine, hydraulic actuator or electric motor is transferred using belt or chain transmission. Promising trends are direct drive of a work tool when hydraulic or electric motor is built inside a work tool. Normally, a surface miner work tool has one common drive, however component work tools may be equipped with several drives (each of cutting units has a separate drive unit and a transmission).

Depending on the complexity of suspension, a surface miner work tool may travel by various paths. In case of a simple travel path, a work tool moves up and down ascending into transport position and descending into operating position. Complex travel path is provided by a miner work tool moving up and down and its inclination in against a frame in vertical plane. Functional capability of a work tool inclination enables forming flat even work site of an open-cut with a slope towards settling sump for water drainage. Random travel path performed in designs of boom miners enables a work tool to travel freely and in any direction in horizontal and vertical planes, ascend, descend, turn in plan, telescope a boom which makes deep selectivity of mineral mining possible [3, 5].

By arrangement, a surface miner work tool is often a single-section one representing screw or drum on which rock destruction tools are mounted as per a certain scheme. A work tool also may consist of two and more identical cutting units as with miners by Alpine Westfalia company (model VASM) [5].

In order to increase efficiency of rock formations destruction process, a work tool may be outfitted with destruction process intensifiers, for example, ones that ensure vibrational movement of a work tool or the capability of its reciprocating offset (in relation to a miner boom), active rotor work tool design equipped with impact devices has been developed [5].

4 Conclusions

Surface miners are effective mining machinery which introduced into mining a great deal of innovative process solutions for mining of solid mineral deposits. Undoubtedly that manufacturing companies carry on improvement and development of designs and process parameters of surface miners and the scope of application of these machines at open-cut mining and in construction expands.

References

1. Wirtgen surface mining for selective limestone mining in the North Caucasus / Russia. Zement-Kalk-Gips Int. **67**, **10**, p. 18. (2014).
2. A.Iu. Cheban The application of milling machines in construction and mining or construction materials // Bulletin of Pacific national university. **3**. pp 105-108. (2012).
3. E.E. Shvabenland Concerning the potential of continuous milling machines with the development deposits by open-cut mining // Rational subsoil development. **1**. pp 54-60. (2014).
4. M. Pikhler, Ui.B. Pankevich Directions of development and results of the application of miners by Wirtgen Surface Miner at open-cuts and open-casts of the world // Mining industry. **3**. pp 42-45. (2000).
5. A.R. Mattis and others Blastless technologies for open-cut mining of solid minerals. Novosibirsk: Publishing house of Siberian Branch of Russian Academy of Sciences, p. 337. (2007).
6. *Iu.B. Pankevich*, G. Khartman Experience of surface miners by company Wirtgen on plaster quarries of the world // Mining industry. **1**. pp 4-9. (1997).
7. Drebenshtedt K., Pessler C. Calculation methods for mining machines. Materials of the international conference "Forum of miners - 2006". D: National mining University. pp. 26-32. (2006).
8. K.N. Trubetskoi, M.G. Potapov, K.E. Vinitskii, N.N. Melnikov and others. Open-cut mining. Bureau of Mines. p. 590. (1994).

Transportation of steam-water mixture during development of steam-water fields.

Alexander Shulyupin[1] and *Natalia* Varlamova[1*]

[1]Mining Institute of Far eastern branch of Russian Academy of Sciences, Khabarovsk, Russia.

Annotation. The main problems of transporting the steam-water mixture revealed during steam-water fields development: pulsations of parameters at low flow rates and considerable hydraulic resistance at high flow rates. The domestic experience of the calculation of steam-water mixture pipelines, oriented to the dispersed-ring flow, is considered. Based on the analysis of domestic and foreign experience in the steam-water fields development, new challenges have been identified that make it urgent to develop scientific principles for the transport of steam-water mixture at low rates, taking into account modern concepts of the stability of steam-water flows.

1 Introduction

It is impossible to imagine modern life without the fuel and energy complex. Lighting, heating, heating food, water supply – almost any sphere of activity is carried out by energy consumption. For many regions where fuel is imported, as well as in view of the search for environmentally friendly solutions for energy production in the world as a whole, the question of using alternative energy sources, which include geothermal resources, is relevant. Geothermal energy is considered to be renewable, does not depend on weather conditions, it is not a source of carbon dioxide emissions into the atmosphere, and is available around the clock.

More than 80 countries use geothermal resources as a direct source of heat in heat pumps, balneological basins, heating systems, etc. [1]. More than 20 countries produce electricity from geothermal sources. [2]. In such countries as Costa Rica, El Salvador, Iceland, Kenya, the Philippines, the share of geothermal energy is 15-22% of the total electricity production. In future, this method is considered to be possible to produce about 8.3% of the total world electricity, which allows to meet the needs of 17% of the world population [3].

The greatest development of geothermal energy in Russia has received in the far East, where there are 5 geothermal electric power station (GEOENERGY). The source of heat carrier for this electric power station is the steam-water field, whose production wells bring to the surface a steam-water mixture. The largest field is Mutnovskaya (Kamchatka), which provides more than 80% of domestic electricity generation on geothermal resources. At the Mutnovsky field, the transportation of the heat carrier from the production wells to the geo-power Station is carried out through the pipelines of the steam-water mixture [4]. This work

* Corresponding author: vnn-dvgups@mail.ru

is devoted to problems of the specified type of transport of the heat carrier in the fields paramerters.

2 Problems of transportation of steam-water mixture and their solutions

At the beginning of development of steam-water fields widely used separate scheme of transporting heat transfer fluid separation was carried out near the mouth of wells, steam pipelines were delivered to the station, water was poured to the terrain. The tightening of environmental requirements and the widespread introduction of reflux to maintain reservoir pressure, have caused the need to transport not only steam, but also water to a common place of use. In such cases, it is advisable to transport the steam-water mixture in the form in which it is obtained from the wells, and the phase separation is carried out at a common separation station [5]. In this regard, at the end of the last century, the scheme of two-phase transportation began to be actively implemented [6-9]. Interest in two-phase transportation remained in the following years [10-13].

Operation of steam-water mixture pipelines revealed two major problems: the pulsations of flow parameters, typical for low flow rates, and a significant hydraulic resistance, which is characteristic for high flow rates. That's why, at the design stage of pipelines there was a problem of choosing the optimal diameter, which would provide minimal hydraulic losses and would provide a pulsation-free mode of transportation. Also there was a methodical problem - to perform the calculation of the pipeline in terms of the interrelation of pressure drop and flow, while taking into account the dependence of the flow rate of wells from wellhead pressure.

There is very little information in the world practice about the specific ways of solving these problems. Consider in more detail their decision in the development of domestic fields. A. N. Shulyupin and A. A. Chermoshentseva developed a computer program MODEL for calculation of steam-water mixture pipelines [4]. Initially, the program was conceived as a simplified version of a larger product based on a complex model of the dispersion-ring flow [14]. But the ease of use, and most importantly sufficient accuracy, identified in the course of emergency calculations, predetermined its widespread use.

The program involves the calculation of the pressure drop by the value of the parameters at one node point, that is applicable for short pipes. Long pipes should be divided into short design areas (up to 200 m). The state parameters are calculated by the equations IAPWS-IF 97 [15]. The calculated recommendations are focused on the dispersion-ring flow. For the calculation section of the program determines the steam quantity according to the principle of the possibility of pulsation-free transportation under the condition

$$x > \frac{1}{1+1{,}6\sqrt{\rho'/\rho''}}, \tag{1}$$

where x - mass flow steam content; ρ' and ρ'' - density of water and steam.

Then a recommendation is issued on the choice of diameter for non-pulsation transportation according to the condition

$$D \le 0{,}252\left(\frac{4G}{\pi\rho}\right)^{0,4}, \tag{2}$$

where D is the inner diameter of the pipe; G is the mass flow rate of the mixture; ρ is the mixture density determined by the homogeneous model.

After entering the selected diameter, the pressure drop is calculated

$$\Delta p = \Delta p_f + \Delta p_m , \tag{3}$$

where Δp – total pressure drop; Δp_f – pressure drop due to friction; Δp_m – total pressure drop in local resistances.

The differential pressure on friction is calculated by the formula obtained as a simplified generalization of the results of numerical implementation of the model of the dispersion-ring flow [14]

$$\Delta p_f = \frac{0{,}02\rho(v - v_b)^2 L}{2D} - , \tag{4}$$

where ρ, v – the density and velocity of the mixture corresponding to the homogeneous model; v_b – the velocity of the friction surface, equal to the critical velocity of the saturated water; L -the length of the design pipeline.

The pressure drop in local resistances are calculated according to the formula

$$\Delta p_m = \frac{1{,}4\rho v^2}{2} , \tag{5}$$

where ζ is the total coefficient of local resistance for a single-phase flow.

Usually, the pressure at the endpoint of the pipeline (at the inlet to the group station separator) is set as a constant parameter. The flow parameters of the mixture depend on the wellhead pressure. Practice has shown that to find a solution taking into account the influence of wellhead pressure is most convenient by graphic method, when the base of the well performance chart, reflecting the dependence of the flow rate of wellhead pressure, using the calculations of the pipeline to build a graph of the performance of the wellhead–pipeline, reflecting the dependence of the flow rate on the pressure at the outlet of the pipeline [4]. Further steps are already being taken on the basis of this timetable.

With the use of the MODEL program, most of the existing pipelines at the Mutnovsky field were calculated [4], a lot of modifications of pipelines were reasonably made (replacement of U-shaped compensators with bellows, removal of non-functional shut-off valves, etc.), the calculation of two pipelines at the Pauzhet field was performed. In all cases, estimates of pressure drops and costs were confirmed.

In the work [11], 6 different variants of calculating the friction pressure drop are compared with experimental data obtained from the Lahendong geothermal field, Indonesia for horizontal pipes with a diameter of 10-22" (0.254–0.559 m). It is noted that in the experimental data the main one is the dispersion-ring flow regime. It is found that the homogeneous model gives one of the closest results to the experiments, especially for large diameter pipes. Note that the formula (4) is also close to the classical homogeneous model – only a correction for the velocity of the friction surface is introduced. The introduced correction (for the conditions of the Mutnovsky field t is about 6 m/s) reduces the calculated drop, but for large pipelines this reduction can be compensated by an increased value of the friction coefficient. In the formula (4), the coefficient of friction adopted 0.02. Unfortunately, in [11] there are no recommendations for determining the coefficient of friction, but indicates its dependence on the roughness. If it define the friction coefficient as a function of roughness, for pipes of small diameter it is most probably to be more than 0.02, and for pipes of large diameter – less than 0.02. Thus, it can be argued that the calculation of the considered recommendations is consistent not only with domestic experience, but also does not contradict the experimental data presented in [11].

3 New challenges in transportation of steam-water mixture

As noted, the low transport speeds of the mixture are associated with the risk of pulsations. In this regard, it is not recommended to combine the flows from different wells into one main pipeline – disconnection of one well can significantly reduce the flow rate and cause pulsations. However, the construction of individual pipelines from each well requires increased costs. Unsurprisingly that there are attempts to combine the flows from wells [11]. Note that the Mutnovsky field also has a similar experience. In this regard, the question of the scientific justification of such a decision, including the justification of the possibility of transportation with low flow rates, becomes relevant.

Transportation of the steam-water mixture under the condition (5) is associated with significant hydraulic losses. This raises the question about the possibility of liberalization of the conditions tailored to the specific circumstances. And the reasons for this are. For example, by designing a pipeline from a well 053 of 2240 m long, and given the absence of ascending sections, a diameter exceeding the limit (5) was deliberately chosen to reduce hydraulic losses. Subsequent exploitation confirmed the validity of this decision.

Experience shows that well over time reduces productivity. In a pipeline originally designed for a specific flow rate, the flow rate can fall over time to a value that does not meet the condition (5). The question of adjusting this condition arises again.

All of these points require a new approach to the definition of the conditions of non-pulsation transportation, which makes it possible to expand the calculation method in the area of lower flow rates. Note that the formula (4) is fundamentally oriented to high flow rates, i.e. a radical change in the calculation method is required.

A new view on the problem of stability of steam-water flows is proposed in [16]. It is shown that the stability of the operation mode of the well is directly related to the flow conditions in the system of ground transportation of the coolant. The condition of stable operation of the well is formulated as

$$\frac{\partial \Delta p_{int}}{\partial G} + \frac{\partial p_w}{\partial G} > 0 \quad , \tag{6}$$

where Δp_{int} is the internal pressure drop in the well (or the sum of the pressure drops on friction, convective acceleration and gravity), Δp_w is the wellhead pressure determined by the flow in the ground transportation system.

Since the wellhead pressure is the sum of the pressure at the outlet of the pipeline (in the group station separator) and the pressure drop in the ground pipeline, taking the outlet pressure constant, independent of the well flow, the condition (6) is written in the form

$$\frac{\partial \Delta p_{int}}{\partial G} + \frac{\partial \Delta p_{tr}}{\partial G} > 0 \quad , \tag{7}$$

where Δp_{tr} is the pressure drop in the ground transportation system.

According to new ideas, the pipeline is a factor in stabilizing the operation of the well. With a spontaneous increase in the flow rate at the wellhead, the pressure drop in the pipeline increases, the wellhead pressure increases, the development of instability is extinguished. With a spontaneous decrease in the flow rate at the wellhead, the pressure drop in the pipeline is reduced, the wellhead pressure is reduced, causing a compensatory increase in the flow rate.

The new approach imposes even more stringent requirements for combining flows from different wells. The main pipeline is less dependent on the flow rate of one well, which reduces the second component of the left part (7), reducing the stability of the system. In this case, the stability of the system will be determined by the positive value of the first term. Moreover, a spontaneous disturbance, triggering the mechanism of instability can rely on a favorable for the development of instability disturbance in the main pipeline.

Special attention is paid to transportation at low flow rates, at which friction does not play a significant role. In the presence of upward sections, the pressure drop in the transportation system will be significantly dependent on the gravity component, which can decrease with increasing flow. In this case, the negative value of the second term of the left part (7) is possible, it means the pipeline will play a destabilizing role.

Thus, new challenges point to the relevance of the development of scientific foundations of transportation of steam-water mixture with low flow rates, taking into account modern ideas about the stability of steam-water flows. On the developed scientific basis, it is required to create a calculation method, which must necessarily take into account the gravitational component of the pressure drop in the pipeline.

4 Conclusion

1. Despite the rapid development of alternative energy sources, as well as of interest in many countries in this matter, today there is very limited information about the methods of calculation of two-phase transport systems of the coolant.

2. The existing methods of calculation of pipelines for the transportation of steam-water mixture are designed for high flow rates typical for the dispersion-ring mode.

3. Known methods of calculation are limited to taking into account hydraulic resistances due to friction and local resistances.

4. New challenges point to the need to develop a method of calculation of steam-water mixture pipelines for low flow rates, taking into account the gravitational component of the pressure drop.

The study was performed with financial support RFBR, research project No. 16-05-00398.

References

1. Lund, J.W., Boyd, T.L. Direct utilization of geothermal energy 2015 worldwide review. Geothermics 60, 66-93. (2016). doi:10.1016/j.geothermics.2015.11.004.
2. Bertani, R. Geothermal power generation in the world 2010–2014 update report. Geothermics 60, 31-43. (2016). doi:10.1016/j.geothermics.2015.11.003.
3. Bertani R. Geothermal energy: an overview on resources and potential // International Geothermal Days. – Slovakia, (2009).
4. SHulyupin A.N. Voprosy gidravliki parovodyanoj smesi pri osvoenii geotermal'nyh mestorozhdenij. – Vladivostok: Dal'nauka. 262 s. (2011).5. Earth sciences. – Paris: UNESCO, 12. (1973).
6. Lee K.C., Jenks D.G. Ohaaki geothermal steam transmission pipelines // Proceedings, 11-th New Zealand Geothermal Workshop. P. 25–30. (1989)
7. Wigly D.M. Separation plant and pipework design – Ohaaki steam field // Proceedings, 11-th New Zealand Geothermal Workshop.P. 19–24. (1989).
8. Delnov Y., Shulyupin A. Geothermal power generation in Kamchatka, Russia // Geothermal Resources Council Transactions. – Portland. **20**. P. 733–736. (1996).
9. Zhao H.D., Lee K.C., Freeston D.H. Proceedings, World Geothermal Congress 2000. Kyushu-Tohoku. P. 3349–3353. (2000).
10. Ghaderi I. Comprehensive comparison between transmission two-phase flow in one line and two line separately for 50 MWe power plant in Sabalan, Iran. In: Proceedings of the World Geothermal Congress. Bali, Indonesia, no. 2501 (2010).
11. Rizaldy, S.J. Zarrouk. Pressure drop in large diameter geothermal two-phase pipelines // Proceedings 38th New Zealand Geothermal Workshop. – New Zealand. P. 1–5. (2016).
12. A. Garcia-Gutierrez, J.I. Martinez-Estrella, R. Ovando-Castelar, A. Vazquez-Sandoval, C. Rosales-López. Thermal Efficiency of the Los Humeros Geothermal Field Fluid Transportation Network. Proceedings World Geothermal Congress 2015 Melbourne, Australia, 19-25 April 2015, no. 25007. (2015).

13. H.S. Cheik, H.A. Ali. Prefeasibility design of single flash in Asal geothermal power plant 2x25 MW, Djibouti. Proceedings World Geothermal Congress 2015 Melbourne, Australia, 19-25 April 2015, no. 25030. (2015).
14. SHulyupin A.N., CHermoshenceva A.A. Model' dispersno-kol'cevogo potoka v geotermal'noj skvazhine // Dinamika geterogennyh sred v geotekhnologicheskom proizvodstve. Petropavlovsk-Kamchatskij: Izd-vo KGARF. S. 23-35. (1998).
15. Aleksandrov A.A. Sistema uravnenij IFPWS-IF 97 dlya vychisleniya termodinamicheskih svojstv vody i vodyanogo para v promyshlennyh raschetah. CH. 1. Osnovnye uravneniya // Teploehnergetika. № 9. S. 69-77. (1998).
16. Shulyupin, A.N. Steam-water flow instability in geothermal wells. International Journal of Heat and Mass Transfer 105, 290-295. (2017). doi:10.1016/j.ijheatmasstransfer.2016.09.092.

The tasks of hydraulics of steam-water wells in the development of steam-hydrothermal fields

Alla Chermoshentseva[1*] and *Alexander* Shulyupin[2]

[1] Kamchatka State Technical University, Petropavlovsk-Kamchatsky, Russia
[2] Mining Institute of Far Easten Branch of Russian Academy of Sciences, Khabarovsk, Russia

Abstract. Actual problems and prospects of practical development of geothermal fields are noted in the present work. The mathematical models for the calculation of flows in steam-water geothermal wells are described. The author's models cover the whole range of possible flow conditions.

1 Introduction

The use of the deep heat of the Earth is not currently an exotic direction of non-traditional energy [1, 2]. In the main provisions of the "Energy Strategy of Russia for the period up to 2020" [3], the potential ability of non-traditional energy, including geothermal, to abundantly provide domestic demand for the country is noted.

Now in Russia, five geo-power Stations with a total installed capacity of 81.2 MW are operated, three of which are located in Kamchatka (74 MW). Intensively developing geothermal energy requires improving the efficiency of the use of available resources and equipment [1, 4-6]. The prospects for the development of geothermal fields for electricity generation are obvious. In addition, geothermal fluids are a source of valuable chemical components and compounds used in balneology for heating residential and industrial premises [7, 8].

In the development of geothermal fields, unresolved problems constrain the development of this direction. They are fundamental for the appropriate stages of development of fields. The relevance of the direction as a whole determines their high practical importance. Most of these problems relate to the system of production and transportation of the heat-carrier.

Prospects for the practical development of geothermal fields are associated with deposits, the heat-carrier of which is a mixture of water and steam. Of particular importance is the need to develop reliable methods for calculating steam-water flows [9, 10], which requires appropriate scientific and methodological foundations for the calculation of steam-water flow in the wellbore.

Some of the problems relate to the measurement of flow parameters of steam-water wells. These measurements are used to calculate the reserves of the field, its development and ground equipment are designed. The difficulty lies in the need to measure two independent parameters characterizing the steam-water mixture. Traditional single-phase

* Corresponding author: allachermoshentseva@mail.ru

hydraulic methods are not acceptable in this case. A combination of methods or special approaches is required.

One of the important points in the development of geothermal fields is the calculation of the flow of steam-water mixture in the production wells necessary to determine the parameters in the reservoir by wellhead parameters for the calculation of reserves and reverse calculation for the design of field development.

The complexity of the processes of dynamics of the steam-water mixture, which are interrelated with thermodynamic processes, often does not allow obtaining simple solutions to the problems. Mathematical modeling of these processes helps significantly [7].

The development of modern science is impossible without mathematical modeling and computational experiment, which in some cases can reduce the number of field experiments or completely replace them. A qualitative mathematical model is usually more accessible and convenient for research than a real object. The model allows learning how to manage the object by testing different options. Mathematical modeling of steam-water flows is widely used in the development of geothermal fields for the calculation of flows in wells.

2 The development of the model assumptions

Hydraulically wells are circular vertical or inclined pipes of telescopic design. The difficulties of modeling the flow in wells are associated with the presence of steam-water flow. A wide range of rate steam content suggests the possibility of all the main structures of gas-liquid flow.

All currently known methods and models for the calculation of steam-water flows in wells the integral method for describe are use. In this method uses to describe the averaged flow parameters over the cross section of the stream. The models differ in the quantity and type of empirical dependences used to close systems of equations [7, 11].

At present, the most famous and widely used is the HOLA model, developed by G. Björnson in 1987 [12]. It based on the numerical solution of one-dimensional equations of continuity, motion, and energy. The model assumes the possibility of single-phase flow and different regimes of two-phase flow, also takes into account the phase sliding. In the same year, A. N. Shulyupin developed a model of WELL similar in ideology. Coordination of the calculated results with experimental data is presented in [13]. A significant disadvantage of the first version of the program WELL was the loading of the model and data using punch cards. Perfection and development of computer technology required modification of the program. Further study and research of the dynamics of the steam-water mixture, as well as the application of various approaches to the description of flow, led to the creation of new models based on the developed WELL.

By the end of the last century, heat transfer with surrounding rocks was taken into account when modeling the flow in wells [14]. Different researchers in their models considered only radial heat flow. Specifically for the research of the vertical component of the heat flow, the authors developed the WELL-2 model in 2005. The WELL model was taken as a basis. Heat losses in the energy equation were calculated by solving the problem of two-dimensional heat transfer with the surrounding rock in cylindrical coordinates [15]. The changes affected using equations of state. The original model was intended for the wells of the Pauzhetsky field, where the limit of up to 25 bar for the equations of state of saturated steam and water was sufficient. The new model was developed with a focus on the deeper and higher enthalpy wells of Mutnovsky field with a pressure at a depth of 100 bar. New equations of state IFPWS-IF 97 were introduced into the model [16]. This made it possible to use it for the Mutnovsky field. The developed model was used for research. The result showed that to take into account the thermal losses in the surrounding rocks it is enough to take into account the radial heat flow by means of introducing the coefficient of unsteady heat transfer [7].

Next, the quantity of considered two-phase mixture flow regimes and the criteria of their existence are changed in various modifications of the model. On the basis of the accumulated experience of developing mathematical models for calculating the flow of steam-water geothermal wells for practical problems of hydraulics, models of WELL-4 and its modification of WELL-4C, WELL-4G, WELL-4CG were developed [17].

3 Model WELL-4

The basic model of WELL-4 consider. To describe the flow in the well, stationary hydrodynamic equations are used. The flow is assumed to be quasi-stationary (i.e., over time, the heat flow on the walls, which is included in the energy equation). The basic one-dimensional equations of continuity, movement and energy were obtained using the integral method and the two-speed model for two-phase flow

$$dG = 0\,, \tag{1}$$

$$\rho''\varphi v'' dv'' + \rho'(1-\varphi)v' dv' + \frac{(v''-v')}{\pi R^2} G dx = -dp - \frac{2\tau}{R} dz - (\rho''\varphi + \rho'(1-\varphi)) g dz\,, \tag{2}$$

$$dh + gdz + de = dq\,, \tag{3}$$

where G – mixture mass flow rate, ρ'', ρ' – steam and water density, φ – true volumetric steam content, v'' и v' – steam and water speeds, x – mass rate steam content, p – pressure, R – radius of the well, τ – shearing stress at the wall, g – gravitational acceleration module, z – coordinate along the pipe axis oriented upwards, h – specific enthalpy of the mix, dq – change of the specific energy of the flow by means of the heat flow from the well walls. e – specific kinetic energy determined as

$$e = (xv''^2 + (1-x)v'^2)/2. \tag{4}$$

The direction of calculations in WELL-4 is carried out from the wellhead to bottom-hole. The integration step is entered, it is possible to change the calculated diameter of the well, taking into account its telescopic construction.

The model describes single-phase (pure water) and two-phase (steam-water) flow. For two-phase flow three regimes are provided: flow with low steam content (structures with continuous liquid phase), the flow of the transition structure and the flow with a high steam content (structures with a continuous gas phase).

In the event of the flow with a minor steam content (bubble and assembly structure), the well-known formula is used for determining the speed of the steam, applied successfully for the corresponding structures

$$v'' = 1{,}2w + 0{,}35\sqrt{2gR(1-\rho''/\rho')}\,, \tag{5}$$

where w - reduced speed of the mix, determined also as the speed corresponding to the homogeneous model.

The emulsion structure is transient between the assembly and dispersed-ring structures, and in the event of high speeds is transient between bubble and dispersed-ring structures. This formula is used to determine the steam velocity

$$v'' = w + v_c(1 - w/v_s)\,, \tag{6}$$

where v_c - critical speed of the movement of saturated water and v_s - is the speed of steam determined by formula (4).

The existence of the conditions with a continuous gas phase associated with the dispersed-ring structure requires meeting two conditions.

First, the steam content should be sufficient for the formation of the flow core.

$$\beta > 0{,}8\,, \quad (7)$$

where β - volume consumed steam content.

Second, the speed in the core should be sufficient for holding the liquid film on the wall, which requires the predominance of the inertia forces of the steam core over the gravity forces; i.e., the Froude number should be above one:

$$\mathrm{Fr} > 1\,. \quad (8)$$

Moreover, the conditions in which the specified constraints were not met but the steam speed has reached the critical speeds of saturated water are considered transient.

The choice of formulas determining the true speed of one phase or the sliding factor (ratio of gas and liquid speeds) is the principle issue for the description of gas and liquid flows.

The known formula of Z.L. Miropolsky is used to determine the slip coefficient in the dispersion-ring flow

$$s = 1 + \frac{13{,}5(1 - p/p^*)(1 - \mathrm{M}^2)}{\mathrm{Fr}^{5/12}\,\mathrm{Re}^{1/6}}\,, \quad (9)$$

where s - sliding factor, Fr, Re and M - the Froude, Reynolds, and Mach numbers, p^* - pressure at the critical point ($22{,}115 \cdot 10^6$ Pa).

Determining the shearing stress at the pipe wall is also important. In the proposed model, the following formula is used for all conditions of the two-phase flow:

$$\tau = \xi(\rho'' v''^2 \varphi + \rho' v'^2 (1 - \varphi))/8\,, \quad (10)$$

where ξ - coefficient of friction.

The following formula [7] is recommended for determining changes in the specific energy of the flow.

$$dq = \frac{-\Delta T(z) 2\pi\lambda}{G \ln(1 + \sqrt{\pi a t / R^2})} \quad (11)$$

where $\Delta T(z)$ - difference in the current temperature of the heat carrier and its initial temperature equal to the temperature of the rock massif at the infinite border, λ - coefficient of heat conductivity of the surrounding rocks, t - is the time from the beginning of the operation of the well, a - conductivity temperature of the surrounding rocks.

4 Practical use

To substantiate the project for reconstruction of the A-2 well of the Mutnovsky geothermal field, the forecast of this well's performance was performed using the WELL-4 model. The planned reconstruction was to be installed from the mouth to a depth of 1200 m inside the existing casing with an internal diameter of 0.225 m of the liner with an internal diameter of 0.16 m. By the computer program WELL-4, according to the measurement of flow parameters at various stages of wellhead pressure, the pressure at a depth of 1200 m was determined with the existing wellhead design. Then, using the inverse calculation, the wellhead pressure was calculated at the new inner diameter. The result of the reconstruction confirmed the predictions [18], which is a confirmation of the model's effectiveness in the conditions of the Mutnovsky field.

5 Modification of the WELL-4

All known models are based on the equations assuming the invariability of the mass flow of the heat carrier along the flow channel, these recommendations are correct only until the border of the feed zones, below which the rate changes.

The thermal aquifer sequence of Russian geothermal fields consists of rocks characterized by fracture- vein-type permeability. In the process of drilling the well intersects 1 to 7 feed zones 1 to 300 m thick. The wellbore in the feed area consists of pipes, the walls of which are perforated for fluid supply from the thermal aquifer sequence. In some wells drilled at the stage of field exploration and the wells being operated, the supply occurs through the uncased part of the wellbore.

The WELL-4G model was developed for the calculation of flows in the feed area based on WELL-4 [17]. The well feed is modeled as one zone stretched from the upper border of the feed zones to their lower border. It was supposed to be a linear change in the flow rate with depth. In this case, the equations of continuity, motion, and energy have the form

$$dG = (G / L)dz\,, \qquad (12)$$

$$\begin{aligned}\rho''\varphi v''dv''+\rho'(1-\varphi)v'dv'+\frac{(v''-v')}{\pi R^2}Gdx+\frac{(v''x+v'(1-x))}{\pi R^2}dG = \\ = -dp-\frac{2\tau}{R}dz-(\rho''\varphi+\rho'(1-\varphi))gdz\end{aligned}\,, \qquad (13)$$

$$dh + gdz + de + edG / G = 0\,, \qquad (14)$$

where L – thickness of the feed area.

In the case of nonlinearity of the flow rate change with depth in the feed zone, instead of (12), another dependence can be introduced without changing the rest. Other calculation formulas are taken from WELL-4.

This model was realized in the same way as WELL-4, with a slight change in the input block. In particular, one of the initial values is the current specific enthalpy of the mixture. In the base model, the initial value was almost the measured specific enthalpy of the inhibited flow of the mixture, which required an additional unit to calculate the current enthalpy at the wellhead. It is logical to use the new model together with the previous model: make a calculation from the well- head to the upper border of the feed zones with the WELL-4 model, and make the calculations below it with the WELL-4G model.

Table 1 presents information on operating the wells 24, 042, 037 Mutnovsky field parameters.

Table 1 The main characteristics of the wells of the Mutnovsky field

Well	Wellhead pressure, bar	Flow rate, kg/s	Mix enthalpy, KJ/kg	Deph, m	Feed zone, m
024	7,9	15,5	1020	1300	1000-1300
042	8,4	69,9	1200	1860	1355-1837
037	9,6	21,5	1200	1771	1339-1669

Figure 1 shows the calculated pressure profiles in the functioning wells 24, 042 and 037 of the Mutnovsky field of steam hydrotherm.

Points A in the curves correspond to the upper border of the feed zone. Points B correspond to a transfer from the single-phase current to the two-phase one.

- Well 24: in the feed zone the heat-carrier is in the single-phase (point B in the curve is above point A).
- Well 042: in the feed zone, there are both sections with single-phase (below point B) and dual-phase flows (the section from A to B).
- Well 037: in the feed zone the heat-carrier is in the double-phase (there is no point B in the curve).

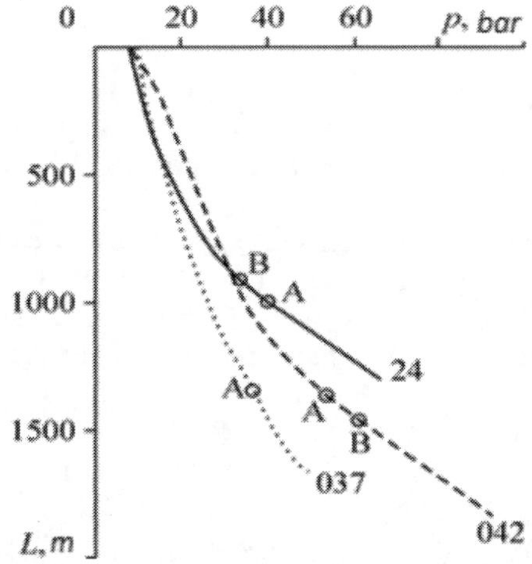

Fig. 1. Calculated distribution of pressure in depth in operating wells 24, 042, and 037 of the Mutnovsky steam hydrothermal field.

Determination of the pressure profile in the bore of the operating well intersecting the feed zone and its comparison with the data of stationary conditions will allow for studying the filtration flows in the bottom-hole zone having a significant impact on well productivity.

Directional drilling is actively used today to improve the efficiency of the intersection of the productive aquifer. The structure of the slant well consists of three sections: the upper vertical, middle deviated (with the set change of the inclination angle), and lower inclined (with a constant inclination angle). The WELL-4C [19] is based on the WELL-4 model and is designed for the calculation of flows in slant wells.

In Eqs. (2) and (3) introduced amendment: module of gravitational acceleration is multiplicatied on the cosine of the well deviation angle from the vertical. Respectively, a package responsible for determining the specified angle is introduced in the implementation of the model in the software.

The WELL-4GC is based on the model WELL-4G and is designed for the calculation of the flows in the feed area of the slant wells. In Eqs. (13) and (14) introduced amendment: module of gravitational acceleration is multiplicatied on the cosine of the well deviation angle from the vertical. The tilt angle is constant and in the implementation of the model is entered in the source data. The other formulas and dependences are identical to WELL-4 G.

Thus, to calculate flow in the feed zone vertical steam-water wells developed a model of WELL-4G. To calculate steam-water flows in inclined wells at constant mass flow rate is proposed model WELL-4C. To calculate in the feed zone inclined steam-water wells is proposed model WELL-4GC.

6 Conclusion

In the present work, existing problems, approaches and methods used in modeling steam-water flow in geothermal wells are investigated. The described WELL-4 computer program developed and implemented by the authors and the WELL-4-G (with variable flow) modifications, WELL-4-C (with the possibility of insertion into the inclined section and the previous section with a set angle) WELL-4-GC allow to cover the full range of possible tasks associated with the calculation of flow in wells in the development of steam-hydrothermal field.

The study was funded by RFBR according to the research project № 16-05-00398 a.

References

1. R. Bertani, Proc. World Geothermal Congress. **01001**. (2015).
2. J.W. Lund, T.L. Boyd, Geothermics 60:66-93. doi:10.1016/j.geothermics (2015).
3. Main provisions of the "Energy strategy of Russia for the period up to 2020", Thermal power engineering, **1**, 2-8, (2002).
4. M.F. Armenta, M.R. Montes, L.M. Alcala, Proc. World Geothermal Congress, No. 22021, (2015).
5. H. Gudmundsdottir, M.T. Jonsson, Proc. World Geothermal Congress, **22071**, (2015).
6. M.T. Fathaddin, H.A. Jibriela, I.M. Azmi, Proc. World Geothermal Congress, **25002**, (2015).
7. A.N. Shulyupin, I.I. Chernev, Georesources, **1(43)**, 19-21, (2012).
8. J.W. Lund, D.H. Freestone, T.L. Boyd, Proc. World Geothermal Congress (2010).
9. V.I. Dobrohotov, O.A. Povarov, Thermal power engineering, **1**, 2-11, (2003).
10. A.A. Chermoshentseva, A.N. Shulyupin, Mathematical modeling of steam-water flows in the elements of geothermal equipment, (Kamchatka St. Tech. Univ., 2011).
11. A.N. Shulyupin, A.A. Chermoshentseva, Technical Physics, **58**, (2013).
12. G. Bjornsson, Report LBL-23546. Lowrence Berkeley Laboratory, (1987).
13. A.N. Shulyupin, Volcanology and seismology, **4**, 25-31, (1991).
14. A. Palachio, Geothermics, **19 (4).** 311–328, (1989).
15. A.A. Chermoshentseva, Mathematical Models and Computer Simulations, **18 (4).** 61-76. (2006).
16. A.A. Alexandrov, Thermal power engineering, **9.** 69-77. (1998).
17. A.N. Shulyupin, A.A. Chermoshentseva, Mathematical Models and Computer Simulations, **9 (1).** 127–132. (2017).
18. I.I. Chernev, A.N. Shulyupin, Mining inf. and analytical Bulletin, № 4, 103-107, (2013)
19. M.H. Mubarok, Y.D. Cahyono, S. Patangke, E.E Siahaan, Proc. World Geothermal Congress, No. 25013. (2015).

The system of main categories of integral assessment of mining technologies

Gennadii Sekisov[1*]

[1]Mining Institute of Far Easten Branch of Russian Academy of Sciences, Khabarovsk, Russia

Abstract. The evaluation of rationality and efficiency of mining technologies of mining operations and the development of solid commercial minerals deposits as a whole is given. It is considered and justified as big and rather topical problem. Methodological framework of its solution is represented by a system of evaluation categories the base of which is represented by criteria, indices and parameters of mining technologies. Original component of evaluation system is a some kind of sub-system of evaluation criteria. Indices for each type of criteria representing in some extent their content are represented as a central unit, the parameters of mining technologies are represented as components of evaluation system, and subsequently substantial expression of criteria, and indices. **Key words:** Mining technologies, mining operations, solid commercial minerals deposits, deposits development, parameters of mining technologies, evaluation categories, evaluation criteria, evaluation indices and evaluation parameters.

1. Introduction

Reliable and timely evaluation of the development and use of mineral resources and products has always be playing an important part, but especially great significance in our country it gains currently: first, in connection with the transfer of economics to market relations, second due to the absence so far of genuine advanced methodological frameworks of evaluation: rationality and efficiency of mining operations; development of deposits and their use as a whole. This is a major and complex issue which for it to be solved requires the accomplishment of special and extensional researches and analysis. Therefore, in this work we justify and represent one of the most essential components of original scientific and methodological frameworks of mineral deposits development — system of evaluation categories.

It is unrealistic to count on the creation in rather short terms of comprehensive methodological frameworks of this evaluation with plurality of types of mineral deposits themselves and social and economical, organizational and process, technical and other conditions of their development. Since we deal with a very branching and complex scientific and production system, on given stage this refers to only original frameworks of aggregative evaluation of efficiency of the development of one of the main and the most

[*] Corresponding author: adm@igd.khv.ru

wide-spread types of mineral features — ore deposits and their major components. Originally, we base upon the system of evaluation categories being put forward which includes criteria, indices and parameters in hierarchical population and interrelation. As often as not in mining literature and, especially, in practice they are put on the same level being considered as synonyms of "rational" and "comprehensive" concepts. While these conceptual and terminological categories are different in sense. In respect to specific conditions, the most comprehensive justification of "comprehensive subsoil development" category was provided by academic M. I. Agoshkov [1].

We provide substantive definitions of the following ones of major conceptual and terminological categories in the field of subsoil mining and industrial development:

Development of deposits – production and economic acquirement of mineral deposits and mineral components being contained in them, concentrated within natural or residue mineral formation of industrial or sub-industrial significance.

Rational development of deposit – reasonable and feasible production and economical acquirement of industrial concentration of mineral deposit (or deposits) and mineral components being contained in them within natural, natural and residue mineral formation.

Effective development of deposit generally – resultant production and economical acquirement of industrial concentration of mineral deposit (or mineral deposits) and mineral components being contained in it (in them) within a specific natural, natural and residue formation in case of following social, ecological and legal regulations.

Optimised development of deposit – the most advantageous in integrated expression (economical, ecological, social and legal) production acquirement of industrial concentration of mineral deposit (or mineral deposits).

Comprehensive development of deposit (mineral feature) in the own sense – acquirement of mineral deposits concentrated in industrial and economical quantities within specific mineral formation or mineral feature.

2. Evaluation criteria for the development of mineral deposits and mining operations

Generalising and analysing economical and technical literature, the author has determined essential discrepancies in interpretation and definition of concepts of "criterion" categories although some authors, specifically, academic T.S. Khachaturov defining "efficiency of social production" category, proceed from wide interpretation of "criterion" and "index" concepts [2]. Unfortunately, quite a number of economists and representatives of some other specialities identify "criterion" concept with "index" concept. Considering widely used literature, primarily reference one, various interpretations and definitions of these terms are noted. Similar approach we applied as well to establish the essence of related categories, in particular – "index" or "indices", "parameter" and "parameters", at that versus "criterion" concept [3-5].

The analysis of interpretations and definitions of categories "criterion", "index", "parameter", allows deducing that "criterion" (if broadly defined) – this is a kind of general standard of some activity. Subsequently, it is necessary to judge from the fact that criterion was understood to be the most general principle, specific feature of evaluation of some human's activity, and index was to be understood as one or another (mathematical, symbolic, verbal, etc.) expression of this activity level as a whole, its separate aspect or a row of aspects simultaneously, i.e. the expression of criterion level.

Famous classical expression may serve as satisfactory evidence of feasibility of such approach: "first try then trust".

Hereafter, we judge by the fact that criterion is, essentially, a general specific principle of evaluation of one or another activity, specifically, deposits development.

Having made specific researches, we put forward a general system of evaluation criteria based on the assumption of their range, including evaluation criteria for the development of deposits, including the gold ore deposits, such as:

integrated, generalizing;

Quasi-integrated (nearly integrated) criteria, i.e. close by its range to integrated criterion, but other than such (completely);

Sub-integrated (sub-generalizing);

aggregative; nodal and differential (elemental).

As an original criterion of the development of deposits and mining operations, "deposit development rationality" is put forward. From the viewpoint of this criterion, the development may be generally rational and irrational.

"Efficiency" should be referred to integrated criteria, "optimality" — to quasi-integrated criteria representing various but incomplete combinations of sub-integrated criteria, for example, "social and ecological rationality of gold or deposits development".

3. Evaluation indices of deposits development mining operations

In general meaning [6, 7] and in specified interpretation, the index of some activity, including, naturally, the development of deposits and mining operation, is a certain expression by the form of a particular level of activity. In relation to a criterion, the index is a kind of means of its specific expression in one or another informative form: mathematical, symbolic, signalling, etc. Great variety of indices types, even more quantity of types, appear from the given above system of evaluation criteria. Their full disclosure is possible only within the frames of some sort of a systems, the development of which — special, wide and detailed survey. In this work, only general rank system of evaluation indices is stated.

In accordance with the above established ranking of evaluation criteria we distinguish following rows of indices:

1. Integrated indices expressing integrated development evaluation criteria;
2. Quasi-integrated indices reflecting quasi-integrated criteria;
3. Sub-integrated indices expressing sub-integrated criteria;
4. Aggregative indices expressing aggregative criteria;
5. Differential (elemental) indices expressing accordingly differential criteria.

From the perspective of categorization of evaluation itself, it is feasible to distinguish indices of three various kinds: quantitative evaluation indices, qualitative evaluation indices and combined (quantitative and qualitative) evaluation indices.

High categories evaluation criteria indices are given below.

The following may be referred to integrated indices:

1. Rationality indices.
2. Efficiency indices.
3. Optimality indices.

Similar and equivalent to the criteria system, also systems of types of quasi-integrated, sub-integrated, aggregative and differential (elemental) indices are distinguished.

Quasi-integrated indices by its gradation and description are analogues to quasi-integrated criteria. Comprehensiveness of the development of deposits and other mineral features should be referred to one of the most important quasi-integrated indices of evaluation.

It is proposed to use rationality and efficiency indices as sub-integrated ones: economical; ecological; social; process; technical, etc. expressed by relevant factors.

Currently, quantitative evaluation of the level of rationality is rather handicapped due to the following reasons:

– first, the content itself of "rationality of development" category is not expressed quite clearly (definitely) from the perspective of its quantitative expression;

– second, even if it is granted that clear expression of its content is accomplished, at current stage it is rather encumbering to evaluate it. Since it is still impossible to bring down to a common denominator, adequately and in terms of quantity, such its major aspects as economical, ecological, social, etc. likewise so far the situation roughly is with reliable, authentically integrated quantitative evaluation of development efficiency.

In these conditions it is possible to evaluate the rationality of mineral deposits development in terms of quality. The rationality index of deposit development (*Po*) is put forward as integrated index of such evaluation. Whereby, five levels of rationality of deposits development are distinguished which may be established as a result of its qualitative evaluation:

– Rather high degree of rationality (Rrh);

– High degree of rationality (Rh);

– Mid level of rationality (Rm);

– Not high level of rationality (Rnh);

– Low level of rationality (Rl), i.e. rationality of development is on the edge of irrationality.

It is proposed to refer such its level to rather (or exclusively) high degree of rationality when

$$R_{gen} = R_{rh} > R_h > R_m \rightarrow max, \quad (1)$$

It is possible when the level of major components (quasi- and sub-integrated indices) are rather high (50% and more) and high.

It is feasible to refer to high degree of rationality its level when

$$R_h \leq R_{rh} \geq R_{h,}, \quad (2)$$

This, in turn, is possible when all its major large components have high level or predominately high one, and the rest part (lesser) — not lower than mid one.

Mid level of rationality is its level with

$$R_m \leq R_h \leq R \leq R_{rh} \quad (3)$$

This is possible under condition when all its components have mid level or generally mid level and only its small part – not high and high.

Pursuant to the above stated approach, we refer rationality to not high level when

$$R_{nhs} \leq R_m \leq R_h, \quad (4)$$

and to its low level when

$$R_l \leq R_{nh} \leq R_m, \text{ as per } R_l > Ir, \quad (5)$$

where I_r – the level of irrational deposits development

Since at present it is impossible to strictly express the level of rationality in terms of quantity, its establishment may be accomplished based upon, primarily, expert evaluation and rank correlation. Likewise, it is proposed to determine the level of quasi- and sub-integrated rationality.

No less various appear indices reflecting criteria of efficiency which is to be represented by such its components as economical, ecological, legal, process, technical efficiency.

Specifically, it is proposed to use qualitative and quantitative expression of efficiency factor. At that, as its integrated characteristic, it is feasible to use an index expressed in rank scale and changing five discrete gradations:

1. Rather high efficiency or the highest efficiency *($\sum$ Erh)*, i.e. when

$$\sum E_{rh} \rightarrow max; \quad (6)$$

2. High efficiency $(\sum Eh)$, i.e. when

$$\sum E_h < \sum Э_{вв} \; ; \quad (7)$$

3. The efficiency of mid level $(\sum Em)$, i.e. when

$$\sum Эс \approx \frac{Эв+Эн}{2}; \quad (8)$$

4. Low efficiency (El), i.e. when

$$\sum E_l \leq \sum E_m \leq \sum E_h \quad (9)$$

5. Rather low efficiency $(\sum Erl)$, i.e. when

$$\sum E_{rl} < \sum Э_н \rightarrow min \quad (10)$$

Efficiency, in particular the efficiency of deposits development (EDD), may be considered as rather high if its main components (economical, social, ecological, etc.) are rather high or the majority of components refer to rather high level and the rest to high level.

Other levels of E_{DD} are established as follows:

– high level of its components if of the same high level or from the combination of components of rather high, high and mid levels;

– mid – from components of mid level or from the combination of high, mid and, possibly, one component of low level;

– rather low – from components of rather low or predominately low level.

In case of different, including in case of significant discrepancies in levels of separate components, the level of full E_{DD} is established based upon average weighted evaluations approach. It is suggested that, in turn, level of each component of E_{DD} may be established by one of known mathematical and statistical method, including rank correlation method or via expert evaluation.

In order for preliminary or approximate quantitative evaluation of E_{DD} effectiveness, quasi-integrated index is brought in, conventionally called as a index of net effectiveness (E_{DD}) expressed by the dependence of general view:

$$Э_{ом} = \sum_{j=1}^{n} R_j - \sum_{j=1}^{m} S_j - \sum_{k=1}^{i} D_k \quad (11)$$

where P_j – reduced (both by time factor and by effect) positive result obtained from the production and the use of j-products with effective deposit development (in monetary or natural expression);

S_j – reduced value of j-type of expenditures for obtainment of positive results and prevention of negative consequences related to complete cycle of effective development of deposit;

D_k – reduced value of k-damage caused by negative consequences of deposits development or mining operations;

n – quantity of positive results obtained from effective deposit development or mining operations;

m – quantity of types of expenditures carried ut in case of effective deposit development or mining operations;

i – quantity of types of damage (negative consequences) associated with deposits development or mining operations.

Integrated indices of efficiency of deposits development (E_D) and mining operations may be expressed by the following dependence of general view:

$$Э_о = \sum_{1}^{t} P_П + \sum_{1}^{m} P_{пm} + \sum_{1}^{n} P \rightarrow max \quad (12)$$

where P_n – level of efficiency of the development of natural mineral resources and mining operations within deposit and used raw materials over a period t_i years;

P_{nm} – level of efficiency of the development of natural and residue features and use of obtained mineral raw materials over the period t_i years;

P_m – level of efficiency of the development of residue features and use of their mineral raw materials over the period t_i years;

m – quantity of natural and residue mineral features formed during the development of original natural deposit or by mining operations;

n – quantity of residue features formed in the process of the development of original natural deposit and formed at that natural and residue features, and also in case of using their mineral raw materials.

Table 1. General categories of parameter of the system of mining operations and technologies of ore deposits development

Attributes		Categories of parameters and their types
Item	Description	
1	Parametrical purpose	Mining technologies: geological exploration, mineral features development, deposits mining, mining and development (mainly overburden mining), mining operations, mineral secondary development, processes
2	Level of parameters	Integrated, sub-integrated, aggregative, sub-aggregative, nodal, link, elemental
3	Performance (efficiency and expenditures)	General efficiency, economical, ecological, power, social, process, technical, geomechanical, etc.
4	Production space	Territoriality: process, geometrical, geomechanical, geological, geophisical, physical and geographical, etc.
5	Production time	Duration: process preparation. Operational stage, process impact on production process entities and as a whole mining technologies, mineral products production
6	Quantity	Process quality: physical, mathematical (including geometrical), labour-intensive, consumptive, consumable, range, etc.
7	Quality	Process quality: innovativeness, selectiveness, reliability, productivity, small waste, resource-saving, economical, ecological, power, social, adaptability, versatility.
8	Demand	Needs, demand, consumption, realization
9	Application conditions	Process and working conditions: Physical and geographical, common geological, mining and process, ecological, economical, power, infrastructure, technical, legal.
10	Informativeness and information	Process informativeness: its integrity, operational efficiency, authenticity, reliability, completeness, usefulness, valuation.
11	Movement	Movement: intensity, consistency, mobility, variability, stability, safety, etc.
12	Infrastructure common-process	General mining mining and process infrastructure: transport, power, communication, information, construction, etc.

4. System sub-complex of mining technologies parameters

As a rule, values characterizing main properties of certain features — systems, process, events — are referred to parameters [8,9].

In respect to problem set-up by us, parameters are values reflecting main properties of one or another mining technology, its components. Quantitative definition of parameters, i.e. values of given technical system and its main components is provided for being accomplished on the following stages of scientific research since it is rather volumetric sub-problem. On this stage, volumetric definition of the system of mining technologies parameters themselves is accomplished in respect to peculiarities of the development of gold ore deposits of Russian Far East. In relation to methods of the development of gold ore deposits and to methods of the development of ore deposits as a whole, a specific, specifically – hierarchical system of given parameters is being formed by us from the perspective of their volumetric expression. At that, keeping to the system of main attributes given in the table 1. Within which general categories and types of parameters are reflected.

Criteria, indices and parameters are considered and justified as the most essential components of general evaluation system. All these taken all together are feasible to be referred to theoretical justification methodological frameworks of parameter system complex volumetric evaluation. Process is a basic system as main feature of parameter system complex being represented by us. In the capacity of peculiar parameters of ore deposits as one of main features of mining technologies from the perspective of substantivity of features given in fig. 1.

Parameters of substantial features – geological and production parameters of deposit – quantitative characteristics of mineral deposits banks and their natural properties, to a significant extent predetermine not only conditions but also performance of deposit mining, extraction and mineral raw materials processing and half-finished products.

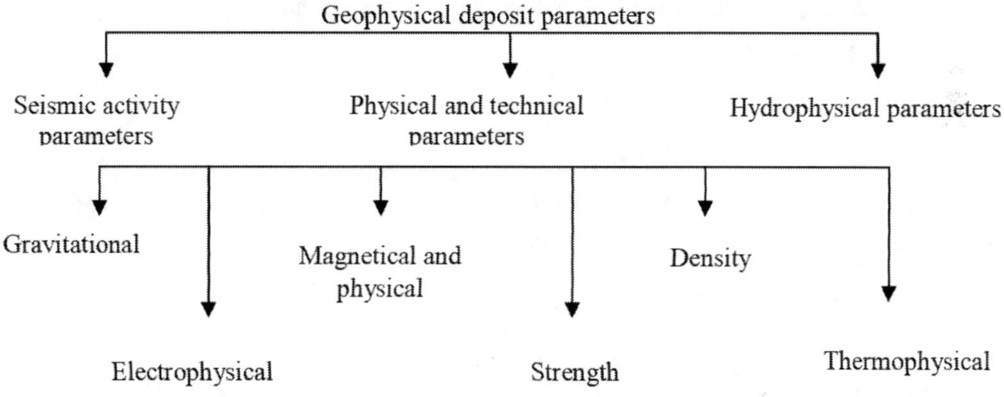

Fig. 1. Main geophysical parameters of ore deposits and built-up rock formations

Geochemical parameters of solid minerals deposits and, primarily, – gold ore ones – is, mainly, the quantity of types of commercial and harmful components of deposits, and as a whole and its components; their quantity and combination of components; the content of components in commercial mineral(s); chemical composition of rock formations building up geological bodies of deposits.

The following are referred to main categories of *geophysical parameters* of deposit as development target: physical and technical properties of rock formations and masses, seismic activity of mining and geological features, Hydrophysical properties and some other.

5. Results

It is feasible to refer the following to main results of the above-stated:

1. Parametrical categories of mining technologies evaluation as well as hierarchical system of mining technologies comprehensive evaluation consisting of criteria, indices and parameters are represented systematically.
2. Sub-system of evaluation criteria is determined and represented by original component block of this system.
3. Sub-system of indices and their composition, criteria level, the dependence of original types are determined for their quantitative expression.
4. Sub-systems of indices of mining technologies quantitative evaluation, and in a certain degree and mining operations are established. At that, they are represented in hierarchic sequence.
5. Original definitions of main concepts — "criterion", "index", and "parameter" as well as basic categories, such as rationality, effectiveness, optimality and comprehensiveness are corrected and reflected in accordance with modern conditions of mining operations.

6. Conclusions

1. For the first time, hierarchical system of evaluation categories is proposed in complex relationship of components.
2. It is possible and feasible to reliably accomplish the use of substantively established categories of the evaluation of mining technologies and mining operations in the shape of hierarchical system of criteria, indices and parameters as scientific and methodological means of justification of rational modernization of traditional, and most significantly – creation and effective practical use of innovative mining technologies.

References

1. Mikhail Agoshkov. Publishing house of Academy of Sciences of the USSR Research Institute of Comprehensive Exploitation of Mineral Resources. (1986).
2. Tigran Khachaturov. Academy of Sciences of the USSR, Economics division, 2nd edition. Nedra, (1987).
3. Gennadii Sekisov, Nikolai Zykov. The development of mineral features and evaluation method. – M.: "Mining book" Publishing house. pp 383-391. (2012).
4. Statistical criteria. Great Russian Encyclopedia, page 65.
5. Criteria of oil and gas geological exploration efficiency. Geological dictionary. "Nedra" Publishing house. **2**. p 375. (1978).
6. Economic parameters! Great Russian Encyclopedia. "Sovetskaia Entsiklopedia" research publishing house. **26**. p 598. (2014).
7. Index of reserves exploration extent. Geological dictionary. "Nedra" Publishing house. **2**. p 107. (1987)
8. Parameters. Geological dictionary. "Nedra" Publishing house. **2**. p 71. (1987).
9. Parameter. Great Russian Encyclopedia. – M.: "Sovetskaia Entsiklopedia" research publishing house. **25**, p 302. (2014).

Modelling the development of drilling operations in iron ore quarries (based on Kachkanar MPP)

Andrey Regotunov[1*] and *Vladimir* Antonov[1]

[1]Institute of Mining, Ural Branch of the Russian Academy of Sciences, Ekaterinburg, Russia

Abstract. The necessity of application of quantitative models of drilling operations development in iron ore quarries for the justified planning of drilling volumes and the relevant fleet of drilling rigs has been shown. For this purpose, experimental regression studies of the patterns of formation and development of drilling operations in the transition period of the modern market economy have been carried out using the example of the Kachkanar Mining and Processing Plant, which has typical characteristic features and dynamics of changes in mining production in Russia. Based on the analysis of the annual technical and economic indices of Kachkanar MPP, conclusions have been drawn about the characteristics of the depression period of drilling production in the 1990s and its subsequent growth, coupled with qualitative changes in the fleet of drilling rigs and productivity thereof. In order to establish the relevant quantitative relationships and development forecasts of the drilling operations at the plant, a nonlinear regression model for the distribution of the length of wells drilled per year has been created, depending on the calendar time, the number of drilling rigs and their utilization factor. The model has a high determination factor of 0.9. It is adequate to the errors of the initial indices, it is statistically reliable, and in this confidence interval, it expresses regularities in changes in the annual length of the drilled wells. According to the regularities, interpretation of transient processes and forecast estimates of the development of drilling operations have been provided.
Key words: forecast, extraction, ore, drilling operations, drilling rigs, indices, length of wells, regression, model.

1 Introduction

Forecasting and planning of drilling volumes and an associated fleet of drilling rigs is directly determined by the state of mining in Russia. Over the past 35-40 years, it has changed significantly due to the transition of industrial infrastructure and management to a market economy. The moderate increase in production volumes that began in the 1980s since the early 1990s had severely disrupted. The crisis in metallurgy in 1993-2000 had led the iron ore industry in Russia to the same uncertainty. In paper [1] transitional periods of

* Corresponding author: pochta8400@inbox.ru

decline in volumes of mining production (1991-1994), unstable stabilization (1995-1998) and, since 1999, - relative growth are noted.

To predict drilling operations in iron ore quarries and management thereof, it is necessary to disclose and understand the patterns of their formation and development to the stage of modern manning and use of drilling rigs. In [2-4], devoted to this problem, the optimization of technological operations of mining production has been considered, and in the studies [5-10] a technique has been proposed and the results of calculation of the optimal number of drilling rigs in quarries have been presented. The results of the minimization of costs for drilling operations have also been stated. However, these studies and conclusions are fragmentary and do not give any idea of these regularities.

In this article, the results of the corresponding experimental regression studies are presented, which make up the shortcomings noted. They are based on the data of annual technical and economic indices of mining enterprises, in which the dynamics of drilling operations have the characteristic features typical for the transition period.

2 Analysis of annual drilling performance at Kachkanar MPP

EVRAZ KGOK "Kachkanar MPP", OJSC, located in the Urals, has been among the iron ore enterprises that have the most characteristic drilling performance dynamics. According to the collection "Technical and Economic Indices of Mining Enterprises" [12], periodically published by the Institute of Mining of the Ural Branch of the Russian Academy of Sciences, the following annual drilling parameters have been sampled at this plant: L-the length of drilled wells per year (drilling volume), K– calendar time rig use factor, n– average number of rigs.

Based on the Kachkanar MPP, let's note features in the distribution of indices similar for the plants. Over the years, since 1970-1980, the number of drilling rigs available at enterprises has decreased by a factor of 1.5-2. At the same time, the factor of their use increased from 0.5 to 0.8. The length of the drilled wells for the year in the 1991-1998 time interval has sharply decreased, and then it has risen slowly to recovery and some advance of the previous indices for the 1980s.

The average number of drilling rigs does not give a complete picture of their impact on drilling results since the factor of their use for the calendar time of operation of the enterprise changes significantly. Then the presence of only working rigs, expressing their conditional quantity by the product of nK should be taken into account. For example, with the list structure of 20 rigs and the factor of their use equal to 0.7 it is conventionally believed that during the annual working hours, drilling of wells has been performed by 14 rigs. In view of this, the efficiency of the use of rigs is best characterized by the ratio L/nK, which is the specific length of the drilled wells per year by a single rig with a conditionally full time of its operation.

A typical distribution of drilling parameters based on the example of Kachkanar MPP is shown in Fig.1. Here during the observed time interval, the drilling rig fleet was reduced with a relevant increase in the factor of their utilization, and the specific annual length of the wells drilled by one working rig after 2000 has only increased.

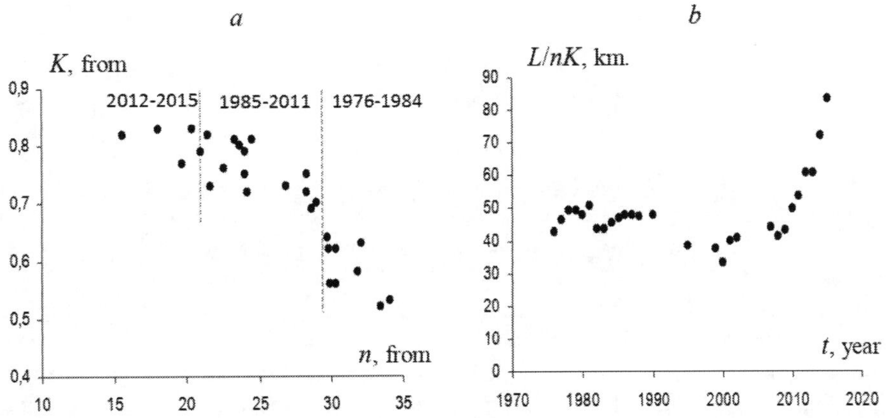

Fig. 1. The calendar distribution of the utilization rate of drilling rigs at the Kachkanar MPP by their number (a) and the specific length of the drilled wells by one machine with the conditionally full time of its operation (b).

The discrete distribution of the length L by the years of drilling and the number of working machines *nK* at the Kachkanar MPP is shown in Fig. *2, a*. The error of the reproducibility of the fixed values *L* is estimated at 5%. It only qualitatively reflects the calendar change in the drilling operations. This is not enough to establish links and quantitative predictive assessments of its development. Therefore, to achieve this goal, this distribution has been studied by the experimental regression modelling.

3 Modelling the distribution of the length of drilled wells

The distribution of the annual length of the drilled wells *L* in the space of a changing calendar time *t* and the number of working drilling rigs *nK* has been modelled by the method of non-linear functional factorial regression [12-13]. Preliminary, taking into account the root-mean-square error of the L discrete values at the Kachkanar MPPσ_K=43 km, the lower R^2_l =0.85 and the upper R^2_u=0.94 allowable interval of the determination coefficient of the sought model have been calculated with a confidence probability of 0.95.

The general view of the model is formed from the phenomena of the experimental distribution of the length *L* in the *t coordinates, nK*. It is noted that the monotonous and long-term change in mining production, caused by the balanced development of geo-resources, develops exponentially over time. The factor of depression and subsequent recovery of drilling operations, limited in time and associated with market restructuring, is expressed by an asymmetric quasi-Gaussian function. The influence of the number of working rigs *nK* on the length of the drilled wells is described by the power function. According to the previous section, it has been taken into account that its index can vary linearly over time. Summarizing the effects of the above factors, we represent the model in the following generalized form

$$L = A_1 e^{\frac{t}{\tau}} + A_2 e^{-\left[\frac{(t-t_o)\left(1+e^{\frac{t-t_o}{\varepsilon}}\right)}{\tau_1+\tau_2 e^{\frac{t-t_o}{\varepsilon}}}\right]^2} + A_3 (\mathrm{nK})^{a(t-1975)+b} + B \quad (1)$$

where: τ – time relaxation of long-term changes in drilling production; t_o, τ_1, τ_2 – the time of the greatest decline in production, the steepness of its decline and, accordingly, the rise in the depression period; ε - time parameter of transition from recession to recovery; a, b– degree parameters.

In this expression, the optimal coefficients A_1, A_2, A_3, B *have* been calculated by the method of least squares, and the functional parameters by the parabolic vertex approximation method using the M3T (τ), M9T (t_o, τ_1, τ_2, ε), M5T (a, b) [14]. The model for the distribution of the annual length of the drilled wells at the Kachkanar MPP, obtained after optimization in the confidence interval of its mean square error of 44.5 km, is presented in a specific mathematical form

$$L = -2{,}35\cdot 10^{34} e^{-\frac{t}{27}} - 463{,}4\, e^{-\left[\frac{(t-1998)\left(1+e^{\frac{t-1998}{4{,}1}}\right)}{3{,}5+11{,}3\, e^{\frac{t-1998}{4{,}1}}}\right]^2} + \quad (2)$$

$$+1{,}4384\cdot 10^{-29}(nK)^{23{,}58-0{,}03(t-1975)} + 1199 \pm 44{,}5$$

A graphic view of the model is shown in Fig. *2 b*.

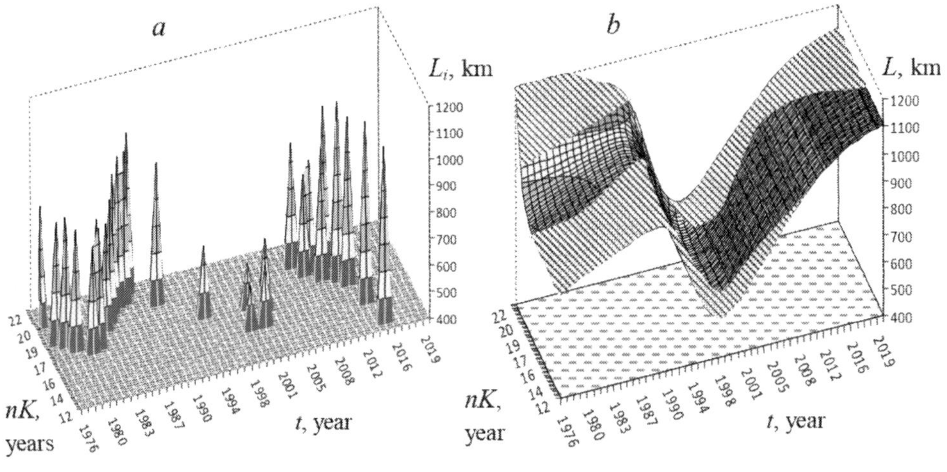

Fig. 2. The distribution of the length of the drilled wells by the time of mining and the number of machines used, reflected in the annual reports (a) and the regression model (b).

Due to the fact that during the time the number of working drilling rigs changed gradually, model changes in the annual length of the drilled wells are shown in the limited intervals of changes in the *nK* argument corresponding to the actual availability of drilling rigs at the enterprise. The model determination factor 0.90 respectively, enters the allowable range, and the deviations of the regressions from the *Li* discrete values have the properties of normal distribution and homoscedasticity. Thus, the presented model reflects the distribution of the annual length of the drilled wells at the MPP, adequately to the error of discrete values in the initial data, that is, they are statistically reliable, and in the indicated confidence interval they express regularities in its changes.

4 Results and discussion

In the model, the time parameters of changes in the annual length of the drilled wells have been determined identically, due to the long-term development of natural iron ore resources and the depressive influence of transitional economic processes. The annual length of the drilled wells, taking into account the initial design decisions for the development of natural resources, has been increasing exponentially with a relaxation interval τ = 27 years. The time-limited depression in drilling operations, starting from their relatively stable state in 1990 to the date of minimum t_o =1998, occurs along the left slope of the right-sided asymmetric function with a steepness index of τ_1 =3.5 years. The subsequent increase in drilling operations then follows the right slope of this function with a steepness index τ_2=11.3 years. According to the transitional parameter ε =4.1 years it has been determined that starting from 2005, the slope of the rising rate passes with a difference of no more than 10% in the relaxation interval of the above function.

The length of wells drilled per year increases with the number of working drilling rigs nK in terms of the exponential function. Its exponent at the Kachkanar MPP is determined by the indexes a=-0.03, b=23.58, which are in linear dependence of the time. In 1988-1991, when the number of drilling rigs varied from 18 to 22, the exponent was 23.1, and in 2015, with 13 working rigs, this index has a reduced value of 22.4.

The decrease in the index of working rigs nK in the iron ore quarry recently, despite the growth of the K factor, occurs due to a significant reduction and qualitative change in the fleet of the existing drilling rigs. In this case, the length of wells drilled per year increases, that is, drilling productivity increases. The given regularities in the distribution of the annual length of the drilled wells are explained by the structural and qualitative changes in the fleet of drilling rigs and the improvement in the organization of operation thereof. Thus, as a result of the modernization of the drilling rigs SBSh-250MNA-32 [15-16], carried out by the "Rudgormash", OJSC manufacturing plant for the last 15 years, drilling productivity has increased by 15-20%. In addition, new high-performance SBSH 250/270-60 KP frame-platform drilling rigs have recently been used at enterprises. The organization of their repair improved and reliability of operation increased [17]. After 2000, an investment program for the replacement of a drilling rigs fleet started to be implemented at the MPP. Along with improved domestic samples, foreign are rigs are purchased, for example, American D75KS and SKS, which have higher drilling productivity [18]. As a result of modernization, modern domestic and imported rigs of various improved types are used at Kachkanar MPP [19-22].

Currently, the period of depression in the drilling operations is coming to an end. The effect of the depressive function in these models becomes very small. The annual length of the drilled wells is largely determined by the volume of mining operations, balanced by natural resources and the needs of the metallurgical complex. Obviously, in the long term, the annual drilling length at the Kachkanar MPP is approximately L = 1.200 km (B = 1.199 km).

5 Conclusions

The research revealed the quantitative pattern of formation and development of drilling operations in the iron ore quarries has been revealed using the example of a large mining enterprise (Kachkanar MPP) in the transition period of the modern economy. The reliability of the created model describing the change in the annual length of the drilled wells is marked by a high index of their determination of 0.9. The set of characteristic parameters of the complex development of drilling operations determined by the model shows the high cognitive efficiency of the method of nonlinear functional factor regression used in the studies. The noted regularity allowed to give an interpretation of the dynamics of drilling operations in the transition period and to evaluate their development in the short term.

The research has been carried out within the framework of the Government Task 007-00293-18-00. Topic No.0405-2018-0015, and also with additional attraction of economic and contractual means and financing under the competitive project No.0405-2018-0001. Project No.18-5-5-10.

References

1. V.L. Iakovlev, A.I. Kuklin, A.I. Pavlov, Mining information-analytical bulletin, **2**, 105. (2004)
2. L. F. Riehs, C. L. Henrique, R. D. Pacheco, L. F. Sartori Piran, Resources Policy, **52,** 427 (2017)
3. M. Rosienkiewicz, E. Chlebus, J. Detyna, Applied Mathematical Modelling, **49**, 87 (2017)
4. N. Bilim, A. Çelik, B. Kekeç, Journal of African Earth Sciences, **134**, 564 (2017)
5. W.A. Hustrulid, *AIME, Ch. 18 Drill Evaluation by R.H. Heinen*, (1979)
6. M.A Sayed, Bulletin of the faculty of Engineering, Assiut University, **29**, 255 (2001)
7. M. M. Elbeblawi, M. A. Sayed, M. T. Mohamed, E. I. Abdelrasoul, Journal of engineering sciences, **42**, 557 (2014)
8. T. B. Afeni, Mining Science and Technology, **19**, 73 (2009)
9. K. Žarko, Č. Miodrag, T. Dražena. Rudarski fakultet Prijedor, **9**, 35 (2013)
10. V. Krúpa, M. Kruľáková, E. Lazarová, M. Labaš, K. Feriančiková, L. Ivaničová, Measurement, **117,** 165 (2018)
11. Technical and economic indices of mining enterprises over 1990-2012, 361 (2013)
12. V.A. Antonov, Mining information-analytical bulletin, **10**, 17 (2017)
13. V.A. Antonov, Mining information-analytical bulletin, **23**, 90 (2017)
14. V.A. Antonov, Izvestiia of SFedU. Technical sciences, **7 (192)**, 133 (2017)
15. A.V. Tokarenko, E.V. Gudenkov, Mining magazine, **3**, 76 (2013)
16. V.V. Zenin, Ugol, **10 (978)**, 50 (2007)
17. V.D. Butkin, A.V. Gilev, Selection and rational operation of drilling tools and rigs in quarries, 236 (2010).
18. Iu.I. Lel, Mining information-analytical bulletin, 6, 199 (2011)
19. S.P. Reshetniak, Mining industry, **5(87)**, 18 (2009)
20. Iu. S, Alfereve, D.V. Kiselev, Iu.A. Klimashevskii, E.V. Dokukin, Mining magazine, **5**, 23 (2017)
21. R.I. Sukhov, A.S. Regotunov, Kazakhstan Mining magazine, **10**, 10 (2013)
22. R.I. Sukhov, V.S. Bolkisev, A.S. Regotunov, Mining equipment and electromechanics, **5**, 46 (2014)

MEGALOPOLISES GEOTECHNICS

Vladimir Makarov[1*] and *Valery* Makishin[1]

[1]Far Eastern Federal University, Vladivostok, Russia

Annotation. Necessity of formation of the uniform concept of maintenance of development of the megalopolises, combining building of aboveground and underground complexes is shown. Interaction of elements of geotechnical system «a megalopolis - an underground construction» with a view of maintenance of ability to live of underground constructions of megalopolises is considered. Classification of conditions of development of megalopolises by stage of use of the underground space, added with criteria of reliability, safety and an economic optimum is developed. Conditions of maintenance of ecological compatibility and infrastructural appeal of megalopolises, as centres of cultural and commercial development are considered. Necessity of formation of the general scientific discipline «Megalopolises Geotechnics», covering a complex of methods of aboveground and underground building of megalopolises taking into account reliability and safety of functioning, ecological compatibility, infrastructural appeal, and also an economic optimality answering to criterion is formulated.

1. Introduction

Development of megalopolises in modern conditions demands a comprehensive approach considering both the characteristics aboveground, and underground building, bound among themselves principles of reliable and safe functioning at the optimum economic decision Such approach in case of a megalopolis locating in island terrain where the aboveground space is circumscribed by the connatural factor - a coastal line is especially important. If thus the terrain has a mountainous relief without use of underground space a big city it is simple cannot be developed effectively [1].

Thus, classification of conditions of building by principles of a sustainable development [2] becomes the important factor of regulation of use of underground space of megalopolises. Among the most important requirements it is necessary to note peak efficiency of building of a construction (its reliability, safety, ecological

---

* Corresponding author: vlmvv@mail.ru

compatibility, and also infrastructural appeal). Other principle it is maximum an effective utilisation of underground space (compactness of a locating of underground constructions in relation to each other and terrestrial surface at observance of the first principle). Both first conditions should answer also to a principle of the minimum cost of building.

Definition of the specified principles shows that the existing contradiction between requirements of the minimum cost and peak efficiency leads to a problem of optimization, which should be resolved in each specific case with application of classification and mathematical approaches.

2. Interaction of aboveground and underground building complexes

To one of the major aspects of operation of underground constructions of different function is maintenance of their interaction with objects a terrestrial surface. For maintenance of it process carrying out of auxiliary developments of different function - transport (escalators, ladders, foot tunnels, motor transportation inclined developments for maintenance of ability to live of objects of welfare, industrial appointment, a warehouse economy) is necessary; power supply (cable tunnels and chinks); ventilations; water drains and water removals.

At designing and building of city underground constructions, it is necessary to consider that the underground construction is operated in the conditions of its interaction with a city infrastructure on the terrestrial surface, existing underground communications of different function both mountain-geological and geomechanical conditions of a soil and rock massif. Underground objects in a combination to objects on a terrestrial surface provide building of a complex infrastructure that in modern social and economic conditions is a necessary condition of functionality of a city infrastructure as a whole. Thus, as a rule, optimum conditions of formation of underground transport and foot paths with the minimum height of movings on a vertical [3, 4] are provided.

By researches, it is positioned that limiting extent of transport-ventilation and other auxiliary openings of various section confines their admissible component within 15-20 % from volume of an underground construction and size of aerodynamic resistance of a ventilating network of an underground construction.

The lobe of these openings can be increased to 30-37 % in case of continuous development on the area and in depth of a soil and rock massif of geotechnogenic underground space [5]. It follows from this that economically effective volume of an underground construction assumes its continuous development for the purpose of optimisation of technological and logistical communications between a megalopolis and underground constructions.

At the initial stage of formation of uniform geotechnological system each separate underground construction can depend in full or in part on maintenance from a terrestrial surface. It, taking into account an admissible lobe of developments of auxiliary appointment, assumes building of shallow underground constructions with rather small volumes.

At a substantiation of a configuration of underground object it is necessary to consider that for high-grade functioning of an underground construction its maintenance with fresh air according to the positioned norms is necessary. Feature of underground constructions is necessity of constant illumination of developments and air conditioning in them. The large underground object is the complex power consumer demanding presence of powerful systems of constant and emergency power supply. Appointment, the dimensions of an underground construction, hydro-geological conditions of a soil and rock massif and a profuseness of deposits demand building of powerful system of the water drain and water removal. In this connection one of priority problems of Megalopolises Geotechnics is the establishment of laws and communications in geotechnical and geotechnological systems, making system «an underground construction - a city infrastructure» [6].

Gradual development of an underground infrastructure assumes dilating of developments on the area and in depth of the given mountain tap taking into account influence of objects of a terrestrial surface and geomechanical characteristics of a soil and rock massif.

In process of development of this system there is an affiliation by a series of the located underground constructions of different function in the uniform system, allowing to tolerate from a terrestrial surface to underground developments not only objects of welfare, technological, industrial and power appointment, but also to incur functions of maintenance of objects of a city infrastructure.

Development of underground geotechnological logistical systems of underground space makes essential impact on formation of an infrastructure of a megalopolis. Conduction to underground developments of objects of material and power supply allows to make rational layout decisions of placement of buildings and the constructions most of different function in the most necessary places for a city, including in the conditions of the extremely constrained building, to reconstruct a transport network taking into account constantly increasing transport streams, to create and develop underground transport, both the railways, and undergrounds, to optimise parametres of development of city engineering networks, to provide decrease in consumption of fuel and energy resources and to promote decrease in negative influence on city ecology [3, 4, 6].

Modern explorers notice that on the one hand, the underground space of megalopolises and cities-companions by their consecutive development on the area and in depth of hills, affiliation of close located underground complexes in uniform supersystem practically has no restrictions.

But, as well as any complex system, it as becomes to vulnerable and subject system failures [5]. In this connection carrying out of the researches defining at a modern level of development technicians and technologies of optimum design data of a uniform geotechnological complex taking into account their area, number of storeys, appointments, completeness of use, influence on formation, maintenance and development of objects of a city infrastructure of a megalopolis for the purpose of rational geographical demarcation of underground space and specialisation of underground constructions is necessary.

The major part of strategy of transferring from traditional aboveground building to underground is geographical demarcation of underground space, applicable for building. For this purpose it is necessary in reacting documents on working out of the general layout of a city (on long-term prospect) to lead a condition of conducting works on a complex assessment of development of underground space along with a traditional complex assessment of city lands and terrains adjoining to it.

The complex assessment of development of underground space is manufactured on the basis of the detailed analysis of underground space of a city. Characteristic regions precipitate out, considering not only functional features of the yielded field of a city, but also geological, hydro-geological and other conditions in which building and operation of the future underground constructions will be carried out.

3. Completeness of use of underground space of the Vladivostok City

The research made within the framework of complex development of underground space of the Vladivostok City, consists in working out of procedures of definition of rational parametres of underground constructions taking into account their interference and means of maintenance of the working space. The procedure should solve the questions bound with three last from above-stated directions of procedure of structurization are space-planning, constructive and technological decisions.

Definition of the minimum dimensions of pillars between openings it is manufactured taking into account functional features of rocks of constructions erected in the rock massif, i.e. completeness of use of underground space should not depend on that in the congenial, normal or preferable regions building of those or other constructions [3] is supposed. Finally completeness of use of underground space will be defined by optimum distances between openings (safety pillars) which are in turn defined by stress concentrations in pillars and strength properties of rocks.

Completeness of use of underground space on conditions of a stressed state of rocks can be divided into three conditional categories (σ_c – ultimate strength of containing rocks on axial compression):

1. Safe – at which dimensions safety it is whole get out of a condition that stress concentration in them will not exceed $0.5\sigma_c$;
2. Centre stage of danger – at which strains in pillars are in limens $0.5\text{-}0.8\sigma_c$;
3. Dangerous conditions – at which strains are in limens $0.8\text{-}0.9\sigma_c$.

Definitions of the minimum dimensions of safety pillars it is whole, providing the set safety factor can be carried out with use of a design procedure of strains in a rock massif around interacted openings [7]. In this case the concept of quotient of completeness of use of underground space is introduced:

$$K_{\Pi\Pi} = \frac{V_{\Pi C}}{V_0},$$

where $V_{\Pi C}$ – volume of an underground construction,

V_0 – total amount applicable for building of underground space.

The check point augmentation at invariable values V_0 has the big economic value as descends at the expense of involving in economic circulation of additional volumes of underground space.

4. Classifications of conditions of development of megalopolises by conditions of interaction of aboveground and underground parts of a city

The considered problems demand introduction of classification of conditions of development of megalopolises on stage of use of the underground space, answering to the above resulted principles. Three cores of a case of interaction of aboveground-underground development of a megalopolis precipitate out. First, it is conditions of local placement of the underground construction bound exclusively with any one object, placed on a terrestrial surface. Secondly, conditions independent of functioning of objects of aboveground building of "an Underground City». And thirdly, conditions «Integrated aboveground-underground» developments of the megalopolis representing a uniform complex. Each of the considered divisions has it's one the advantages and disadvantages. So, local building of underground constructions provides peak efficiency of operation of objects of aboveground and underground building at minimisation of extent of auxiliary openings. But thus use of underground space of a megalopolis not effectively, and frequently and simply is not considered. Also efficacy of building for a city as a whole is not considered.

The concept of an "Underground City» means possibility of independent coexistence of a aboveground part of a megalopolis and its underground part. Advantages of such variant of development is possibility of effective operation of an underground part of a megalopolis, relative simplicity of building and minimisation of costs on maintenance of communication with a surface. Disadvantages consist in necessity of carrying out of additional underground communications and auxiliary openings, difficulty of maintenance of a reliable communication of underground and aboveground objects.

The variant «Integrated aboveground-underground» megalopolis developments in many cases is preferable, allows to aggregate advantages of first two considered cases. However, it has such negative sides, as complication of technical decisions of building.

5. Discussion

The resulted classification of conditions of a sustainable development of megalopolises in aggregate with the specified principles developments and criteria of reliability and safety of functioning, ecological compatibility, infrastructural appeal, and also criterion of an economic optimality compound a basis of a scientific discipline «Megalopolises Geotechnics» which urgency of formation is confirmed by modern practice [8].

Main ideas have been discussed at the International Geotechnical Conference "Geotechnical Challenges in Megacities" (Moscow, 7-10 June 2010), where the many practical decisions had been done [9]. The article is widening the Conference recommendation and introducing the approach to the "Megalopolises Geotechnics" formation.

6. Conclusion

Problems of interaction are relevant for geotechnics of megacities. Interaction of foundations, effect of new buildings and constructions on underground structures along with an effect of new underground structures on existing buildings and networks – all that problems are already within the range of interests of geotechnical engineers and must attract even more attention in the practise. Systems approach to these problems dictates the necessity to both geofailures and geological risks assessment in urban planning. The latter is often related with construction on problematic soils and rocks.

Many megacities were formed on the island territories. The need to take in account the surface limit and mountain like form of relief causing difficulties for geotechnical engineers. Another feature of prolonged human activity in built-up areas is pollution of the environment, giving rise to geoecological problems of construction on contaminated soils. Preservation of hydrogeological situation in course of underground space development, water pumping and other man-caused effects lies within the same range of problems.

The complex of the practical geoengineering decisions can be achieved on the way of the "Megalopolises Geotechnics" formation.

The research is supported by a grant from the Ministry of Science and Higher Education of the Russian Federation No. RFMEFI58418X0034.

References

1. Makarov V., Anikeev V., Pushkarev I., Sadardinov I., & Mirobian A. Procedia Engineering, **165**, 497-503. (2016). DOI: 10.1016/j.proeng.2016.11.725
2. Porfiryev, B. N., AND Bobylev, S. N. Studies on Russian Economic Development, **29 (2)**, 116-123. (2018).
3. The Management on complex development of underground space of big cities V.A.Ilyichev, G.E.Golubev, A.V.Zamaraev, A.N.Skachko, O.I.Ignatov, V.G.Budanov, O.N.Korotkov. (2004). http://rykovodstvo.ru/exspl/31572/index.html (in Russian)
4. Korchak, A. V. Problems, directions and paths of problem solving of development of underground space of megalopolises / A. V. Korchak//the Mining information-analytical bulletin.. **5**. pp. 89-98. (2014) (in Russian)
5. Mining sciences. Development and conservation of bowels of the Earth / the Russian Academy of Sciences, АГН, the Russian Academy of Natural Sciences, МИА; under the editorship of K.N.Trubetsky. Publishing house AGN. 478 with. (Mining Sciences. Development and Conservation of Mineral Resources of the Earth/RAS, AMS, RANS, IEA; Under General Editorship of K.N. Trubetskoy.-M.: Academy of Mining Sciences, Publishing House. 478 p.). (1997) ISBN 5-7892-0008-7 (in Russian)
6. Makishin, V.N. Methodology of a substantiation of rational parametres of transport developments of underground constructions / V.N.Makishin, A.A.Fatkulin, B.F.Lesovsky. - Vladivostok: Publishing house DVGTU. 157 p. (2007) (in Russian)
7. Fotieva N.N., A.N.Kozlov Fix parallel openings in seismic countries. Nedra. 231 p. (1992). (in Russian)
8. Makarov V. Island Megalopolises: Tunnel Systems as a Critical Alternative in Solving Transport Problems. Engineering, **4 (1)**, 138-142. (2018). doi.org/10.1016/j.eng.2018.02.001
9. "Geotechnical challenges in megacities". Proceedings of the International geotechnical conference. V.1, M.: Edited by V.P. Petrukhin, V.M. Ulitsky, I.V. Kolybin, M.B. Lisyuk, M.L. Kholmyansky. 392 p. (2010)

Concerning the principles of development of mineral resources in the polar zone of Russia by small enterprises

Vladimir Gavrilov[1*], *Yevgeniy* Khoyutanov[1] and *Natalia* Batugina[1]

[1]Mining Institute of the North, Siberian Branch, Russian Academy of Sciences, Yakutsk, Russia

Abstract. Increasing the efficiency of developing deposits of solid minerals in the polar zone of the northeast of Russia is an objective necessity. Unlike traditional territories, the peculiarities of the region predetermine the use of new approaches and methods in studying the feasibility of developing various deposits, including the widely represented relatively small deposits. Compared to large ones, medium and small mining enterprises operating or created on their basis have fewer opportunities to protect their interests. Their design and actual technical and economic indicators are generally worse than those of more powerful economic entities. The decrease in the scale effect can be achieved in various ways. Reduction of risks in the implementation of investment projects is possible through more thorough study and subsequent additional training, as a rule, complex in structure and heterogeneous in quality deposits involved in operation. A set of basic and private principles of different levels and values has been formed, which should guide small and medium-sized enterprises in the planning, design and development of deposits in the polar zone of Russia. Its use in the preparation and decision-making allows at least partially to reduce the impact of complicating and costly production factors. **Keywords.** Development of deposits, principles, solid minerals, small enterprises, the polar region, geotechnology, decision-making.

1 INTRODUCTION

Long-term analysis of the theory and practice shows that the extreme combination of natural climatic, mining-geological, technological, logistical and environmental conditions in the polar region of Russia significantly complicates or makes unacceptable by a combination of criteria the main activities related to the effective extraction and processing of mineral wealth. Really, only a small part of the reserves of solid minerals (SM) is processed. Basically, these are deposits of high-value raw materials with short technological chains and low-tonnage output of finished products. The imperfection of many aspects of the subsoil use mechanism in the region leads to irrational approaches to the development of subsoil. A consequence is the full or partial non-fulfillment of projects

* Corresponding author: gvlugorsk@mail.ru

for the development of many types of SM widely migrating from one program document to another over years, the inability to achieve the planned indicators, the low demand for geo-resources, the lack of research to identify and explain the reasons for such a situation.

In the polar zone of Russia, various SM deposits are being developed or planned for development. They can be large (iron or apatite ores in Murmansk region, precious metal and polymetallic ores in Norilsk industrial hub, gold ore in Chukotka), developed by large enterprises over decades; and there may be middle and small deposits, for example, placers, which are worked out by small artels, including during only a few seasons.

The first of these, as a rule, have multiple opportunities to defend their interests. The latter have techno-economic indicators are generally much worse than more powerful business entities. The possibilities of applying modern types of equipment and technologies, methods of organization and management in accordance with modern trends due to a number of reasons, primarily financial, are more limited. Without special support measures, for example, using public-private partnerships [1, 2], their normal functioning under extreme polar conditions is greatly complicated. Limited in comparison with large enterprises, the opportunities for small and medium-sized enterprises require the use of a synthesis of general and private approaches to identifying reserves for effective development of subsoil in the region. In this case, universal approaches should be based on general knowledge and the patterns of their production [3, 4, etc.], and individual – apply taking into account the specific specificity of each individual deposit based not only on clarifying general knowledge, but also by obtaining new, resulting from the features of the fields and the supply chains that they form.

This combination allows, when implementing investment projects in the mineral and raw materials sector, to achieve a level of profitability acceptable for relatively normal operation of enterprises, taking into account the interests of investors, federal, regional and local government authorities. Two main things should be remembered. First: there is a stable, often unjustified, view of the uniqueness and richness of the region's mineral and raw materials base (MRB) and constantly announced hope for high and stable demand in the country and the world for its SM. Second: the current state and prospects for the development of the territory are associated primarily with the raw material sector and are actually non-alternative. For the region, more economical types of production cannot be offered by science and practice yet. The efficiency of a job in the mining sector is 3-5 or more times higher than in other sectors.

Adaptation of common solutions for real mining and geological conditions, properties of mined minerals and formed chains of creation of value added allows one to more accurately consider and evaluate the feasibility of involving profitable SM deposits in the region. The foregoing allows one to state that the solution of interrelated and complementary tasks on the analysis of existing principles for the development of mineral resources and on the synthesis of a complex of those that can be rationally used to increase the efficiency of small-scale mining enterprises is urgent. There is a need to formulate the necessary and sufficient conditions for real involvement of regional MRB in effective and environmentally safe operation.

2 THEORY

Reduction of risks in the implementation of investment projects is possible through more thorough study and subsequent additional preparation for operation of deposits, as a rule, complex in structure and heterogeneous in quality. At the same time, following a unified approach to business taxation, regardless of where it is based in Moscow or the Arctic Circle, is an erroneous understanding of the principle of equal competitiveness or "the conscious displacement of the remnants of economic presence from the pioneer development regions" [5]. It is necessary to revise the approaches used to select geotechnologies and related processes. Priority should be given to those of them that allow for the most complete consideration of local climatic, mining and geological and cryological features of deposits that are able to react quickly to changes in the conditions for the development of their individual sites.

The scientific literature covers a wide range of issues related to the Russian polar zone [6, 7] and the world [8-10], the ecological efficiency of technologies for the development of SM deposits [11], the development of individual industries in the regions [1, 12], the transport and logistics infrastructure [13], the combination of geotechnologies for a sustainable development of subsoil [14], geotechnological preparation of deposits with small reserves [15], the association of small and medium-sized gold mining enterprises to increase the efficiency of their operation [16], etc. However, the copying of research results and best practices of the mining sector to the conditions for the development of specific deposits or their groups in the polar regions of Russia is in most cases difficult or has a short-term positive impact. Standard approaches, as a rule, do not provide additional competitive opportunities, because have already been widely used in other regions with more comfortable conditions for solving similar problems.

Despite the attempts made, the whole complex of constantly replenishing specific features of the development of remote, inaccessible deposits in the polar zone has not been fully taken into account. It is necessary to develop and implement a set of measures aimed, if not to complete overcoming of objective difficulties, then at least to partially neutralize them.

3 RESULTS AND DISCUSSION

The work on the justification of effective geotechnologies for mining deposits in the conditions of cryolithozone based on the development of new principles for the rational development and conservation of explored reserves and predictive resources has been carried out at the Institute of Soil Science of the SB RAS for many years. Much attention is paid to improving the methods for assessing the qualitative and quantitative structure of reserves, the extent and nature of the impact of variability of the geological environment and external factors on the efficiency of development of the extraction and processing processes. Serious influence on the formation of ideas in this case was provided by Batugin [3, 17].

First, the authors consider a small part of the known and new, general and particular principles applicable to the innovative development of the deposits of the polar zone, synthesized from the conceptual provisions developed by the world and Russian science. The main attention will be concentrated on their use by enterprises of medium and low power. Such an accent is due to the fact that in these organizations, in comparison with large business, have limited opportunities for self-adjustment of principles in the preparation of local strategies and development programs because of organizational and financial reasons.

The general principle of nature management: maximum possible advance study of the natural environment with a high level of accounting for its features and patterns in strategic and tactical projects and actions.

Ecological and economic principle of nature management: maximum full utilization of extracted minerals. According to statistics, up to 98% of all extracted raw materials are not used, forming waste. This has an extremely negative impact on the economic efficiency of nature management, growing costs of storage and disposal of waste, reclamation of disturbed land.

Principles of economic policy and economy of natural resources: it is important to avoid the widespread view of the homogeneity of any resource, the reserve of which is strictly determined by its volume. The analysis should proceed from a wide range of qualitative characteristics of mineral resources, which are determined primarily by the geological features of the deposits, the impact of which has not been sufficiently studied. In turn, economic and technological conditions determine which properties of the resource are most attractive for development and subsequent use [18].

Principle of systemology: dynamic stability and development of mining systems are possible only in an environment that has a certain combination of favorable and unfavorable parameters of subsystems "georesource", "technological transformation of georesource into products and potential resources", "result, product, effects of different origin" 19].

Principles of information stage of development of natural resources. The information stage of development plays a paramount role in complex studies on the problems of the development of mineral resources, unique objects and territories [20]. In its course, a deliberate accumulation of data on a set of conditions that ensure the inclusion of undeveloped subsoil blocks in operation is conducted. The principles and concepts formulated within the framework of this group are as follows: the research potential aimed at informational development of a new territory should be more proportional to their resource potential, rather than the concentration of industrial production [20]; the whole technological chain of development of new territories and subsoil plots should be linked by modern means of computerization and automation; the development of territories and subsoil, starting from the information stage, should be carried out taking into account the interdisciplinary integration of scientific forces and cooperation between industries in accordance with targeted scientific and technical programs. Now, adding a few less general but important principles of informational development of the territory and subsoil [21]: principle of analogy, which means that the methodology for studying a new territory or subsoil area is selected taking into account the studied analogous objects (in the absence of such objects or the lack of a degree of their analogy, without changes); principle of successive approximation, realized by the breakdown of the process of information development (including prospecting and evaluation of subsoil) at the stage, in each of which the study is conducted with increasing detail; principle of selective detailing, which consists in an additional study of individual sections; principle of necessary and sufficient accuracy, detail and complexity of studying the territory and subsoil as a whole. The categories of "necessary and sufficient" accuracy have practically no valid solutions to date, although research is being done in this direction. For example, there is a way to assess the level of knowledge of coal quality indicators based on the coefficient of spatial information variability and information stability [22]. With the application of methodological recommendations, a quantitative assessment of the reliability of coal deposits is carried out [23].

Principles of the stage of industrial development of natural resources: continuous improvement of the organization, requiring constant improvement of the existing management structure, its adaptation to the changing conditions of production and social relations; transition from the principle of "mineral wealth for a known technology" to "technology for extracted mineral wealth, creation of integrated quality management systems for mineral raw materials, from the field to the end user.

Principles of preparing a geo-resource for effective development, use and conservation: the need to apply scientific knowledge already at the early stages of prospecting and exploration of deposits in order for their subsequent development to be cost-saving; natural heterogeneity of the most important indicators of productive bodies, their zonality, power, properties, discontinuity cannot be established with necessary and sufficient accuracy at the exploration stage, their identification and correct recording in mining and processing systems of mineral raw materials are possible only on the basis of improving and implementing a representative advanced and operational testing; the need for proper use and interpretation of the results of geological testing of stocks and laboratory analysis of samples for minerals and elements to increase the reliability (for example, based on [17]).

Principles of improving and developing new effective geotechnologies, mining equipment, technological processes and operations. 1. In the mineral and raw materials sector, when choosing the directions for improving equipment and technologies, it is necessary to take into account: the growing rates of depletion of the deposits of most geotechnologically-utilized types of SM; uncertainty of their substitution with alternative scenarios; inadequate attention to fundamental research projects and forecasted areas of development of mining equipment and technologies. 2. A rational combination of two or more technologies can create a synergistic effect leading to a significant improvement in the general schemes of field development. 3. Special conditions developed in the systematic lag in the modernization and adaptation of mining equipment and geotechnologies. This is due to the continuous increase in the depth of development in the fields with associated other, also deteriorating, mining and geological conditions; forced transition to mining of deposits of another geological type or in new, as a rule, worse conditions. This principle has not been sufficiently investigated, which is manifested in mistakes in the justification of the strategies for scientific and technical policy of nature management.

Clarification of the above principles, their addition and subsequent ranking by importance for each of the development objects is planned to be performed using various foresight methods [24] (Delphi [25], scenario building, SWOT analysis, road maps), including taking into account possible options for the future development of Yakutia.

4 CONCLUSIONS

The reduction of risks in the implementation of investment projects for the development of fields in hard-to-reach polar regions by medium and small mining enterprises is possible with the use of the considered set of basic and particular principles.

The adoption of rational solutions in the planning, design and development of complex in terms of structure and heterogeneity in the quality of deposits should be carried out on the basis of the most accurate and reliable knowledge of the basic qualitative and quantitative characteristics of mineral raw materials in the subsoil and the technological streams formed during extraction.

The use of research results by stakeholders is important for resource-saving, economically acceptable and environmentally safe extraction and primary processing of mineral raw materials, which is competitive in the market, and allows increasing reliability and validity of solutions for the development of enterprises based on the territory of the region.

References

1. I. P. Glazyrina, S. M. Lavlinskii, I. S. Kalgina, Geography and Natural Resources, **4.** (2014).

2. Status and prospects of development of public-private partnerships in the context of the complex development of mineral resources; Ed. A. E. Kontorovich, S. M. Nikitenko, E. V. Goosen (Kemerovo, Sib. Izdat. Gruppa, 2015).

3. S. A. Batugin, V. L. Yakovlev, The patterns of development of mining (Yakutsk, YSC SB RAS, 1992).

4. G. V. Sekisov, Vestnik ZabGU, **7.** (2016).

5. V. P. Orlov, Problems of Subsoil Use (2000-2006) (Moscow, Geoinformmark, 2007).

6. K. N. Trubetskoi, Yu. P. Galchenko, G. V. Kalabin, Engineering Physics, **4**. (2016).

7. E. I. Zakharov, Izvestiya TulGU. Sciences of Earth, **2.** (2012).

8. G. Duhaime, N. Bernard, P. Fréchette, M-A. Maillé, A. Morin, A. Caron, The Mining Industry and the Social Stakes of Development in the Arctic (Canada Research Chair in Aboriginal Comparative Condition, 2003).

9. Circumpolar Information Guide on Mining for Indigenous Peoples and Northern Communities URL: http://www.sdwg.org/wp-content/uploads/2016/04/Circumpolar_Mining_Guide_Draft-Apr_2011.pdf (Date of access: 14.06.2018).

10. D. Avango, L. Hacquebord, U. Wråkberg, Journal of Historical Geography, **44**, (2014).

11. G. V. Sekisov, N. I. Grekhnev, V. G. Romanov, N. V. Zykov, Mining Journal, **3.** (2007).

12. A. I. Tonkikh, Mining informational and analytical bulletin, **11.** (2009).

13. O. A. Tatarinova, Mining informational and analytical bulletin, **6.** (2013).

14. D. R. Kaplunov, V. A. Yukov, Mining Journal, **11.** (2015).

15. A. G. Mikhailov, Mining informational and analytical bulletin, **2.** (2015).

16. I. A. Kradenykh, Mining informational and analytical bulletin, **spec. iss. 21.** (2016).

17. S. A. Batugin, E. D. Chernyi, Theoretical Basis of Sampling and Reserves Estimation (Novosibirsk, Nauka. Sib. predpriiatie RAN, 1998).

18. A. Endres, I. Kvener, Economics of Natural Resources (2nd ed., Saint-Petersburg, Piter, 2004).

19. S. A. Batugin, S. M. Tkach, N. S. Batugina, V. L. Gavrilov, Mining informational and analytical bulletin, **9.** (2012).

20. K. P. Kosmachev, V. P. Mosunov, Yu. S. Nikulnikov, A. A. Sysoev, Methodological problems of specific Sciences, **55.** (1984).

21. L. I. Chetverikov, Theoretical Basis of Mineral Resources Exploration (Moscow, Nedra, 1984).

22. N. N. Grib, P. Yu. Kuznetsov, Coal (Ugol'), **9,** 186 (2016).

23. T. Rogova, S. Shaklein, Reliability of the reserves of coal deposits. Quantitative assessment and monitoring (Saarbrucken, Lambert Academic Pablishing, 2011).

24. M. Conway, An Overview of Foresight Methodologies URL: http://www.forschungsnetzwerk.at/downloadpub/An-Overview-of-Foresight-Methodologies1.pdf (Date of access: 19.06.2018).

25. J. Caron, S. Durand, H. Asselin, Journal of Cleaner Production, **119.** (2016)

Internal forces analysis of gas permeable shelter with emphasis on short-delay blasting

Aleksandr Leshchinskiy[1], *Evgeniy* Shevkun[1], *Aleksandr* Lovtsov[2] and *Olga* Kostiunina[2*]

[1]Pacific National University, Transport and Power Engineering Faculty, Khabarovsk, Russia
[2] Pacific National University, Engineering Faculty, Khabarovsk, Russia

Abstract. A mathematical model of transformable gas permeable blasting shelter made from worn out tires of heavy trucks bound together with chains, ropes or cables is proposed. Large displacements and unilateral constraints are specific features of this model. An algorithm was developed to calculate the displacements of tires and internal forces in connecting elements of blasting shelter subjected to dynamic loading. The results of the numerical simulation of simple shelter are presented for assessment of the adequacy of the mathematical model. This algorithm can be used to determine the parameters of shelter such as mass, number of tires, diameter of connecting elements which ensure its safe operation under any sequence of explosions.

1 Introduction

Blasting is a fast and efficient way of rock fragmentation. It is commonly used not only in the mining industry but also in civil engineering, construction and reconstruction of roads and railway lines. Position of the Far East region provides favorable conditions for domestic and international transportation. There is a direct railway access to several major border-crossing terminals. At the same time the density of railway network remains quite low in the regions of Siberia and the Far East. It curbs economic and industry growth in certain regions [1]. Prospects for the development of transport and communication conditions in the Far East are considered by Antonov [2].

Rock fragmentation by blasting is accompanied by flyrock which is a potential source of numerous hazards to people, surrounding objects and environment. Flyrock accidents in surface mining are analysed by Kecojevic and Radomsky [3]. A new mass explosions technology for destruction of semi-rocky and rocky soils preventing flyrock was developed at Pacific National University [4, 5, 6]. It is based on the use of blast holes with air-deck in overdrill along with transformable gas permeable blasting shelter made from worn out tires of heavy trucks bound together with chains, ropes or cables. Experiments carried out on the sites of federal highway #58 "Amur" proved predictability, effectiveness and estimated safety of the developed technology. An example of blasting shelter for a mass explosion conducted on 12.04.07 is shown in Figure 1, *a*. Explosions were made row by row (rows 1-

* Corresponding author: 006934@pnu.edu.ru

3) with 0.02 seconds (s) delay (Fig. 1, *b*). Two additional rows 4 and 5 were added to ensure safety on the village side.

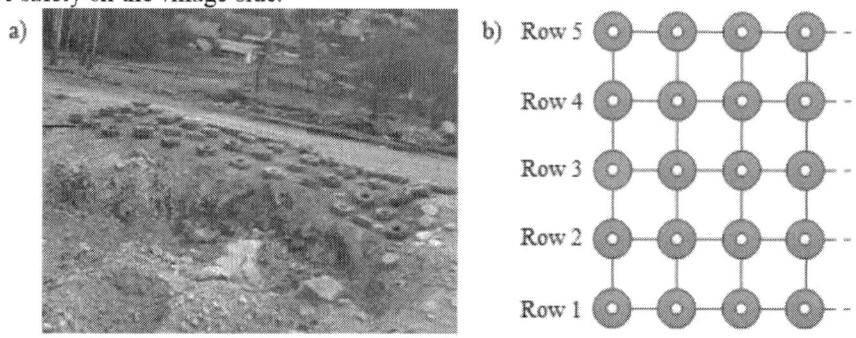

Fig. 1. (a) View of blasting shelter for mass explosion on 12.04.07; (b) Fragment of the model

Currently, mining industry tends to point blasting instead of multiple row blasting. It was caused by two factors: technological development in blasting (shift from detonating chord to non-electric detonation system etc.) and new requirements implemented to reduce seismic impact on nearby settlements. The environment friendly method of rock fragmentation by blasting is proposed in the latest researches. It combines the use of blast holes of special design [4, 7] with gas permeable blasting shelter [8] and point blasting with long-period delays [9-13]. However, the existing computational techniques do not allow to estimate the behavior of the shelter under various sequences of explosions.

An algorithm was developed for the analysis of the transformable gas permeable blasting shelter [14]. Numerical experiments showed that maximum flight height of the tires depends on delay intervals and detonation sequence of the blast holes. Employed mathematical model was verified by full-scale experiment carried out on 12.04.07 on the site of federal highway #58 "Amur" near Teploozersk village (Fig. 1, *a*). In present research the mathematical model is modified by adding frictional forces acting on the tires which are in contact with the earth's surface.

2 Model and methods

The dynamic spatial model of blasting shelter is represented by concentrated masses connected using elastic weightless incompressible elements. The problem solving involves the combination of two types of nonlinearity: geometrical nonlinearity associated with large displacements and unilateral contact which emanates from the fact that contact region changes due to deformation of the elements. Contact region is defined by a set of "switched on" elements: elastic elements under tensile stresses and supports with compressive interaction forces between surface and tires (Fig. 2).

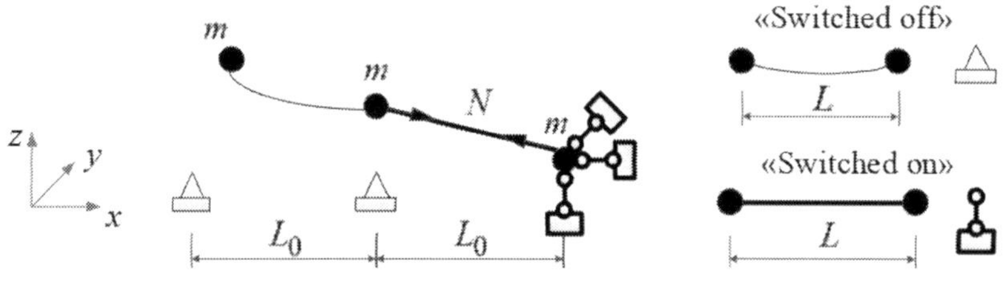

Fig. 2. "Switched on" elements and supports

Time discretization techniques are used for dynamic contact problems and the corresponding contact problem is solved at each time step using the appropriate numerical method [15-17]. There are many papers on nonlinear dynamics of geometrically nonlinear systems. The conceptual bases are given by Crisfield and Oller [18-20].

The shelter is exposed to short-term loads which are idealized by instantaneous impulses. Thus, the problem leads to investigating the behavior of the system under the action of its own weight and instantaneous impulses, applied at different times to different masses. We denote this load as $\mathbf{F}(t,x,y)$. Equation of motion can be written as

$$\begin{aligned} &\mathbf{M}\ddot{\mathbf{z}}(t)+\mathbf{R}(t)\mathbf{z}(t)=\mathbf{F}(t,x,y), \\ &\mathbf{z}(0)=0, \\ &\dot{\mathbf{z}}(0)=\dot{\mathbf{z}}_0, \end{aligned} \tag{1}$$

where $\mathbf{z}(t)$, $\dot{\mathbf{z}}(t)$, $\ddot{\mathbf{z}}(t)$ are nodal displacements, velocities and accelerations, $\mathbf{M}$ – diagonal mass matrix, $\mathbf{R}(t)$ – dynamic stiffness matrix at the moment t.

Only nonnegative vertical displacements are considered due to unilateral contact between surface and tires

$$\mathbf{w}(t)\geq 0, \tag{2}$$

$$\mathbf{z}(t)=\begin{Bmatrix}\mathbf{u}(t)\\ \mathbf{v}(t)\\ \mathbf{w}(t)\end{Bmatrix},$$

where $\mathbf{u}(t)$, $\mathbf{v}(t)$ are horizontal displacements of masses, $\mathbf{w}(t)$ – vertical displacements of masses (Fig. 3, *a*).

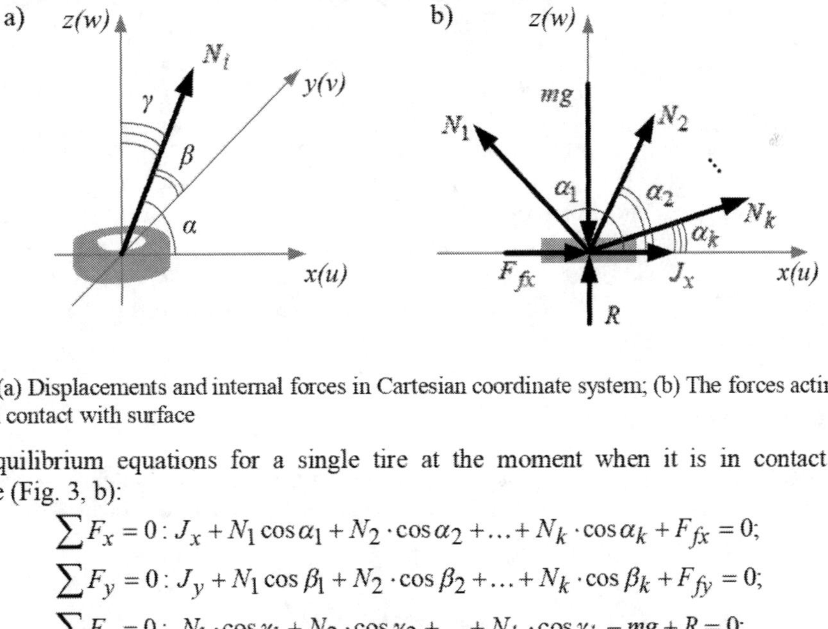

Fig. 3. (a) Displacements and internal forces in Cartesian coordinate system; (b) The forces acting on a tire in contact with surface

The equilibrium equations for a single tire at the moment when it is in contact with surface (Fig. 3, b):

$$\sum F_x=0:\ J_x+N_1\cos\alpha_1+N_2\cdot\cos\alpha_2+\ldots+N_k\cdot\cos\alpha_k+F_{fx}=0;$$

$$\sum F_y=0:\ J_y+N_1\cos\beta_1+N_2\cdot\cos\beta_2+\ldots+N_k\cdot\cos\beta_k+F_{fy}=0;$$

$$\sum F_z=0:\ N_1\cdot\cos\gamma_1+N_2\cdot\cos\gamma_2+\ldots+N_k\cdot\cos\gamma_k-mg+R=0;$$

where N_i – internal force in the elastic element, J_x, J_y – inertia forces, R –constraint reaction force perpendicular to the surface, F_{fx}, F_{fy} – the x and y components of actual friction force.

If the tire is in contact with surface the actual friction force F_f does not exceed the force of kinetic friction:

$$F_f \leq \mu \cdot R, \tag{3}$$

$$F_f = \sqrt{F_{fx}^2 + F_{fy}^2},$$

where μ is the coefficient of friction.

When the tire is detached from the surface: $w(t) > 0$ and $F_f = 0$. If $F_f < \mu \cdot R$ there is no horizontal movement of the tire and constraints are imposed to it in xy plane. Otherwise $F_f = \mu \cdot R$ and sliding occurs between the tire and surface.

When the system experiences deformation a set of "switched off" elastic elements is defined by current length of each element L (Fig. 2). If current length of the element due to deformation is less than initial length L_0, then this element is not taken into consideration because it doesn't affect stiffness of the system:

$$L < L_0 \tag{4}$$

Various time integration strategies can be used to solve equation (1) with the Newmark methods being very popular. Dynamic equilibrium is considered at the end of the time step:

$$\mathbf{M}\ddot{\mathbf{z}}_{n+1} + \mathbf{R}_{n+1}\mathbf{z}_{n+1} = \mathbf{F}_{n+1}.$$

The solution of this equation can be obtained using a "predictor-corrector technique" [19]. Equation (5) provide an incremental "predictor step" $\Delta\mathbf{z}$:

$$\Delta\bar{\mathbf{F}} = \bar{\mathbf{R}}_n \Delta\mathbf{z} \tag{5}$$

where $\Delta\bar{\mathbf{F}} = \left(\mathbf{F}_{n+1} - \mathbf{F}_n\right) + \mathbf{M}\left(\frac{4}{\Delta t}\dot{\mathbf{z}}_n + 2\ddot{\mathbf{z}}_n\right)$, $\bar{\mathbf{R}}_n = \mathbf{R}_n + \frac{4}{\Delta t^2}\mathbf{M}$, $\mathbf{R}_n$ – static tangent stiffness matrix.

Having solved (5) for $\Delta\mathbf{z}$, the incremental displacements, velocities and accelerations at step $n+1$ are obtained. Then Newton-Raphson corrector is employed.

Nodal coordinates of the masses change as the system experiences deformation and current length of the elastic elements is calculated at each time step. Inequalities (2) – (4) are analyzed to define contact region when Newton-Raphson at step $n+1$ converges. If contact region at step $n+1$ differs from step n "predictor-corrector step" should be repeated. Otherwise we move to the next step.

3 Numerical simulation

Simple model of blasting shelter (Fig. 4) was taken for numerical simulation. The shelter consists of 5 tires weighing 280 kg each bound together with 6.2 mm cables. A single blast hole is located under the central tire m_C.

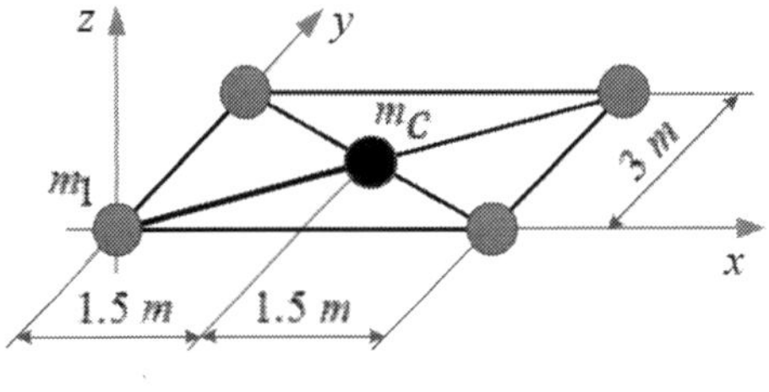

Fig. 4. Blasting shelter

Initial velocity V_0 of the central tire after explosive detonation set to 6 m/s. If the tire is not a part of the shelter its flight height equals $h_{max} = V_0^2 / 2g = 1.83\ m$. The coefficient of friction for rocky soil $\mu = 0.3 - 0.5$. Considering the most unfavorable conditions μ equals 0.5. Time integration step set to $2 \cdot 10^{-4}\ s$.

According to the simulation results the maximum flight height of the shelter is 0.708 m. It takes 0.25 s for the central tire to reach this altitude (Fig. 5, *a*). The maximum internal force in the diagonal elastic element is 17.7 kN (Fig. 5, *d*).

The numerical results depend on the time step Δt. The values of maximum vertical displacement of the central tire and the maximum internal force in the diagonal elastic element depending on integration step size are shown in Table 1. The analysis shows that the displacement converges with a decrease in the integration step size and tensile force requires more detailed consideration. The values of internal force in the diagonal elastic element for $t = 0.16\ s$ are also presented in Table 1. The percent error is defined by the expression $\delta = \frac{N(j) - N(r)}{N(r)} \cdot 100$, where $j \in 1 \ldots r$, $N(j)-$ internal force corresponding to integration step size in the row j of Table 1, $N(r)-$ tensile force corresponding to the minimum value of integration step size in the range of values considered in this study. In accordance with the calculation results the convergence of the tensile force can be observed in a certain range of values Δt. Further decrease of the integration step size leads to increase in the percent error.

Table 1. The dependence of numerical results on integration step size

j	Integration step size $\Delta t \cdot 10^{-4}$, s	Maximum vertical displacement of the central tire, m	Maximum internal force in the diagonal elastic element, kN	Internal force in the diagonal elastic element for t=0.16 s, kN	δ, %
1	5	0.705	17.668	9.265	24
2	2.5	0/707	17.687	8.284	11
3	2	0.708	17.704	7.461	0.1
4	1	0.708	17.704	7.456	0.06
5	0.625	0.708	17.701	7.521	0.9
6	0.5	-	-	7.452	-

The analysis of movement of the tires shows that the central tire pulls the corner tires of the blasting shelter when it goes up and the friction forces between the corner tires and the surface arise as a result. In figure 6 the functions of the actual friction force $F_f(t)$ and the force of kinetic friction $\mu \cdot R(t)$ acting on a corner tire are drawn. When these two lines are coincident the corner tires slide on a surface.

The system is symmetric so the absolute values of corner tires displacements are equal. The corner tire detaches from the surface for the first time at $t = 0.08\ s$. It hit the surface at $t = 0.13\ s$ and internal force in the diagonal elastic element at this moment is zero (the element is "switched off"). The corner tire rests on a surface until diagonal element "switches on" (Fig. 5, *b*, *c*). The central tire continues to move upward and it leads to a sudden change of contact region at $t = 0.15\ s$. Elastic elements located along the perimeter are "switched off" throughout observation period.

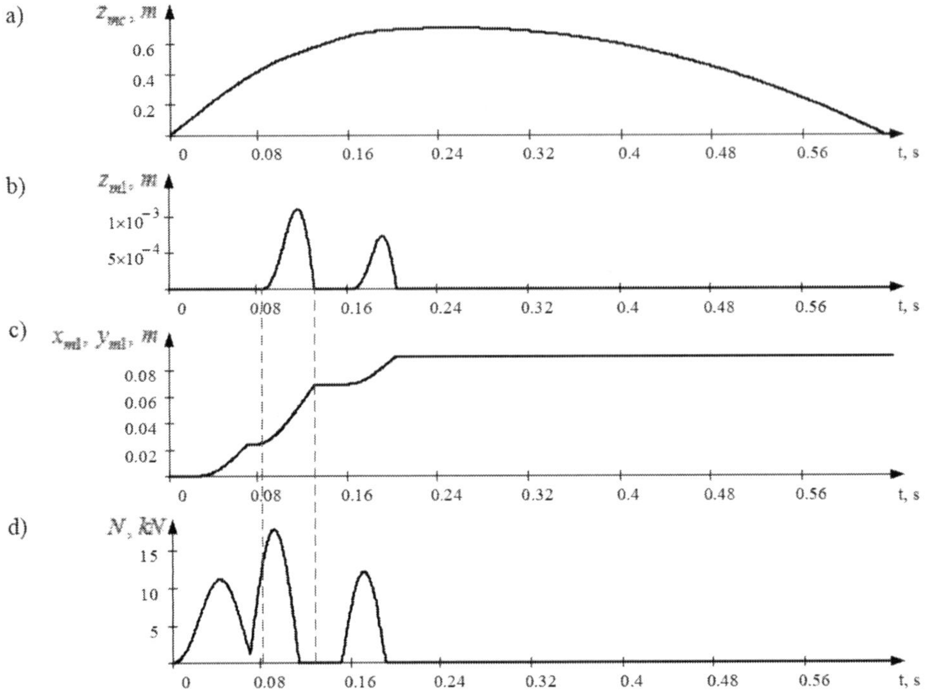

Fig. 5. (a) Vertical displacements of the central tire, (b) Vertical displacements of the corner tire, (c) Horizontal displacements of the corner tire, (d) Internal force in the diagonal elastic element as function of time

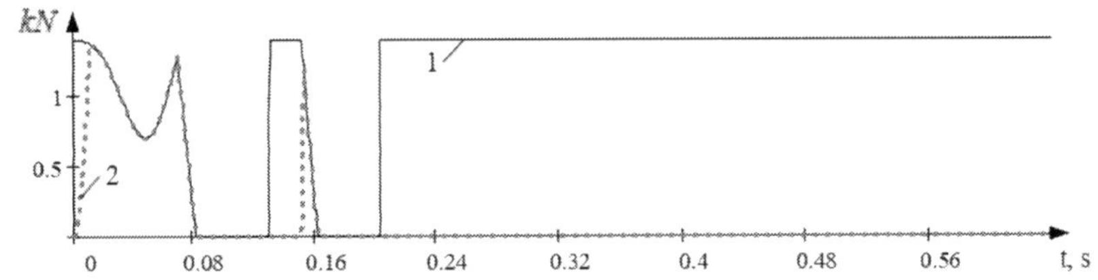

Fig. 6. 1 – Function of the force of kinetic friction $\mu \cdot R(t)$; 2 – Function of the actual friction force $Ff(t)$

4. Conclusion

The dynamic spatial model of gas permeable blasting shelter is represented by concentrated masses interacting with unilateral foundation and by elastic weightless incompressible elements. The contact of mass with a surface in vertical direction is represented by non-penetration condition and in the horizontal directions is governed by Coulomb friction. Time discretization technique is used to obtain numerical solution of the problem. Contact region is defined at each time step.

The algorithm was developed for nonlinear dynamic analysis of transformable gas permeable blasting shelter. The behavior of blasting shelter under various sequences of explosions can be investigated using proposed algorithm. The results of numerical simulation of simple shelter are presented for assessment of the adequacy of the mathematical model.

References

1. Strategiya razvitiya zh.–d. transporta RF do 2030 goda. Utv. rasporyazheniem Pravitelstva RF ot 17 june 2008 g. # 877-r.
2. S. L. Antonov, Proektirovanie razvitiya regionalnoy seti zh.–d., **4,** 8. (2016).
3. V. Kecojevic, M. Radomsky, Safety Science, 43, 739. (2005).
4. E. B. Shevkun, A. V. Leschinskiy, A. A. Piotrovich, Vestnik TOGU, **2**, 93. (2014).
5. E.B. Shevkun, A.V. Leschinskiy, pat. 2317521 RF, patentoobladatel TOGU.
6. E.B. Shevkun, A.V. Leschinskiy, I. M. Urenev, G. P. Vagina, pat. 2310812 RF, patentoobladatel TOGU.
7. P.K. Singh, M.P. Roy, R.K. Paswan, Md. Sarim, S. Kumar, R. R. Jha, JRMGE, **8**, 225. (2016).
8. A.V. Leschinskiy, E.B. Shevkun, N.K. Lukashevich, Vzryivnyie rabotyi pod ukryitiem v transportnom stroitelstve (Yurayt, Moskva, 2017).
9. N. Changping, J. Sjöbergb, D. Johanssona, N. Petropoulosa, IJRMMS, **100**, 250. (2017).
10. W. Bangbiao, C. Rong, X. Kaiwen, IJRMMS, **80**, 12. (2015).
a. Momeni, M. Karakus, G.R. Khanlari, M. Heidari, IJRMMS, **77**, 89. (2015).
11. E. B. Shevkun, A. V. Leschinskiy, Yu. A. Lyisak, A. Yu. Plotnikov, GIAB, 4, 272, (2017).
12. Yu. A Mityushkin., Yu. A. Lyisak, A. Yu. Plotnikov, A. V. Ruzhitskiy, E. B. Shevkun, A. V. Leschinskiy, GIAB, 4, 341. (2015).
13. A.D. Lovtsov, O.A. Kostiunina, Procedia Structural Integrity, **6C**, 122. (2017).
14. T. A. Laursen, Computational Contact and Impact Mechanics (Springer-Verlag, Heidelberg, 2002).
15. I.E. Lyuminarskiy, Raschet uprugih sistem s odnostoronnimi svyazyami (MGIU, Moskva, 2006).
16. K. Rolf, W. Mirjam, 2012. Applied Numerical Mathematics, **10**, 1393. (2012).
17. M. A. Crisfield, Non-linear finite elements analysis of solids and structures. Volume 1: Essentials (John Wiley and Sons, 1991).
18. M. A. Crisfield, Non-linear finite elements analysis of solids and structures. Volume 2: Advanced topics (John Wiley and Sons, 1997).
19. S. Oller, Nonlinear Dynamics of Structures (Springer, 2014).

PARTICULARITIES OF SOLVING THE PROBLEM OF SUSTAINABLE DEVELOPMENT OF CHROMITE UNDERGROUND MINING AT DEEP HORIZONS BY MEANS OF COMBINED GEOTECHNOLOGY

David Kaplunov[1], *Dosanbay* Bekbergenov[2*] and *Gulnar* Djangulova[2]

[1]Institute of Comprehensive Exploitation of Mineral Resources, Moscow, Russia.
[2]Mining Institute after D.A. Kunayev, Almaty, Kazakhstan.

Annotation. The paper covers the particularities of development of Don chromite field and improvement of applied technologies of underground mining using uncontrolled caving. The weakest point of the technology of uncontrolled caving and any other similar structures is mining operations under the caved mass that considerably affects the stability of mine workings and their safety during the whole period of stoping. This issue is one of the most important and urgent, especially at the transition of mining operations to deeper levels where technological particularities of the minefield are getting more complicated. Therefore, the Project Development and Substantiation of Parameters of Combined Geotechnology for Safe and Sustainable Development of Chromite Mineral Resource Mining for underground mining will be a new technological solution ensuring the replacement of exhausted mineral resources in the Republic of Kazakhstan in the conditions of transition of mining operations to deeper levels. Further development at the combined and sequential underground mining operations using the system of uncontrolled caving and the system of horizontal layers with stowing is essential. Key words: geotechnology, development system, sustainable development, chromite mining.

1 Introduction

The mining industry in Kazakhstan takes one of the leading positions and Kazakhstan is one of 15 leading mining countries and the second chromite mining nation after the SAR [1-3].

The most significant representatives that widely apply the caving systems are ore miners of Donskoy mining and processing plant developing thick ore deposits of the fields confined within the southern edge of Kempirsay massif that includes over 160 fields,

* Corresponding author: kdbekbergen@mail.ru

occurrences and chromite mineralization. At that, the prevalent part of ore deposits, 93% is mined underground [4-6].

In the conditions of scarcity and impossibility to replace mineral reserves, of special significance at their development is the task of sustainable use of subsoil that requires the improvement and application of new options of underground mining technology.

As the second working develop in the mentioned conditions and the scope of preproduction and temporary working increases and more confined stopes are created, there arises a need in more rational control of processes, in reliable structural elements of the development system, selection of methods and means of mine working support and finally in effective and safe system of development of ore deposits.

Therefore, it is necessary to understand the essence of these complications and to make reasonable choice of process charts based on regularities of mining technical processes associated with the development and to determine their parameters in order to exclude accidents and ensure rational subsoil management.

2 Materials and Methods

The technology of uncontrolled caving at Donskoy mining and processing plant is mainly applicable in the medium and low-stability rocks which is very typical of the Don chromite fields. Their distinguishing feature is high fracturing of the rock mass, various fracture systems and spatial orientation. All these eventually stipulate low strength of ore bodies and the host rock mass. In this connection, one of the major and most urgent issues at the underground development of chromite fields is the provision of stability and reliability of operation of development and temporary workings during the whole period of operation.

The ores of the Kempirsay field are classified as easily concentrated. There are two types of ore grades [7]: high-grade with over 45% of Cr2O3, and run-of mine ore. The raw ore is mined underground (94.1%) at the mines Molodezhnaya and the Tenth Anniversary of Kazakhstan Independence (DNK), as well as using open-cast mining (5.9%) at the Yuzhnyi open pit.

The chromite ore is mined underground using the system of uncontrolled caving (Figure 1) which is class 2 of the development system according to Prof. V.R. Imenitov's classification.

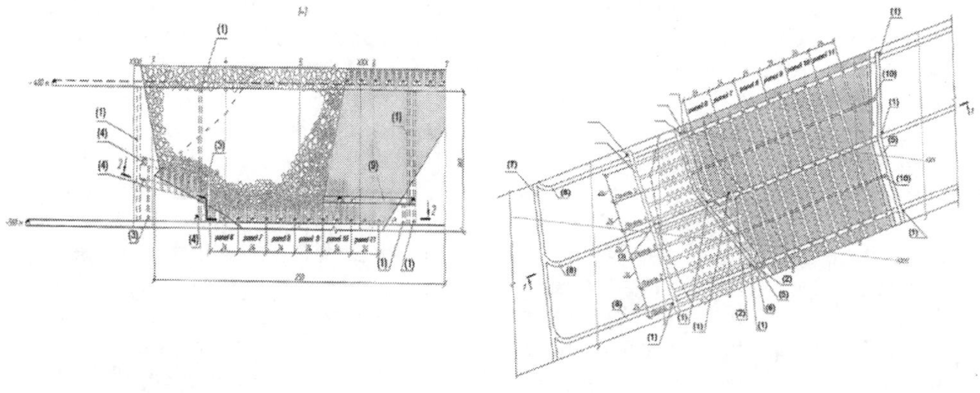

1 – VMR (Ventilation manway raise); 2 – Ore chute; 3 – Rock chute; 4 – Prefabricated air level; 5 – Air level; 6 – Storage drift; 7 – Haulage ort; 8 – Ort; 9 – Air orts; 10 – Household ort; 11 – Panel
Fig.1. Development of Don chromite reserves using the technology of uncontrolled caving

This development technology of uncontrolled caving is characterized by gradual separation of the ore from the rock mass under the proper weight and mine pressure without forced explosion [8]. In the low part of the block, under the caved ore mass, the network of drawing workings is laid that are separated from each other by small pillars. Above the draw horizon, a horizontal exposure is formed using an explosion – an undercut onto which the mass is caved under the force of the proper weight and pressure of overlying rock by means of its cracking in the conditions of unstable fractured ores.

There are four stages of the process in the system of uncontrolled caving:

- Formation of a blasted slot and beginning of undercut formation;
- Initiation of 0uncontrolled caving;
- Development of uncontrolled caving of ore and host rocks;
- Uncontrolled caving reaches the surface.

The mining geological conditions most favorable for the application of the system of uncontrolled caving are: possibility of caving of the surface and overlying stratum of host rocks, and considerable fracturing of the ore body.

The technology of uncontrolled caving is of great interest due to its cheapness and simplicity.

The case history of the technology at underground mines of the company De Beers (South Africa) revealed that the cost of mining is commensurate with the open-cast mining cost. Of special interest is the practice of those working with this technology and belonging to the authors [9], as well as the description of experience of efficient operation of several foreign mining companies: diamond mines "Finch", "Premier" and "Koffiefontein" (South Africa), copper-nickel mine "North Sparks" (Australia), copper-molybdenum mine "Henderson" of the company "Climax" (USA) and mine "El Teniente" (Chili).

The development system of uncontrolled caving has long been considered as the main mining technology of chromite ores at the mines of Donskoy mining and processing plant. The effectiveness of this system is due to the solving of geomechanical tasks related to the stability of mine workings at the stope sill, the reduction in the abutment pressure that tends to increase as the second working begins, the selection and substantiation of optimal design of the stope sill, the sequence of development of adjacent production units and timetable of ore drawing.

The mining geological and mining technical conditions of chromite mining are classified as very complicated in view of instability and high fracturing of ores and host rocks. The engineering geological description of the field developed within the DNK mine is characterized by the hardness factor of host rocks around 8÷10, and ores, 6÷8 due to the structural weakening and volumetric scale effect. In fact, the hardness factor of ores and rocks does not exceed 4-6.

These properties of the rock mass create certain conditions for the application of the development system using uncontrolled caving. However, the instability and tendency to uncontrolled caving are the source of constant manifestation of mine pressure during all mining operations at the mines of Donskoy mining and processing plant. In most cases, despite double additional supports, the supports of development headings and stopes are destroyed and stop working long before the end of second working. In fact, stump pillars, which are the elements of the stope sill, do not serve as such because they are located in a kind of "loose condition because of initial intensive fracturing of the rock mass and dynamic impact of explosions on it. Thus, these pillars as if "lie" on the support of stopes with all their weight and the weight of the whole thickness of caved rocks and overlying rocks and deform them practically pressing them in the ground. Therefore, to excavate the remained ore mass it is necessary to pay extremely undesirable additional expenses for timber relieving (sometimes even repeatedly). These measurements negatively affect the timing of mining operations and rhythm of the mine activity.

Therefore, it is possible to state that despite all the advantages such as maximum productive capacity and the lowest cost, the application of technology of uncontrolled caving of host rocks at the mines of Donskoy mining and processing plant is rather a forced necessity than an effective technical solution of rational sustainable mining in the absence of any other decent option.

In addition to the above-listed major problems, there are less urgent issues arising right at the realization of development and second working in a block, such as combat against inrushes of rocks (ore) at the road heading, premature inrush of overlying rocks into draw points from so-called "tubes" resulted from uneven mineralization and consolidation ability, humidity and looseness of ores, issues related to loss of time at the elimination of overhung boulders, difficulties with ore drawing timetable, unsatisfactory accuracy of losses evaluation and ore dilution for the developed block.

The following institutes and laboratories have performed works with regard to the selection of technological options of field development, finding ways to overcome negative sides of mining geological conditions and increase the overall efficiency of underground chromite mining: Mining Institute after D.A. Kunayev, Mining Institute of Ural Department of the Russian Academy of Sciences, Ust-Kamenogorsk Institute "Kagiprotsvetmet", Institutes VIOGEM and VSEGINGEO (All-Russian Research Institute of Hydrogeology and Engineering Geology) as well as the resources of the research group of the plant. This research has brought expected results, and complied with the requirements of that time. At the present time, further chrome mining and increase in production capacities at the mines of Donskoy mining and processing plant take place due to the involvement of deeper deposits. Transition to deeper levels results in harder mining technical conditions, mainly geomechanical and technological ones.

A new technical solution by the Kasgiprotsvetmet Institute was the application of the mining technology with consolidating stowing at the transition of mining operations to deeper levels of the DNK mine. The project [10] corrected the stripping scheme, the order of preparation and development of reserves of the Millionnoye and Almaz-Zhemchuzhina fields. The application of the combined geotechnology suggests the separation of the ore field in order to develop one part of reserves using horizontal descending layers with stowing (Figure 2) and another part of the field, using the development system of uncontrolled caving.

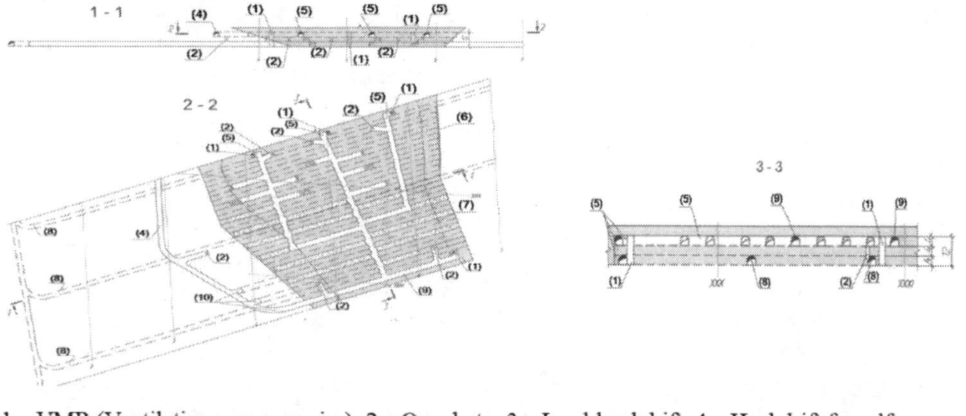

1 – VMR (Ventilation manway raise); 2 – Ore chute; 3 – Load-haul drift; 4 – Haul drift for self-propelled equipment; 5 – Stope drift; 6 – Contour of reserves mining using the primary layer; 7 – Contour of reserves mining using the final layer; 8 – Ort; 9 – Cut ort;10 – Cross slits stope drift

Fig. 2. Development of Don chromite reserves using the system of horizontal layers with stowing according to the Kazgiprotsvetmet project

There is a positive experience of road heading in the stowing, for example, at underground mines of JSC GMK Norilskiy Nickel.

According to the results of performed research and published works in scientific foreign editions [11] as well as the review of the practice at the mines of Donskoy mining and processing plant, despite the development of lower levels of the chromite field of Donskoy mining and processing plant in the evolving conditions, it is quite realistic to transit to artificial stope sills. And currently in line with the Kazgiprotsvetmet project this type of the stope sill is necessary. To tackle the issues of designing artificial concrete stope sills, the staff of the Mining Institute after D.A. Kunayev developed the design of artificial concrete stope sill on a stope sill-support basis ensuring sufficient reliability, stability and operability of scraping drifts. The developed design of the stope sill represents the main designed structural elements, sequence of road heading of service and main workings and the order of erection of concrete support [12].

The Kazgiprotsvetmet Institute has developed the project of transition of mining operations to the lower levels of the mines of Donskoy mining and processing plant using the geotechnology combining the application of development systems using uncontrolled caving and horizontal layers with stowing.

Up to the present no study of such kind of combined geotechnology has been done. Guided by the significance of this direction for mining production, in line with the Decision of the National Academic Council of the Republic of Kazakhstan with regard to the program *Grant Financing of Scientific Research* for the years 2018-2020, the Mining Institute after D.A. Kunayev began the research work on the subject: *Development and Substantiation of Parameters of Combined Geotechnology for Safe and Sustainable Development of Chromite Mineral Resource Mining at Deep Levels.*

3 Results and discussion

The principles and strategies of sustainable development ensure the long-term existence and development of production with due account for the environment based on the targeted controlled economic growth. For this very reason many authors in their practice connects the geotechnological issues with the environmental problems of the mining industry [13].

At the same time, the measures aimed at the formation of the general policy of sustainable development of mining and smelting industry become more significant in line with current requirements of this field. An important condition for successful realization of the Plan of Actions of the World Summit on Sustainable Development (Johannesburg, 2002) related to the mining and smelting branches of industry is the principles formulated regarding the issues of sustainable development of the mining industry of the Republic of Kazakhstan [14]:

First of all, the satisfaction of the main needs of the current and future generations and conservation of existing ecosystems and restoration of the disturbed ones. Development of mineral resources using environmentally clean and safe mining processing technologies with due account for the balance of the government, society and representatives of mining business.

In view of the mentioned solution to the problem, the development of combined geotechnology for safe and sustainable development of chromite mining requires certain transition from setting the task of application of underground geotechnology depending on the economic growth to setting the task and providing economic development, substantiation of physical and technical parameters of the combined geotechnology. Such model includes technological solutions that ensure the replacement of exhausted mineral resources in the Republic of Kazakhstan; in this particular case, it is based on the development and substantiation of parameters of combined geotechnology in the conditions of transition of mining operations to deeper levels at simultaneous and consecutive underground mining operations using the development system of uncontrolled caving and horizontal layers with stowing.

An adequate complex of processes for sustainable development of the combined geotechnology depends on a large number of factors.

The analysis of research by contemporary authors testifies the availability of various approaches to the identification of factors of sustainable development of underground mining geotechnology.

A combination of factors affecting the extent of sustainability of development of underground mining geotechnology by various kinds of ores is divided into the management level, time factor management and technical maintenance.

It is worth noting that majorly the factors of sustainable development are backbone, and only when the impact of these factors is comprehensively accounted for, it is possible to achieve the goals of sustainable development of underground ore production using the combined geotechnology.

The methodology of development and substantiation of parameters of the combined geotechnology for safe and sustainable development of chromite mining at deep levels must meet the safety requirements within the proposed combined geotechnology, requirements of technological properties and cost effectiveness, i.e. profitability of the development and combined geotechnology at the development of reserves at the transition of mining operations to deeper horizons.

4 Conclusion

The above conditions and factors will contribute to the realization of the new project idea related to the combined geotechnology of chromite mining at deep levels, namely the creation of designs with the stable reinforced-concrete platforms with artificial stope sill made of concrete consolidating stowing for producing blocks. In this way the platform is created for the system of uncontrolled caving capable of bearing high mine pressure and ensuring increased stability of haul crosscuts as well as the safety of draw and haulage workings during the whole period of block development for independent ore transportation.

Based on the foregoing, the problem of sustainable development of chromite mining at deep levels at the companies of Donskoy mining and processing plant requires the development and substantiation of parameters of the artificial stope sill of the system with uncontrolled caving at deep horizons of the DNK mines on the basis of the study of reasons for the rock mass caving around development and transportation workings at deep levels of mines belonging to Donskoy mining and processing plant.

Along with this, another important and urgent issue is the forecasting of risks to ensure sustainable ore mining.

References

1 Ibragim Edilbaev. Revival (1995-2005), (2004).
2 Guidelines on Claculation of Loads on the Support of Mine Workings in the Conditions of Mines of the Donskoy mining and processing plant. (Karaganda - 2002).
3 Leonid Zherebko, Lyazzat Shamganova, Gulnar Jangulova. Specific Character of Don Chromites Development in Complex Mining and Geological Conditions. The 16th International Multidisciplinary Scientific Geoconference. Exploration and Mining. Mineral Processing. Sgem 30 June - 6 July (2016).

4 Dosanbai Bekberenov, Gulnar Dzhangulova. Geomechanical Nature and Technological Particularities of Application of the Uncontrolled Caving Technology at Deep Levels of the DNK Mine of Donskoy mining and processing plant / The International Scientific Technical Conference "Problems and Ways of Innovative Development of Mining and Smelting Industry", Tashkent 27-29 November (2014).
5 Ibragim Edilbaev, Marat Bitimbaev, Leonid Zherebko. Development of the Caved Zone in the Superincumbent Rock at the Development of Ore Deposits Using the System of Uncontrolled Caving at Fields of Don Chromites. Scientific and Technical Provision of Mining Production. Collection of Proceedings of the Mining Institute after D.A. Kunayev, No 68, Part I. Almaty, (2004).
6 Kabden Nugmanov, Leonid Zherebko. Stress Condition of the Mass in the Zone of Road Heading, Scientific and Technical Provision of Mining Production. Proceedings of the Mining Institute after D.A. Kunayev, Almaty, V.65 (2004).
7. Dosanbai Bekbergenov, Ayan Toktarov. Problems of Development of Deep Horizons of Chromite Ores of Kempirsay Mass. (Collection of proceedings of the branch Mining Institute after D.A. Kunayev, volume 84, Almaty, 2013).
8 Evgeniy Kuzmin, Uzbekova. Uncontrolled Caving of Ore at the Underground Ore Mining: Tutorial. - M.: Press of the Moscow Mining University, (2006).
9 Stefan Nadolski, Mike Samuels1, Bern Klein, Craig J.R. Hart. Evaluation of Bulk and Particle Sensor-Based Sorting Systems for the New Afton Block Caving Operation. Minerals Engineering 121 (2018).
10 "Kazgiprotcvetmet", Project, volume 2, book 3, Ust-Kamenogorsk. (2003).
11 Rafiee, Ataei, Khaloo Kakaie, Jalali, Sereshki, Numerical Modeling of Influence Parameters in Cavabililty of Rock Mass in Block Caving Mines. International Journal of Rock Mechanics and Mining Sciences, 105 (2018).
12 Gulnar Dzhangulova, Leonid Zherebko, Larisa Pivovarova. The RK Patent 79933. The Method of Erection of the Artificial Stope Sill at the Development of Unstable Ore Deposits at Great Depth (27.01.2012).
13 Yuriy Galchenko, Leonid Burtsev. Environmental Component of Sustainable Development as Applied to Underground Ore Mining. Materials of the International Symposium "Miner's Week – 98" (Seminar 4, MGGU, 1998).
14 David Kaplunov, Yukov. The Principles of Sustainable and Environmentally Balanced Subsoil Development Based on the Combined Geotechnology, (Mining Magazine No 11, 2015).

Concerning the capability of monitoring the condition of coal bed by tomography based upon the representation of object model using Chebyshev polynomials

Larisa Nazarova*[1], *Leonid* Nazarov[1], *Maxim* Protasov[1], *Anton* Panov[1] and *Petr* Konichek[2]

[1] The Institute for Problems of Integrated Development of the Subsurface named after Academician N.V. Melnikov of RAS, Moscow, Russia
[2] Institute of Geonics of the Czech Academy of Science, Ostrava-Poruba, Czech Republic

Abstract. 3D geomechanical model of "Vorkutinskaia" mine, which allowed describing the development of stressed and deformed state of coal bed rock mass, has been developed and realized in numbers. Using empirical dependences of longitudinal waves velocity V on mean stressσ established by laboratory testing data, 3D distribution of V has been plotted. First arrival time tomography task, based upon the approximation of a bed geometrical model by Chebyshev polynomials, in order to give appraisal to the applicability for bed velocity structure reconstruction, has been analysed. With the use of data on dynamic events, registered by permanent stations, the object tomography has been prepared and 3D field of elastic waves velocities in highlighted part of area under study. It enables essential opportunity of stress state, if functions $V(\sigma)$ for rocks building up the massif are known. **Key words:** coal-rock massif, 3D model, tomography, Chebyshev polynomials, stress field reconstruction

1 Introduction

The safety of mining is based upon two major elements: optimal technology of deposit development called upon ensuring not only ceiling volume of production, but as well the level of stresses not exceeding the critical one [1]; rock mass state operation monitoring system [2-4] is designed for the collection of seismoacoustic and deformographic and other direct or indirect information of geomechanical fields, according to which analysis findings, preventive measures on the mitigation of emergency occurrence risks may be taken as well as changes in mineral wealth mining may be introduced. Generally accepted criterion of such information appraisal does not exist since all monitoring systems function in the conditions of uncertainty and, as a rule, use statistical methods of the detection of spatio-temporal regularities of distribution of stochastic fields parameters which directly or indirectly characterize the degree of fracturing in the elements of the object under control [5-10]. In [11], it is noted that long-and mid-range forecast and seismic risk appraisal

* Corresponding author: lanazarova@ngs.ru

(SRA) are used in mining operations planning. However, general opinion is that short-range SRA currently is not sufficiently reliable, despite significant efforts. Although, it was proven that a this is a hard task, it is necessary to take efforts on SRA technology improvement and further.

There are several directions of this subject development trends. For example, forecast of mine bumps may be improved by means of the integration of seismic, deformation and mining-engineering data with geomechanical modelling of rock mass trend. Precisely such approach was realized in [12,13] when the appraisal of rock mass condition was performed using stochastic and deterministic data within the frames of geomechanical model: regressional dependences between the parameters of seismic emission and integral characteristics of stress field have been prepared for various sections of the object under control. This allows verifying the model in terms of forensic analysis [14] forecasting the number and total energy of man-triggered events on the distribution of stresses, calculated in accordance with the long-range mining plan. In [15,16], new approach to modeling of current stressed and deformed state of a deposit mined was theoretically justified by the data of passive and/or active tomography and experiments on the determination of empirical dependence of longitudinal waves velocity on average stress.

With the use of data from [17], detailed 3D model of the mined section of "Vorkutinskaia" mine has been prepared in the first part of the work. In the second part, the capability of the application of tomography based upon the approximation of coal bed geometry by Chebyshev polynomials [18] has been justified.

2 Geomechanical model of "Vorkutinskaia" mine section

2.1 Object geological structure

"Vorkutinskoe" deposit represents brachy-synclinal fold with the dimensions of 30×12 km. Geological structure of the deposit is represented by alternating sandstones, aleurolites and coals with thicknesses from 3 to 100 m. Tectonically, the mine field is simple therefore discontinuities were not considered in geomechanical model. "Troinoi" bed (average thickness 2.85 m) is the upper commercial bed.

2.2 Geomechanical model description

In order to prepare the 3D model of the object under study, mining plan was digitized. The dimensions of computational area D are 840×1570×330 m, the upper border is located at the depth of 590 m from daylight surface, the thickness of coal bed is 3 m. Finite element discretization of the area (about 500 thousands) was performed in accordance with the structural features of coal bed and mined-out space configuration. Discretization pitch by horizontally is 10 m, vertically (z-axis) — 0.5 m in the bed, and gradually increases by moving away of it.

The properties of rocks (density ρ, Lamé parameters λ and μ), building the massif are given in Table 1.

Table 1. The properties of rocks.

Rock	ρ, kg/m^3	λ, GPa	μ, GPa	A, m/s	B, m/s	C
Enclosing rocks	2,200	20	20	4,212	1,498	0.592
Coal	1,500	15	10	2,698	484	0.779

Information on horizontal stresses in natural field borrowed from [19]: in the neighbour of "Vorkutinskaia" mine, there is a faulting geodinamic mode, lateral repulse factor in the direction of mining operations development is q_y=0.6, in orthogonal direction — q_x=0.4. The deformation of medium is described by the system including: equilibrium equations (1), Hook's law (2) and Cauchy equations (3)

$$\sigma_{ij,j}+\rho g\delta_{iz}=0 \quad (1)$$

$$\sigma_{ij}=\lambda(\varepsilon_{xx}+\varepsilon_{yy}+\varepsilon_{zz})\delta_{ij}+2\mu\varepsilon_{ij} \quad (2)$$

$$\varepsilon_{ij}=0.5(u_{i,j}+u_{j,i}) \quad (3)$$

where σ_{ij} and ε_{ij} – components of stresses tensors and deformations ($i, j = x,y,z$); u_i — of shifting; δ_{ij} — Kronecker delta. System (1)(3) with relevant boundary conditions on ∂D is realized through finite-element technique using original code [20].

2.3 Calculation results

On fig. 1a, by way of example, the distribution of horizontal stresses σ_{yy} in section x=420 m (negative values correspond to compression) is shown. Expectedly, in the neighbour of a mine face, there is stress concentration.

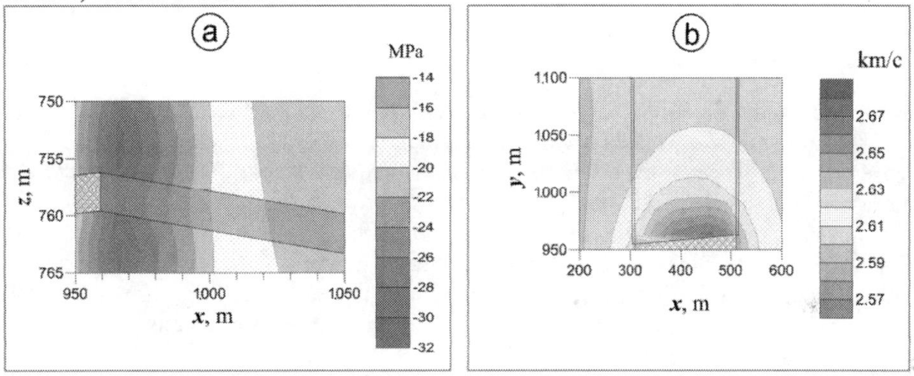

Fig. 1. Isolines of horizontal component of stress tensor σ_{yy} (MPa) in vertical section x=420 m (a); the distribution of longitudinal waves V in a bed mid-section (b).

Laboratory experiments [21] for the determination of seismic waves distribution velocity in loaded rock samples allowed establishing the dependence of longitudinal waves velocity V on average stress σ

$$V(\sigma)=A-B\cdot\exp(C\sigma/\sigma_0), \quad (4)$$

where empirical constants A, B and C are given in Table 1, σ_0=5 MPa. On fig. 1b, the lines of level of velocity V in a bed mid-section (crosshatched zones – mined-out space), calculated by (4) where $\sigma=(\sigma_{xx}+\sigma_{yy}+\sigma_{zz})/3$, are represented. Obtained distributions $V(x,y,z)$ were being used further on as a precise model in tomography.

3 Tomography based upon the representation of coal bed model by Chebyshev polynomials

3D model of the medium containing coal-rock massif is represented in the shape of a set of layers L_j (j layer index) divided by surfaces Z_j. In each L_j, the function of slowness (value reciprocal velocity) $S_j(x,y)$, which depends only on lateral coordinates x and y defined. Function S_j is determined through Chebyshev polynomials of zero-third order

$$S_j(x,y)=C_0+C_1x+C_2y+C_3xy+(2x^2-1)C_4+(2y^2-1)C_5+ y(2x^2-1)C_6+x(2y^2-1)C_7+(4x^3-3x)C_8+(4y^3-3y)C_9, \quad (5)$$

where coefficients C_C are various for each bed. Surfaces Z_j are described by Chebyshev polynomials similarly to (5) with own expansion factors. The use of Chebyshev polynomials for the parametrisation of the model allows obtaining significant calculative advantages: travel-time information and their derivatives by parameters determining ray are calculated analytically.

3.1 Point-to-point ray tracing by bend method

We suppose that in each layer, the ray represents the length of a straight line, described depicted by coordinates of points of interception with the upper (x_j,y_j,z_j) and lower $(x_{j+1},y_{j+1},z_{j+1})$ surfaces,. Note that z_j depends on (x_j,y_j) through the representation of Chebyshev surfaces. Thus, in order to trace a ray satisfying Fermat's principle, it is necessary to minimize objective functional

$$T=\sum T_j(x_j,y_j,x_{j+1},y_{j+1}) \qquad (6)$$

by variables (x_k,y_k) with fixed position of a source and receiver. The search of minimum T is performed by non-linear conjugate gradient method. For that end, analytical forms for the information of travel-time T_j and derivatives of T_j along current trajectory are depicted.

3.2 Numerical studies

In order to study the applicability of 3D tomography algorithm created based upon Chebyshev polynomials, tresses (p. 2), and with the use of (4) precise distribution $V(x,y,z)$ were calculated. Then, the part of the bed, planned for mining, and, accordingly, for monitoring, was selected and transferred to terms of Chebyshev polynomials.

It should be noted that the average velocity in the bed is about 2,600 m/s, in enclosing rocks — 3,800 m/s, the bed has a rugged relief. Therefore, the three-dimensional task, standard methods of rays tracing and tomography based upon them are inapplicable. Exactly, for these reasons, waves tracing method with the approximation of Chebyshev polynomials functions is used here [18].

At "Vorkutinskaia" mine, accepted the following observation system for registration of seismic events (fig. 2): 5 receivers with spacing of 10 meters are located along board gates. Let 50 sources are identified in a bed.

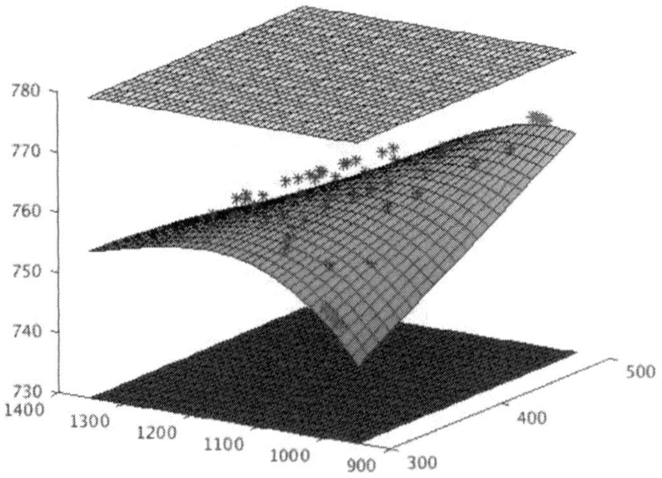

Fig. 2. Coal bed relief and observation system.

Tomography results for such observation system are represented in fig. 3. Initial velocity model for coal bed was selected as homogeneous (V=2,600 m/s). It is apparent that tomography gives good result, almost identical to precise model in highlighted area.

4 Conclusions

Tomography, based upon the representation of geomechanical object model using Chebyshev polynomials in respect to the task of monitoring coal bed by seismic data, has been studied. Such representation allows:

- briefly describing bed relief and distribution of velocities in it;
- correctly solving the task of tracing in contrast thin bed;
- substantively reducing the number of required parameters compared to standard grid model representations what determines solution reliability even in conditions of lack and/or noisiness of seismic information.

Authors express gratitude to data processing center of Novosibirsk State University for computational resources provided. The work has been prepared with the assistance of Russian Scientific Fund, project No.16-17-00029.

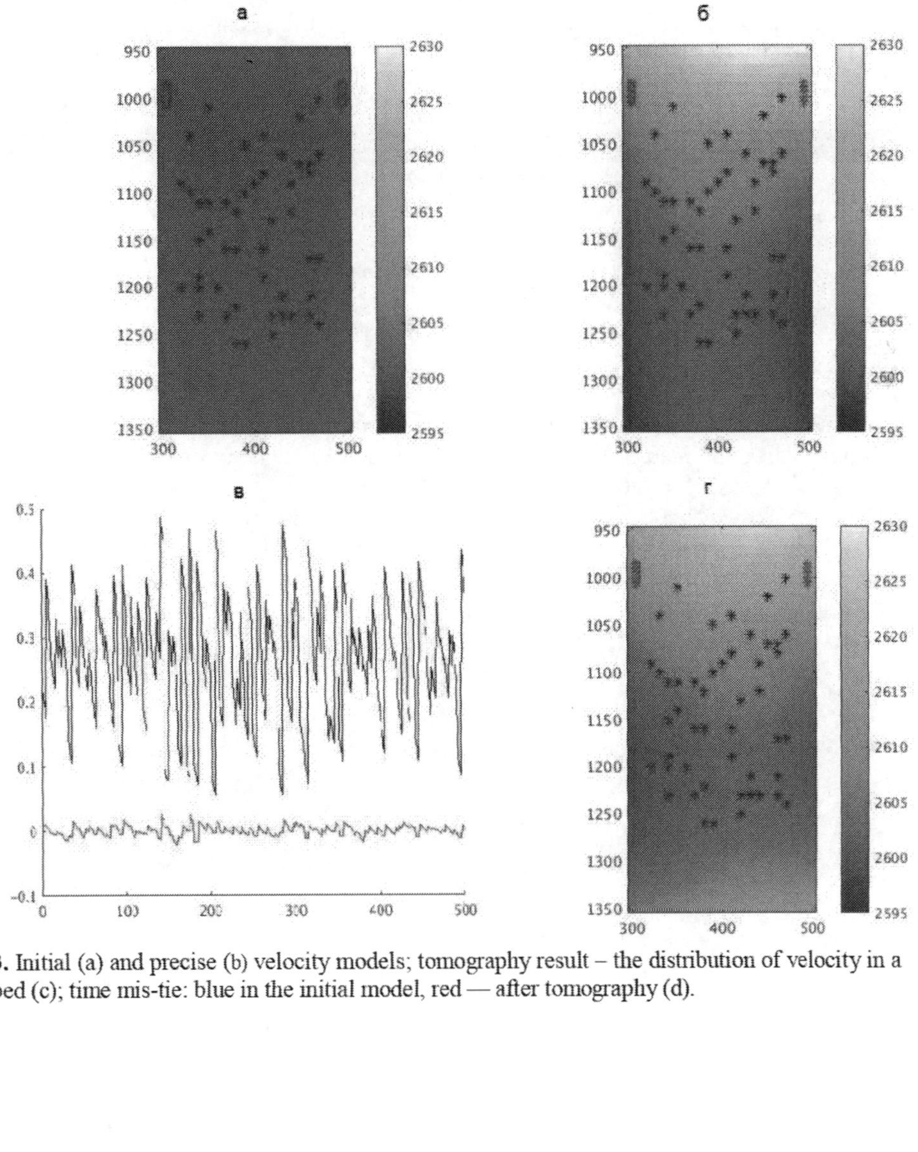

Fig. 3. Initial (a) and precise (b) velocity models; tomography result – the distribution of velocity in a coal bed (c); time mis-tie: blue in the initial model, red — after tomography (d).

References

1. D. M. Bronnikov, N. F. Zamesov, G. I. Bogdanov, *The development of ores at great depths* (M.: Nedra, 1982)
2. V. N. Zakharov, *Seismoacoustic prediction and control of condition and properties of rocks during coal deposits development* (m.: National Mining Research Center – A.A. Skochinsky Institute of Mining, 2002)
3. L. Zhenbi, Zh. Baiting, International Journal of Computer Science Issues, **9**, 24-28 (2012)
4. M. R. Hudyma, R. K. Brummer, Proceedings of the First Canada-US Rock Mechanics Symposium, **2**, 1,423–1,430 (2007)
5. V. N. Oparin, A.P. Tapsiev, V.I. Vostrikov and others, Journal of Mining Science, **4**, 3-22 (2004)
6. V. N. Oparin, A.P. Tapsiev, V.I. Vostrikov and others, Journal of Mining Science, **5**, 3-25 (2004)
7. V. N. Oparin, A.P. Tapsiev, V.I. Vostrikov and others, Journal of Mining Science, **6**, 5-22 (2004)
8. V.S. Kuksenko, Physics of the solid state, **47**, 788-792 (2005)
9. I. Iu. Rasskazov, S.V. Tsirel, A.O. Rozanov and others, Journal of Mining Science, **2**, 29-37 (2017)
10. I. Leslie, Mining Magazine, **April**, 38-40 (2013)
11. R.J. Durrheim, A. Cichowicz, R. Ebrahim-Trollope, et al., Deep Mining, 249-261 (2007)
12. M. Al Heib, Int. J. of Geosciences, 3, 834-846 (2012)
13. L.A. Nazarod, L.A. Nazarova, N.A. Miroshnichenko and others, Journal of Mining Science, **6**, 3-12 (2011)
14. L.A. Nazarova, L.A. Nazarov, V.L. Shkuratnik, et al., SAIMM Symposium Series, **1**, 593-604 (2017)
15. L.A. Nazarova, V.N. Zakharov, V.L. Shkuratnik, et al., Procedia Engineering (2017)
16. L.A Nazarova., L.A. Nazarov, M.I. Protasov, Journal of Mining Science, **4**, 12-21 (2016)
17. A.V. Panov, L.A. Nazarovs, P.V. Nikolenko, A.P. Averin, Deformation and destruction of materials with defects and dynamic codnitions in rocks and mine roadways, 182-186 (2017)
18. D.A. Nekludov, M.I. Protasov, Seismic tomography technologies, 2, 32-38 (2016)
19. [Electronic media]. – Access mode: www.dc-app3-14.gfz-potsdam.de
20. P.G. Djadkov, L.A. Nazarova, L.A. Nazarov, Geol.&Geofiz, 12, 2001–2010 (1997)
21. V.L. Shkuratnik, P.V. Nikolenko, A.E. Koshelev. Phys. technical problems of development of commercial minerals, **5**, 48-53 (2016)

GEOINFORMATIONAL SUPPORTING OF GEOTECHNOLOGIES

Viktor Alenichev[1*]

[1]Institute of Mining, Ural Branch of the Russian Academy of Sciences, Ekaterinburg, Russia

Abstract. The use of complex geotechnical systems, helping for a comprehensive exploitation of mineral resources, is possible only in combination of physical and technical methods of the physical, chemical, biohydro-metallurgical and other special geological technologies of extraction of useful components from natural deposits and man-made formations. To substantiate the parameters of geotechnologies and of extraction indicators of the useful component, information is needed about the modes of processes in the rock mass characterized by certain properties. The creation of a geodatabase, including the characteristics of the material composition and technological properties of the mineral in various fields, will allow to form a database of technological knowledge for the selection of the method of extraction of minerals from the depths and enrichment schemes based on the classification of situations. The bank of technological knowledge should be formed as a multi-level structure that takes in account the physical phenomena and effects accompanying the extraction of a useful component from the depths, technological methods of mining, geotechnological methods and processes of raw material extraction, its processing and enrichment, modes of technological processes. **Key words:** geoinformation supporting, complex field, geotechnology, extraction, geodata, formation, knowledge bank.

1 Introduction

Technologies of development of solid mineral deposits are based on the use of processes: physical-and-technical, physical, physical-and-chemical, chemical, biohydrometallurgical and other or their combinations. To choose an effective method of extracting a useful component directly from the subsoil, information is needed about the impact of the mining environment on the regime of a specific technological process. The traditional mining process, based on the physical-and-mechanical properties of the mineral, is associated with the extracting from the entrails of the earth some crushed or friable rock mass and its subsequent transportation to the points of further processing, storage and stockpilling. This technology, preserving the qualitative characteristics, physical and chemical properties, texture-and-structure features of the mineral in the extracted raw materials, is associated

* Corresponding author: alenichev@igduran.ru

with high costs for the transfer of rock mass to the mobile state, transportation of waste rock into dumps and ores to the points of bringing it to the state, a satisfying a customer.

2 Research of the question

The way out of this situation is the use of mining-technical systems focused on the full cycle of geotechnological processes of extraction and processing of raw materials and providing an increase in the field geopotential [1, 2]. The conversion to complex geotechnical systems in the present time can be carried out only when combination of physical-and-technical methods with physical, chemical, biohydrometallurgical and other special geotechnologies of extraction of useful components from natural deposits and man-made formations. The possibility of using optimal technologies and complex geotechnical systems is determined by the results of a system analysis of geological, mining, hydrogeological, geochemical conditions, texture-and-structure characteristics and the material composition of the mineral [3, 4]. At the same time, it is necessary to revise The State Balance taking into account the development of various industry and energy branches, modern market requirements, international standards for estimating of mineral resources and reserves [5]. The "Guidelines for the preparation of feasibility studies of conditions" should identify the features of research on the phases and stages of exploration, taking into account the type of mineral raw material and the complexity of the field [6].

Minerals are mainly presented in the form of compounds that have arisen as a result of certain geological processes: magmatism, metamorphism, destruction and sedimentation. Minerals located in the Earth's entrails are characterized by certain physical properties, chemical composition and internal structure [4]. The number of stable natural compounds – minerals – does not exceed three thousand. By chemical composition the minerals are divided into simple substances: sulfides, halides and oxygen compounds. Most minerals have a crystalline structure and only some of them are amorphous. The physical properties of minerals are very diverse and are predetermined by the chemical composition and internal structure. The choice of geotechnological method of extraction depends on the texture-and-structure features of the mineral and the enclosing rocks, which are associated with the size, shape and relative position of grains, aggregates and inclusions [7, 8].

The basis of a particular geotechnological method is physical, chemical, biochemical, electrochemical, thermal, photo, electromagnetic, radioactive and other processes, and often their combinations. Therefore, to assess the parameters of geotechnologies and of extraction indicators of the useful component, we need information about the flow regimes of a particular process in the rock mass characterized by a complex of properties. The classification of physical properties of rocks is based on their reaction to external fields or influences, in interaction with which specific properties are shown. In accordance with this postulate, the following classes of properties are distinguished: density, mechanical, thermal, electromagnetic, radiological, mining-technological and others.

When using geotechnologies of selective extraction, detailed information about the physical properties, chemical and mineralogical composition of the mineral, the processes of its transfer to the mobile state, morphology, texture-and-structure features of ores, the minimal industrial matter of mineral components in the block, the technological effectiveness of the mineral and the onboard content. On the basis of technological studies of the productive strata and of the host rocks, additional indicators are being established which characterize a certain geotechnology and are necessary for effective impact on the mountain massif and for extraction of the useful component from the depths.

The most popular physical and technical techniques of the mining industry are mainly associated with the development of solid mineral deposits by open, underground and combined methods. Physical and physic-chemical geotechnological methods of extracting the useful component from the subsoil are: heap and underground leaching, underground dissolution, underground smelting, underground gasification, borehole hydro production. These methods allow extracting the useful component directly from the subsoil, excluding traditional methods of mining operations. Heap and underground leaching are used while the extraction of precious, non-ferrous, rare and radioactive metals, underground dissolution – by the operation of salt deposits and by the creation of underground tanks. Underground smelting is widely used by the extraction of sulfur. Underground gasification, based on the transfer of minerals into gaseous state, is used by the extraction of coal, shale, peat, heavy oil and bitumen. Boreholes hydraulic mining, classified as a processing technology, used in the extraction of useful component (components) from the productive mass by its hydro-mechanical destruction. The implementation of geotechnologies for selective extraction of only useful components from the subsoil requires the study of physical, physical-and-chemical and many other processes, as well as the search for new qualitative innovations in mining practice. Along with the generally accepted characteristics of the field the creation of geoinformation support of physical and chemical geotechnologies should use the concept of "physical and geological environment", including geological, hydrogeological and geothermal characteristics of the deposit, physical and chemical properties of minerals and host rocks. A detailed study of the "physical-geological situation" gives a perspective view for the ability of the mineral to move into a mobile state by means of erosion, dissolution, leaching, combustion, melting, sublimation, etc. The development of geodatabases describing the characteristics of the physical composition and technological properties of minerals in various fields will allow forming in the future a database of technological knowledge for selecting the method of extraction of minerals from the depths and enrichment schemes based on the classification of situations. The realization of this measure will lead to a reduction of the search area, to reasonable consideration of the characteristics of the object of research, to making conclusions about the expected results and improving the quality of design decisions [9, 10].

Herewith, situations can be structured according to various technical and technological features: physical and mechanical properties of rocks and minerals, content of useful and harmful components, grades and types of ores, natural varieties, characteristics of the distribution of minerals by size, morphology, etc. As a criterion of structuring contrast technological properties, a rational complex of methods and apparatus of ore preparation and enrichment cycles can be used. The classified situations, together with non-standard situations, stored in the form of statistically accumulated data, will allow to expand the knowledge base used to select the mode of extraction, processing of poor and difficult-to-rich raw materials and man-made waste, which will provide further justification for the promising internal structure of the projected geotechnology of mining of mineral resources and technologies for its processing.

The bank of technological knowledge should be formed as a multi-level structure that accumulates knowledge in the following areas: physical phenomena and effects, reflecting the technology of extracting a useful component from the depths; technological method (methods) of mining, geotechnological methods and processes of extraction of raw materials, its processing and enrichment, modes of technological processes in the extraction of a useful component of their subsoil. The procedure for the creation of such knowledge bank requires the solution of complex problems, the main ones are: the rationale for a universal conceptual framework for describing the subject area of the investigation of extraction of useful component from the subsoil, the improvement of modules of knowledge representation, the creation of effective methods and means of forming knowledge bases, and creation of information support and special software. In essence, the bank should be a computer system designed to classify, accumulate, store and issue knowledge at the request of the user, control the truth of the postulates (regularities) used. The difference between the bank of technological knowledge and presently known systems of storage and issuance of technological data is that it is possible to extract new, previously unknown for specific conditions technological solutions that are not stored in a suitable form for instant use, but can be synthesized as a result of the interaction of the in the bank existing knowledge. When creating geoinformation support for complex fields and deposits, objective geodata about texture-and-structure characteristics and morphological features of the mineral, as well asabout the chemical composition and physical properties of the rock mass are needed [11, 12].

3 Summary

1. Foundation of physical-and-technical geotechnologies constitute explosive, mechanical, hydraulic, cryogenic, chemical, gravitational and other effects on the rock mass, due to which destruction of rocks, strengthening and stabilization of mine workings and massifs, transportation of minerals to the enrichment points and of waste rocks to the dump are carried out.

2. For the successful implementation of physical, physical-chemical and other new geotechnologies designed for extracting the useful component directly from the subsoil, it is necessary to accelerate researching activities on the creation of the geoinformation support system and of the knowledge bank, based on the optimal modes of the relevant processes.

References

1. Kaplunov D.R., Ruban A.D., Rylnikova M.V. Integrated development of mineral resources united by geotechnologies. NIIK Nedra XX1 LLC. 304 p. (2010)
2. K. N.Trubetskoy, G.L. Krasnyansky, V.V. Khronin, V.S. Kovalenko. Designing of open-cast mines. 3rd ed. Higher education. 694 p. (2009)
3. Doroshenko M.V., Bashlykova T.V. Technological properties of minerals. Heat-and-power engineer. 296 p. (2007).
4. Bashlykova T.V. Rational development of mineral resources. **5 - 6.** 4 - 14. (2015)
5. Ilin O.V. Subsoil use of the XX1 century. **2 (71).** 112 - 115. (2018)
6. Ivanov S.M., Annenkova T.E., Casanov O.V., Kushnarev P.I. Subsoil use of the XX1 century. **2 (71).** 106-110. (2018)
7. Trubetskoy K.N. Mine surveying and subsoil use. **3.** 22 - 29. (2010)
8. Borisov F.F., Alenichev V.M. Mining Information Analytical Bulletin. **10.** 256 - 262. (2015)
9. Sukhanov V.I., Alenichev V.M. Mining information-analytical bulletin. **11.** 320-329. (2015)
10. Alenichev V.M., Alenichev M.V. Problems of subsoil use. **1 (8)**: peer reviewed. sci. articles. Ekaterinburg: IGD UB RAS. 12 - 19. (2016) //trud.igduran.ru.
11. Alenichev V.M., Sukhanov V.I. Mining Inform. and Analyt. Bulletin. **8.** 5 - 15. (2016)
12. Larionov A.N., Kalinichenko L.S., Ryazanov M.A. Rational development of subsoil. **5 - 6.** 51 – 52. (2015)

Choosing the method of excavating vein reserves and mineralized zones on the basis of simulation modeling

Konstantin Strukov[1], *Irina* Nikiforova[1*], *Vladimir* Lavenkov[1] and *Yuri* Iun[1]

[1]Institute of Complex Exploitation of Mineral Resources of the Russian Academy of Sciences. Moscow, Russia

Abstract For the effective involvement of off-balance ores into development, a simulation modeling of gross and selective mining variants at "Kochkar" deposit reserves has been carried out by the Monte Carlo method, provided that a new type of geological resource – mineralized rocks has been involved in mining, and an economic evaluation of competing options has been made. **Keywords:** "Kochkar" deposit, underground mining, excavation method, selective and gross mining technologies, Monte Carlo method, ore veins, mineralization zones, poor ore raw stock.

1 Introduction

An analysis of negative trends in the state of the mineral wealth sector of Russia shows that their main reason is the wide use in the mining sector of the concept on extensive subsoil use, which led to a significant deterioration in the quality and primary specifications of the mineral wealth of most mineral deposits. To the full, this refers to the development of veined gold ore deposits in the South Urals, for which the uneven nature of the distribution of minerals in the solid mass of the deposit predetermined the initial selective excavation of rich sites and the abandonment of ore reserves with low quality characteristics.

According to current forecasts, given the existing scale of production, the availability of gold reserves for underground mining in the region is 20-30 years, open - 30-40 years [1]. However, taking into account the increase in the production capacity of mines and the retirement of the facilities of existing enterprises, the supply of the sector will drop sharply, which will bring the region closer to the critical level of the state of the ore gold resource portfolio.

The generalization of the experience of exploitation of gold deposits, developed by "Yuzhuralzoloto Group of Companies", OJSC ("YuGK", OJSC), specializing in the extraction and treatment of gold-bearing ores and metallurgical manufacturing of commercial gold, allowed justifying the concept and identify the principles of preserving the production potential of the mining engineering system in the conditions of reducing volume of reproduction of the resource portfolio and falling quality specifications of the mined ore, as well as increasing mining depth. In order to prevent the liquidation of the

* Corresponding author: nikiforova495@mail.ru

enterprise and, as a consequence, the emergence of a set of socioeconomic and environmental problems, fundamentally new geotechnological solutions have been adopted aimed at changing the technology of mining and treatment of gold-bearing ores with subsequent disposal of waste in the developed subsoil area [2-6]. That will ensure the stabilization of not only the economic, but also the geomechanical and ecological situation [7].

One of the principles of the new concept is full disclosure of the potential of the resource portfolio, rejecting of the initial selective testing of the richest sites of deposits and transition to joint development of gold ore veins and adjacent mineralized zones. The fulfillment of this condition allows, by increasing the vertical capacity of ore deposits, to increase the intensity of the cleaning works, to apply more productive development systems with mass breaking of ore, to involve hard sites of deposits, with a small and highly variable gold content in the ore, and also to move to deeper horizons, where "traditional", previously used technologies become less profitable, or unprofitable.

To date, the reserves of vein deposits and mineralized host rocks that were previously abandoned during the initial development due to the economic inexpediency of mining have not been considered as promising for involvement in the development.

Analysis of the gold content in veins and in host rocks, applied to the conditions of development of the "Kochkar" deposit, showed that on the worked out horizons there are promising sections of veins and mineralized rocks, in terms of size and techno-economic conditions suitable for development. These reserves include mineralization zones in the hanging and lying sides of gold ore veins (*tabashka*), as well as in quartz-flooded granites. The thickness of such ores reaches 10-14 m, with an average gold content of 3 g / t. In connection with the growth of prices for precious metals, the cleaning-up of these mineralization has become profitable. For their work out, a system has been introduced for the development of sub-floor drifts with borehole breaks, which has made it possible to increase labor productivity at the stoped excavation and to reduce the prime cost of the extracted ore mass [8-10].

In turn, the need to take into account the specifics of the technological schemes for processing ore raw materials of rich veins and mineralized rocks required a set of research works aimed at justifying the parameters of high-performance development systems, choosing the preferred option for mining rich ore reserves and cutting mineralized zones and identifying the economic feasibility of applying various methods of excavating ore veins and mineralized zones: a comparison of the selective working of the vein with gross simultaneous extraction of all the stocks of the mineralized zones and veins included in the chamber contour.

2 Methodology

To select the best option for excavating ore in areas including veins and mineralization zones, a simulated Monte Carlo modeling of gross and selective excavation was performed. The comparative evaluation was carried out according to two criteria: the mass of extracted gold and the resulting specific profit per 1 ton of recoverable reserves for the options considered.

During the simulation, variants of the stepwise selective excavation of rich ores with their separate extrusion onto the surface for self-processing of the extracted ore mass by cyanidation with different gold contents were considered. Also, the options with gross development of ore veins and mineralized zones with varying gold content in veins and mineralized rocks were considered.

As initial data for the construction of the probability model, the gold extraction parameters were used depending on the base gold content in the ore, provided that the extracted ore mass was processed by cyanidation at the Artem completed enrichment cycle factory. Economic indicators of ore mass redistribution at this factory were adopted during the modeling. A general view of the probabilistic model, which reflects the dependence of the recovery index on the gold content, is shown in Fig. 1 and is described by the expression:

$$\varepsilon = 100 \cdot \left(\frac{\alpha}{|\alpha| + d} + \delta \right), \tag{1}$$

where ε – extraction parameter, %; α – gold content, g / t; δ – sporadic error of the model; d – model's parameter.

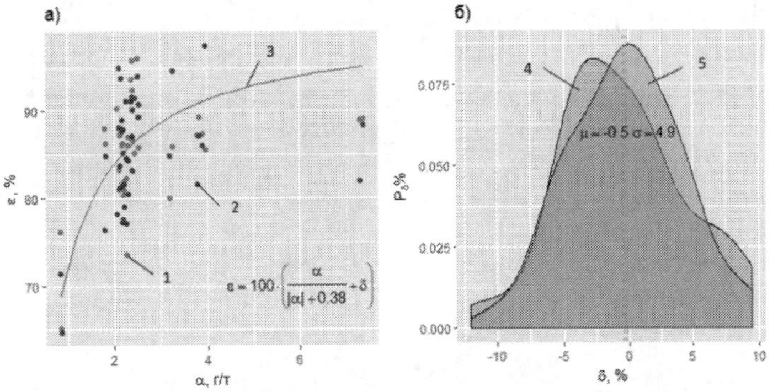

Fig. 1: *a* - Regression model of the dependence of gold recovery in leaching on the gold content in ore: 1 - observational data; 2 - model-generated data; 3 - graph of the regression model; *b* - error distribution density of the regression model: 4 - according to observation data; 5 – software-generated results

In Fig. 1, position 1 shows the points obtained from the data of actual observations, position 3 - the results of model processing obtained by regression analysis. Fig. 1b shows the error distribution density for the actual data (4), the values of its mathematical expectation (μ) and standard deviation (σ) are presented, on the basis of which the sporadic error was programmed (5). Based on the regression model (3) and the error distribution density (5), recovery indices were generated depending on the gold content (2).

In the future, the regression model of gold recovery from the extracted ore mass was used to programmatically generate the gold recovery parameter from the extracted ore mass by cyanidation taking into account the error, which included the identification of the mass of recovered gold per 1 ton of extracted ore mass for the options of selective and gross working of veins and mineralized rocks .

This experiment assumed the generation of initial values of gold content in veins and mineralized rocks, their combination, the generation of the recovery index for each combination according to the formula (1), the calculation of the specific gravity of extracted gold for two options of working out (Me_{gr} and Me_{sel}) according to formulas (2) and (3), as well as their difference (ΔMe),calculated according to (4). This algorithm was implemented in the model 100 times, and for each combination of gold content in the vein and mineralized rocks, the probability P_{me} was calculated that the mass of extracted gold for the selective production option would be higher than the mass for the gross extraction.

$$Me_{вал} = \frac{1 - П_{кам}}{1 + P_{кам}} \cdot \bar{\alpha} \cdot \varepsilon_{\bar{\alpha}}, \text{ g / t;} \tag{2}$$

$$Me_{gr} = 1 - L_{cham} / 1 + D_{cham} * \bar{\alpha} * \varepsilon_{\bar{\alpha}} / \text{g / t;}$$

$$Me_{\text{сел}} = \frac{V_{\text{жила}} \cdot \alpha_{\text{жила}} \cdot \varepsilon_{\alpha_{\text{жила}}} \cdot \dfrac{1-П_{\text{ПШ}}}{1+P_{\text{ПШ}}} + V_{\text{МП}} \cdot \alpha_{\text{МП}} \cdot \varepsilon_{\alpha\text{МП}} \cdot \dfrac{1-П_{\text{кам}}}{1+P_{\text{кам}}}}{V_{\text{жил}} + V_{\text{МП}}}, \text{ g / t}; \quad (3)$$

$$Me_{sel} = V_{vein} * \alpha_{vein} * \varepsilon_{\bar{\alpha}\ vein} * 1 - L_{sfl} / 1 + D_{sfl} + V_{mr} * \alpha_{mr} * \varepsilon_{\bar{\alpha}\ mr} * 1 - L_{cham} / 1 + D_{cham}$$

$$\Delta Me = Me_{sel} - Me_{gr}, \text{ g / t}, \quad (4)$$

where $\bar{\alpha}$ – weighted average content of gold in the ore block, g /t; α_{vein} and α_{mr} – gold content in vein and mineralized rocks, g / t; $V_{vein} = l * h * m_{vein}$ and $V_{mr} = l * h * m_{mr}$ – volume of vein and mineralized rocks in the block, respectively, m^3; h and l - height and width of the block, m; $\varepsilon_{\bar{\alpha}}$ –extraction of gold from ores in gross mining, unit fraction; $\varepsilon_{\bar{\alpha}\ vein}$ and $\varepsilon_{\bar{\alpha}\ mr}$ – extraction of gold from vein and mineralized rocks in selective mining, unit fraction; L_{cham}, D_{cham}, L_{sfl} and D_{sfl} – loss and dilution of ore under systems of chamber and sub-floor drifts, unit fraction.

Weighted average content of gold in the ore block:

$$\bar{\alpha} = \frac{V_{\text{жила}} \cdot \alpha_{\text{жила}} + V_{\text{МП}} \cdot \alpha_{\text{МП}}}{V_{\text{жила}} + V_{\text{МП}}}. \quad (5)$$

$$\bar{\alpha} = V_{vein} * \alpha_{vein} + V_{mr} * \alpha_{mr} / V_{vein} + V_{mr}$$

Similarly, the probability (P_{pr}) was calculated that, with the selective extraction option, the profit would be higher than with the gross mining, respectively. The profit for selective and gross options of mining was found by the formulae:

$$\text{Pr}_{\text{уд сел}} = Me_{\text{сел}} \cdot Ц_{\text{золота}} - (C_{\text{ПНР ПШ сел}} + C_{\text{ПНР кам сел}} + C_{\text{закл сел}} + C_{\text{обог сел}}), \quad (6)$$

$$Pr_{sp\ sel} = Me_{sel} * P_{gold} - (C_{POD\ sfl\ sel} + C_{POD\ cham\ sel} + C_{fill\ sel} + C_{enr\ sel})$$

$$\text{Pr}_{\text{уд вал}} = Me_{\text{вал}} \cdot Ц_{\text{золота}} - (C_{\text{ПНР кам вал}} + C_{\text{закл вал}} + C_{\text{обог вал}}), \quad (7)$$

$$Pr_{sp\ gr} = Me_{gr} * P_{gold} - (C_{POD\ cham\ gr} + C_{fill\ gr} + C_{enr\ gr})$$

$$\Delta\ Pr = \Delta\ Pr_{sp\ sel} - \Delta\ Pr_{sp\ gr} \quad (8)$$

The cost of preparatory-development operations (PDO) for a system of sub-floor drifts in selective mining was defined as:

$$C_{\text{ПНР ПШ сел}} = \frac{C_{\text{ПНР ПШ}} \cdot V_{\text{жила}}}{V_{\text{жила}} + V_{\text{МП}}} \cdot \frac{1-П_{\text{ПШ}}}{1+P_{\text{ПШ}}}, \quad (9)$$

$$C_{POD\ sfl\ sel} = C_{POD\ sfl} * V_{vein} / V_{vein} + V_{mr} * 1 - L_{sfl} / 1 + D_{sfl}$$

where $C_{POD\ sfl}$ – cost of PDO for a system of sub-floor drifts per 1 ton of mined ore mass, rubles/t.

Cost of PDO for a chamber system in selective mining was defined as:

$$C_{\text{ПНР кам сел}} = \frac{C_{\text{ПНР кам}} \cdot V_{\text{МП}}}{V_{\text{жила}} + V_{\text{МП}}} \cdot \frac{1-П_{\text{кам}}}{1+P_{\text{кам}}}, \quad (10)$$

$$C_{POD\ cham\ sel} = C_{POD\ cham} * V_{mr} / V_{vein} + V_{mr} * 1 - L_{cham} / 1 + D_{cham}$$

where $C_{POD\ cham}$ – cost of PDO for a chamber system, per 1 ton of mined ore mass, rubles/t.

Cost of filling operations accounting for underfilling factor and loss/dilution in gross mining option is assessed as:

$$C_{fill\ sel} = (1 - L_{cham}) * (1 + D_{cham}) * (1 - L_{sfl}) * (1 + D_{sfl}) * k_{uf} * C_{fill} \qquad (11)$$

where C_{fill} – cost of filling operations per 1 ton of mined ore mass, rubles/t.

Cost of enrichment per 1 ton of ore in block:

$$C_{обог\ сел} = \frac{V_{жила} \cdot (1 - П_{ПШ}) \cdot (1 + P_{ПШ}) \cdot C_{обог\ жил\ 1m} + + V_{МП} \cdot (1 - П_{кам}) \cdot (1 + P_{кам}) \cdot C_{обог\ МП\ 1m} +}{V_{жила} + V_{МП}}, \qquad (12)$$

$$C_{enr\ sel} = V_{vein} * (1 - L_{sfl}) * (1 + L_{sfl}) * C_{enr\ vein\ 1t} + V_{mr} * (1 - L_{cham}) * (1 + L_{cham}) * C_{enr\ mr\ 1t} + / V_{vein} + V_{mr}$$

where $C_{enr\ vein\ 1t}$ – cost of enrichment of 1 ton of ore from vein, rubles/t; $C_{enr\ mr\ 1t}$ – cost of enrichment of 1 ton of ore from mineralized rocks, rubles/t.

Cost of PDO for chamber system in gross mining:

$$C_{ПНР\ кам\ вал} = C_{ПНР\ кам} \cdot \frac{1 - П_{кам}}{1 + P_{кам}}. \qquad (13)$$

$$C_{POD\ cham\ gr} = C_{POD\ cham} * 1 - L_{cham} / 1 + D_{cham}$$

Cost of filling operations accounting for underfilling factor and loss/dilution in gross mining option is assessed as:

$$C_{fill\ gr} = (1 - L_{cham}) * (1 + L_{cham}) * k_{uf} + C_{fill} \qquad (14)$$

Cost of enrichment per 1 ton of ore in block:

$$C_{enr\ gr} = (1 - L_{cham}) * (1 + L_{cham}) * C_{enr\ vein\text{-}mr\ 1t} \qquad (15)$$

where $C_{enr\ vein-mr\ 1t}$ – cost of enrichment of 1 ton of averaged ore from veins and mineralized rocks, rubles/t.

3 Results and Discussion

As the initial data for geotechnological modeling, the extraction and enrichment indicators achieved at the "Kochkar" underground mine (Table 1) were adopted.

Table 1. Initial data of geotechnological modeling

Parameters	Symbol	Unit	Value
Block length	l	m	60
Block height	h	m	60
Loss in sub-floor drift systems	L_{sfl}	unit fraction	0.11
Dilution in sub-floor drift systems	D_{sfl}	unit fraction	0.10
Loss in chamber systems	L_{cham}	unit fraction	0.15
Dilution in chamber systems	D_{cham}	unit fraction	0.15
Gold price	P_{gold}	rubles	2,480
Underfilling factor	k_{uf}	unit fraction	0.95
Cost of filling mixture per 1 ton of mined ore mass	C_{fill}	rubles/t	382.80
Cost of preparatory-development operations for a system of sub-floor drifts per 1 ton of mined ore mass	$C_{PDO\ sfl}$	rubles/t	1,121.92
Cost of preparatory-development operations for a chamber system per 1 ton of mined ore mass	$C_{PDO\ cham}$	rubles/t	518
Cost of enrichment of 1 ton of averaged ore from veins and mineralized rocks	$C_{enr\ vein\text{-}mr\ 1t}$	rubles/t	700
Cost of enrichment of 1 ton of vein gold	$C_{enr\ vein\ 1t}$	rubles/t	700

Parameters	Symbol	Unit	Value
Cost of enrichment of 1 ton of mineralized rocks	$C_{enr\ mr\ 1t}$	rubles/t	700

Fig. 2 presents in the form of diagrams the results of calculating the probability of preferred selective extraction based on the maximum profit criteria and the maximum gold extraction for various values of gold content in ore and zones adjacent to the vein as the thickness of veins and mineralized rocks varies. Taking into account the influence of the contrast of the gold content in the ore mass on the factors of maximum profit and maximum extraction of gold, the areas of preference for the use of selective excavation have been identified (dark color selection in the graph).

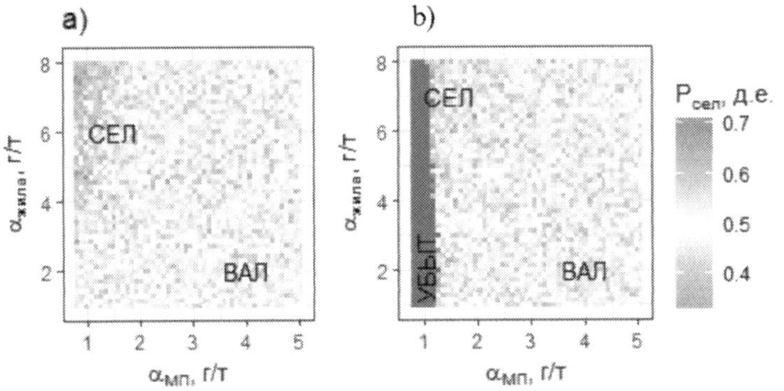

г/т	g/t
СЕЛ	SEL
ВАЛ	GR
УБЫТ	LOSS
д.е.	unit fraction

Fig. 2. Diagrams of the probability of preferring the use of selective technology of mining in comparison with the gross method by the criterion: *a* - recoverable gold; *b* - maximizing the corporate profit

An analysis of the data in Fig. 2 allows drawing a conclusion that for ranges of gold content in veins from 1 g/t to 8 g/t and in mineralized rocks from 0.7 g/t to 3 g/t, respectively, the preferred technology is gross extraction, both by the criterion of maximum metal extraction and by the criterion of profit.

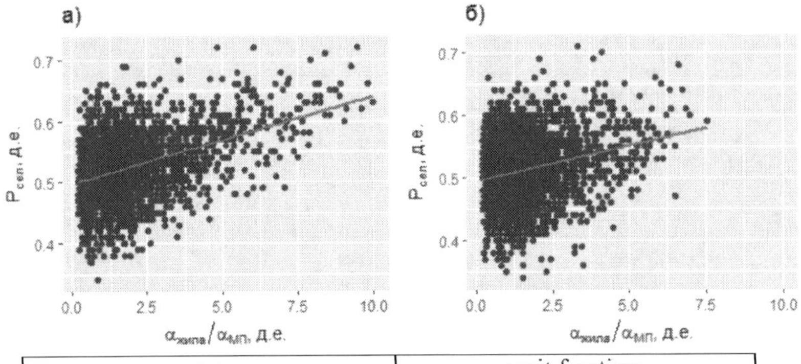

д.е.	unit fraction

Fig. 3. Probability of the effectiveness of applying selective extraction method depending on the contrast of the contents by the criterion: *a* - recoverable gold; *b* - maximizing the corporate profit

The influence of the contrast of the average gold content in the vein and the adjacent mineralized zone on the area of preferential application of selective extraction for different values of the vein and mineralized rock thickness by these factors is shown in Fig. 3. An analysis of the diagrams presented in Fig. 3, allows concluding that as the contrast of the content of ores increases ($\alpha_{vein}/\alpha_{mr}$), the probability of effective application of the selective extraction method also increases.

4 Conclusions

On the example of "Kochkar" deposit, for the first time for the conditions of involving a new kind of geological resource - mineralized rocks of vein gold deposits - at the final stage of exploitation of balance reserves, the efficiency of gross and selective excavation of gold-bearing ore mass was compared on the basis of geological testing of ore veins and mineralized zones. As shown by the comparison of the options performed, with the current level of gold prices and the existing difference in production costs for gross and selective excavation, the difference in the processing of rich and poor ores is not compensated by a change in production costs in the extraction methods under consideration. When the content of gold in veins and adjacent mineralized zones is up to 8 and the thickness of the veins is up to 1 m, gross excavation of ore is preferred at the final stage of underground mining of vein gold deposits in the South Urals. Selective excavation is expedient only at high contrast of gold content in the vein and adjacent mineralized zones - above 8, vein thickness more than 1.5 m and thickness of mineralized zones less than 10 m.

References

1. Cencerning the state and use of mineral resources in the Russian Federation in 2015: State report / Moscow: the Ministry of Natural Resources and Ecology of the Russian Federation, 2015. URL: http://www.mnr.gov.ru/upload/iblock/c50/2015_msr.pdf.
2. K.N. Trubetskoy, Development of resource-saving and resource-reproducing geotechnologies for the integrated development of mineral deposits. Moscow: IPKON RAS. 2014. - p. 196.
3. N.N. Melnikov. Jubilee session of the Department of Geology, Geophysics, Geochemistry and Mining Sciences of Russian Academy of Sciences "Development of new scientific directions and technologies for the development of the earth interior". FSUE National Science Center for Mining Production. Skochinsky Institute of Mining. 24-26. (2000)
4. D.R. Kaplunov, D.N. Radchenko. Mining Magazine. **11.** 52-59. (2017)
5. V.N. Kalmykov, K.I. Strukov, G.P. Konstantinov, R.V. Kulsaitov. Bulletin of the Nosov Magnitogorsk State Technical University. **14 (3).** 13-20. (2016)
6. D.J. Packey. Resour. Policy. 37 (1). 104–108. (2012)
7. V.N. Kalmykov et al. Eurasian mining. **2**. 12-15. (2017)
8. K.I. Strukov, V.V. Fedoseev, R.V. Berger, M.V. Rylnikova. Mining Magazine. **6**. 5-8. (2017)
9. M.V. Rylnikova, I.L. Nikiforova, V.A. Yezhov, S.N. Plotnikov. Mining Magazine. **9.** 35-40. (2017)
10. K.I. Strukov, M.V. Rylnikova, I.L. Nikiforova. Mining Industry. **2**. 46-49. (2018)

The reasoning of mining methods parameters toward development of the apatite-nepheline ore deposits based on results of forecast of massif stress state

Ilya Bagautdinov[1*], *Anton* Kuranov[1], *Nikita* Belyakov[1] and *Anatoly* Streshnev[2]

[1]Mining University, Saint-Petersburg, Russia.
[2]'Apatit', Kirovsk, Russia.

Abstract The article describes the experience in which the argumented parameters of the sublevel caving system in the conditions of mining operations take into account the Khibiny apatite-nepheline deposits. The stress state of the massif under the given conditions is determined by the parameters of the initial gravitational-tectonic field stresses, which have a complex distribution and redistribution for each individual isolating situation, which, in turn, is determined by the zone of the bearing pressure from the overlying stages. The necessary consideration of the factors indicated determined the order of the research: the parameters of regional field stress should be determined using the global numerical model of the deposit; after this, in order to justify the parameters of mining methods in local areas, it is necessary to make local numerical simulation for each area. All calculations were performed using Simulia Abaqus software, which implements the finite element method. Thus, the article presents numerical simulation of the stressed state of the massif of areas of "Plato Rasvumchorr" and "Apatitovii Tsirk" deposits and further describes the order of intersection of the results with the help of a global numerical model to a local numerical model of stope block at + 310 m at the Rusvumchorr mine. There are fixed minimum pillar sizes according to the accepted mining method in numerical simulation results using multiscale models. **Key words:** stress, numerical simulation, workings, pillars, rock massif

1. Introduction.

Existing empirical methods for calculation of pillars and the local numerical models for estimation of massif stress state unable made whole assessment of the mining-and-

[*]Corresponding author: Bagautdinov_II@pers.spmi.ru

geological and mine technical situation in the region which was formed by the time of planning mining operations.

The regional numerical model of "Apatitovii Tsirk" deposit which development by Rasvumchorr mine was created for estimation of initial stress state of rock massif [1].

Geomechanical safe order of preparing of haulage level +310m by Rasvumchorr mine to excavation and reasoning primary main technical parameters of mining method was selected using local numerical model. The results of calculation the massif stress state according global numerical model was taken as boundary conditions for local numerical model.

Thus, the paper includes calculation order of safe pillar sizes and process of creating numerical models.

2. The procedure of experimental research.

"Apatitovii Tsirk" deposit is part of great apatite-nepheline field on South-western ore field by the Khibiny rock massif. Whole length of the ore body is 2350 meters. Dip azimuth is changing from 35^0 to 40^0 for north-western part and from 60^0 to 70^0 for south-eastern part of the deposit.

Rocks and ore on the deposit is strongly and coherent strongly with high elastic properties and sharp-crack fracture with considering their deformational and strength parameters.

The initial stress state of the rock mass on the deposit is gravitational-tectonic, with significant horizontal component.

The global numerical 3D model of "Apatitovii Tsirk" deposit was created in CAE SIMULIA Abaqus software [2]. According geological model was built wireframes of body ores, tectonic joints and their spatial location.

The analytical model of regional numerical model is shown in Fig.1. The boundary conditions of numerical model was performed as follow: the lateral faces of numerical model unable to displacement along to OX axis; the lower face unable to displacement along to OY axis; the end faces unable to displacement along to OZ axis; the top face of numerical model is free.

The field stress in massif is forming by own weight of rocks in vertical direction (gravitational component of field stress) and additionally by tectonic joints in horizontal direction (tectonic component of field stress).

Tuning of software Simulia Abaqus solver was adjustment as such a way as all stress in massif was compressive only. Therefore, any displacements of model top face (Earth surface) due to own weight of rocks were banned.

Whole solid space of the numerical model was separated in 482.000 three-dimensional solid elements of first order with tetrahedral form. The finite element mesh was created with differentiate by sizes. There were small finite elements around ore bodies and large finite elements at approach to boundary of numerical model. Linear sizes the faces of elements had size 30 meters around ore bodies and 50 meters on boundaries.

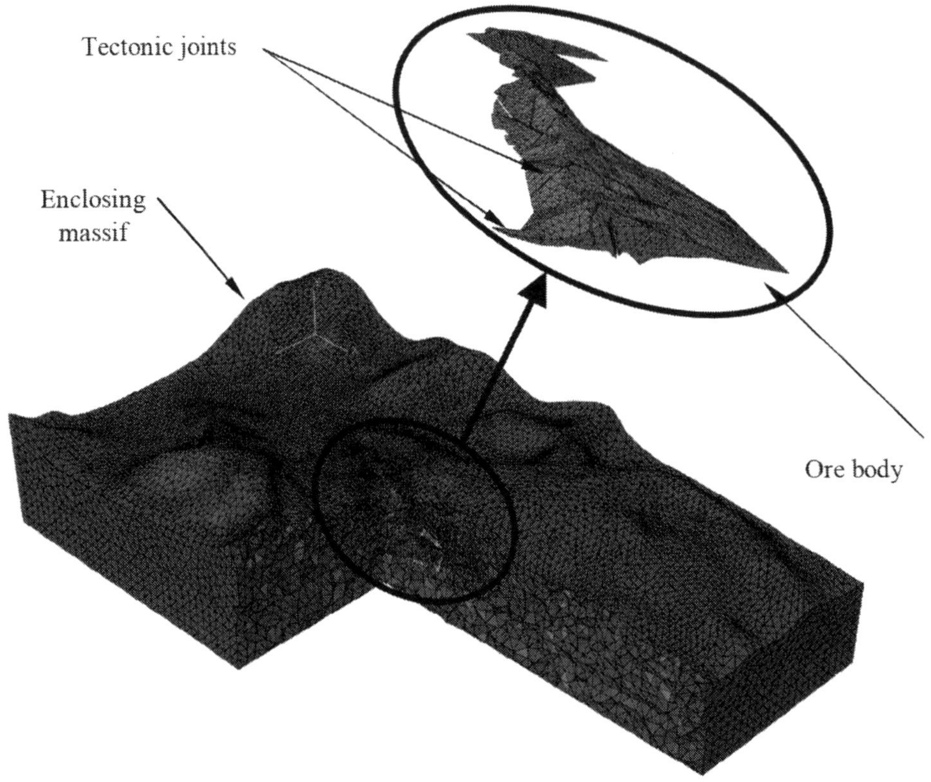

Fig 1. Analytical model with finite element mesh

The behavior of ores and rocks in enclosing massif is describing by elastic model. Deformation characteristic of massif in global numerical model was equal deformation characteristic of ores and rocks. For reason of some limits in calculation capacity and big scale the created numerical model was accepted some assumptions regard changing exist geometry of tectonic joints to some simple shapes. However, created tectonic joints were similar to exist in nature. The fragment of calculation results of the maximum principal stresses in the ore deposit at the time of the development of overlying levels is shown in Fig. 2.

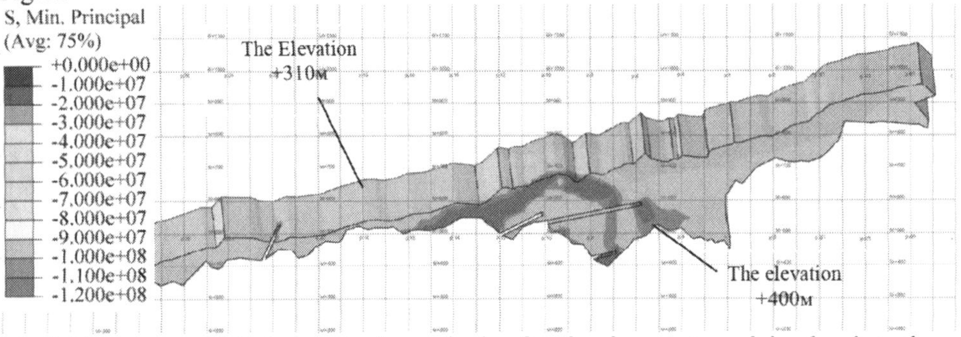

Fig 2. Maximum principal stress in the ore body after development overlying levels to the elevation +400m, Pa

Analysis of the results revealed irregular distribution of maximum principal stresses in ore body both along depth and plan. The main reasons of these results are complex Earth surface and gravitational-tectonic character of distribution stresses.

The local numerical 3D model of stope block was created in CAE SIMULIA Abaqus software. Therefore, we can forecast of stress state of massif according extraction front and other mining activity by now and near future.

As boundary conditions for local numerical model was taken the stresses of global numerical model accordance same points [3].

Thus calculated components of stress tensor for local numerical model have following details: σ'_{xx} = 62.5 MPa; σ'_{yy} = 41.5 MPa; σ'_{zz} = 10.5 MPa.

Furthermore the numerical model has following geometric details: 3500 m × 2000 m × 2000 m (according with direction main axes OX, OY, OZ). Beside the local numerical model has some features of stope block 7/10:

- relative space location of the three stope levels +310m, +325m, +375m in stope block 7/10 according elevations and angle of slope ore body;
- boundaries of share zone with linear angle parameters such as angle of slope β and angle of fracture β'; an angle of slope β determines boundaries of share zone from the side of hanging wall against low boundary of waste area;
- influence exist front extraction on stress state of massif around stope block 7/10 on level +310m.

The analytical model of local numerical model is shown in Fig.3. There are main principles of sublevel caving system presented in Fig.3.

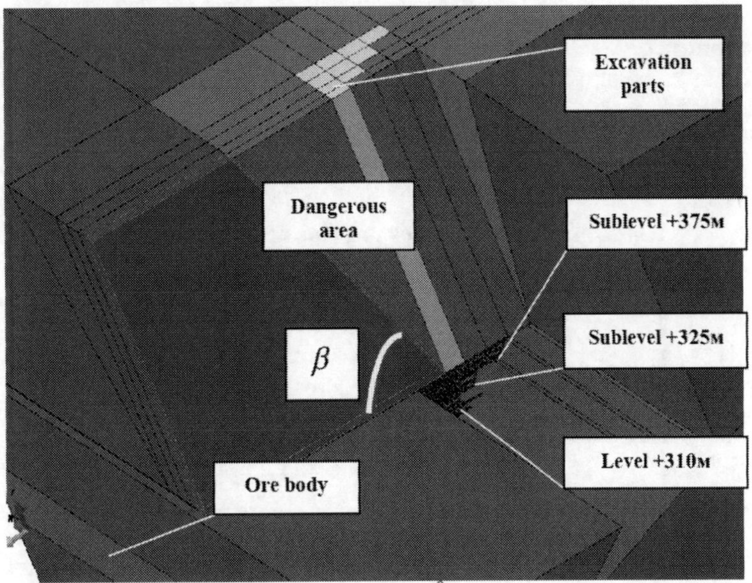

Fig 3. Analytical model: β - angle of slope, $\beta = 70^0$

3. Results and discussion.

Geomechanical safe sizes of pillars, height sublevel and development order of stope block was calculated by numerical simulation.

The fragment of calculation results of the minimum principal stresses in pillars wide of 13.2m (3.3D, D – span of stope working) for stope level +310m with gate order development is shown in Fig. 4.

On Fig.4 you can see that displacement from hanging wall to bottom is reason for common reduction of stresses in pillars of stope level [4].

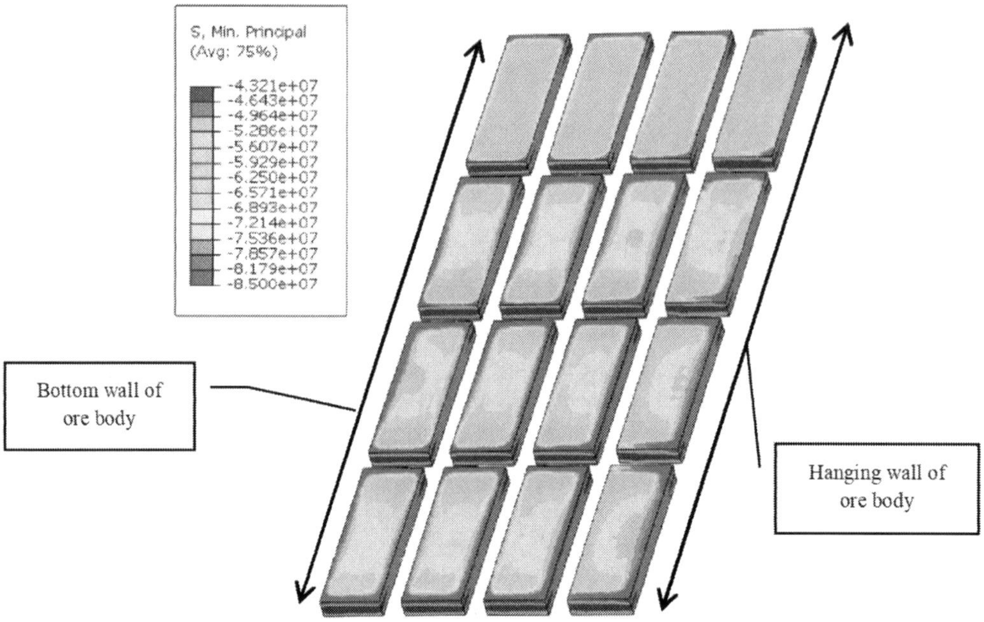

Fig 4. Maximum principal stress in pillars wide 13.2m for stope level +375m, Pa (gate order development)

Similar results we have using numerical modeling of pillars for width 11.2 m (2,8D) and 15.2 m (3,8D).

On Fig. 5 you can see chart of strength pillar dependence on it geometric sizes. Additionally in Fig.5 shown chart with maximum principal stress in pillars accordance geometric sizes.

Safe size of pillar determined as intersection both charts with considering coefficient of safety.

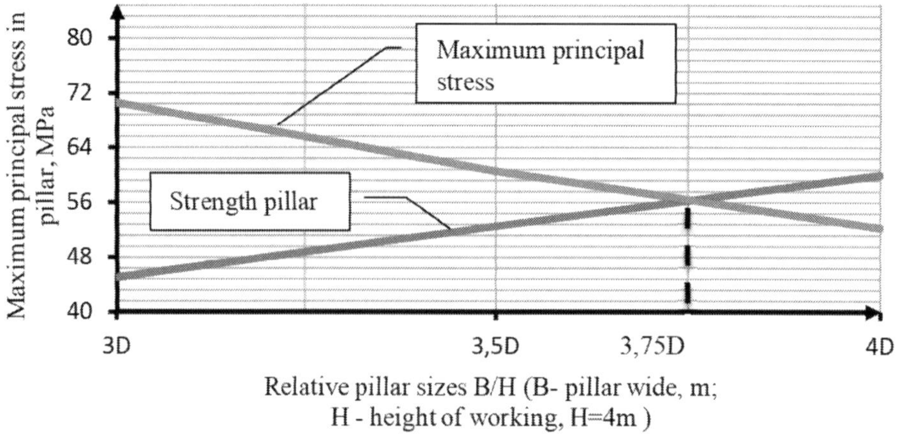

Fig 5. Determining of safe size pillar for stope level +375m with considering coefficient of safety

4. Conclusions.

The method of creating the local numerical models based on calculation results of global numerical models is important tool, which can use for geotechnical calculations with considering extraction fronts and other mining activity in deposit.

The big scale numerical model allows to take into account complex surface of mining region. Besides using numerical models may help with establish dependency degree of stress state massif from landform.

Estimation impact of tectonic joints on stress state massif around stope workings should be made based on results of geodynamic zoning using created big scale numerical model.

Along path from hanging wall of ore body to bottom wall is observed common reduction of stresses in pillars.

In this way safe pillar sizes from point of view geomechanics are determined according to results of numerical simulation using multi scale numerical models.

Reference.

1. Trout, P. L. Formulation and application of new underground mine scheduling models. PhD thesis, The University of Queensland, Brisbane, Queensland, Australia, 344pp. (1997)

2. Beck, D. A. and Duplancic, P. Forecasting performance and achieving performance indicators in high stress and seismically active mining environments. In Y. Potvin and M. Hudyma (eds.), Controlling Seismic Risk, Proceedings of the Sixth International Symposium on Rockburst and Seismicity in Mines, Perth, Western Australia, Australia, 9–11 March, pp. 409–417. (2005) Perth, Western Australia, Australia: Australian Centre for Geomechanics.

3. Kuranov, A.D. and Bagautdinov, I.I. Experience of geomechanical research and calculations toward ore deposit development in complex mining conditions. Geomechanics and Geodynamics of Rock Masses – Litvinenko (Ed.). Taylor & Francis Group, London, 1037-1042 pp. (2018)

4. Lunder, P. Hard rock pillar strength estimation: An applied empirical approach. MASc thesis, University of British Columbia, Vancouver, British Columbia, Canada, 166pp. (1994)

Multifractal analysis of electron images of fossil coal surface

Vasiliy Malinnikov[1*], *Valeriy* Zakharov[2], *Denis* Uchaev[1], *Dmitry* Uchaev[1] and *Olga* Malinnikova[2]

[1]Moscow State University of Geodesy and Cartography, Department of Space Monitoring and Ecology, Moscow, Russia
[2]Institute of Comprehensive Exploitation of Mineral Resources Russian Academy of Sciences, Moscow, Russia

Annotation. The multifractal spectra of electronic images of fossil coal surfaces obtained from outburst-hazardous zones, outburst-nonhazardous (quiet) zones and outburst zones are studied. It is shown that the structural organization of coal surface elements can be represented by a multifractal with a corresponding $f(\alpha)$-spectrum of fractal dimensions, which is suitable to determine quantitative difference in microstructures of coal specimens from outburst-hazardous and outburst-nonhazardous beds. Coals from outburst-hazardous and outburst-nonhazardous beds differ in the direction of skewness of multifractal spectra characterizing their surface structure. If the $f(\alpha)$-spectrum is skewed to the right, then the analyzed coal specimen belongs to an outburst-hazardous bed, and on the contrary, if $f(\alpha)$ is skewed to the left, then it can be concluded that the coal specimen most likely belongs to an outburst-nonhazardous bed.

1 Introduction

Currently, fossil coal is widely studied using the scanning electron microscopy (SEM) for direct multi-scale visualization of micro- and nanostructures. SEM microscopes have some advantages over light microscopes [1, 2]:

- great depth of focus, which allows analyzing both polished sections and relief surfaces;
- high image magnification: the ultimate magnification of the up-to-the-minute optical microscopes is limited to 2,000 times while SEM microscopes offer magnification to 1,000,000 times;
- sharp contrast imaging of multi-phase specimens since the image contrast in scanning electron microscopes is conditioned by an atomic number of a substance.

Supermolecular structure parameters estimated by means of scanned coal surface images (SCSI) are used to analyze processes of coal formation and the related methane release factor of coal, to calculate coalbed methane release during mining and gas drainage, to predict gas-dynamic events in coal mines etc. Naturally, the researchers search and

* Corresponing author: malinnikov@mail.ru

develop new methods and techniques for automated interpretation of SCSI with the maximized correction for the probabilistic spectral and spatial structure of bituminous coal. The analysis of fractal and multifractal properties of SCSI is a promising trend in the geophysics inasmuch as it takes into account the fact that the energy spectrum of scanned coal surface images decreases as the spatial frequency increase by power law [3]. Such spectrum is invariant under multiplicative change in scale. The ideal fractal images possess no such notions as a typical size or a characteristic spatial frequency: what happens in a spatial-frequency interval can occur at any scaled frequency range. On the other hand, real image can be assumed as a fractal only within a certain range of spatial frequencies. As a rule, this range is very limited and hardly detectable due to instability of energy spectrum estimates obtained from the images, and owing to the influence of a transfer function of the specific image visualization system. For example, in electron microscopy imaging of microstructures at a tolerable contrast, the frequency response characteristic defines frequency bands visualized at the maximum and minimum contrast. This being the case, the spatial spectral features and typical scales fail to describe real characteristics of an image. It should be taken into account that the transition to a new scale for some discrete image is only possible in case of scale decrease [4].

The purpose of this work is to demonstrate results of multifractal analysis of images of fossil coal surfaces, aimed to identify differences in the spatial structure of outburst-hazardous and outburst-nonhazardous coals.

2 Methodology of research

The methodology of research included five successive stages.

1. Visual interpretation of original electronic images to pre-identify objects, to eliminate noises, blur and degradation, to highlight boundaries of irregularities, ill-defined regular components etc.

Images of scanning electron microscopy (SEM images) of coal specimens are usually characterized with a high level noise from a number of sources [5]. The paper [6] shows that the Gaussian distribution can be used to approximate noise in SEM images, which appreciably complicates further interpretation due to considerable difference between the values of characteristics obtained using images with and without the Gaussian noise.

In this paper, the multi-scale filter for improving the image quality in the Chebyshev moment domain was used to suppress the noise of SEM images of coal specimens [7]. This filter has already demonstrated its effectiveness in suppressing noise in coal specimen images obtained by SEM [8].

2. Exploratory analysis of investigated SEM images, including:

- assessment of statistical law and characteristics of grey levels in SEM images.
- formation of two-dimensional spatial Fourier spectra and their informative features for analyzed images or sub-images.

During the image analysis, we used power spectrum or power spectrum in log–log axes to calculate the informative features of spatial spectrum (SS) listed below:

- aspect A characterizing elongation of SS: $A = l_{max} / l_{min}$, where l_{min} and l_{max} are shortest and longest sides of a outlined spectrum, respectively;
- SS connection parameter, which is equal to one for simply connected spectra and to zero for multiply connected spectra;
- slopes of one-dimensional sections of SS in different directions;
- minimum, maximum and average values and dispersion;
- central moments;
- variation coefficient, excess and asymmetry.

3. Checking the hypothesis about a multifractal nature of coal surface SEM images or sub-images.

The hypothesis about a multifractal nature of grey-level SEM images of coal surface was checked using a number of criteria:

– The energy spectrum of the images decreases as the spatial frequency increases by power law. Such spectrum is invariant under multiplicative variation in scale [9].
– Statistical distribution of grey levels for the images is a near to log-normal [10].

4. Calculation and analysis of multifractal characteristics for investigated SEM images or sub-images.

This stage involves calculation of multifractal spectra $D(q)$ и $f(\alpha)$ and derived characteristics by means of local-global multifractal analysis of digital images [11]. Evaluation and analysis of multifractal characteristics of the investigated images were processed in GEO-PC environment developed at MIIGAIK. Further, informative multifractal characteristics were investigated in order to estimate their usefulness to differ images of coals obtained from outburst-hazardous and outburst-nonhazardous beds, as well as coals taken from outbursts (Table 1).

Table 1. List of the most informative multifractal characteristics.

Multifractal characteristic	Calculation formula
$\Delta_q D$ — order parameter	$\Delta_q D = D_1 - D_q \quad (q >> 1)$
α_{min} — minimum value of Hölder exponents α	$\alpha_{min} = \min\limits_{1\le i\le W, 1\le j\le H} \alpha_{i,j}$, where W and H — are width and height of the analyzed image or its local region
α_{max} — maximum value of Hölder exponents α	$\alpha_{max} = \max\limits_{1\le i\le W, 1\le j\le H} \alpha_{i,j}$ where W and H — are width and height of the analyzed image or its local region
$f(\alpha_{min})$ — fractal dimension of the image subset characterized by α_{min}	$f(\alpha_{min}) = \inf\limits_{q\in R}\{\alpha_{min}q - \tau(q)\}$
$f(\alpha_{max})$ — fractal dimension of the image subset characterized by α_{max}	$f(\alpha_{max}) = \inf\limits_{q\in R}\{\alpha_{max}q - \tau(q)\}$
$\Delta\alpha$ — range of Hölder exponent values α (width of $f(\alpha)$)	$\Delta\alpha = \alpha_{max} - \alpha_{min}$
S_L — area of a figure under the left branch of $f(\alpha)$	$S_L = \int_{\alpha_{min}}^{\alpha_0} f(\alpha)d\alpha$
S_R — area of a figure under the right branch of $f(\alpha)$	$S_R = \int_{\alpha_0}^{\alpha_{max}} f(\alpha)d\alpha$
S_f — area of a figure under $f(\alpha)$	$S_f = S_R + S_L$
r — relative asymmetry coefficient for $f(\alpha)$	$r = \dfrac{\alpha_0 - \alpha_{min}}{\alpha_{max} - \alpha_0} = \dfrac{\Delta_L\alpha}{\Delta_R\alpha}$
R — normalized asymmetry coefficient for $f(\alpha)$	$R = \dfrac{\Delta_L\alpha - \Delta_R\alpha}{\Delta_L\alpha + \Delta_R\alpha} = \dfrac{\Delta_L\alpha - \Delta_R\alpha}{\Delta\alpha}$

5. Analysis of results of multifractal analysis of coal surface SEM images.

The final stage of the proposed procedure consists in an expert analysis of the obtained multifractal parameters of coal surface SEM images.

3 Source Data

In this study, imaging of coal specimen surfaces was performed using scanning electron microscopes JEOL JSM 5910-LV and Jeol-6610-LV. The spatial resolution of the microscopes is more than 10 and 100 nm for secondary and reflected electrons, respectively. Low-energy secondary electrons are used in imaging surface topography. Natural-shape coal specimens were placed in a work camera via a gate. In the mode of registration of secondary electrons, the work camera was vacuumized (with >10-6 mm hg vacuum). Secondary electrons were recorded by a standard detector, which a type of a sweeping-field photomultiplier tube connected to scintillator.

As the source data for this study was IPKON collection of coal specimens from the Zapolyarnaya mine (Vorkuta) and the Kirov mine (Leninsk-Kuznetsk), obtained from outburst-nonhazardous zones, outburst-nonhazardous zones and outburst zones. For our research, we analyzed microstructure of coal surface in the images magnified 1,000 times which showed coal grains with a characteristic size from .5 to a few microns. Methane can desorb from such grains, diffuse and flow in fractures as free gas [12].

4 Research Results

Results of Fourier analysis demonstrated that the different irregularities in the coal surface microstructure are described by spatial spectra with different shape and statistical characteristics. At the same time, the spectrum curve for all analyzed specimens can be approximated by a power function, which proves the hypothesis about the fractal nature of the analyzed images.

The statistical analysis of the analyzed images shows that their grey-level distributions have peaks and a distinct positive asymmetry. Approximations of grey-level distributions showed that they could be described by a log-normal law with 95% confidence level. This important outcome is a consequence of a multifractal structure of analyzed coal surface images.

Thus, both tested criteria showed the feasibility of multifractal analysis and the possibility of correct interpretation of results of the multifractal analysis. On this basis, for each image, multifractal spectra $D(q)$ and $f(\alpha)$ with characteristics derived from these spectra were calculated. Then, for each multifractal characteristic, it was analyzed the possibility of each multifractal characteristic to serve as a disciminative feature to differentiate outburst-nonhazardous and outburst-nonhazardous coal beds using images of coal specimens from these beds. As a result of the analysis, it was revealed that the normalized asymmetry coefficient R for $f(\alpha)$-spectrum has the greatest deciphering potential. After calculation of R values for all analyzed coal specimens, we found that the surface structure of outburst-hazardous coal specimens can be characterized by a fractal dimension spectrum with positive skewness whereas the surface structure of outburst-nonhazardous coal specimens is characterized by a symmetrical fractal spectrum (with negative skewness close to zero) (figure 1). This result is verified for nearly 80% of all test coal specimens.

For each of our groups of coal specimens, ranges of the most likely values for R were also obtained (shown as dashed lines in Fig. 1). As can be seen from Fig. 1, approximately 60% of all estimates for R obtained for coals from outburst-hazardous zones, outburst-nonhazardous zones and outburst zones fall into non-overlapping ranges [–0.845, –0.505],

[−0.005, 0.215] and [−0.300, 0], respectively. Thus, we can assume that the asymmetry of fractal dimension spectra characterizing coal surface structure can be used as a criterion for verification of coal susceptibility to gas-dynamic events.

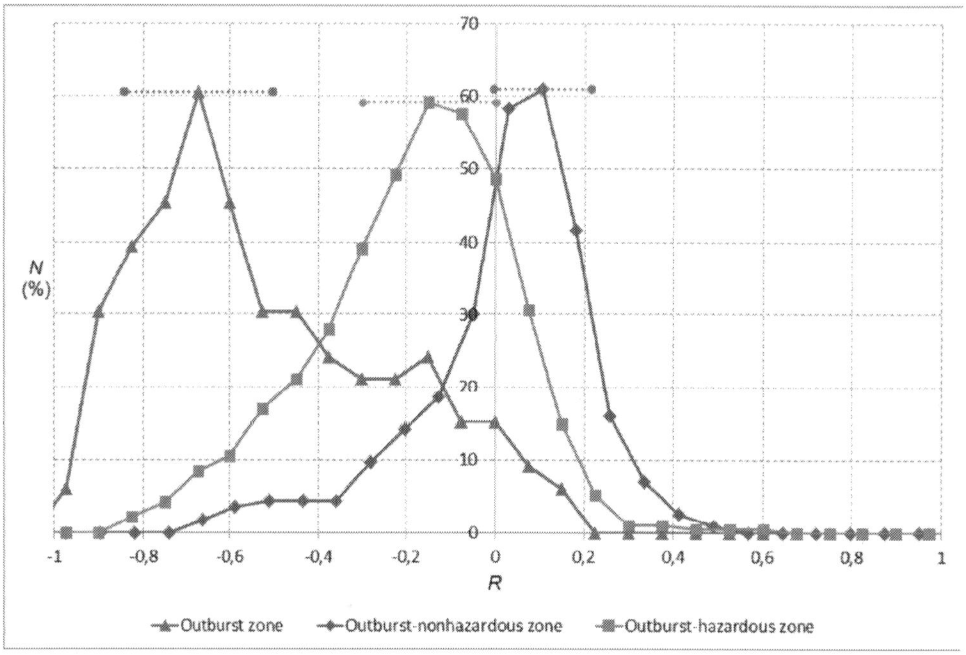

Fig. 1. Number N (%) of estimates for R, calculated for analyzed coal specimens. The distance between centers of bins and the bin width for each histogram were chosen so that about 60% of all estimates for R fall into the ranges of the most likely values for R.

5 Conclusions

Based on the implemented research, it has been concluded that the multifractal approach is efficiently applicable to investigations of fossil coal damage and can be used to estimate the connection between the fractal properties of damaged coal structure and coal susceptibility to gas-dynamic events. As our investigations have shown, there is a certain relationship between the degree of coal susceptibility to outbursts and the asymmetry of multifractal spectra calculated using coal specimen images. The obtained criterion relations can be used in the preliminary prediction of coal susceptibility to gas-dynamic events by the structural factor; however, these relations are advisable to refine through analyzing a wider class of different scale electron images of coal surface.

The development of the technique for exploratory multifractal analysis, as well as the technique and algorithm for calculating multifractal characteristics, was conducted in the framework of the state task of the Ministry of Education and Science of the Russian Federation (No. 5.6470.2017/8.9). The remaining studies were supported by the Russian Science Foundation, Project No. 16-05-00816.

References

1. W. Zhou, Z.L. Wang, *Scanning microscopy for nanotechnology: techniques and applications (BINOM: Laboratory of Knowledge, 2014).*
2. L. Chen, J. Xu, J. Chen, Science China Earth Sciences. **58**(10), 1768–1778 (2015).

3. V.A. Malinnikov, O.N. Malinnikova, V.N. Zakharov, Procs of XXVII International School of Academician Khristianovch «Deformation and destruction of materials with defects and dynamic phenomena in rocks and excavations», 158–162 (2017).
4. B.N. Grudin, V.S. Plotnikov, N.A. Smolyaninov, Optoelectronics, Instrumentation and Data Processing, **46**(3), 215-221 (2010).
5. D.V. Nesterenko, Computer Optics, **35**(2), 166–174 (2011).
6. N. Marturi, S. Dembélé, N. Piat, Scanning. **36**(4), 419–429 (2014).
7. Dm.V. Uchaev, D.V. Uchaev, V.A. Malinnikov, Proc. SPIE, **9875**, 987512 (2015).
8. O.N. Malinnikova, V.A. Malinnikov, Dm.V. Uchaev, D.V. Uchaev, Procs of XXVII International School of Academician Khristianovch «Deformation and destruction of materials with defects and dynamic phenomena in rocks and excavations», 134–139 (2016).
9. M. Schroeder, Fractals, Chaos, Power Laws: Minutes from an Infinite Paradise («RHD», 2001).
10. S.V. Bozhokin, D.A. Parshin, Fractals and multifractals («RHD», 2001).
11. V.A. Malinnikov, D.V. Uchaev, Dm.V. Uchaev, Izvestia vuzov «Geodesy and aerophotography», 4, 64–68 (2010).
12. O.N. Malinnikova, V.N. Odintsev, V.A. Trofimov, Methane/GIAB, SI 11, 189-204 (2009).

Estimation of the thermo-hydrodynamic parameters of the coolant in the Koshelev geothermal system

Dmitry Mamaev[1*]

[1]Research Geotechnological Center, Far Eastern Branch of the Russian Academy of Sciences, Petropavlovsk-Kamchatsky, Russia.

Abstract. The Koshelev geothermal system is located in the southern part of the Kamchatka Peninsula. On the surface, the system manifests itself by steam-hydrotherms at two thermal fields. Estimates of the predicted power from different data sources vary significantly, because the parameters of the coolant at great depths are not currently known and are presumable. Based on the available geological and geophysical data, a numerical three-dimensional thermo-hydrodynamic model of the geothermal system has been developed. The model was calibrated according to thermometry data in exploratory wells. The calibrated model adequately describes the distribution of thermo-hydrodynamic parameters of the coolant in the system. It can be used to evaluate the coolant parameters during the development of the thermal resources of the Koshelev geothermal system. In the course of computational experiments, the developed model is used to obtain the distribution of the thermo-hydrodynamic parameters of the hydrothermal coolant in the geothermal system in the natural state and during the development of thermal resources.

1 Introduction

The Koshelev geothermal system is located in the southern part of the Kamchatka peninsula, 15 km east of the coast of the Sea of Okhotsk. It belongs to the Pauzhetko-Kambal-Koshelev geothermal region, which is part of the South Kamchatka geothermal province [1]. The Koshelevsky volcanic massif is composed of five volcanic structures of different ages, which are partially overlapping: Western, Valentine, Central, Eastern (Figure 1) and Ancient. The latter is buried under more young buildings.

On the surface, the geothermal system is manifested by unloading steam hydrotherms at two thermal fields. The Lower Koshelev thermal field is located on the slope of the Western volcano at elevations of 700-800 m. The field area along the 20°C isotherm at a depth of 0.5 m is about 38,000 m^2. The maximum temperature of thermal manifestations is 127 °C [2].

Upper Koshelev thermal field located in an erosion crater Valentin at elevations 1100 - 1300 m [2]. The area of the field along the isotherm of 20°C at a depth of 0.5 m is about

* Corresponding author: ghtio@mail.ru

303,000 m^2. The most typical form of thermal manifestations are powerful jets of superheated steam at a flow rate of 0.1-0.4 kg/s, at a temperature of 120-153°C [2]. There are more than 40 such jets, and the number of smaller ones is hundreds.

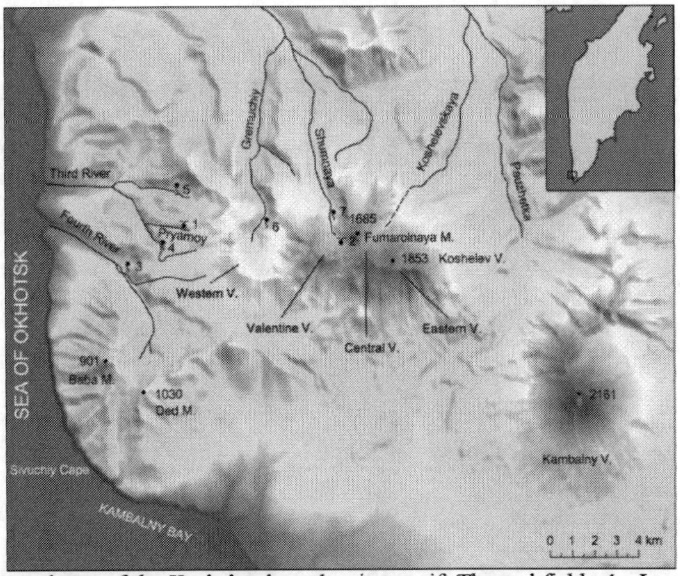

Fig. 1. Overview scheme of the Koshelevsky volcanic massif. Thermal fields: 1 - Lower Koshelev; 2 - Upper Koshelev. Thermal springs: 3 - Sivuchinskie; 4 - Promezhutochnye; 5 - Skazka; 6 - Caldernye; 7 - Shumnye.

The table contains estimates of the forecast geothermal resources of the Lower Koshelev thermal field and the Koshelev geothermal system as a whole, according to data from various sources.

Table 1.

Data source	Natural heat transfer, MW		Forecast electric power, MW			
			The whole of the Koshelev system		Lower Koshelev field	
	The whole of the Koshelev system	Lower Koshelev field	According to the heat transfer	According to the reservoir thermal energy	According to the heat transfer	According to the reservoir thermal energy
[2]	306	105	–	–	30	–
[3]	–	–	–	87	–	–
[4]	–	–	144	215±64	48	100±30
[5]	–	–	943		–	–

Estimates of the predicted power for different data sources vary significantly. It should be noted that calculations based on the thermal energy of the reservoir are approximate. The characteristics of the reservoir and coolant at great depths to the present time are largely presumed [4]. In this paper, the distribution of thermo-hydrodynamic parameters of the coolant in the Koshelev geothermal system is established using a model calibrated according to actual data.

2 Materials and methods

A set of available data on the geological structure of the Koshelev geothermal system was used to construct its numerical three-dimensional thermo-hydrodynamic model. Calibration of the developed model, performed on the basis of exploratory drilling data, ensures the adequacy of the model to the object of investigation. In the course of computational experiments, the developed model is used to obtain the distribution of the thermo-hydrodynamic parameters of the hydrothermal coolant in the rocks of the geothermal system in the natural state and during the development of thermal resources. For the thermo-hydrodynamic calculations of the parameters of the geothermal fluid, the HYDROTHERM program [6] was used, which is capable of performing numerical three-dimensional modeling of heat and mass transfer processes in a porous medium in the temperature range 0-1200°C and pressures of 0.05-1000 MPa, which allows using it for supercritical thermodynamic conditions.

The modeling area is a rectangular parallelepiped measuring 17 km in latitudinal direction, 6.8 km in meridional and 9.3 km in vertical direction (Fig. 2). The lower boundary of the modeling area is the horizontal plane at a depth of 7400 m below sea level. The upper limit of the models is set so that the maximum in the terrain of 1853 m falls into the modeling area - this is the plane at an altitude of 1864 m above sea level.

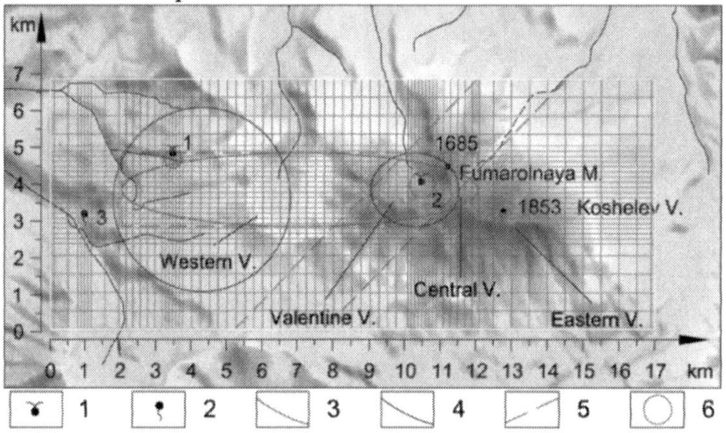

Fig. 2. Modeling area and computational grid. 1 - thermal fields with steam jets (1 - Lower Koshelev, 2 - Upper Koshelev); 2 - thermal springs (3 - Sivuchinskie); 3 - boundary of the magma chamber; 4 - boundaries of subvolcanic intrusions; 5 - boundaries of tectonically disturbed zones; 6 - thermally withdrawing channels.

Three-dimensional numerical thermo-hydrodynamic models of the Koshelev geothermal system are constructed on the basis of analysis of available geological and geophysical data [2, 3, 7-13] and physical properties of rocks [14-18]. The structure of the models in the section is shown in Fig. 3.

The boundary conditions are defined as follows: on the lower and lateral boundaries of the modeling area, there is no flow of heat and fluid; on the day surface - a constant temperature of 10°C and a constant atmospheric pressure. The region of the magma chamber is given by a boundary condition with a constant temperature of 900°C. The initial distribution of rock temperature is given by an average geothermal gradient of 30°C/km. The initial distribution of fluid pressure in the array is hydrostatic. The simulation time is 40 thousand years. Spatial sampling of the model is made by means of a rectangular grid measuring 71×28×55 cells with an irregular step of 100-500 m. In total, the model contains 109,340 computational blocks.

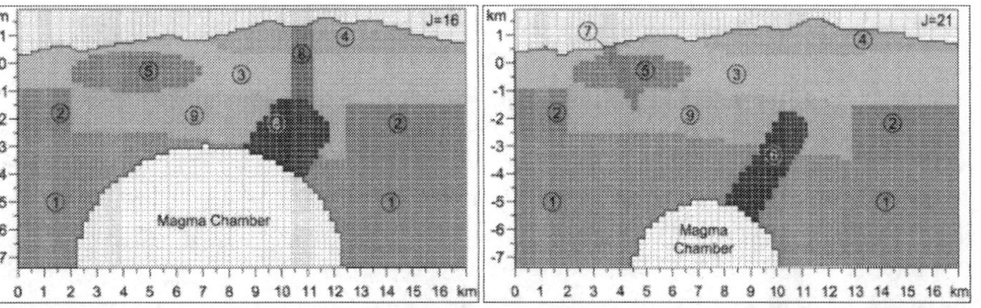

Fig. 3. Rock domains and computational grid. Left - slice through the center of Upper Koshelev field, on the right - through the center of Lower Koshelev field.

A number of numerical calculations of models with different values of rock permeability were made. The model was calibrated using four-dimensional unconditional optimization by the deformable simplex method (Nelder-Mead method). Data for the temperature in the five wells at the Lower Koshelev field were used to calibrate the models. The target function is the average linear deviation between the actual bottomhole temperature and the temperature of the calculating block corresponding to the spatial position of the bottomhole in the model:

$$f(k_3, k_5, k_7, k_9) = \sum_{i=1}^{5} |T_i - T_{0i}|/5 \tag{1}$$

where T_i – calculated bottomhole temperature; T_{0i} – actual bottomhole temperature; k_3, k_5, k_7, k_9 – permeability of rock domains 3, 5, 7 and 9, respectively.

3 Results and discussion

In Fig. 4 shows the distribution of the temperature and the phase state of the coolant in the calibrated model. The average deviation in the wells is 30.9°C, or 15.9%.

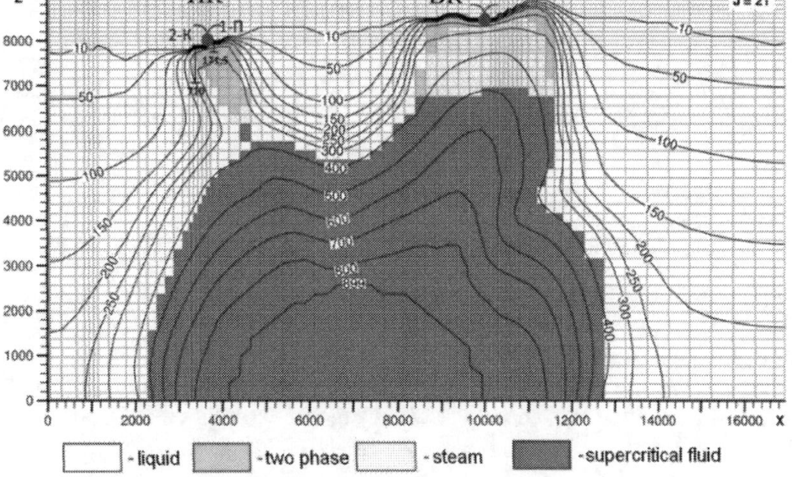

Fig. 4. Distribution of temperature (°C) and phase state of the coolant.

The calibrated model can be used to study the spatial distribution of the thermo-hydrodynamic parameters of the coolant and their changes during the development of the thermal resources of the Koshelev system with the help of numerical thermo-hydrodynamic modeling.

The results of calculating the calibrated thermo-hydrodynamic model at the final time of simulation adequately represent the distribution of thermo-hydrodynamic parameters in the system in the natural state. It used as initial conditions for determining the parameters of the coolant in the operation of the geothermal circulation system. The scheme for the development of the geothermal circulation system assumes the drilling of 6 wells with a depth of 1000 m: 3 production and 3 injection wells. In the calculations, the temperature of the injected coolant is assumed equal to 80°C. The specific enthalpy of the injected coolant corresponding to this temperature and the pressure at the bottom of the injection well is assumed to be 346 kJ/kg.

Thermo-hydrodynamic modeling of system operation at different values of coolant flow was carried out. Dynamics of the total thermal power of the geothermal circulation system and its impedance during operation for 30 years are shown in Fig. 5 and 6 respectively.

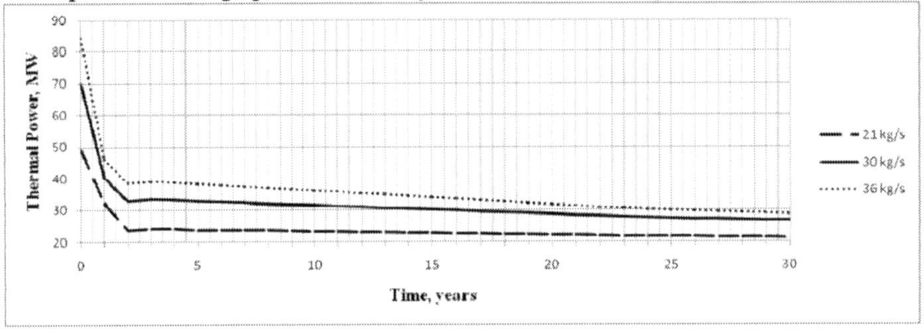

Fig. 5. Change in thermal power of the geothermal circulation system during the operating time for different values of the coolant flow rate.

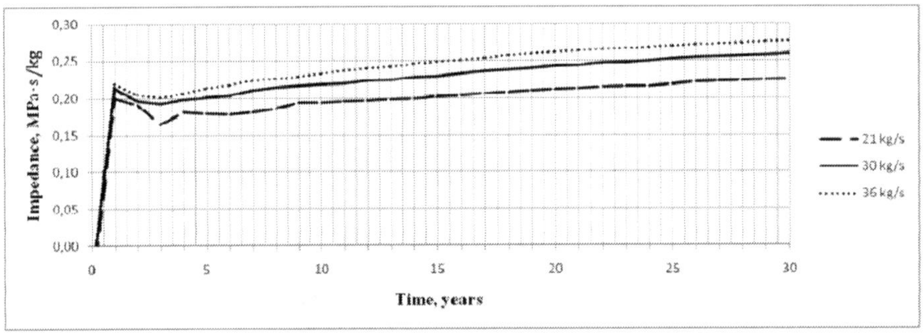

Fig. 6. Change in the impedance of the geothermal circulation system during the operating time for different values of the coolant flow rate.

During the first three years, the geothermal circulation system enters a stable operating mode. The thermal power slowly falls to 26 MW by the end of the service life with a total flow rate of 30 kg/s. As with increasing and decreasing the flow, there is a corresponding slight change in the values of thermal power and impedance. This indicates the stability of the geothermal circulation system with small changes in the coolant flow rate.

4 Conclusion

The model of the Koshelev geothermal system, developed and calibrated on the basis of the available geological and geophysical data, adequately describes the distribution of thermo-hydrodynamic parameters of the coolant. It can be used to study the parameters of the coolant and their changes over time in the development of thermal resources.

References

1. V.V. Averyev, Hydrothermal process in volcanic regions and its relation to magmatic activity, 118 (Moscow, 1966).

2. E.A. Vakin et al., Hydrothermal Systems and Thermal Fields of Kamchatka, 58, (Vladivostok, 1976).

3. A.I. Pozdeev et al., Volcanology and Seismology, **3**, 32 (2008).

4. V.M. Sugrobov et al., Materials of the International Field Kuril-Kamchatka Seminar, July 16 - August 6, 2005 (Petropavlovsk-Kamchatsky, 2005).

5. V.M. Sugrobov, Hydrothermal Systems and Thermal Fields of Kamchatka, 267 (Vladivostok, 1976).

6. K.L. Kipp et al., Guide to the revised ground-water flow and heat transport simulator: HYDROTHERM - Version 3, 160 (Reston, 2008).

7. P.P. Blukke, et al., Detailed search for deep zones of the Lower Koshelev field of steam-hydrotherms to provide the coolant for the first line of the Koshelev geothermal power plant with the capacity of 94-100 MW with preliminary exploration of its central part in 1989-1995. Book 1, 193, (Termalny, Kamchatka region, 1989).

8. V.L. Leonov, Structural conditions for localization of high-temperature fluids, 104, (Moscow, 1989).

9. M.I. Zubin, Geophysical fields and deep structure by geophysical data, Long-living center of endogenous activity of South Kamchatka, 172, (Moscow, 1980).

10. V.L. Leonov, Volcanology and Seismology, **5**, 32 (2001).

11. A.V. Kiryukhin et al., Hydrogeology of Volcanoes, 395 (St. Petersburg, 2010).

12. N.E. Litasov, The active volcanoes of Kamchatka, 384 (Moscow, 1991).

13. M.V. Pisareva, Volcanology and Seismology, **2**, 52 (1987).

14. V.V. Shanina et al., Geoecology. Engineering geology. Hydrogeology. Geocryology, **5**, 459 (2013).

15. V.V. Shanina et al., Water: Chemistry and Ecology, **1**, 3 (2015).

16. F.A. Yanovsky, Volcanology and Seismology, **5**, 77 (1989).

17. N.B. Dortman et al., Physical properties of rocks and minerals (petrophysics). Handbook of Geophysics, 527 (Moscow, 1976).

18. V.N. Kobranova, Physical properties of rocks, 470 (Moscow, 1962).

Method of placer mining by directional well drilling

Viktor Bagazeev[1], *Niyaz* Valiev[1], *Konstantin* Kokarev[1], *and Vasiliy* Startsev[*1]

[1]Federal State Budgetary Educational Institution of Higher Education "Ural State Mining University" Yekaterinburg, Russia

Abstract. The study aims at justifying the method for the development of placers by directed drilling of wells. A description is given of the opening of the site, the receding and oncoming scheme of the digging excavation, drilling and expansion of the wells, the main hydraulic unit with remote control. The physics-mechanical substantiation of technological parameters is given: hydro-jet erosion, the shape and size of the cleaning chambers, the adaptation of hydrotransport calculation methods. With jetting, the required flow rate and pressure of water to break rocks is based on the laws of soil mechanics. The parameters of the clearing space and the shape of the chambers are taken in accordance with the magnitude of the vault of natural equilibrium. As a determining factor in the calculation of gravity hydrotransport, the consistency of the pulp is adopted with observance of the condition necessary for the bearing capacity. To adapt the methodology for calculating the hydraulic lifting of pulp to the conditions of well development, the coefficient of dilution is taken as the main design parameter. Modeling of jetting jet in the laboratory conditions was carried out, dependences of decrease in axial pressure were obtained at removal of the face from the nozzle.

1 Introduction

A promising trend set to increase the gold reserves is to involve placer accumulations and deposits hard-to-reach by traditional technologies in the development using the method of hydraulic mining by horizontal directional well drilling (HDD). Horizontal directional drilling makes it possible to place a high-capacity hydromonitor barrel at the well bottom, to increase the productivity of hydraulic washing-out, to clean the placer bedrock, to exclude the impact of sediments on the well and the collapse of barren rocks (peats).

The research aims at justification of the method and technological parameters of hydraulic placer mining by means of directional well drilling on the basis of physical and mechanical properties of rocks.

The research focuses on the mechanisms of hydraulic washing-out, hydraulic transportation, and the stability of rock outcrops.

The justification of the method and its technological parameters is of crucial importance now when the placer reserves are being depleted while the mining depth is being increased.

* Corresponding author : Vasiliy.Startsev@m.ursmu.ru

The implementation of the method will make it possible to work out unmined areas of deep depressions and bedrock cuts as well as placers and man-made massifs having a small volume of rock mass and a high gold content.

The scientific novelty of the study is protected by the patent as a method of hydraulic well drilling of useful minerals [1].

The leading developers of hydraulic well drilling technologies in Russia are academician V. Zh. Arens, professor N. I. Babichev, V. L. Kolibaba, A. V. Pankov, A. S. Khrulev, E. I. Cherney, D. N. Shpak . Problems of hydraulic well drilling are also solved in the works of V. P. Drobadenko, G. A. Nurok, N. G. Malukhin, V. F. Khnykin, S. S. Shavlovskiy, and others.

2 Theory and experimental work

The mining complex of hydraulic mining by horizontal directional well drilling includes (Fig. 1):

- drilling equipment: drilling rig, drilling tool set, equipment used to prepare, deliver, and reclaim drilling mud, control location systems;
- washing equipment: hydromonitor barrel, pump and pipes to supply water under high pressure;
- pulp lifting equipment: soil pump, hydraulic elevator, air lift.

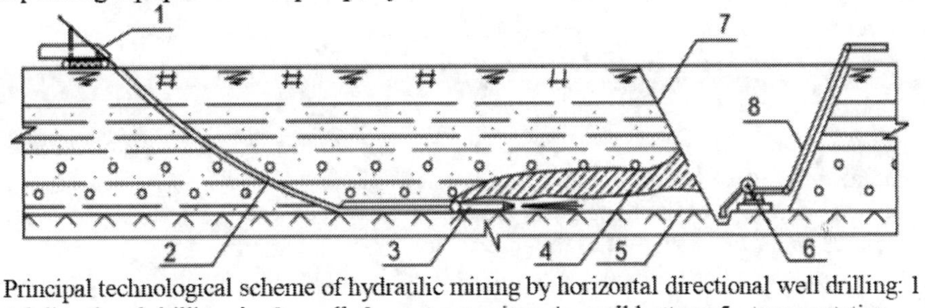

Fig. 1. Principal technological scheme of hydraulic mining by horizontal directional well drilling: 1 - horizontal directional drilling rig; 2 - well; 3 - water monitor; 4 – well bottom; 5 - transportation section of a well ; 6 - soil pump or hydraulic elevator; 7 - working trench; 8 – pulp line

Production wells are assumed to be drilled by a typical horizontal directional drilling rig used to lay pipelines [2]. Taking into account the diameter of the hydromonitor barrel, wells are drilled in stages:

- drilling of a pioneer well having a diameter of up to 200 mm (entering a massif at an angle of 28 - 30° or up to 40°);
- reverse washing-out of a pioneer well up to 400 - 450 mm.

The well accommodates a pressure conduit ending with a hydromonitor barrel. The hydromonitor barrel connected with a pipeline by means of a ball joint or a pressure hose is supposed to be manipulated in the horizontal plane.

As the devices identical in design, a typical hydromonitor barrel with two nozzles used to wash out the rocks under water and a typical hydromonitor barrel GMD (ГМД) -250 used for opencast workings are considered to be the most appropriate ones. It is advantageous to use a remote control cable stretched along the pipeline.

The lifting equipment is typical and its choice is determined by the mining depth.

To determine the technological parameters, it is necessary to provide a special physical and mechanical justification of hydraulic washing-out and stopes formation as well as the adaptation of hydraulic transportation calculation methodologies.

The mechanism of hydraulic fracturing of clay rocks is outlined as follows: the rock mass is solid, isotropic and impermeable; the loading condition with an impinging water jet is quasistatic; the destructive load is applied to the area of a circle whose radius is somewhat smaller than the radius of a water jet; the load distribution is uniform. When the destruction takes place, the value of pressure P_p, MPa, is taken as calculated according to the Prandtl-Reisner formula [3]approximated for loamy rocks within the internal friction angle φ = 10 - 35° by the expression:

$$P_p = c \cdot ctg\varphi \cdot 0.00025 \cdot \varphi^{2.6}, \quad (1)$$

where c is the soil adhesion, MPa.

The destructive load corresponds to the head of the water jet at the well bottom H_{min}, m:

$$H_{min} = K_3 \frac{P_p}{2g \cdot \rho} = 55 \cdot P_p, \quad (2)$$

where K_3 = 1.05 - 1.1 is the safety factor;

g = 9.81 m/s^2;

ρ = 1.0 t / m^3 is the water density.

The required amount of water [4]:

$$q_т = \frac{7.05\sqrt{H \cdot E \cdot \rho}}{(f \cdot H - H_{min})}, \quad (3)$$

where $q_т$ is the discharge of water , m^3, having the head H, m, to wash out 1 m^3 of rocks;

f is the coefficient of the relief of axial pressure of the water jet due to the increase in the distance from the well bottom to the nozzle l.

The analytical expression of the coefficient f was derived on the basis of analogy with S. S. Shavlovskiy's studies of the opencast workings [5] and the results of the axial pressure measurements under laboratory conditions (Fig. 2).

$$f = \frac{P_l}{P_0} = exp \cdot (-d_n \cdot l^{0.78} \cdot \sqrt{H}), \quad (4)$$

where P_l, P_o are the values of the axial pressure of the jet at a distance l and in the immediate vicinity of the nozzle, kPa;

d_n is the diameter of the nozzle of the hydromonitor barrel, m;

H is the water head in front of the nozzle, m

l is the distance from the nozzle to the well bottom, m.

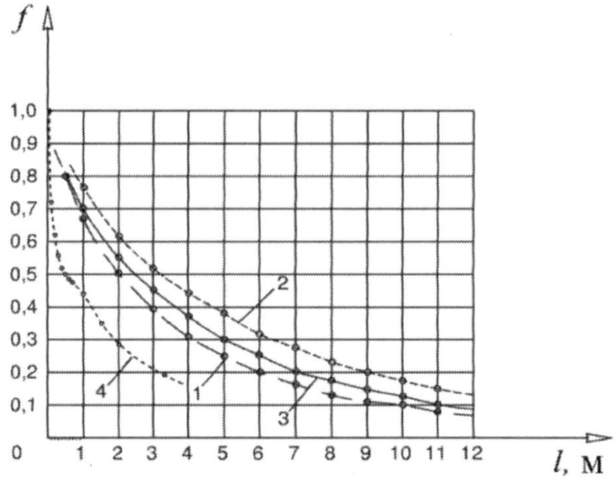

Fig. 2. Dependence of the coefficient of the relief of the axial pressure of the medium-pressure jet on the distance to the nozzle: 1, 2, and 3 are based on S.S. Shavlovskiy's measurement results, while [5], 4 are based on the model results;

1 - $d_н\sqrt{H}$ = 0.4; 2 - $d_н\sqrt{H}$ = 0.28; 3 - $d_н\sqrt{H}$ = 0.34; 4 - $d_н\sqrt{H}$ = 0.0032

The features which can be distinguished as specific to slushy gold-bearing placer mining are as follows:

- the lack of a stable roof and a well- defined boundary line between the sands and peats;
- the presence of coarse clastic material in placer deposits;
- the possibility of the roof collapse.

At the same time, the parameters of the exposed stope (length, width, area) are not limited to the conditions of the required long-term stability. Under these conditions, the corresponding spatial shape of the stope washed out by a water jet will look like ½ part of the spherical sector with its top at the nozzle of the hydraulic monitor placed in the well. In the plan a stope has the form of a sector.

In accordance with the hypothesis of the dome of natural equilibrium [6] and the formula of Borisov A. A.

$$L_d = (2.0 \div 3.0)h_d)\, L_{\text{пр}} = (2{,}0 \div 3{,}0)h_{\text{св}}, \tag{5}$$

where L_d, h_d are the width and height of the dome of natural equilibrium, m.

If the average value of the parabolic arch $h_{\text{св}}$ is taken to be equal to the thickness of the productive placer deposits M, m, the width of the stope $B_{\text{к}}$, m, appears to be as follows:

$$B_{\text{к}} = \frac{3}{2}M \,. \tag{6}$$

In case of hydraulic placer mining, it is advisable to take the pulp consistency expressed as the solid to liquid ratio (Т:Ж) as the determining factor for the calculation of the hydraulic transportation of rocks,. The solid to liquid ratio (Т:Ж) is the amount of water, m^3, per 1 m^{3}of the solid in the pulp. According to A. I. Kuprin's formula [7], the gravity flow of the pulp will have a sufficient load-bearing capacity at the ratio:

$$\text{Т}:\text{Ж} \geq \frac{\rho_{\text{тв}}}{\rho_{\text{тв}} - \rho_{\text{п}}} \cdot \frac{\text{К}_{\text{т}} \cdot J}{f_{\text{л}} - J}, \tag{7}$$

where $\rho_{\text{тв}}$, $\rho_{\text{п}}$ are the densities of solid particles and the pulp, t / m^3;

$\text{К}_{\text{т}}$ = 0.07-0.14 is the coefficient of the energy transfer to the flow of liquid;

$f_{\text{л}}$ = 0.34-0.47 is the coefficient of friction of a solid moving along the chute (gutter).

On the basis of generalization of theoretical propositions and experimental studies, B. E. Fridman [8] developed a methodology for the calculation of the pressurized hydraulic transportation by a hydraulic elevator. In the methodology the main design parameter is the pulp density.

This technique has been adopted to the hydraulic well drilling. It takes the coefficient of the pulp dilution in the hydraulic elevator [9] as its main design parameter:

$$\text{К}_{\text{рп}} = \frac{n_1 + 1}{n_2 - n_1}, \tag{10}$$

where n_1, n_2 is the liquid to solid ratio in the feed pulp (before the hoist of the hydraulic elevator) and in the pulp after the hoist of the hydraulic elevator.

3 Results and generalization

The research results in the justification of the method of placer mining by directional well drilling which excludes the surface damage.

Analytical expressions of technological development parameters are derived from physical and mechanical properties of placer rocks. The parameters of the washing-out and the stopes are based on the laws of loamy rock mechanics. The methods for the calculation of gravity and pressurized hydraulic transport are adapted to the conditions of hydraulic well drilling.

4 Conclusion

The implementation of the method opens bright prospects for the development of unmined small placers and accumulations of alluvial gold in man-made mineral formations.

References

1. V. A. Startsev, N. G. Valiev, V. K. Bagazeev, I. L. Zdorovets, D. I. Simisinov, Patent of the Russian Federation No. 2640611, МПК 21C 45/00 (2006.01), E21C 41/16 (2006.01), E21C 41/26 (2006.01) Method of hydraulic well drilling of mineral resources / FSBEI HE USMU. No. 2016121266, applied 05/30/2016; published 01/10/2018. **Bul. 1.**
2. Method of horizontal directional drilling in subsurface utility engineering. STO NOSTROY-15. 124 p. (2011).
3. N. A. Tsytovich, Soil mechanics. Moscow: Vysshaya Shkola, 288p. (1984).
4. V. K. Bagazeev, N. G. Valiev, D. I. Simisinov, Physical and mechanical justification of hydraulic fracturing of rocks in the process of hydraulic well drilling. Gornyi zhurnal. **12.** (2015).
5. S. S. Shavlovskiy, Fundamentals of jet dynamics in the process of rock mass destruction. Moscow: Nauka, 173 p. (1979).
6. M. V. Rylnikova, O. V. Zoteev, Geomechanics. - Closed Joint-Stock Company "Publishing House" Ore and Metals ", (2003).
7. A. I. Kuprin, Gravity pipeline transportation. Moscow: Nedra, 244 p. (1980).
8. B. E. Fridman, Hydraulic elevators. Moscow: Matgiz, (1960).
9. V. K. Bagazeev, V. I. Shok, Justification of the parameters for the selection of a hydraulic elevator for the development of alluvial deposits, Izvestiya vysshikh uchebnykh zavedeniy. Gornyi zhurnal. **8.** P. 4-10. (2016).

Combined method of opencast and underground mining of valuable ore

Niyaz Valiev[1*], *Vyacheslav* Berkovich[1], *Vladimir* Propp[1], and *Konstantin* Kokarev[1]

[1]Federal State Budgetary Educational Institution of Higher Education "Ural State Mining University" Yekaterinburg, Russia.

Abstract. The efficiency of the combined opencast and underground mining of deposits directly depends on the overburden amount within the open pit boundaries. The amount of overburden to be removed, in its turn, is determined by the angles of the open pit slopes. At the same time, the influence of the open pit slope angle on the amount of the rock mass within the open-pit contours increases in an arithmetic progression with the mining depth. This fact confirms the direct relationship between the open pit slope angle and the efficiency of the mining operations from both an economic and ecological points of view. The article considers the technology for mining steeply inclined deposits of valuable ores by means of underground chambers mined out in advance with consolidating stowing to support unworked sides of an open pit called "well". The chambers which can be not only vertical but also inclined at an angle of 60-80^0 serve to create an artificial fence (a ring concrete support). The work of the fence supporting the open pit side is markedly different from that of retaining walls of any type since the consolidating stowing mixture used to create the fence provides an ideal contact of the concrete wall with the enclosing rocks, thus making it impossible for an artificial fence to overturn. The proposed combined mining technology is expected to reduce considerably the number of the overburden operations, to minimize the damage caused by mining operations to the environment, and to increase the intensity of mine workings.

1 Introduction

In our country and abroad mining practice makes an increasingly wider use of combined (simultaneous or consecutive) opencast and underground mining of one and the same mineral deposit and therefore ensures its complete extraction and protection of subsoil with lower costs of mining mineral raw materials with simultaneous decrease of harmful influence on the environment. The combined mining method aims at increasing the efficiency of both opencast

* Corresponding author: science@ursmu.ru

and underground mining operations by means of the combined use of their most advantageous technological characteristics.

The efficiency of opencast mining is known to depend directly on the overburden amount within the open pit boundaries. The amount of overburden to be removed, in its turn, is determined by the angles of the open pit slopes. At the same time, the influence of the open pit slope angle on the amount of the rock mass within the open-pit contours increases in an arithmetic progression with the mining depth. Thus, there is a direct relationship between the open pit slope angle and the efficiency of the mining operations from both economic and ecological points of view. The open pit slope angle is the result of stripping and mining methods used in open-pit workings. In general, it is reflected in two values of the open pit slope angle: constructive and stable ones. Each of these angles must be as high as possible for a given deposit. However, it is required that the constructive value angle should not exceed the stable value one. Vertical open pit slopes are certain to be an ideal condition which can be created due to the combined method of mining. Using the advantages of the underground mining, it is possible to lay the foundation for surface mining of deposits with steep open pit slopes.

2 Theoretical propositions

There is a tendency for the combined development of steep ore deposits to use a technology with unworked open slopes equal to 70-90 °. Academician V.V. Rzhevskiy called this technology "unbenched" mining or "well" mining [1]. For the application of this mining method it is important to maintain the stability of steep open pit slopes. In order to mine the deposits with the slope angles ranging from steep to vertical, it is necessary to solve two major problems. The first one is "how to extract a useful mineral and what means to use to bring it to the surface"; and the second one is "how to maintain stability of the open pit slopes at the angles close to vertical".

The article describes a method which makes use of underground chambers mined out in advance with consolidating stowing to support unworked slopes of an open pit called "a well". They can be not only vertical but also inclined at an angle of 60-800 (the angles are determined by the stability condition of the enclosing rocks [2]). The essence of the method boils down to the following (Fig. 1).

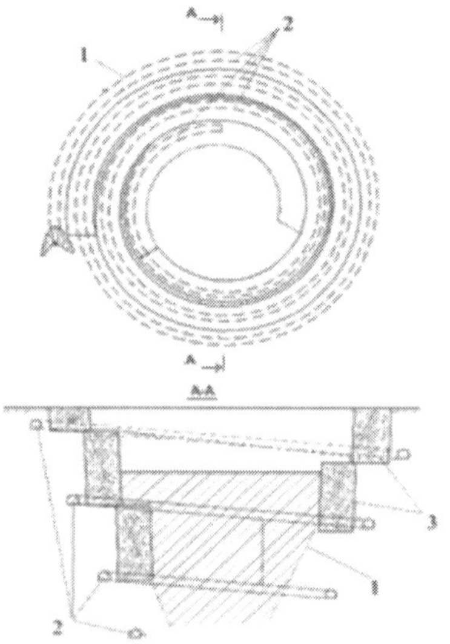

Fig. 1. Unbenched mining of the ore body:
1 – working zone contours; 2 - spiral stoping; 3- ring support

The upper part of the deposit is worked out to the optimum depth by means of a traditional technology of the opencast mining. At the same time, an underground open pit is prepared by means of an unbenched mining technology to clean up the pit reserves of ore. For this purpose, outside the contours of the working zone of a deposit, a spiral stoping with an inclination allowing the use of self-propelled equipment is sunk downwards. In parallel with the spiral stoping along the contours of the working zone of the deposit, the worked-out space is formed for the construction of the advance support (a well) by means of sequential working of the primary and secondary chambers. For this reason, the spiral stoping is used to drive drilling and haulage workings to drill and blast chamber reserves. The blasted rock is transported by the self-propelled equipment through the spiral working to the surface. After the chambers are worked out and cleaned up, they are stowed with consolidating mixture. Secondary chambers are worked out after the consolidating stowing develops its normative strength (Fig. 2).

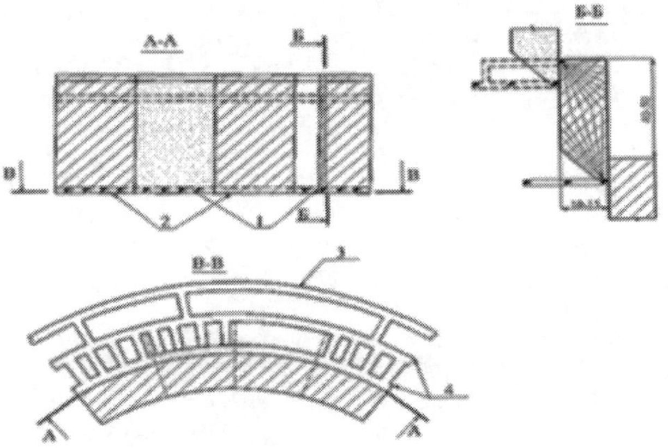

Fig. 2. Technology for construction of the ring support: 1-primary chambers; 2-secondary chambers; 3-spiral ramp; 4-drilling and haulage workings

Ventilation of the preparatory-development workings is carried out by means of the plenum system with the air flowing through the ventilation holes drilled from the surface. After the construction of the first turn of the ring support, the reserves inside the well are worked out by means of an opencast mining directed from the center to the periphery (to the "well" support). Along with these works, the "well" supports are built up by excavating a trench from the surface to the depth of the first turn of the ring support to transport the bulk of the ore reserves by means of the underground method. The ore reserves under the trench are worked out for last. In this case, the ore is also hauled to the surface along the spiral working. Thus, it is possible to work out a steep ore body limited in plan more intensively thanks to a continuous excavation of ore reserves in the central and peripheral parts.

For the construction of a ring support along with the chamber-level mining method described above, it is possible to use dipping flat-back cut-and-fill system. The choice of the mining method depends on the properties of the enclosing rocks and their stability. The combined method of ore mining makes it possible to reduce significantly the overburden amount and the area of the mined-out land.

It should be noted that the work of the artificial fence (concrete ring support) supporting the open pit side is markedly different from that of retaining walls of any type since the consolidating stowing mixture used to create the fence provides an ideal contact of the concrete wall with the enclosing rocks, thus making it impossible for an artificial fence to overturn.

To give a more accurate assessment of the effect of the artificial fence surcharge weight on the stability of the open pit slopes, a series of calculations was made according to V.G. Zoteev [3]. The calculations were made for the open pit sides having slope angles of 50 ° and 70 °. The open pit depth in both cases was 200 m. The angle of internal friction of the massif and adhesion in it were assumed to be equal to 26° and 0.6 MPa respectively while the massif density amounted to 2.75 t / m3. In the calculations the height of the artificial fence varied from 0 to 200 m in increments of 20 m. The thickness (normal density) of the concrete wall was assumed to be the same throughout the height of the side. At the same time, the stability coefficient was calculated for each height of the fence with different thickness set equal to 10, 20 and 30 m. The results of the calculations in the form of the graphical representation of the dependence of the stability coefficient on the height and thickness of the artificial fence are shown in Fig. 3.

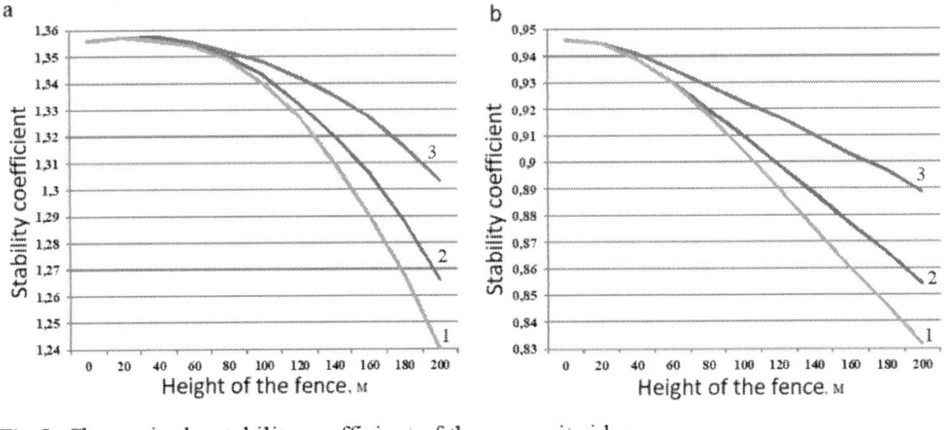

Fig.3. Change in the stability coefficient of the open pit sides:
a) with a slope angle of 50 °; b) with a slope angle of 70 °;
1,2,3- the fence thickness set equal to 10m, 20m and 30m respectively

With the increase in the height and thickness of the artificial fence, the stability coefficient decreases according to the non-linear law. The convex character of the graphs and the divergence of the branches corresponding to different thicknesses of the fence indicate that the intensity of the decrease in stability grows with the increase in height and thickness of the fence. With the increase in the angle of inclination of the open pit side, the decrease in the stability coefficient during the construction of the fence throughout the height of the exposure can reach 12% or more.

The artificial fence to be erected has height limitations determined by the strength of the stowing, the angle of the inclination, and the thickness of the fence to be constructed, as well as the rate and the ratio of the initial stresses (the higher the ratio between horizontal and vertical stresses, the greater the ratio between the width of the fence and its height). In order to overcome the limitations, in principle, an artificial fence should be erected in the form of several isolated benches (Fig. 4). Each of the benches can have its slope and thickness determined in accordance with the engineering and geological characteristics of the rock massif. Since a harder and stronger rock massif will support each part of the fence, the stress concentration zones will be separated in space. In addition, the proposed technology makes it possible to increase the scope of work aimed at the construction of an artificial fence, since the fence on each bench can be erected independently of the other ones.

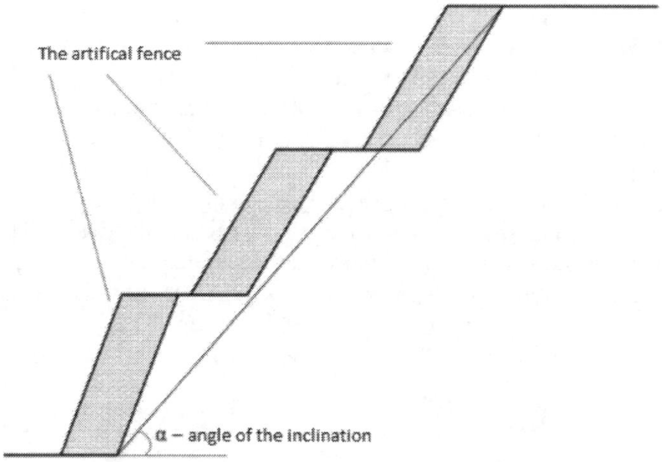

Fig. 4. Structure of an artificial fence

3 Results and discussion

The basic concepts of the proposed technology were used to plan the development of Shemurskoye and Novo-Shemurskoye deposits of copper-sulphide ores on the eastern slope of the Northern Ural. Located at the distance of 2 ÷ 5 km from the nature reserve "Denezhkin Kamen", they have a grim prospect to be mined in a traditional way in view of the great ecological danger. The proposed mining technology with a minimum damage to the earth interior makes it very promising.

In the northern and central parts the ore zone of Shemurskoye deposit has thickness up to 200 m. The angle of the ore zone dip reaches 50°. The central part of the deposit is exposed to the surface. The harmful impurities in the ores are lead, arsenic, antimony and fluorine. The balance reserves of C1 + C2 category copper ore amount to5.1 million tons.

The plan of Shemurskoye deposit development is carried out in a traditional way. The open pit is 120 m deep, 630 m along the strike, and 370 m wide. The rock mass output is 10720 thousand m3, including overburden accounting for 8920 thousand m3. The average stripping ratio is 1.7 m3 / t. The open pit mine production is 800 thousand tons of ore per year. The mine life is 8.5 years.

When various parameters of the support (artificial concrete fence) were compared, the ring support with an upper mark of 750 m, an outer diameter of 100 m at the top, an external diameter of 50 m at the bottom (630 m), a height of 120 m, and a thickness of 14 m appeared to be the most optimal (Fig. 5).

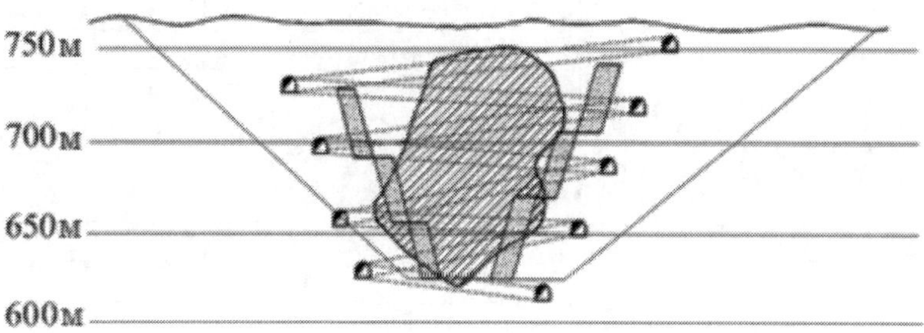

Fig. 5. The alternative method of mining of Shemurskoye deposit with steep unworked sides

The technology of the ring support construction makes it possible to work simultaneously with seven chambers. Before the beginning of the opencast mining, at least two turns of the ring support must be ready inside the well. It means that 14 chambers must be worked out by the underground method of mining and that it will take 3 years to do this.

Novo-Shemurskoye deposit is expected to be developed in a similar way. An alternative to the project is supposed to be a combined mining method. The central part of the deposit, which is 240 m above the horizon, is worked out by an open pit with the depth of 190 m. The rock mass output will be 31.0 million m^3. The south-eastern section of the deposit below the mark of 200 m is mined by means of a new unbenched technology (Fig. 6). Moreover, the time schedule of the preparation, construction, and operation of the mine makes it possible to construct the underground "well" up to the mark of 50 m during the time of the open pit mining (up to the abslute mark of 200 m). The parameters of the "well" will be approximately the same as those at the Shemurskoye deposit.

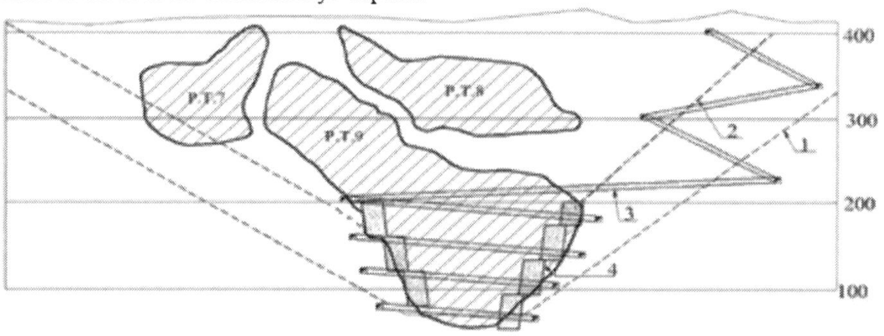

Fig. 6. Combined method of Novo-Shemurskoye deposit mining: 1 – project-oriented open pit contours; 2 -new open pit contours; 3 - inclined ramp; 4 - ring support.

The costs associated with the construction of the ring concrete support include the cost of driving an inclined ramp and underground mining of the chambers with consolidating stowing. The main technical and economic indicators associated with the mining of Shemurskoye and Novo-Shemurskoye deposits by means of the unbenched mining technology in comparison with the project-oriented traditional mining method are given in the Table. As can be seen from the table, the rock mass output is 5-8 times less when deposits are mined by means of a new technology than when they are extracted in a traditional way. This happens due to the reduction of the overburden operations, which, in its turn, minimizes the damage to the land, the cost of mining operations and solves many environmental problems of mining.

Table 1. Comparison of technical and economic indicators associated with the development of Shemurskoye and Novo-Shemurskoye deposits

Indicators	Deposit			
	Novo-Shemurskoye		Shemurskoye	
	Alternative		Alternative	
	project-oriented	proposed	project-oriented	proposed
Total rock mass output, million m^3 including	247.0	31.0	10.7	2.1
overburden	238.7	27.6	8.9	-
ore	8.3	5.5	1.8	2.0
Total cost of 1 m^3 of rock mass, rub	1026.6	254.5	193.0	180.0
Total costs of a deposit mining, million rubles	8521.0	1403.0	347.5	360.0

4 Conclusion

1. The conducted researches and feasibility study of the economic efficiency evaluation of the proposed combined opencast-underground method of mining have shown that the proposed technology makes it possible to reduce the costs of mining, to minimize the damage caused by mining to the environment, and to increase the intensity of the deposit development.

2. The development and justification of the proposed technology will require a new approach to the opencast and underground mining operations, the use of new constructive solutions, as well as an in-depth study of the mechanism of interaction of the system open pit side-ring support from the geomechanical point of view.

References

1. Rzhevskiy V. V. Problems of the mining industry and the complex of mining sciences. V. V. Rzhevskiy. Scientific publishing house:Moscow Mining Institute. 240 p. (1991).
2. Inventor's certificate 1745931 Russian Federation. Method of mining mineral deposits / V.M. Berkovich [and others]. No. 4827141/03; claimed 21.05.90; published. 07.07.92, Bul. 25.
3. Zoteev V.G., Zoteev O.V., Tarasov E.B. The method of slip surfaces construction for the calculation of stability of enclosing dams accumulating liquid industrial waste. "Izvestiya. Vuzov". Mining magazine. **3.** P. 33-38. (2005).

Development of an electronic control module and research of dynamic and power characteristics of an electromagnetic hammer for destruction of boulders.

Konstantin Vasin[1,2*], *Yerik* Yedygenov[1], *Nakhypbek* Aldiyarov[2] and *Vladimir* Voronin[3]

[1]D.A. Kunaev Institute of mining, Almaty, Kazakhstan
[2]K.I. Satpayev Kazakh National Research Technical University, Almaty, Kazakhstan
[3]Pacific National University, Khabarovsk, Russia

Abstract. The developed electronic control module of an electromagnetic hammer for destruction of boulders is described and studied in the article. The possibility of using an inductive slot sensor for both controlling electromagnets and for measuring dynamic and energy characteristics is demonstrated. During the research of the possibility of controlling the impact energy of an electromagnetic hammer, the advantage of the method of adjusting the height of the lifting of the armature in relation to the method for controlling the magnitude of the electric current of the electromagnets is clarified.

1. Introduction

Currently, in the mining industry, the mechanization of technological processes is carried out by high-impact machines, mainly hydrohammers, which have a significant drawback associated with the multiple conversion of energy from one type to another.

The research shows that the creation of electromagnetic hammers is perspective way of developing such machines [1,2,3]. The main advantages of electromagnetic hammers in comparison with other types of impact machines are: the simplicity of construction and the absence of high precision parts; transformation of electrical energy directly into the kinetic energy of a rectilinearly moving peen; the possibility of transmitting electricity over significant distances, which is a very important factor in the creation of a number of machines; great opportunities in increasing reliability and efficiency.

* Corresponding author: kvas2500@mail.ru

2. The Theory and Methods of Experimental Research

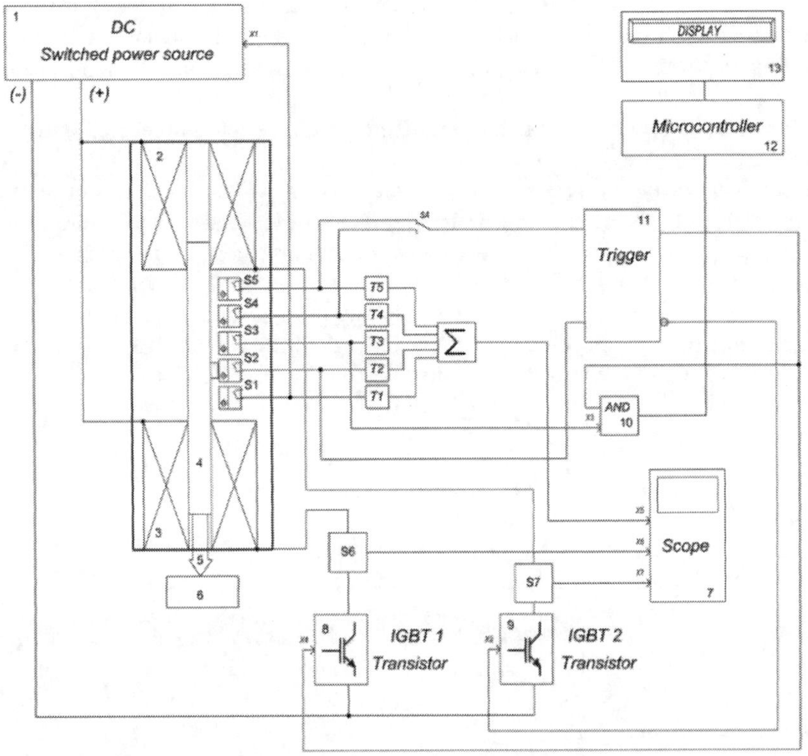

Fig. 1. Electronic control module for electromagnetic hammer.

The developing an electronic control module for electromagnets is very important in developing an electromagnetic hammer. The scientific novelty of this research is the use of an inductive slot sensor for both controlling electromagnets and for measuring dynamic and energy characteristics. Figure 1 shows the circuit diagram of the electronic control module of the electromagnetic hammer.

The simplest electromagnetic hammer has two electromagnets. No.2 is the electromagnet of idling. He draws the armature (No.4) up. No.3 is the electromagnet of the working stroke. He pulls the armature (No.4) down. The armature transfers the energy of impact through the peen (No.5) to the collapsible boulder (No.6). The armature (No.4) has a protrusion for the inclusion of inductive gap sensors as the idle and working motion. In this electronic control module, we have applied five inductive slot sensors (S1-S5). The sensor S1 is used to turn on the electric power of the electromagnets when the peen (5) is pressed against the broken boulder (No.6) and to cut off the electrical supply of the electromagnets when the oversize collapses and the peen (No.5), together with the armature (No.4), fall into the lower position. In this case, the sensor S1 is triggered and the signal X1 turns off the switchable electric power supply (No.1) with a direct current. The sensor S2 is triggered when the electromagnetic hammer is pressed against the next boulder and the peen (No.5), together with the armature (No.4), ascends to the working position. The signal from the sensor S2 switches the trigger control device (No.11) which generates the signal X2 and opens the second power IGBT transistor (No.9). This power transistor turns on the electromagnet of idling and the armature (No.4) rises. The upper position of the armature (No.4) is controlled by the sensors S4 and S5, signals from which through the switch SA, switch the trigger device (No.11). Sensor S5 operates at full idling, i.e. when lifting the armature to full height. Sensor S4 operates at the middle of idle and shortens the height of the armature, thus reducing the working stroke and, accordingly, the impact energy. At the output of the trigger device, the signal X2 is turned off and through the second power transistor (No.9) disconnect the electromagnet of idling (No.2), and the signal X4 is turned on and opens the first power IGBT transistor (No.8). The first power transistor (8) turns on the electromagnet of the working stroke (No.3) and the armature (No.4) accelerates and strikes the peen (No.5). The sensor S2 is triggered and the duty cycle is repeated until the boulder is destroyed.

The S3 sensor is used to measure the speed of an armature. The inductive slot sensor remains in the on state until the armature protrusion travels in the sensor slots 23 mm apart. To isolate the signal from the sensor S3 during the working movement of the armature, the "AND" element of the module (No.10) is applied. The signal is emitted to the microprocessor controller (No.12), where the signal duration is measured, followed by calculating the armature (No.4) speed and the impact energy of the peen's strike (No.5).

The speed of movement of the armature is calculated by the formula:

$$V = \frac{l}{t}; \tag{1}$$

V – the speed of the armature movement (m / sec);
l – the distance of the armature movement (m);
t – the time of the armature movement (sensor signal time) (sec).

The impact energy of the peen is calculated by the formula:

$$E = \frac{m * V^2}{2}; \tag{2}$$

E – the energy of a single impact of the peen transmitted by the armature (J);
m – the mass of the armature (kg);
V – the speed of movement of the armature (m / sec).

Calculated by the microprocessor controller (No.12) energy of a single impact of the peen is displayed on the display, indicated on the diagram (No.13).

In addition to measuring and calculating the signal by the sensor S3, the signals of all inductive sensors are emitted to attenuators, where they are scaled in level and then are emitted through the mixer to a multichannel electronic oscilloscope in the form of the common signal X5. Simultaneously with the signal X5, signals X6 and X7 from the sensors S6 and S7 are emitted to the multichannel electronic oscilloscope carrying information about the change in the electric current in the coils of the working and idling, respectively.

On the oscillogram, the signals from the sensors S1-S5 are separated in time and differ in amplitude. For this reason, they are easily distinguishable. In addition to illustrating the general nature of the armature movement when idling and running, the oscillogram allows to measure the signal duration from the S3 sensor for subsequent manual calculations.

3. Results and discussions

Investigations of the electronic control module for electromagnets were carried out on an experimental electromagnetic hammer, developed in the laboratory "Destruction and Delivery of Rocks" of the D.A. Kunaev Institute of Mining. The electronic control module of the electromagnetic hammer was investigated in three operating modes of the hammer:

1) The work of an electromagnetic hammer with a current of electromagnets is 80 amperes. The total stroke of the idle and working motion, i.e. switching on the signal from the sensor S5;

2) The work of an electromagnetic hammer with a current of electromagnets is 100 amperes. The total stroke of the idle and working motion, i.e. switching on the signal from the sensor S5;

3) The work of an electromagnetic hammer with a current of electromagnets is 100 amperes. Incomplete stroke of idle and working motion, i.e. switching on the signal from the sensor S4.

Figures 2 and 3 show oscillograms of the first two operating modes of an electromagnetic hammer. The upper graph shows the signals from the inductive sensors S2-S5. Sensor S1 is not used, because the experimental hammer is mounted on the stand, and the hammer strikes the metal bar. The average graph shows the electric current's change during the operation of the idling electromagnet, i.e. when the armature moves up. The lower graph shows the change in the electric current when the electromagnet of the working stroke is operating, i.e. when the armature moves down.

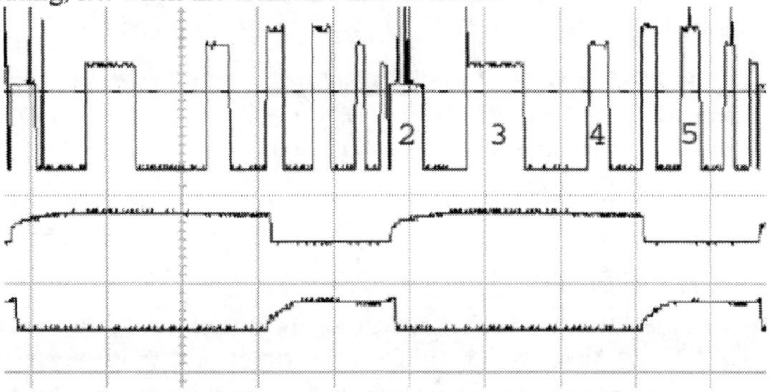

Fig. 2. Operating mode of the electromagnetic hammer No. 1 (80A, S5).

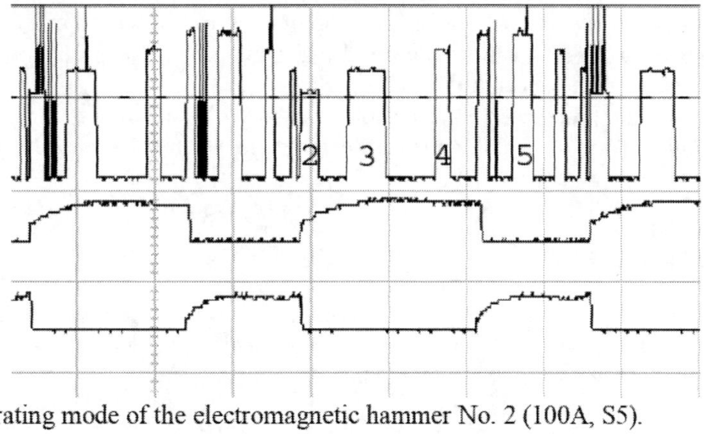

Fig. 3. Operating mode of the electromagnetic hammer No. 2 (100A, S5).

Figure 4 shows an oscillogram of the third mode of operation of an electromagnetic hammer when current of electromagnets is 100 amperes and an incomplete idling and working stroke of the armature. The armature does not go up to the S5 sensor. Therefore, the upper graph shows the signals from the inductive sensors S2-S4.

On the oscillograms of the digital oscilloscope, the periods of the full cycle and the duration of the signal of the sensor S3 were measured during the working stroke of the armature. Then, using formulas 1 and 2 for l = 23mm and m = 90 kg, the armature velocities and the impact energy of the peen were calculated for all three operating modes of the electromagnetic hammer. The results are shown in the table 1.

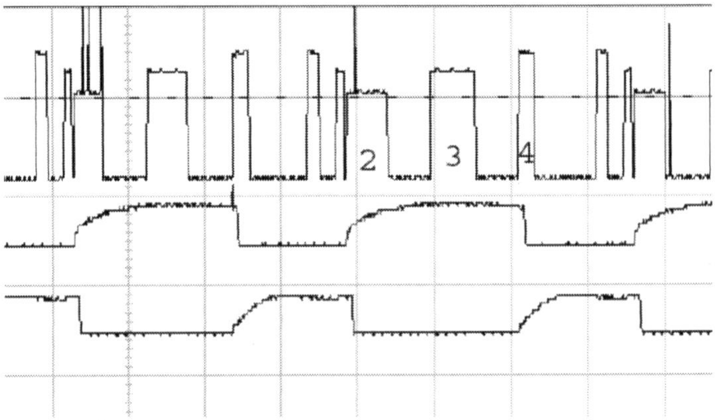

Fig. 4. Operating mode of the electromagnetic hammer No. 3 (100A, S4).

Table 1— The results of measurements and calculations.

Mode No.	Period, msec	Number of hits per minute	Duration of S3, msec	Speed, m/sec	Hit energy, J
1	500	120	8	2,875	372
2	380	158	6	3,833	661
3	370	162	8	2,875	372

The results of calculations show that energy of impact of an electromagnetic hammer can be managed both by change of force of an electric current in electromagnets, and by shortening of a working course of an armature. In this case, the method of shortening the working stroke of the armature makes it possible to substantially increase the frequency of impacts at the same impact energy. This is important in the destruction of boulders. In addition, there is no need to use a bulky and expensive voltage regulator of voltage for powering an electromagnetic hammer.

The use of a microprocessor controller for measuring of the duration of the signal from the S3 sensor with subsequent calculations allows you to monitor the impact energy in real time.

4. Conclusions

Therefore, the developed electronic control module of the electromagnetic hammer for the destruction of boulders allows you to change and explore the operating modes of the hammer and can be used for testing new models.

References

1. Electromagnetic hammers. Ed. A.T. Malova, N.P. Riashentseva, Novosibirsk, "Nauka", (1979).
2. E.K. Yedygenov, V.I. Lyashkov, N.N. Neborokova, Electromagnetic strike machine for the destruction of rocks // Scientific and technical support of mining. 80. (2010).
3. Yedygenov E. Lyashkov V. Electromagnetic rock breaker for non-explosive rocks breaking // The 20th International Symposium on Planning of Mining and Selection of Equipment MPES 2011 Almaty (2011).

Hydraulic borehole mining method possible application at Middle Larba alluvial gold field

Vikror Rochev[1*]

[1]North-Eastern Federal University. M.K. Ammosov, Technical Institute (division) Neryungri, Sakha Republic (Yakutia), Russia.

Abstract. Gold on placer deposits is mainly extracted in two ways: open and underground. But the open method of mining has a great negative impact on the environment, and the mine method is characterized by high labor intensity, capital and operating costs. As an alternative to these methods of development of alluvial gold, the hydraulic borehole mining method can be used. The use of a well as a hydro-mining creates favorable opportunities for ensuring nature protection and safe operation. When applying the method, the well does not have hydraulic extractions: stripping, this allows preserving the cultural layer of the soil in its integrity; explosive, loading works and use of motor transport, excluding dustiness and gas pollution of the atmosphere; the hard and harmful work for people's health is eliminated. The method of well drilling requires further serious scientific, technical and developmental studies to increase its reliability and expand the field of application. Alluvial deposits of gold are the most favorable objects for hydraulic borehole mining under their mining and geological conditions. Recently, new gold placers have been identified and are being surveyed at a depth of more than 40 m. With a high content of metal in them, these placers are very promising for mining by the hydraulic borehole mining method. The reserves of gold in these fields reach several tens of tons. Against the backdrop of a reduction in inventories for open and underwater mining, large technical difficulties, significant capital investments and operating costs for underground mining of thawed buried placers, the advantages for gold mining by the hydraulic borehole mining method are increasing. Based on mentioned above, the study of the application of the hydraulic borehole mining method in the conditions of Siberia and the Far East is a timely scientific task. **Key words**: Southern Yakutia, hydraulic borehole mining, alluvial gold, gold-bearing area, thickness of a seam, ore occurrences, concentration.

1. Introduction.

In the system of geotechnologies, hydraulic borehole mining is one of the main. With this method a production is carried out through wells at the same time with the help of energy of water and air, by means of which destruction of rocks, delivery of destroyed rocks to the well and lifting them to the surface [1,2,3] are carried out.

* Corresponding author: viktor-rochev74@mail.ru

The most obvious objects for the application of hydraulic borehole mining are alluvial gold deposits. Here the sands are represented by easily disintegrated, disconnected or loosely connected rocks. A feature of the northern regions of the country is the presence of permafrost on placers, where there is a predominantly stable roof of the productive formation, which allows the extraction of large amounts of sands from the well without collapse.

Due to the depletion of surface reserves, large technical difficulties, significant capital investments and operational costs for the development of placer deposits in Siberia and the Far East, the prospect of gold mining by the hydraulic borehole mining method in rich deep-buried placers is increasing. [6]

Based on above mentioned, the study of the hydraulic borehole mining method application in conditions of permafrost is a timely scientific task.

In the field of hydraulic borehole mining studying, a great contribution was made by the following researchers: Arens [1-4], D.N. Shpak [13], P.M. Tupitsin [9], B.S. Fedorov [10], and many others. In addition, the method of mining through wells can be successfully used in the development of coal [12] and ore deposits [1,2,3,4,7,11], titanium-magnetite sands [5,6], as well as in the construction of underground reservoirs (for storage of oil, etc.) [8] and refrigerators [8].

To apply the hydraulic borehole mining method in conditions of permafrost, we were asked to conduct laboratory experimental studies to further use the results obtained at the alluvial gold deposit Middle Larba.

2. Field characteristics.

The Middle Larba field is located in the Tyndinskii district of the Amurskaia oblast, 62 km north-west of the district center of Tynda. The Baikal-Amur Mainline is 35 km to the south, the AYAM highway and the Tynda-Neryungri railway are 30-43 km to the east. The area of the object covers the valley of the upper course of the Middle Larba River from the mouth of the Arbat stream to the mouth of the Khorojchikan stream and is 53.86 km^2.

Geographically, the relief of the area is medium-mountainous intensely dissected. The soils are represented by sandy-argillaceous rocks with inclusions of gravel. Alluvial gold laying depth is 5m. The temperature of frozen soils is -3^0 C.

Alluvial gold of the Middle Larba laying is located within the floodplain. The thickness of the loose sediments ranges from 2.0 to 6.7 m, an average of 4.4 m. The sediments of the Middle Larba are in a long permafrost state. Water bearing tabetisols are observed near the bed only.

The gold-bearing layer is confined to the lower horizons of the pebbles and the weathering crust of the bedrock. The thickness of the layer fluctuates from 0.4 to 2.8 m, an average is 1.0 m. Sands have high washability.

The placer bedrock consists of biotite granites, and less often gneisses broken up to the state of crushed stone and gravel, mostly to large blocks.

3. Experiment technique.

Based on the above data for the application of the hydraulic borehole mining method in laboratory conditions, we carried out experimental studies on the rocks brought from the given field with experimental unit presented on Fig. 1.

The simulated formation consisted from fine-grained sand, clay and gravel. Magnetite was used as a mineral. The roof and the soil of the reservoir corresponded to the maximum original conditions.

A sandy-argillaceous gravel mixture with a moisture content of up to 20% was placed in an experimental unit. Then, the mixture was frozen with the help of columns placed to a certain depth, into which the coolant (chladone-22) is pumped under pressure from the freon freezing station. Freezing of the rock was carried out to a temperature of -35^{0}C and then melted to -3^{0}C.

A pulp-lifting tube (6) was built into the casing on the cotter pins (5) and then a supply pipe (7) with a smaller diameter was placed inside. At the top of the supply pipe were two nozzles: one was supplied with an water solution, and another with a certain pressure air. In the lower part there was a hydromonitor for washing of frozen rock. The supply pipe was adjusted in depth and rotated by 360^0. The experiments were carried out with the diameter of the hydromonitor nozzles of 4,5; 5,0; 5,5 and 6 mm (8), water pressure from 0,03 to 0,07 mPa. In addition, an airlift with a diameter of 14 mm was used to create a vortex flow and a rapid entrance of the pulp and its passage through the pulp pipe.

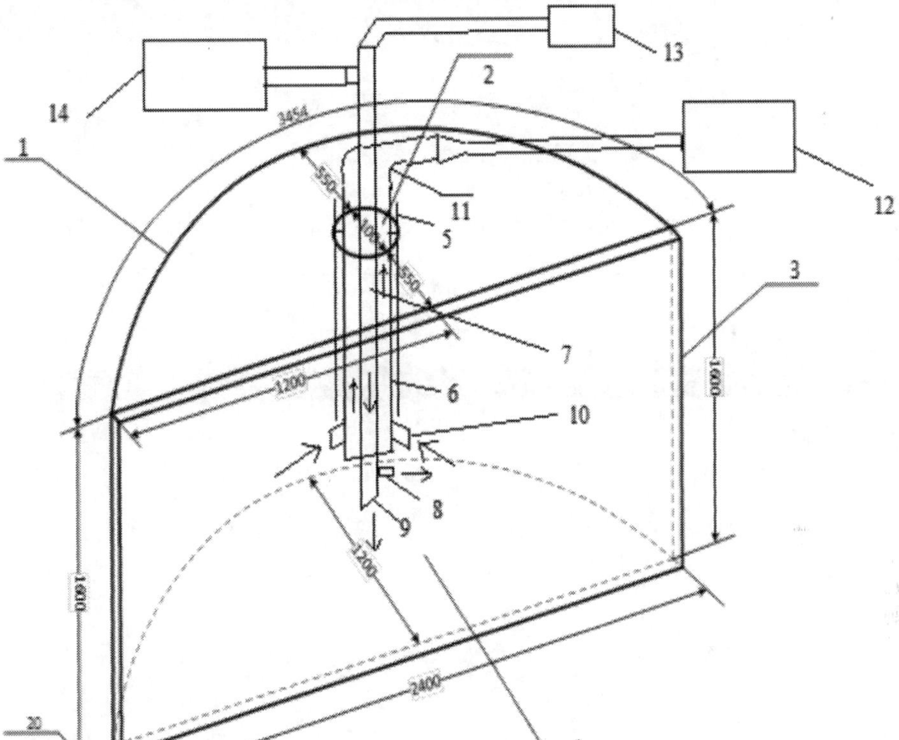

Fig. 1. Laboratory experimental unit for studying the processes of hydraulic borehole mining of send layout. 1 - removable lid, 2 - production well, 3 - plexiglass for observation, 4 - chamber, 5 - casing, 6 - pulp tube, 7 - water-air supply pipe, 8 - jetting nozzle, 9 - airlift, 10 - pulp input, 11 - head, 12-mud pump, 13 - centrifugal surface pump, 14 - compressor

In theory, increasing the diameter and pressure of water will increase the washing rate and increase the efficiency of the chamber [1]. However, the increase in efficiency is limited by the productivity of the pressure device. When minerals are mined from the top down, favorable conditions for the inflow of the slurry to the pressure device are created. A conical sump is formed around the hydroelevator inlet, and the jet acting on top will not interfere with the flow of the slurry.

A centrifugal surface pump (13) was used to supply water. Creation of vortex motion of the destroyed rock in the borehole and increase in water velocity was carried out with the help of a compressor (14). The mixture was removed to the surface by means of a mud pump (12).

4. Results and discussion.

As a result of experimental studies it was found that the depth of destruction (Fig. 2) and the radius of destruction (Fig. 3) depends on the time. With the increase in time, the aforementioned indicators will increase.

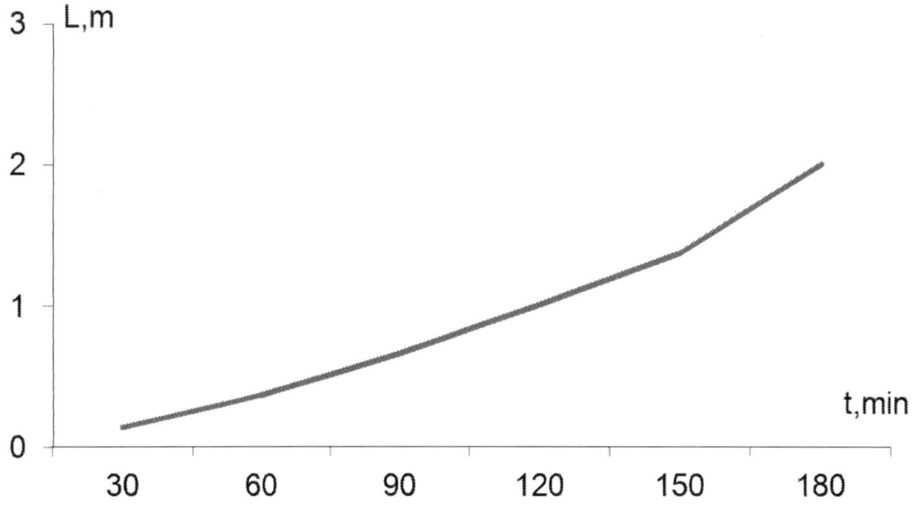

Fig. 2. Dependence of the destruction of frozen rocks depth on time.

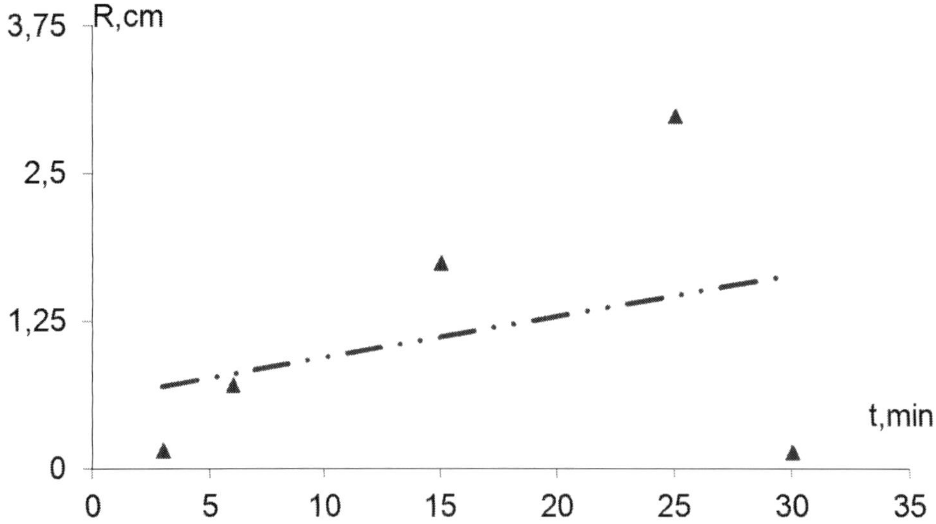

Fig. 3. Dependence of the radius of destruction of frozen rocks on time.

The effectiveness of hydraulic breaker on the diameter of the nozzle study has been carried out (Fig. 4). It can be seen that the radius of the destruction depends on the washing efficiency and the diameter of the nozzle of the hydromonitor. The radius of destruction increases with an increase in the diameter of the nozzle.

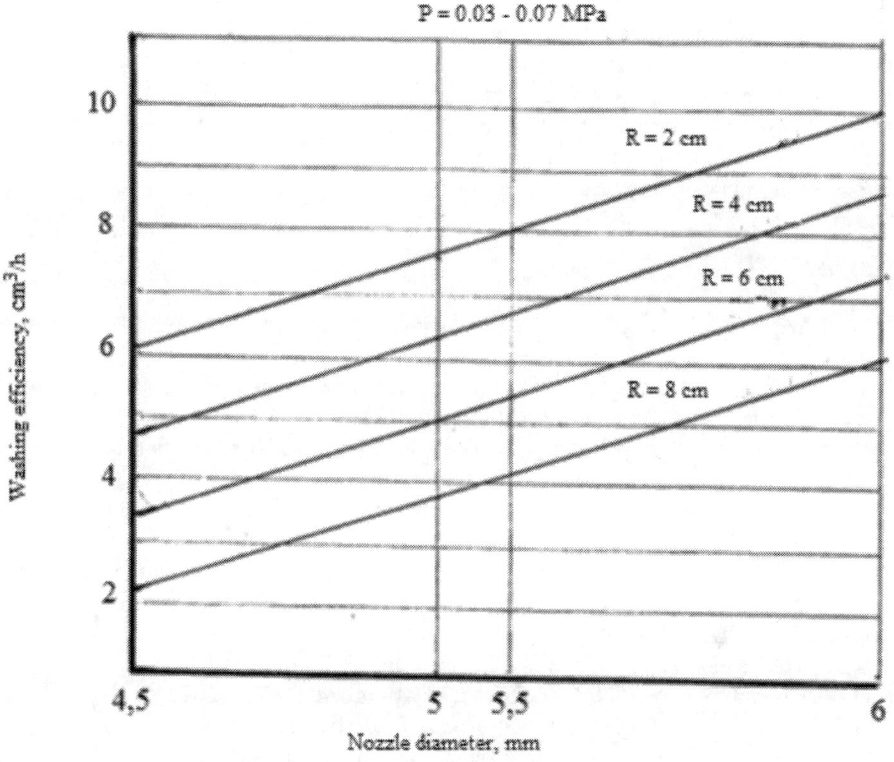

Fig. 4. Dependence of the productivity of a hydraulic breaker on the diameter of the nozzle.

The final result of the experimental studies was that a sandy-clay-gravel mixture and magnetite was elevated to the surface.

On the basis of experimental studies, we calculated all the parameters for using the hydraulic borehole mining method at the Middle Larba field.

Drilling of rocks to gold-bearing sands will be carried out by drilling rig URB 2A2 with casing pipes of 286 mm in diameter and with a coupling. Well-to-well distance is 40 m.

It is proposed to develop the deposit with a mining projectile of the SGS-5 type with a washing by hydromonitor and a hydroelevator lifting of the gold-bearing slurry. Pulp is transported through the pipeline to the device PGSH-II-50 for further enrichment.

In addition, the diameter of the lifting pipe is 140 mm, the diameter of the transported pipe is 113 mm. For air supply, a compressor of the brand XAC47 with a capacity of 1.6 to 5 m^3 / min will be used. As a feed pump, Grundfos CR series is proposed, with a capacity of up to 180 m^3 / h. Water with a temperature of + 4^0C will flow to the pump from a borehole drilled near it, and settle after being raised in 2 sump-settlers with a volume of up to 216 m^3. The water pressure on the nozzles of the hydraulic monitor with a diameter of 20 mm is 3 mPa, the water pressure at the face is 0.6 mPa

To raise the hydrevlic fluid, an additional suction pump 8НДв will be used.

5. Conclusions

As it was revealed from the analysis of geological materials there are a lot of gold deposits suitable for development by hydraulic borehole mining method in Siberia, and most of them are unsuitable for development by traditional methods (open and underground), because of their inefficiency, and can be proceeded by hydraulic borehole mining only.

The hydraulic borehole mining method still requires huge scientific, technical and experimental study to improve its reliability and expand the field of application.

In the permafrost zone, sufficiently thick layers of frozen soils are widely distributed, where hydraulic borehole mining is possible. Such a development method will result in minor disturbances in the environment, provide safe working conditions and will allow for the automation of the processes of washing and rising of soil to the surface, and also achieve high technical and economic performances.

References

1. V. J. Arens, B.V. Ismagilov, D.N. Shpak, Hydraulic borehole mining of solid minerals. 229 p. (1980)
2. V. Zh. Arena, O.M. Gridin, E.V. Kreinin, V.P. Neber, M.I. Fazlullin, A.S. Khrulev, G.Kh. Hceyan, Physical and chemical geotechnology. (2010).
3. V.Zh. Ahrens, N.I. Babichev, A.D. Bashkatov et al., Hydraulic borehole mining of minerals. (2011).
4. V. Zh. Arens, G. Kh. Khcheyan, A. A. Khrulev. The Markscheider Herald. **3**, 13-17. (2010).
5. Yu.P. Bobkov, S.I. Kulyapina, V.I. Mesitsi et al., Problems of Mountain Thermophysics. The II all Soviet Union scientific-techn. Conf. On mining technical thermophysics. 131-132 (1981).
6. V.P. Drobadenko, N.G. Malukhin, A.L. Wilmis, Gold mining. **155**. 17-19. (2011).
7. O. V. Zamyatin, V. M. Mankov, Journal of Mining. **4**, 22-26. (2011).
8. G.P. Kuzmin, A.V. Yakovlev, Underground reservoirs in frozen soils. (1992).
9. P.M. Tupitsyn, Device for underground mining of minerals by hydraulic method. - Author. cert. of USSR № 58591, (1936).
10. B.S. Fedorov, P.A. Konovalov, Building Materials. (1974).
11. Kh. A. Sergeevich, Technology of hydraulic borehole mining of gold from buried permafrost placers (2002).
12. S.M. Shorokhov, Development of placer deposits and design basics (1963).
13. D.N. Shpak, Scientific and Technical Review. **4**, 37. (1982).

Principles of the non-Euclidian model application to the problem of dissipative mesocracking structures of highly compressed rock and massifs modelling

Mikhail Guzev[1] and *Vladimir* Makarov*[1]

[1]Far-Eastern Federal University, Vladivostok, Primorskiy krai, Russia.

Abstract. New experimental results such as "zonal disintegration" around deep openings and "reversible deformations" of highly compressed rock samples cannot be described correctly from contemporary rock mechanics, which is based on the principals of classical Continuum Mechanics theory. A new approach to rock mechanics mathematical models consists of the application of non-Euclidian modelling to the problem of the description of anomalous experimental results. This leads to the formation of the "Geomechanics of Highly Compressed Rock and Rock Massifs" – a new branch of the existing theory of Geomechanics – in which framework a radical rise in geodynamical phenomena forecasting can be achieved. Principles of the geomechanics of highly compressed rock and rock massifs are discussed in this paper. The effectiveness of the application of non-Euclidian modelling to the anomalous experimental effects observed in research is demonstrated on two hierarchical geomedia block levels such as rock samples and rock massif around underground openings.
Keywords: reversible deformation, zonal disintegration, non-Euclidian modelling, high compression.

1 Introduction

Research on the deformation and failure relationships of rock massifs, which are in conditions of high compression – exceeding the strength by one half – has shown two noteworthy experimentally determined abnormal phenomena: the reversible deformation of rock samples [1-3] and the zonal character of the deformation and failure of the rock massif around underground openings [4, 5] (Fig. 1).

It is impossible to explain both of these new phenomena from the standpoint of existing classical representations of geomechanics. Attempts at providing a theoretical description of these separate effects, undertaken by a number of authors, have not led to success [6-10]. A general theory which would allow us to explain all of the available experimental data from a uniform position does not currently exist.

At the same time, representations in other areas of science regarding thermodynamically nonequilibrium processes have been developed for open energy systems and demanding the

* Corresponding author: vlmvv@mail.ru

statement of the local irreversibles [11]. Gauge theory methods of representing a defective continuum have been applied to the mechanics of continuous media, allowing a transition to a new class of models in which the defective media has been replaced by an equivalent continuous one via the introduction of a non-Euclidian (curved) model space [12].

Localization of the defect appearance processes, in turn, has led to the mesomechanics representation in which attention has been focussed on the processes of formation and development of the various types mesostructures in a continuum media with defects, which are the areas prepared for the occurrence of macrodefects [13-15]. Moreover, a rock massif under the conditions found at great depths can be represented as an open nonequilibrium thermodynamics system of defective media [16, 17]. All of the positions are complicated by the hierarchical character of defects existing in a rock massif according to its hierarchical block structure [18].

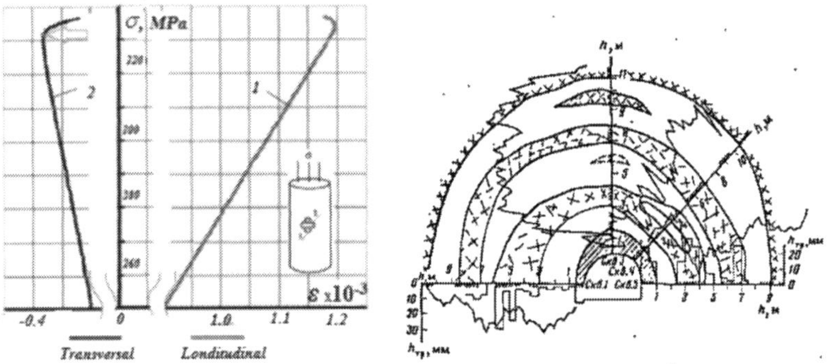

a b

Fig. 1. Anomalous phenomena of highly compressed rock massifs: a - Reversible deformation of rock samples [3], b - Zonal character of destruction near deep underground openings [4].

Therefore, the development of a new model of a rock massif, which is under the conditions found at great depths, has demanded the creation of a theoretical base corresponding to the new requirements, discovering new, abnormal from the point of view of the existing theory, the phenomena of deformation and destruction of rock.

2 Non-Euclidian model to simulate anomalous geomedia phenomena

Our approach uses a non-Euclidian model of the zonal failure phenomena description on the hierarchical level of the mining opening scale of the geomedia block [19]. The model describing the zonal structure of the stress field around a cylindrical opening is considered in a stationary formulation and is assumed to be a plane loosed by a circular hole. For the non-Euclidean model, the classical equations of equilibrium and the boundary conditions remain true:

$$\frac{\partial \sigma_{ij}}{\partial x_j} = 0\,. \tag{1}$$

In this model, the stress σ_∞ is specified at infinity under the condition that $\sigma_r = 0$ at the opening boundary $r = r_0$, where r_0 is radius of the opening. In accord with the linearity of equation (1), it is possible to present the required solution as the sum of the classical Σ_{ij} field and an additional field T_{ij}:

$$\sigma_{ij} = \Sigma_{ij} + T_{ij}\,. \tag{2}$$

The T_{ij} field is reflected by the parameter of incompatibility R. In a plane-strain approximation, for deep openings with a circular cross-section, we have a R-function

$$\frac{R}{2} = 2\frac{\partial^2 \varepsilon_{12}}{\partial x^1 \partial x^2} - \frac{\partial^2 \varepsilon_{11}}{\partial x^2 \partial x^2} - \frac{\partial^2 \varepsilon_{22}}{\partial x^1 \partial x^1}, \tag{3}$$

that characterizes the incompatibility of the components ε_{ij}. In polar coordinates, we have the following expression for the stress components [19]:

$$\sigma_r = \sigma_\infty\left(1 - \frac{r_0^2}{r^2}\right) + (\gamma + \beta - \alpha)\frac{\partial^2 R}{\partial r^2} + \frac{\gamma}{r}\frac{\partial R}{\partial r},\quad \sigma_\theta = \sigma_\infty\left(1 + \frac{r_0^2}{r^2}\right) + (\gamma + \beta)\frac{\partial^2 R}{\partial r^2} + \frac{\gamma}{r}\frac{\partial R}{\partial r}, \tag{4}$$

where α, β, γ are constant coefficients.

This model (1) – (4) can describe the periodical character of the stress field around a cylindrical opening. The results of the boundary problem solution give the formulas for the stress components:

$$\sigma_{rr} = \sigma_\infty\left(1 - \frac{r_0^2}{r^2}\right) - \frac{E}{2\left(1-\nu^2\right)\gamma^{3/2}}\cdot\frac{1}{r}\times\left[aJ_1\left(\sqrt{\gamma}r\right) + bN_1\left(\sqrt{\gamma}r\right) + cK_1\left(\sqrt{\gamma}r\right)\right];$$

$$\sigma_{\varphi\varphi} = \sigma_\infty\left(1 + \frac{r_0^2}{r^2}\right) - \frac{E}{2\left(1-\nu^2\right)\gamma}\times\left[aJ_0\left(\sqrt{\gamma}\cdot r\right) + bN_0\left(\sqrt{\gamma}\cdot r\right) - cK_0\left(\sqrt{\gamma}\cdot r\right)\right] + \tag{4}$$

$$+\frac{E}{2\left(1-\nu^2\right)\gamma^{3/2}}\cdot\frac{1}{r}\left[aJ_1\left(\sqrt{\gamma}\cdot r\right) + bN_1\left(\sqrt{\gamma}\cdot r\right) + cK_1\left(\sqrt{\gamma}\cdot r\right)\right],$$

where r is the radius-vector, which characterizes the zones positions.

After carrying out the calculations with the failure criteria [20], it was determined that the characteristics of the zones depend slightly on the values of the deformational parameters of the model: E, the elastic modulus of the rock, and ν, the Poisson ratio. The effectiveness of the model has been demonstrated using wide cases of rock massif properties for solid rock (σ_c = 150 MPa, where σ_c is rock strength) and for weak rock (σ_c = 15 MPa). Algorithms and programs have been developed to calculate the expressions of the stresses, and the criterion function $K(r)$ [5].

Here, the results of the analysis of zonal failure in the Nikolaevskij ore mine (Dalnegorsk, Russia) are presented. The forecasted depths of the failure zones are shown in Table 1. The correlation between the theory and the experimental research has been estimated by comparing the results of in situ measurements of the radial displacements near the openings (Nikolaevskij ore mine) with the model predictions (Fig. 2). It was determined that the quantitative difference between the forecasted and measured data was no more than 47%, and the full quality satisfaction.

A comparison of the analytical and experimental results for soft rock also shows good correspondence. The rock analyses can be summarized by stating that the basic factor influencing the zonal failure structure parameters is the value of the stress level in the rock massif. With increasing depth, the number of failure zones increases and their radial width also increases.

The anomalous character of the strains in rock samples (Fig. 1a) can also be described by the non-Euclidian model [21]. Extensive natural and experimental studies carried out in laboratories worldwide allow us to make conclusions about the inhomogeneous space-time allocation of longitudinal and lateral deformations of local fields of the sample (Fig. 3).

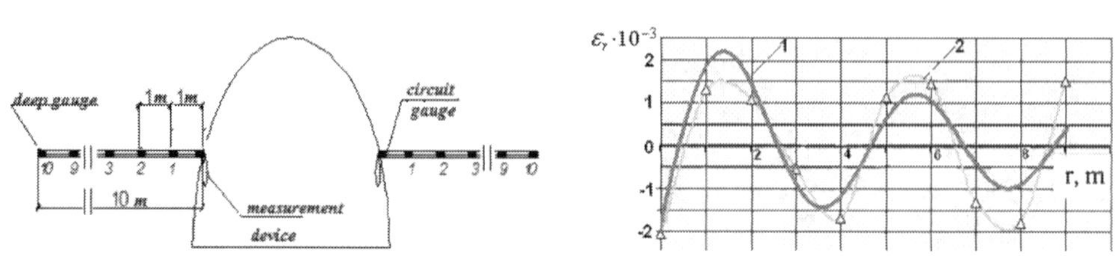

Fig. 2. Theoretical (1) and experimental (2) radial deformations near the opening (right) and research station (left)

Table 1. Depth of zonal structure forecasting

Zone Destruction Number	**I**	**II**	**III**	**IV**
Critical relative stress of zone formation	1.3	2.3	2.9	3.3
Depth of zone appearance (m)	520	920	1160	1320

The phenomenon of the localisation of macroscopic deformations is the motivating force behind the further application of the non-Euclidian model to the study of rock sample behaviour allowing for a dilation of the possibilities for the theoretical description of the rock massif medium.

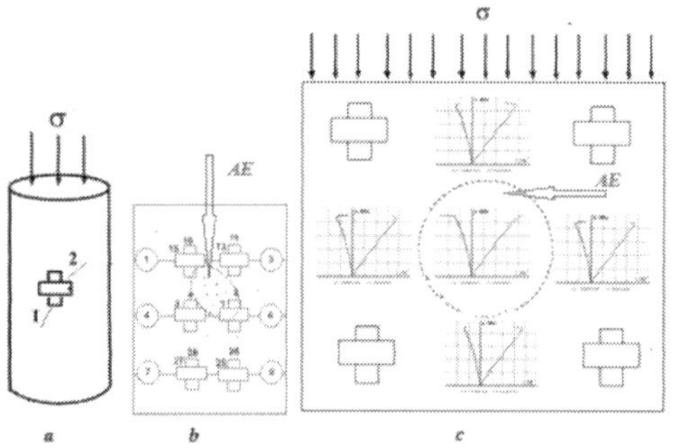

Fig. 3. Mesocracking structure of the source type:
a – Scheme of loading (1, 2 – linear deformation gauges), b – Source area location (1-9 – AE sensors), c – Reversible deformation location in the near-source area (AE – acoustic emission)

The boundary problem for the allocation of a field of stresses σ_{ij} in a cylindrical sample is formulated in a stationary statement (1). The solution is written in the form (1). Let the external force $\boldsymbol{P} = (0,0,P)$ act on the sample ends. Then for balance it is compensated by the force of internal stresses, and the boundary conditions are satisfied at the ends of the sample ($S_{\pm}$) and on its lateral surface (Γ):

$$\int_{S_{\pm}} \sigma_{ij} n_j dS = 0, \quad \int_{\Gamma} \sigma_{zj} n_j dS = \int_{\Gamma} P dS \,. \tag{5}$$

The demands (5) are satisfied on the condition that

$$\Sigma_{zz}\Big|_{z=\pm h} = P, \quad \Sigma_{zr}\Big|_{z=\pm h} = 0, \quad \Sigma_{z\varphi}\Big|_{z=\pm h} = 0,$$

$$\Sigma_{rr}\Big|_{r=R} = 0, \quad \Sigma_{r\varphi}\Big|_{r=R} = 0, \quad \Sigma_{rz}\Big|_{r=R} = 0, \tag{6}$$

$$\int_0^R r dr \int_0^{2\pi} d\varphi T_{iz}\Bigg|_{z=\pm h} = 0, \quad \int_{-h}^{h} dz \int_0^{2\pi} d\varphi T_{ir}\Bigg|_{r=R} = 0 \;, \tag{7}$$

where R is the radius of the sample and $2h$ is its height. The conditions (6) coincide with the classical conditions locally at each point. It is seen from (7) that the components T_{ij} are the self-balanced fields.

To construct T_{ij} the stress function g must be introduced:

$$T_{ij} = \delta_{ij}\Delta g - \frac{\partial^2 g}{\partial x^i \partial x^j}. \tag{8}$$

It is well known that the classical theory demands $\Delta\Delta g = 0$, where Δ is the Laplace operator. Then T_{ij} satisfies the compatibility conditions. The non-Euclidian model results in the equation

$$\Delta g = -\gamma g. \tag{9}$$

with a phenomenological parameter γ. In the cylindrical coordinate system, we have the following expression for the stress components (8):

$$T_{rr} = \frac{1}{r}\frac{\partial g}{\partial r} + \frac{1}{r^2}\frac{\partial^2 g}{\partial \varphi^2} + \frac{\partial^2 g}{\partial z^2}, \quad T_{r\varphi} = -\frac{\partial}{\partial r}\left(\frac{1}{r}\frac{\partial g}{\partial \varphi}\right), T_{rz} = -\frac{\partial}{\partial r}\frac{\partial g}{\partial z},$$
$$T_{\varphi\varphi} = \frac{\partial^2 g}{\partial r^2}, T_{z\varphi} = -\frac{\partial}{\partial z}\left(\frac{1}{r}\frac{\partial g}{\partial \varphi}\right), T_{zz} = \frac{\partial^2 g}{\partial r^2} + \frac{1}{r}\frac{\partial g}{\partial r} + \frac{1}{r^2}\frac{\partial^2 g}{\partial \varphi^2}. \tag{10}$$

The common solution of the field components T_{ij} is expressed through a Fourier series. The model numerical calculation was carried out for the experimental research using rock sample characteristics: ν=0.26, E=1.7*10^4 MPa, h=5 cm, R=2.5 cm. The calculation results for the deformations at the same experimental conditions show a full qualitative coincidence of the analytical and experimental results, and the maximum quantitative difference of the strain values does not exceed 23% [21].

3 Geomechanics of highly compressed rock massifs: subject and method.

The novelty of the theory is provided by the subject of the research, which does not have an analogue in existing theoretical conceptions. Additionally, the new theory demands, besides new conceptual definitions, the introduction of our own method of research.

The subject of the Geomechanics of Highly Compressed Rock and Rock Massifs is the rock objects, which hierarchical block massifs generate at each level of the geomedia, and the similar wave jointing mesostructures of the contrast type, which are areas of preparation for macrodefects of the following hierarchical level.

The method of the Geomechanics of Highly Compressed Rock and Rock Massifs consists of the representation of hierarchical block geomedia by a system of non-Euclidian models and the replacement of mesodefective structures in each hierarchical level of the defective media by a model of the continuous media in which the presence of defects is replaced with a condition of curvature of the model space (a condition of incompatibility) [21, 22], and the occurrence of macrodefects is defined by a system of strength criteria [20]. In this case, the macrodefects of the bottom hierarchical level are mesodefects of the neighbouring upper hierarchical level of the geomedia.

4 Principles of the Geomechanics of Highly Compressed Rock and Rock Massifs

The basic consequences of the experimental data analyses and the non-Euclidian approach to modelling may be formulated as the following scientific positions. The rock mass represents hierarchical block medium with crystal-like formations (minerals) located at the lowest level, and the highest level is limited to the scale of the objects considered within the crust. Dissipative mesocrack structure (DMS) is considered as a set of areas of active (increase) and passive (decrease) defect formation. At each hierarchical level, it is reasonable to consider three scale levels of defects, i.e. micro- meso- and macrodefects, which makes difficult the analysis of such media and construction of adequate mathematical models. Macrodefects of the lowest hierarchical level can be considered mesodefects at the next higher hierarchical level (macro-meso-transition).

Conception of "high compression of rocks and rock mass" characterizes formation and development of the dissipative mesocrack structures (DMS) of the corresponding hierarchical level. So, subject of the geomechanics of highly compressed rocks and rock mass is the rock phenomena generated by hierarchical rock mass at each level of geomedium, Method of geomechanics of 'highly compressed rocks and rock mass' is presentation of hierarchical block geomedium by a system of non-Euclidian models. Thus, we replace mesodefective structures at each hierarchical level of the defective medium by the models of the continuous medium. Therefore, main principle of geomechanics of highly compressed rock and rock mass is the principle of non-Euclidian hierarchy of geomedium, i.e. replacement of block hierarchical medium by continuous hierarchically structured medium with the non-Euclidian metrics, where transition between structural levels (models) is defined by condition criteria corresponding to the meso-macro-meso process. The conditions of compatibility of elastic deformations are not satisfied in each point in the defective medium, while conditions of compatibility of general deformations for the corresponding hierarchical level of blocks remain satisfied (the monolithic block principle).

5 Some Conditions of Geomechanics of Highly Compressed Rock and Rock Massifs applicability

Some conditions of the geomechanics of highly compressed rocks and rock mass applicability must be added to the principles described above. These are the condition of the geomedium being far from thermodynamic balance, the condition of transference from one structural level of the hierarchical block medium to another, and the condition of criteria satisfaction [21-22].

6 Formation of a new scientific branch of Geomechanics

Representations regarding the reliability of the forecast of geodynamic phenomena are based on reliable precursors of the formation and development stages of the source type mesocracking structures in a massif of the corresponding hierarchical level [25]. The formation of zonal type structures around deep mining openings (zone disintegration) forces us to reconsider the principles describing the interaction of the lining with a rock massif and the technology employed in its construction [26].

7 Conclusion

The Geomechanics of Highly Compressed Rock and Rock Massifs is a new scientific branch of general Geomechanics in which, on the basis of considering a rock massif under the conditions found at great depths from the standpoint of nonequilibrium thermodynamics and gauge theories of the defective media, models describing mesocracking structures at all levels of the block hierarchy are developed, with the designation of a method of transitioning from the mesostructure of each lower level to similar structures of upper levels through the corresponding meso-macro process.

Acknowledgements

The authors thank Prof. V. Odintsev for his important recommendations in the process of writing this paper. We thanks moreover PhD L. Ksendzenko for her income to the models calculation and PhD A. Golosov for the experimental research results. The research is supported by a grant from the Ministry of Education and Science of the Russian Federation No. RFMEFI58418X0034.

References

1. Seldenrath T. R., Gramberg J. Mech. Prop. of Non-Metallic Materials, Butterworths, pp. 79–102. (1958)
2. Guzev M. A., Makarov V. V. Deformability and failure of highly stressed rocks around openings. 231 P. (in Russian) (2007)
3. Guzev M. A., Makarov V. V., Ushakov A. A. J. Mining Sci., **6**. 3–13. (2005)
4. Shemyakin, E. I., Fisenko, G. L., et al. Journal of Mining Science. **3**. 3–15. (1986)
5. Vladimir, V., Makarov, M., A. Guzev, V. N. Odintsev, L. S. Ksendzenko. Journal of Rock Mechanics and Geotechnical Engineering. **8 (2)**. 164–169. (2016)
6. Shemyakin, E. I., Fisenko, G. L., Kurlenya, M. V., Oparin, V. N., et al. Journal of Mining Science. **1.** 3–8. (1987)
7. Shemyakin, E. I., Fisenko, G. L., Kurlenya, M. V., Oparin, V. N., et al. Proc. of the USSR Academy of Sciences. **289 (5).** 1088–1094. (1986)
8. Qian, Q. H., Zhou, X. P., Yang, H. Q., Zhang, Y. X., Li, X. H. Theoretical and Applied Fracture Mechanics. **51 (2).** 129–138. (2009)
9. Reva, V. N., Tropp, E. A. Proceedings of Physics and the Mechanics of Rocks Failure, VNIMI Publishing: Saint Petersburg, pp. 125–130. (in Russian) (1995)
10. Chanyshev, A. I. Intense-deformed condition of a rock massif, Novosibirsk: Mining Inst. pp. 3–8 (in Russian) (1988)
11. Prigogine, I., Kondepudi, D. Modern thermodynamics. Pub. 1998 by John Wiley & Sons Ltd, 506 P. (1998)
12. Kadich, A., Edelen, D. The gauge theory of dislocation and disclination, M: World, 168 P. (in Russian) (1987)
13. Panin, V. E., Grinjaev, J. V., et al. Structural levels of plastic deformation and destruction, Novosibirsk: Science, 255 P. (in Russian) (1990)
14. Panin, V. E., ed. The physical mesomechanics and computer designing of materials: in 2 v., Novosibirsk: Science, **1,** 297 P.; **2**, 320 P. (in Russian) (1995)
15. Lockner, D. A., Byerlee, J. D., Kuksenko V., Ponomarev A., Sidorin A. Nature. **350 (7)**. 39–42. (1991)
16. Guzev M. A., & Myasnikov V. P. Mech. solids, **33(4),** 131-144 (in Rus) (1998)
17. De Grotto, C., Masur, P. Nonequilibrium thermodynamics. World, 524 P. (in Russian) (1964)
18. Sadovski, M. A. Natural blocky of rock, Proc. AS USSR. **247 (4).** (in Rus) (1979)
19. Guzev M. A. J. Rock Mechanics and Geotechnical Engineer. **6(3).** 180–185. (2014)
20. Odintsev, V. N. Tension destruction of a massif of brittle rocks. IPKON, RAS, 166 P. (in Russian) (1996)
21. Guzev M., Makarov V., Ksendzenko L. ISRM Regional Symposium "EUROCK 2015: Future Development of Rock Mechanics", EUROCK 2015 & 64th Geomechanics Colloquium, Salzburg, Austria. pp. 979–984. (2015)
22. Goldin, S. V. Physical Mesomechanics, **8 (1).** 5–14. (in Russian) (2005)
23. Guzev, M. A., Odintsev, V. N., & Makarov, V. V. Tunnelling and Underground Space Technology, **81**, 506-511 (2018). https://doi.org/10.1016/j.tust.2018.08.018.

Prognosis of lithological variability of the water-blocking strata based on the results of complex interpretation of seismic and electrical data

Igor Sanfirov*[1], *Yuri* Stepanov[1], *Tatiana* Baybakova[1] and *Alexander* Taynitskiy[1]

[1]Mining Institute of the Ural Branch of the Russian Academy of Sciences, Branch of the Federal State Budgetary Institution of Science of Perm Federal Research Center of the Ural Branch of the Russian Academy of Sciences, Perm, Russia.

Abstract: This article deals with a comprehensive geophysical approach intended to ensure safety of mining operations at the Verkhnekamskoe field of potassium and magnesium salts. As an example, the article presents the results of additional detailed seismic and electrical studies performed on one of the mine fields of the Upper Kama potash-magnesium salt deposit (UKPMSD) within the site designed for mining operations. Based on the obtained results the structure in the form of a "dome" was identified. According to the data of the complex interpretation, the area of an indefinite lithology (most likely, associated with the replacement of salt deposits) was identified in the upstructure part.

1 Introduction

Seismic and electrical exploration methods take the lead within the complex of geophysical methods intended to ensure safety of mining operations at the Upper Kama potash-magnesium salt deposit (largest in Europe). The most popular seismic studies are studies that use the multifold reflection surveying procedure (MRSP) with the use of interference detection systems. As for the electrical exploration methods, the most popular are direct current methods in combination with the assessment of variability of natural polarization (NP): vertical electrical sounding (VES), symmetrical electrical profiling (SEP) and NP.

Seismic methods are mainly intended to be used for the determination of features of the physical and geological structure of the productive pool and its surrounding deposits. Electrical exploration methods allow to identify and assess negative hydrogeological processes in oversalt deposits. Such a combination of interpretative possibilities is relevant to existing and abandoned sites of a mine take of the salt mine. A complex interpretation of seismic and electrical exploration data allows to localize dangerous areas characterized by low elastic properties of the deposits in question and increased intensity of hydrogeological processes in the oversalt strata. Further such sites are considered as monitor objects

* Corresponding author: Sanf@mi-perm.ru

controlled by using various types of monitoring and design of specialized mining engineering activities.

Combined interpretation conclusions for the sites with wave field problems within the salt strata are based on the laws of changes in propagation velocities of elastic waves and apparent electrical resistance in the similar intervals of the geologic section of oversalt deposits. For example, low values of propagation velocities of longitudinal elastic waves at high apparent resistances can often be caused by the presence of dry fractured intervals of the geologic section, whereas in case of low apparent resistances such values can be caused by similar but already watered areas. Such methodological approaches of a complex interpretation of geophysical data can be used for an advanced prognosis of geological structure features that influence safety of the designed mining operations.

2 Materials and methods

Seismic studies are carried out using the technique of non-explosive seismic exploration of a high resolution with the use of the interference detection system for the common depth point. The elastic vibrations were recorded by the seismic recorder "IS128", the number of active channels was 64, the sampling interval was 0.2 ms and 0.5 ms, the record length was 1500 samples, the magnification was 64 Db. The distance between the receiver points was 2 and 8 m, the distance between the source points was 2 and 8 m as well. The 64-channel asymmetric variable system is used as the detection system.

Seismic materials were processed using SPS-PC and Express-OGT systems with the use of original programs of the active Seismoacoustics Department of the MI UB RAS. The materials were interpreted in the original complex developed by the Seismoacoustics Department of the MI UB RAS.

Electrical exploration studies in the lines of seismic profiles are carried out using the vertical electrical sounding (VES). The sounding was carried out using a symmetrical four-electrode device fitted with the equipment of AMS-1 resistance method made with the help of employees of the Geophysics Department of Perm State University and LLC Scientific Production Enterprise "Intromag".

As an operating device, the symmetric four-electrode Schlumberger device was used. The distance between the transmitter electrodes and the center of the measuring device was sequentially increased with a progression of 1.25. The maximum AB separation was 1000 m, the MN detector line length was 5 m and 25 m. The sounding interval was 100 m.

The program IPI2Win developed by the Geophysics Department of Moscow State University was used for the interpretation of VES data.

3 Results and discussions

Additional detailed seismic and electrical exploration studies have been carried out on one of the mine fields of the UKPMSD within the site designed for mining operations.

According to the previous studies [1], a number of the most dynamically expressed seismic events is usually noticed on time sections (see Fig. 1, a) within the UKPMSD, which correspond to the target seismic reflecting horizons (RH) associated with the lithophysical boundaries. The selected reflecting horizons are associated with the section of the lower part of the salt-marly strata, where the presence of the first continuous formations of rock salt (CFoRS*/salt "mirror") is possible; the location of the formation top E (Et) and sylvinite zone (Sil.), as well as with the marking clay (MC).

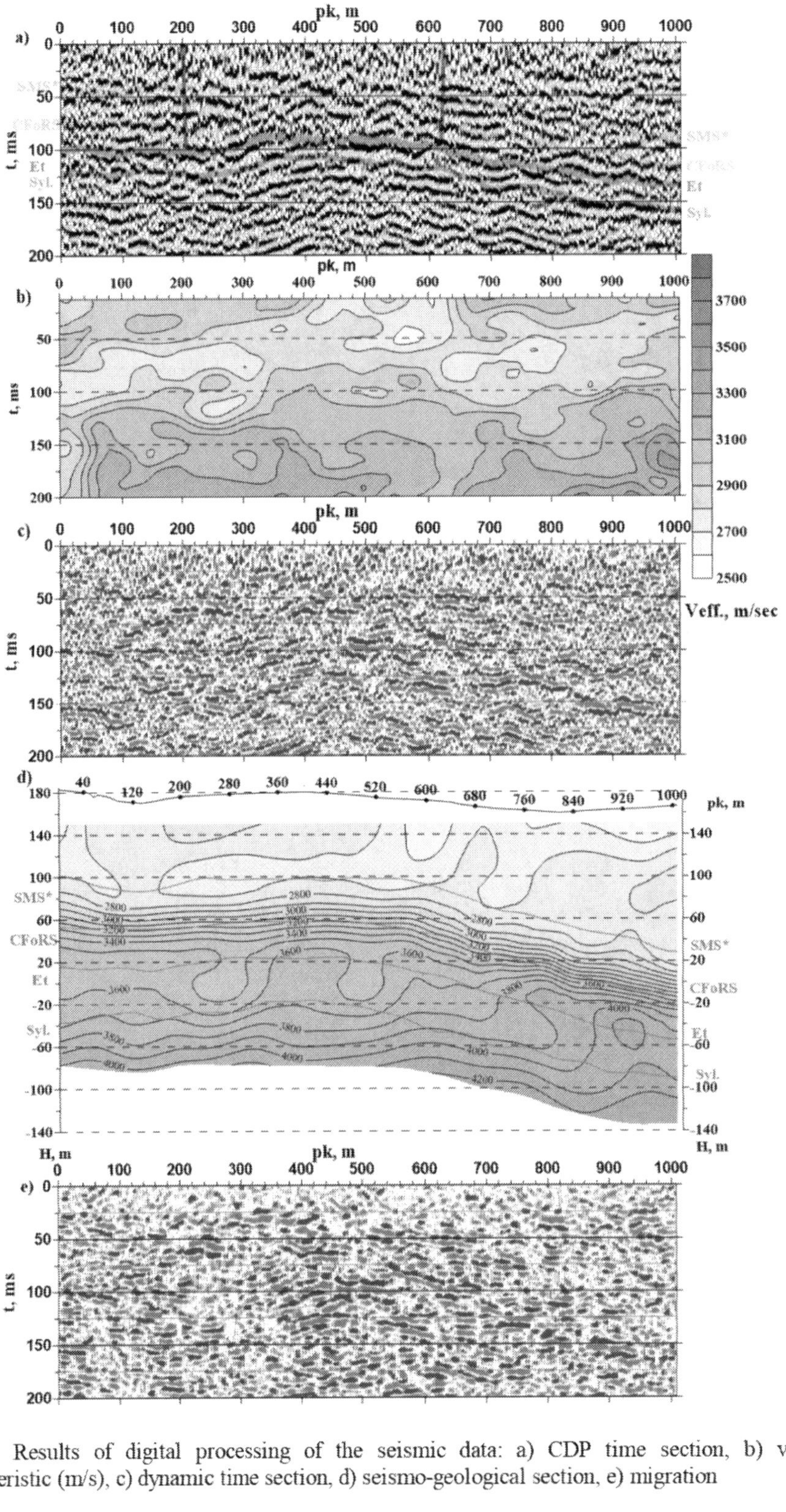

Fig. 1. Results of digital processing of the seismic data: a) CDP time section, b) velocity characteristic (m/s), c) dynamic time section, d) seismo-geological section, e) migration

Due to the presence of a negative vertical velocity gradient showing in the speed characteristics, RH SMS* was additionally introduced above RH CFoRS* in the section of the salt-marly strata (see Fig. 1, b). Such velocity boundary can be caused by vertical lithological differentiation in the salt-marly strata.

According to the results of seismic studies, a wave field problem in the form of a "dome" is identified in the section between the observation surface and the top of the sylvinite zone (200-620 m). In this range there is an increased velocity alternating differentiation both vertical and lateral in the velocity section (see Fig. 1, d) (in comparison with other parts of the profile). The reduced values of the elastic wave propagation velocities in the section of the salt-marly strata and carnallite zone on separate pieces of the profile line and partially within the limits of this problem are separated by a high-velocity formation in the salt-marly strata (CFoRS*—Et). Such areas are observed in the beginning (0-260 m) and end parts (550 m and >) of the profile. According to the depths, this formation corresponds to the possible position of the rock salt formations in the lower part of the salt-marly strata. Within the areas with low-velocity sections one can also see the increased attenuation of the intensity of the total wave field which is mostly obvious on the dynamic time sections (see Fig. 1, d). There is an increased intensity of scattered [2] waves (see Fig.1, e) on the "dome" borders, which may indicate the presence of certain vertically oriented contacts of deposits that differ in elastic properties.

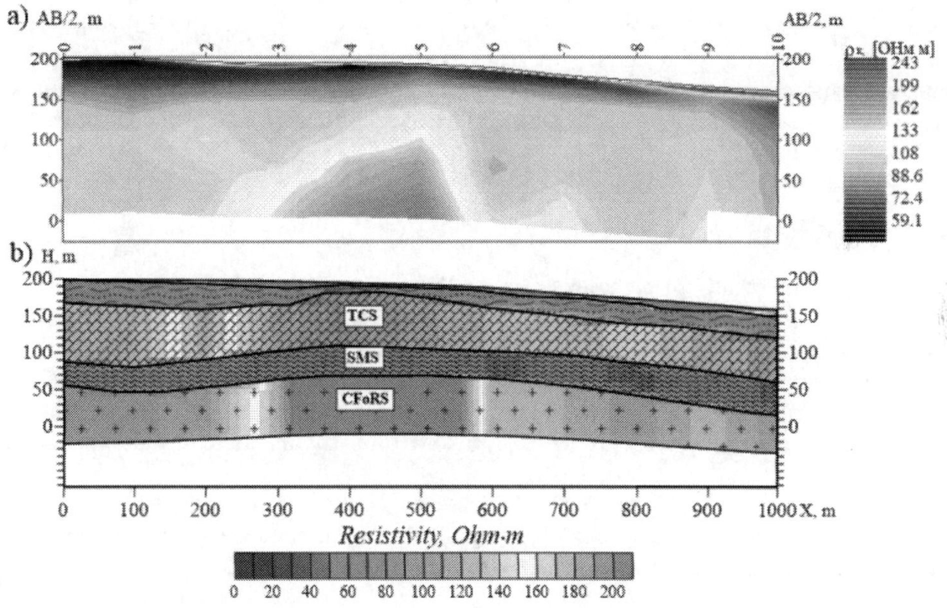

Fig. 2 Sections of apparent resistivity (AR) a) pk (Heff), b) geoelectric section

The section of apparent resistivity is based on the VES data, whereas the geoelectric section is based on the quantitative interpretation results (see Fig. 2).

On the section one can see a vertical variability of AR; there is an insignificant change of the section in the lateral direction. There are four formations in the oversalt deposits on the geoelectric section. Top of the fourth formation corresponds to the depths of occurring of the rock salt, which is the indicator horizon, immerses in the direction from west to east from the mark of 50 m to -10 m respectively. According to the interpretation results, the specific electrical resistance (SER) is 110-150 Ohm*m.

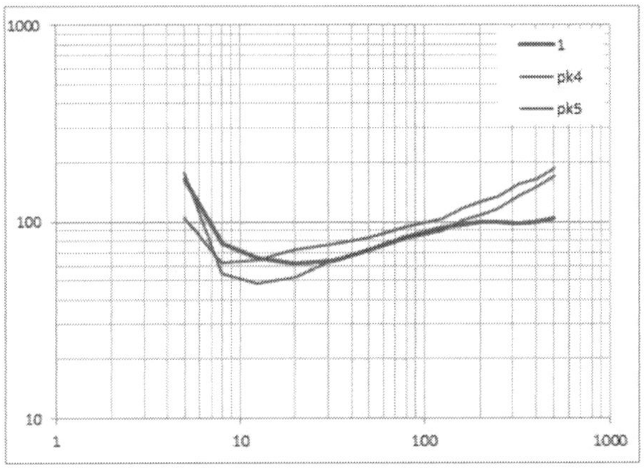

Fig. 3 Faired VES curve (1) and pk 4 and pk 5 VES curves

The analysis of VES curves shown in Fig. 3 indicates a significant difference between the faired VES curve and the apparent resistivity curves at the pickets 4 and 5 in the asymptotic part of the curve.

According to the results of electrical studies, the areas of increased resistances at the level of maximum separation were identified matching with the positive structural form selected in accordance with the seismic data. It should be noted that the features of the apparent resistance curves in the upstructure part can be interpreted as those caused by the lithological replacement in the depth range of the rock salt formations in the lower part of the salt-marly strata.

4 Conclusions

According to the results of complex interpretations of seismic and electrical data of the depth ranges of the upper formations of rock salt and in the upper part of the carnallite zone, (RH SMS*-CFoRS-Et) the area of an indefinite lithology was identified in the upstructure part. The replacement of the salt deposits with rocks with slightly lower elastic properties is assumed. At the same time the replacement contacts and salt deposits show a significant decrease of elastic properties (up to 10 %) and an increased dissipation of reflected waves.

The received results confirm the interpretative capabilities of the examined complex of geophysical methods and the appropriateness of a combined seismic and electrical exploration, including the areas designed for mining operations.

References

1. I.A. Sanfirov, A.I. Babkin, A.G. Yaroslavtsev, G.Yu. Priyma, K.B. Fatkin Seismic studies on conditions of potash deposit mining. / Geophysics. **5**. P. 53-59. (2011).

2. I.A. Sanfirov, A.G. Yaroslavtsev, A.A. Zhikin, K.B. Fatkin, T.V. Baybakova Determination of shallow natural and technogenic heterogeneities of rock mass in different implementations of the interference field of reflected waves / Geophysics. **5**. P. 56-63. (2017).

Nonlocal Criteria for Brittle and Quasi-Brittle Fracture of Geomaterials and Rocks

Sergey Suknev[1*]

[1]Mining Institute of the North, Siberian Branch, Russian Academy of Sciences, Yakutsk, Russia

Abstract. Nonlocal criteria are used for prediction materials and rock mass failure near stress concentrations (pores, faults, openings, excavations). A common property of nonlocal fracture criteria is the introduction of the intrinsic material length characterizing its microstructure, which allows one to describe the size effect in conditions of stress concentration. At the same time the scope of their application is limited to cases of brittle or quasi-brittle fracture with a small fracture process zone. To expand the scope of the criteria for cases of fracture with a developed fracture process zone, it is proposed to abandon the hypothesis of the size of this zone as a material constant, associated only with the material structure. New fracture criteria are proposed, which are the development of the average stress criterion, and point stress criterion, and which contain a complex parameter that characterizes the size of the fracture process zone and accounts not only for the material structure, but also plastic properties of the material, geometry of the sample, and its loading conditions. Expressions are obtained for the critical pressure in the problem of the formation of tensile cracks under compression in the samples of geomaterials with a circular hole. The calculation results are in good agreement with the experimental data on the fracture of drilled gypsum plates.

1 Introduction

Nonlocal fracture criteria are based on the concept of the formation of a fracture process zone in the material, in which there is local redistribution of stresses, whereas the main material is deformed elastically until fracture [1–8]. Typical representatives of such quasi-brittle materials are geomaterials (concrete, gypsum), and rock. A general property of these criteria is the introduction of intrinsic material length that characterizes its structure, which allows describing the size effect in conditions of stress concentration, thereby expanding the scope of application in comparison with conventional criteria. Pointing to the general property of the nonlocal criteria, Taylor [9] proposes to consider them as partial expressions of some general "theory of critical distances". Recently, there have been a large number of papers that, in order to describe the fracture of materials with openings, appeal to the theory of critical distances. In 2008, a special issue of the Engineering Fracture Mechanics was devoted to this theory [10].

* Corresponding author: suknyov@igds.ysn.ru

The scope of application of nonlocal criteria is, first of all, the brittle fracture of materials with openings. To describe ductile fracture, additional information is needed on the inelastic behavior of the material in the fracture process zone or on the material constants characterizing its plastic properties. An intermediate position between brittle and ductile fracture is occupied by quasi-brittle fracture region. Paper considers the possibility of expanding the scope of application of nonlocal criteria for cases of fracture with a developed fracture process zone.

2 Fracture criteria

Below, quasi-brittle fracture is understood as a sudden propagation of an unstable crack (also characteristic of brittle fracture), accompanied by the formation of a large fracture process zone. The size *d* of the fracture process zone is not associated with the size of the crack, as is customary in fracture mechanics, but with the characteristic size d_0 of the material structure. We talk about brittle fracture if $d = d_0$ and ductile fracture if $d >> d_0$. The redistribution of stresses within the limits of d_0 is not related with the plastic (in the macroscopic sense) deformation of the material. The plastic properties of the material begin to appear when $d > d_0$, and the larger d with respect to d_0, the more they manifest themselves. Taking this into account, we represent *d* in the following form:

$$d = d_0 + \beta L_e . \quad (1)$$

Here L_e is the size of the stress concentration zone, β is the dimensionless parameter that characterizes the plasticity of the material. For brittle materials, $\beta = 0$; for plastic materials, $\beta >> 1$. In the case where $\beta \sim 1$, the material is characterized by moderate plastic properties.

For ductile fracture, the critical stress does not depend on the size of the opening, so the size of the damage zone is proportional to the size of the opening and, accordingly, to the size L_e (under unchanged boundary conditions). In brittle fracture, on the contrary, the size of the damage zone does not depend on the size of the opening and is determined by the structure of the material.

During compression, the behavior of the failure stress characterizing the formation of tensile cracks at the opening has the form shown in Figure 1. For small values of L_e, the material does not feel the stress concentration and collapses as a smooth sample when the applied pressure reaches the compressive strength C_0. After reaching the critical size of the opening, the failure pressure p_c decreases, asymptotically approaching the tensile strength of the material T_0 in the case of brittle failure and the stress T_s ($C_0 > T_s > T_0$) in the case of ductile fracture.

This behavior of the failure stress agrees with the current understanding of the real solid as containing primary, inherent defects. Accordingly, small artificial defects, whose sizes are comparable with the sizes of the microstructural components of the material, do not affect its strength until their sizes reach a certain (critical) value.

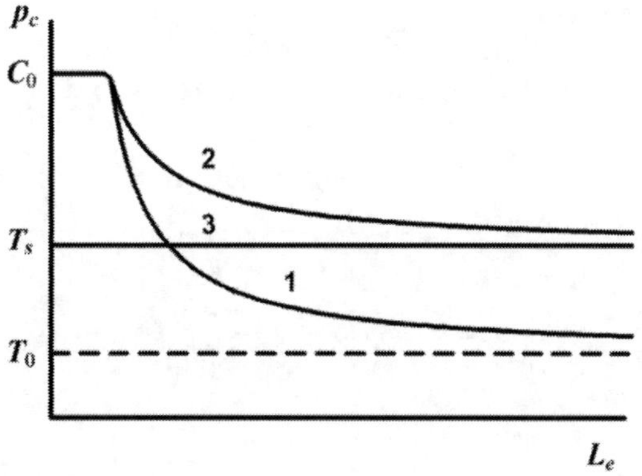

Fig. 1. Fracture diagrams: 1 – brittle fracture; 2 – quasi-brittle fracture; 3 – ductile fracture.

2.1 Average stress criterion

The most well-known nonlocal criterion is the average stress criterion (ASC), or integral criterion [1–3], which has the form

$$\langle \sigma_e \rangle_d < \sigma_0 , \tag{2}$$

where $\langle \sigma_e \rangle_d$ is the equivalent stress averaged at the distance d along the dangerous cross section and characterizing the internal stressed state of the deformed body:

$$\langle \sigma_e \rangle_d = \frac{1}{d} \int_0^d \sigma_e(x) dx . \tag{3}$$

For brittle materials, the size of averaging of d is assumed as a material constant, which characterizes the structure of the material: $d = d_0 = \text{const}$, the material strength σ_0 is also assumed to be a constant.

Critical pressure for the sample with a circular hole [11]:

$$p_c = \chi C_0 \frac{(1 + 2d / l)^3}{1 + d / l}, \quad l > l_c . \tag{4}$$

Here $\chi = T_0 / C_0$, l is the hole diameter, and l_c is the critical diameter of the hole.

To describe the quasi-brittle fracture, we determine the size of averaging according to equation (1), in which the size of the stress concentration zone has the form

$$L_e = \frac{\sigma_e}{|\text{grad}\, \sigma_e|} . \tag{5}$$

The normal stress distribution σ_y along the axis of loading has the form [12]:

$$\sigma_y = \frac{p}{2}\left(3\frac{a^4}{x^4} - \frac{a^2}{x^2} \right) . \tag{6}$$

Here p is the applied compression stress (pressure), and a is the hole radius. The origin is chosen at the center of the hole, and the stress p is considered to be positive. The size of the stress concentration zone calculated by equation (5) with account for equation (6) is $L_e = l/10$.

The critical pressure is determined by substituting equation (1) into equation (4) with account for the obtained estimate for L_e:

$$p_c = \chi C_0 \frac{(1+2d_0/l+0{,}2\beta)^3}{1+d_0/l+0{,}1\beta}. \tag{7}$$

As $l \to \infty$, we have:

$$\frac{T_s}{T_0} = \frac{(1+0{,}2\beta)^3}{1+0{,}1\beta}. \tag{8}$$

2.2 Point stress criterion

The application of the ASC requires the implementation of an integration procedure, which, in some cases, causes certain difficulties, especially when solving asymmetrical problems. Therefore, along with the ASC, the point stress criterion (PSC) is widely used [3]. In this criterion, the integration is replaced by calculating the equivalent stress σ_e at some point that is located at the distance d from the maximum point. The strength criterion takes the form

$$\sigma_e(d) < \sigma_0. \tag{9}$$

The parameter d is also considered to be a material constant that does not coincide with the identical parameter in the integral criterion. The critical pressure for the sample with a circular hole is determined by the equation [11]:

$$p_c = \chi C_0 \frac{2(1+2d/l)^4}{3-(1+2d/l)^2}, \quad l > l_c. \tag{10}$$

Equation (1) is substituted into equation (10), and, with account for the estimate for L_e, we have:

$$p_c = \chi C_0 \frac{(1+2d_0/l+0{,}2\beta)^3}{1+d_0/l+0{,}1\beta}. \tag{11}$$

As $l \to \infty$, we have:

$$\frac{T_s}{T_0} = \frac{2(1+0{,}2\beta)^4}{3-(1+0{,}2\beta)^2}. \tag{12}$$

3 Experimental verification and discussion

The applicability of the developed quasi-brittle fracture criteria can be verified by the experimental data on the tensile cracks initiation under compression in the samples of geomaterials with circular holes.

We test samples made of gypsum cement of various compositions. One series of samples is made from a gypsum cement with high (higher than 90 %) content of hemihydrate gypsum (Gypsum 1), and the second one from the gypsum cement having a low (within 60–70 %) content of hemihydrate gypsum in the initial composition (Gypsum 2). The peculiarities of sample preparation and the experimental procedure are given in [13, 14].

The samples from Gypsum 1 demonstrate brittle fracture. The formation of tensile cracks at the contour of the circular hole is sudden for all the investigated diameters of holes, and the crack length at the time of initiation is 5 to 6 cm. With an increase in the hole diameter, there is a decrease in the critical pressure at the time of crack initiation, and further development of cracks leads to failure of the sample in the form of cleaving into two pieces.

The formation of tensile cracks at the contour of the circular hole in the samples of Gypsum 2 have a different character for small and large holes. Their formation on the contour of a small diameter (up to 5 mm inclusive) is sudden, and the samples demonstrate brittle fracture as in the samples of Gypsum 1. The formation and propagation of cracks on the contour of a hole of large diameter (10 mm and larger) occurs gradually, which is typical for ductile fracture. After the formation of the remote cracks located at a distance from the hole, the opening of primary tensile cracks decreases, their growth stops, and they no longer affect the further fracture of the sample.

3.1 Average stress criterion

Figure 2 shows the experimental data (solid circles) on the load at the moment of tensile cracks initiation from the hole as a function of its diameter, obtained from the samples of Gypsum 1, and the results of the calculations of the critical pressure (curve 1), carried out according to equation (7) for $\beta = 0$. The size d_0 is 1.1 mm and comparable to the size of the largest pores. The dashed line is calculated according to the conventional approach.

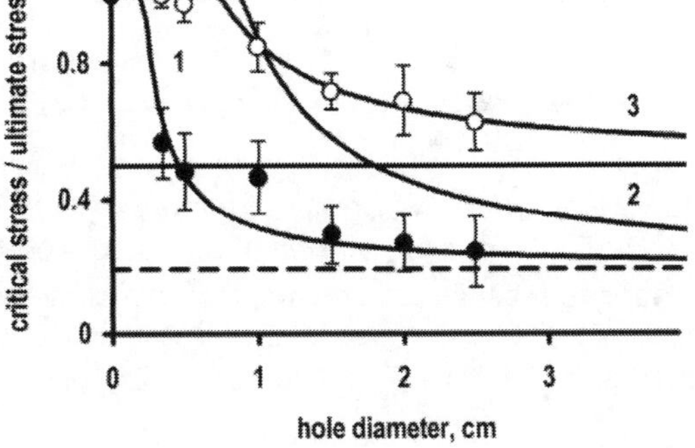

Fig. 2. Relationship of critical pressure with hole diameter. Calculated by the modified ASC.

The experimental data (open circles) and the calculation results for Gypsum 2 for $\beta = 0$ (curve 2) and $\beta = 2,5$ (curve 3) are also given in figure 2. The size d_0 is 4.5 mm in the first case and 2 mm in the second case. In accordance with equation (8), the stress T_s is equal to T_0 in the first case and $T_s = 2{,}7T_0$ (solid straight line) in the second case.

Figure 2 illustrates a significant size effect, i.e. the effect of the diameter of the hole on the local strength of the material. When it decreases, the critical pressure increases, reaching the compressive strength limit. When it increases, the critical pressure asymptotically approaches the tensile strength T_0 for Gypsum 1 and the stress T_s for Gypsum 2. This behavior is well described by the modified ASC, in which the averaging size d is determined by equation (1).

3.2 Point stress criterion

Figure 3 illustrates the experimental data (solid circles) on the value of the load at the moment of tensile cracks initiation from the hole as a function of its diameter, obtained from the samples of Gypsum 1, and the calculation results for the critical pressure (curve 1) according to equation (11) for $\beta = 0$. The size d_0 is 0.55 mm. The dashed line is calculated according to the conventional approach.

The experimental data (open circles) and the calculation results for Gypsum 2 for $\beta = 0$ (curve 2) and $\beta = 1$ (curve 3) are also given in figure 3. The size d_0 is 2.25 mm in the first case and 0.6 mm in the second case. In accordance with equation (12), the stress T_s is equal to T_0 in the first case and $T_s = 2{,}7T_0$ (solid straight line) in the second case.

The experimental results are well described by the modified PSC, in which the size d is determined by equation (1).

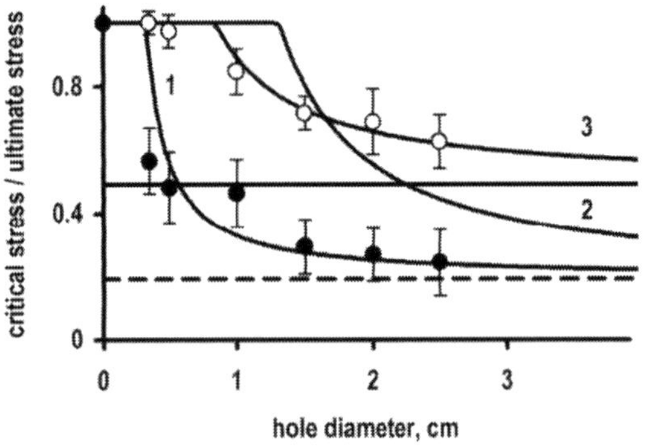

Fig. 3. Relationship of critical pressure with hole diameter. Calculated by the modified PSC.

As can be seen from figures 2, and 3, the experimental data obtained for brittle (Gypsum 1) and ductile (Gypsum 2) crack propagation are equally well described by the modified nonlocal criteria that take into account the change in the size of the fracture process zone in accordance with equation (1). The plasticity parameter β is zero in the first case, and it is determined by the form of the nonlocal criterion and depends on the ratio T_s / T_0 in the second case. The stress T_s is determined by the asymptotic form of the dependence $p_c(L_e)$ and equals $2{,}7T_0$ for the criteria considered.

4 Conclusion

On the basis of the analysis of existing nonlocal fracture criteria, whose distinguishing feature is the introduction of the additional material constant characterizing its structure (intrinsic material length), it is shown that the scope of their application is limited to cases of brittle or quasi-brittle fracture with a small fracture process zone. To expand the scope of application of the criteria for cases of quasi-brittle fracture with a developed fracture process zone, it is proposed to abandon the hypothesis of the size of the fracture process zone as a material constant associated only with its structure. The structural parameter underlying the nonlocal criteria combined into the "theory of critical distances" should be considered as a material constant only in one particular case, which is brittle fracture. For quasi-brittle materials, this parameter reflects not only the structural features of the material, but also plastic properties of the material, geometry of the sample, and its loading conditions.

The proposed approach is used in the development of new (modified) average stress criterion, and point stress criterion. The applicability of the developed quasi-brittle fracture criteria is verified on the problem of the tensile cracks initiation under compression in the samples of geomaterials with circular holes. It is shown that the modified criteria describe well the experimental data on the fracture of quasi-brittle materials.

The research was supported by the Russian Foundation for Basic Research under grant number 18-05-00323.

References

1. H. Neuber, Kerbspannungslehre, Grundlagen für genaue Spannungsrechnung (Springer, Berlin, 1937).
2. V.V. Novozhilov, J. Appl. Math. Mech. **33**, 201 (1969).
3. J.M. Whitney, R.J. Nuismer, J. Compos. Mater. **8**, 253 (1974).
4. B.J. Carter, E.Z. Lajtai, Y. Yuan, Int. J. Fract. **57**, 3, 221 (1992).
5. D. Radaj, S. Zhang, Eng. Fract. Mech. **50**, 1, 111 (1995).
6. S.E. Mikhailov, Eng. Fract. Mech. **52**, 4, 731 (1995).
7. Z. Yosibash, A. Bussiba, I. Gilad, Int. J. Fract. **125**, 3-4, 307 (2004).
8. D. Taylor, P. Cornetti, N. Pugno, Eng. Fract. Mech. **72**, 7, 1021 (2005).
9. D. Taylor, The theory of critical distances: a new perspective in fracture mechanics (Elsevier, Oxford, 2007).
10. Critical distance theories of fracture, Eng. Fract. Mech. **75**, 1695 (2008).
11. S.V. Suknev, J. Appl. Mech. Tech. Phys. **56**, 6, 1078 (2015).
12. S.P. Timoshenko, J.N. Goodier, *Theory of elasticity* (McGraw-Hill, New York, 1970).
13. S.V. Suknev, V.K. Elshin, M.D. Novopashin, J. Min. Sci. **39**, 5, 460 (2003).
14. S.V. Suknev, J. Min. Sci. **44**, 1, 43 (2008).

In-situ rock mass stress-state measurements in scales of mineral deposits: problem-solving

Alexander Balek[1*] and *Anatoly* Sashourin[1]

[1]Mining Institute of Urals Division of Russian Academy of Sciences, Yekaterinburg, Russia.

Abstract: This paper describes the new method for in-situ measurements of the most large-scale stress fields that are comparable to dimensions of deposits and mining leases. The method considers open cast mines and collapse zones of underground mining as disturbing cavities. The calculation of stresses is carried out thru measured deformations for the conditions of plane stressed state from the measured displacements of survey marks on the earth surface. GPS or GLONASS technology is used for the measuring. The results of in-situ study deformations of pit edges Pipe Udachnaya Mine (Sakha Republic, Russia) have been presented as an example of practical application of the method.

1 Justification of in-situ stress state measurements technique

The most of modern deformative methods of the stressed state rock mass in-situ measurements are based on calculation of the stresses according to Hooke's law from the measured deformations caused by purposeful stress field distortions. Ordinarily for this purpose we use the redistributions of stresses resulted of the excavated from the rock mass stressed cavities with known shape and size such as boreholes, crevices, mine workings, worked-out areas, etc. It is assumed that in calculations of the stresses there can take place elastic deformation of continuous homogeneous isotropic medium in a uniform stress field [1-3].

Mining Institute of Urals Division of Russian Academy of Sciences for in-situ measurements of the most large-scale (among geomechanics objects) stress fields that are comparable to the capacity of deposits and mining leases considers open cast mines and collapse zones of underground mining as disturbing cavities [4]. The elastic deformations of the surrounding rock mass are determined by the measurements of displacement of the survey marks which were pre-installed in the field of the future of elastic deformation in the affected zone of the taken out cavity, but out of the trough of rock mass gravitational settling [5-6].

As far as in-situ measurements are aimed at identifying the tectonic stresses in the horizontal plane, the calculation of stresses is carried out thru measured deformations for the conditions of plane stressed state from the measured displacements of survey marks on the earth surface. GPS or GLONASS technology is used for the measuring [7-12]. Open

* Corresponding author: balek@igduran.ru

cast mines and collapse zones are approximated by elliptical holes which are taken out in a stressed thin plate.

The displacements, according to the superposition principle, caused by the excavation in elastic isotropic plane of elliptical hole, can be represented as the difference between the displacements, caused by loading of plane with a hole, and the displacements caused by loading of the plane without hole.

Numerical values of displacement vectors for survey marks on the earth surface are determined by analytical dependences which are developed on the basis of the complex variables. The solution of combined equations, that are composed on combinations of measured displacement vectors of survey marks, allows to evaluate unknown parameters of the stressed state.

The difficulty lies in the fact that not all of the required unknowns are independent physical parameters. Having data about the geometry of the disturbing cavity and about displacements of the surrounding massif with an arbitrarily large number of measured vectors and with any of their locations it is impossible to determine:

- the values of the principal stresses σ_1 and σ_2 without knowing Young's modulus of the massif (or vice versa);
- the ratio of the principal stresses σ_1 and σ_2 without knowing the value of the Poisson's ratio of the massif (or vice versa).

The proposed technique is aimed to resolve this problem.

Analysis of the equations, which are linking changes in the geometry of the disturbing cavity with the displacements of surrounding massif, shows that the influence of the Poisson's ratio μ at solving equations does not exceed few percent. If the changes characterize the rocks this influence is insignificant and the Poisson's ratio is in the range $\mu = 0.2 \div 0.3$. Adopting Poisson's ratio as the predetermined value we can find the remaining unknown parameters by the follow two methods.

Method 1. By setting different Young's modulus E for each vector of measured horizontal displacements the solution of the two equations system determine the ratio of principal normal stresses σ_2 / σ_1 (σ_2 corresponds to the maximum compression) and their orientation by the angle α (that is axes azimuth σ_2).

A comparative analysis of the received data variation for different values of the Young's modulus E is performed after finding the initial massif stresses of all measured vectors. As a result, the value of the modulus and ratios of the principal normal massif stresses σ_2 / σ_1 and their orientation are taken so (appropriate to this modulus) that the difference between the standard deviation of the calculated angles α and their arithmetic means is minimized.

The method is very remarkable for comparative simplicity and the ability to find all the unknown parameters through rigorous analytical solutions. However, it has the drawback due to the nonlinearity of basic relationships. Regardless of the reference point position and a vector of its displacements with increasing of default values of the Young's modulus the calculated direction of major axes of the stress tensor σ_2 quickly approach to the azimuthal direction of the point. The final result appears to be dependent on the uniformity of the metering points distribution around the cavity.

And with this method Young's modulus of the surrounding massif is taken the same for all metering points. This assumption requires justification too.

Method 2. By selection of pairwise combinations of measured vectors (i.e. solution of system of the four equations) the three unknown parameters are determined: the ratio of the principal stress σ_2 / σ_1, the direction of their axes through the angle α, and the massif Young's modulus E (defined as a fraction of maximum stresses: E / σ_2).

The calculations are carried out in two stages in order to facilitate the task and make the analysis of the results more convenient. Initially is defined the set of possible values of E

and σ_2 / σ_1 over a range of axes azimuths σ_2 from 0° to 180° by the solution of two equations system for each of the measured vectors. These equations satisfy the criterion of compressive stresses existence subject to the condition of maximum value σ_2.

Further calculations are reduced to finding such pairwise combinations of displacement vectors, which have the same direction of the principal stress axes at the same values of σ_2 / σ_1 and E. It is obviously that such task as this has no analytic solution. It can only be solved numerically because the task represents finding of the three curves crosspoint in three-dimensional space or finding of four independent nonlinear functions $f(\alpha, \sigma_2 / \sigma_1)$ and $f(\alpha, E)$ for each pair of vectors in two dimensional space.

In practice this solution is more convenient to carry out graphically. It is possible to draw the curves on the computer screen and to identify common angles α with a given precision for the crosspoints of these functions on two-dimensional graph by 3 axes: $\alpha - (\sigma_2 / \sigma_1)$ and $\alpha - E$. In this case angle α is set to a specific level of precision (for example: ± 5°) and with really possible range of relationships between the principal stresses, for example: $1 \le (\sigma_2 / \sigma_1) < 5$.

2 Examples of practical application of technique

According to the in-situ measurements analysis of "Udachnaya" pit edges deformations Mining Institute of Urals Division of Russian Academy of Sciences was first who obtained experimental data about the initial stresses field and the relative characteristics of the massif deformation around deposit "Udachnaya pipe" on the bases of several kilometers. For this purpose there were used the results of instrumental GPS monitoring carried out by LLC "GENESIS CENTER" for the period 1997 to 2009. There were analyzed vectors of horizontal displacements of those bench marks that reflect the deformation of massif from pit excavation and those bench marks which are not affected by local deformation symptoms caused by "not geomechanical" factors: subsidence of edges and separated rock blocks, the processes of freezing, thawing, swamping etc. According to the analyzed results the legitimacy of analytical methods and of models of continuous elastic medium mechanics was substantiated for the displacement vectors of 29 bench marks which are shown in Figure 1.

On the 3D model of the earth's surface there is the outer boundary of "Udachnaya" pit with the size 2000 to 1900 meters and the boundary of its bottom with the size 600 to 260 meters. The excavated from 1997 to 2009 rock volumes (about 40 million cubic meters) were approximated by a cavity in form of cylinder with the vertical depth h = 160 meters and with ellipse base. The ellipse semi-axes are: a = 300 meters; b = 130 meters (Fig. 1). The coordinate X-axis is directed from west to east, and coincides with a major semi-axis a. The Y-axis is directed from south to north and coincides with the minor axis b.

The bearing strata is represented by sandy, clay and algal limestones and dolomites, which are weak and rather weak, with ultimate strength on uniaxial compression of 10 to 35 MPa and 4 to 10 MPa, respectively. According to degree of disturbances the rock massif is divided as follows:

- by eastern edge – the disturbance is very high (more than 12 joint systems);
- by south edge – the disturbance is high (10 - 12 joint systems);
- by south-west and west edges – the disturbance is medium (8 - 10 joint systems);
- by the northern edge – low disturbance with 6 - 8 joint systems.

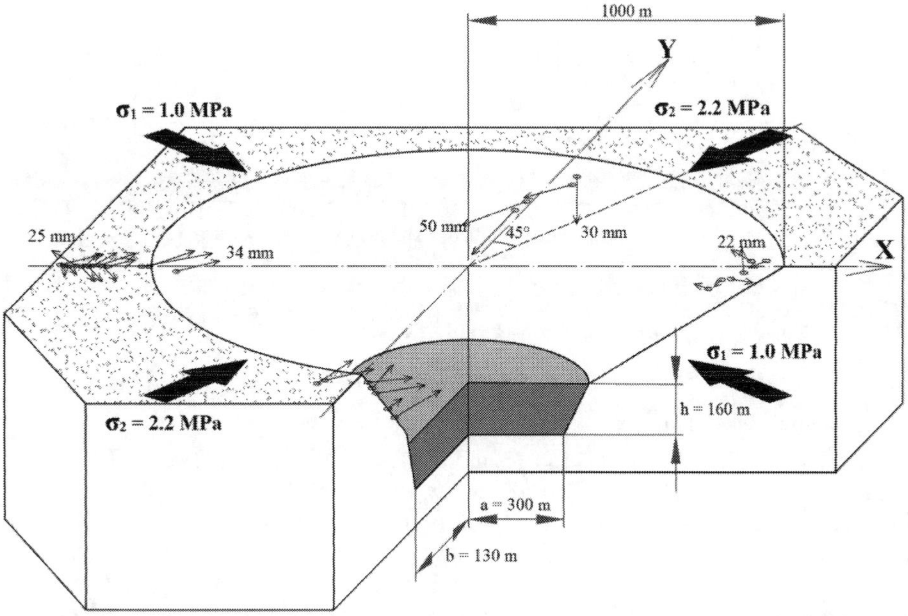

Fig. 1. Scheme for in-situ measurements and for calculation of the initial stresses of massif around the excavated volumes of pit "Udachnaya" in the period from 1997 to 2009. These volumes were approximated by an elliptical cylindrical cavity.

The calculations performed with the both the above-stated methods for the following boundary conditions: a = 130 meters, b = 300 meters, μ = 0.3, σ_1 = -1 MPa.

Solution of equation systems by the first method (i.e. with equation roots σ_2 / σ_1 and α by enumeration of possibilities of Young's modulus E) gave us 29 pairs of equation roots satisfying the above-stated criterion. Comparative analysis of the obtained angles α showed that the minimum standard deviation (that was 52°) is reached at E = 1700 MPa.

The results of solving the same equations system by the second method (finding the equation roots E and σ_2 / σ_1 by carrying out comparative analysis of the pairwise combinations of 29 bench marks) showed that in the range $1 \leq (\sigma_2 / \sigma_1) < 5$ with an accuracy of ± 5° common azimuths of axis σ_2 are available in 48 crosspoints of the considered functions $f(\alpha, \sigma_2/\sigma_1)$ and $f(\alpha, E)$.

Obtained by the first method summary diagram which represents the calculated ratios σ_2/σ_1 and orientation of the principal axes of the massif initial stresses is shown in Figure 2.

A similar diagram for the obtained by the second method results is shown in Figure 3 with the values of Young's modulus E, defined for the relevant sections of the massif.

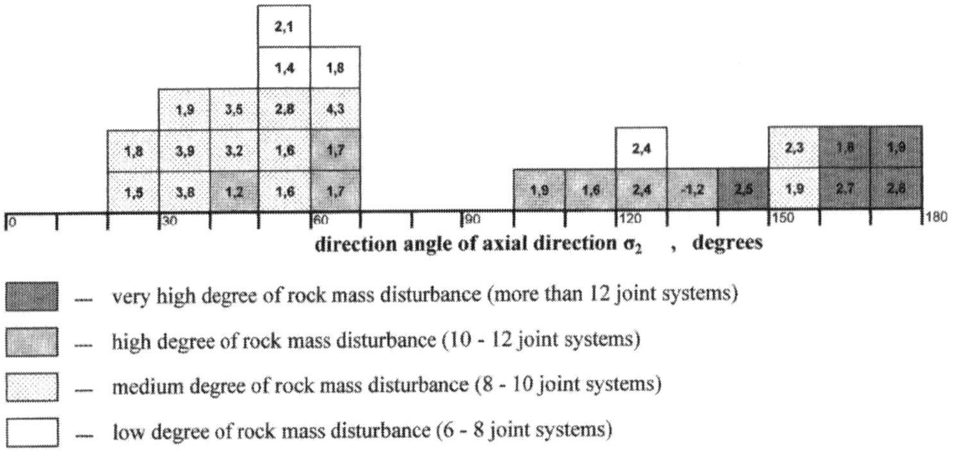

Fig. 2. Diagram with the calculation results of ratio σ_2/σ_1 and orientation of the principal axes of the tensor of rock massif stresses in the vicinity of the pit "Udachnaya" for each of the 29 bench marks at $\sigma_1 = -1$ MPa and the massif Young's modulus $E = 1700$ MPa. On the horizontal axis in the azimuthal range 0° - 180° with step 10° there are azimuths of axial direction σ_2, and on a vertical axis there is identification frequency (about the relevant direction) with the calculated values of σ_2/σ_1 and degree of the massif disturbance in the area of one of the bench marks.

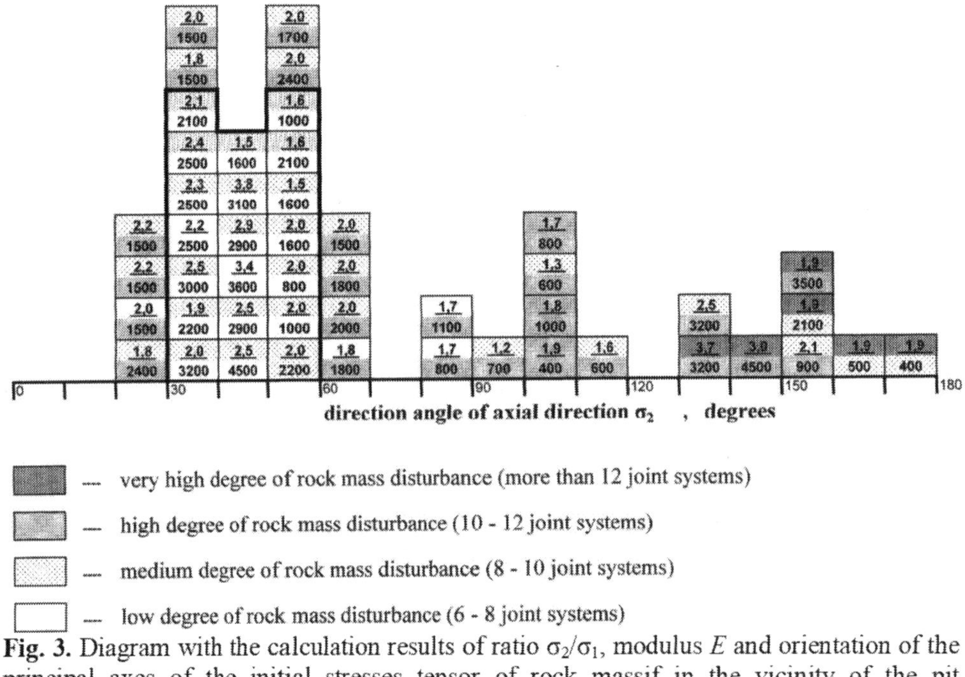

Fig. 3. Diagram with the calculation results of ratio σ_2/σ_1, modulus E and orientation of the principal axes of the initial stresses tensor of rock massif in the vicinity of the pit "Udachnaya" for each of the 48 pairwise combinations of the bench marks. On the horizontal axis in the azimuthal range 0° - 180° with step 10° there are azimuths of axial direction σ_2. On a vertical axis there is identification frequency (about the relevant direction) with the calculated values σ_2/σ_1 (represented in the numerator), modulus E (represented in the denominator as fracture of σ_1) and degree of the massif disturbance in the area of one of the bench marks (represented by filling).

3 Discussing the results and conclusions

In both methods were obtained qualitatively identical results. Prevailing ratio of principal stresses of the surrounding rock massif σ_2/σ_1 was close to 2.2, and in the distribution of the calculated directions of the of maximum compression axis σ_2 appeared two statistical samples: prevailing sample at the north-east direction and less clearly defined at the south-east direction.

For the analysis of independence of the variables by criteria «χ^2» there are $n = 29$ of calculated differences between axis azimuth σ_2 and direction angle defined by elliptic coordinate of the appropriate bench mark. These differences were compared to the expected frequency of their uniform distribution with the arithmetic mean difference of 67°. The obtained $\chi^2 = 381$ for $(n - 1) = 28$ degrees of freedom even for a significance level of 0.05% is 6.5 times higher than the table value $\chi^2 = 59.3$. This indicated significant independence of identified directions of axes σ_2 by the uneven distribution bench marks around the pit.

Comparative analysis indicated statistical significance of both samples of calculated axis azimuths σ_2, because the mean square error of difference S_d between sets of north-eastern and south-eastern directions was: for the 1st method $S_d = 7.6$ and for the 2nd method $S_d = 7.8$. Received mean square errors are less than the differences between the mean values of azimuths (which are 80° - 95°) for more than an order of magnitude.

This does not allow to exclude samples of southeast direction from consideration and therefore to make a clear conclusion about prevailing direction of the greatest compression σ_2 in the initial stresses field of the surrounding rock massif. The problem was solved by using the calculated modules of elasticity of different parts of the massif determined by the 2nd method and geological data about their disturbance.

The diagram in Figure 3 shows that northeastern sample belongs to areas with medium and low disturbances. The highest disturbance concentrates in the extreme south-eastern sector. This gave grounds for advancing a working hypothesis that the southeastern sample (that is more statistically "weak" and with more variability) is causal inference of the deformation processes of fractured sections of the massif. Hierarchically block medium directed chaotic shifts of rock blocks and bench marks.

By the criterion of rock disturbance in the installation site of a benchmark the whole array of the 2nd method from $n = 48$ calculated data was divided into $m = 2$ statistical samples:

- 1st sample: 20 pairs of bench marks installed in areas with extremely low and medium degree of disturbance (6 - 10 joint systems) with an arithmetic mean value of the modulus $E = 2345\ \sigma_1$;
- 2nd sample: 28 pairs of bench marks in areas that include high and very high disturbance (more than 10 joint systems) with an average modulus $E = 1621\ \sigma_1$.

The variance between samples S^2_B with the degree of freedom $(m - 1) = 1$ was $S^2_B = 6290112\ \sigma_1$, and variance within sample S^2_W with degrees of freedom $(n - m) = 46$ is not exceeded 972,101 $S^2_W = \sigma_1$. The ratio between the variances $F = S^2_B / S^2_W = 6{,}47$ is more than table value of Fisher criterion $F = 4.05$ for the significance level of 5%.

This allowed to accept the hypothesis of different sources that form variability the samples and therefore to consider only the results obtained by 20 pairs of bench marks in areas with low and medium disturbance for determining the parameters purpose of the initial stress field (in Figure 3 are shown in bold line). According to the given bench marks the calculated axis azimuths σ_2 are concentrated in a narrow sector from 31° to 60°. Calculated by the remaining bench marks azimuths are uniformly distributed in the sector from 22° to 171° (Fig. 3).

Thus, as a result we obtained the following generic parameters.

The measured parameters of the rock mass stressed state of deposit Udachnaya Pipe (for assumed 95% confidence interval for the Poisson's ratio 0.3) are:

- the ratio between principal initial stresses of the rock mass is

$$\sigma_2/\sigma_1 = 2.2 \pm 0.3 \quad (\text{with } S = 0.6); \tag{1}$$

- the azimuth of principal direction σ_2 is

$$45^\circ \pm 4.4^\circ \quad (\text{with } S = 10.1^\circ); \tag{2}$$

- Young's modulus of the rock mass is

$$E = 2345\ \sigma_1 \pm 410\ \sigma_1 \quad (\text{with } S = 931\sigma_1), \tag{3}$$

Where S is the standard deviation of relevant parameter.

The offered technique was first that allowed to obtain experimental data about initial stress-strain state and the elastic characteristics of the rock massif at the most large bases (for space it is about 5 - 10 km, for time it is about 10 years), and to consider statistical different degree of significance of in-situ measurements results caused by hierarchical block structure the deformed rock strata.

References

1 Kodama J., Miyamoto T., Kawasaki S., Fujii Y., Kaneko K. et al. International Journal of Rock Mechanics & Mining Sciences. **63**. P. 1–11. (2013).
2 Figueiredo B., Cornet F. H., Lamas L., Muralha J. International Journal of Rock Mechanics & Mining Sciences. **72**. P. 37–48. (2014).
3 Wittke, W., Sykes R., Semprich S., Plischke B. Rock Mechanics, Theory and Applications with case histories, Springer-Verlag, Springer; 1 edition, 1076 p (1990).
4 A.Ye. Balek. Rock mechanics– a challenge for society: Proceedings of the ISRM regional symposium EUROCK 2001. P. 807-812. (2001).
5 Geertsma J. The Homogeneous Case. Verh. Con. Ned. Geol. Mijnbouwk. Gen. **28**. p. 43-62. (1973).
6 Sainoki A., Mitri H. S. International Journal of Rock Mechanics & Mining Sciences. **66**. P. 19–29. (2014).
7 C. Liu, J. X. Gao, X. X. Yu, J. X. Survey Review. **47**. P. 79-86. (2015).
8 Yan Bao, Wen Guo, Guoquan Wang et al. Journal of Surveying Engineering. **144**. (2017).
9 Zanutta, A., Negusini, M., Vittuari, L. et al. Journal of Geodynamics. **110**. P. 31-42. (2017).
10 Ludovico Biagi, Florin Calin Grec, Sensors. **16**. P. 21-40. (2016).
11 Yigit C.O., Coskun M.Z. Yavasoglu H. et al. Measurement. **91**. P. 398-404. (2016).
12 Mohtarami E., Jafari A., Amini M. International Journal of Rock Mechanics & Mining Sciences. **67**. P. 43–56. (2014).

Management of geomechanical processes - the basis for selecting the optimal technologies for the development of mineral deposits

Anatolii Kozyrev[1], *Iuliia* Fedotova[1 *], and *Eduard* Kasparyan[1]

[1]Mining Institute KSC RAS, Lab. of Geomechanics, Apatity, Russia

Annotation. When carrying out mining operations, a rock massif responds to technological impacts in the form of developing strains and fractures. Under certain conditions, this response occurs as dynamic and gas dynamic destructions of rocks with intensive release of energy, which creates threats to security and disrupts a working technology. To substantiate the optimal mining technology, which would be maximum adequate to the specific geological and geomechanical conditions of a deposit development, it is necessary to organize a comprehensive monitoring of geomechanical processes in rock massifs. The paper considers general principles to organize the geomechanical monitoring under conditions of the hierarchically-blocked rock massifs in the gravitational-tectonic field of the natural stress state. The authors give main recommendations for managing geomechanical processes in the rock mass for various mining methods.

1 Introduction

When carrying out mining operations, a rock massif responds to technological impacts in the form of developing strains and fractures. Under certain conditions, this response occurs as dynamic and gas dynamic destructions of rocks with intensive release of energy, which creates threats for security and disrupts a working technology. The only way to overcome the dangerous increase of accidents' frequency during mineral deposits mining is to increase the level of knowledge about processes and events in mining operations and on this basis to develop rules and techniques that ensure their effectiveness and safety.

The rock massifs form special physical media, the state of which is determined by three components: rock properties, structural features and natural stress state. Since for all these components the degree of heterogeneity is high enough, in general for rock massifs the heterogeneity is substantially higher than for any other artificial or even natural materials [1]. The significant heterogeneity of the rock massifs determines the extremely wide variability of rock characteristics and makes it necessary to apply specific methods to study the properties and patterns of their variation, depending on the considered volumes, regimes of force impacts, the time of impact, etc. [2]. A characteristic feature of mineral deposits is an extremely wide variety of mining and geological conditions, which makes each mine or

* Corresponding author: fjulia@mail.ru

open pit (or their various parts) unique both in the form of ore bodies developed and in mining methods. From there, the management of geomechanical processes in the rock mass implies a purposeful regulation of the nature and parameters of the processes. In this case, the researchers strive to achieve the development of processes in a desired safe direction, and to keep their parameters within predetermined safety values. To do this, in some cases engineering solutions are used to reduce the stress concentrations in the elements of the mining methods; in others it is necessary, on the contrary, to increase the concentration of stresses and create conditions for the destruction of rocks in some areas of the rock mass. The geomechanical processes management also includes the measures to target changes in the properties of the rock mass, both to improve its stability and effectively destroy the rock mass.

For each deposit the monitoring systems, as well as the methods of controlling geomechanical processes, have their own specifics [2-6]. At that, knowledge of the patterns of changes in these processes, depending on the influencing factors, allows not only predicting the development of negative events and developing the measures to ensure the efficiency and safety of mining operations, but also, with due regard for them, choosing the optimal mining techniques to reduce the cost of prevention and the elimination of the consequences of accidents.

2 The organization of integrated monitoring

Monitoring of changes in rock mass parameters is an integral part of modern subsoil use. It is organized for mining provinces to obtain the necessary information for the development of measures to ensure the safety of mining operations and any facilities, both underground and on the surface [7-11].

Since mining-engineering systems in general are integrated multicomponent structures, the organization of geomechanical monitoring in them should be based on systemic principles that reflect the change in the state in space and time of the rock mass and various engineering facilities built during the construction or the exploitation of deposits.

Taking into account the general regularity of the hierarchically-blocked structure of the rock mass from the point of view of the scale of strains and destructions, it is possible to identify regional and local levels of geomechanical processes and, accordingly, regional and local monitoring of geomechanical processes is needed. At that, the development of geomechanical monitoring systems should include the following provisions [12]:

- the structure of geomechanical monitoring should be multicomponent and multilevel (hierarchical) and should correspond to the structural features of the monitored rock mass (the entire hierarchy of its constituent blocks), geomechanical zoning results, and classes of controlled objects;
- geomechanical monitoring should be integrated with the use of different observation methods in accordance with the features of controlled objects;
- the structure of geomechanical monitoring should be optimized in terms of economic and technical costs for its organization;
- a unified geomechanical monitoring system should be open and allow for painless connection of newly emerging control objects and application of new observation methods.

Based on the experience of performing geomechanical monitoring at various mining enterprises of the Kola Peninsula, a single integrated system for geomechanical monitoring of rock mass is represented as a network of testing sites consisting of separate observation points located both on the day surface and in underground excavations [12]. Moreover, the observation points are laid within the limits of the mining areas in such a way that their minimum number within each controlled boundary of the structural blocks is not less than two. If possible, the observation points are also constructed directly into the structural heterogeneities, which are the boundaries of the structural blocks. Hence it follows that the minimum number of points on each edge of the structural block must be four. The total number of observation points in the monitoring system can reach several tens [12].

On the daily surface, it is reasonable to use geodetic and seismic observations and, consequently, to develop networks of observation points of two types over the entire area of the deposits. In underground conditions it is reasonable to organize three kinds of observations: inclined-deformometric, seismic and geodetic, and, accordingly, to develop three types of monitoring networks for the rock mass state at different levels of structural blocks [12].

Also, within the rock mass blocks indicated on the basis of geomechanical zoning, it is necessary to measure the parameters of the operating stress fields using direct or indirect methods.

It is very important that all observational networks and testing sites to be strictly interconnected by applying a single coordinate system, agreed time frames for measuring cycles and methodologically uniform methods of processing the information received.

3 Creation of geomechanical models

The analysis of the monitoring data provides for the indispensable development of a general model of the rock mass, containing a developed (planned to develop) deposit and the monitored processes. The models are determined by the degree of detail of the available information for the rock mass and the processes under consideration. However, even in the case of a clearly insufficient amount of information, the models, even primitive ones, should be developed, otherwise monitoring becomes unproductive [13].

The next stage of the model design is the compilation of particular geomechanical models of separate earlier identified areas of the rock mass. In this case, an actual mining and engineering situation is superimposed on an engineering-geological model and all changes in the natural state of the rock mass, caused by mining at a certain time, are recorded.

During monitoring, the level of knowledge about an object of observation is increased, and as a result, models are usually complicated and their adequacy to the really developing processes in the rock mass is increased.

4 Management of geomechanical processes

From the point of view of geomechanics, there are no great differences in the development of geomechanical processes, both in the construction and operation of underground and surface structures. This makes it possible to apply uniform methodological approaches to ensuring safety and solving geomechanical problems in mining operations by both open and underground methods [13].

4.1 Open-pit mining operations

The only but very significant feature of rock mass in open mining operations is a very high proportion of near-surface, mostly weak rock layers, often representing a soil mass. In the soil mass, stratification is most often found from structural heterogeneities, and the stressed state is caused exclusively by the gravitational field. To a certain extent, the same applies to dump mass, which also represents actually artificial bulk soils, but, unlike typical soils, it has significant heterogeneities in the fractional composition.

As the depth of mining increases, the proportion of soil masses decreases; the marked features for open pits lose their significance and rock mass, from the geomechanical point of view, becomes more and more identical to the rock mass with which miners deal during the construction of underground structures and mining enterprises. For dumps, these features remain, and they have to be taken into account when solving geomechanical problems.

Technological parameters of open mining operations are largely determined by the development of geomechanical processes (strains, displacements and fractures) in the rock mass of the deposits being developed. At that, the main geomechanical problems in open mining are to ensure the slope stability and stability of benches and dumps.

Based on the very long service life of the open pit walls, especially in the upper benches, the open pit walls are almost analogous to the capital excavations of underground structures, in terms of the development of geomechanical processes. Therefore the methods of ensuring their stability can be carried out in analogous methods - either by regulating (usually decreasing) acting stresses in the contour rock mass, or by a purposeful change in the properties of the rocks composing the benches and walls.

Thus, the management of geomechanical processes in open-pit mining is to justify the optimal structures of the open pit walls, in terms of view of safety and efficiency, to determine the angles of slopes and the height of the benches, parameters of the safety and transport berms, as well as the angles of slopes and the height of the dumps.

4.2 Underground mining operations

Geomechanical processes in the rock mass during capital and preparatory excavations are developed in various ways depending on the geological conditions, i.e., all the determining factors: properties, structural features and natural stress state of the rock massif, as well as shapes of sections and sizes of underground excavations. Based on this, the management of geomechanical processes during driving of the excavations is carried out by choosing the optimal shapes and sizes of the cross sections of the excavations, their optimal orientation relative to the natural stress field of the rock mass and the prevailing systems of structural heterogeneities, and also by choosing sustainability measures and, in particular, necessary types of supports and the calculation of their parameters. For the room mining methods, the stability of the pillars and the roof of the rooms must be maintained for the set time. For the backfilling mining methods, the strength of the material-filler and the permissible deformations of the covering rocks should be ensured. For the caving mining methods, the stability of the covering rocks must be initially ensured, and then their collapse with specified parameters.

Depending on the strain and strength characteristics of the rocks directly adjacent to an excavation's contour, the rocks on the contour of excavations can be deformed or destructed. Due to the high velocity of deformation (approaching the velocity of sound), elastic deformations around the excavation's face have time to develop before the construction of the support, and as a result often go unnoticed. At the same time, the previously driven parts of the excavation, if they are fixed with very rigid support, can be exposed to huge destructions caused by additional, even insignificant, elastic deformations. To avoid this, rigid permanent supports are always constructed with some delay from the face. At that, it is important to emphasize that this lag of the permanent support must be carried out not only for the purpose of protecting it from the effects of blasting operations, if drilling is performed by drilling and blasting (a very common opinion), but also from the indicated effect of elastic deformations. In fact, the choice and justification of the lag of the permanent support installation from the face is an element in the management of geomechanical processes in the near-boundary rock mass.

Thus, the safety and efficiency of underground mining will be ensured only in a case of correct choice of methods controlling the state of the surrounding rock mass, when the stresses and deformations arising therein do not exceed the permissible values.

4.3 Combined mining

Combined (open and underground) mining is an independent mining system of extracting minerals within a single deposit with its inherent features, advantages and disadvantages. The principles of spatial and temporal combination of open and underground developments within a single field are used as a basis for distinguishing individual varieties of combined mining methods. It is assumed that the main feature of the space-time combination of open and underground mining operations is their carrying out in zones of mutual geomechanical influence.

During open mining in the collapse zones, the most dangerous form of deformation of the undermined rock mass is funnelling. A decisive moment is the moment when the caving funnels come to the surface. In this connection it is necessary to carry out underground mining operations under the bottom and in the open pit walls according to schemes and parameters that would ensure minimum conditions for the formation of caving funnels. But if the caving funneling cannot be avoided, then during combined mining of the deposit, the obligatory condition of underground operations is to ensure the exit of a caving funnel to the surface in the shortest possible time.

Experience shows that the formation of voids, hanging and funnels is largely determined by a mining method. So the probability of funneling increases with a one-stage ore block breaking in the clamp in comparison with a two-stage block breaking into the compensation space. This is explained by lower values of the soil conversion factor (1.06-1.26) against (1.35-1.4) with a two-stage breaking. Another important condition, which also makes it possible to significantly reduce the funnel formation in the collapsed rock mass, is the uniform release of the broken ore in the area of the blocks. The most convenient is the simultaneous uniform release of broken ore from all the blocks that undermine a part of the open pit.

However, for geological and other objective reasons, it is not always possible to implement the above recommendation. Therefore, in each specific case it is necessary to choose such a scheme for working out the blocks and releasing the ore, which would ensure the most uniform movement of the collapsed rock mass. For example, for conditions of steeply-inclined deposits, it is recommended to maintain a uniformly sequential order of ore release from as many holes as possible, with a minimum distance between them and small doses of release. At that, the velocity of release should be of the order of 4-5 t / m^2 per day. It is reasonable to start the release from the footwall of the ore deposit (at large reservoir thickness), and in the direction of strike the "ore - rock" contact should be kept flat, with an angle of inclination to the horizon of the order of 50°. In other conditions, optimal schemes and production modes may be different.

5 Conclusions

The organization of a single system for the integrated monitoring of geomechanical processes in rock massifs during the development of mineral deposits should be aimed at monitoring the state of different-scale faults. The analysis of the obtained data will allow predicting the development of changes in the state of the hierarchically-blocked environment based on the created multi-scale geomechanical models, taking into account specific mining and geological conditions in the gravitational-tectonic natural stress state field.

General recommendations for managing the geomechanical processes in rock massifs when using various mining methods require the detailed consideration, taking into account the specifics of concrete deposits. It should be noted that the management of the geomechanical processes is the basis for the justification of the optimal mining technology.

References

1. M.V. Rats, Heterogeneity of rocks and their physical properties. M., Nauka (1958).
2. Methods and systems of seismic deformation monitoring for mining-induced earthquakes and rockbursts. Integration projects of SB RAS, Novosibirsk: SB RAS, **24 (1).** (2009).
3. B.V. Vlasenko, V.P. Potapov, V.A. Fedorin, Mining physical-engineering monitoring. Mining physical-engineering bulletin, **10**, 108-113. (2001).
4. V.A. Il'ichev, P.A. Konovalov, N.S. Nikiforova Particularities of geomonitoring when constructing underground facilities, Foundations, basis and mechanics of soils, **4**, 20-26. (1999).
5. Yu.S. Isaev, S.N. Mulev, Experience in creation of geodynamic monitoring systems in mining enterprises, Mining Journal, **10**, 22-24. (1999).
6. E.V. Kasparian, A.V. Kozhukhovskii, I.Yu. Rozanov, Experience in organization of monitoring of slope stability, Mining Journal, Izvestia VUZov, **5**, 67-74. (2015).
7. A. Kozyrev, V. Panin, Iu. Fedotova, Mining-Induced Seismicity in the Kola Peninsula Mines. Journ. of Civil Eng. and Arch., ISSN 1934-7359, USA. Volume 7, **7 (68)**, 897-906. (2013).
8. S. Cesca & all, MINE, Mining Environments: continuous monitoring and simultaneous inversion, Geotechnologien science Report, Vol.18, 3-13. (2011).
9. C. I. Trifu, F. T. Suorineni, Use of microseismic monitoring for rockburst management at Vale Inco mines, Controlling Seismic Hazard and Sustainable Development of Deep Mines. C. Tang (ed.), Rinton Press, NY, 1101-1114. (2009).
10. G. Nosenzo, B. E. Whelan, M. Brunton, D. Kay and H. Buys, Continuous Monitoring of Mining Induced Strain in a Road Pavement Using Fiber Bragg Grating Sensors, Photonic Sensors Vol. 3, **2**, 144–158. (2013).
11. T. Urbancic, C. Trifu, Seismic Monitoring of Mine Environments, Mapping and Monitoring the Mine Environment, Proceedings of Exploration 97: Fourth Decennial International Conference on Mineral Exploration, A.G. Gubins (ed.), 941–950. (1997).
12. A. A. Kozyrev, E. V. Kasparian, Iu. V. Fedotova, Monitoring of natural and mining-induced processes during mining operations in highly-stressed rock massifs, Monograph, LAP Lambert Academic Publishing RU, Saarbrucken, Deutschland 161 p. (2016). ISBN 978-3-659-97138-9
13. A.A. Kozyrev, V.I. Panin, I.E. Semenova, Yu.V. Fedorova, V.V. Rybin, Geomechanical support of engineering solutions in mining in highly-stressed rock massifs. Physical-engineering problems of minerals development, SB RAS, **2**, 46-55. (2012).

THE ASSESSMENT METHODOLOGY OF CHAMBER MINING SYSTEM ELEMENTS SUSTAINABILITY

Vladimir Asanov[1*], *Anton* Evseev[1], *Valeriy* Toksarov[1] and *Nicholas* Beltiukov[1]

[1]Mining Institute of the Ural Division of the Russian Academy of Sciences - Division of Federal state budget institution of science of the Perm Federal research center of Ural Division of Russian Academy of Sciences, Perm, Russia

Abstract. The forecasting method of roof treatment works stability in the silvinit seams mining of the Verkhnekamsk deposit of potassium salts, based on the strain properties determination of its constituent species in field conditions is examined. **Key words:** roof, interchamber intacts, salt rocks, stresses, strains, resistance

1 Introduction

One of the important conditions to ensure the safe conduct of underground mining is to control the stopes and development workings roof stability. Instrumental measurements allow adjusting the parameters of the industrial stocks project and take additional positive measures of working protection, if necessary. Compliance with this approach is especially important in the development of water-soluble ores, which safety testing relates to the need of the undermined strata continuity (of impermeable intact) preservation [1, 2].

The object of the study is the Verkhnekamsk deposit of potassium and magnesium salts (VKDPM). Leaving the hard interchamber intacts, the silvinite seams ore mining are being developed by chamber mining system. The drifting of development workings and stopes is carried out by the combine method. Productive strata of the deposit, located at depths of 250-450 m, is lamellar mass of silvinite and stone-salt rocks, often separated by clay-anhydrite layers. Silvinite layers RfII and AB an average power of 5.0 and 3.5 m are respectively separated from each by other technological interlayers (RfII -AB) with a capacity of 2.5–8.0 m. The roof of the working layers (especially RfII) overlain by having low resistance clay-salt rocks (cake) layers. With the strength characteristic of red silvinite 23.0 MPa [2] the strength of the silvinite seam RfII can be 1.5 - 2.0 times below the average values for the field. Strength and deformation clayey material properties are directly in 10-15 times lower than the properties of the rock salt layers and silvinite [3].

So at the opening of Polovodovsky block VKDPM reserves it is identified, that the clay material content in productive strata is 17-25%, and clay layers capacity reaches up to 0.5–0.7 m. With the low strength and deformation properties are correspond to the first layer of interlayer RfI - RfII, where there is the highest content of clay material (table. 1).

* Corresponding author: ava@mi-perm.ru

Table 1. The salt rocks properties study results of productive strata 1 YUVP of "EuroChem – UKK"

Layer, stratum	Rock	Clay content	Compressive resistance	Tensile strength	Modulus of deformation	Elastic modulus
		%	MPa	MPa	GPa	GPa
RfI-RfII (cakes)	rock salt	39.6	1.23	0.21	0.0567	0.11
Rf2lr1	red silvinit	19.8	13.73	1.44	0.49	5.29
Rf2lr2	rock salt	22.5	15.48	1.39	0.55	6.50
Rf2lr3	red silvinit	7.7	15.00	1.72	0.56	6.04
Rf2lr4	rock salt	15.7	19.00	1.83	0.83	6.77
Rf2lr5	red silvinit	7.6	19.02	1.34	0.64	7.08
Rf2lr6	rock salt	15.7	23.00	1.78	0.76	8.03
Rf2lr7	red silvinit	4.0	22.12	1.52	0.72	7.31
Rf2-Rf3	rock salt	11.1	23.62	1.76	0.95	9.06

2 The roof rocks stability analysis

When working out the south-eastern sections reserves of the deposit, characterized by a high content of clayey material in the mined strata roof, the problem to ensure the conservation of the roof stopes had intensified. So when working out the reserves on 9-11 eastern panel of the mine BKPRU-2, camera width 5.5 m, where the clay content of material on average is 12-15 %, there was intensive balmstoning in the roof, leading to violation of technological process and creates hazardous conditions for miners. A number of cameras failed to finish until the end for sinking of heading machine because of the danger of their "gripping" and a high probability of roof collapse.

Despite the presence of roof bolting, examination of exhaust chambers showed that, a small roof appear within the first weeks after the sinking, and after 1.5-2.0 months marked the collapse of the roof rocks of a large area and partial destruction of the interlayer RfII–AB.

With the aim of establishing the main regularities of deformation processes development in water treatment chambers depending on the geological conditions of mining operations conducted complex long-term field studies of the deformation nature of the system development structural elements, which included measurement of deformations for interchamber intacts, roof extraction chambers and interlayers.

Contour and deep bench mark system was the design of the metering station.

The central part of the mine at each metering station had a 4 deep bench mark with a length of 0.5, 1.0, 1.5 and 2.0 m, and two sets of contour bench marks for monitoring the chambers roof and walls convergence on both layers (Fig.1a). The study results analysis showed that there is intense stratification of seam roof rocks for RfII with the subsequent clay layers collapse, within the first 1.0 to 2.5 months after the advancement. All layers of technological interlayers RfII – AB are involved in this deformation.

During the 54 days of observation the seam roof deformation RfII in chamber No. 137 (9 VP) was about 200 mm, there was observed uniform character of deformation of the roof across its power (Fig. 1, b).

At the approach of mining operations to the metering station is to increase the speed of rocks soil deformation on the layer AB to 5.5 – 6.0 times. The rate of deformation for soil rocks extraction chamber is stabilized at the same level In the future. In the north-eastern part of the mine field mine BPKRU-2, where the clay content is much less, the bulk of the deformation accounts for the marginal layer of the seam roof RfII, while strains an order of magnitude less [4].

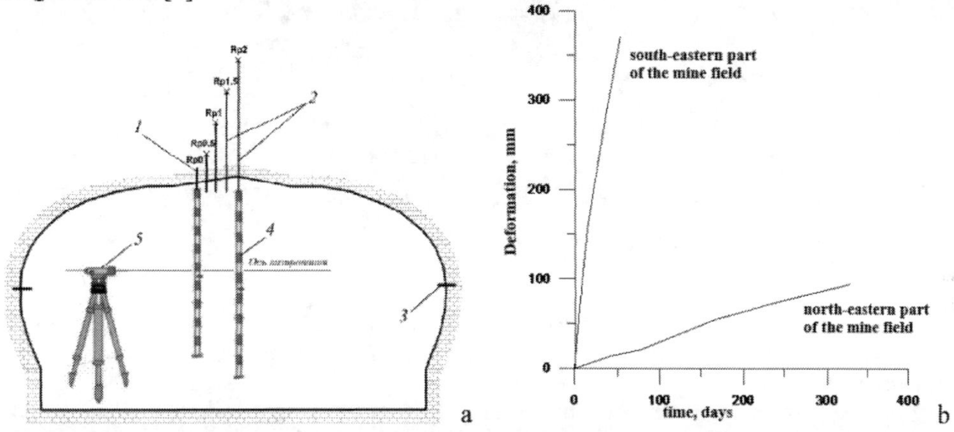

Fig.1. The measurement scheme (a) and the subsidence roof layer Kr (b) graph

The implermented studies complex have shown that it is necessary to know the physico-mechanical properties of rocks for the outcrops stability justification, taking into account structural and their structure textural features, which can be evaluated to the strength and deformation properties study of adjacent strata in mass.

The loading circuit layer methods of the rock hydraulic load devices with various designs is widely used for their definition: flat metal spinning units, downhole hydraulic jack.

3 Experimental test procedures

The field studies complex to determine the deformation modulus of rocks in-circuit mass using Goodman's hydraulic jack[5], which represents a logging sonde designed to measure deformation of the borehole walls under the action of the applied load (Fig. 2) are made for a substantiation the criteria of roof rocks sustainability for extraction chambers.

The borehole walls loading is carried out by two movable metal plates at each end which are mounted the displacement sensors. During the strain indicators field measurements, the following procedure is carried out. Hydraulic jack is placed in the borehole at the required depth. Then using the hand-pump working fluid is supplied into the cylinder of the measuring device (to the level of pressure 15 – 20 MPa). During the experiment check of strains every 2 to 5 MPa. After the loading is done stepwise unloading. Each measuring point was performed 2 – 3 the measurement cycle in the "load-unload". The measurements are performed using 0.3 m distance from the contour of the outcrop. The maximum well depth is 3,5 – 5,0 m.

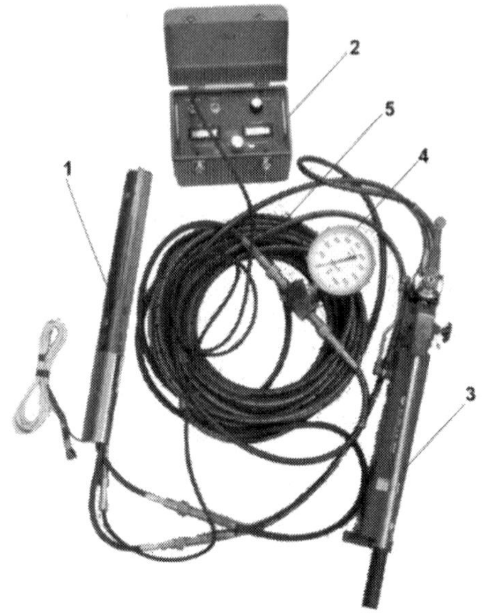

Fig. 2. Equipment set for the modulus of deformation salt rocks determining in field conditions: 1 - hydraulic jack; 2 - a portable device for measuring deformation; 3 - hand pump; 4 - manometer; 5 - pressure hoses

As a result of the experimental studies graphs "pressure (Q), radial strain (εd)" which are determined by the deformation parameters of the mass. There are three characteristic stages (Fig. 3) In the borehole walls deformation diagrams.

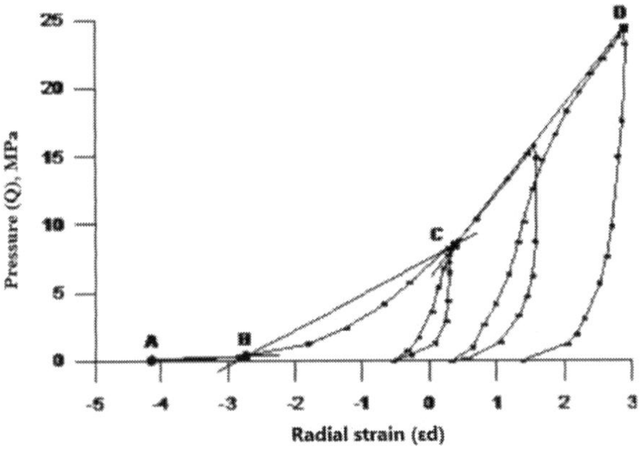

Fig. 3. The borehole walls deformation graph

In the first stage, the movement of the pressure plates to contact with the borehole wall (segment AB) is being. The second stage is timed (segment BC) to bulging-in borehole walls jack irregularities. The third stage corresponds to the complete contact of the pressure plates with the walls of the bore and the deformation of the circuit layer of the rock. The modulus of deformation is calculated by the tangent at the site of a full contact of the pressure plates with the walls of the well.

Studies of the deformation properties of rock prisms pre-loading them according to the stress condition that exists in the mass (Fig. 2) was also undertaken for method development. The prism loading of is carried out by hydraulic jack with a maximum force of 100 t. Prior to testing, the sample surface is levelled using a layer of salt chips with a thickness of 1 – 2 cm. The chips are compressed under load, compacted and ensures uniform transmission of pressure from the plate to the block. Then was the drilling of the measuring wells, in which hydraulic jack was installed, carried out and loading block was made from inside. The studies of the mechanical properties on samples with a diameter of 50 mm. were performed in parallel laboratory.

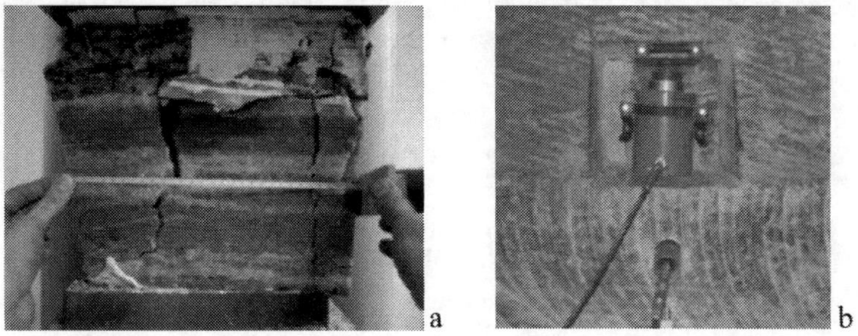

Fig.4. Determining the salt rocks modulus of deformation in laboratory (a) and field conditions (b)

As a result of researches it is established that the modulus of deformation of salt rock, defined in field conditions ranges from 0.5 to 2.7 GPa and depends on the stress level. With the increase of stresses in the mass marked increase in the modulus of deformation. The laboratory studies results tend to give higher (10-20 %) values of the designated indicators (table. 2).

Table 2. The modulus of deformation of rocks formations KrII-KrI and KrII (mine workings BKPRU-2)

Designation rock, seam	Determining method	The modulus of deformation the rock, GPa		
		11VP	9VP	20ZP
Rock salt, RfII-RfI	Laboratory tests	1,260	1,360	1,560
	Mining measurement	0,933	1,061	1,260
Red silvinite, RfII	Laboratory tests	1,350	1,530	1,360
	Mining measurement	1,421	1,363	2,087

The modulus of elasticity is on average 4 to 5 times higher than the modulus of deformation. These values correlate well with values of deformation properties of salt rocks determined in the laboratory on small sized samples. This suggests that the scale effect influence in the salt rocks is insignificant [4].

4 Discussion of results

The large amount of field research carried out in the mines of "Uralkalii" OJSC has shown that the deformation modulus varies in the interlayer rocks RfII – AB within a fairly wide range from 0.7 to 3.0 GPa In the depth interval of 0.5 – 1.5 m from the contour of the outcrop noted its growth from 1.2 to 2.0 GPA. There is an increasing of the modulus of deformation up to 2.5 GPa in the middle of interlayers (in the depth interval of 2.7 – 3.2 m), and then, it is reducing to 0.9 GPa, when approaching the overlying soil layer of the AB.

Simultaneously with the evaluation of deformation properties in the marginal layer, the voltages measurement based on the use of acoustic emission memory effects, is well manifested in quasiplastic salt rocks [4] was carried out. Studies have found that in the middle of interlayers RfII – AB the magnitude of the horizontal stress varies from 5.0 to 7.5 MPa. Near the roof of the reservoir RfII and in the soil layer of the AB voltage is reduced to 1.5 MPa, which indicates the destruction of the marginal layer of rocks and confirms the results of studies on the nature of deformation and destruction of rocks interlayers at high clay content. Telemetric examination of the walls of the measuring wells also confirmed the presence of zones of high fracture and delamination in the top layer RfII and in the soil layer AB.

The joint studies analysis of the deformation nature and roof treatment fracture of stopes and the determination results for the modulus of deformation in field conditions has led to the conclusion that the critical value of the modulus of deformation near the outcrop in which there is intense stratification and the roof rocks collapse of the treatment chamber with the existing parameters of mining is 1.1 GPa (Fig. 5).

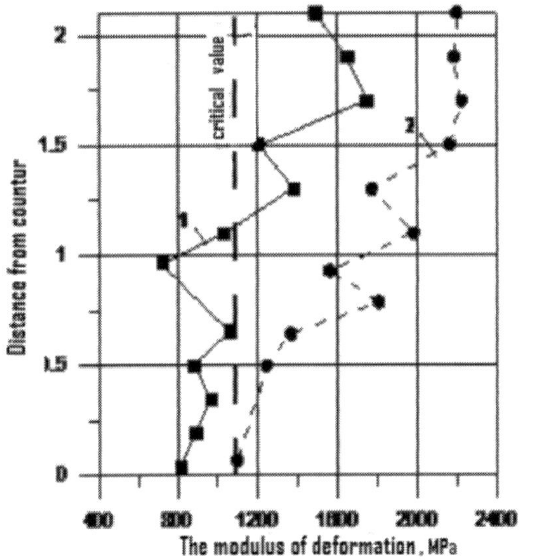

Fig. 5. The dependence of the rocks interlayers modulus deformation with distance from the contour of the exposure (the mine BKPRU-2): 1 – extraction chambers roof is unstable (9 VP); 2 – extraction chambers roof is steady (20 ZP)

So when the width of the extraction chambers more than 5.5 m and in the modulus of deformation less 1.1 GPa the separation process of roof extraction chambers with the consequent loss of its stability is dramatically intenses. When the module of deformation is from 1.5 to 2.5 GPA salt rocks of interlayers are rocks of medium resistance, which contain no more than 5 – 6% of clay material, which allows to develop reserves of chambers without the technological complications, the term steady state of the roof extraction chambers is approximately one year. When the module of deformation is more than 2.5 GPa of the host rock are characterized by high homogeneity (absence of clay "cakes" are working in the roof layers), have a high roof resistance, providing exposure on large areas (with a width of chambers up to 16 m) for up to several years

5 Conclusions

Thus, the modulus of deformation measured in full scale, is a comprehensive indicator of the salt mass structural heterogeneity, allowing it to be used for the assessment of the species sustainability. On the research basis the sustainability assessing methods of the rocks roof extraction chambers during the mining of silvinit seams in the mines VKDPM are developed. The stopes control stability task(safety of mining operations) during the mining of silvinite seams can be divided into two stages. The first step is to assess the degree of danger (risk) discontinuity waterproof strata to identify potentially dangerous areas for development opportunities emergency. Geological survey documentation obtained in the course of operational intelligence, the acts of inspection stopes and development workings, the data on land subsidence, the results of geophysical and seismological research is initial information to evaluate the undermined massif state of potentially dangerous areas. In the second phase, selected dangerous sections of the mine field runs comprehensive monitoring of load-carrying elements of the development system and the water column. In these areas, studies of physico-mechanical properties of the main lithological types of salts rock shall be performed, the evaluation of the stress-strain state and destruction in time for Interchamber intacts and interlayers are assessed. According to the results of experimental studies the intact deformation models, displacement of rocks and earth's surface when working out of the working layers clarifies. The implementation of multivariant calculations of stress-strain state of the salt mass in the process of mining for specific mining-geological conditions allows to evaluate the stability degree of chamber systems load-carrying elements and the danger of discontinuity VZT [6, 7, 8]. This technique can be adapted for use in similar conditions.

References

1. Federal norms and rules in the field of industrial safety "Safety regulations for the mining and processing of solid minerals": approved by the order of Rostechnadzor dated 11.12.2013, no 599.
2. Guidelines for the protection of mines from flooding and protection of undermined objects in the conditions of the Verkhnekamsk deposit of potassium salts. Perm (2014).
3. 3. A. A. Bariakh, V.A. Asanov, I.L. Pankov. Physico-mechanical properties of salt rocks of the Verkhnekamsk deposit of potassium / tutorial. (2008).
4. 4. V.A. Asanov, V.N. Toksarov, A.V. Evseev, N.L. Beltiukov Behevioral of working roof on the southern flank of the Verkhnekamsk deposit of potassium salts. Phys. Tech. Probl. Mineral resources. No.1 (2012).
5. "Suggested Method for Estimating the In-Situ Modulus of Deformation of Rock Using the NX-Borehole Jack", by Heuze, F.E., ASTM Geotechnical Testing Journal, GTJODJ. **7 (4).** (1984).
5. 6. V.A. Asanov, A. A. Bariakh, V.N. Toksarov, A.V. Evseev. The geomechanical assessment method for interchamber intacts long-term stability// Scientific and technical support of mining production (collection of scientific works. D.A. Kunaev Institute of Mining Engineering), 87. (2015).
7. Analysis of time-to time variation of load on interchamber pillars in mines of the Upper Kama potash salt deposit Baryakh A.A., Lobanov S.Y., Lomakin I.S. / Journal of Mining Science. **51 (4).** Pages 696-706. (2015).
8. A.A. Bariakh, I.A. Sanfirov, A.K. Fedoseev, A.I. Babkin, A.A. Tsayukov, Seismic and geomechanical prediction of the waterproof strata state at the potash mines // Physical-technical problems of mining. **6**. pp. 10-22. (2017).

Critically Stressed Areas of Earth's Crust as Medium for Man-caused Hazards

Andrian Batugin[*1]

[1]National University of Science and Technology «MISiS», Mining Institute, Moscow, Russia.

Abstract. Despite advances in rockburst studies, suddenness of major geodynamic events is reported in a number of cases. Phenomenological tectonophysical model is suggested to explain some geodynamics phenomena. Prof. Petukhov I.M. suggested a concept: the Earth crust's critical stress condition is developed due to horizontal compressive forces and entrains rock strata from the sub-surface to a certain depth. The conditions that induced earthquake in 2013 at Bachat coal field in south west Kuzbass are considered in terms of critical stress developed in the top layer of the Earth crust. Estimates show that the size of the critical stress zone, produced presumably by interaction of huge (over 100 km) crustal blocks is at least 10km. Whereas critical stress zone is located in the top part of Earth's crust, mining operations in the pit including blast operations was making a direct impact on this area. Shallow occurrence of critical stress area and its size can provide insight into why mining works brought about induced earthquake with hypocenter at the depth of several kilometers. The conclusion has been made that regional areas of critical stress within rock massif developed as a result of crustal blocks interaction create hazard medium for mining. Key words: rockburst, induced earthquake, earthquake source, critical stress condition, crustal blocks, Bachat earthquake, reactivation of faults.

1 Introduction

Geodynamic safety issue remains to be high on the agenda in various mining areas of the world [1-9]. Significant success has been achieved in Russia in solving rock burst problem, methods of forecasting and prevention have been developed, geodynamic mechanisms have been learned. However, many of the recent severe geodynamic events can not be explained based on established conceptions on geomechanical processes inside the rock massif and onset of its stress strain behavior [9]. As an example, consider Bachat earthquake happened in June 2013. According to [10] its hypocenter was located below pit excavation at the depth of approximately four kilometers. In the meantime such disciplines as rock geodynamics, seismology and geomechanics put forward ideas that critical stress condition is related not only to local rock volumes but to large segments of Earth's crust [11-15].

The idea that significant amount of Earth's crust is in critical stress state was suggested by I. M. Petuchov in his works and was further supported in the book [11]. This work

* Corresponding author: as-bat@mail.ru

substantiates the hypothesis that mechanism and occurrence of Bachat earthquake were related to release of critical stress zone in the Earth's crust resulting from nature tectonic process and man-caused impact.

2 Theory and method

2.1 Insight into critical stress state of lithosphere.

Critical stress state theory in geomechanics is used to explain mechanisms of rock burst, rock behavior in abutment zones, pit slope stability, etc. [11, 15-16]. In situ studies show that impact on the edge of coal seam, such as insertion of shearer's cutting bar, explosion, some other impact, brings about immediate changes in critical stress area, reallocation of loads, their transfer depthward [11, p. 27]. Coal seam in this area maintains the ability to accumulate potential energy of elastic compression. Critically stressed zone is considered as similar to that which is ultimately saturated with potential energy. If, in case of impact on this area, the seam releases the load slowly, this results in brittle failure of coal manifested by rock bumps and bursts. To eliminate these consequences or make them less intensive the cutting rate/shearer advance rate (for example) needs to be slowed down.

Academician Ye. I. Shemyakin defined the area in front of the maximum bearing pressure zone as prefracture zone. Within prefracture zone which as such is in critically stressed state, rock deformation occurs along newly developed surfaces of weakening - through relocation of rock slabs, some of which may be in the state of elastic strain.

Making comparison between processes in coal seam, relying upon geodynamics postulates, was put forward the idea that lithosphere as such is in critically stressed state [11]. Critical stress is developed in the Earth's crust due to horizontal compressive forces and entrains shallow rock strata in the first place. If horizontal stress is great enough, progressively deeper strata of lithosphere acquire critical stress condition. Specific feature of this critically stressed stratum (by analogy to face adjacent part of coal seam) is the presence of elastic areas (volumes) inside. In such conditions a wide range of deformations resulting from man-caused impact may be manifested inside the rock massif - creep, rock bumps, naturally occuring earthquakes.

In seismology basic concepts describing mechanism of earthquakes in the Earth's crust [17], use in one way or other the concept of critical stress state of the stratum in focal area. Some works [12,14] bear direct indications that there are areas in the Earth's crust which state is close to critical.

2.2 Identifying mining risk bearing areas by geodynamic zoning

Mining practice gives evidence that adverse geodynamic events are manifested in particular zones. By mid- 1970s it became obvious that probability of adverse geodynamic events in mines is dependant, among other factors, on natural forces that are responsible for stressed condition and structure of the rock massif. The idea of upfront identification of areas hazardous for mining activity based on the knowledge of tectonic stress areas and current rock massif block structure has found its reflection in geodynamic zoning method [18].

According to the concept of geodynamic zoning, zones of danger in the area of the deposit are formed due to the interaction of geodynamically active crust blocks (modern tectonic blocks) of various hierarchical ranks, the formation of which is associated with global geodynamic processes and the divisibility of lithospheric plates.

2.3 Evaluation of rock massif stressed state by tectonophysical methods

Evaluation of the stress condition of the massif is the key to understand the conditions for occurrence of geodynamic phenomena [1-8, 19-27]. Tectonophysics methods here have an important advantage over instrumental methods because they give a possibility to investigate the stress condition of large volumes of rock massif based on analysis of rock slabs shifting along current faults [18, 28]. Many tectonophysical methods offer graphic solutions in which the displacement vector on the indiscriminately oriented fault plane should be positioned in a specific manner on the stereogram thereby showing the relationship between the displacement direction and orientation of the primary stress axes. Moreover, it follows on the ideas of the hierarchy of the stress fields [29] that the movement along the fault of the specified size can be induced only by the stress field of a certain rank. The stress field causing the movement along the fault remains active in some areas which includes this dislocation and which is also regional towards it.

Since present tectonic crustal movements are reflected in the relief of the Earth's surface, and the block boundaries are detected by indicators of relief, then it is assumed that the method of geodynamic zoning reveals currently active structural elements and stress field of the Earth's crust. Using the data on blocks displacement along their boundaries with tectonophysics methods makes it possible to study present stress fields [18].

3 Geodynamic position of Bachatsky coal pit

Block composition of Kuzbass is represented in Figure1. Bachat coal deposit is located at the south west edge of Kuzbass next to intersection of two present-day regional tectonic zones II-II and I-II-IV.

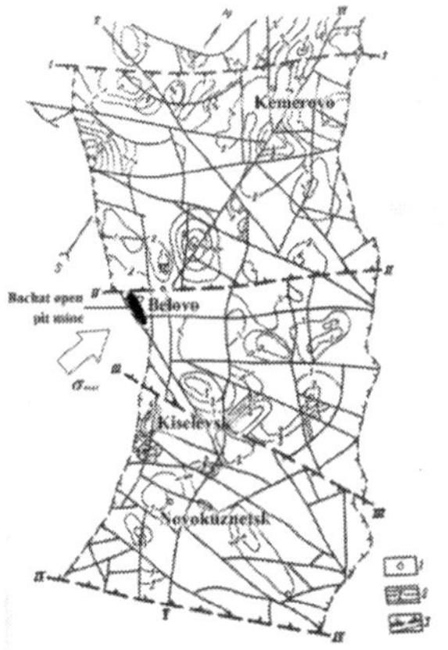

Fig. 1. Scheme of Kuzbass block composition [18]. 1- towns, 2- isoline of moving of earth surface; 3- boundary of block (2nd and 3rd ranks)

Fault zone I-II-IV stretches to the north-west and aligns with major boundary thrust fault separating Kuzbass from Salair mountain range located to the south–west. Fault zone II-II strikes transversally to key Kuzbass structures, stretching to the north-east. Geologists point out high structural complexity of Bachatsk coal deposit stretching along 1-1 boundary line. The deposit is divided into Northern and Southern parts and is located within one of tectonic plates limited by thrust faults striking in the north-west direction (Figure.2-a). It is assumed that due to Salair's pressure on Kuzbass coal bearing deposits a series of several major thrust faults acquired their shape during one of folding phases. These are Tyrgan,

Bachat and Salair thrust faults, dipping to the south-west at 50-70 degree. There are no detailed studies on the deep structure of this part of Kuzbass, yet regional geologic materials suggest that these thrust faults stretch down to greater depth and undercut Bachat coal pit at a depth mark of several kilometers (Figure 2).

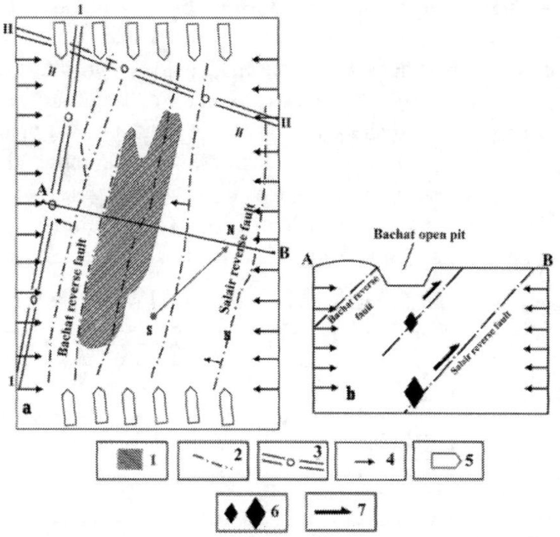

Fig. 2. A simplified tectonophysics scheme of the Bachat open mine area: 1 - Bachat open pit; 2 - tectonic faults; 3- boundaries of blocks of rank II; 4, 5 - direction of maximum stress axis of present stress fields of ranks II and III respectively, defined by tectonophysical method; 6 - aftershock and main shock hypocenters; 7 - direction of the displacement along the reverse faults during the earthquake

Taking into account the direction of relative displacement of the walls of regional fault II-II and reverse fault displacement of the southwest wall of Bachat thrust, it is possible to determine the direction of regional stress in this part of the Kuzbass area: SW – NE (Fig 2).

Bachat coal pit is approximately 300 m deep, about 10km long and its width is 2km. Mining operations involve routine blasts with up to 350 t mass of charge.

The earthquake (M=6) that happened at the turn of the 20st century near Novokuznetsk city is known as the heaviest earthquake in Kuzbass. Throughout the time of instrumental monitoring up to 2013, no earthquakes of over 4.5 magnitude was observed [30]. However, Kuzbass today is one of the mining regions in Russia with technogenic seismicity. Since 2006, an increase in seismic activity is observed in the area of operating deep mines and pits in the western part of Kuzbass (Leninsk-Kuznetsk and Bachat districts). In 2013, there was Bachatsky earthquake of $M_L = 6.1$ which is rated as the strongest mining-induced earthquake in coal mining areas by the nature of the main shock location and the subsequent aftershocks (at Bachatsky open mine). Focal mechanism solutions show that the main shock event was reverse fault. The depth of seismic focus was assessed as 4km [10].

4 Tectonophysical model of Bachatsky earthquake

Present day tectonic stress field in west Kuzbass is partially derived from one of paleotectonic stress fields. The existing thrust faults are represented in surface relief and in present day stress field and go through reactivation process which is manifested by nodal surface orientation of the main shock and aftershocks of Bachat earthquake. At the same time NE striking shears are developed transversal to thrust faults, which is also reflected in the structure of current surface relief. Development of present day tectonic deformations next to and around Bachat coal pit and seismic activation processes allow to draw a conclusion: tectonic stress in the Earth's upper crust has reached its critical state in this region. As Bachat coal pit is located at the intersection of two current major faults in the Earth's upper crust, critical stress state could be reached primarily here.

Tectonophysical studies at rock-burst mines show that the hazard of tectonic rockbursts and mining-induced earthquakes is connected with the possibility of reactivation of large tectonic faults, oriented in a certain way in the massif [31]. On the basis of the approach developed by us, it can be supposed that the mechanism of Bachat earthquake is related to the influence of present regional stress field to the fault planes of major faults, Figure 2.

Under the influence of present regional contraction on tectonic faults existing in the massif, depending on their orientation in space, high shearing stresses may be achieved. The calculated ratio $\tau^* = \tau_n / \tau_{max}$ of shearing stress (τ_n) in thrust fault planes and the maximum shearing stress of this stress field (τ_{max}) are presented in Table 1.

Table 1. Tectonophysical conditions of large faults in the present stress field

Fault	$\tau^* = \tau_n / \tau_{max}$	
	stress field of rank II with orientation of axis σ_{max} to SW-NE	stress field of rank III with orientation of axis σ_{max} to SE-NW
Bachat, Salair thrusts	0.8	0.6
In-plane shear	0.9	0.4

The table shows that the thrusts are located in planes with high shearing stresses, i.e. they are hazardous in manifestation of movements through them in the stress field of rank II. In the field of rank II with SW-NE orientation of axis σ_{max} and vertical σ_{min}, Salair and Bachat thrusts behave as classic-type reverse faults.

Indeed, main shock mechanism is interpreted as a proper thrust fault, i.e. at the time of the earthquake major thrust faults got reactivated in agreement with the tectonophysical model under consideration, Figure 2-b.

5 Discussing the findings

Based on the data available in paper [32, Figure 3], where generalized empiric formula determined by a number of authors provide assessment of earthquake seismic focus size L as function of its magnitude M, we may assume that seismic focus size L of Bachat earthquake of magnitude M_L=6.1, is at least 10km. This assessment correlates with aftershock zone size which according to data [10], covers not just the pit area but the space beyond its boundaries. Let us take into account that the hypocenter of the earthquake main shock was at the depth of 4 km. Considering that the size of seismic focus exceeds 10km, this means that the Earth's upper crust in vicinity of the pit was in critically stressed state from the day surface down to the depth of several kilometers. In this case mining operations in the pit including blasts of up to 350t explosives can be regarded as factor of permanent impact upon this regional critically stressed area. It is discovered that cyclical impact is able to influence for nature of seismic displacement along fault [33]. Due to mining operations and development of large scale voids - mined-out space - equilibrium in shallow part of this area is inevitably disrupted, hence stress redistribution may propagate to much greater depths than mining levels. In this relation the fact that earthquake hypocenter was located at a great depth below the pit could be so explained: actually, mining works were executed within critically stressed area of the region.

Furthermore, considering the relation between tectonic bump focal size *r* and areal size of this event preparation process *R*, *R/r*=10-30 [31, 34] it is possible to estimate sizes of the

Earth's upper crust blocks involved in preparation of induced earthquake. If the size of seismic focus area equals 10km, the relevant event preparation area shall be 100km to 300km, and this is comparable with the size of crust blocks of the 2nd rank, as shown in Figure 1. In this case, conceivably, due to interaction of large blocks of lithosphere, within or at the boundaries of these large plates critically stressed areas are built up. As geologic processes are really slow, energy encapsulated in critically stressed areas have enough time to release in the rock massif and no conditions are developed for dynamic rupture. In case of mining works with high rate of face advance in deep mines and/or mass scale blasts in pits, load increment rate may exceed the rate of load release, thus conditions are created for onset of adverse geodynamic events. Taking into account that critically stressed areas built up due to geodynamic processes embrace first and foremost subsurface layers of the Earth's upper crust, monitoring shall be envisaged as part of mining operations both in deep and surface mines.

4 Conclusions

1. Influence of mining operations on major seismic events with hypocenters at great depth such as Bachats earthquake may be explained using the concept of critically stressed condition of the Earth's upper crust.
2. Analysis of conditions contributing to Bachat earthquake occurrence shows that with seismic focus area size of over 10km and earthquake hypocenter depth of 4 km the Earth's upper crust in vicinity of the pit was in critically stressed state from the day surface down to the depth of several kilometers. In fact, mining operations were carried out within an extensive critically stressed area, and this is the reason why stress equilibrium in this part of subsurface zone was continuously disrupted. Ultimately it triggered dynamic release throughout the critically stressed zone with rock blocks shifting along the major fault.
3. Size of Bachat earthquake preparation area might have exceeded a hundred kilometers, and this is comparable with the size of crust blocks of the 2nd rank, participating in their own tectonic motion. Major upthrust faults and lateral shears in present day stress field coinside with high shearing stress planes, which explains why such direction of displacement when the above upthrust faults and shears got reactivated at the time of earthquake.
4. Under specific conditions triggered both by tectonic processes and man-caused impact it is the upper part of the Earth's crust that first becomes critically stressed, and this creates conditions provoking occurrence of risk bearing geodynamic events in case intensive mining is performed in these areas.

References

1. I. Yu. Rasskazov, B. G. Saksin, V. I. Usikov, M. I. Potapchuk, Gornyi Zhurnal **12**, 13-19. Doi: 10.17580/gzh.2016.12.03 (2016).
2. W. Cai, L. Dou, M. Zhang, W. Cao, J.-Q. Shi, L. Feng, Tunnel. and Underground Space Techn. **80**, p. 232-245 (2018).
3. Y. Yu, D.-X. Geng, L.-H. Tong, X.-S. Zhao, X.-H. Diao, L.-H. Huang, Int. J. Geomech. **18(7),** Art. No. 06018016 (2018).
4. D. Malovichko, G. van Aswegen, R. Clark, J. South. African Inst. of Mining and Metallurgy. **112**, 419-429 (2012).
5. Y.-D. Jiang, Y.-S. Pan, F.-X. Jiang, L.-M. Dou, Y. Ju, J. China Coal Society. **39(2),** pp. 205-213 (2014).

6. T. G. Butler, B. S. Simser, Proceedings of the 9th International Symposium on Rockbursts and Seismicity in Mines, Santiago Chile, 290-296. (Santiago, 2017).
7. Zhang Hongwei, Du Kai, Rong Hai, et al., J. of Liaoning Techn. Univ. (Natural Science), **34(2),** 165-169. Doi:10.11956/j.issn.1008-0562.2015 (2015).
8. S. Lasocki, B. Orlecka-Sicora, G. Mutke, W. Pytel, I. Rudzinski, P. Markowski, P. Piasecki, Proceedings of the Ninth International Symposium on Rockbursts and Seismicity in Mines, Santiago Chile (RaSim-9), 316-324 (Santiago, 2017).
9. A. V. Lovchikov, Forecast and prevention of tectonic rock bursts and earthquakes: measurement of deformations, residual and actual stresses in rocks: First International Symposium on 21-23 of September, 39-49, (Bishkek, 2016).
10. A. F. Emanov, A. A. Emanov, A. V. Fateev, E. V. Leskova, E. V. Shevkunova, V. G. Podkorytova, J. Mining Sc. **50(2),** 224-228 (2014).
11. I. M. Petukhov, I. M. Batugina, Geodynamics of the Earth's Interior. (Nedra Communications, Moscow, 1999).
12. G. A. Sobolev, A.B. Ponomarev, Physic of Earthquake and precursor. (Science, Moscow, 2003).
13. Yu. L. Rebetsky, Active geological and geophysical processes in lithosphere: methods and results of investigation: Int. Conference, **2**, 105-110 (Voronezh, 2006).
14. W. L. Ellsworth, Science, 341, Doi: 10.1126/science.1225942 (2013).
15. A. F. Revugenko J. Mining Sc. **50(3),** 33-40 (2014).
16. L. Xue, Q. Sun, Y. Wang, S. Wang, G. Li, X. Pan, J. of Basic Sc. and Eng. **21(4),** pp. 710-724 (2013).
17. J. Rice, In*: Physics of the Earth's Interior*, 555-649 (1980).
18. I. M. Batugina, I. M. Petukhov *Geodynamic zoning of mineral deposits for planning and explotation of mines.* (New Delhi: Oxford & IBH Publishing Co. Pvt. Ltd.,1990).
19. L. A. Nazarov, L. A. Nazarova, O. M. Usol'tseva, O. A. Kuchai, J. Mining Sc. **50(5),** 831–840 (2014).
20. T.-W. Lan, H.-W. Zhang, L. Sheng, at al, Sc. World J., Article ID 516095. (2015).
21. I. Yu. Rasskazov, D. S. Migunov, P. A. Anikin, A. V. Gladyr', A. A., Tereshkin, D. O. Zhelnin, J. Mining Sc. **51(3)**, 614-623 (2015).
22. A. Batugin, V. Odintsev, Yu Lijiang, K. Kolikov, Ev. Khotchenkov, Proceedings of the 18th International Multidisciplinary Scientific GeoConference & EXPO (SGEM-2018), Albena (Bulgaria), doi.org./10.5593/sgem2018/1.3, 18(1.3), pp. 319-326. (2018).
23. L. A. Puchkov, N.O. Kaledina, S.S. Kobylkin, Euras.Mining, **2**, 3-6 (2015).
24. V. N. Odintsev, N. A. Miletenko, J. Mining Sc., **51(3)** 423-434. (2015).
25. V. N. Morozov, V.N. Tatarinov, J. Int. Nucl. Energy Sc.and Techn., **2(1/2)**, 65–74. Doi: 10.1504/ijnest. 2006. 010648 (2006).
26. A. Batugin, V. Musina, I. Golovko, Proceedings of WMESS-2017, Praga, on September 11-15, 2017. (IOP Conference Series: Earth and Environmental Science, **95** (2017).
27. D. Zhirov, V. Glaznev, A. Zhirova, Proceedings of the 15th International Multidisciplinary Scientific GeoConference & EXPO, Albena. **1**. 97-104 (2015).
28. E. M. Anderson, The dynamics of faulting, 206 (Edinburgh: Oliver and Boyd 1951).
29. D. N. Osokina, J. Geodynamics. 10(2–4), 331–344 (1988).
30. D. V. Yakovlev, T.I. Lazarevich, S.V. Tsirel'. J. Mining Sc. 49(6), 862–872 (2013).
31. A. S. Batugin, I.M. Batugina, Lan Tianwei. J. Liaoning Techn. Univ., Natural Science. 35 (6), 561-565 (2016).
32. E. G. Bugaev, Geodynamic & Tectonophysics. **2(3)**, 244-265. (2011).
33. G. G. Kocharyan, V. N. Kostyuchenko, D. V. Pavlov, Physical Mesomechanics, **7(1)**, 5-21 (2004).
34. P. Xiang, H.-G. Ji, M.-F. Cai, Y.-Z. Zhang, Rock and Soil Mech. **39(2)**, pp. 457-466 (2018).

Determining displacement parameters of "Kletevaya" shaft axis over mined-out space

Albert Zubkov[1], *Sergey* Sentiabov[1] and *Konstantin* Selin[1*]

[1] Federal State Budgetary Institution "Institute of Mining of Ural Division of Russian Academy of Sciences", Yekaterinburg, Russia

Abstract. The development of mineral deposits very often occurs in stages. First, open-cut mining, then underground one. At the same time, the arrangement of shafts is always problematic, when it is necessary to choose between security and economic efficiency. Design solutions always convince us that the main thing is safety, but is it really so? This article is the initial stage in the formulation of a very topical issue, in the opinion of the authors, which consists in developing a method for estimating and forecasting the change in the state of shafts during operation. Experimental work was carried out at "Gaiskoe" underground mine, where seven shafts are located in operation, located 100-200 meters from the collapse zone, the depth of the trunks is more than 1300 meters. The main purpose of the research was to determine the wandering from the vertical axis of the trunk at different depths. **Key words**: stress-strain state, deformation, rock shift, control of linear parameters.

1. Brief description of the mine

"Gaiskoe" copper-pyrite mine began operating in 1959. The mine is represented by several isolated ore mines of complex lenticular and vein-like forms, differing in mineral composition. The ore-bearing formation is represented by albitophyres, tuff breccias, tuffs of basic and mixed compositions, intensively schistose and silicified.

On the basis of the mine was built "Gaiskoe mining and processing works", where 76% of the copper reserves of the Orenburg region are concentrated. In order to accelerate the involvement of rich ore in exploitation and increase the capacity of the enterprise, the development of the mine was carried out in a combined way, with the combination of open-pit and underground mining.

The first stage of the pit No.1 was commissioned in 1963 with a design capacity of 2 million tons. Simultaneously with the construction of the pit on its northern side, the construction of an underground mine was performed. The extent of the horizons at the first stage of the autopsy (with a deck height of 60 meters) was 170-440 meters.

The extent of the horizons at the second stage of the autopsy was 440 - 685 meters. When designing the second stage of the mine, it was planned to increase the production capacity several times by using self-propelled equipment.

* Corresponding author: skvekb@mail.ru

The opening was carried out by vertical shafts of the mines "Ekspluatatsionnaia", "Kletevaya" and "Zakladochnaia", access ramp and ventilation shafts. The height of deck was 80 meters, the width of chambers was 20 meters, haulage level was passed at a elevation of 685 meters from the surface. In the intermediate horizons (510, 590, 670 meters), ore from blocks was delivered using self-propellered LHD to ore passes, then the ore was transferred to production level (685 meters) and transported to the shaft in wagons.

The third stage of the mine opening provided for the penetration of intermediate horizons at 750, 830, 910 meters and a haulage level of 936 meters. Opening of the reserves was performed via vertical mine shafts " Ekspluatatsionnaia", " Zakladochnaia", "Novaia", "Sredniaia ventiliatsionnaia", "Iuzhnaia ventiliatsionnaia", "Severnaia ventiliatsionnaia" and access ramp.

In 2006, a large-scale reconstruction of "Gaiskoe mining and processing works" OJSC underground mine shafts was commenced. Since 2012, the priority reconstructed objects of the underground mine are shafts of three mines - Kletevaya, Skipovaia and Novaia.

The "Kletevaya" mine shaft was put into operation in 2013 after reconstruction. The purpose of "Kletevaya" mine shaft is the descent of people to the elevation of 1390 meters, ensuring the operation of mine drainage, supply of compressed air and water to the mine.

2. Measuring method

At mines, there is a widespread sinking of shafts in the zone of influence and mined-out space, which includes the pit and caving area for underground mining. At the Gaiskoe mining and processing works all eight shafts turned out to be at the distance of 140 to 480 meters (on the average — 200 meters) from the side of pit No.1 and the possible caving area for underground mining operations with a diameter of 1,400 meters for a half a century of operation (Figure 1).

In the geomechanical calculation of the deviation of the axis of the shaft from the position of conductors in relation to the axis, several difficulties arose:

1. The displacement of the rock massif on mined-out space occurs in the zone of residual stress [1].

- gravitational component causes a deformation of the expansion of mined-out space;
- tectonic component causes its compression;
- time variable astrophysical component [2] compresses or expands mined-out space with an average period of 11 years.

2. Deep mine shafts rarely pass in one stage. "Kletieva" mine shaft in the Gaiskoe underground mine were taking place in 4 stages since 1959 to 2012. At this time, the massif in the mining zone underwent a variety of deformations, both in magnitude and sign.

In these conditions, the geomechanical calculation of the deviation of shafts axis toward mined-out space is very problematic. It is easier to measure it instrumentally. It was done by the experimental method.

The proposed control method assumes the determination of the deviation of shaft from the vertical axis, caused by the theoretically possible displacement of the rock mass in the direction of the pit mined-out space.

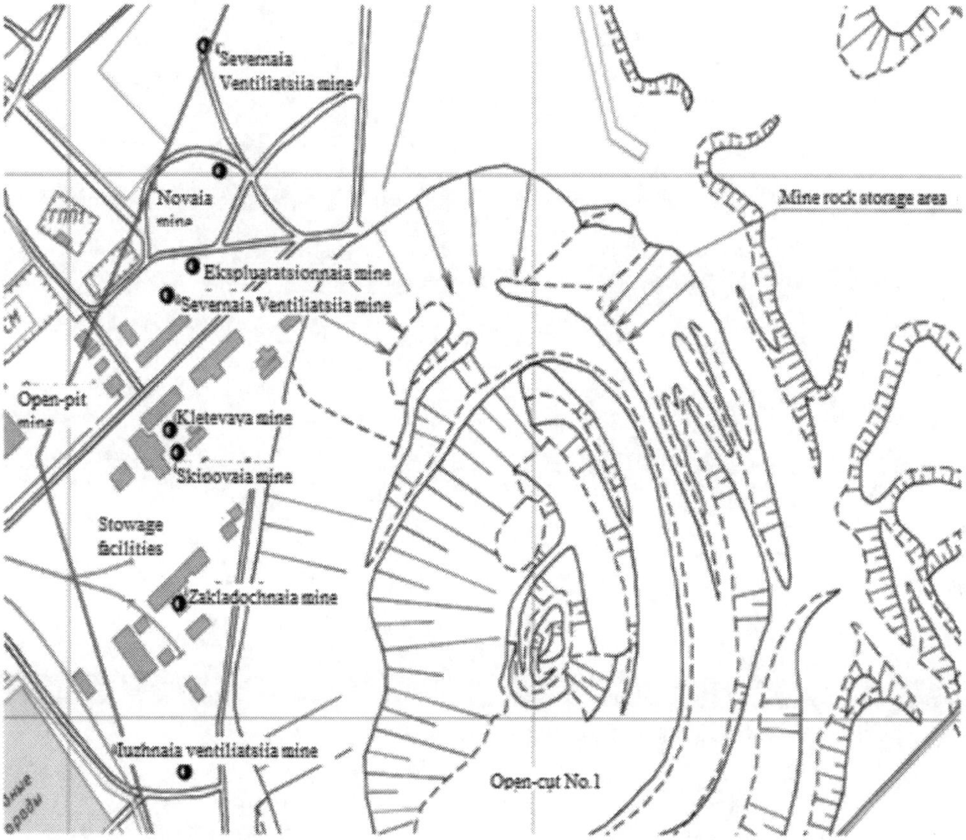

Fig. 1. Industrial site plan

The basis of the method is the determination of the mutual position of the observation stations (OS) located in the shaft of the "Kletevaya" mine, at horizons: 0; - 260 m; - 440 m; -752 m; -990 m. The mutual position control is carried out by determining the change in the linear parameters of the position of laser plumb spot on a target. The laser plumb function is performed by the precision tachymeter "Sokkia NET 1200" with a telescope at the zenith point. The function of the target is performed by an observation table made of PMMA, on which shroud sheet is fixed (Fig. 4).

The tachymeter is mounted on the console, which is attached to the H-beam of a shaft manway. The device is located under the target, consistently, on the horizons: - 990 m; - 752 m; -440 m, -260 m while the target is set at horizons: -752; - 440 m; - 260 m; 0. The scheme of tachymeter is shown in Fig. 2, 3, 4. Tachymeter is set to operating position (zenith angle of 0°), after which the spot of cross-section is marked on the target. Then tachymeter is rotated by 90° and the position of cross-section spot is again marked on the target. In total, the above-described operation is performed 4 times, then, 4 contours of the cross-section spot remain on the target with a 90° tachymeter rotation pitch. The center of the cross-section is determined by 4 contours of the spot of the cross-section. Then, linear parameters of the center position in relation to joints of the elements of the shaft furniture, in the target setting area (Fig. 4), are determined with a tape ruler. Measurements of linear parameters of the device position are performed in a similar manner.

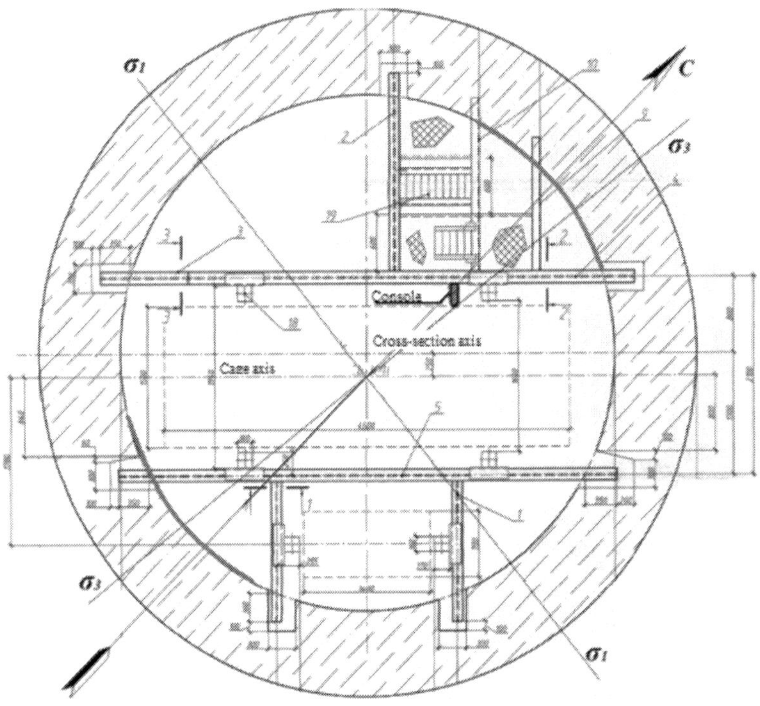

Fig. 2. The observation station of the tachymeter (console).

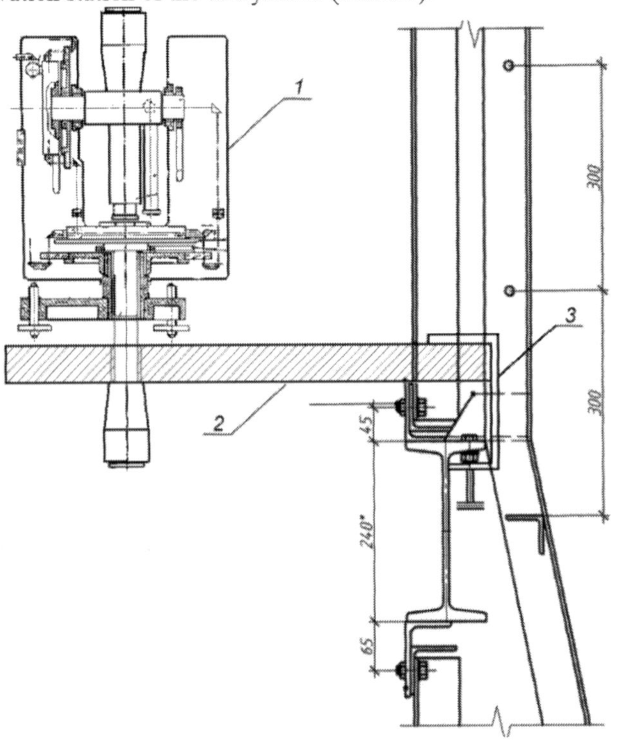

Fig. 3. Scheme of tachymeter installation, where: 1 - tachymeter; 2 - console; 3 - console attachment.

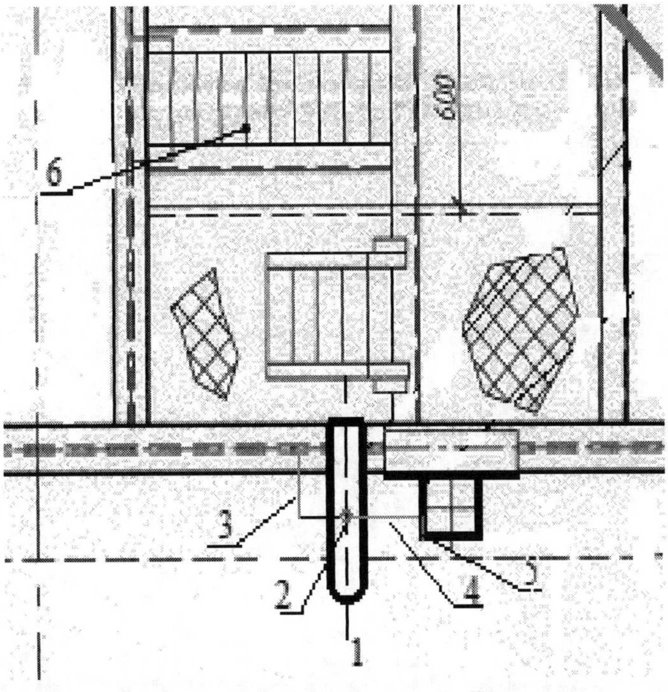

Fig. 4. Scheme for the measurement of linear parameters the position of central axis of instrument (axis of the plumb line), 1 - console; 2 - plumb line axis; 3 - distance from the axis of plumb to the center of the bunton; 4 - distance from the axis of plumb to conductor; 5 - distance from the axis of plumb to conductor surface; 6 - manway.

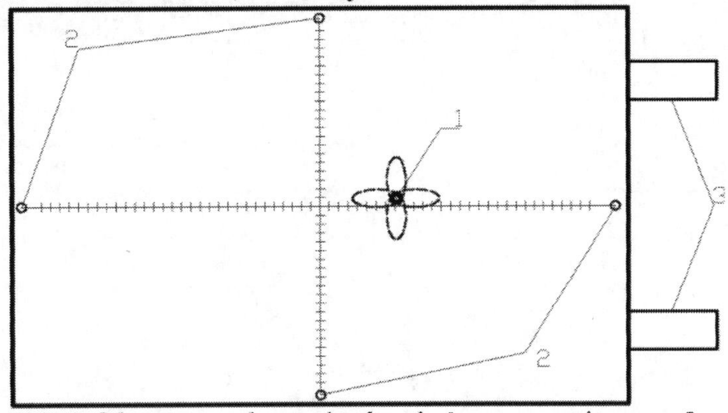

Fig. 5. The position of the spot on observation bench, 1 - cross-section spot; 2 - marks for cross-section spot position reference; 3 - bench attachment.

3. The results for monitoring parameters of "Kletevaya" mine shaft axis displacement and rock massif at Gaiskoe field

In accordance to the condition that shaft slope changes, measured linear parameters of plumb axis position relative to the reinforcement elements change. Changes in the linear parameters on the OS of all horizons may be summed to obtain the total deviation of the shaft from 0 to -990 m.

According to the paragraph No.324 [3], permissible deviation of the shaft axis is calculated by the formula:

$$\Delta U\rho = (50 + 0.15H), \text{ mm}$$

H - shaft depth (in area under calculation), m.

For "Kletevaya" shaft at the area of 990 m, the maximum deviation will be 198 mm.

The results of three observations series for determining the position of shaft axis are given in Table 1 and in Fig. 6.

Table 1. The results of observations

Stage / Date	June 2014 X / Y (mm)	October 2014 X / Y (mm)	July 2017 X / Y (mm)
-10/-260	266 / 7	398 / 26	431 / 2
-260/-440		307 / 5	330 / 1
-440/-752	190 / 39	208 / 75	233 / 49
-752/-990	-	45 / 4	60 / 11

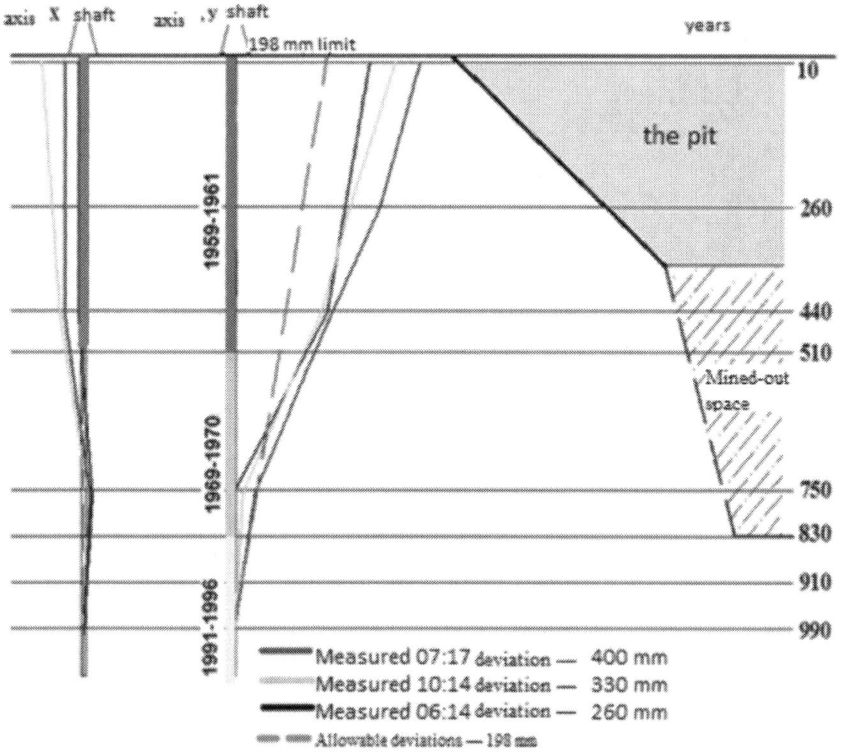

Fig. 6. The results of determining the position of "Kletevaya" shaft axis

As a result of the studies, the general trend of the shaft axis wandering was confirmed. The total inclination of the axis towards mined-out space was 400 mm (along the "Y" axis), with an allowable deviation of 198 mm (in accordance with paragraph No.324 [3]). The measurements were carried out according to the methodology developed by the Institute of Mining of the Ural Division of the Russian Academy of Sciences, which was not certified at the moment. The proposed method of stage measurements, unlike traditional methods [4], allows us to quickly detect the deviation of shaft axis from the allowable limits at any intervals. The proposed theoretical methods allow calculating massif deformation around the caving area when limited conditions are gravitational and gravitational-tectonic stresses which are constants for the whole period of the enterprise's operation [5].

It has been established for now that the residual stresses of the rock mass form gravitational ($\sigma^{п}_{\gamma H}$), tectonic ($\sigma^{п}_{T}$) and the astrophysical component ($\sigma^{п}_{Af}$) which is time-dependent, with a cyclicity of 11 years [1]. This circumstance fundamentally changes the idea of the regularities in the development of the processes of displacement of the rock massif into mined-out space, which the world science has not considered.

The deviation of the shaft axis towards mined-out space was formed under the following conditions:

1. Up to a mark of -510 m, the shaft was conducted during 1959-1961, when there was no open-pit and no underground excavations were in progress, $\sigma^{п}_{AF} = 0$;
2. Deeping from -510 to -830 m was carried out in 1969-1970, during the period when cleaning works in the pit and underground were conducted. The displacement of the shaft axis during this period occurred within the framework of the factors of the change in the parameters of the stress-strain state, as a reaction to the formation of mined-out space, when the sums of gravitational ($\sigma^{п}_{\gamma H}$) and tectonic ($\sigma^{п}_{T}$) stresses act in the array, when $\sigma^{п}_{Af} = 0$. At the surface, where $\sigma^{п}_{\gamma H} = 0$, the displacement was smaller than at the depth;
3. 3. Deeping from -830 m to -1150 m was carried out from 1991 to 1996, when $\sigma^{п}_{Af}$ increased from zero to a maximum;
4. In 2014 mined-out space at the mine was formed up to the mountains (-830 m). As a result, the displacement of the shaft axis in the interval -750 ÷ -830 m was minimal, and at an interval of -510 ÷ 440 m the displacement increased, closer to the surface it should become larger, which was confirmed during the experiment.

4. Conclusions

The main conclusions obtained during the work are as follows.

1. The change in the parameters of the stress-strain state of the rock massif in time radically changes the notion of the regularities in the development of the processes of displacement of the rock mass to mined-out space, which necessitates a change in the approaches to the assessment of the location and parameters of the actual state of shafts and other capital mine workings.

2. Instrumental measurements were made of the deflection of the shaft of "Kletevaya" mine from the vertical axis at various horizons, the value of which at the surface reached 400 mm, with permissible 198 mm for the depth 990 m.

3. An analogous situation with the inclination of the shafts should be observed in the fields where they start or have started underground mining, which is produced at depths of more than 500 m.

4. The proposed method of stage measurements, unlike traditional methods, allows us to quickly detect the deviation of the shaft axis from the allowable limits at any intervals.

References

1. A.V. Zubkov, Lithosphere, **5**, 146 - 151 (2016).
2. A.V. Zubkov, K.V. Selin, S.V.Sentyabov, Lithosphere, **6,** 116 - 129 (2015).
3. Instructions for the production of surveying work, RD 07-603-03.
4. I.I. Dobkin, V.B. Lebedev, M.N. Galinskaya, E.D. Platonov, Yu.V. Anufriev, Mine surveying for the installation and operation of mine hoisting equipment, 221 (1983).
5. A.D. Sashurin, Movement of rocks in the mines of ferrous metallurgy, 268 (1999).

The work was carried out within the framework of the state task (subject №0405-2015-0012).

Experimental substantiation of acoustic method for determination of "coal bed - country rock" contact relations based on boundary inverse problem solution

Vladimir Shkuratnik[1], *Petr* Nikolenko[1], *Larisa* Nazarova[1*] and *Leonid* Nazarov[1]

[1]Institute of Complex Exploitation of Mineral Resources of the Russian Academy of Sciences, Moscow, Russia.

Abstract. The article describes the acoustic method developed by the authors to determine the contact conditions between coal seam and host rocks on the basis of boundary inverse problem solution. Justification of the applicability of the method is made by means of physical modeling, in which samples from the model material were subjected to step-by-step uniaxial loading with simultaneous ultrasonic sounding using a multichannel scheme. In this case, various friction conditions were set for different samples at the "sample-plate of the press" interface. Based on the results of sounding, the distribution of velocities in the illuminated domain of the sample was reconstructed using tomographic processing. The array of data obtained as a result of acoustic tomography was used as input data to solve the inverse problem. As a result of the solution of the inverse problem, it was shown that the appearance of slip regions at the boundaries of the sample generates tensile stresses, the magnitude of which reaches 0.8 from the axial load. The developed method makes it possible, on the basis of acoustic measurements, to identify areas with a reduced coefficient of friction between the coal seam and the host rocks, the presence of which may be the reason of dangerous dynamic phenomena.

1 Introduction

Combine excavation of coal is accompanied by a rapid change of stress fields in the rock mass, which sometimes leads to dangerous dynamic phenomena (sudden outbursts of coal and gas, rock burst, etc.) [1,2]. To monitor the state of the coal massif and forecast such phenomena, mines are equipped with microseismic monitoring systems [3,4,5]. However, modern methods of interpreting data from seismic observations in mines do not allow to detect heterogeneities in the massif with linear dimensions of several centimeters [6, 8, 7, 9]. Such heterogeneities include also thin contact zones of coal seams with host rocks, and their acoustic properties differ insignificantly [10]. Meanwhile, the presence of areas with reduced coefficient of friction may be one of the reasons for the destruction of the bottom-hole part of the seam, followed by the coal and gas bursts [11]. Such arias introduce

* Corresponding author: lanazarova@ngs.ru

qualitative changes into the stress-strain state of the coal massif and can be observed by acoustic methods of mine geophysics [12, 13].

In the present work, the method for determining contact conditions of a coal seam with host rocks is developed on the results of physical modeling. The method is based on the solution of the inverse boundary problem by seismic tomography data with the use of the empirical dependence of the longitudinal waves velocity on the mean normal stress.

2 Experimental setup

An epoxy resin filled with muscovite was used as model material. The content of mica was selected so that the acoustic and deformation characteristics of the resulting composite, as well as its density (1600 kg / m3) were close to those for coal.Five cylindrical (diameter 38 mm, height 76 mm) and eight rectangular samples were made (length X = 100 mm, height Y=140 mm, thickness 40 mm).

Cylindrical samples were subjected to uniaxial compression by a standard procedure [14], as a result of which the static Young's modulus $E = 9.2$ GPa and the Poisson's ratio $\nu = 0.21$ (average values) were determined, which corresponds to the same parameters for coals[15,16,17]. At each loading step an ultrasonic time-pulse method with UD2M-PN flaw detector was used to determine the velocity of longitudinal waves V. The processing of the obtained data by the original technique [18, 19] made it possible to obtain an experimental dependence

$$V(\sigma) = A - B\exp(-\alpha\sigma / \sigma_0), \quad (1)$$

where σ-mean normal stress, A = 3041 m/s, B = 671 m/s, $\alpha = 0.87$ и $\sigma_0 = 3$ MPa.

During main experiments, a rectangular sample was subjected to mechanical loading along the Y axis using a GT 2.0.8-2 loading machine. In this case, special gaskets $l$were placed on the contact areas of the sample with loading plates P_1 and P_2 (Figure 1) to simulate zones with low friction. On the vertical surfaces of the plate, twelve acoustic piezoelectric transducers S_1, ...S_{12} (resonant frequency 100 kHz) were fixed. These transducers can function as receivers or emitters. The signals were recorded by means of ADC E20-10 with a sampling rate of 5 MHz coupled with PC.

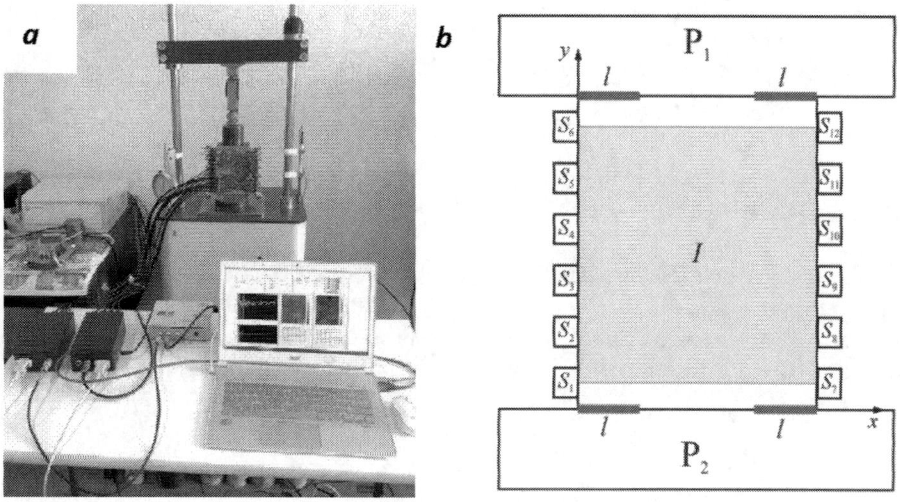

Fig. 1. General view of the laboratory setup (a) and loading scheme (b) (P_1, P_2 - steel plates, S_i - acoustic transducers, I – illuminated area, l – length of reduced friction zones).

The sample was loaded discretely with a step of 2 MPa. At each stage of loading k, sounding was carried out: the S_m transducer generated a pulse, which was received by six receivers at the opposite end. The absolute accuracy of recording the first arrival of P-wave was 0.2 μs. On the first arrival, the time t_{mn}^k of the path of the longitudinal wave between S_m and S_n was determined. Matrix $\{t_{mn}^k\}$ become an input data for tomography. Database $\{t_{mn}^k\}$ was formed as a result of two experiments:

- press plates are rigidly fastened to the sample (l = 0);
- at the edges of the contact surfaces "sample-plate" there are areas with low friction of length l =25 mm.

The values t_{mn}^k were calculated from the accumulation of ten signals.

3 Reconstruction of speed distribution of longitudinal waves

Tomography of the sample was carried out using the original iterative procedure [18, 19], the stability of which is ensured by the expansion of the solution at intermediate steps in a specially constructed basis [20]. In Figure 1, the illumination domain I is highlighted for the selected observation system $\{S_m\}$. Figure 2 shows the longitudinal velocity distributions $V(x,y)$ in region I under different conditions of contact of the loading device with the sample and the magnitude of the vertical load P.

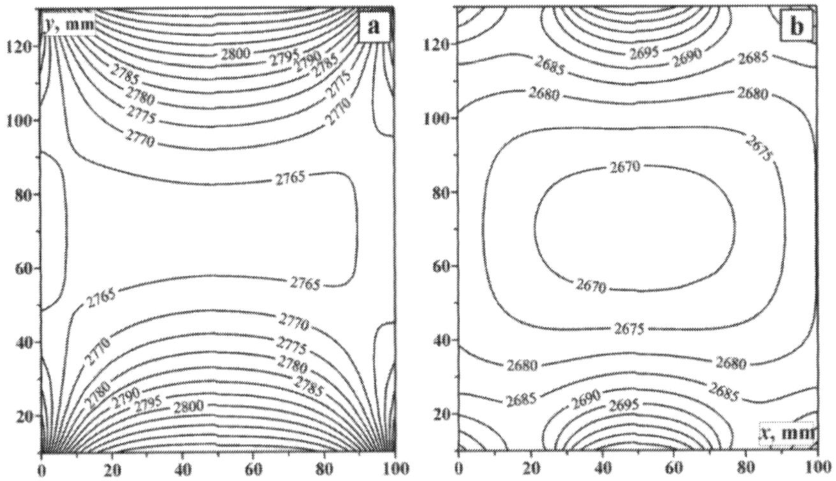

Fig. 2. Lines of the velocity level of longitudinal waves V (m/s) in a sample reconstructed as a result of tomography: P = 6 MPa, l = 0 (a); P = 6 MPa, l = 25 mm (b).

The main reason for the asymmetry of isoline patterns is in chaotic arrangement of the aggregate particles in the sample matrix. As an example, Figure 3 shows the stress curves in the horizontal section of $y = y_0$, calculated from (1) and found by the tomography of the velocity V (Figure 2) for different values of P and contact conditions (solid lines l = 25 mm, dashed lines l = 0)

$$\Sigma(x) = \sigma(x, y_0) = \frac{\sigma_0}{\alpha} \ln \frac{B}{A - V(x, y_0)} \tag{2}$$

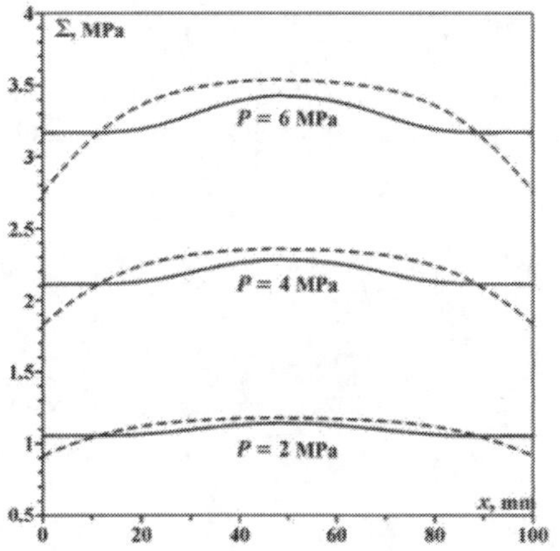

Fig. 3. Mean normal stress σ via x in the cross section y=y_0 (y_0=0.2Y, positive values correspond to compression).

In experiments, the magnitude of the load was set so that the deformation of the sample occurred in an elastic regime, therefore, σ is practically proportional to P. It is the dependences of $\Sigma(x)$ that are the input data for the inverse problem considered below.

4 Determination of shear stresses on the boundaries of the sample based on the solution of the inverse problem

We assume that on horizontal boundaries the distribution of vertical stresses is given

$$\sigma_{yy}(x,0) = \sigma_{yy}(x,Y) = P \tag{3}$$

and according to the data of the tomography $\sigma(x, y_0) = \Sigma(x)$ is known. It is necessary to find shear stresses σ_{xy} for $y = 0,Y$. For this we use the method developed in [21].

Figure 4 shows the results of the solution of the inverse problem: the distribution of tangential stresses $\sigma_{xy}(x,Y) = F(x)$ at the upper boundary of the sample at different load values and contact conditions. As expected, the function $F(x - 0.5X)$ is odd, in the neighbourhood of the points $x = l$ and $x = X - l$, the concentration of stresses occurs due to a sudden change in the contact conditions of the plates with the sample.

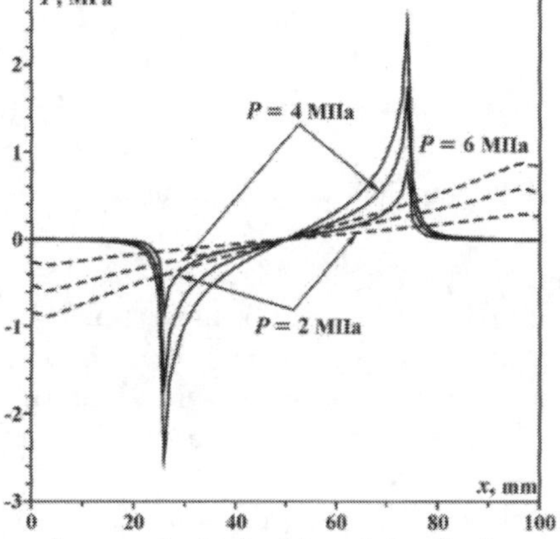

Fig. 4. The stresses σ_{xy} at the boundary $y = 0$, obtained by solving the inverse problem. Solid lines correspond to $l = 25$ mm, dashed lines – to $l = 0$.

5 Stress state of the sample at the various conditions of contact

Let us analyse the stress field in a sample under different contact conditions on horizontal boundaries. For this, within the frames of a linear-elastic model, which includes:
equilibrium equations

$$\sigma_{ij,j} = 0; \tag{4}$$

Hooke's law

$$\sigma_{ij} = \frac{E}{1+\nu}\left(\frac{\nu}{1-2\nu}\varepsilon\delta_{ij} + \varepsilon_{ij}\right) \tag{5}$$

and the Cauchy relation

$$\varepsilon_{ij} = 0.5(u_{i,j} + u_{j,i}). \tag{6}$$

We solve the following boundary-value problem in the domain $D=\{0 \le x \le X,\ 0 \le y \le Y\}$ (Figure 5)

$$\begin{gathered}\sigma_{xx}(0,y) = \sigma_{xy}(0,y) = \sigma_{xx}(X,y) = \sigma_{xy}(X,y) = 0,\\ \sigma_{xx}(x,0) = P \quad \sigma_{xy}(x,0) = F(x),\\ \sigma_{xx}(x,Y) = P \quad \sigma_{xy}(x,Y) = -F(x),\end{gathered} \tag{7}$$

where ε_{ij} - components of the strain tensor (i,j = x,y), $\varepsilon = \varepsilon_{xx} + \varepsilon_{yy}$, u_i - displacements, summation is performed over repeated indices, and the function $F(x)$ is the result of solving the inverse problem. The system (4)-(7) was solved by the finite element method using the original code [22].

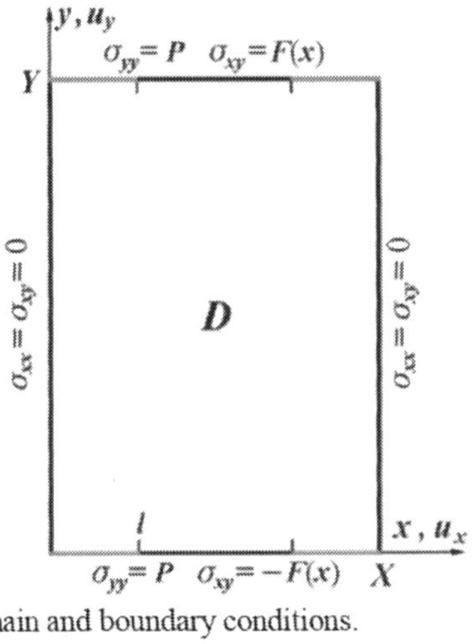

Fig. 5. Scheme of calculation domain and boundary conditions.

Figure 6 shows isolines of horizontal stresses σ_{xx} (related to P) for l = 0 and l = 25 mm, respectively. It turned out that if the surface of the sample and the loading plate (in actual conditions - the coal bed and host rocks) have good adhesion, then in the central part D there is an extensive zone of horizontal tensile stress. Stress in that zone is insignificant (about 0.02P) and does not exceed strength limitsfor coal even at P = 25-30 MPa (which corresponds to lithostatic pressure at a depth of 1 km).

If the edges of the horizontal boundaries are areas of slippage, then in their vicinity there arise areas of horizontal tension stress reaching 0.8P. These areas will be the most likely foci of destruction and sudden bursts.

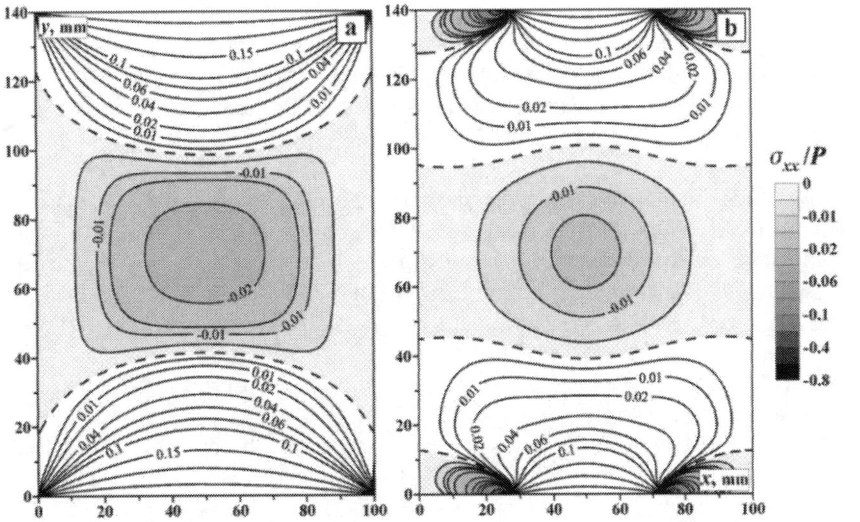

Fig. 6. Isolines of horizontal stress σ_{xx} for different sizes of slip zones: $l = 0$ (a); $l = 25$ mm (b).

Conclusion

A method has been developed that makes it possible to determine the conditions at the contact of a coal seam with host rocks. It is based on the solution in the geomechanical model of the inverse boundary problem by seismic tomography data using empirical dependences of the speed of elastic waves on the stresses obtained under laboratory conditions. The method can be used to reconstruct the stress-strain state of rock mass, changing during the excavation of reserves, as well as to obtain information about the geological conditions of coal seams occurrence, necessary for the prediction of the outburst hazard.

The study was supported by the Russian Science Foundation, project no. 16-17-00029.

References

1. J. Seidle, Fundations of coalbed methane reservoir engineering. PennWell Books. 416 p. (2011).
2. V. Bondarenko, I. Kovalevska, K. Ganushevych, Progressive Technologies of Coal, Coalbed Methane, and Ores Mining. CRC Press. 534 p.
3. V.N. Zakharov, Sejsmoakusticheskoe prognozirovanie i kontrol' sostoyaniya i svojstv gornykh porod pri razrabotke ugol'nykh mestorozhdenij. M.: IGD im. A.A. Skochinskogo, 172 s. (2002).
4. L. Zhenbi, Zh. Baiting, Microseism monitoring system for coal and gas outburst // Int. J. Computer Sci. Issues. V. **9(5)**. N 1. P. 24 – 28 (2012).
5. Al Heib, M. Numerical and Geophysical Tools Applied for the Prediction of Mine Induced Seismicity in French Coalmines // Int. J. of Geosciences. V. **3(4).** P. 834 – 846. (2012).
6. D. Palmer, High resolution seismic reflection surveys for coal // Geoexploration.. V. **24**. 397-408 (1987).
7. H.Jr. Henson, J.L. Sexton, Premine study of shallow coal seams using high resolution seismic reflection methods // Geophysics. V. 56(9). P. 1494 – 1501 (1991).

8. J. Cocker, M. Urosevic, B.A. Evans, High Resolution Seismic Survey to Assist in Mine Planning / Proceedings of Exploration 97: Fourth Decennial International Conference on Mineral Exploration. Edited by A.G. Gubins. P. 473 – 476. (1997).
9. M. Khukhuudei, U. Khukhuudei, High Resolution Seismic Reflection Survey for Coal Mine: fault detection / American Geophysical Union, Fall Meeting. Abstract NS43A-3856 (2014).
10. K.V. Mironov, Spravochnik geologa-ugol'shchika. M.: Nedra. 363 s. (1988).
11. A. L. Karchevsky, L. A. Nazarova, V. N. Zakharov, L. A. Nazarov, Stress state estimation in coal bed under random conditions in contact zone with enclosing rocks based on inverse problem solution. Gornyi Zhurnal. **N 11**. P. 37-40. (2017).
12. M.E. Everett, Near-surface applied geophysics. Cambridge University Press. 403 p. (2013).
13. W.M. Telford, L.P. Geldart, R.E. Sheriff, Applied Geophysics. 2nd Edition. Cambridge University Press. 744 p.(1990).
14. GOST 28985-91. Metod opredeleniya deformatsionnykh kharakteristik pri odnoosnom szhatii.
15. V. L. Shkuratnik, P. V. Nikolenko, A. E. Koshelev. Stress dependence of elastic P-wave velocity and amplitude in coal specimens under varied loading conditions. Journal of Mining Science. **V.52(5)**. P. 873-877. (2016).
16. Fiziko-tekhnicheskie svojstva uglej i gornykh porod Kuznetskogo bassejna: Spravochnik / G.G. SHtumpf, Yu.A. Ryzhkov, V.A. SHalamanov i dr. M.: Nedra,. 447 s. (1994).
17. Fizicheskie svojstva gornykh porod i poleznykh iskopaemykh (petrofizika): Spravochnik geofizika. Red. N.B.Dortman. M.: Nedra, 526 s. (1976).
18. L.A. Nazarova, V.N. Zakharov, V.L. Shkuratnik et al. Use of tomography in stress-strain analysis of coal-rock mass by solving boundary inverse problems, Procedia Engineering, P. 1048 – 1055. (2017).
19. L. A. Nazarova, L. A. Nazarov, M. I. Protasov. Reconstruction of 3D stress field in coal–rock mass by solving inverse problem using tomography data. Journal of Mining Science. **V.52(4).** P. 623-631. (2016).
20. M.J. Woodward, D. Nichols, O. Zdraveva et al. A decade of tomography // Geophysics, V.73(5). P. VE5-VE11. (2008).
21. A. L. Karchevsky. Determination of the possibility of rock burst in a coal seam. Journal of Applied and Industrial Mathematics. **V. 11(4)**. P. 527-534. (2017).
22. L. A. Nazarova, L. A. Nazarov. Dilatancy and the formation and evolution of disintegration zones in the vicinity of heterogeneities in a rock mass. Journal of Mining Science. **V.45(5).** P. 411-419. (2009).

The role of modern geodynamic movements in the formation of geomechanical problems in subsoil use

Anatoly Sashourin[*1], *Andrey* Panzhin[1], *July* Konovalova[1] and *Vladimir* Ruchkin[1]

[1]Institute of mining of UB RAS, Ekaterinburg, Russia

Abstract. The paper describes the role of geomechanical processes and phenomena in the development and manifestation of hazardous situations in the development of mineral deposits, threatening the working personnel and the surrounding population. The studies of the Ural Scientific School of geomechanics functioning in the Institute of Mining of Ural Branch of RAS have established that the basis for the formation of hazardous geomechanical processes and phenomena are the fundamental properties of the rock mass - hierarchical blocking and constant mobility caused by the recent geodynamical movements. The constant mobility of hierarchical block medium of rock mass under the influence of trend and short-term recent geodynamical movements gives rise to the phenomenon of the secondary structuring. The recent geodynamical movements concentrate at the borders of the secondary structures and result in the deformations exceeding by 2 - 5 times the intrablock and integral deformations and representing the main source of the hazardous impact on the mining objects and on the surrounding population. The technology of the geological environment diagnostics for the forecast of hazardous mining situations is aimed at the identification of the boundary zones of secondary structures, the estimation of the hazard level of deformation processes and the adoption of proactive measures.

1 Introduction

The modern society satisfies its needs in raw materials at the expense of open-cut and underground mining. And throughout the history of mining some of the constraints determining the mining efficiency and, at the same time, some of the threats to the safe mining are the intrinsic geomechanical processes and phenomena in the rock mass. First of all they include: rock pressure, rock movement, emissions of coal and gas, rock bumps and technogenic earthquakes appeared in the wake of rising the degree of technogenic impact on the rock mass during mining of solid minerals and especially of hydrocarbons. The founder of the Russian geomechanical science, I. M. Bakhurin (1946), said in the middle of the last century that "we'd be about right if we said that the entire history of mining, the entire history of finding the best research systems is a history of struggle with the rock

* Corresponding author: sashour@igd.uran.ru

movements", understanding by rock movements at that period all the rock mass deformation processes accompanying the development of mineral deposits to whose research the geomechanical science is devoted [1].

The studies of geomechanical processes and phenomena made over the past nearly a century period have fully confirmed the correctness of those views. In fact, all the geomechanical processes and phenomena carrying a threat to the mining safety and impeding the mining efficiency as well as violating the safety and comfort of the surrounding population are connected, directly and sometimes indirectly, with deformation processes caused by interference in the natural environment. The comprehensive role of deformation processes in destabilization of the rocks in mines of different purposes and in generating the emergency situations has predetermined that the geomechanics studying the regularities of deformation processes in the rock mass under the conditions of its natural occurrence and in the fields of mining influence has the status of the key in ensuring the safe and efficient mining. Its practical use requires the knowledge of the basic factors involved in generating the emergency situations and the regularities of their interaction.

Until recently, in the mining science and practice of mining works the rock mass was considered and, in many cases, is being considered as a stable, equilibrium environment being in a homogeneous stress state generated by its own rock weight and tectonic forces arising from the geodynamical processes undermined in the past geological periods. The destabilization occurred only from technogenic mining activity. Such a simplified approach to the rock mass of the upper part of the lithosphere as an integral part of the Earth does not reflect its endogenous and exogenous processes and phenomena and leads to serious catastrophes with tragic consequences. Among them, the most large-scale catastrophes should include gas emissions and inflammations with mass mortality of people in coal mines, rock bumps and technogenic earthquakes, mine floods whose fresh examples have occurred in the past decades: Mine Sokolovskaya (SSGPO, 2005), mine Magnetitovaya (VGOK, 2009), mine Mir (ALROSA 2017). Along with such large-scale catastrophes, the development of mineral deposits is constantly accompanied with other smaller-scale emergency events. The negative consequences caused by discrepancy of projected processes and phenomena taken in the mining designs to feasible ones determine the relevance of the geomechanical studies over the years, growing as the mining production expands and the mining descends down to deeper depths.

The studies of the Ural school of geomechanics, established in the Institute of mining of Ural Branch of RAS have allowed on the basis of the confirmed data on the fundamental properties of the rock mass, such as the hierarchical blocking and constant mobility, to identify a number of basic regularities of processes and phenomena proceeding in the rock masses in their natural occurrence and in the reach of the technogenic mining activity. The studies have been conducted in mining enterprises with open-cut and underground developments of solid mineral deposits in the Urals and Kazakhstan as well as in the mineable hydrocarbon fields in Western Siberia and during the construction of the subway in Ekaterinburg [2].

2 Methods of study

The magnitude of the processes and phenomena proceeding in the mining and entering the circle of interests of the geomechanics leads to an integrated approach to the implementation of the research. Acquisition of baseline information is carried out by experimental methods under field conditions. Their following interpretation is performed with the use of the theoretical methods and the computer modeling using the apparatus of continuum mechanics.

The research has established that the behavior and the state of rock mass, as in its natural state and in the area of mining activities, are determined by its stress-deformed state formed with two fundamental properties of the rock mass - its hierarchical block structure and recent geodynamical movements, supplying it with continuous mobility [3-9]. In accordance with this, the methods of research are focused primarily on the study of the structural features of rock mass, which is produced by geophysical methods of remote sensing [10, 11]. The hierarchical block environment plays a determining role in the behavior of rock mass and the stability of mining objects, creating serious problems in the predictive assessment of the reliability and safety of mining structures of various sizes. So, for example, the preparatory mines require to estimate the behavior of the volumes of rock mass with the dimensions reckoned by the tens meters, and the open-pit side requires to estimate the behavior of rock mass with the dimensions of the hundreds of meters and sometimes kilometers. This task is solved with the electrometry methods in various versions, including the electrotomography, the method of georadar sensing and spectral profile shooting.

The stress-deformed state of rock mass in natural conditions is generated under its own rock weight and tectonic activity which is currently characterized by the recent geodynamical movements divided conditionally into trend and cyclical short-term movements. In the area of mining influence the natural factors are added by technogenic factor which disturbs the natural dynamic balance creating a secondary stress-deformed field around the mines depending on their parameters. The recent geodynamical movements are studied with the geodetic methods by monitoring the changing of coordinates of geodetic signs of observing stations on the earth's surface by the method of satellite positioning systems (GNSS). The measurements performed by the complex GNSS have showed its high efficiency to meet the challenges of the geomechanics, and its application enables not only the discrete measurements, but also the regular monitoring of deformations and stresses occurring in the earth's crust [12, 13]. The spatial coordinate binding of geodynamical polygons is performed by the double-system (GPS-GLONASS) and double-frequency receivers of geodesic class whose potential and actual accuracy corresponds to 3-5 mm+1 ppm in plan and 5-7 mm+1 ppm in height.

During the first stage the current spatial coordinates of geodynamical polygons are determined, while the centering and orientation of geodesic constructions are carried out using the permanent GNSS stations (CORS) of the Ural region and the global network IGS with coordinate referencing to the International Terrestrial Reference Frame (ITRF) [14].

The trend movements of geodynamical polygons and observing station benchmarks are determined on the basis of the analysis of the spatial coordinate increments (vectors) ΔX, ΔY, ΔZ in the intervals between repeated measurement cycles, and the cyclic short-term movements are determined by the continuous measurements [15]. The interval deformations received as a result of instrumental measurements using the mathematical apparatus of continuum mechanics are converted into a vector and tensor representation of deformation field with the allocation of the main components of the deformation tensor [16].

During the second stage the program of instrumental observations includes the determination of high-altitude observing station benchmarks and soundings of courses between the benchmarks as well as the spatial coordinating using the complex of satellite geodesy and the special tasks for diagnostics of geodynamical activity of the investigated territory. As a result of instrumental observations in the vertical plane the benchmark subsidences and the dip deformations as well as the dip moveout velocities are determined; in the horizontal plane the movement vector parameters and the horizontal tension-compression deformations as well as their velocities are determined.

The results of instrumental measurement of movements and deformations are visualized by building a vector field of horizontal displacements (Figure 1) and tensors of horizontal deformations of regular geomechanical network model built by the Finite Element Method (FEM). The finite element network boundary and form change reflecting the investigating area deforming nature is also analyzed. The key elements of FEM model are points of the geodynamical polygon applications and the observing station benchmarks for which the spatial movements between performed repeated cycles of measurements have been determined.

3 Research results

The results for the monitoring of recent trend geodynamical movements in the open-pit mines of chrysotile asbestos for the period from 2006 to 2017 in the form of offset vectors are presented in Figure 1 «a». The total values of offsets caused by the recent geodynamical movements and by the changing of the open-pit mine parameters, as it appears from Figure, are different and vary in the different areas of the open-pit mines from 10 - 15 to 60 -70 mm. The directivity of offset vectors is also inhomogeneous. In many respects it is technogenic factor whereby the directivity inside the open-pit mine prevails, however the observed variation in their directions and values indicates a significant influence of the initial geodynamical movements.

The inhomogeneity and mosaic structure of stress-deformed state contrasts especially in the field of horizontal deformations submitted in Figure 1 «b» in the form of deformation tensors obtained by the Finite Element Method (FEM) using the offset vectors as boundary conditions.

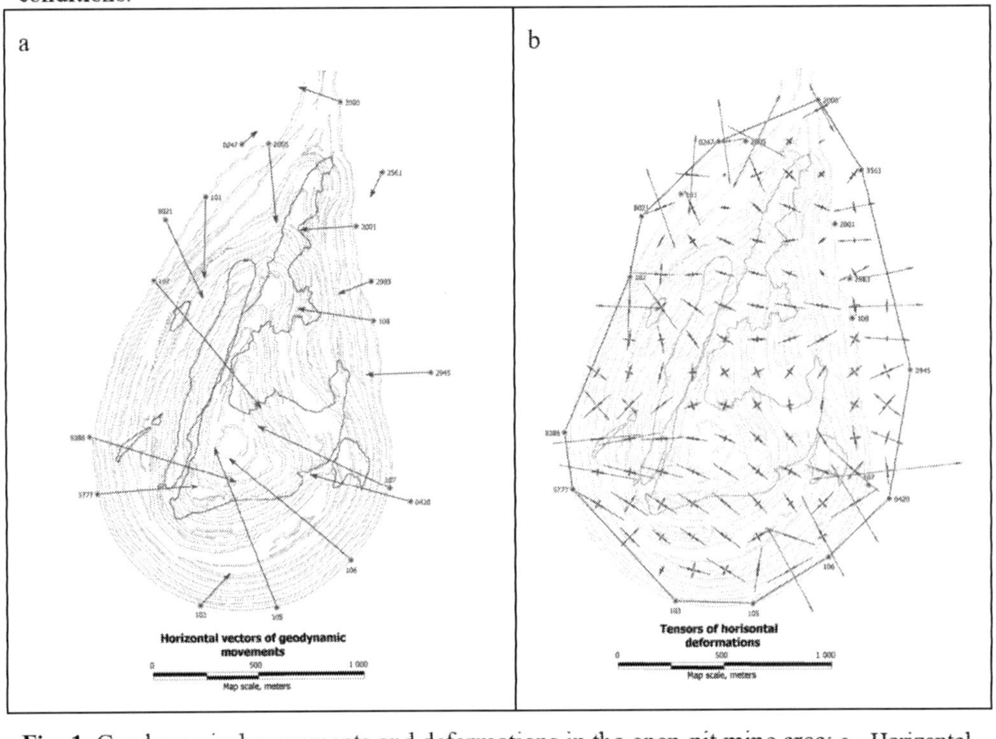

Fig. 1. Geodynamical movements and deformations in the open-pit mine area: a - Horizontal vectors of recent geodynamical movements in the open-pit mine area for the period from 2006 to 2017; b - Tensors of horizontal deformations caused by recent geodynamical movements in the open-pit mine area at same time

By the pit outline prevail the tensors with various signs of deformation – tension and compression differing in values. In the central part the tensors are more homogeneous and presented by the minor compressions in both main directions.

4 Discussion of results

The interpretation and analysis of the research results for the stress-deformed state at the various enterprises which develop the mineral deposits in open-cut and underground ways under various geomechanical conditions have revealed the basic regularities of the development of deformation processes in rock mass which play a key role in the safety and efficiency of mining. The research has established the following:

- - the hierarchical block environment of rock mass, being in constant mobility under the influence of trend and short-term cyclic recent geodynamical movements, is subjected to the secondary structuring;

- the consequence of secondary structuring of hierarchical block mass is a discrete mosaic structure of the stress-deformed state, and the recent geodynamical movements concentrate on the borders of secondary structures, causing the deformations which exceed intrablock and integral movements by 2-5 times (Figure 2);.

- the mosaic structure of the stress-deformed state is amplified by interblock counter movements on the borders of secondary structures, creating the zones of concentration of compressive stresses and the zones of depression where rock mass can be completely freed of compressive stresses (Figure 3);

- abnormal deformations in the mosaic stress-deformed state caused by the secondary structuring and constant mobility are the main source of the hazardous impact on technogenic mining objects in the depression zones [17] (Figure 4) as well as in the areas of concentration of compressive deformations.

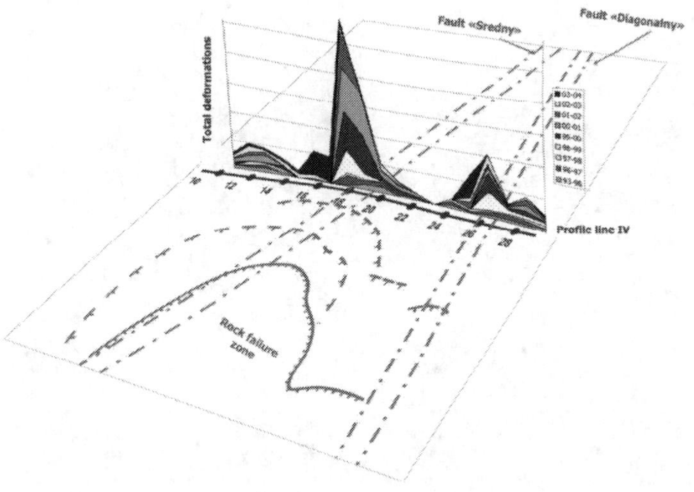

Fig. 2. Abnormal values of deformations on the borders of secondary structures

Thus, the performed research has identified the mechanism and regularities of the development of catastrophic deformation processes in rock mass, as reflected in the shear fractures caused by the mosaic structure of the stress-deformed state with a wide range of values of the components of the stress and deformation tensors in the hierarchical block environment under the influence of recent geodynamical movements, which establishes the key status to the geomechanics in ensuring the safety and efficiency of mining. The technology of the geological environment diagnostics for the forecast of hazardous mining conditions allows for the identification of boundary zones of secondary structures, the estimation of the hazard level of deformation processes and the adoption of proactive measures.

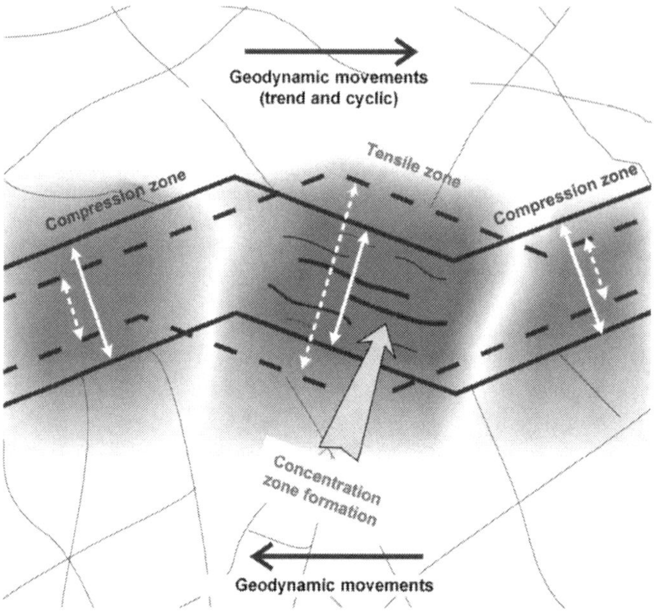

Fig. 3. Diagram of development of mosaic structure of the stress-deformed state under the influence of interblock movements of secondary structures

Fig. 4. Formation of the caving zone over the developed deposit in depression zone

5 Conclusion

The formation of the stress-deformed state of the upper part of the lithosphere during mining is determined by the fundamental properties of rock mass – hierarchical block structure and constant mobility under the influence of recent geodynamical movements, and their combination results in open non-linear system self-organizing in space and time under the influence of the phenomenon of secondary structure.

The phenomenon of secondary structure forms a mosaic structure of the stress-deformed state, determining the uneven impact of deformation processes in the mining objects.

The main source of emergency and catastrophic events at the mining objects are the interblock movements between secondary structures whose parameters exceed the intrablock and integral movements by 2-5 times.

The key to ensuring the safety and efficiency of mining is diagnostics and predictive assessment of the impact of the geomechanical processes and phenomena on mining objects on the basis of identification of boundary zones of secondary structures and determination of the parameters of the movements and deformations in these zones, taking into account the ratio between the object sizes and the parameters of secondary structures.

Thanks

In the preparation of the paper the authors have used the results of the multi-year fundamental and applied research performed by a team of geomechanics of the Institute of mining of Ural Branch of RAS at the mining enterprises of the Ural Region and Kazakhstan, wherefore the author team expresses its deep appreciation to their colleagues as well as to engineers and technical workers of mining enterprises who took the liveliest participation in solving the geomechanical production problems

References

1. 1. I. M. Bakhurin, Rock Movement Under Influence of Mining (Gostopizdat, Moscow, 1946) 229 pp. [in Russian].
2. S. V. Usanov and Yu. P. Konovalov, Problemy nedropolzovaniya, **4 (11)**, 104 (2016). [Mining Problems, No. 4 (11), 104 (2016)]. DOI: 10.18454/2313-1586.2016.04.104
3. S. V. Usanov, Yu. P. Konovalov and O. D. Zheltysheva, Gorny zhurnal, 1, 36 (2012).
4. M. A. Sadovsky, "Natural Rock Lumpiness", Dokl. Akad. Nauk SSSR, **4**, Vol. 247, 829 (1979).
5. M. A. Sadovsky, "Hierarchy from Dust to Planets". Earth and Universe. **6.** 4 (1984).
6. Yu. O. Kuzmin, Modern Geodynamics and Estimation of Geodynamical Risk at Mining, (Agenstvo ecologicheskikh novostei, Moscow, 1999) 220 pp. [in Russian].
7. A. D. Sashourin. Mining Magazine, **1**, 29 (2012); Eurasian Mining. **1**, 17 (2012).
8. A. A. Panzhin. Mountain Information and Analytical Bulletin. **9**, 313 (2010).
9. Yu. P. Konovalova. Mountain Information and Analytical Bulletin, **7**, 269 (2010).
10. V. V. Melnik and A. L. Zamyatin. Mining Informational and Analytical Bulletin. **4**, 226 (2005). Dokl. Akad. Nauk SSSR, No. 4, Vol. 247, 829 (2005) [Sov. Phys. Dokl., No. 4, Vol. 247, 829 (1979)].
11. V. V. Melnik. Mine Surveying and Mining, **6**, 56. (2014).
12. Gao Jing-Xiang, Hu Hong. Procedia Earth and Planetary Science. **1(1).** P. 1081-1088. (2009).
13. A. A. Panzhin, Mining Magazine, **1**, 39 (2012).
14. A. A. Panzhin. Herald of the Magnitogorsk State Technical University named after G. I. Nosov, **1 (49)**, 22 (2015).
15. Fengshan Ma, Haijun Zhao, Yamin Zhang, Jie Guo, Aihua Wei et al. Journal of Rock Mechanics and Geotechnical Engineering. **4 (1)**. P. 82–87. (2012).
16. B. T. Mazurov, A. A. Panzhin, A. A. Silaeva. Geodesy and cartography. **3**. 35 (2016).
17. V. I. Ruchkin. Mountain information and analytical bulletin. **5**, 45 (2013).

The assessment of the impact of natural and anthropogenic factors on the current stress-strain state of rock massif of burst-hazardous ore deposits in the East of Russia

Igor Rasskazov[1*], *Boris* Saksin[1], *Vitaly* Usikov[1] and *Alexander* Sidlyar[1]

[1]Mining Institute of Far eastern branch of Russian Academy of Sciences, Khabarovsk, Russia.

Abstract This paper does a case study of endogenous deposits located in the zone of influence of convergent (converging, colliding) boundaries of tectonic plates in the zone of the continent-ocean transition, where there is increased present seismic and geodynamic activity, covers the issues of the ratio of natural and anthropogenic components of the stress-strain state (SSS) in rock massifs. It has been shown that, despite the confinement of the studied deposits to a single rock structure, as well as the proximity of their common neotectonic position, the SSS of the rock massifs containing them varies significantly. Based on a comprehensive analysis of geological and geophysical data with the use of space data about the Earth (GRACE mission data for calculation of stress fields and radar images of the Earth's surface), the kinematics of the process of modern deformation of the upper part of the Earth's crust within the studied area is specified. It is established that the peculiarity of the region is the pronounced contribution of the natural component to the general present SSS of rock massifs of impact-threatening deposits. On a real example, the possibility of a predictive assessment of SSS has been shown, which is important at the stages of designing and conducting underground mining operations in severe mining-geological and burst -hazardous conditions. **Keywords:** Rock bursts, mine seismicity, geodynamics, remote sensing of the Earth, stress-strain state, tectonic structure, forecast

1 Introduction

During underground development of subsoil in severe mining and geological conditions and at great depths, one of the most difficult and topical problems is preventing rock bursts and other dangerous geodynamic phenomena. The difficulty of solving this problem is due to the complexity of the geological structure of the deposits, the high heterogeneity of the properties and the state of the rock masses composing them, which is further strengthened as a result of mining rocks [1, 2]. At a certain stage of mining, the risk of manifestation of such formidable phenomena as mining-tectonic impacts and man-made earthquakes with catastrophic consequences increases. A sign of increasing geodynamic risk is mine

* Corresponding author: rasskazov@igd.khv.ru

seismicity, which reflects the process of redistribution of stresses in a rock massif and its structural reorganization under the influence of intensive mining [3-6].

In the conditions of dangerous and burst-prone deposits of Far Eastern region, it is established [1] that the stress-strain state (SSS) of rock massifs is significantly influenced by the heterogeneity of the regional natural background, on which the stress field formed in the natural-technical system is superimposed. Accounting for the SSS of the rock massif, especially at the stage of mining design, ensures effective and safe mining of deposits.

All this requires further improvement of methods and tools for studying the geological environment under conditions of intense man-made impact, obtaining and rethinking new experimental data obtained in regions with different geodynamic regimes.

2 Analysis of the geodynamic position of the burst-hazardous deposits of the East of Russia

In the East of Russia, a significant number of ore deposits (tin, tungsten, polymetals, noble metals, etc.) are concentrated within the relatively young Pacific orogenic belt, which position is controlled by the zone of interaction of the Pacific and Asian tectonic plates and the global East Asian system of shifts, having north-eastern orientation [7]. According to the regionalization scheme for the territory of the USSR in terms of the absolute value of the maximum tangential stresses in the Earth's crust (Gzovskiy, 1975), the area under investigation is characterized by a sufficiently high level of stress equal to 70 ± 35 MPa, which is inherent in active modern orogenic regions. Long-term geodynamic studies on the ore deposits of the region have revealed the specificity of the manifestation of rock pressure in the described territory, which is proposed to be attributed to a particular type of Edge-Asian type of model regions [1].

Territories of this type are located in the zone of influence of convergent (converging, colliding) boundaries of tectonic plates, where as a result of long-term and still continuing interaction of plates, a shift-magmatic regime of formation of the structure of ore fields is created and increased modern seismic and geodynamic activity is observed. As a result, in the upper horizons of the Earth's crust, as a rule, anomalous deformation is established here, due to the fact that excess stresses acquire a non-gravitational nature, and the main interactions and structural-material transformations of the geological environment are carried out due to mechanical energy. For such conditions, the effect of a very uneven stress state of sections of rock massifs and an increased activity of neotectonic processes are characteristic. This cannot but leave an imprint on the local geodynamic system, and, consequently, on the character of the dynamic manifestations of rock pressure. This is confirmed by the results of full-scale studies in the conditions of burst-hazardous deposits in the East of Russia, in the massifs of which horizontal compressive stresses (due to the action of tectonic forces) are exceeded over vertical (gravitational), which is largely determined by their geodynamic position within the tectonically active Amur plate (Table 1).

The authors will examine in more detail the Nikolaevskoye, Yuzhnoye and Vostok-2 deposits located within the Sikhote-Alinsky segment of the Pacific orogenic belt in the area of its articulation with the Central Asian orogenic belt. They are confined to the zone of active interaction between the Amur and Okhotsk tectonic plates of the next hierarchical level. These are postmagmatic deposits attributed to vein, stockwork and skarn types. They consist of a composition of predominantly steeply dipping ore bodies. Despite the proximity of their common neotectonic position, the nature and parameters of the stress field of these deposits are significantly different, and the burst hazard manifests itself at different depths.

Table 1. Parameters of stress fields in rock massifs of some bump-hazardous deposits of Far Eastern region

Deposit	Digging depth, m	Maximal stress, MPa	Ratios of primary stresses (σ_1:σ_2:σ_3)
Nikolaevskoye	860	70-150	2.5:1.5:1
Yuzhnoye	500	95	1.8:1:1
Vostok-2	500	50-70	2.5:1.3:1
Perevalnoye	600	50	2:1:1
Hinganskoye	550	50-70	3:1:1
Antei	700	50-100	2.5:1.4:1
Darasunskoye	700	27	1.7:1.3:1
Irokindinskoye	200	11	1.3:1.2:1

3 Methods of research and principles of complex interpretation of data

In [8, 9], the authors outlined the rationale for the principles of complex interpretation of geophysical fields by various methods. The basis of the applied approach is the assertion that the natural SSS of various depth levels of the Earth's crust is reflected in regional physical fields, and its upper part - in the morphological structures of the daytime relief and other neotectonic elements. The method is approved on materials on the Amur tectonic plate.

To clarify the neotectonic structure of the upper part of the crustal section in the region under investigation, the authors used various methods of interpreting satellite geodetic data [10, 11], results of an instrumental assessment of the stress-strain state of rock massifs of bump-hazardous ore deposits, generalizing geological, geophysical and neotectonic materials on the region. Processing and subsequent visualization of satellite images were carried out using the original programs and software environment ArcGis 9.3. The construction and direct visual study of computer volumetric 3D models of relief was carried out using GTOPO_30 altitude matrices, using Google Earth online service and original programs. The construction of a more detailed neotectonic scheme of the Dalnegorsk-Kavalerovo ore region and its surroundings was based on the morphostructural analysis of the relief and the concept of horizontal motion of tectonic flows [12]. Initial materials were digital relief models (DRM). They are interpreted using the altitude matrices srtm30 and srtm03 with visualization in the form of maps with tinted washing in color, black and white and in the form of anaglyphic images. Refinement of modern deep features of the territory was carried out according to satellite gravimetry (GRACE project).

Full-scale geomechanical studies included: instrumental measurements of parameters of properties and states of rock massifs, their seismoacoustic control and visual observations of the manifestations of rock pressure.

4 Results

Data on the stress state of deep horizons of the Earth's crust were obtained from the interpretation of the materials of the GRACE mission. It is established that the most burst-hazardous deposits of the region (Yuzhnoye and Nikolaevskoe) are located between regional anomalies of compression and stretching. Vostok-2 deposit is located in the contour of the regional positive anomaly. The fundamentally different geodynamic position of the studied deposits with respect to regions with unequal intensity of the present compression of the upper part of the Earth's crust (local linear anomalies of intense compression, associated with activated fault structures) was described by the authors in [9, 13].

On the example of Dalnegorsky ore region, a large-scale neotectonic scheme was constructed (Fig. 1). It is based on the interpretation of the morphostructural analysis of space materials (the srtm03 altitude matrix), performed within the .framework of the concept of horizontal motion of tectonic flows [12]. When studying the modern structure of the upper part of the Earth's crust of Sikhote-Alin and surrounding territories, an important contribution of fold-thrust dislocations is noted [7, 14, 15, etc.]. This is also evidenced by the high values of the main horizontal stresses at all the investigated deposits (see Table 1), as well as the data of GPS measurements [16].

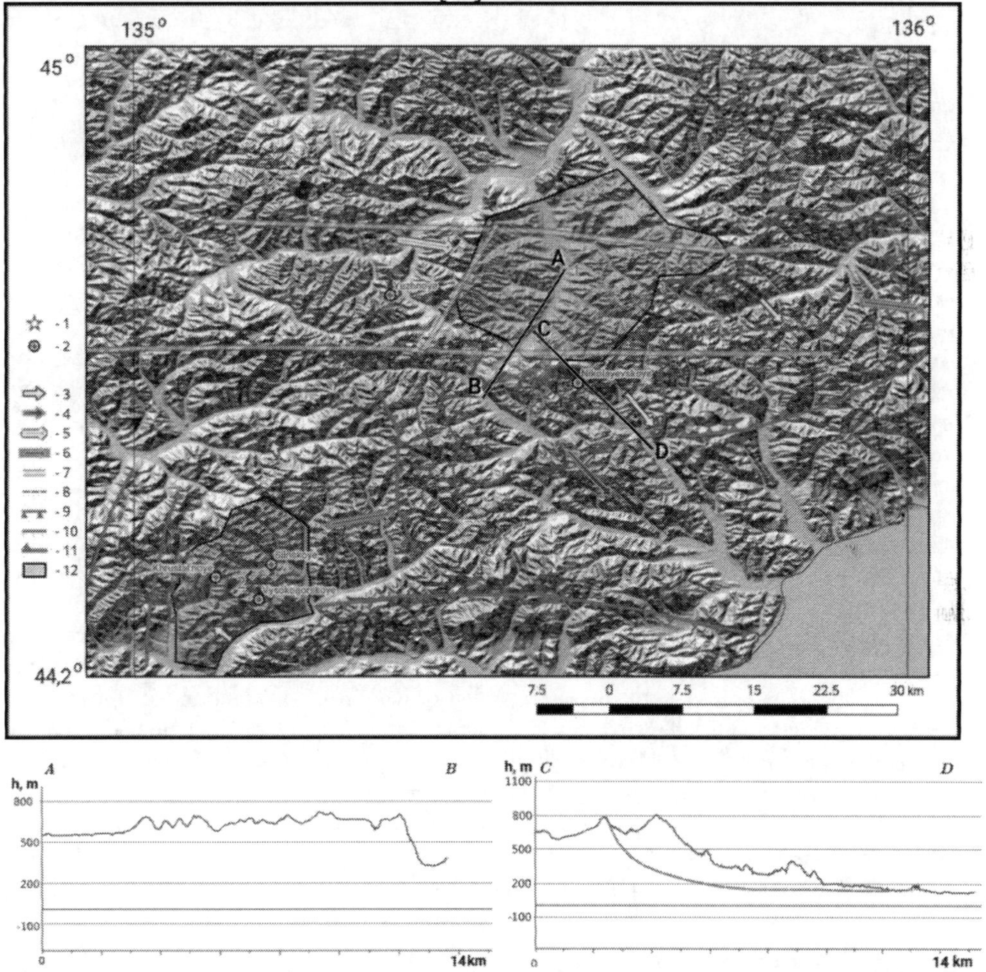

1 - earthquake in the area of Kamenka community in 2016; 2 - deposits; 3 - displacements of tectonic scales relative to each other; 4 - directions of the movement of tectonic flow of the early stages; 5 - vector of modern stresses; 6 - modern directions of the movement of tectonic flow; 7 - linear stress zones; 8 - tectonic disturbances (without separation); 9 - contours of the rear borders of the covers; 10 - lines of detachment of covers from the underlying layers; 11 - thrusts; 12 - tectonic blocks (1 - Verkhnenikolaevsky, 2 - Silinsky); AB and CD - the position of the relief surface profiles. The red line shows the projection of tectonic disturbance No. 1 on the vertical plane.

Fig. 1. The main structures of Kavalerovo-Dalnegorsk ore cluster, identified by DEM based on the SRTM03 altitude matrix

The morphostructural scheme of Kavalerovo-Dalnegorsk ore region (see Figure 1) reflects features of the tectonic cover of this territory, which manifest themselves as areas of congestion and tearing of tectonic scales and blocks formed as a result of uneven horizontal displacements. With its help, it was possible to establish a clear connection (according to the latest arc fault) between Nikolaevskoe deposit and the epicenter of the earthquake located 30 km south-eastwards with a magnitude of 4 points on March 25, 2016, which played the role of a trigger and coincided with a strong destructive geodynamic phenomenon on deep horizons of Nikolaevsky mine [13].

It follows from the scheme that the most burst-hazardous Nikolaevskoe and Yuzhnoye deposits are located in the peripheral zone of the Verkhnenikolaevsky hard block of the upper part of the Earth's crust. The block is clearly distinguished by the features of the morphology of the relief and is composed of early Cretaceous terrigenous-sedimentary deposits, and is also expressed by a positive anomaly of the gravitational field. However, on this the generality of the neotectonic position of the named objects ends. Thus, Nikolaevskoe deposit is located in the zone of transition from a rigid block to an active tectonic structure of the type of reversed overthrust. In this area, a tear-off zone (structure 1 in Figure 1 and CD section), which affects the instability of the entire system, was outlined in the rigid block. If this field is confined to the rear zone of tearing of smaller tectonic blocks (scales), then the Yuzhnoye deposit is to the areas where they are clustering. The dragging effect is expressed particularly clearly at the eastern edge of a large limestone plate (olistolith), held by Nikolaevskoe deposit (Fig. 2).

Other deposits of the Kavalerovo ore region (see Figure 1) are localized within the Silin neotectonic block, which is characterized by the predominant tearing of tectonic scales and lower average relief heights in comparison with neighboring areas. The level of tectonic stress within this block is predicted to be lower than in the neighboring block, which is confirmed by the absence of signs of burst hazard during underground mining.

Analysis of the geomechanical situation at Nikolaevsky deposit showed that in recent years the movement of the mining front towards the eastern edge of the olistolite has led to an increase in the number and intensity of the dynamic manifestations of rock pressure. Their most part is confined to the geodynamically active section of the rock massif, within which there is structural reorganization with the opening of tectonic disturbances and movements along the boundaries of tectonic blocks [13]. Widely manifested at Nikolaevsky and Yuzhny mines has been mine seismicity, which is a reaction of the natural geoenvironment to anthropogenic activity [3-5]. Anthropogenic impact in conditions of high tectonic disturbance of the rock massif can lead to the spontaneous occurrence of foci of trigger seismicity with unpredictable energy [3]. At Nikolayevskoye deposit, the centers of mine seismicity are mainly confined to the ribbed zones of a large plate of ore-bearing limestones (especially its eastern part), where shocks and shifts of smaller blocks occur during the process of rebuilding the rock massif.

5 Conclusions

A comprehensive analysis of geological, geophysical and geomechanical data with the use of materials from the remote study of the Earth from space (GRACE mission data for the calculation of stress fields and radar images of the Earth's surface) made it possible to significantly refine the kinematics of the process of modern deformation of the upper part of the Earth's crust within the boundaries of the Edge-Asian model region . It is established that the peculiarity of the region is the significant contribution of the natural component to the total current SSS of rock massifs of impact-threatening deposits.

The fundamentally different position of the deposits of the Kavalerovo-Dalnegorsk ore region (located in the continent-ocean transition zone) has been revealed in relation to the tectonic structures that are unevenly activated in the upper part of the section during the current stage of development of the area and are formed as a result of uneven horizontal displacements.

The interdependence of the geodynamic position of deposits with the level of stresses in rock massifs is established, which must be taken into account when designing and conducting mining operations in severe mining-geological and burst-hazard conditions.

This research was supported by Russian Science Foundation grant No.16-17-00018.

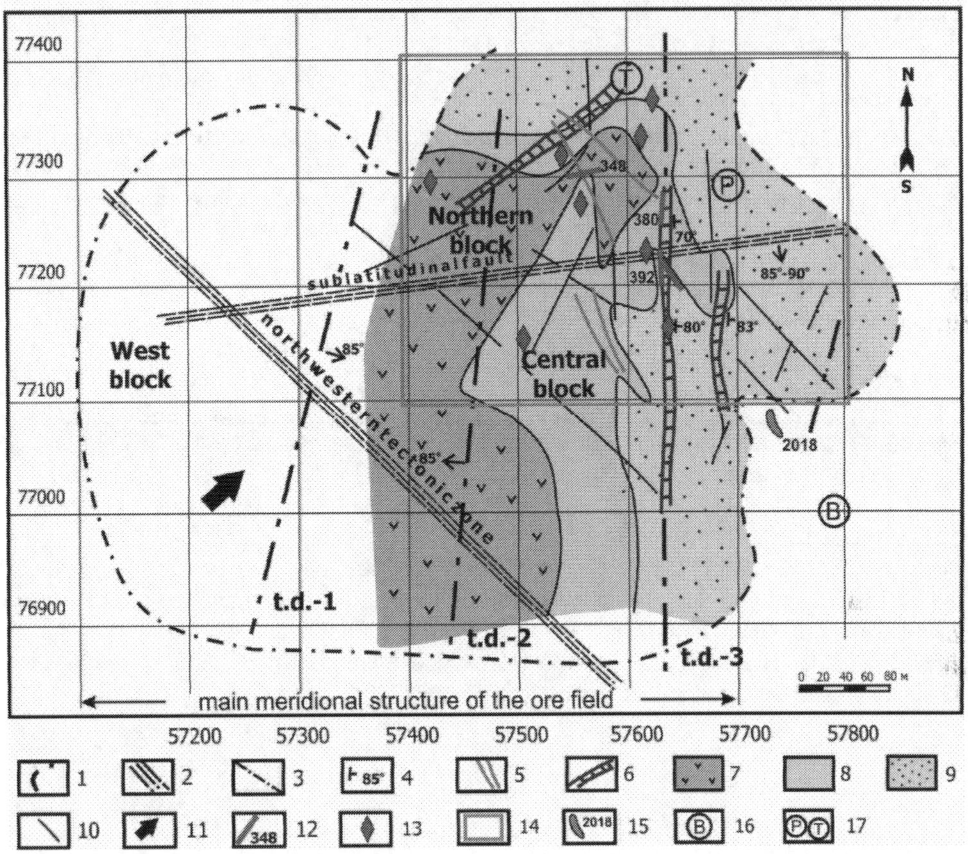

1 - projection of the outline of the olistolite on the horizontal plane; 2-6 - geological exploration data: 2 - block-forming faults and their names; 3 - other tectonic disturbances and their numbers; 4 - established elements of occurrence; 5 - dikes of medium and basic composition; 6 - gorges in the body of limestones, filled with metasomatites; 7-10 - geodynamic regions formed by the morphological features of the olistolith and the modern stress field: 7 - relatively geodynamically stable (raised or prihrebtovy) area; 8 - a hollow-slope region with intermediate geodynamic activity; 9 - steeply oblong geodynamically active peripheral region; 10 - lineaments of the topography of the roof of the olistolite; 11 - direction of the modern horizontal movement of the ore-bearing geoblock and the main horizontal compression of the rock massif; 12 - location of uncovered cracks associated with dynamic manifestations on 03.24.2016; 13 - position of large dynamic manifestations of rock pressure in 2016-2017; 14 - general contour of geoacoustic monitoring studies, 15 - new steeply falling zone of anomalous acoustic activity (2018) near the eastern edge of the olistolith (horizons - 340 - minus 400 m); 16 - location of the mass explosion (2017); 17 - places of manifestation of dynamic events caused by the explosion in 2017.

Fig. 2. Block structure and geodynamic zoning of limestone body of Nikolaevsky deposit

References

1. I. Iu. Rasskazov, Monitoring and management of rock pressure on mines of Far Eastern region. – Moscow: Mining book. p. 329. (2008)
2. Butra, J., Kudeiko, J. Cuprum. **61 (4).** P. 5–20. 17. (2011)
3. V.V. Adushkin, S.B. Turuntayev, Technogenic seismicity - induced and triggered. - Moscow: IDG RAS, 2015.
4. Li, T., Cai, M. F., Cai, M. International Journal of Rock Mechanics and Mining Sciences. **44 (8).** pp. 1149-1171. 2007
5. Marcak, Henryk, Mutke, Grzegorz. J. Seismol. **17 (4)**. 1139-1148. (2013)
6. Paige E. Snelling, Laurent Godin, Stephen D. McKinnon. International Journal of Rock Mechanics and Mining Sciences. **58.** 166-179. (2013)
7. Geodynamics, magmatism and metallogeny of the East of Russia. Vladivostok: Dalnauka. p. 572. (2006)
8. Present Day Stress Strain State in the Upper Crust of the Amurian Lithosphere Plate / I.Iu. Rasskazov, B.G. Saksin, V.A. Petrov, B.F. Shevchenko, V.I. Usikov, G.Z. Gilmanova // Izvestia, Physics of the Solid Earth. **50 (3).** 444–452. (2014)
9. B.G. Saksin, I.Iu. Rasskazov, B.F. Shevchenko. Physico-technical problems of mining. **2**. 53-65. (2015)
10. B.F. Shevchenko, B.G. Saksin, G.Z. Gilmanova, M.M. Dovbnich. GIAB: Far East. **OB4.** 49-60. (2013)
11. O.V. Rybas, G.Z. Gilmanova, The exploration of the Earth from Space. **6.** 45-52. (2011)
12. V.I. Usikov. Pacific Geology. **6.** 14-33. (2011)
13. I.Iu. Rasskazov, B.G. Saksin, V.I. Usikov, M.I. Potapchuk. Mining magazine. **12.** 13-19. (2016)
14. V.P. Nechaev. Pacific Geology. **6.** 91-101. (2015)
15. V.Iu. Zabrodin, O.V. Rybas, G.Z. Gilmanova, The fault tectonics of the mainland part of Russian Far East. - Vladivostok: Dalnauka. pp. 132. (2015)
16. V.G. Bykov, V.A. Bormotov, A.A. Kokovkin et al. Bulletin of Far Eastern Division of Russian Academy of Sciences. **4.** 83-93. (2009)

Software and hardware improvement for the Streltsov ore field geodynamic testing area

Andrei Gladyr[1*], *Vladimir* Miroshnikov[1] and *Alexander* Konstantinov[1]

[1]Mining Institute of Far Easten Branch of Russian Academy of Sciences, Khabarovsk, Russia

Abstract. For reliable forecasting and prevention of dangerous geodynamic phenomena in the Streltsov ore field area, the geodynamic testing area has been created. The geodynamic testing area includes modern measuring complexes that have different measurement methods and provide effective registration, and in-depth analysis of a wide range of geophysical wave fields changing in space and time. In the presence of anthropogenic seismicity, with the complication of the geomechanical and hydrological conditions of mining, the constant development of methods and means used at the geodynamic test site is required. It is necessary to install additional elements of a multi-level geomechanical monitoring system at sites in which dangerous geodynamic events are predicted and evidence of their reliable precursors. The modernization and integration of the measuring complexes of the geodynamic range into a multi-level geomechanical monitoring system allows to expand the controlled volume of the rock massif, to increase the quantity and quality of the information obtained about the state of the rock massif, which makes it possible to identify regional and local harbingers of dangerous dynamic phenomena and to justify a set of effective measures to reduce the level geodynamic risk improve the quality of forecast evaluation of geomechanical condition.

1 Introduction

At present, according to many experts, the influence of human activity on the earth surface state becomes more and more noticeable. Urbanization of territories, along with technogenic impact in the process of mining and processing of minerals, often leads to a significant deterioration of the overall environmental situation of the environment, climate change, activation of seismic processes [1-2].

In this connection, the role of creating specialized geodynamic testing areas on which complex (geodetic, geomechanical, engineering-seismological, engineering-seismometric, geophysical, etc.) observations of natural, technogenic and technogenic-induced geodynamic processes affecting the state of the techno-natural system is increasing. What is especially important for the areas that are subject to active man-caused impacts in the form of underground mining of minerals, construction of large engineering structures, reservoirs, etc. Emergency incidents in these areas can lead to catastrophic consequences for the

[*] Corresponding author: rush3112@mail.ru

population as well as for the infrastructure of the facilities themselves and for the environment [3-5].

2 Materials and methods

In increasingly complex mining conditions, more in-depth complex geodynamic studies are needed, including: assessment of the geodynamic, seismic situation and stress-strain state of rock massifs, both within the Transbaikal block of the Amur plate, and in the area of the PAO "PPGHO" fields; studying regularities of geodynamic fields and processes in the field of technogenic influence of mining operations using seismic, seismoacoustic, geodetic, seismic and other methods to obtain fundamentally new information on the stressed-deformed state of the geological environment, as well as the identification of regional precursors of dynamic phenomena of different energy levels in mountainous arrays [6].

To solve the set tasks, a comprehensive program of R & D was developed in the period 2010-2016, the purpose of which was the creation of a multi-level complex geodynamic monitoring system ("Geodynamic testing area") in the Streltsov ore field.

At present the following systems are part of the geodynamic testing area [7]:

- geoacoustic monitoring system "Prognoz-ADS" (ASKGD) ("Antey" deposit, 34 geophones);
- a system of microseismic monitoring "Prognoz-S" (ASHSGD) (PUR No. 1, 12 seismic pavilions);
- system of deformation monitoring "SMD" (cages, places of possible deformations);
- laser scanning system "FARO FOCUS - 3D" (PUR No. 1, observation stations at 11, 12, 13 horizons);
- core drilling machine DIAMEK-232 (the core disk exploration method is the base one);
- portable impact control device "Prognoz-L" (local assessment of the state of the edge parts of the rock massif);
- laser deformometer (industrial site of PUR No. 1, IV horizon of mine No. 1);
- nanobarograf (industrial site of PUR No. 1, IV horizon of mine No. 1);
- regional network of GPS observations (the territory of Southern Transbaikalia);
- analytical program complex of mathematical modeling of the stress-strain state of rocks Sigma GT and ARGIS-10 (regional forecast of geodynamic activity);
- spatial model of deposits and mine workings based on the software product MINEFRAME (solving geotechnological problems and creating mining documentation).

At present, the team of Institute of Mining FEB RAS is actively working to modernize the geodynamic testing area facilities in order to improve its characteristics and improve the quality of the received information.

From 2016 to the present, the following works were planned and performed:

- expansion of the observation network for geomechanical processes;
- development of a special version of the elements of the system for working in explosion-proof design;
- modernization of software and hardware monitoring systems to increase the information and convenience of work;
- development of means for integrating the measuring means of the geodynamic test site into a single integrated observation network.

3 Results and discussion

At the "Antey" deposit, related to the Streltsov ore field, 14 additional receiving transducers of the "Prognoz ADS" geoacoustic system were installed in the mine workings on the XIII and XIV horizons and the number of data transmission channels from 5 to 9 ports was expanded. On the XIII horizon, there is an underground apparatus room in the chamber opposite the mouth of the 6a-1302 ort and the gangway 6a-1301.

The installed sensors form several additional combinations of the receiving antenna and in combination with the existing geophones allow us to cover an even larger part of the minefield of the Antey deposit, including the riskiest areas of the massif near blocks 6a-1006, 6a-1110-1102, 6a-1202 -1212, 6a-1312-1302 and 6a-1412, where dynamic manifestations of rock pressure have recently been observed.

For geomechanical monitoring under gas conditions, a set of geoacoustic equipment "Prognoz-ADS" in explosion proof design was developed. It included an explosion-proof switching and distribution unit consisting of cabinets and terminal boxes fixed to the frame. All elements on the explosion-proof frame with IP54 degree of protection are manufactured and certified by GORELTECH. In the main section (control cabinet) there are: an industrial computer with low power consumption (included in one network with the operator's computer, serves as an intermediate server for data collection of the system) complete with an external explosion-proof monitor and keyboard, a board with an RS-485 interface from MOXA; converter IMS-21GA-LX; the device of a filtration of sound channels and 4 devices of protection of information channels MOXA; power supply and time synchronization in the FPC model PSUt4.

Some elements of "Prognoz-ADS" developed in explosion-proof design and installed on an underground mine are shown in Fig. 1.

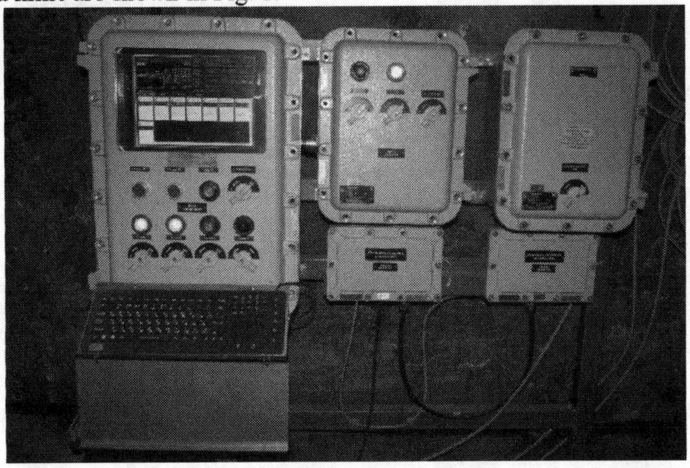

Fig. 1 - Explosion-proof switching and distribution unit "Prognoz-ADS" automated system (PB Ex d I Mb)

The network of microseismic observations in the area of the Streltsov ore field has been expanded by increasing the number of seismic pavilions of the microseismic system "Prognoz S" installed in underground workings on horizons V to XIV of the First Uranium Mining Management (PUR-1) from 8 to 12, which ensures greater reliability of the measuring complex and better accuracy of determining the parameters of seismic events in the controlled area of a rock mass with a volume of more than 12 km^2.

To improve the efficiency and reliability of the microseismic system "Prognoz S", the technical and software tools have been modernized, including: the replacement of a central computing unit with a more modern one, built on the basis of the Intel NUC DE3815TYBE processor module; equipment of data acquisition modules by a specialized node - the watchdog timer WD021, which excludes "freezing" of equipment; development of special software for monitoring the state of the seismic station and the software shell of the operator's workplace with enhanced capabilities.

The software used in the system of microseismic monitoring can be divided into service, providing continuous collection of microseismic monitoring data and analytical, allowing to analyze and process the registered data (Fig. 2) [8-11].

To manage the process of interrogating the analog-to-digital converter, preliminary processing and subsequent transfer of information on the results of microseismic monitoring to the data collection server, software is developed and applied at each of the seismic stations. This software includes the following programs: the primary data server and the graphical interface of the data collection server.

Primary data server

The program is designed to collect data from primary converters of the microseismic monitoring system, preliminary processing, subsequent transmission of observation data and can be used as part of real-time geodynamic monitoring systems. To increase the speed and stability of the work, this program works in text mode.

The program provides the following functions:

- continuous data collection;
- the organization of a circular data buffer on an external medium with the ability to select data for an arbitrary time interval;
- digital filtering, decimation, compression of collected data;
- continuous transmission of processed data through a computer network for long-term storage and additional analysis.

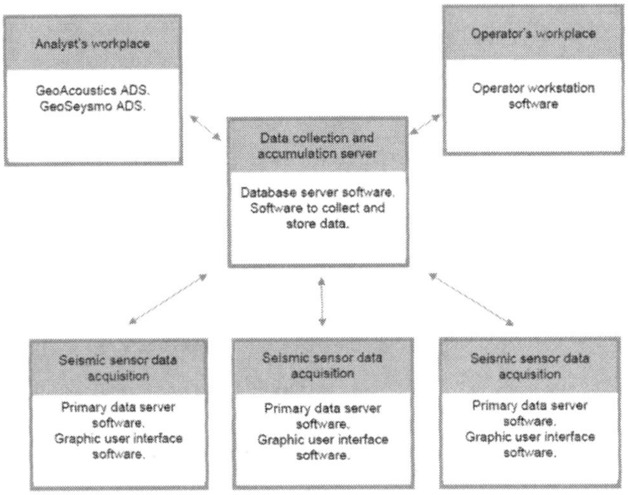

Fig. 2. Block diagram of the microseismic monitoring system software

Graphical interface of the primary data server

This program is designed to display information on the functioning of the primary data server of the microseismic monitoring system in real time and can be applied at each of the established seismic stations of the system to obtain up-to-date information on the process of collecting and transmitting the data of microseismic observations.

The program provides the following information:

- the operating status of the device, which displays information about the program version, the serial number of the analog-to-digital converter module, the number of active channels, the current sampling frequency, the operation modes of the program clock, the operation modes of the built-in digital filters.

- graphical display of the digitized information coming from the analog-digital conversion module in real time.

- the protocol of the program, in which the protocols for sending the decimated data to the data collection server and isolating the potential microseismic pulses are displayed separately.

Simplified, the GUI of the primary data server is the "wrapper" for the primary data server. With this program, you can manage the primary data server and monitor its status. The program has three working screens: the first screen displays information about the device, mode of operation, status; The second screen contains graphs that show incoming signals from the sensor in real time; The third screen displays the protocol of the primary data server operation, including service and diagnostic information.

On the communication lines, the microseismic monitoring data is fed to the data collection and accumulation program.

Data collection and accumulation program

This program is designed to collect data from primary converters of the microseismic monitoring system, integrated into a single information network. The program performs preliminary processing of information on observations and is applied in the real-time monitoring system. To increase the speed and stability of the work, this program works in text mode.

The program provides the following functions:

- obtaining data (including decimated data) from the devices of primary converters, united in a single information network, through connection to data servers on each of the primary converters;

- detection in the received data of impulses on the set parameters;

- loading of the detected impulses of detailed data from all connected primary converters and recording to an external medium for long-term storage.

Workplace of the operator of the microseismic monitoring station

This program interacts in real time with a program of sensor interrogation, preliminary processing of information and long-term storage of the data of the microseismic monitoring system, obtaining from it the necessary information for the generation of graphical information reports.

The program of the operator's automated workstation for the microseismic monitoring system allows:

– to expand the possibilities of visual display of the processes of information collection and processing;

– analyze the performance of each seismic post individually and the system as a whole for at least the last seven days;

– evaluate the performance of data channels in real time with memory for the last not less than seven days;

– to observe in real time the process of downloading registered microseismic pulses from seismic stations;

– view graphically the downloaded microseismic pulses registered previously.

Simplified, "Operator's seat of the microseismic monitoring station" is the "shell" for all service programs of the microseismic monitoring system. With this program, you can manage the operation of data collection modules and monitor the status of the process of collecting, transmitting and storing monitoring data for a long time.

The most important task for the generalization of experimental data characterizing the manifestations of seismoacoustic activity and technogenic seismicity in the conduct of underground works is the close integration of the applied systems into a single integrated observational network.

The integration process includes the following functionally completed steps [12]:

- integration into the information exchange system (using interfaces and protocols USB, RS-485, RS-485, SHDSL with a transmission rate of up to 15.2 Mbps);
- maintenance of common time management on all systems entering the geodynamic range (the implemented technical solutions ensure the accuracy of the system clock on each of the data collection servers is not less than 15 ms) [13];
- realization of accumulation of monitoring data for geodynamic polygon systems in a single database;
- implementation of the ability to display and analyze monitoring data of geodynamic polygon systems using a single software package.

The main integrating platform of the geodynamic testing area is the software complex "Mineframe", intended for the complex solution of wide mining-geological and technological problems [14]. The three-dimensional graphic core, which is the basis of the software of the system, allows you to work with models of objects in a multi-window mode and provides a wide range of control options for displaying and editing objects of mining infrastructure.

4 Conclusions

To solve the problem of preventing dangerous manifestations of rock pressure in the underground mines of PAO "PPGHO" in the Streltsy ore field area, a multi-level system of complex geodynamic monitoring (complex geodynamic testing area) is created, which includes a number of complementary methods and technical means that ensure effective registration and in-depth analysis of the wide spectrum of spatially and time-varying parameters of geophysical wave fields (geoacoustic, microseismic, deformation ets).

The modernization allows to expand the controlled volume of the rock massif, improve the quality and quantity of information received and improve the quality of the predictive estimates of the geomechanical state.

Integration of geodynamic testing area systems into a single network of information exchange allows:

- manage the operation of any of the subsystems used from any workplace of the operator included in a single network, including those located in the surface complex;
- remote control and adjustment of acoustic emission, microseismic and deformation subsystems through a single Internet gateway;
- consolidate measurement data of geodynamic polygon subsystems on a single data server for centralized storage, processing and analysis, with the ability to automatically synchronize with remote data servers.

Integration of systems into a single network of providing the exact time allows us to trace with sufficient accuracy the correlation dependence between the indications of subsystems of microseismic, geoacoustic and deformation monitoring.

Joint interpretation of geoacoustic and microseismic monitoring data, correlated with geological and mining technical models, provides high reliability of forecast estimates.

The application of a multilevel system of geodynamic monitoring makes it possible to increase the quantity and quality of the information obtained about the state of the mountain massif, which makes it possible to identify regional and local harbingers of dangerous dynamic phenomena and to justify a set of effective measures to reduce the level of geodynamic risk in the development of mineral deposits in complex geological and shock hazard conditions.

The work was supported by the Russian Science Foundation (agreement No. 16-17-00018).

References

1. A.J. Mendecki, Seismic monitoring in mine. London: Chapmen and Hall (1997)
2. L. P. Ishukova, Uranium deposits of the Streltsov ore field in Transbaikalia (2007)
3. Aleksander J. Mendecki, Richard A. Lynch & Dmitriy A. Malovichko. Routine Micro-Seismic Monitoring in Mines. Australian Earthquake Engineering Society. 2010 Conference Perth, Western Australia, 33 (2010)
4. P.L. Swanson, Rockbursts and Seismicity in Mines – RaSiM5, South African Institute of Mining and Metallurgy, 11 (2001)
5. I. Yu. Rasskazov, B. G. Saksin, V. A. Petrov, B. A. Prosekin, Physical and technical problems of mineral development, **3**, 3 (2012)
6. I. Yu. Rasskazov, A. V. Gladyr, P. A. Anikin, V. S. Svyatetskiy, B. A. Prosekin, Mining Journal, **8**, 9 (2013)
7. A. V. Gladyr, D. S. Migunov, P. A. Anikin, G. A. Kalinov, V. I. Miroshnikov, Geomechanical and geotechnological problems of subsoil development in the North: Mining Information Analytical Bulletin, **11**, 329 (2017)
8. I. Yu. Rasskazov, A. V. Gladyr, S. V. Deykin, D. O. Zhelnin, V. I. Miroshnikov, Certificate of registration of the computer program RUS 2014613073 (2014)
9. I. Yu. Rasskazov, A. V. Gladyr, S. V. Deykin, D. O. Zhelnin, E. A. Katella, Certificate of registration of the computer program RUS 2014613552 (2014)
10. A. V. Gladyr, Certificate of registration of the computer program RUS 2017660983 (2017)
11. A. V. Gladyr, Certificate of registration of the computer program RUS 2017660981 (2017)
12. A. V. Gladyr, Mining Information and Analytical Bulletin, 6, 220 (2017)
13. D. O. Zhelnin, A. V. Gladyr, E.A. Katella, S. V. Deykin, Problems of complex development of georesources Materials of the V All-Russian Scientific Conference with the participation of foreign scientists dedicated to the 30th anniversary of the Institute of Mining of the Far Eastern Branch of the Russian Academy of Sciences and The 100th anniversary of his birth. corr. RAS E. I. Bogdanova, 49 (2013)
14. S. V. Lukichev, O. V. Nagovitsyn, I. E. Semenova, O..V. Belogorodtsev, Mining Journal, **8**, 53 (2015)

Justification of safe and effective working conditions for pit reserves of "Pioner" gold deposit

Marina Potapchuk[1 *], *Gennady* Kursakin, *Viktor* Krukov[1] and *Mikhail* Lomov[1]

[1]Mining Institute of Far eastern branch of Russian Academy of Sciences, Khabarovsk, Russia

Abstract. Further prospects for the development of "Pioner" gold deposit are associated with the development of deep-seated ore bodies by an underground method. When assessing the bump hazard of proposed technological solutions for stripping and working out the pit reserves of the deposit and substantiating the safe working conditions, some comprehensive studies were carried out, including an analysis of the geological and geomechanical conditions of deposit's development, detailed study of the physical and mechanical properties of rocks and ores, investigation of the fracture tectonics of the deposit and stress and strain state of the rock massif of the deposit. The performed modeling of the geomechanical processes of the rock massif with the use of numerical methods made it possible to identify the safe parameters of the guarding safety pillars under the pit bottom, and also to identify areas and elements of rock structures characterized by a minimum margin of stability according to the criteria of the acting maximum compressive and tangential stresses after complete mining of ore bodies. Recommendations were developed on the rational order of mining of ore bodies and on effective ways to protect and maintain mine workings.

1 Introduction

"Pioner" gold deposit is located in Amur region, Zeya district. The deposit is represented by a series of numerous small lenticular bodies of irregular shape, nonpersistent in the fall and strike. Ore bodies fall to the north-north-west at an angle of about 60. The average horizontal thickness is from 2.0 to 6.8 m. The tectonic structure of the deposit is determined by a combination of faults in the orthogonal and diagonal systems.

The formation of the structures of localization of ore-magmatic systems is due to the factors of difference in their internal structure and composition of magmatic rocks. There are five systems isometric in plan view and one oblong. Three of them (I, II, III), being the earliest ones, belong to the diorite-granodiorite assemblage, two (V, VI), occupying the average age-related position, are represented by the subalkaline-granite kindred (according to the existing ideas, the Jurassic age) and one (IV) - by fluidolite. The isolated systems

* Corresponding author: potapchuk-igd@mail.ru

occupy a certain spatial position and correspond to a specific structure. Within these structures, 15 mineralization zones are shown, as shown in Fig. 1.

Currently, the development of the Bakhmut ore zone is under way, the upper part of the deposit (N-E Bahmut) has largely been worked out by quarries, further development of deep-seated ore bodies 2-9, 1-5, 1-4 and 10-24 by underground method is planned. The structure of the ore-rock block of the deposit is very heterogeneous due to the combination of heterogeneous rocks (diorite porphyrites, microdiorites, granodiorite-porphyry, fluidolites), metasomatites, different morphology fault structures and breccias. The fracture of the submeridional and north-western strike is divided into blocks of high order [1, 2]. Two blocks are completely outlined: the first and the second. The first block is composed mainly of diorite porphyrites, which with depth are replaced by fluidolites of pelitic texture. The stability of rocks is reduced in a series of pseudomorphic metasomatites (apodiorites, apogranites) – berezitized rocks – near-ore (automorphic) metasomatites. Their ratio is 30:65:5, respectively. There are two areals with a relatively high density of disjunctive breaks. The second block is composed mainly of pelitic fluidolites, which have a sufficiently high strength. The stability of rocks is reduced in a series of pseudomorphic metasomatites (apodiorites, apogranites) – tourmaline rocks – quartz-tourmaline (automorphic) metasomatites. Their ratio is 25:65:10, respectively. There are four areals of disunctive breaks with relatively high density. The following sections of increased fracturing are identified: falling on the central part of the ore body 1-4, with cracks often being of vein material (quartz, carbonates, tourmaline, carbon minerals), with a thickness of tenths of a millimeter, less often up to 1-5 mm, and adjacent to the drilling chamber, but largely resulting from blasting. On the basis of generalization of the obtained data, the studied section of the deposit was divided into three groups according to the degree of stability, namely: relatively stable rocks that coincide spatially with altered diorites and granites (apodiorites, apogranites) (at least 30% of the total volume), medium rock stability, berezitized and tourmalineized varieties occupy 60-65%, and unstable rocks, berezites, quartz-feldspar-chlorite and quartz-tourmaline metasomatites, quartz kalishaspatites - up to 10%.

Based on the mining and geological conditions of deposit's development and in accordance with the Rules for engineering operation of minery developing deposits of non-ferrous, rare and precious metals [3] for underground mining of ore bodies, a system was proposed using sub-floor drifts with open worked out space and breaking of ore by deep wells with a diameter of 65 mm with a height of 30-40 m, a sub-floor being 15 ÷ 20 m.

The transition to an underground method for the development of ore reserves is associated with a number of complicating factors, such as the impact of borrow cut on the stability of mine workings in the near edge part of the massif, the presence of sites with various rock disturbances, and the complex tectonic structure of the deposit. In order to obtain reliable geomechanical information necessary for the development of recommendations on the rational order of mining ore bodies, and effective methods of protection and maintenance of mine workings, a set of studies was carried out, including geodynamic zoning of the deposit with the identification of its tectonic structure, study of the properties and stress state of the rock massif, as well as the study of stress distribution in the rock massif as the deposit was being mined.

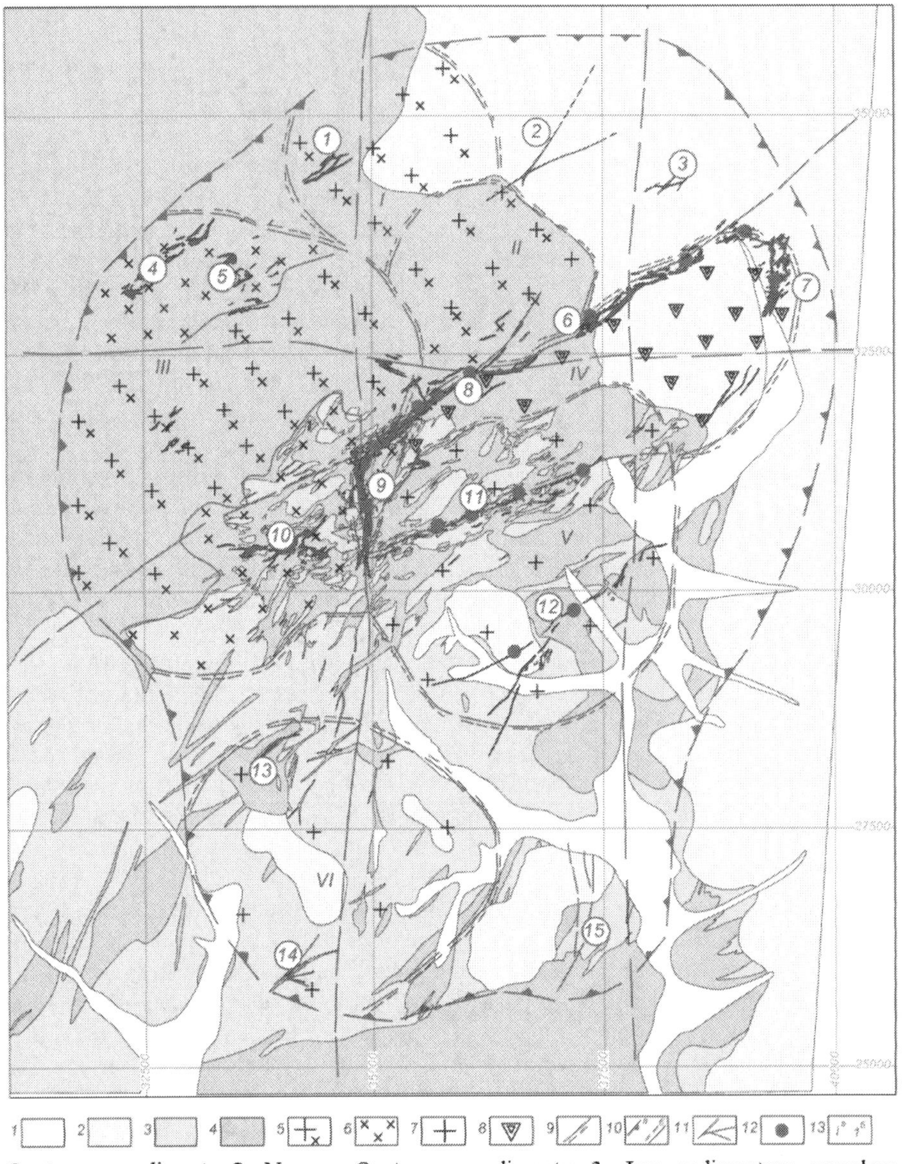

1. Quaternary sediments 2. Neogene-Quaternary sediments; 3. Jura sedimentary complex; 4. Cretaceous intrusive formations; 5. Granodiorites, granodiorite-porphyry, quartz diorite porphyrites; 6. Gabbro-diorites, diorite porphyrites, microdiorites; 7 Subalkaline granites, granite-porphyry; 8. Fluidolites, eruptive and explosive breccias; 9. Boundaries of ore-magmatic systems. 10. Faults: a) bordering the Pioneer arch, b) linear structures traced and distinguished by interpretation, geophysical features; 11. Ore bodies and mineralization zones; 12. Ore posts; 13. Objects: a) structures of localization of ore-magmatic systems; b) ore bodies and zones of mineralization: 2 – z. Dump, 3 - z. Original, 4 - o.z. Asteriks, 5 - o.z. Asteriks 2, 6 - o.z. Bahmut, 7 o.z. Erosion, 8 - o.z. Intermediary, 9 – o.z. South, 10 - o.z. East, 11 - o.z. Andreevskaya and Contacting, 12 - o.z. Nikolaevskaya, 13 - o.z. Babaevskaya, 15 - o.z. Propsective

Fig. 1. Structure of ore-magmatic systems of ore field Pioneer

2 Materials and methods

In order to study the regularities in the process of anthropogenic impact in the distribution of initial stresses in the rock massif (in the pillars and the main structural elements of the mining system), the computational method (the finite element method) used widely for solving various geomechanical problems was applied [4, 5, 6].

The applied method was implemented with the help of the FEM software package consisting of three modules: FEM1 (for the volumetric problem - FEMV1), FEM2-3 (FEMV2-3), and FEM4 (FEMV4), allowing solving the problems of the theory of elasticity and plasticity by the finite element method, both in plaane and in volumetric formulations. The program allows the assignment of homogeneous and non-homogeneous massifs with any kind of heterogeneity: ore bodies with different angles of inclinations, workings of different configurations [7].

When assigning loads along the boundaries of the finite element model, the results of an experimental evaluation of the stress state parameters using the acoustic-emission memory effect [8, 9] were used, and also the fact that the natural stress field of the upper (upland) part of the massif is determined by the influence of the modern relief and at a depth below the bottom of the valleys, tectonic forces that form the tectonic component of the stress tensor begin to act. Therefore, the depth and upland parts of the massif were differentiated according to the level of the initial stresses, supposing that the gravitational stress field, described in accordance with the well-known hypothesis of Dinnik [10], acts in the upper part of the deposits.

To obtain objective information about the properties and state of the rock massif used in the modeling, the geodynamic zoning of the deposit site was carried out, during which its tectonic structure was studied, including the separation of the boundaries of tectonic blocks by geological and morphometric methods, and the physico-mechanical characteristics of the entire lithological complex were found.

3 Results and discussion

An analysis of the geodynamic position of "Pioner" deposit in the modern structure of the Amur plate showed that it is confined to the region of compression with a shift [11]. According to the results of the morphometric analysis of the relief from satellite geodesy data and seismic survey materials [12-15], it was established that the compression mode is typical for the deposit area and the vector of the current main horizontal compression at the site is directed south-west-westwards.

To assess the stress state, a method based on the well-known Kaiser effect was also used, according to which rocks have the ability to store, keep and reproduce the information on the stresses experienced [9]. According to the results of the research, it is established that the Kaiser effect is manifested in 30% of the samples studied, sampled at a depth of more than 80 m, in which the spread between the measured and calculated values of σ_0 to σ_k does not exceed 30%. The obtained results indicate that the stress state at the depths of sampling (80-200 m) is close to hydrostatic, due to the weight of the overlying rock mass.

The generalization of the set of data obtained made it possible to substantiate the parameters of the stress field in the region of the deposit, which is characterized by a gravitational type up to a depth of 320 m, while horizontal compressive stresses oriented in south-westwards and 1.5-2 times higher than the vertical stresses predominate below.

The evaluation of the stressed-strain state of "Pioner" rock massif was carried out in three stages: first, a change in the level of stresses at individual points in the massif was studied with a decrease in the height of the safety pillar during the mining of ore bodies; further, the degree of influence of tectonic structures on the distribution of stresses in the elements of rock structures and the surrounding rock massif was established, and at the final stage the assessment of the applied development system from the position of bump hazard was carried out.

Studies of the geomechanical state of the ore-rock massif and ore bodies 10-24 and 10-24-1 and pillars decreasing during the process of mining under the bottom of the quarry have shown that a complex anthropogenic stress field is formed characterized by the presence of both discharge areas (mainly in the quarries and separate areas under the bottom of the quarry), and the appearance of zones of increased stress concentrations in the marginal parts of the massif. Prior to the beginning of mining, two regions of high stress concentrations were identified: at the interface between the bottom and the lower side of the quarry along the southeastern border of the quarry, and in the deep part of the deposit in the northwestern part of the quarry below the horizon of 140 m. The level of maximum compressive stresses in these sections exceeds 12.5 MPa, the tangential stress level is 5.5 MPa. Unloading zones are observed mainly along the bottom and sides of the quarry, the magnitude of the maximum tensile stresses here does not exceed 0.5 MPa.

As the ore bodies are being worked out and the height of the guard pillars decreases, a redistribution of stresses and a formation of new areas of increased stresses occur. The site, which is formed on the boundary of the enclosing massif and the worked out space at 155 ÷ 165 m mark, at the final stage of mining the ore body 10-24 is characterized by the maximum values of compressive and tangential stresses, reaching 19 and 6.8 MPa, respectively. In the safety pillars of ore bodies 10-24-1 and 10-24, the stresses amount to 15 MPa.

For the geomechanical justification of the limiting parameters of the safety pillars, the theory of maximum normal compressive stresses and the Coulomb-Mohr criterion was used. The calculation of the stability of the most stressed sections according to these criteria showed (see Figure 2) that stability is maintained only at a height of safety pillars beinbg 20 m for profile lines 486, 482 and 23 m for profile line 484.

The results of volumetric modeling of successive stages of ore blocks mining in the following sequence: block 1-5, block 2-9 showed that before the beginning of mining of ore bodies, the maximum stress concentrations are observed along the northwest and southeast boundaries of the quarry at the interface between the bottom and the lower slope.

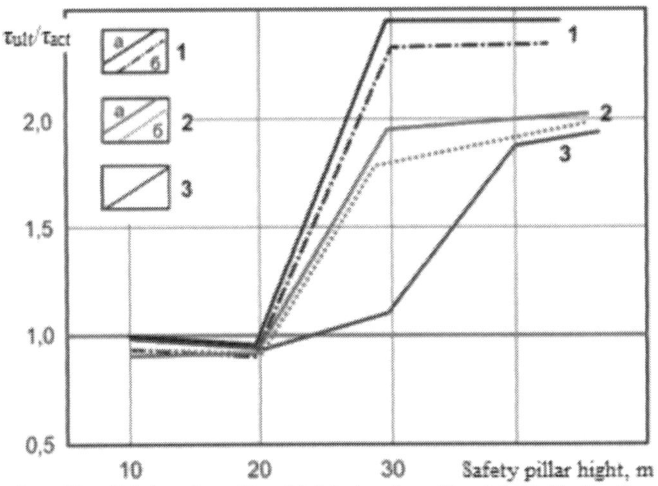

1 - for conditions of profile 486 (*a* - for pillar 10-24, *b* - for pillar 10-24-1); 2 - for conditions of profile 482 (*a* - for pillar 10-24, *b* - for pillar 10-24-1); 3 -for conditions of profile 484 for pillar 10-24

Fig. 2. Dependence of the change in the ratio of the acting tangential stress in the rock mass to the ultimate shear strength of rocks (τ_{ult}/τ_{act}) on the safety pillar height.

The values of the maximum compressive stresses in these sections do not exceed 15 MPa, the shear stresses are 10 MPa, respectively.

Working out part of the ore reserves of the upper sub-floor (115 ÷ 130 m) in the area of the ore block 1-5 leads to a redistribution of stresses and formation of an area of increased rock pressure in the vicinity of the bottom of the lower sub-floor, with a tangential stress and medium pressure level reaching 13 MPa, indicating an increased probability of destruction of this section of the massif in a static form (Fig. 3).

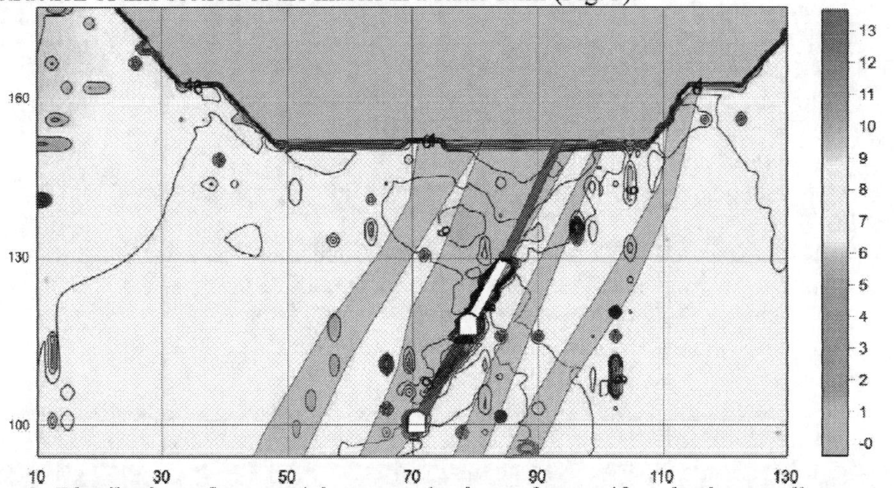

Fig. 3. Distribution of tangential stresses in the rock massif at the intermediate stage of mining ore block 1-5

At the final stage of mining of ore block 1-5, the voltages are concentrated in the lower part of the safety pillar and at the section below the sub-floor drift of horizon 100 m. The maximum compressive stresses increase by 40%, and reach 20 MPa, but the tangential stresses decrease, compared to the intermediate stage of 2 MPa and constitute 10 MPa.

Partial mining of the ore sub-floor 115 ÷ 130 m of block 2-9 leads to the formation of an additional area of increased stresses in the rock mass along the sides of the chamber's exhausted space, but the stress level does not increase significantly, only by 0.5 MPa from the stresses to the beginning of working. After the complete excavation of this ore block, the maximum stresses are concentrated at the section of the rock massif below the sub-floor drift of horizon 100 m. The values of the maximum compressive stresses are 16.5 MPa, tangential – 10.8 MPa.

3.1 Conclusions

The performed geomechanical calculations and modeling results allowed establishing the safe parameters of the guarding safety pillars formed during mining of the reserves of ore bodies 10-24, 10-24-1, excluding the development of deformation processes of the surface with the facilities of the mine's site located, which are 20-23 m .

The main elements of the mining structures of the applied development system will remain stable by the criteria of the acting maximum compressive and tangential stresses at all the considered stages of mining of ore bodies 1-5 and 2-9. But the minimum reserve of stability, calculated by the Coulomb-Mohr criterion, will have an ore massif of the upper sub-floor (above 130 m) of ore body 1-5 before the start of the mining and ore body 2-9 during mining.

Based on the results of geomechanical studies, recommendations were developed on the rational order and order of mining of ore bodies, effective ways to protect and maintain mine workings, as well as the parameters of the pillars and other structural elements of the mining system.

References

1. Petrographic Code. Magmatic, metasomatic, metamorphic and impact formation. Ed. 3, St. Petersburg: Publishing house VSEGEI. p. 200. (2009).
2. G.I. Tugovik, Fluid-explosive structures and formation mineralization. Moscow: Nauka. p. 193. (1989).
3. Rules for engineering operation of minery developing deposits of non-ferrous, rare and precious metals, Moscow: Nedra. - p. 109. (1981).
4. A. Houmat Int. J. Numer. And Anal. Meth. Geomech **37, 11**, 1552. (2013).
5. S. Shnorhokian, H. Mitri, D. Thibodeau, Int. J. of Rock Mech. and Mining Scien. **66**, 13. (2014).
6. I.M. Petukhov, A.M. Linkov, V.S. Sidorov, Calculation methods in the mechanics of rock bumps and blows: Guideline. Moscow: Nedra. p. 256. (1992).
7. O.V. Zoteev, Higher education bulletin of Mining Journal. **5**, 108. (2003).
8. T. Kanagawa, M. Hayashi, H. **Report 375017** Centr. Res. Inst. of Electric Power Ind. (CRIEPI) (1976).
9. V.S. Yamshchikov, V.L. Shkuratnik, K.G. Lykov, Physical and technical issues of mineral wealth. **2**, 23. (1990).
10. I.A. Turchaninov, E.V. Kasparian, Fundamentals of the mechanics of rocks - Leningrad: Nedra. p. 488. (1989).
11. B.G. Saksin, I.Iu. Rasskazov, B.F. Shevchenko, Physical and technical issues of mineral wealth. **2,** 53. (2015).
12. V.I. Usikov, Tectonics, Magmatism and Geodynamics of the East of Asia: VII Kosygin Readings: Proceedings of the All-Russian Conference Dynamics and Structure of Tectonic Flows. **328.** (2011).
13. K.G. Levi, S.I. Sherman, V.A. Sankov, Map of modern geodynamics of Asia. Siberian Division of RAS. (2007).
14. S.V. Ashurkov, V.A. Sankov, A.I. Miroshnichenko, A.V. Lukhnev, A.P. Sorokin, M.A. Serov, L.M. Byzov, Geology and Geophysics. **2,** 299. (2011).
15. I.Iu. Rasskazov, Pacific Geology. **25**, 104. (2006).

Geomechanical justification of security measures in the development of the South Khingan deposit of manganese ore

Maksim Rasskazov[1*], *Marina* Potapchuk[1], *Gennady* Kursakin[1] and *Denis* Tsoy[1].

[1]Mining Institute of Far Easten Branch of Russian Academy of Sciences, Khabarovsk, Russia

Abstract. The paper presents the results of geomechanical studies on the assessment of the potential rockburst hazard of the rock massif of the South Khingan deposit of manganese ore at the stage of development. Geodynamic zoning has been performed, mining and technical, mining and geological conditions of field development have been studied, and parameters of physical and mechanical properties of enclosing rocks and ores have been determined. Numerical simulation methods have been used to estimate the stress state of a rock massif at various stages of the deposit development. The tendency of the lower part of the South Khingan deposit to rockburst has been established. The complex of effective organizational and technical security measures has been substantiated in the development of this field. **Keywords:** rocks, physical and mechanical properties, stress-strain state, geomechanical justification, geodynamic zoning, development system, rockburst hazard, monitoring system.

1 Introduction

In the underground mining of solid mineral deposits, the prevention of mountain impacts, sudden releases of coal, rock and gas, and other dynamic phenomena are among the largest and most complex problems[1-2].

Rockbursts, representing a process of brittle fracture of the ends or edge parts of the massif, occur with a certain combination of geological and mining conditions. One of the reasons for the origin of mountain impacts is the unstable release of a part of the energy inflow, which is formed by decreasing the potential energy of the rocks around the excavations. The changes occurring in the rock massif under the influence of mining work provoke activation of the hidden focus of the seismic event, formed mainly due to the energy of deformations generated by the natural unevenness of the stress field [3-4]. It is known that the most effective solution to this problem is achieved at the design and construction stages of mining enterprises when it is possible to provide in advance a set of regional and local preventive measures that sharply reduce the danger of dynamic manifestations of rock pressure.

The object of the study is the South Khingan deposit of manganese ores located in the Oktyabrsky district of the Jewish Autonomous Region [5]. Its geodynamic position and

* Corresponding author: rasm.max@mail.ru

confinement to tectonically active structures in the southern part of the Bureinsky middle massif at the junction of the major modern morphological elements of the region: The Middle Amur Riftogenic Depression and the Bureinsky Range, which often predetermines the deposit's inclination to rockbursts during the underground mining operations, have given rise to the geomechanical research.

2 Mining-geological and mining-technical conditions of field development

The main reserves of the South Khingan deposit are concentrated in the "Poperechniy" section, which is a linear ore-bearing block with a width of 40 to 80 m (on average - 50 m) and a length of 240 m, containing a steeply dipping ore horizon, broken by transverse and longitudinal faults into a higher-rank blocks. Ore bodies are represented by steeply dipping low-lying strata, which can be traced along the strike for hundreds of meters. The deposits are in most cases sustained, but there are crimps, bulges as well as pinches (Figure 1).

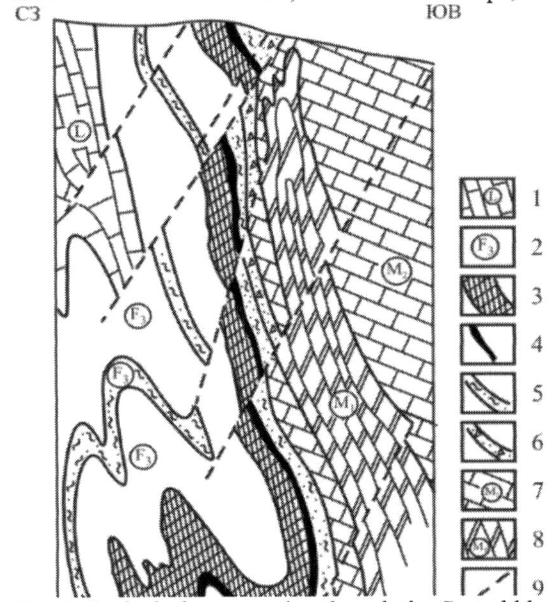

1-6 – Terrigenous-carbonate deposits of the Lower-Middle Cambrian; 7-8 – terrigenous-carbonate deposits of the Upper Proterozoic - Lower Cambrian period: layered and schistous dolomites with interbeds of various shales (7); siliceous dolomites, dense dolomites (8); 9 – tectonic disturbances

Fig. 1. Geological cross-section through the Central block of the "Poperechniy" section of the South Khingan deposit

The terrigenous-carbonate deposits of the Lower-Middle Cambrian period are: black limestones with clusters of carbonaceous-argillaceous and siliceous-argillaceous shales in the basement (1); clayey, carbonaceous-argillaceous and siliceous-argillaceous shales with interlayers of dolomitic sandstone (2); iron-ore layer, magnetite, magnetite-hematite microquartzites with breccia lenses (3); layered brownite-rhodochrosite-hausmanite ore (4); dolomite sandstones with siliceous-argillaceous shales (5) and sedimentary dolomite breccias (6).

The deposit is supposed to be developed by an underground method. The opening scheme provides for the penetration of two inclined trunks, the location of which is to be determined taking into account the results of the research, and coupled slopes. For the extraction of reserves, it is planned to use a development system with ore storage (with an ore body capacity of up to 3 m) and systems with sub-level drifts (with an ore body capacity of 2.5-3.0 m and above), with borehole or longhole stopping. The following parameters of the development systems are provided: the length of the excavation blocks is up to 60 m; the width of the excavation blocks is determined by the thickness of the ore body, but not less than 1.0 m; the height of the excavation blocks is determined by the height of the levels and amounts to 60-70 m. When using the development system by sub-level drifts, the excavating unit is divided into sub-stages with a height of up to 15-17 m.

3 Methods and results of the research

Faults and tectonic structure of the deposit, play an important role in the formation of impact and the manifestation of technogenic seismicity [6, 7]. Therefore, to assess the stressed state of the upper part of the lithosphere in the area of field development, its geodynamic zoning is performed, including: a complex morphostructural interpretation of different-scale topographic maps and a volumetric analysis of the modern relief of the earth's surface; target analysis of radar images (srtm30, srtm3) with the construction of appropriate neotectonic maps; generalization of information on the regional natural background of the stress-strain state of the geo-environment [11].

Calculation methods using GPS measurements and remote sensing of the earth's crust allowed for the reconstruction of the natural stress field in the region of the South Khingan deposit, which in the upper (upland) part of the massif has been formed under the action of gravitational forces, and in the lower part of the massif - from a depth of 300 m it is complicated by the additional action of tectonic forces. In the massif of the deep part of the deposit, the effect of an unequal component stress field is predicted, in which the horizontal compressive stresses prevail, the largest of which is oriented in the northeasterly direction and is 2.5-3 times greater than the gravitational component of the weight of the superincumbent rock [12].

The analysis of physical and mechanical properties of rocks and ores of the South Khingan deposit have shown that almost all of them (with the exception of layered and schistose dolomites) are quite strong (the strength of uniaxial compression varies from 82.3 to 174 MPa) and highly elastic (Young's modulus is in the range of 72.5-98.6 GPa). The results of the research show that the majority of the rocks in the "Poperechniy" section of the South Khingan deposit are those with high strength and rigidity. A negative factor is also the fact that all the rocks (F1, F2, F3) within which the cleaning dredging are planned are almost twice as strong as the surrounding M2 and L. This will contribute to a higher level of stress in the areas of the massif of mountain rocks, where the main mining works will be conducted.

Also, an unfavorable factor is the large variation in the strength and elastic parameters of the rocks within the groups, due to the diversity of their lithologic composition. For example, the increase in tensile strength is naturally correlated with the enrichment of rocks with silica (quartz and its modifications - F_2 and M_1 horizons), which characterizes these rocks as very fragile and capable of breaking down in a dynamic form. The brittleness index, which is the ratio of the compressive strength of rocks to the ultimate tensile strength, for all types of rocks and ores of the deposit exceeds 15.

The density of the studied section of rocks as a whole exceeds the corresponding reference characteristics of analogous metamorphosed differences. This circumstance is explained by the anomalous sulfidization of rocks of the entire section, visually observed by us when viewing the range.

Thus, the results of the studies performed give grounds to consider the rocks and ores of the South Khingan deposit as potentially dangerous.

To study the patterns of redistribution of initial stresses in a rock massif under the influence of mining operations, the finite element method (FEM) was implemented in a flat and volumetric formulation of problems [11, 12].

The stress-strain state of the South Khingan deposit rock massif was evaluated in two stages: first, the change in the stress-strain state (SSS) in the rock mass was studied with depth as the deposit was worked and the mining was reduced, and then the redistribution of stresses in the area cleaning blocks with allocation of zones of increased stresses in the elements of development systems. Based on the results of the simulation, it has been established that a complex technogenic stress field was formed during the development of the deposit. With the deepening of mining operations in individual areas of the massif, which include: non-ore areas, tectonic disturbance zones, interlevel and other pillars and marginal parts of the massif, normal and tangential stresses increase, which is especially noticeable in the lower part of the deposit. In the level -20 ... -120 m and below (depth 300 m and more), the value of the first main normal stresses σ_1 in the entry and arch pillars when working overlying reserves exceeds 90 MPa. To assess the stresses in the structural elements of the proposed development systems (sub-level drifts and ore storage), a volumetric modeling of the stage-by-stage treatment of the cleaning unit was performed.

Analysis of the results of numerical simulation of SSS at various stages of the cleaning unit treatment showed that when the deposit was worked out by the system with sub-level drifts, the highest level of stress was noted in inter-chamber pillars, mainly in the marginal parts from the sub-level drift. As the cleaning unit is being developed, a direct proportional increase in the magnitude of the maximum compressive and tangential stresses, the magnitude of which reaches a maximum (up to 85 MPa and 40 MPa, respectively) occurs when the unit is fully developed (Fig. 2). The complete development of the cleaning unit leads to the formation of an area of increased stresses in the central part of the overhead track on the side of the developed chamber (the main normal and tangential stresses reach 65 and 35 MPa, respectively), which increases the likelihood of dynamic manifestations of the rock pressure when it is reclaimed.

Using a system with ore stoping, the concentration of stress is observed in the inter-chamber ends, as well as in the marginal parts of the ore unit, even at the intermediate stage of the cleaning unit (at a stope height of 20 m). An increase in the stope height leads to an increase in the first principal σ1 and tangential stresses in the pillars by 20-25%, the values of which reach 70 MPa and 45 MPa (Figure 3), respectively. In addition, the most loaded are the inter-chamber pillars on the side of the spent cleaning unit.

The increase in the impact hazard in the area of inter-chamber towers is predicted with the complete development of the unit (complete release of the stoped ore from the cleaning unit), when the values of the first principal σ1 (85 MPa) approach the compressive strength, shear stresses (55 MPa) to the shear strength limit. Increasing the depth of mining will increase the potential impact of these elements of the development system.

Based on the analysis of the obtained results of SSS numerical simulation and the experience of working out the deposits with similar mining geological conditions, it can be concluded that the system of sub-level drifts is less dangerous since only marginal parts of the inter-chamber pillars are loaded. The development of ore reserves using the ore shrinkage system leads to the formation of a large number of potentially rockburst-dangerous ore inter-chamber pillars, which creates the conditions for dynamic manifestations of rock pressure. In general, the results of the simulation show that starting from a depth of 300 m (below the -20 m mark), some sections of the massif (pillars, free zones, tectonic disturbance zones) are characterized by an increased stress level, which approaches the rock strength limit. This circumstance indicates the possibility of the manifestation of rock pressure in a dynamic form while working out the lower part of the South Khingan deposit. The obtained results of the research had served as the basis for its classification as prone to rockbursts and made it possible to justify and put into the project a set of measures for the prevention of dangerous geodynamic phenomena, protection and maintenance of mine workings in rockburst-prone areas.

a *b*

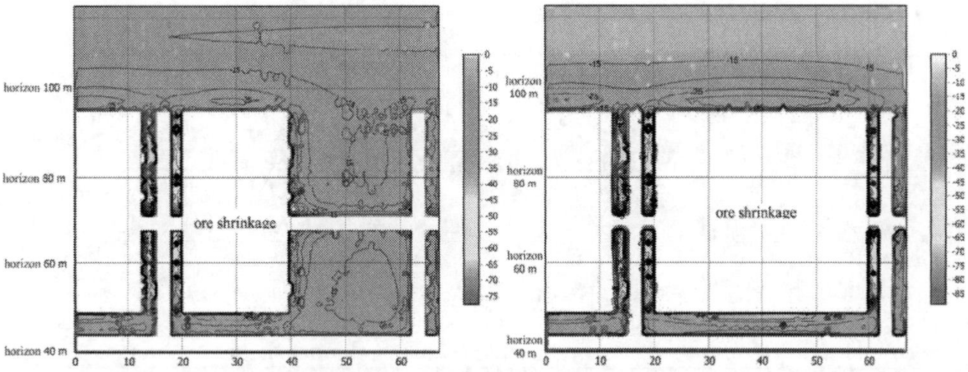

Fig. 2. Distribution of the first main stresses in the rock massif: *a* -when working out 50% of the cleaning unit; *b* - with the complete cleaning of the cleaning unit in the level 20 ÷ 100 m a (in the projection on the inclined plane)

a *b*

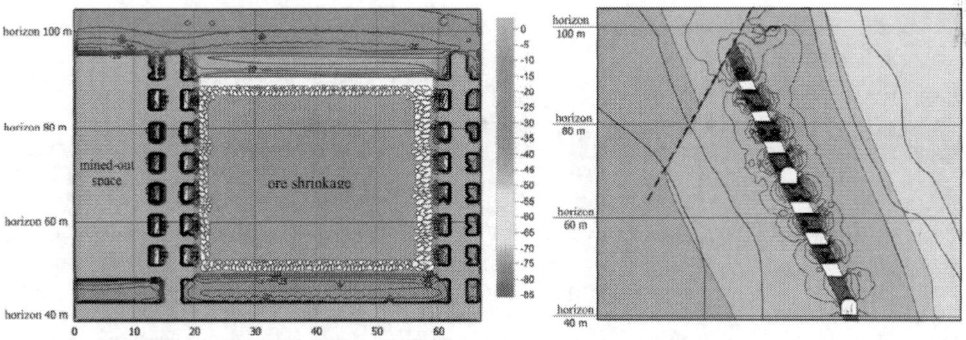

Fig. 3. Distribution of the first major stresses in the rock massif at a stope height of 40 m: *a* - projection on the inclined plane; b - cut into the cross of the strike of the ore body along the line of the interchamber pillar

4 Recommendations for the Prevention of Rockbursts

The main principle of safe mining of deposits prone to rockbursts is to ensure the order and methods of mining, in which excessive concentrations of stresses are eliminated in the rock massif and the harmful effect of rock pressure around mine workings is reduced. Mining operations should develop systematically from one flank of the field to another or from the center to the flanks, leaving the minimum number of lanes, followed by their reclamation. This principle may be implemented through the use of special preventive measures, the choice and justification of which is based on the regional and local forecast of the rockburst hazard [1, 12-13].

The results of the analysis and geomechanical justification of the South Khingan deposit made it possible to propose the following recommendations on the expediency of applying a set of the most promising organizational and technical security measures for the development of this field (Table. 2).

Table 2. The complex of measures to prevent rockbursts

Measures to prevent rockbursts	Content of measures
Tracing of workings	When opening and building deep horizons in tectonically tense rock massifs, a mandatory preventive measure should be the tracing of excavations, which contains the norm of observing reasonable requirements not to place the workings production in places of increased rock pressure and orient them along the direction of action of the greatest main stresses.
Giving the polygonal or elliptical shape of the cross-section (or face) of the working	In the case of an unfavorable arrangement of the workings (especially when oriented perpendicular to σ_{max}), it is advisable to give to their cross-section a polygonal or elliptical (for vertical workings) shape.
Camouflet blasting	When carrying out the workings in high-stressed areas of the massif that are classified as "Dangerous", a protective zone created by camouflage blasting or by cutting the discharge slots in a plane perpendicular to $\sigma_{ma.}$ must always be kept ahead of the moving face.
Creating the discharge slots	With a persistent high impact hazard, it is advisable to form discharge slots in the sides of the workings, oriented along the direction of action $\sigma_{max.}$
Drilling of discharge wells	When the shearing stresses are higher than the permissible values, during the development of mining operations, a discharge zone is created by drilling the ring of the discharge wells, while orienting it if possible in a plane perpendicular to the action of the maximum stresses. The angle of inclination of wells in the unloading zone can be determined by the greatest destruction of working.
"Standage" of workings	The standage of workings can be used in potentially dangerous areas of the massif, only when the process of relaxation of dangerous stresses in the contour zone takes a short time (not more than 5-10 days).
Fixing the workings	The most responsible capital workings with a long service life can be fixed with a constant and amenable support.

Along with the justification of the effective parameters of the shockproof measures, it is recommended to consider the use of special technical means for assessing the impact hazard, including the "Prognoz L" device, and "Prognoz-ADS" automatic seismoacoustic system for controlling the rock pressure and others [14].

5 Conclusions

1. The South Khingan deposit of manganese ores is characterized by a complex tectonic structure, its massif is composed of strong and highly elastic rocks. A large amount of silica in rocks and ores characterizes them as very fragile, capable of accumulating significant potential energy and breaking down dynamically.

2. Mathematical modeling of the stress-strain state had established its regular change with the depth of the deposit development. Starting at a depth of 300 m, an elevated stress level is observed in some parts of the massif (interlevel pillars, uninhabited areas, tectonic disturbance zones), approaching (and in some cases exceeding) the shear strength of the

rocks. A high potential rockburst hazard is also represented by a decreasing ore embankment under a mining development system with ore storage.

3. According to the results of geodynamic zoning, the current position of the fault zones limiting the main blocks of the site has been revealed and their position relative to the regional stress field has been shown. The obtained data are the basis for making design decisions when designing the optimal direction of horizontal excavations and the places for laying vertical ones

4. The geomechanical conditions of the South Khingan deposit and the revealed patterns of stress distribution in the elements of development systems give grounds for its attribution to those prone to rockbursts.

5. To conduct mining operations in the rockburst-prone areas of the South Khingan deposit, a set of measures to prevent dangerous manifestations of rock pressure has been justified.

References

1. I.Yu. Rasskazov, Monitoring and management of rock pressure on mines of Far East region. Mountain book. pp. 329. (2008)
2. I.M. Petukhov, P.V. Egorov, B.S. Vinokur, Preventing rockburst in mines. Nedra, (1984)
3. Cook N. G. W. Rockbursts: prediction and control. Institution of Mining and Metallurgu, pp. 1-9. (1984)
4. Salamon M. D. G. Rockbursts: prediction and control. London: Institution of Mining and Metallurgy. pp. 11-36. (1984)
5. G.I. Arkhipov, Far Eastern ferrous metallurgy: iron ore raw materials base and development opportunities. Institute of Mining of FED RAS. (2005)
6. Marcak Henryk, Mutke Grzegorz. J. Seismol. 17 (4). 1139-1148. (2013)
7. Paige E. Snelling, Laurent Godin, Stephen D. McKinnon. International Journal of Rock Mechanics and Mining Sciences. **58.** 166-179. (2013)
8. V.I. Usikov. Tectonics, magmatism and geodynamics of the East of Asia: the 7th Kosygin readings: Materials of the All-Russian Conference. – Khabarovsk city: Institute of tectonics and geophysics of FED RAS. 328-331. (2011)
9. I.Yu. Rasskazov, B.G. Saksin, M.I. Potapchuk, V.I. Usikov. Physical and technical problems of mining. **1.** 13-22. (2014)
10. O.V. Zoteev. Izvestiya vuzov. Mining Magazine **5.** (2003)
11. Instruction on the safe management of rocks in mines and non-ore deposits, objects of construction of underground structures, inclined and dangerous by rockbursts] State Scientific and Technical Center for Safety in Industry of the Gosgortechnadzor of Russia, (2000)
12. I.M. Petukhov, P.V. Egorov, V.Sh. Vinokur, Preventing rockburst in mines. Nedra. pp 231. (1984)
13. Control of rock pressure in tectonically tense massifs. Kolskii Scientific Center. **1.** pp 159. (1996)
14. M.I. Rasskazov, P.A. Anikin, D.S. Migunov, A.A. Tereshkin, A.V. Gladyr, D.I. Tsoi, M.A. Lomov, A.S. Fedianin, E.K. Pul. Innovative trends in the design of mining enterprises: geomechanical support for the design and support of mining operations: collection of research papers of the 8th International scientific and practical conference. St. Petersburg Mining Univ. 95-101. (2017)

THE EVALUATION OF THE INFLUENCE OF TECTONIC STRUCTURE OF NIKOLAYEVSKY DEPOSIT ON ROCK-BURST HAZARD OF ROCK MASS

Aleksandr Sidliar[1*], *Boris* Saksin[1], *Pavel* Anikin[1] and *Mikhail* Lomov[1]

[1]Institute of Mining of Far Eastern Branch of the Russian Academy of Sciences, Khabarovsk, Russia

Abstract. Using the joint analysis of seismoacoustic monitoring data and geodynamic features we have determined the consistent patterns of industrial stress field formation in deep aquifers of Nikolayevsky polymetallic deposit. This analysis allowed us to explain the hypothesis about the mechanism of hazardous geodynamic phenomena in the complex structure rock mass. Results of the research on the geostructural features of the Nikolayevsky scarn-polymetallic deposit allowed us to improve its geodynamic zoning. As the result, we have discovered major heterogeneity of the geological environment of the ore field, which defines its modern structure, as well as features of the strain-stress state of the rock mass (olistolith).

1. Introduction

The problem of rock-bump hazard in the underground mines of Far Eastern region has been relevant for decades. Rock pressure, which is inevitable during underground exploitation of deep aquifers, and high tectonic stress level of the rock mass pose a great threat to worker's lives, interfere with the normal work process during mining operations, thus reducing efficiency of mining production [1, 2]. Solution to the problem of rock pressure management is yet to be discovered, because of, to a large extent, insufficient study of the nature and mechanisms of geodynamic processes and phenomena appearing in the rock mass under the influence of multiple natural and industrial factors.

One of the most hazardous deposits in the Far East is Nikolayevsky polymetallic deposit, located in the East Primorye. It is characterized by its complex geological structure and confinedness to the tectonically active regions of the crust [3].

Monitoring the deposit between 2011 and 2018 registered more than 250 dynamic occurrences of rock pressure in various forms: from pressure burst to rock and rock-tectonic bumps. Thanks to continuous employment of geomechanical monitoring and timely taken stress-relief actions at the Nikolayevsky deposit, the amount of dynamic occurrences in the

[1] Corresponding author: alex-igd@mail.ru

later years has significantly reduced, compared to the period of 2011-2012 (see fig. 1).

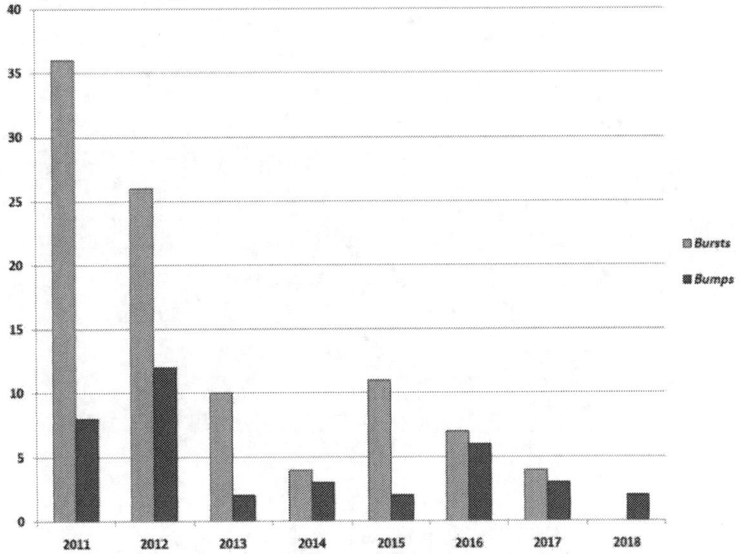

Fig. 1. Distribution of dynamic occurrences of rock pressure (bursts and bumps) at the Nikolayevsky deposit, 2011-2018.

2. Analysis of seismoacoustic monitoring results

Unlike other deposits in the Far East, evaluation and rock-bump hazard control at the Nikolayevsky deposit is conducted by instrumental measurements of seismoacoustic activity of the rock mass with the help of digital geoacoustic monitoring system "Prognoz-ADS" and local control tool "Prognoz-L", as well as systematic visual monitoring of dynamic occurrences of rock pressure in the mines.

"Prognoz-ADS", automated rock pressure control system, which is used to estimate rock-bump hazard level in the region, provides effective collection of acoustic signals, their digitization, processing and transmission by digital communication lines to the central computer, which controls the measuring and computing system [4, 5].

Currently, the monitoring network includes 21 digital receiving transducers (DRT) mounted in the wells within the mines on aquifers -323, -360, -380, -390, -406, -420, -433.

In total, the geoacoustic monitoring system "Prognoz-ADS" has registered more than 17,000 AE events between 2011 and 2018. Retrospective analysis of the results of seismoacoustic monitoring, conducted from 2011 to 2018, showed that most of them were limited to mining sites, as well as fault network, especially the active subvertical fault TN3 (fig. 2). Sites of the mine field at the southern and north-western flanks of the deposit show the most seismoacoustic activity.

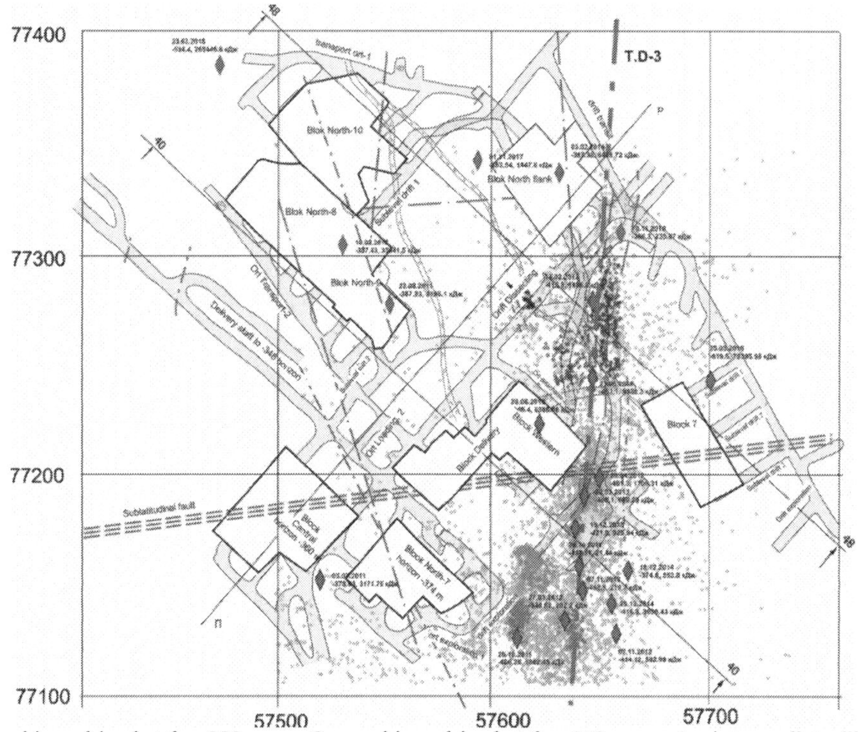

1 – workings drived at the -390 m aq.; 2 – workings drived at the -360 m aq.; 3 – intermediate dike; 4 – stripped space of the stope a) at the -390 m aq.; b) at the -360 m aq.; 5 – blocking sub-lateral fault; 6 – tectonic disturbances a) of the second order; b) other; color shows source zones

Fig. 2. Map of acoustic activity of the Nikolayevsky deposit mass according to seismoacoustic control data in 2011-2018.

Analysis of major dynamic occurrences reveals the similar picture. Between 2011 and 2018 Prognoz-ADS system registered 29 shocks at the deposit. Majority of them was limited to the fault network, especially to the active subvertical submeridional fault TN3 with the accompanying fracture set, specifically at the sites of its interception with subvertical dike bodies of diabase porphyrites, predominantly in the north-west direction (fig. 2). This structure represents the area of increased fracture density up to 50-60 m wide, with scattered sulphide mineralization and separate crushing zones.

It is worth mentioning that most of the shocks registered within the tectonically active TN3 fault initiated as the result of seismic influence, both natural and industrial (after blasting works).

For example, a strong geodynamic event was registered on March 25, 2016, at 12:57 AM near the Vostok-1 ore deposit (block 40). It was accompanied by the powerful acoustic occurrence, shaking of the whole mine field and seismic vibrations recorded in the overground mining complex. As the result of this event, the lining and walls of the mine at the aquifers -375, -390, -406, -420 m were damaged. The volume of destroyed rock mass inrush exceeded 400 m^3. During the first hours after the geodynamic event, acoustic activity of the rock mass more than two times exceeded the average level. As noted earlier [6, 7], the time of the described event matched the time of the earthquake of the 4th magnitude in Kamenka village at the Japanese sea coast. The described geodynamic event resulted in the appearance of open faults and fractures dozens of meters long in the rock mass. There were cases of soil elevation for 10-12 cm along the fault borders, and faults themselves coincide with elements of deposit tectonic structure (fig. 5).

Therefore, at the Nikolayevsky polymetallic deposit, which had the most of hazardous geodynamic events in the recent years, we revealed the pattern for nature-industrial stress field formation, resulting to a large extent from rock mass tectonic structure and complex geometry of the mined-out area. The majority of registered AE events and major dynamic occurrences (shocks) is limited to the fault network, especially to the area of the active subvertical fault T3 with accompanying fracture sets, subvertical dike bodies, predominantly in the north-west direction. Analysis of experimental data allowed us to explain the hypothesis about the mechanism of hazardous geodynamic phenomena in the complex structure rock mass, according to which these events are limited to the areas in boundary state of stress at the borders of tectonic blocks and initiate as the result of a natural or industrial seismic influence.

3. Geodynamic features of the deposit

We contrasted geomechanical monitoring data with the features of Nikolayevsky deposit tectonic structure, where the large complex limestone slab (olistolith) plays a major role. In the area of study we registered zones of abnormal increased seismoacoustic activity of two types: areal (near-isometric) and more local (elongated) (fig. 3).

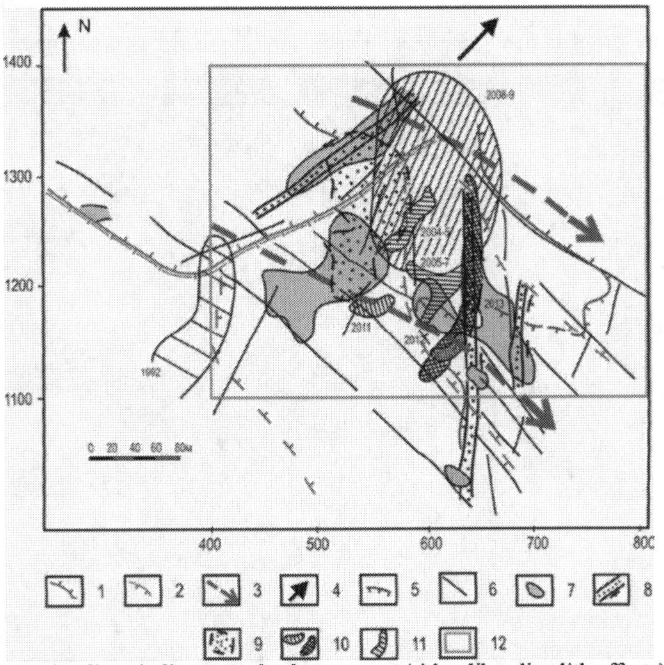

1-2 – axial water-parting lines in limestone back geometry (ridge-like olistolith offsets): 1 – primary, 2 – secondary; 3 – direction of ore body subsidence; 4 – direction of modern horizontal movement of ore hosting geoblock and main horizontal rock mass compression at the aquifer -320 m; 5 – area of sharp drop of absolute marks of limestone olistolith back; 6 – back geometry lineaments (conjectural neotectonic faults); 7-8 – sites of lower metasomatite increased capacity: 7 – isometric; 8 – linear (chasms in limestone bodies filled by metasomatites) and direction of their fall; 9 – location of the lower rim of resurgent feeder for upper metasomatites; 10 – results of acoustic monitoring between 2004 and 2013 (different hatching shows anomalies discovered in the specified years); 11 – region of rock pressure dynamic occurrence registered in 1992; 12 – outline of geoacoustic monitoring

Fig. 3. Interpretation results of the maps showing modern morphology of large limestone slab back, as well as lower metasomatite bodies in the Nikolayevsky deposit

We began to analyze the data by constructing summary olistolith block graph (fig. 4), where we put our predecessors' views [8, 9] on block boundaries, which were based on geological data, and our suggested geodynamical zoning of the studied part of the object (at the modern stage of development), based on different morphological expressiveness of regions with uneven modern geodynamic condition.

Moreover, the figure shows lithological irregularities which, as monitoring process showed, influence the distribution of seismoacoustic abnormalities.

Morphology of olistolith outline on the graph leads to a conclusion that it is defined by faults of the following directions (starting from the most important): submeridional, north-west, and north-east. This is confirmed by the results of geological prospecting work, according to which the olistolith body is split by a series of faults that are a part of the main meridional structure of the ore field, secant in relation to faults in other directions. Three blocks of the first order are formed by the intersection of sub-lateral fault and north-western tectonic zone (see fig. 4). They are split into smaller blocks by faults in other directions. Most of the faults are subvertical. Comparison of fig. 3 and 4 shows that there is no distinct connection between the abnormal zones discovered by monitoring and blocks of the I order. We can only note the confinedness of lateral-oriented abnormal zones to sub-lateral fault separating Central and Northern blocks. At times seismoacoustic anomaly is controlled by the faults of another rank (for example, TN3 and other smaller faults in north-west direction), which are most likely limiting blocks of a higher order (not shown on the graph).

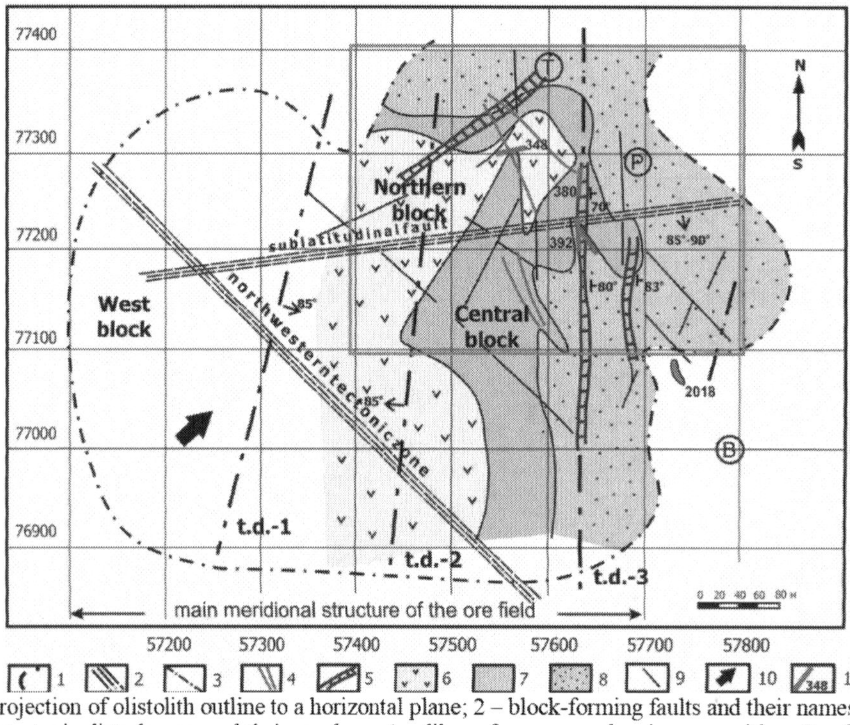

1 – projection of olistolith outline to a horizontal plane; 2 – block-forming faults and their names; 3 – other tectonic disturbances and their numbers; 4 – dikes of average and main composition; 5 – chasms in limestone bodies filled by metasomatites; 6-8 – geodynamical regions formed by olistolith morphological features and modern stress field: 6 – relatively stable geodynamic region; 7 – slightly sloping region with intermediate geodynamic activity; 8 – steep lateral geodynamically active region; 9 – olistolith back geometry lineaments; 10 – direction of the main horizontal rock mass compression; 11 – location of fractures and cracks created by the earthquake on 03.25.2016.

Fig. 4. Ore hosting limestone olistolith body block structure of the Nikolayevsky deposit and its modern geodynamic zoning

Consistent patterns of interaction between natural and industrial systems can be indicated more distinctly by the joint analysis of results of local monitoring and complex research aimed at discovering currently active block-tectonic forms. As you can see on fig. 4, location of abnormal areas is controlled by the intersection of primary and secondary ridge-like olistolith body offsets. Within the studied area the main water-parting line is of north-west direction, along the modern horizontal compression, and the secondary ones are oriented perpendicular to it. The largest areal anomaly, discovered in 2008-2009, is located at the commissure of the main water-parting line with two secondary ones, and also at the joint of divergent zones, where the absolute marks of olistolith back drop sharply, and where the vertical channel for upper metasomatites is fixed. Therefore, areal anomalies conceal the most stressed areas of a relatively stable geodynamic olistolith region.

In the south-west direction from the main water-parting line we first encounter local anomalies mainly of north-east, less often north-west and sub-lateral orientation that are likely to be limited to the crevasses in olistolith body, which appear under the influence of mining works along the direction of primary and secondary water-parting line orientations. Incidence rate of linear local anomalies increases in the regions with intermediate geodynamical activity and in active lateral ones. The latter ones are characterized by occurrence of elongated linear local anomalies limited to fault zones in meridional direction. These crevasses in olistolith body are formed during premineral and mineral stages of the geological development of the region, because they are cured by subvertical bodies of wallrock metasomatites. These crevasses are distinguished by the fact that acoustic emmission anomalies in them are registered even at significant depths. The lateral olistolith region, currenlty geodynamically active, is also characterized by the presence of chasms in its back geography, cured wallrock metasomatites and increasing degree of its absolute mark drop. All of the above indicates the increased small-block nature of olistolith body structure towards the lateral area (subvertical eastern olistolith rim), which determines increased rock-bump hazard in the south-west direction occurring downward the mine.

Therefore, geodynamics of the territory of the region results from confinedness to the concealed deep submeridional fault, shifts in which defined the elements of the tectonic structure of the deposit. It is characterized by its block composition, main elements of which are subvertical sub-lateral fault and north-west tectonic zone, breaking the ore field into three main structural blocks: central, northern and western. Within the boundaries of the deposit we can also see elongated subvertical ruptural post-mineral disturbances in the submeridional direction (fig. 4) [10].

Under these conditions we need more thorough and complex geodynamic research, including the following: identifying the consistent patterns of current strain-stress state formation of the rock mass and evaluation of geodynamical and seismic conditions based on it, as well as the strain-stress state of the rock mass; investigating the role of natural and industrial influence in conjunction with strain-stress state of the rock mass in order to lay the basis for measures for managing rock pressure and decreasing geodynamic risks.

4. Conclusions

1. We revealed the features of acoustic activity occurrence in the rock mass of Nikolayevsky deposit at the stage of formation of potentially rock-bump hazardous zones. The majority of the registered AE events was limited to the fault network, especially to the active subvertical submeridional fault TN3 with the accompanying fracture set, specifically at the sites of its interception with subvertical dike bodies, predominantly in the north-west direction. We revealed the correlation between the dynamic occurrences of rock pressure and seismic waves from natural (earthquake) and industrial (technological blasts) sources.

Seismic waves act as a trigger and initiate destruction in the dynamic form of rock mass sites in boundary state of stress.

2. After studying the conditions and reasons for minihg tectonic impact on March 25, 2016, at the Nikolayevsky polymetallic deposit, which caused sizeable damage to the mines at various aquifers, appearance of faults and fractures dozens of meters long with cases of soil elevation for 10-12 cm, we established that dynamic occurrences of rock pressure and industrial seismicity at the deposit is determined, to a large extent, by complex geodynamic processes in form of realignment and self-organizing of the mass in nature-industrial stress field, which are accompanied by various displacement along the tectonic disturbances of different scale.

3. The main irregularity of the geological environment of the ore field, which defines its modern structure and features of strain-stress state of the rock mass, is a large inclined stratified limestone slab (olistolith). Any displacement in this rigid slab results in activation of previously formed structures, and morphology of its back highlights the occurring movement and determines strain-stress state of various parts of the plate. The results of studying geostructural features of Nikolayevsky scarn-polymetallic deposit helped us specify the graph of its geodynamic zoning, including the construction of summary limestone olistolith block graph, which allows us to give more reliable interpretation to seismoacoustic monitoring data highlighting the geomechanical processes in the exploited rock mass and determine potentially rock-blast hazardous areas.

References

1. M. Henryk, G.Z. Mutke. Journal of Seismology. **17(4)**. 1139–1148. (2013).
2. P.E. Snelling, L. Godin, S.D. McKinnon. International Journal of Rock Mechanics and Mining Sciences. **58**. 166-179. (2013).
3. I.Iu. Rasskazov, B.G. Saksin, V.A. Petrov, B.F. Shevchenko, V.I. Usikov, G.Z. Gil'manova. Izvestiya, Physics of the Solid Earth. **50 (3).** 444–452. (2014).
4. I.Iu. Rasskazov, M.I. Potapchuk, G.A. Kursakin, Iu.I. Bolotin, A.V. Sidliar. M.I. Rasskazov. Mining information-analytical bulletin. **4.** 96-102. (2012).
5. M.I. Potapchuk, I.Iu. Rasskazov, G.M. Potapchuk, A.V. Sidliar. M. I. Rasskazov. Problems of complex exploration of geological resources: materials of the Fourth All-Russian scientific conference with foreign participants (Khabarovsk, September 27-29, 2011). In 2 vol. – Khabarovsk city, "Institute of Mining" of the FED RAS. 1. 110-117. (2011)
6. I.Iu. Rasskazov, B.G. Saksin, V.I. Usikov, M.I. Potapchuk. Mining Magazine. **12.** 13-19 (2016)
7. B.F. Shevchenko, B.G. Saksin, M.I. Potapchuk. MIAB. **S21**. 182-191. (2016)
8. B.G. Saksin. I.Iu. Rasskazov // Problems of complex exploration of geological resources: materials of the Fourth All-Russian scientific conference with foreign participants (Khabarovsk, September 27-29, 2011). – Khabarovsk: "Institute of Mining" of the FED RAS. **1**. 125-130. (2011)
9. L.I. Sosnovskii. L.I. Sosnovskii // Proceedings of Irkutsk State Technical University. 4. 32-33. (2006)
10. K.G. Levi. Irkutsk: IEC SB RAS. (2007)

Determination of the distortions introduced by the compression algorithm in the results of the measurement of the local impact control device "Prognoz-L".

Dmitry Migunov[1*] and *Alexey* Grunin[2]

[1]Mining Institute of Far Easten Branch of Russian Academy of Sciences, Khabarovsk, Russia
[2]"Pacific National University", Khabarovsk, Russia

Abstract. The article discusses the application of the u-law algorithm from the G.711 standard, which performs lossy compression. A theoretical error introduced by this algorithm into discrete samples of digitized acoustic emission signals is determined. The effect of compression on signal parameters was also determined. During the analysis, the data from the local "Prognoz-L" device were used. "Prognoz-L" performs rockburst hazard assessment. The data were processed in an automatic mode to isolate pulsed signals containing acoustic emission. The processing of the obtained sample of pulse signals made it possible to determine the basic statistical characteristics of the errors introduced by the compression algorithm into the values of the signal parameters. The possibility of using this algorithm for preliminary processing of acoustic emission signals is established with the aim of keeping rock pressure monitoring in low-power acoustic-emission devices.

1 Introduction

An effective method for predicting dangerous dynamic manifestations of rock pressure in a mining facility is the geoacoustic method. This method has a high resolution (in comparison with the microseismic method) and allows to observe the evolution of the destruction points formation process directly from the initial stages [1]. The method consists in recording with piezo accelerometers (sensors) acoustic emission signals in a rock mass in the frequency range 1-20 kHz and a dynamic range of more than 80 dB. This frequency range of signals has a large attenuation coefficient, so that to monitor the entire volume of mining requires either a large number of sensors spaced a distance of several kilometers or sequential monitoring by one sensor in the most dangerous places. In this case, the amount of data received from one sensor at a sampling frequency of 40 kHz (twice the maximum signal frequency) and the resolution of the ADC of 16 bits to provide a dynamic range of more than 80 dB will be 78 kb/s or more than 6.4 GB/day. To store such large amount of data on local low-consumption devices for continuous monitoring of

* Corresponding author: dimisi@mail.ru

rockburst hazard is a big problem. To reduce the amount of data, there are various methods, for example, saving data containing only acoustic emission or preserving only the parameters of signals without saving the waveform itself [2]. However, these methods exclude careful processing of the entire volume of received data using new algorithms for the isolation of acoustic emission signals. One way to solve this problem is to compress the data stream before saving. In this paper, we consider the method of data companding by the G.711 u-law algorithm and its effect on the parameters of acoustic emission signals.

2 Determination of theoretical error

G.711 is the standard containing algorithms for companding audio data, mainly used in telephony. The compression ratio can vary depending on the problem being solved. However, the compression ratio 2 described in the standard is most often used, as it has the most optimal relationship between the compression ratio and the insertion distortions. There are two main algorithms u-law and a-law in the standard. Both algorithms are logarithmic, but u-law is optimized for hardware implementation. Data compression using the u-law algorithm is performed according to the following equation [3]:

$$y = sign(x) \cdot \frac{Ln(1 + \mu|x|)}{Ln(1 + \mu)}, \mu = 255 \quad (1)$$

where x – the initial (input) value of the sample signal;
y – the corresponding μ-type sample.

The formula (1) is not directly applied because of the high resource intensity of calculating the logarithm. To reduce the resource intensity in the standard, a special approximation is presented, which can be easily calculated by hardware [4].

This approximation ensures the calculation of the μ-type sample using only 128 bytes of RAM, 7 logical operations, 2 addition operations and 1 conditional operator. Such a low resource capacity allows applying the algorithm in energy-efficient devices based on 8-bit microcontrollers without digital signal processing modules with battery power supply or with limited power supply through long communication lines. For example, using the low-consumption STMicroelectronics processor STM32F051 [5] and the clock frequency of 2 MHz, the algorithm allows processing up to 55 thousand samples per second, which corresponds to the input bandwidth of 22 kHz, while the power consumption is only 4 mW.

The drawbacks of this method of compression include the loss of data, which causes distortion of the reconstructed waveform. Figure 2 shows a theoretical error over the entire range of input linear samples.

Fige 1 shows that the algorithm error for most initial sample values does not exceed 5%, but in the region of small values it can reach 100%. Such a high error value is not something that is not acceptable since its contribution to the total error budget is much less than the contribution of noise introduced by the ADC (for example) in this range of values. As an example, we take the popular low-power ADC, designed for battery-powered devices AD7988-5BRMZ from Analog Devices [6]. This ADC has a bit capacity of 16 bits and at a reference voltage of 5 V provides a minimum noise. The maximum deviation of the measurement result is 4 least significant bits (LSB). Substituting this value in the formula 2 where (x-x') = 4 for the whole range of values, we get an estimate of the maximum noise effect of the ADC on the error of the measurement result. The combination of the error graphs introduced by the ADC and the companding algorithm, Fig. 2, shows that the contribution of the companding algorithm in the region of small values is less by a factor of 4.

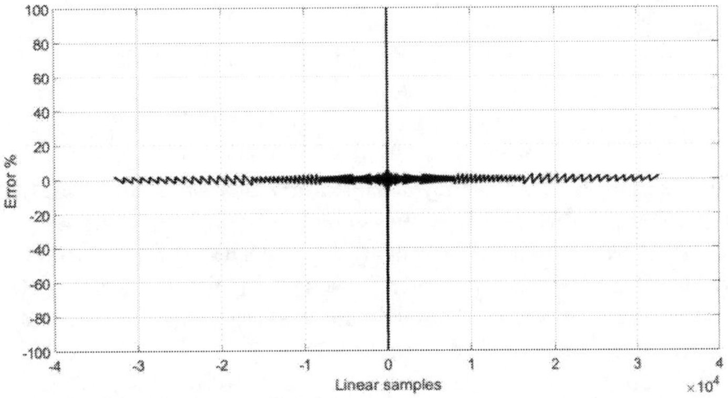

Fig. 1. Theoretical value of companding error.

Error introduced by the algorithm is calculated using the formula:

$$err = \frac{x - x'}{x} * 100\%, \qquad for\ x \neq 0 \tag{2}$$

where x – the initial (input) value of the sample signal;
err - error of the value obtained after sequential applying of the forward and reverse companding algorithm;
x' - value of the sample signal after sequential applying of the forward and reverse companding algorithm.

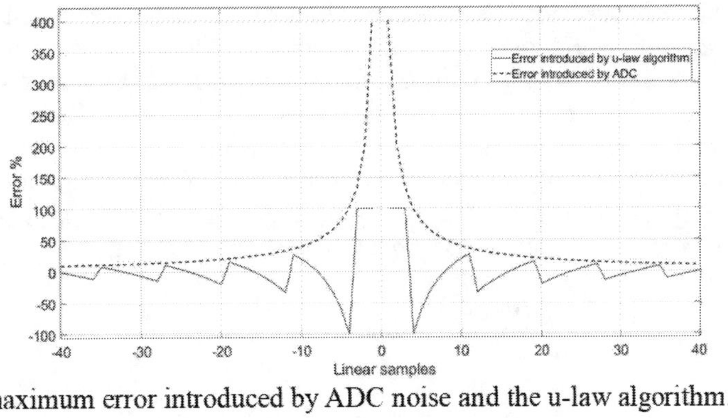

Fig. 2. The maximum error introduced by ADC noise and the u-law algorithm.

3 Algorithm influence on the macro parameters of acoustic emission

As was shown above, the u-law algorithm influences the shape of the acoustic emission signal. However, only the signal waveform in practice is not used when evaluating the impact parameters. The shape of the original wave changes significantly both during propagation through the material and during transformation by the sensor, so the signal coming from the sensor is very remotely similar to the original signal [7]. The shape changes lead to the fact that the correlation parameters of the received signal with certain fracture characteristics are used to estimate the fracture parameters. These parameters of the final acoustic emission signal are called macro parameters. The most widespread macro parameters are [8]: the time of the first crossing of the threshold level, the count of threshold crossings, the time to the amplitude peak, peak amplitude, signal energy, signal duration, signal strength (area under the MARSE envelope), average signal frequency, signal peak frequency, the frequency centroid of the pulse, the RMS value, the average signal level. In the presented work, the following set of macro parameters was chosen: maximum amplitude, signal energy, signal strength (MARSE), RMS amplitude value, fundamental frequency and frequency centroid signal. The choice of these macro parameters was based on how often they are used in various acoustic-emission monitoring systems of rock mass, as well as the maximum influence on them of the distortions introduced into the waveform by the companding algorithm. To determine the effect of waveform distortion on the macro parameters, statistical methods were used over a large sample of signals recorded at various mines in the Russian Federation. The registration was carried out with the help of the local rockburst monitoring device "Prognoz-L" [9].

As a result of field experiments using this device, more than 2.2 GB (6 hours) of data were obtained. The data is recorded in files in the "wav" format with a duration of 10 minutes, the recording was made at various mines with different intensity of acoustic emission pulses. Later on, only data related to acoustic emission signals were extracted from the array of files with measurement results. To isolate these signals, the automatic processing algorithm STA / LTA was used [10]. As a result, 5338 fragments containing acoustic emission signals and a small number of drill noise signals were isolated.

In the future, for each signal from the 5338 samples, the value of the maximum amplitude was calculated. In general, the sample has the following parameters: the maximum amplitude of the weakest signal was 22 units of the least significant digit of the input ADC (LSB), the maximum value of the amplitude of the strongest signal was 32767 LSB. The largest number of signals (3976) has a maximum amplitude of less than 327 LSB. That also indicates that all samples of these signals also have an amplitude not exceeding 327 LSB. This range in accordance with Figure 2 is one of the cases in which the u-law algorithm introduces the maximum distortion. Thus, by calculating the error introduced into the macro parameters by the companding algorithm on the signals from this range of amplitudes, we obtain its maximum estimate. The method used to determine the statistical characteristics of errors introduced by the algorithm into macro parameters consisted in the following. The initial values of macro parameters were calculated for the acoustic emission signals extracted from the total volume of data. The calculation of macro parameters was performed according to the following formulas [11]:

$$A = \max_{i=1}^{n}(|x_i|) \tag{3}$$

$$W = \sum_{i=1}^{n}(x_i^2) \tag{4}$$

$$S = \sum_{i=1}^{n}|x_i| \tag{5}$$

$$RMS = \frac{1}{n} \times \sum_{i=1}^{n}(x_i^2) \tag{6}$$

$$\bar{f} = \frac{\sum_{i=1}^{k}(f_i \times \omega_i)}{\sum_{i=1}^{k}(\omega_i)} \tag{7}$$

where, A- maximum amplitude; W- signal energy; S- signal strength (MARSE); RMS- RMS amplitude value; $\bar{f}$ - frequency centroid signal; x – is the discrete value of the signal amplitude; n – is the number of signal points; f – is the discrete value of the harmonic amplitude of the signal; k – is the number of harmonics of the signal in the frequency domain, ω – is the discrete value of the harmonic frequency of the signal.

The forward and backward companding algorithm was applied to the original signals successively. For the signals processed in this way, the macro parameters were recomputed. Formula 2 uses the values of the macro parameters before and after applying the algorithm and calculates the errors introduced. Based on the error values of each macro parameter, their probability density function, mathematical expectation, minimum, maximum value and mean square value, as well as the interval in which most of its values are located, were calculated.

As a result of calculation according to the algorithm proposed above, statistical error estimates for each macro parameter were obtained for all 5338 signals. The graphs of the probability density function are shown in Figure 3.

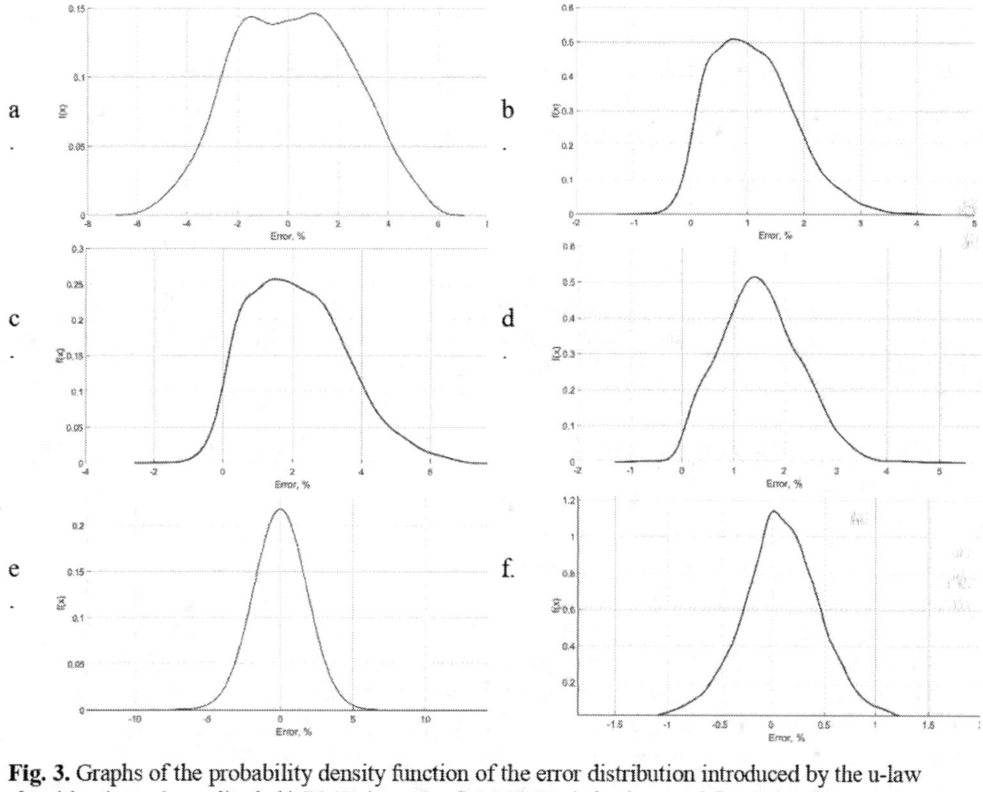

Fig. 3. Graphs of the probability density function of the error distribution introduced by the u-law algorithm into: a) amplitude b) RMS c) energy d) MARSE e) fundamental frequency f) spectral centroid of acoustic emission signal realizations.

From the graphs obtained, only the error distribution of the frequency of the fundamental harmonic has the normal distribution form. Therefore, it is not permissible to apply the rule of three sigma to estimate the interval in which in which most of values are located. In order to simplify the calculations, the calculation of the boundaries of these intervals was made numerically based on the probability density graphs. An interval was chosen in which the error values are located with a probability of 0.993. This probability is based on an analysis of the average number of signals in one "wav" file and the assumption that in an average file, no more than one signal should have an error beyond the computed interval. Statistical estimates of errors in the calculation of macro parameters are presented in Table 1.

Table 1.Statistical estimates of errors introduced into macro parameters by the u-law algorithm.

Macro parameter	Minimum error value, %	Maximum error value, %	M(x)	Standard deviation	The interval of the most probable error values with a probability of 0.993
Maximum amplitude	-12,4	12,5	0,3	2,4	[-5,7 ; 6,9]
RMS amplitude value	-0,8	16,9	1,16	1,03	[-0,3 ; 3,4]
Signal energy	-1,7	31	2,3	2	[-1,8 ; 8]
Signal strength (MARSE)	-22,4	5	1,4	1,6	[-0,6 ; 4,1]
Fundamental frequency	-86,2	40,6	0	1,97	6δ*
Frequency centroid signal	-6,8	1,85	0	0,62	[-2,15 ; 2]

*- for the normal distribution, the probability of falling into the interval 6δ is 0.9973.

Thus, to reduce the error in calculating the integral macro parameters using data processed by the u-law algorithm, it is necessary to add an offset of -1.55% for the RMS amplitude value, -3.1% for the signal energy, -1.75% for MARSE function. The intervals of the most probable error value (probability 0.993) are [-1.85; 1.85] [-4.9; 4.9] [-2.35; 2.35], respectively.

4 The contribution of the u-law algorithm distortions to the total error budget

The distribution characteristics of the maximum amplitude error generally agree with the analytical calculations shown in section 2. The distribution has zero expectation and the interval of the most probable error value from -5.7 to 6.9 percent. This interval has the maximum spread among the errors introduced by the algorithm into macro parameters. If we take the maximum error introduced by the algorithm in 6.9%, then before the amplitude value of the 58 LSB signal is reached, errors introduced by the noise of the ADC will prevail in the error budget. However, after overcoming the amplitude of the signal of this magnitude, the main contribution to the distortion of the parameters will be due to errors in the u-law algorithm. To determine the criticality of the error introduced by the companding algorithm in signals with an amplitude of more than 58 LSB, it is necessary to estimate the practical limits of the error in determining the macro parameters in problems of controlling rock pressure by the acoustic-emission method. Consideration of this issue is beyond the scope of this article and requires a separate study. However, if we consider such a parameter of polycrystalline rocks as the absorption of acoustic vibrations with a wavelength much greater than the average crystal size, it can be noted that even for different types of granites this value can differ by 15%. [12]. At the same time, in the conditions of a mining enterprise, the acoustic emission signal can pass from the source to the recorder through a variety of rocks with an even wider spread of the values of this coefficient. Therefore, depending on the location of the receiving sensor, even if the constant distance from the source (the point of destruction) is maintained, the amplitude of the received signals can differ by tens of percent, which is much larger than the error introduced by the companding algorithm.

5 Conclusions

The results of the research allow us to conclude that the application of the u-law algorithm for preliminary processing of geoacoustic data is justified. At a compression ratio of 2 and a low resource intensity, it contributes only a small contribution to the overall budget of errors in determining the parameters of the initial pulses of acoustic emission. The companding algorithm described above could be used to upgrade the "Prognoz-L" local control device in order to increase the autonomy of its operation and increase the number of stored measurement results.

References

1. I.Yu.Rasskazov, P.A. Anikin, A.Yu. Iskra, D.S. Migunov, ISSN 0236-1493,Rezul'taty geoakusticheskogo kontrolya udaroopasnosti na rudnikakh Dal'nego Vostoka (Results of geoacoustic impact control in the mines of the Far East), **7**, 104-111. (2008).
2. I.Yu.Rasskazov, G.A. Kalinov, K.O. Kharitonov, D.A. Kulikov, D. S. Migunov, ISSN 0236-1493, Sovershenstvovanie tekhnicheskikh i programmno-metodicheskikh sredstv geoakusticheskogo monitoringa udaroopasnogo massiva gornykh porod (Perfection of technical and program-methodical means of geoacoustic monitoring),**6**, 119-125, (2007)
3. Douglas Lyon, Ph.D, JOURNAL OF OBJECT TECHNOLOGY, The u-law CODEC, **7**, 17-31, (November-December 2008).
4. ITU-T Recommendation G.711, PULSE CODE MODULATION (PCM) OF VOICE FREQUENCIES.
5. STM32F051. Datasheet - production data (Jan-2017).
6. AD7988-1 Datasheet - production data Patent 6,703,961 (July-2017).
7. Dr. Pollock A. Physical Acoustics Corporation (PAC), METALS HANDBOOK, **17**, (1989).
8. GOST 27655-88 (Russian state standard), Akusticheskaya emissiya. Terminy, opredeleniya i oboznacheniya (Acoustic emission. Terms, definitions and notation), (1988).
9. I. Yu. Rasskazov, D. S. Migunov, P. A. Anikin, A. V, Gladyr', A. A. Tereshkin, D. O. Zhelnin, Journal of Mining Science, New-Generation portable geoacoustic instrument for rockburst hazard assessment, **51**, 614–623, (2015).
10. W.F. Freiberger, Quarterly Appl. Math, An approximate method in signal detection, **20**, 373-378, (1963).
11. S. A. Bekher, ISBN 978-5-93461-613-8, Osnovy nerazrushayushchego kontrolya metodom akusticheskoy emissii(Basics of non-destructive testing by acoustic emission method), (2013).
12. O.I. Babikov, Moskva: Gosudarstvennoe izdatel'stvo fiziko-matematicheskoy litterature, Ul'trazvuk i ego primenenie v promyshlennosti (Ultrasound and its application in industry), (1958)

Evaluation of geomechanical conditions at Magnezitovaya mine when undermining natural and anthropogenic objects

Roman Krinitsyn[1*], *Arkady* Avdeev[1] and *Sergei* Khudyakov[1]

[1]Institute of Mining of the Ural Division of Russian Academy of Sciences. Yekaterinburg, Russia.

Abstract: Development of Satka magnesite deposit is carried out in complicated geomechanical conditions, in particular, mining works are conducted under the Bolshaya Satka river, in the zone of influence of the existing and developed quarries. Underground workings undermine a number of surface objects of industrial and civil purpose: railways, bridges, buildings and structures. There was an urgent need to conduct special studies of geomechanical conditions at the mine in order to assess the feasibility of working off part of the safety pillars located under the railway tracks of "Russian Railways", PJSC and the river bed. In the process of research, the main physical and mechanical properties of rocks have been identified, and the initial natural stresses of the rock mass have been measured. An increased level of horizontal initial stresses in the zone of safety pillars at the horizons + 180. + 260 m was detected. Based on the results of the research, a conclusion was made on the need to develop some engineering measures to ensure the safety of underground works. It was recommended, to ensure the stability of the rocks massif, to apply the option of extracting the treatment chambers with the formation of the ceiling of the vaulted form. It is proposed to monitor deformation processes in mine workings and on the daylight surface. To this end, the installation of an automated control system for rock pressure is recommended. It is necessary to provide for the establishment of long-term observation stations for controlling rock pressure and displacement. According to the results of the research, "Guidelines for the safe control of mining operations at bump-hazardous Satka magnesite deposit" have been developed in accordance with the requirements of the Federal norms and regulations in industrial safety. **Key words:** physical and mechanical properties of rocks, initial stresses of rock massifs, full-scale measurements, laboratory tests, slit discharge, safety pillars, undermining of industrial and civil objects.

1 Introduction

Satka deposit of magnesites (the Southern Urals) is represented by a series of ore bodies of stratum and lenticular form, along a strike length of 1.3-3.6 km, along a dip of 100-500 m. The azimuth of the strike of ore bodies is $45 \div 55^0$, the angles of incidence southeastwards

* Corresponding author: roman_krinicyn@mail.ru

vary within 10÷850^0 averaging 30÷45^0, the normal thickness of ore bodies being in the range of 8÷30 m.

The lithological section of the deposit is represented by a complex of stable sedimentary and metamorphosed rocks: magnesites, dolomites and shales of coaly-carbonaceous composition. Rocks of the diabase type, confined to the dyke veins of a steep fall with a thickness of up to 20 cm, have a minor distribution. Magnesite deposits are sometimes complicated by karst cavities. Active faults of the northeastern strike (azimuth 60^0) are low-amplitude faults.

In general, the entire sedimentary stratum is disintegrated by single systems of cracks of two types. The first type includes multidirectional cracks with vertical or close to them incidence angles up to the first tens of meters, often without aggregate, sometimes uncovered. The second type includes interstitial cracks located in magnesites strictly parallel to their bedding and characterized by a considerable extent along the strike and dip of ore bodies. The angles of incidence of cracks in the second group vary from 15÷20^0 to 45÷50^0, clearly repeating all the bends of the ore bodies.

The main development system for the extraction of ore at the deposit is currently a chamber-pillar system with dry backfilling of the worked-out space. The length of the mining block along the strike of the ore body at horizons +340 m, +260 m, +180 m with the chamber-pillar development system, taking into account the inter-block pillars is 240 m. The parameters of the chambers and pillars during the work-out of the treatment blocks, according to the recommendations of Institute of Mining of the Ural Division of Russian Academy of Sciences (IM UD RAS), have undergone several changes. For example, the width of the cleaning chambers increased from 8 m to 12-13 m, which led to an increase in the completeness of the excavation of magnesite and, consequently, to a decrease in losses and dilution of ore.

Since 2016, preparatory work is underway for the mining of the mine field by a sub-floor-chamber system with hardening backfilling of the worked-out space. The magnesite will be produced at horizon +100, and at horizon +180 the extension of the block No. 3 will be finalized with simultaneous preparation of the blocks No. 4, 5, 6 [1].

In the process of penetrating the buggy gangway and associated mine workings at horizon +180 m, some manifestations of rock pressure in the form of peeling and shooting with sharp sound and flying up to 3 m of slab-shaped pieces of rock with a transverse dimension of 100-300 mm and a thickness of 10-20 mm were noted. Given the tendency of the deposit's rocks to brittle fracture, the strength and elastic properties, the presence of high stresses in the rock massif, and on the basis of the calculated forecast of a bump hazard by the rock pressure laboratory, Satka deposit was classified as bump-hazardous, beginning from horizon +180 m (depth from the surface 250 m) and below. To monitor the stability of rocks IM UD RAS installed observation stations at the mine, consisting of deep frames, profile lines, photoelastic sensors. To automatically control the deformations of the rock massif, "Massiv 2" station was used. Continuous authors' supervision of the parameters of chambers and pillars during the cleaning of the treatment block was exercised, as a result of which the blocks ODB No.1, No.2 and No.3 were worked out without emergency situations [2, 4].

A significant part of the reserves of minefield I of Magnezitovaya mine is located under a number of protected surface infrastructure: tracks of "Russian Railways", PJSC and departmental tracks, the Bolshaya Satka river, railway and automobile bridges, industrial buildings and facilities, as well as private single-floor residential buildings. At present, the management of the mine plans to develop parts of the safety pillars located under the Bolshaya Satka river and the tracks of "Russian Railways", PJSC. In these areas, in addition to potential bump hazard, there is a risk of negative development of deformation processes, which can affect the stability of the daylight surface (Fig. 1).

- border of hazardous subsidence zone;
- projection of ore bodies of the horizon +100 m.

Fig. 1. Plan of the industrial site of Satka deposit

Therefore, there appeared an urgent need to conduct special studies of the geomechanical conditions of mine field I of Magnezitovaya mine in order to assess the possibility of a partial development of safety pilars located under the railways of "Russian Railways", PJSC and the river bed.

2 Methodology

In the course of the research work, instrumental measurements were carried out by the method of slot discharge, according to the methodology of the IM UD RAS, of the initial stresses of the rock massif near the location of blocks No. 4 and 5 at horizons +260 m and +180 m.

In places of full-scale measurements of the rock mass, samples of the main lithological varieties were taken, prepared and tested for strength properties. A total of six samples of the main lithological varieties from the operational blocks No. 4, and 6 were selected; samples being core and lump. In the course of the research, thin sections were made, through which petrographic diagnostics of the tested rocks was carried out in order to identify the possible dependence of the physical and mechanical properties of rocks on their petrographic composition, structure and texture.

In order to develop some engineering measures to ensure the safety of underground operations, mathematical modeling of anthropogenic stresses in the roof of chambers of various configurations by the method of finite elements was made.

3 Results

According to a petrographic analysis of rock samples, it was established that the samples under study are calcareous dolomite, magnesite, diorite. The main minerals are dolomite, magnesite, amphibole, calcite, plagioclase, epidote, pyroxene, quartz. Hydrothermal mineralization is represented by two samples: dolomite with hydrodermal vein and hydrothermally altered diorite. Hydrothermal mineralization strengthens the original rocks. Four samples are not affected by hydrothermal processes, the crystal sizes in these samples reach 2-6 mm, which significantly reduces the strength properties of the rocks. Lenticular, non-uniform and layered textures of the majority of the presented samples characterize the anisotropy of strength properties of rocks.

According to the results of finding of the strength properties of rocks, it has been established that in the naturally dry state the tensile strength at compression of the dolomite is 120 MPa, in the water-saturated state - 118 MPa (Table 1). For limestone, these figures are 131 and 96 MPa, respectively, for magnesite - 56 and 47 MPa; for diorite - 137 and 126 MPa. The tensile strength of rocks varies from 5 to 35 MPa for dry rocks and 4-29 MPa for water-saturated rocks. It is generally seen that rocks with water saturation reduce the strength properties slightly, i.e. are waterproof. The rock strength varies widely from 4 to 14, averaging 7-8. The average clutch of rocks is 19 MPa, the angle of internal friction is 27-34 degrees.

It should be noted that earlier, when developing a project for the use of a deposit (in the 1990s), according to the data of detailed exploration, significantly higher strength parameters of rocks were established: for example, for dolomite 190 MPa for compression, for magnesite 110 MPa. Thus, the strength of magnesite in the investigated sites is half that of dolomite adopted for the project (1.6 times).

Taking into account the coefficient of structural weakening of rocks in the massif (0.4-0.5), the strength of the massif on the sections under study for compression will be 17 ... 55 MPa, averaging 39 MPa; for a tension of 1.5..14 MPa, averaging 5 MPa.

Table 1. Physical and mechanical properties of rocks

Rock name	Air-dried basis					Water-saturated basis				
	σ_{comp}	σ_p	f	φ	C	σ_{comp}	σ_p	f	φ	C
Samples made from core material										
Mica limestone	131.0	6.7	7	-	18	96.4	6.4	6	-	16
Dolomite	106.6	5.6	6	-	17	120.8	5.8	7	-	16
Magnesite	44.4	4.8	4	36	8	42.4	3.7	4	38	8
Samples made from lumps										
Magnesite	67.7	10.5	6	37	13	41.1	6.5	6	22	12
Dolomite	132.2	20.3	10	35	24	115.9	16.4	10	30	25
Diorite	137.3	35.4	14	26	34	126.2	29.4	13	20	38
Note: σ_{comp} - compressive strength, MPa; σ_p - ultimate tensile strength, MPa; f - rock strength according to M.M. Protodiakonov; C - cohesion, MPa; φ - angle of internal friction, deg.										

To estimate the initial stressed state of the rock massif on the investigated areas, processed under the Bolshaya Satka river and railway tracks connecting Bacal city and Berdyaus junction station, in-situ measurements were carried out by the method of slit discharge [3].

Measurements were made at horizons +260 and +180 m in the area of blocks No. 4 and 5. As a result of measurements at horizon + 260 m, the stresses in the massif were: vertical - minus 14.2 MPa, horizontal along the strike of the ore body - minus 23.3 MPa, horizontal crosswise the strike of the ore body is minus 24.5 MPa (the minus sign means compression of rocks). Stresses are gravitational-tectonic. The ratio of the lateral extension is 1.64-1.73.

At horizon +180 m the initial stresses reach the values: vertical - minus 29.8 MPa, along the strike - minus 48.6 MPa crosswise the strike - minus 51.6 MPa.

Such a sharp increase in voltages in 1.3 – 2.1 times on almost adjacent floors can be explained by the fact that the sites of the installation of rock pressure observation stations pass near the spent and filled up Gologorsky quarry, and the workings of horizon +180 m, moreover, are located almost under the bottom this quarry, i.e. are under the influence of increased concentrations of stresses of anthropogenic nature.

It should be noted that in 2006 [2] the specialists of IM UD RAS found that the dry filling applied at Magnezitovaya mine practically does not affect the bearing capacity of the pillars, i.e. lateral pushback is small. The same can be said about the effect of filling the quarry on its contouring workings. Thus, backfilling of a quarry with empty rocks does not reduce the concentration of man-made stresses on the contour. Workings, passable directly near the bottom of the spent quarry, will thus be subject to increased rock pressure and be unstable, especially given the permissible stresses in the massif (55 MPa).

Therefore, when working off the areas located near the contour of the spent quarry, it is necessary to pay close attention to the compliance with the design constructive parameters of the cleaning blocks, to constantly monitor the stress-strain state and deformation processes, and to develop additional measures to ensure the safety of mining operations.

In order to develop technical measures to ensure the safety of underground work, mathematical modeling of anthropogenic stresses in the structural elements of the systems using the finite element method was additionally carried out.

Modeling the roof of the cleaning chambers of various shapes for the variants of the working parameters showed that when forming the overhead roof of the chambers, the rock mass at the base of the vent ort will be destroyed (Fig. 2a). In the rock massif located at the base of the vent ort left and right there will be tensile stresses reaching the value plus 1. .. plus 2 MPa, this will lead to a partial collapse. The use of a polymeric anchor support for fixing these places will be of little effect since these bars work well for stretching, and for cutting they are less effective.

The vaulted roof of the chamber (Fig. 2b) will be more stable. When modeling the vaulted form of roof of the chamber, it was established that the vaulted shape of the ceiling will allow more even distribution of the load and reduce the likelihood of the manifestation of rock pressure.

Thus, with a camera height of 14 m, man-made stresses on the contour of the vaulted chamber are of a compressive nature. In the working walls, the horizontal principal normal stresses reach a level of 0 MPa. The vertical principal normal stresses in the walls are minus 16 ... minus 2 MPa. The level of stresses obtained does not exceed the compressive strength of the massif. The massif will be in a more stable state, in comparison with the overhead form.

a b c

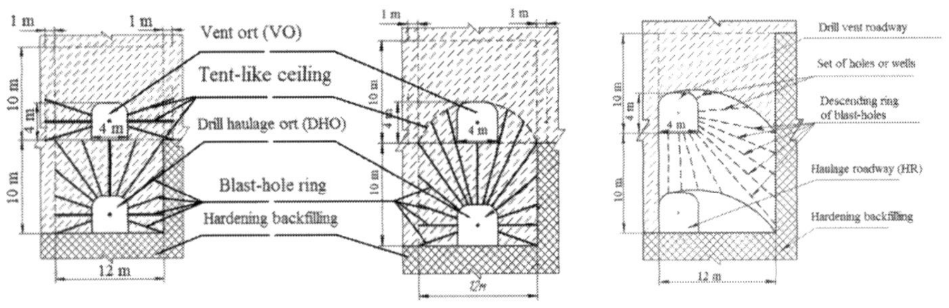

Fig. 2. Options for the chamber with the formation of a) the overhead form of roof of the chambers, b) the symmetrical vaulted ceiling of c) the asymmetrical vaulted ceiling

Based on the results of the modeling, the engineering services of Magnezitovaya mine are recommended for a sub-stage-chamber development system with hardening backfilling of the worked-out space, to ensure the stability of the rock massif, the option of forming chambers with the formation of ceiling of a vaulted shape should be applied.

It should be noted, however, that in the roof of the chamber there is an increased level of stresses, reaching a level of minus 58 .. minus 92 MPa from the action of horizontal forces, minus 5 .. minus 23 MPa from the action of vertical forces. In weak rocks, man-made stresses will exceed the allowable compressive stresses.

Therefore, in low-strength and unstable rocks, it is proposed to give the roof an asymmetrical vaulted shape as an additional measure (Fig. 2c). Thus, modeling the stress of the near-edge massif of chambers of an asymmetric vaulted shape at a chamber height of 14 m showed that when the chamber of the first stage (the under-roof layer) is formed, the horizontal stresses in the roof reach minus 36 .. minus 39 MPa, in the soil - minus 21 .. minus 24 MPa, in the wall - minus 3. minus 9 MPa, the concentration of stresses in the lower corner of the vaulted wall (-51 MPa) is observed. Vertical stresses in the roof are equal to 0 .. minus 3 MPa, in the soil - 0 .. minus 1 MPa, in the wall - minus 30 .. minus 40 MPa, in the lower corner of the vaulted wall there is also an increased concentration of stresses (-78 MPa). Since the stress concentrators in this variant are directed to the soil of workings, the contour massif in the roof and walls of the chamber will be in a more stable state.

4 Conclusions

Based on the above, the authors draw the following conclusions

1. The rocks in the undermined safety pillars of railway tracks and river bed are mainly represented by calcareous dolomite, magnesite, diorite. These rocks are prone to brittle fracture, being anisotropic and waterproof. Strength properties in the investigated areas are much lower than those adopted in the design solutions. The compression strength of the rock massif is on the average 39 MPa; the tensile strength is 5 MPa.
2. An increased level of horizontal initial stresses has been found in the zone of undermining safety pillars at horizons +180 .. +260 m. The coefficient of lateral expansion (ratio of horizontal stresses to vertical ones) is 1.63-1.73. The horizontal stresses acting in this part of Satka deposit are 80-85% or more of the ultimate strength of the rocks of the massif composing the area under investigation.
3. It was found that the level of initial stresses, in addition to natural geological factors, is significantly influenced by closely located old quarries. When testing the sections located near the contour of the spent quarry, it is necessary to monitor the changes in the rock pressure, deformation processes, displacement, and develop additional measures to ensure the safety of mining operations.
4. As technical measures to ensure the safety of underground work, it is proposed to design the ceiling of the cleaning chambers in the form of a symmetrical and asymmetric vault. It is established that in this case the stress concentrators are led to the soil of the excavations, the level of destroying tensile stresses decreases, the contour massif in the roof and the walls of the chambers is in a stable state.
5. Control over the change in the deformations of the massif in the mine workings while undermining the safety pillars is proposed to be carried out with photoelastic sensors, which allows for continuous fixing the change in the stressed state of the massif.
6. It is recommended to install the system "Gradient" ("NTTs Automatika", LLC) for automated control of rock pressure (ASCRP).
7. In addition, provision should be made for the installation of long-term observation stations for controlling rock pressure, shifting in the form of deep benchmarks, profile lines, photoelastic sensors in underground workings and profile lines and benchmarks on the daylight surface.
8. Based on the results of the research, "Guidelines for the safe management of mining operations at Satka magnesite deposit prone to bump hazards" have been developed in accordance with the requirements of the Federal norms and regulations in the field of industrial safety. The Guidelines set out the rules for the safe conduct of mining operations in areas prone to and dangerous for bump hazards, methods for assessing the bump hazard, measures for preventing dynamic manifestations of rock pressure, and methods for calculating the walling and supports of mine workings.

References

1. E.V. Shevlyakov, Improvement of the technology of working out the reserves of Satka deposit within the limits of the floors +180 / +260 m and +100 / +180 m of Magnezitovaya mine / E.V. Shevlyakov, I.S. Shelkovyi // The 5th International scientific and technical conference "Innovative geotechnologies in the development of ore and non-metallic deposits": a collection of reports. – Ekaterinburg city: Publishing house of the USMU. p. 207. (2016)
2. R.V. Krinitsyn, Monitoring of the stressed state and ensuring the stability of the massifs of ores and rocks in the cleaning blocks of Magnezitovaya mine / R.V. Krynitsyn, S.V. Khudyakov // Mining Information and Analytical Bulletin. **7**. 250-256. (2008)
3. A.V. Zubkov, Yu.G. Feklistov, R.V. Krinitsyn et al. Problems of subsoil use. **4 (11).** 41-49. (2016)
4. Iu.P. Shupletsov, Strength and deformability of rock massifs. Ekaterinburg UD RAS. p. 195. (2003) ISBN 5-7691-1428-2.
5. Iu.I. Gurtovoy, A.G. Shadrin. Mine surveyor's bulletin. **4**. 48-51. (2002)
6. R.V. Krinitsyn, S.V. Khudyakov. Eurasian mining. **2**. 16-19. (2017)
7. M.Ye. Chernova, Ya.V. Kuntsyak. Modern scientific research and their practical application. 175-181. (2014)
8. V.A. Petrov, A.B. Leksin, V.A. Sankov, V.V. Pogorelov. Modern Information Technologies in Earth Sciences Proceedings of the International Conference.
9. Siegesmund S., Durrast H. Stone in Architecture: Properties, Durability: Fifth Edition. 97-224. (2014)
10. V.A. Asanov, A.V. Evseev, V.N. Toksarov, V.V. Anikin, N.L. Beltyukov / Eurasian Mining. **2.** 20-24. (2013)

The research of remote earthquakes impact on the intensity of geomechanical processes in burst-hazardous rock massif

Vladimir Lugovoi[1*], *Igor* Rasskazov[1], *Denis* Tsoi[1], *Maksim* Rasskazov[1] and *Aleksandr* Sidliar[1]

[1]Mining Institute of Far Easten Branch of Russian Academy of Sciences, Khabarovsk, Russia

Abstract. Brief summary of works on the detection of pendulum waves intensifying geomechanical processes in rock massif, with the application of high-sensitivity laser measurements, is represented in the article. Peculiarities and technical characteristics of laser deformograph applied as part of comprehensive geodynamic monitoring system in the area of "Streltsovskoe" ore field have been taken into consideration. The results of experimental research for the evaluation of remote earthquakes impact on the condition of rock massif of "Streltsovskoe" ore field, registration of remote earthquakes and their forerunners are given. **Key words:** Deformation field, stress-related waves, comprehensive geodynamic monitoring, high-precision laser measurements, rock-burst hazard

1 Introduction

The occurrence geodynamic activation in large-scale and intense man-made activity which extraction and moving of large volumes of rocks relate to, during solid mineral deposits mining, may be represented by rock-bursts, induced (also in case of impoundment) and other dynamic events, the forecast of which is represented by an actual and simultaneously rather complex problem [1-4]. Reliable forecast of such seismic events and related to them catastrophic phenomena is quite hindered.

The invention, in the end of the previous century, of an event of alternating reaction of rocks to explosion (dynamical) loads gave a great impulse to experimental and theoretical research for the development of fundamentals of pendulum stress-related waves theory [5].

Within the frame of blocks hierarchy model along with fast seismic waves being generated as a result of hard rock destruction, there are slow stress-related waves accompanying the process of elastic energy spread by means of interaction of separate blocks possessing own oscillatory degrees of freedom. Positive results in the registration and record of such waves impact would allow developing new methods for hazardous geodynamic events forecast.

Works on the detection of pendulum waves were being carried out by scientists and specialists from the Institute of Mining of SD RAS and the Institute of laser physics of SD

* Corresponding author: lugovoy@dst.khv.ru

RAS over the period from 2006 to 2011 with the use of high-sensitivity laser measurements applied the Institute of laser physics of SD RAS in order for the study of earthquakes forerunners.

During the research, it was made possible to obtain first results on registration of slow pendulum waves accompanying local earthquakes in Baikal Rift Zone. The processing of data on series of earthquakes allowed not only recording them, but also distinguish features and obtaining quantitative evaluation of such waves spread velocity.

In order to record pendulum waves using laser deformographic method, a row of criteria was developed [6].

Main results of deformographic observations comes down to the following [7]:

- the confirmation of existence of the deformation shaking excitation with oscillatory periods within the band of 0.5–2 hours over 1.5–2 days prior to seismic events obtained;
- the appearance of disturbances in tidal oscillations of the Earth surface over 2-3 days prior to earthquake of class 14 with the focus zone ~530 km away from observation station recorded;
- the origin of quasi-resonant structure of deformation diagrams – with increasing amplitudes of fundamental harmonic as well as noise term commensurable to the amplitude of tidal wave immediately prior to earthquake was observed;
- correlation relationship between diagrams of Earth crustal deformation change and atmospheric pressure takes place;
- within the frames of hypothesis of condition of earthquake aftershock activity by wave pattern if pendulum waves movement, the evaluation of μ-waves spread velocities with the use of laser deformographic and seismological records is performed. For earthquakes of energy classes of K = 14.1÷14.6, the velocity range of these waves is defined by values around 0.5 ÷ 2 m/s.

To lesser extent, works on the detection of stress-related waves were being performed by the Institute of cosmological research and radio wave propagation of FED RAS at Kamchatka Peninsula [8]. In order to solve the task on the research of the properties and efficiency of high-frequency geo-acoustic emission generation in near-surface sedimentary rocks in the conditions of deformation processes activation, the aggregate of observations which consisted of geo-acoustic signals recording system, laser deformograph (one measuring arm), means of meteorological parameters control, data acquisition system were used.

For the purpose of the research of disturbed trend of geo-acoustic emission in on the threshold of strong seismic events at the distances of not more than 200 km over the period of 6-12 hours, laser measurements of the Earth surface deformation alongside with geo-acoustic observations during seismic process activation were performed. As a result, laser deformograph recorded disturbed, from the viewpoint of daily steady diurnal march, deformation process trend – violent oscillations which had high amplitude about 1 μm. The appearance occurred one day prior to strong seismic event which lasted 8 hours. The evaluation of deformation disturbance velocity amounts to 0.1 m/s which corresponds to the velocity of slow stress-related waves.

2 Methods and technical means of research

The problem of industrial disasters risk mitigation in large-scale mining is a burning one for Far Eastern region of Russia where deposits, hazardous by rock-bursts, are mined. Among the enterprises exploiting burst-hazardous deposits is "Priargunsky Industrial Mining and Chemical Union", OJSC. Here, uranium deposits of "Streltsovskoe" ore field (south-east Transbaikal) are developed with underground method [9]. At present, rock massif of ore field is to wide extent disturbed by mining operations. The deposit development is performed by four underground mines where mining operations have reached the depth of 500 – 900 m. At them, the entire spectrum of dynamic occurrences of overburden stress, up to strong and destructive rock-burts [9].

In order to conduct comprehensive research for the study of geo-dynamic and seismic environment at the deposits of "Priargunsky Industrial Mining and Chemical Union ", OJSC by the efforts of a row of academic institutes, multilevel system of comprehensive geo-dynamic monitoring, which merged seismic, geo-acoustic and deformational methods and measuring systems into single measuring network, was created and is being improving [10, 11].

This system includes multi-channel geo-acoustic automated overburden stress control system — "Prognoz ADS", which allows recording and determining the parameters of acoustic-emission events within the frequency band of 0.5 – 12 kHz [12]. Microseismic system "Prognoz S" which records seismo-acoustic events with the energy of up to 10^7 J within the frequency width 10-1000 Hz [13].

In order to control deformation field in a mine field and adjacent rock masses, laser deformograph in used geo-mechanical monitoring system [14].

a *b*

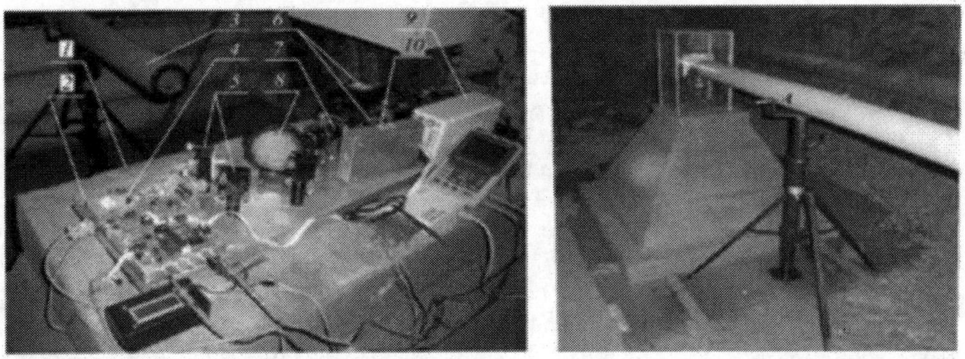

Fig. 1. Laser deformograph installed at geo-dynamic testing area in the vicinity of "Streltsovskoe" ore field: *a* — interferential unit; *b* — angle reflector and a fragment of air-filled beam mode guide: *1* — concrete bed; *2* — optical bench; *3* — beam mode guide; *4* — monitoring system; *5* — piezoceramic units; *6* — laser with a power unit; *7* — collimator; *8* — separating plate; *9* — monitoring system power unit; *10* — resonance amplifier.

Laser deformograph is located inside an underground mine opening at the depth of more than 300 m. Optical part of deformograph is assembled based upon modified asymmetrical Michelson interferometer with 50 m long arm which is targeted at , north-east at the angle of 30°, and frequency-stabilized laser by MellesGriott company. Interferometer is capable of recordng crustal movements with the accuracy of up to 0.1 nm within the frequency width of 0-1000 Hz [10-12].

Deformograph optical components are mounted on two concrete blocks rigidly connected with rock mass. Electrooptical part of interferometer, consisting of frequency-stabilized laser, optimal control system (monitoring and resonance amplifier systems) and other structural and optical elements of laser deformograph (fig. 1*a*), is installed on one of them, on another one — angle reflector. Laser ray was spreading between two concrete blocks on air-filled beam mode guide consisting of coupled pipes with inner diameter of 0.1 m (fig. 1*b*).

In order to ensure required conditions by humidity and temperature, the mine opening is separated, using a separator, into two sections: instrumental section and beam mode guide sections. Air dryer Ballu BDH-25L, set in instrumental section, provides air humidity at the level of 50 – 60%. Temperature mode of instrumental and beam mode guide sections is quite stable with temperature variations of $\pm 2°C$ which allows confident working with interference pattern.

3 Experimental research findings analysis

During testing and commissioning, in the course of laser deformograph implementation in measuring cycle, significant spectrum of earth self oscillations from major spheroidal tone ${}_0S_2$ to another major spheroidal tone ${}_0S_0$ were distinguished by the results of data processing.

Diurnal and semi-diurnal oscillations, the value of which is 23 hours 18 minutes and 11 hours 53 minutes accordingly, were recorded on 14-days deformograph record. Steady registration of diurnal and semi-diurnal oscillations witnesses of the possibility of recording long-period rock massif shift.

It is established that deformograph steadily records rock mass oscillations caused by blasting operations at mines. In order to evaluate the parameters of explosion loads on deformation field in the area of laser deformograph, a range of trial blasts by means of which the directivity graph and its impact on sensitivity of laser deformograph and the degree of signal attenuation in rock massif, was performed.

It is established that various amplitude oscillations, with the interval of about 2 minutes and 3 minutes, are present almost on all sections of deformations records which maybe conditioned by self-frequencies of geo-blocks in the area of deformograph setting. External effects lead to tectonic block oscillations (deformograph recorded their reciprocating movement with the frequency of 2 and 3 minutes).

In order to evaluate earthquake impact on rock massif conditions in the area of mining operations at uranium deposit of "Streltsovskoe" ore field, the development of earthquake and earthquakes forerunners recording procedure, experimental studies with the use of laser deformograph, laser nano-barograph and multi-channel geo-acoustic automated overburden stress control system — "Prognoz ADS" which allows recording and determining the parameters of acoustic-emission events within the frequency width of 0.5 – 12 kHz, were undertaken.

As a case study, the recording results of geo-acoustic system, laser deformograph and laser nano-barograph for the earthquake in Japan on November 16, 2016 with the magnitude of M = 4.7 at the distance of 2524 km away from observation area.

The findings of the analysis of remote earthquakes impact on geo-dynamic activity of "Streltsovskoe" ore field obtained with the use of multi-channel geo-acoustic automated overburden stress control system "Prognoz ADS" are represented on Fig. 2 and 3.

An earthquake duration is depicted as vertical line on fig. 2. It is evident from the figure that the number of events after an earthquake grew twice, and average events energy — nearly thrice.

At that, substantial increase of events number was recorded in the block of mining in which they were observed prior to an earthquake (fig. 3 *a*, *b*).

It should be noted that such situation was observable not always. In other cases, after an earthquake, an increased geo-dynamic activity was observed in neighbouring blocks, i.e. It was spreading across an mine opening. As authors think it, to significant extent, depends on the direction of seismic wave arrival.

Time record of an earthquake moment with laser deformograph and lase nano-barograph is represented on fig. 4 *a*, *b*.

It is evident from the figures that there is a significant distinction of shapes of deformograph and nano-barograph signals in the moment of earthquake. At the same time, it is clearly evident that the stress in the mine opening varies by the action of seismic wave.

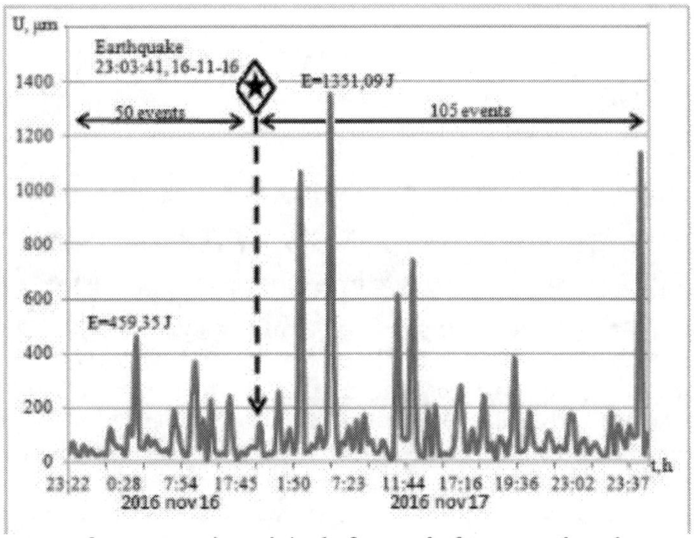

Fig. 2. Diagram of geo-acoustic activity before and after an earthquake

a *b*

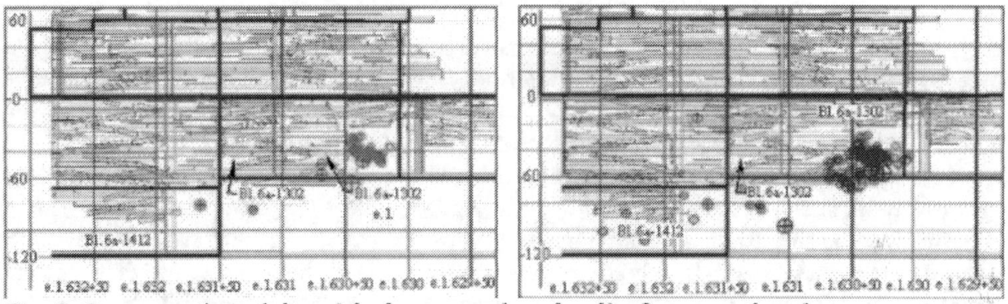

Fig. 3. Geo-acoustic activity, *a*) before an earthquake; *b*) after an earthquake

In order to establish pre- and post-history of seismic events, double diurnal records, diurnal records before earthquakes and after it were studied. Double diurnal records of deformograph (*a*) and nano-barograph (*b*) are represented on fig. 5. Here, unlike the records of an earthquake moment, conspicuous is the fact of high degree of the correlation of results obtained with deformograph and nano-barograph (the difference in the signal polarity is explained by the fact that operating points of Interferometer of nano-barograph and deformograph are selected on various slopes).

a *b*

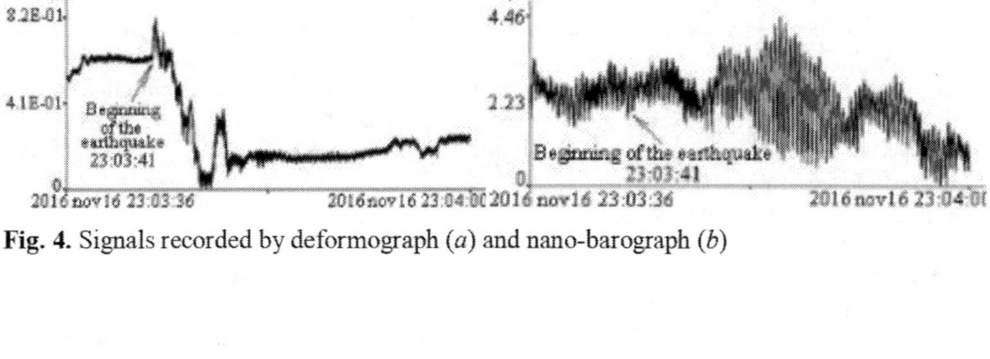

Fig. 4. Signals recorded by deformograph (*a*) and nano-barograph (*b*)

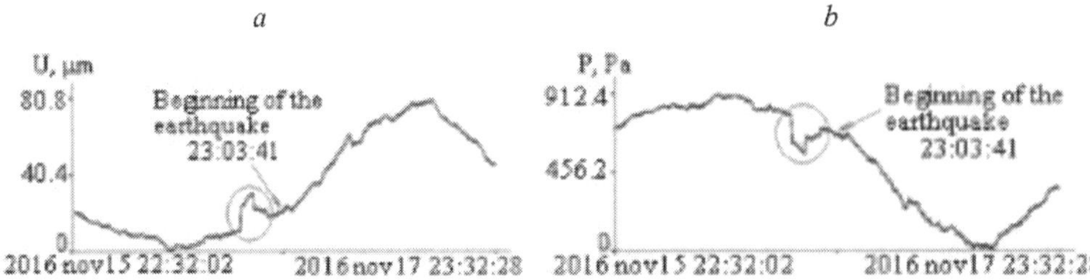

Fig. 5. Record of deformation signals recorded by deformograph (*a*) and nano-barograph (*b*)

The high degree of correlation is connected, apparently, with the fact that deformograph includes air-filled beam mode guide and refraction index variations overlap mechanical oscillations. Same situation is observed in all durable (semi-diurnal, diurnal) records. At that, mechanical oscillations become noise contaminated with variations of refraction index which depends on the stress. The exclusion of refraction index variations impact on deformation measurements is possible to achieve by several ways. First, beam mode guide should be no air-filled, but evacuated to pressures of around 10^{-3} atm and thus minimize the dependence of refraction index on stress and temperature. Second way — installation of one more measuring arm of deformograph orthogonal to the first one which enables obtaining clear differential signal. And the third way is to introduce corrections in post-processing.

In author's opinion, the most advanced is the second way. In case of its realization, it becomes possible not only to increase deformograph resolution by means of interference compensation, but also the possibility of judging of stress-related wave arrival direction emerges which essentially widens the potential of high-precision laser measurements.

One more peculiarity of the results obtained is the presence in each various earthquakes record bay signals of substantive amplitude (on fig. 5 they are encircled).

This may be the reaction of the massif to industrial blasts. Accompanying oscillations with the frequency of 95 Hz, which are steadily recorded in the analysis of blast records, point at that. If to suggest that a blast excited oscillations of block (possibly stress-related wave, local earthquake, block movement) with sufficient power, the block oscillations may shake adjacent medium. In case of accumulated rocks stress such oscillation may cause their destruction. It is fair to assume that the oscillations reached laser deformograph in the shape of bay excitations. It is apparent from figures that deformation goes back almost in the initial point. Otherwise, it would be well to suggest that it's growing somewhere which leads to the destruction and devolution.

4 Conclusions

1. As the researcher' works showed, deformation and wave processes in earth crust are triggering mechanism for various high-energy seismic events (earthquakes, rock bursts, etc.). At the same time, they may be forerunners of these events. Topical becomes the tasks of their detection and recording. The results of research conducted demonstrate that in case availability of required measuring means for recording deformation and wave processes as well as corresponding guidance papers for distinguishing pendulum waves, the opportunity to control dynamic and kinematic attributes of the latter on regional and world-wide basis appears.

2. According to the data of high-precision deformation control at geo-dynamic testing area in the proximity of "Streltsovskoe" ore field it has been established that seismic waves from strong earthquakes have significant impact on deformation field what may be an extract factor initiating dynamic occurrences of overburden stress at "Antei" deposit.

3. It has been experimentally proven that deformation shaking with oscillatory periods within the width of 0.5–2 hours for 1.5–2 days before seismic events exists. Velocity range of pendulum stress-related waves of around 0.5÷+2 m/s was calculated. It should be assumed that, determining and analysing records with rock-bursts registration in the area of "Streltsovskoe" ore field at rather small distances of seismic events from observation area (several kilometres), an opportunity to record stress-related waves, determination of their arrivals time and amplitude-time characteristics of deformational wave-train may be obtained. This will enable significantly increasing the reliability of energy geo-dynamic events in burst hazardous rock in the development of deposits in aid of prevention of catastrophic events.

References

1. V.V. Adushkin S.B. Turuntaev, Technogenic seismic activity – induced and triggered. IDG RAS, 2015.
2. I.Yu. Rasskazov, Monitoring and management of rock pressure on mines of Far East region. Mountain book. p. 329. (2008)
3. Henryk M., Mutke G. Z. Journal of Seismology. 17 (4). 1139-1148. (2013)
4. Paige E. Snelling, Laurent Godin, Stephen D. McKinnon. International Journal of Rock Mechanics and Mining Sciences. **58**. 166-179. (2013)
5. M.V. Kurlenia V.N. Oparin. Journal of Mining Science. **4**. (1990)
6. S.N. Bagaev, V.N. Oparin, V.A. Orlov, S.V. Panov, M.D. Parushkin. Journal of Mining Science. **3**. (2010)
7. V.N. Oparin and others, Methods and systems of seismo-deformation monitoring of induced earthquakes and mine bumps. **1, 2.** Publishing house of Siberian Branch of Russian Academy of Sciences. (2009)
8. I.A. Larionov, B.M. Shevtsov. The problems of comprehensive geo-physical monitoring of Far East of Russia. IKIR FED RAS: Petropavlovsk-Kamchatsky city. (2009)
9. I.Yu. Rasskazov B.G. Saksin, V.A. Petrov, B.A. Prosekin. Journal of Mining Science. **3**. 3-13. (2012)
10. V.A. Petrov, A.B. Leksin, V.A. Sankov, V.V. Pogorelov, I.Yu. Rasskazov, GIS-based 3D geodynamic modelling of Transbaikalia, Russia, Int. Conf. GeoFrankfurt'2014 "Earth System Dynamics". Goethe University Frankfurt a, Main. SDGG Heft 85, Abstract Volume. pp. 471. (2014)
11. I.Yu. Rasskazov, A.V. Gladyr, P.A. Anikin, V.S. Sviatetskii, B.A. Prosekin. Mining Magazine. **8 (2).** 9-14. (2013)
12. I.Yu. Rasskazov, V.A. Lugovoi, G.A. Kalinov, A.V. Gladyr, P.A. Anikin, M.I. Rasskazov and D.I. Tsoi. Proceedings of the 8-th International Symposium on Rockbursts and Seismicity in Mines (Russia, Saint-Petersburg – Moscow. September 1-7, 2013). pp. 121-124. (2013)
13. A.V. Gladyr, V.I. Miroshnikov, Iu.I. Bolotin and others. Mining information-analytical bulletin. **5**. 174 – 180. (2012)
14. I.Yu. Rasskazov, V. A. Lugovoi, B.G. Saksin, D. I. Tsoi, G.I. Dolgikh, S. G. Dolgikh, V. A. Petrov. Journal of Mining Science. **52 (6)**. 1052-1059. (2016)
15. Takemoto Shuzo, Momose Hideo, Araya Akito, etc. Journal of Geodynamics. **41.** 23–29. (2006).

Rock burst hazard assessment of apatite mines on Kola Peninsula by means of local acoustic emission method.

Alexander Rozanov[1*], and *Andrei* Tereshkin[2]

[1]Saint-Petersburg Mining University, Research Center of Geomechanics and Issues of Mining Industry, Saint-Petersburg, Russia

[2] Mining Institute of Far Easten Branch of Russian Academy of Sciences, Khabarovsk, Russia

Abstract. Various acoustic emission (AE) parameters are used to characterize fracture process in rocks. The conventional statistical values are, for example, AE activity, *b*-factor, and distribution of energy. In order to map AE events the location analysis is used. There are also of interest in seismic analysis such kinematic parameters as corner frequency, seismic moment. To describe spatial pattern of hypocenter distribution a so called correlation integral is applied. In this work we apply spectral-correlation analysis to quantify each AE waveform with energy-correlation values. To study the behavior of spectral parameters during fracturing we carried out a series of AE measurements in the areas of extreme mining pressure in the PhosAgro apatite mines. For data acquisition purpose we use a device «Prognoz-L» which has been developed in the Mining Institute of Khabarovsk, Russia, for local stress monitoring.

1 Introduction

Determination of dynamic mechanical properties of rock materials is very important for evaluation of stability of rock structures under dynamic loads. For example, at different scales (seismic scale, oil or gas reservoirs, mine pillars) rock material is subjected to compressive loads at different strain rates. In all cases rock loses its stability not instantaneously, but gradually approaching to its fault. Hence it becomes very important to predict this instability in rock media which starts to develop from micro-fracturing. This problem leads us to the process zone characterization when making observations of fracture process in rocks. For this purpose the AE method is useful.

In terms of microscale faulting fracture is caused by sudden releas of accumulated strain energy when the local strength of material is exceeded. Natural acoustic waves are radiated by grain scale fracturing events. These waves contain information about fractures. AE studies in controlled laboratory conditions provide data of various AE waveforms [1,2].

In view of importance of rock burst monitoring in mines new technologies of geoacoustic systems are widely introduced nowadays. For example, a multifunctional

* Corresponding author: aleksander.rozanov@mail.ru

device of mine local control (Prognoz-L) has been developed in the Mining Institute of Khabarovsk [3]. Now it is successfully used in different mines of Russia, one of them is the appatite mine on Kola Peninsula. The calculations of *b*-value, energy and activity of AE are implemented into the device in online mode. One can carry out the post processing of digitized AE waveforms. In this work the AE waveforms has been subjected to spectral correlation analysis. With the use of spectral correlation algorithm developed on the basis of numerous laboratory data [4,5] the sequence of stick-slip events [6] has been analised in the work [7]. We distinguished three main patterns of AE spectra: narrowband with high correlation (≈1), wideband with low correlation (<0.5), and П-pattern of high correlation (≈1). The AE data acquired under mining conditions have clearly revealed the transition point from stable process of fracturing to unstable process by AE median frequency and standard deviation observations [8].

The aim of this work is to capture an onset of transition process from stable fracturing to unstable fracturing with the use of cross-correlation analysis. The AE data acquired with the device «Prognoz-L» during mining local pressure control in appatite mine have been used.

2 Spectral-correlation method

Historically, Fourier spectral analysis has provided a general method for examining the energy-frequency distribution for a certain time series. As a result, the term "spectrum" has become almost synonymous with the Fourier transform of the data. Although the Fourier transform is valid under the general conditions, there are some important restrictions of the Fourier spectral analysis: the system must be linear; and the data must be strictly periodic or stationary; otherwise, the resulting spectrum will make little physical sense. The far-field seismic phenomena can be approximated by linear systems and so be represented by linear superposition of sinusoid functions adequately. The algorithms of Fourier transform of the signals of finite duration is well developed and the interpretation is quite explicit [9, 10].

The most basic method of data analysis is a limited time window-width Fourier spectral analysis. By successively sliding the window along the time axis, one can get a time-frequency distribution. AE signals are converted from time t to frequency f domain using a Finite Fourier Transform. Let us consider two signals $x(t)$ and $y(t)$ and divide each of them into n_d equal segments $x_k(t)$ and $y_k(t)$ with T duration (time windowing). Then the spectral density estimations $G_{XX}(f)$ and $G_{YY}(f)$ are produced as follows:

$$G_{XX}(f)=\frac{2}{n_d T}\sum_{k=1}^{n_d}\left|X_k(f,T)\right|^2 \tag{1}$$

$$G_{YY}(f)=\frac{2}{n_d T}\sum_{k=1}^{n_d}\left|Y_k(f,T)\right|^2 \tag{2}$$

where $X_k(f,T)$ and $Y_k(f,T)$ are the Finite Fourier Transforms of $x_k(t)$ and $y_k(t)$ respectively:

$$X_k(f,T)=\int_0^T x_k(t)e^{-i2\pi ft}dt \tag{3}$$

$$Y_k(f,T)=\int_0^T y_k(t)e^{-i2\pi ft}dt \tag{4}$$

The cross-spectral density $G_{XY}(f)$ is a complex function of real variable:

$$G_{XY}(f)=\frac{2}{n_d T}\sum_{k=1}^{n_d} X_k^*(f,T)Y_k(f,T) \tag{5}$$

where $X_k^*(f,T)$ is the complex conjugate of $X_k(f,T)$.

Using the cross-spectral estimation one can investigate the frequency dependent correlation between two signals of one radiation event or between two sequent signals of two different radiation events. The so-called coherence function allows us to do that. The coherence function is expressed as a ratio of cross-spectral and spectral densities of two relevant signals:

$$\gamma_{XY}^2(f)=\frac{|G_{XY}(f)|^2}{G_{XX}(f)G_{YY}(f)} \tag{6}$$

Coherence function controls the similarity of two AE signals – the amplitude of the second signal can change relatively to the first signal, but the dominant frequency and the wave shape must remain the same. In this case a value of coherence function tends to 1, and in case the similarity is lost it tends to 0.

The idea of this work is to perform one sequence for all AE signals registered in a certain day of measurements, and then to divide this sequence into the pairs of sequent AE signals. The second step is to apply for each AE pair the cross-correlation analysis according to the equation (6). Finally we obtain the trend of coherence values and compare it with the trend of energy. Coherence value for each AE pair we associate with the dominant frequency of cross-spectral density. Energy estimation is produced from the spectrum of each AE signal according to the equation:

$$EE=\sqrt{\int_{f_-}^{f_+} G_{XX}(f)df} \tag{7}$$

where f_- and f_+ are the lower and upper limits of a valid signal frequency band.

3 Application of spectral-correlation method to AE field data

On the basis of numerous observations of mine AE data we suppose that violation of AE similarity associated with high AE energy is a reliable warning of transition to unstable phase of fracturing which in turn can lead to rock damage. This supposition we test in the mine area which was presumably under high stress conditions. The controlled mine opening was being developed and at the same time excavated area approached too close in the north. On the other hand a substantial geological inhomogeneity of dyke type occurs about 60m to the south. Figure 1 performs a photo of the face of mine opening where AE monitoring has been carried out. AE transducer has been mounted exactly on the wall close to the face of mining.

Fig.1. The face of controlled mine opening and the device «Prognoz-L» installed on the site of AE monitoring.

AE monitoring has been performed over the course of three days. The typical mismatch of AE waveforms (Figure 2b) compared to their correspondence (Figure 2a) is shown in Figure 2. Figure 3 illustrates the cross-correlation calculation of AE waveform similarity respectively. The big circles depict high values of spectral density associated with the dominant frequency of cross-spectrum. One can see that in case of perfect correlation of two AE waveforms the coherence function for big circles approaches 1 (Figure 3a), and when similarity is lost they drop to near zero values of the coherence function (Figure 3b). This is an example of how to detect the similarity violation with the use of cross-correlation analysis.

Then we combine the data of three days of measurements into one sequence and launch continuous cross-correlation and energy analysis. What we obtain is the 3D display in time domain of AE monitoring which detects energy-correlation anomalies evolving in the course of measurements (Figure 4). In the first day (19.03.2018) the AE energies were in general low (almost blue background), only one small green spot of bad correlation and a bit higher energies one can distinguish in the picture. It could be a warning of a small fault zone formation localized in the vicinity of the mining face. The next day of measurements (21.03.2018) we see a significant progression of faulting. The red and green spots of high AE energies and low values of correlation indicate that. And during the last day (23.03.2018) we observe the decay of AE energies without any anomaly to appear (blue background).

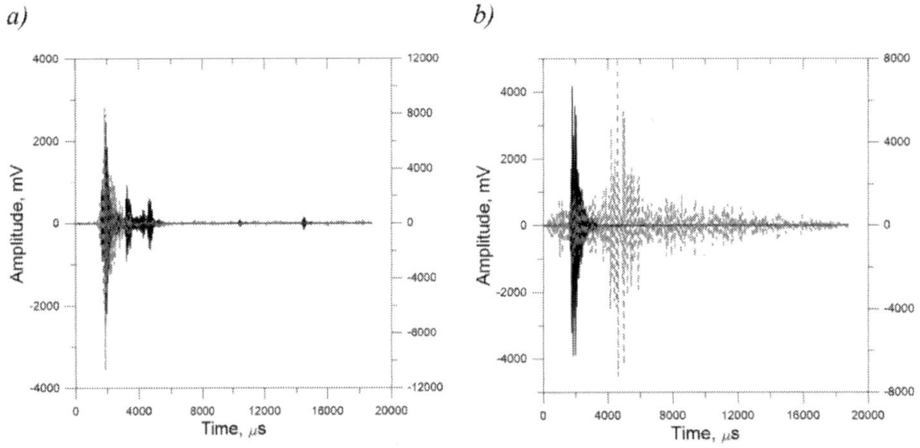

Fig.2. Examples of two pairs of AE waveforms: *a)* – perfect similarity; *b)* – bad similarity.

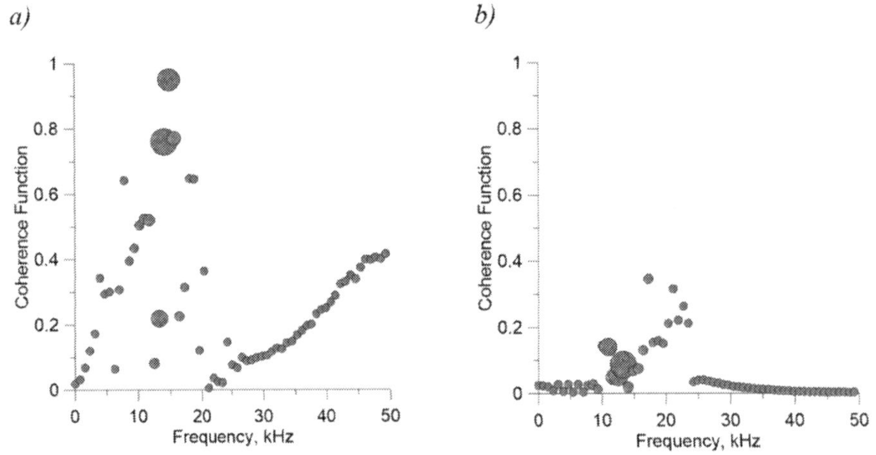

Fig.3. Cross-correlation analysis as an evaluation of AE waveform similarity: *a)* – perfect similarity; *b)* – bad similarity.

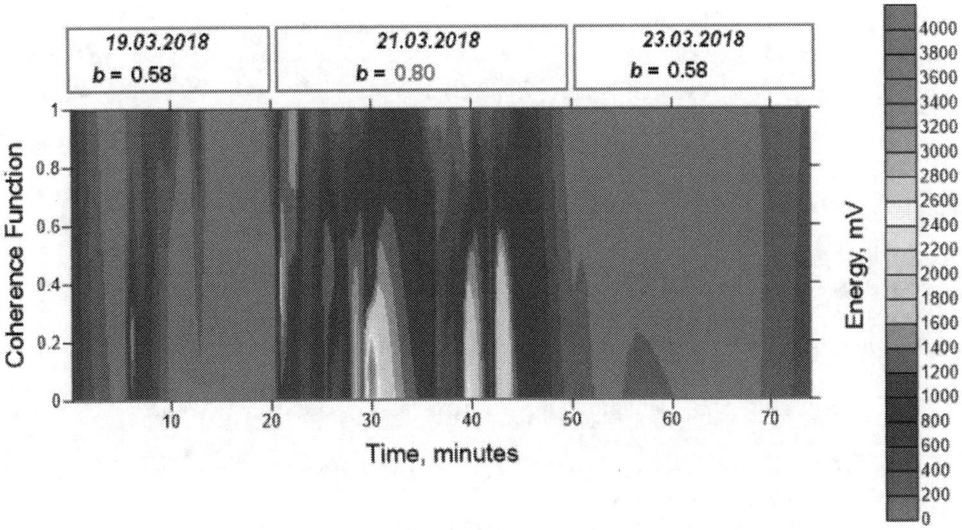

Fig.4. Time domain 3D display of AE monitoring in mine. Red spot is an indication of stress concentration process. The averaged *b*-values for each day are pointed out above.

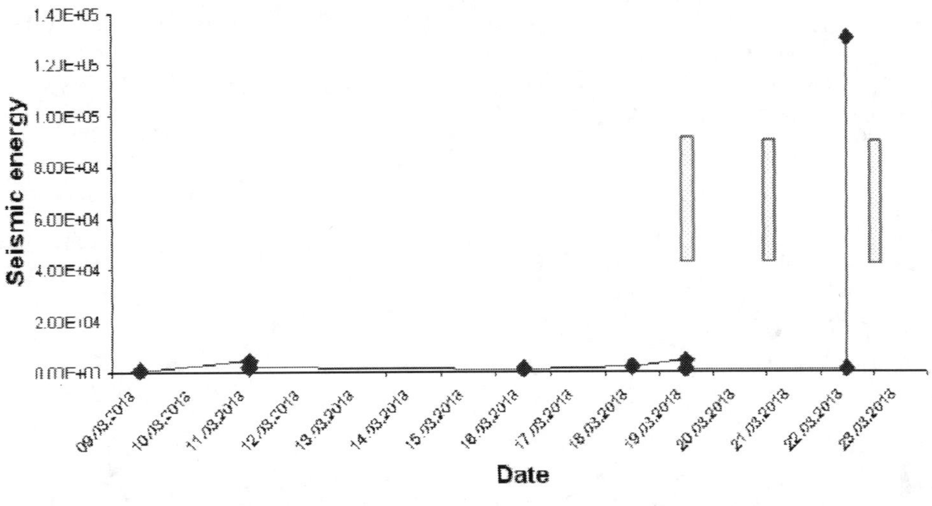

Fig.5. Seismic data obtained in the area where AE monitoring has been conducted. Blue diamonds depict the energy values, and yellow rectangles depict the days when AE monitoring has been carried out.

4 Conclusions

As a result we can make the following conclusions. In order to interpret AE waveform data we make an assumption that similar source mechanisms result in similar AE waveforms. Any change in similarity should produce different values of coherence function. Through the observation of coherence and energy trends we can control a transition to unstable phase of fracturing and take precautions during mining to reduce a risk of rock damage.

In Figure 5 the seismic data for the area of AE monitoring is performed. On 22 of March two strong seismic events have been registered by the seismic acquisition system in the area of AE monitoring. Taking into account that our AE observations (21.03.2018) of intensification of fracturing process have preceded the strong seismic events, we can suggest that it was a process of strain accumulation and afterwards a process of relaxation associated with these strong seismic events.

References

1. Z. Reches, D.A. Lockner, *J. Geophys. Res.*, **99**, B9 (1994).
2. A. Zang, C. Wagner, S. Stanchits, C. Janssen, G. Dresen, *J. Geophys. Res.*, **105**, B10 (2000).
3. I.Yu. Rasskazov, D.S. Migunov, P.A. Anikin, A.V. Gladyr, A.A. Tereshkin, D.O. Zhelnin, *J. Min. Sci.*, **51**, 3 (2015).
4. A.O. Rozanov, *Proc. Int. Geophys. Conf. & Exhibition on Geophysics of the 21st Century*, Extended Abstracts, OS22 (2003).
5. A.O. Rozanov, *Search and Discovery*, **#40897** (2012).
6. K. Aki, P.G. Richards, *Quantitative Seismology. Theory and methods* (2 volumes), W.H. Freeman, San Francisco, 932 pp. (1980).
7. I.Yu. Rasskazov, S.V. Tsirel, A.O. Rozanov, A.A. Tereshkin, A.V. Gladyr, *J. Min. Sci.*, **53**, 2 (2017).
8. A.O. Rozanov, D.N. Petrov, A.M. Rozenbaum, A.A. Tereshkin, M.D. Ilinov, *Geomechanics and Geodynamics of Rock Masses*, pp. 669-672 (2018).
9. J.S. Bendat, A.G. Piersol, *Random Data: Analysis and Measurement Procedures*, Wiley-Interscience (1974).
10. J.S. Bendat, A.G. Piersol, *Engineering Applications of Correlation and Spectral Analysis*, John Wiley and Sons (1980).

Identification method development of focal zones based on seismic monitoring results

Andrey Gladyr[1*] and *Vladimir* Miroshnikov[1]

[1]Mining Institute of Far Easten Branch of Russian Academy of Sciences, Khabarovsk, Russia

Abstract. Method of seismic active zones control is one of approaches of prediction of dynamic mountain pressure manifestations, which is to identify and analyze formation dynamics of seismo acoustic active zones and predict impact hazard based on revealed patterns of changes in geoacoustic activity of rock massif. The key point of this control method is the zone allocation where a potentially dangerous rock destruction source is formed. In the article, we propose an automated method for damage zone identification, based on the preliminary exclusion of background radiation by nonparametric density estimation method, the seismo acoustic active zone allocation by means of probabilistic cluster analysis and damage zone parametrization by characteristic ellipsoid selection. These tasks are part of complex geomechanical monitoring system development and are solved by upgrading and subsequent integration into the interacted software of the considered seismo acoustic active zone monitoring method. Their solution will allow to fully automate predicting dangerous state process of the controlled rock mass, improving forecast quality and significantly reducing the time spent on obtaining final result.

1 Introduction

The kinetic concept of solid destruction, based on the analysis of existing approaches in the forecast of dynamic mountain pressure manifestations, is the basis of most of methods. Many scientists who adhere to this concept found that several characteristic stages are in rock destruction process in which destruction process begins with crack formation with a size of millimeters to cracks with a size of centimeters and of meters, and can end in gaps from tens to thousands of meters, typical of mountain-tectonic shocks, technogenic earthquakes and etc. [1-4].

At present, a number of methodological approaches to dynamic rock pressure manifestations forecast, using different predictive criteria [1, 5, 6]. The author team from the Institute of mining FEB RAS developed a monitoring method of geomechanical rock mass state – the monitoring method of seismo acoustic active zones. This method consists in identification and analysis of acoustically active zones formation dynamics and impact hazard forecast based on established patterns of changes in geoacoustic rock massif activity. Therefore, the key point of the considered method of seismo acoustic active zones

* Corresponding author: rush3112@mail.ru

monitoring is the controlled zone allocation, which can be qualified as damage zone, within which a potentially dangerous rock destruction source is formed.

2 Filtering background radiation

The practice of using cluster analysis methods of seismic monitoring data shows that final clustering result is significantly affected by original data quality, which necessarily contains a certain amount of faulty data. It is recommended to select a data subset in areas with high point density before beginning of clustering. To achieve this goal, we propose to estimate density distribution at each point using a nonparametric density estimate.

At the same time, nonparametric estimates of distribution numerical random variables density and finite-dimensional random vectors were considered in the early stages. In the 1980s, it was possible to construct such estimates in arbitrary environment spaces [7], and then for specific types of non-numeric data [8].

Simplifying, the registered event with coordinates (x_i, y_i, z_i) will be considered as belonging to the region with high density if point number located inside hypersphere centered at the point with coordinates (x_i, y_i, z_i) and radius h, radius h which greater than the threshold value c.

It is necessary to set the threshold value c and filter the seismo acoustic events related to the background radiation upon calculations completion related to calculation of the distribution density estimation for each registered event.

The following method is proposed for automatic selection of value c:

1. Distribution density value is calculated by original input dataset.
2. Median of the resulting distribution is calculated.
3. Resulting distribution is approximated by a given function. Practice of using this method has shown possibility of application as an approximating function of the polynomial of eighth and higher degrees. However, best results were achieved by using as an approximating function of form $f(x) = a_1 \cdot sin(b_1 \cdot x + c_1) + \dots + a_i \cdot sin(b_i \cdot x + c_i)$ with number of terms from six and above.
4. Maximum point's values m1 and m2 of approximating function to the right and to the left of the median are determined.
5. Argument value corresponding to the minimum of approximating function on interval [m1, m2] is determined. Calculated argument value will be required value c.
6. The described practice is illustrated in Figure 1, where points are density distribution values, and continuous line visualizes approximating function.

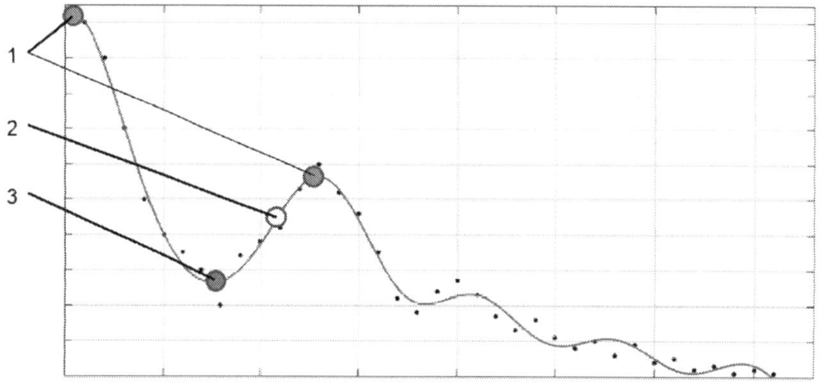

1 – Maximum points of approximating function in the neighborhood of median;
2 – Median of density distribution; 3 – Required threshold value.

Fig. 1. Illustration of a method for automatic selection of threshold density values.

The main task in carrying out qualitative calculations of the weight function is the choice of the parameter *h*. There are several common approaches to choice of this parameter: reference heuristic rules, substitution methods, cross-validation methods, bootstrap methods.

Least-squares cross–validation method is the most suitable of them. It is a fully automatic and data dependent method for selecting parameter h, which is based on principle of choosing parameter h, minimizing integral mean square error of resulting estimate [9-10].

Thus, the developed methodology of background seismo acoustic radiation identification allows, on the one hand, solving problem of data filtering for subsequent separation of seismo acoustic active zones, and on the other hand, to identify recorded seismo acoustic events related to background radiation. It allows assessing background radiation parameters for their further use in predicting dynamic manifestations in the rock massif [11].

3 Damage zones formation

Cluster analysis - a set of mathematical techniques designed for formation of relatively "distant" from each other groups of "close" to each other objects according to distances or relationships (proximity measures) between them [12-14].

Retrospective analysis of considered clustering methods allows us to conclude that use of graph or hierarchical clustering methods seems to be the most appropriate at first glance to solve the problem of damage zones identification based on seismo acoustic monitoring results [15]. This statement is based on idea of physical processes occurring in loading process and subsequent deformation of the rock massif, for example, that destruction process is fractal.

However, these methods of solving clustering problem are related to deterministic mathematical models, as these approaches imply that in clustering process initial set of objects X is divided into several non-intersecting subsets. In other words, these methods are methods of "crisp" clustering.

On the other hand, "fuzzy" clustering methods allow the same object to belong to several (or even all) clusters at the same time, but with different probability. Such probability clustering in many situations is more "natural" than "crisp", because it takes into account random processes inherent in real nature objects and their stochastic description.

Clustering methods are classified by whether the number of clusters is defined or not in advance. In the latter case, clusters number is determined in the process algorithm execution based on distribution of original data. To solve problem of damage zones allocation, C-means algorithm is used, which divides data into known clusters number described by the fuzzy partition matrix. The only difference between the partition matrices for the fuzzy and crisp case is that in the fuzzy partition belonging degree of the object to the cluster takes values from interval [0, 1], and in crisp - from two-element set {0, 1}. Fuzzy partitioning allows you to simply solve the problem of objects located on the border of two clusters. These objects are assigned belonging degrees equal to 0.5.

The spread criterion is used to evaluate quality of fuzzy splitting [16]. Finding the fuzzy partition matrix with the minimum value of the criterion is a nonlinear optimization problem, which can be solved by different methods. The fuzzy c-means algorithm is the most known and frequently used method for solving this problem, which is based on method of undefined Lagrange multipliers [16].

The distance between the object and the cluster center is calculated by standard Euclidean norm, which allows allocating clusters in the form of hyperspheres, in basic algorithm of fuzzy c-means. Diagonal norm allows us to allocate clusters in form of hyperellipsoids oriented along coordinate axes. The Mahalanobis norm allows to allocate clusters in the form of hyperellipsoids. Their axes can be oriented in any direction.

The given clustering algorithm allows us to allocate data clusters for some data sets in form of different geometric shapes: spheres, ellipsoids with different orientations, chains etc. Form of all clusters is the same for clustering algorithm result with a fixed rate. Clustering algorithms how-to impose a structure that is unusual for data, which leads not only to suboptimal, but sometimes to fundamentally wrong results. There are several methods to remove this weakness. The authors recommend using the Gustafson-Kessel algorithm [17].

The Gustafson-Kessel algorithm uses an adaptive norm for each cluster; i. e. there is a norm-generating matrix *Bi* for each *i*-th cluster. This clustering algorithm is optimized not only cluster centers coordinates and the fuzzy partitioning matrix, but also the norm-generating matrices for all clusters. This allows you to select clusters of different geometric shapes.

The number of clusters (c) is the most important parameter in the algorithm. It is quite difficult to choose the right number of clusters for real tasks without any a priori information about structures in data. There are two formal approaches to choosing the number of clusters.

The first approach is based on criteria of compactness and separability of obtained clusters. It is logical to assume that data will be divided into compact and well-separated groups if the number of clusters is chosen correctly. Otherwise, clusters will probably be not compact and well separated. There are several criteria for assessing the compactness of clusters. However, the issue is remained unresolved how to formally and reliably determine the correct choice of the number of clusters for an arbitrary data set. It is recommended to use the index Xie-Beni [19] for fuzzy c-means algorithm [18].

The second approach is suggested to start clustering with a sufficiently large number of clusters, and then sequentially to combine similar adjacent clusters. Various formal similarity criteria for clusters are used.

However, the subtractive clustering method can be considered the most accurate method of determining the number of clusters. One of its advantages is that you do not have to specify the initial number of clusters. The method was proposed By R. Yager and D. Filev in 1993 [20].

The following complex algorithm of damage zones formation is proposed based on results seismo acoustic monitoring of based on above algorithms:

1. Regions with high seismic activity are selected from entire recorded seismic events set using nonparametric density estimation.
2. The number of clusters is determined using the subtractive clustering algorithm.
3. The process of multiple clustering repetition is performed using the Gustavson-Kessel algorithm to obtain a stable result of splitting the initial objects set into the optimal number of clusters by estimating of distribution probability. The highest probability of an event belonging to each cluster is calculated after the process is completed for each case.

The developed algorithm makes it possible to assess the quality of clustering. At the third step of the algorithm, as described above, there is a direct process of objects distribution into clusters. This process is performed as follows: cluster split procedure is performed a large number of times, belonging probability to a given cluster, which it belonged, in the process for each event is saved. Greatest probability (in frequency ratio) of belonging to each cluster is selected for each event are completed after the iterations. Thus, the result is the division of events into probability of belonging to a particular cluster and clustering characteristic, i.e. reference reliability of each object (observation) to each of clusters.

4 Calculation of characteristic ellipsoid parameters

Seismic and acoustic signals location of allows to identify the defects cracking area and to identify the propagation direction of cracks in space. It can be concluded with some assumption that this area will have some ellipsoid form with its main diagonals in space, which characterize length, width and thickness ratio of destroyed layer and orientation of this layer in space. Construction of ellipsoid is based on statistical distribution of crack centers according by location. The coordinate table of recorded events is taken as source data.

We choose select the direction in which the coordinate distribution will be maximal. A projection of event location points on the normal direction will have the maximum dispersion in this case. This normal characterize the length direction of damage zone. In the same way, we determine zone thickness in orthogonal plane to normal, as a direction with a minimum dispersion in the plane. Then we will consider the dispersion of scatter points as a characteristic of zone width in the remaining third orthogonal direction to two previous.

The following practice of ellipsoid construction is proposed. The selected cluster from recorded seismic events set is considered as input data. For further analysis, an origin is moved to damage zone center. The standard event coordinates offsets from the center have different meanings for different directions. We use procedure of successive approximations to determine ellipsoid axes direction. The maximum value of the standard deviation of the coordinate projection on the coordinate axes is revealed. The axis with the maximum deviation is variable. All points of the array in new coordinate system are rotated sequentially around other two axes at some small angle α in positive and negative directions.

We determine such direction of the axis by varying angle α, for which standard deviations of event coordinates will be the maximum. If this value corresponds to a successful variation (rotation), it is taken as the next value of the center point from which the following variations are made. At the same time, new value of coordinate system transformation matrix is determined for each step using same rotation operators.

If central point is the maximum, then we consider the goal reached accurate to angle α. Angle α is reduced to determine a more accurate solution, and then the procedure is repeated until desired accuracy is achieved. The value $\alpha = 10^{-3}$ rad is assumed as a satisfactory accuracy in practical calculations.

Next step second main direction of distribution of seismo acoustic events coordinates is determined. First, the nearest coordinate axis is selected, on which coordinate projections have a minimum spread. Then, rotation operator is calculated around previously inclined axis until the minimum value of mean square spread on the selected coordinate axis is reached. Search procedure is similar to the one described only with the difference that only two variations near center point are determined by rotation operator now. The procedure is performed until the required angle accuracy is achieved. Inverse coordinate system transformation matrix consists of three columns. They are unit vectors of three main orthogonal directions of an ellipsoid corresponding to main standard deviations in original coordinate system.

Main ellipsoid axes are vectors with lengths proportional to the standard deviation. It is believed that no more than 0.27% of values exceeding 3σ, (σ – standard deviation) in the classical Gaussian distribution, however, these values may differ significantly in the case of real data. The region fracture volume estimated as ellipsoid volume with dimensions $z\sigma$, where z- parameter takes into account crack distribution density in selected area.

5 Approbation of the identification method of damage zones based on seismic monitoring results

The seismic monitoring data of the "Prognoz ADS" automated system of mountain pressure control at the "Antey" field for 2011, during which whole range of multi-scale seismic events was registered, were analyzed to check of quality of the developed damage zones identification method.

Was located 14065 seismo acoustic events only in 2011. 38% of the events were related to background radiation based on the results of background radiation filtering. 3 clusters were allocated for the remaining events as a result of cluster analysis. 373 events (21%) were assigned to the largest cluster. The results of characteristic ellipsoid formation describing the selected damage zone are shown in figure 2 (the center of coordinate system is adduced to the center of selected damage zone). 28 events (7.5 per cent) fall outside 3σ, 69 events (18%) – 2σ, 209 events (56%) – 1σ in the figure.

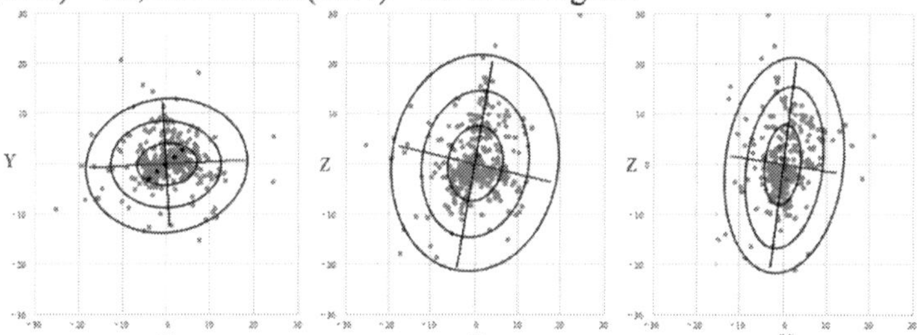

Fig. 2. The result of the damage zone selection algorithm according to seismic monitoring in the projection on orthogonal planes.

Thus, use of the characteristic ellipsoid has allowed reducing number of parameters describing the selected damage zone more than 100 times, while providing ability to control change in volume and shape of a damage zone, both in static representation and in dynamics.

It should also be noted that the proposed method of the damage zones identification is fully customizable and requires a necessary pre-commissioning to adjust parameters in transition to permanent practical use in industrial facilities.

6 Conclusion

It is established by results of existing approaches analysis to dynamic manifestations prediction of rock pressure that the kinetic concept of solids destruction is the basis of most methods. Many scientists who adhere to this concept established that in the process of rocks destruction are several characteristic stages in which destruction process begins with microcracks formation with sizes from millimeters to centimeters and meters, and can be end with breaks from tens to thousands of meters, typical of mountain-tectonic shocks, technogenic earthquakes, etc.

Allocation of the controlled zone, which can be qualified as a damage zone, within which a potentially dangerous source of rock destruction is formed, is the key point of monitoring seismic active zones method, as one of monitoring the geomechanical state methods of rock massif.

Development of mathematical methods capable of statistically sound identification of seismic active zones based on geomechanical monitoring results, as well as the development of a technique for the formation of a characteristic ellipsoid capable of describing geometric properties of identified damage zones with a given accuracy and subsequent integration into associated software of considered monitoring method of seismic active zones will fully automate the process of forecast dangerous state of controlled rock massif, by improving forecast quality, while significantly reducing time spent on obtaining the final result.

References

1. I. Yu. Rasskazov, A. Yu. Iskra, G. A. Kalinov, P. A. Anikin, A. V. Gladyr, M. I. Rasskazov, A. V. Sydlyr, Mining Information and Analytical Bulletin, **8**, 212 (2011).
2. Marcak Henryk, Mutke Grzegorz, J. Seismol, **17**, 4, 1139 (2013).
3. Paige E. Snelling, Laurent Godin, Stephen D. McKinnon, International Journal of Rock Mechanics and Mining Sciences, **58**, 166 (2013).
4. T. Li, M. F. Cai, M. Cai, International Journal of Rock Mechanics and Mining Sciences, **44**, 8, 1149 (2007).
5. V. I. Miroshnikov, A. V. Gladyr, Problems of Complex Development of Georesources V International Scientific Conference, 84 (2013).
6. I. Yu. Rasskazov, S. V. Cirel, A. O. Rozanov, A. A. Tereshkin, A. V. Gladyr, Physical and technical problems of mineral development, **2**, 29 (2017).
7. A. I. Orlov, A nonparametric estimate of the density in topological spaces (Science, 1983).
8. A. I. Orlov, Nuclear density estimates in spaces of arbitrary nature (Perm State University, 1996).
9. A. W. Bowman, Biometrika, **71**, 353 (1984).
10. M. Rudemo, Scandinavian Journal of Statistics, **9**, 65 (1982).
11. A. V. Gladyr, Mining Information and Analytical Bulletin, **21**, 51 (2016).
12. I. S. Enyukov. Methods, algorithms, programs of multivariate statistical analysis: PPCA package (Finance and Statistics, 1986).
13. M. Jambu, Hierarchical cluster - analysis and compliance, Finance and Statistics, (1988)
14. E. Patrik, Fundamentals of pattern recognition theory Sov. Radio, (1980).
15. V. S. Berikov, G. S. Lbov, All-Russian Competitive Selection of Review and Analytical Articles on The Priority Direction «Information and Telecommunication Systems», 26 (2008).
16. J. C. Bezdek, Pattern Recognition with Fuzzy Objective Function, Plenum Press (1981)
17. D. E. Gustafson, W. C. Kessel, Proc. of IEEE CDC, 761 (1978).
18. R. Babuska. Fuzzy Modeling for Control, Kluwer Academic Publishers (1998).
19. X. L. Xei, G. A. Beni, IEEE Trans. on Pattern Anal. and Machine Intell, **8**, 841 (1991).
20. R. Yager, D. Filev, Essentials of Fuzzy Modeling and Control, John Wiley & Sons, (1984).

EXPRESS ASSESSMENT OF ROCKBURST HAZARD IN "NIKOLAEVSK" AND "INTERNATIONAL" MINES BY GEOACOUSTIC METHOD

Andrey Tereshkin[1*], *Pavel* Anikin[1] and *Dmitriy* Migunov[1]

[1]Mining Institute of Far Easten Branch of Russian Academy of Sciences, Khabarovsk, Russia

Abstract. Measurements of parameters of acoustic emission were carried out on a number of burst hazard mineral deposits of the Far East and Transbaikalia. New data on features of phenomena of hazard forms of rock pressure on various sites of the mine are obtained. The technique of carrying out of measurements by "Prognoz L" local monitoring device is developed. Criteria of burst-hazard properties of the rock massif on a number of the mineral deposits, being developed by underground mining, were substantiated.

1 Introduction

Reliable and operational information about a geomechanical condition of the rock massif has an important meaning for the forecast and prevention of hazard phenomena of rock pressure. This information can be received by geophysical methods and measuring tools [1]. Micro seismic (seismoacoustic) and geoacoustic methods are most widely applied. It is possible to carry out both local and regional monitoring of rock massif condition by means of geoacoustic. Scale of monitoring (local or regional) depends on the used technical means. It is based on the experimentally observable and theoretically studied phenomenon of an acoustic emission (AE) which accompanies the process of a mechanical loading and destruction of rocks [2].

2 Design and operating principle of the device

Intensive development of microprocessor and digital technologies has created conditions for development of considerably more perfect geoacoustic portable device ("Prognoz-L") for local monitoring of bust-hazard in Mining Institute FEB RAS in recent years [5]. "Prognoz-L" allows to record a large number of the AE parameters, process and analyze the obtained information. It

* Corresponding author: andrey.tereshkin@bk.ru

considerably increases reliability of operational assessment of a geomechanical condition of the rock massif.

The portable device for local monitoring of burst hard "Prognoz L" consists of two main parts (fig. 1a): primary receiving converter (sensor) and measurement unit. Accelerometer AP99-1000 series (GlobalTest, Sarov) (fig. 1b) was chosen as primary receiving converter for the Prognoz L device by results of the detailed analysis, laboratory and field test.

Observed data and stories of operation of the device is stored on micro-SD flash card. The interface of exchange with cards of this format is realized by STM32F405 microcontroller in the form of the hardware module which allows to read out and record data without using a computing core of the microcontroller.

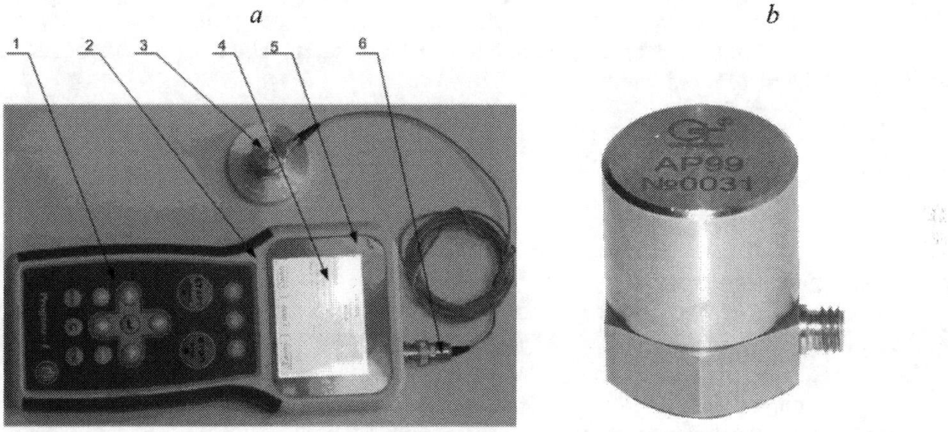

Fig. 1. Prognoz L device: a – measurement unit; b – AP99-1000 accelerometer
1 – keyboard; 2 – device housing; 3 – portable sensor (accelerometer);
4 – liquid-crystal display; 5 – polycarbonate protective glass; 6 – socket connection of the portable sensor

There are four modes of the device. Fist is measurement mode. Second mode is monitoring one; the user and engineering menu are used for the setting of the Prognoz L device.

Setting of the device is performed before the measurement of the AE parameters. Adjusted parameters are date, time, record to digital storage device, volume of sound. There is a possibility of adjustment of frequencies, amplitude and time filter of detection of AE of impulses. User and engineering menu are used for this purpose.

A number of algorithms for selection of the useful AE signals and filtration of technological interference of various nature is realized in the device of local monitoring.

For assessment of a condition of the massif the device of local monitoring "Prognoz-L" uses the following criteria: intensity of AE without visible influence of technological processes N_{AE} and an indicator b (amplitude distribution) [7]. The N_{AE} index of intensity allows to assess the achievement of the ultimate loads in the edge of regional part of rock massif.

The index ***b*** characterizes the instability of process of deformation, increase of quantity of high energy impulses and is calculated by a formula:

$$b = \lg \frac{N^1_{AЭ}}{N^2_{AЭ}} / \lg \frac{A_2}{A_1},$$

where b – a ratio of number of acoustic impulses with a different amplitude (energy); A_1 and A_2 – thresholds (levels of sensitivity of the device); N^1_{AE} and N^2_{AE} – intensity of AE at various thresholds.

Optional AE parameters (spectral characteristics of AE of signals) are used besides the specified criteria (N_{AE} and b) for the assessment of a condition of the massif.

A number of original algorithms for the selection of the useful signals of AE and filtration of technological interferences of various nature is realized in the device of local monitoring. That is extremely important as measurements are very often taken in the composite noise situation of the underground mine.

3 Research burst-hazard

Field tests of the Prognoz L device were carried out in Nikolaevsk, Youzhnyi and Silinsk (JSC Dalpolimetall MMC), Glubokoe (PJSC "PPGHO") and the International mine (Alrosa) underground mines.

Assessment of a geoacoustic condition of the rock massif were carried out by the device of local monitoring in potentially burst-hazard mine caving. Here were obvious visual indications of positive pressure, such as; rupture of timbering, intensive rock cleavage of roof or boards of mine caving, mineral shooting on sites of the massif where the increased level of tension is observed by stationary seismoacoustic control systems [4].

The staff of JSC ALROSA made measurements by the Prognoz L device in April 5, 2018 (mark-779/-862). Most stressed sites of the rocks massif of the mine "International" were measured. The geology of mine course is presented by interlayer of dense, clay and lime dolomite. Rock are unstable, inclined to split. Local showings of oil are noted. Mine course is characterized as I class of gas hazard and as the II group on zones of showings of oil. Mine roof was partly split off during measurements No. 2,3,9.

Twelve measurements were made according to the figure 2 (7 points). Each of them took 5-10 minutes. Results of measurements are given in table 1.

Local sites with high burst hazard degree are defined by results of measurements. Points No. 1,3,4 are included. The category "HAZARD" is appropriated to this site.

Mean values of the measured AE parameters in Avtouklon's face mine "International" (mark -779/-862) are:

- Frequency of impulses is-15113 Hz;
- Duration of impulses - 34,5 ms;
- Amplitude of impulses - 32,5 dB

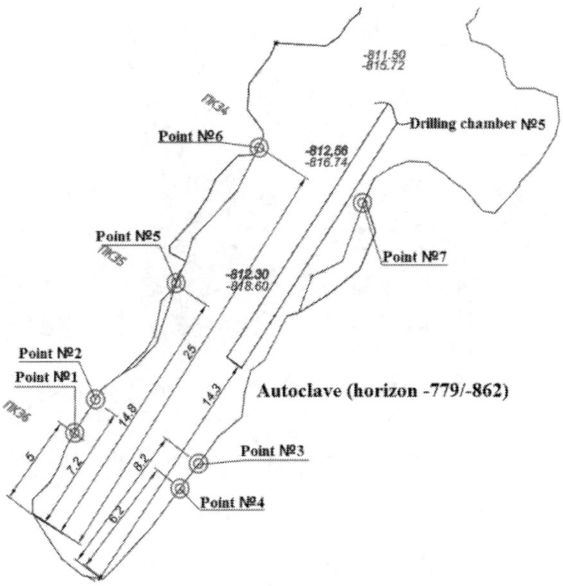

Fig. 2. The scheme of measurement points by the local device on the mine "International"

Table 1. Results of measurement of AE on the mine "International", 5.04.2018.

№	**Time**	**Place**	**t, min**	**Threshold, dB**		**Number of impulses divided by channels**		**Middle activity N_{15}**	**Category of rock burst**
				1	2	N_1	N_2		1
1	10:53	1	10	10	19	119	119	2,975	**HAZARD**
2	11:07	1	10	15	20	95	95	2,375	NOT HAZARD
3	11:23	1	5	15	20	21	21	1,05	NOT HAZARD
4	11:30	1	1	15	20	6	6	1,5	NOT HAZARD
5	11:38	2	10	15	20	47	47	1,175	NOT HAZARD
6	11:55	3	10	15	20	209	209	5,225	**HAZARD**
7	12:07	3	5	17	23	198	198	9,9	**HAZARD**
8	12:17	4	5	10	19	235	235	11,75	**HAZARD**
9	12:24	4	5	15	20	103	103	5,15	**HAZARD**
10	12:35	5	5	15	20	45	45	2,25	NOT HAZARD
11	12:46	6	5	10	19	9	9	0,45	NOT HAZARD
12	12:55	7	5	10	19	9	9	0,45	NOT HAZARD

The high acoustic activity in situ "the North - 8" on marks from-540 m on - 240 m (fig. 3) was recorded from December 1 to December 10, 2017 on the Nikolaevsk mine (JSC Dalpolimetall MMC) ASCRP. Measurements by the local device were performed in mine caving (northern branch, North-8 block, horizon of-375 m) from December 5 to December 7, 2017 year.

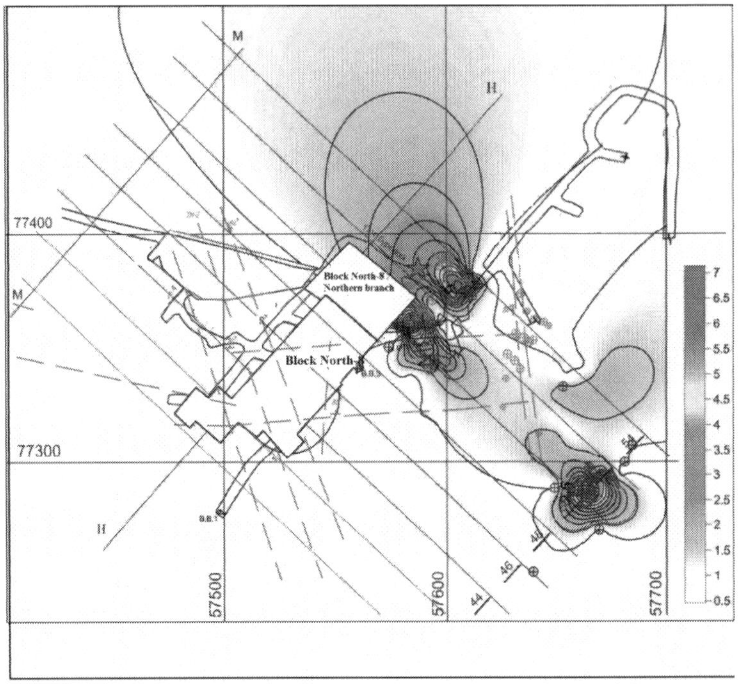

Fig. 3. The map of density of a spatial distribution of the centers of AE of the events recorded on the Nikolaevsk field by the automated Prognoz ADS control system from December 1 to December 10, 2017[3].

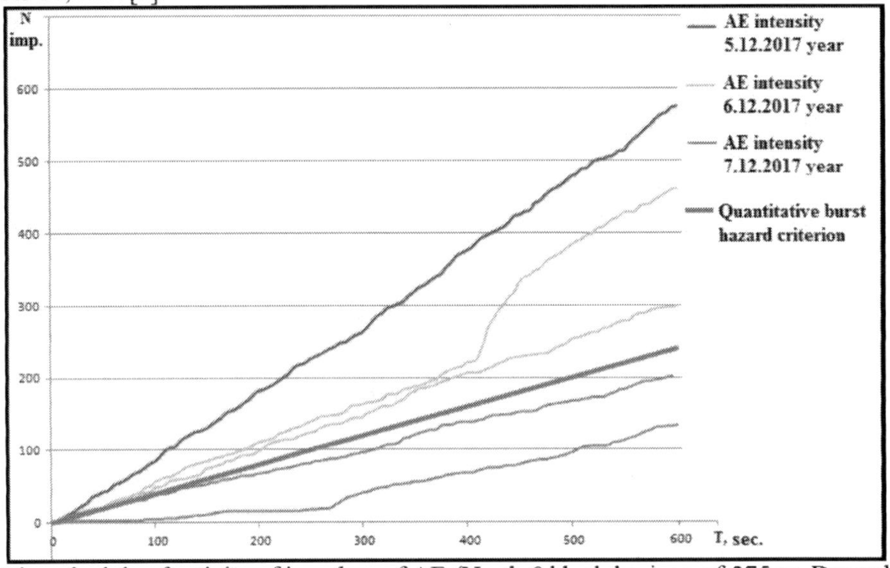

Fig. 4. The schedule of activity of impulses of AE (North-8 block horizon of-375 m, December 5-7, 2017).

Maximal intensity was identified by the device at 10 h 15 min. (5.12.2017), fig. 4. The device indicated "HAZARD" [8, 9]. According the ASCRP indication and "Prognoz-L" data, the mine staff made the decision to close mine caving. After 2 days break the intensity of AE decreased and the device indicated "SAFETY", fig. 6 (9 h 50 min. 7.12.2017 measurement).

Mean values of main AE parameters (North-8 block, -375 m horizon):

- Frequency of impulses -10856 Hz;

- Duration of impulses - 17 ms;
- Amplitude of impulses - 28,0 dB

4 Conclusions

On the basis of use of the modern microprocessor and digital technologies. The geoacoustic portable device for local monitoring of burst hazard potential "Prognoz L" is developed and introduced. "Prognoz L" allows to record and determine the AE parameters in the conditions of technological noise.

Burst hazard research is carried successfully out by the local device in a complex with the automated control system of rock pressure (ASCRP). Research data are verified by numerical modeling of a spatial distribution of the centers of AE events.

The efficient algorithms and the software providing the registration AE signals are developed. They are applied for mine conditions. The following parameters of AE signals are determined: intensity of an acoustic emission; base frequency of impulses of AE; duration, amplitude and the relative power characteristic of AE-events, etc.). The efficient algorithms and the software provide profound analysis of results of geomechanical monitoring.

The measurements of the generated impulses with the given parameters of frequency, amplitude and duration are performed. The results of measurements allow to estimate attenuation of a wave of signals for the distance up to 10 meters from the receiving converter of the device. These measurements were made for different types of rock deep horizons of the "Nikolaev" mine.

The representative volume of the experimental data for development of preliminary criteria of burst hazard of the deep horizons of the Nikolaevsk mine is collected.

5 References

1. Instructions for the safe conduct of mining operations in mines and non-ore deposits, facilities for the construction of underground structures, inclined and dangerous for mountain impacts (RD 06-329-99). GP NTC on safety in the industry of Gostehnadzor of Russia (2000).
2. G.I. Kulakov, G.E. Yakovitskaya FTPRPI. **2**. (1993).
3. A.A. Tereshkin & M.I. Rasskazov / Geomechanics and geodynamics of rock masses / European Rock Mechanics Symposium // Assessment of burst-hazard rock massif by geoacoustic method, P. 1627. **2**. (2018).
4. I.Ju. Rasskazov, B.G. Saksin, M.I. Potapchuk & P.A. Anikin./ Geomechanics and geodynamics of rock masses / European Rock Mechanics Symposium // The researches of burst-hazard on mines in Russian Far East, P. 153. **1**. (2018).
5. I. Y. Rasskazov, D. S. Migunov, P. A. Anikin, A. V. Gladyr, A. A. Tereshkin, D. O. Zhelnin. Journal of Mining Science. **51(3)**. P. 614-623. (2015).
6. A.S. Voznesensky Control systems for geomechanical processes. Publishing house MGGU, (2002).
7. Methodical instructions on seismoacoustic and electromagnetic methods for obtaining criteria for the degree of burst-hazard. VNIMI. (1986).
8. A.O. Rozanov, D.N. Petrov, A.M. Rozenbaum, A.A. Tereshkin & M.D. Ilinov / Geomechanics and geodynamics of rock masses / European Rock Mechanics Symposium // Acoustic emission precursor criteria of rock damage, P. 669. **1**. (2018).
9. I.Yu. Rasskazov, B.G. Saksin, V.A. Petrov, B.F. Shevchenko, V.I. Usikov, G.Z. Gil'manova. Izvestiya, Physics of the Solid Earth. **50 (3)**. PP. 444–452. (2014).

Geodynamic rock condition, mine workings stabilization during pillar recovery at the level +320 m of the Yukspor deposit of the Khibiny Massif

Svetlana Zhukova[2*], *Pavel* Korchak[2], *Anatoly* Streshnev[2] and *Igor* Salnikov[1]

[1]Kirovsk branch of JSC «Apatit», Department of the Technical Director, Kirovsk, Russia
[2]Kirovsk branch of JSC «Apatit», Department of rockburst forecasting and prevention, Kirovsk, Russia

Abstract. The article represents the results of seismic monitoring during pillar recovery at the level +320m of the Yukspor deposit. The number of seismic events with an energy of more than 106 J is sharply increasing in the affected area of overlying rock base. This reflects fracture intergrowth and gradual rock failure due to stress redistribution. Mining operations cause new fractures and interstices, which in turn lead to residual stress redistribution through the formation of new defects. During the rock failure, stresses in the affected area become stable. If fracturing and timely rockfall of overlying rock base do not occur during excavations, thereafter, sudden caving poses a hazard by an underground air strike and can be a threat to the objects on surface. As for stability of mine workings, provoked gradual rock fall does not pose a threat, since this leads to consequent and constant reduction of the mountain base, and, therefore, support pressure, hanging walls become more stable. Underground seismic monitoring based on continuous seismic registration, local monitoring through various geophysical measurements and mathematical modeling of stress-strain rock condition improve the operational safety under difficult geodynamic conditions.

1 Introduction

Up to date, Kirovsk branch of JSC "Apatit" is one of the world's largest enterprises for the production of apatite concentrate. Ore is extracted from three mines: Kirovsky (the Kukisvumchorr and the Yukspor deposits), Rasvumchorr (the Apatite Circus deposit) - underground mining, and Vostochny (the deposits of Plateau Rasvumchorr, Koashva, Nyorkpakhk) - open-pit mining. The exploitation of apatite-nepheline ores at Kirovsky and Rasvumchorr mines is carried out using an underground method of induced sublevel caving. Over 89 years, approximately 2 billion tonnes of ore had been mined, 4 billion tonnes of mined rock had been transported, ground relief had changed significantly, and the connections of tectonic elements had been broken.

* Corresponding author: Szhukova@phosagro.ru

To increase the production capacity in the 2000s, it was settled to make a second slot at the level + 320m of the Yukspor deposit. In 2013, Mining Institute of Kola Science Center corrected previously developed regulations on pillar recovery for the level + 320m, substantiating optimal position and dimensions of sublevel joint sections. The technology for rockburst conditions, the methods of rock monitoring and mine workings stabilization were proposed. Figure 1 illustrates the layout of joint section of the Yukspor deposit.

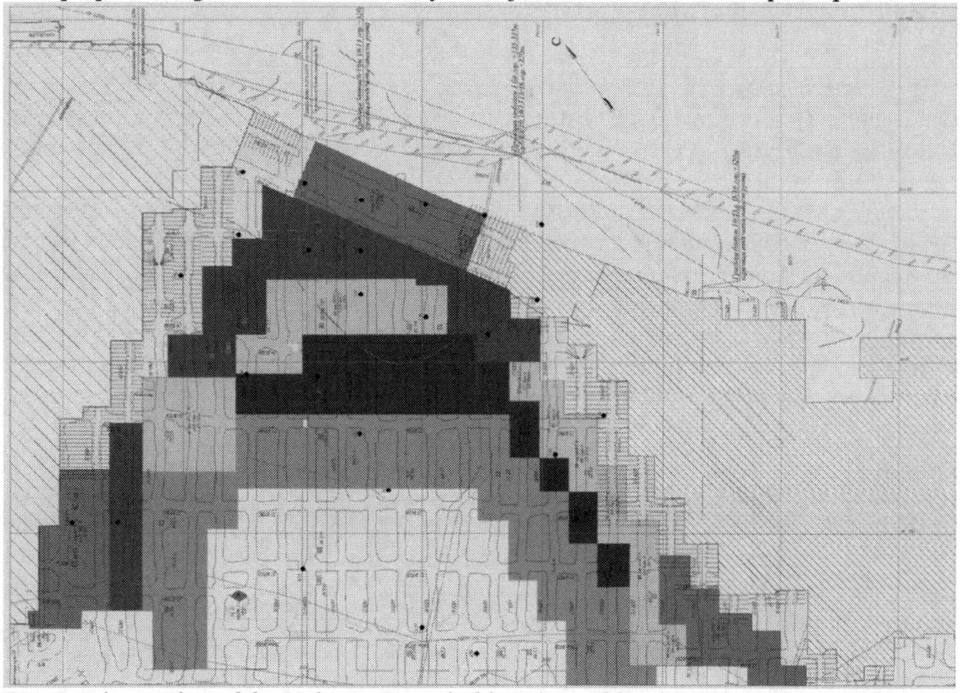

Fig. 1. Joint section of the Yukspor Deposit, bl. 10/13, sublevel + 358m, level + 320m.

The loss of stability of mine tunnels and pillars is mainly caused by the following reasons:

- pillar mining in the affected area and rock pressure increase cause dynamic rockfall, strainburst on the outline of the excavations and pillar bursting;
- seismic induced dynamic events of different origin;
- in highly fractured rocks, as well as in oxidized areas, rockfall occurs under natural and man-made broken conditions. If there are no timely arrangements for safe mining, the formation of gradually increasing kettlebacks begins;
- geological inclusions in the rock mass (dikes, veins, etc.) cause the rockfall along the outline.

The development of three sublevels of the level + 320m of the Yukspor deposit showed that all of the above factors are relevant for the conditions of the mines of Kirovsk branch of JSC "Apatit" and significantly affect the safety of mining operations.

To prevent negative effects of rock pressure, the Department of rockburst forecasting and prevention was established in Kirovsk branch of JSC "Apatit". The following forecasting tools for geomechanical events are used:

- Seismic monitoring - continuous registration, interactive editing of seismic data, analysis of local rock-bump hazard based on continuous seismic monitoring.
- Local monitoring - visual inspection, disking method, control of borehole walls' destruction, various geophysical measurements.

• Mathematical modeling of stress-strain rock condition - detailed numerical local models of stress-strain rock condition of the researched areas based on mining, geological and geomechanical data analysis.

The results of rock condition monitoring are provided below.

2 Seismic monitoring

Underground seismic monitoring is carried out by geophysical monitoring group (GMG) of the Department of rockburst forecasting and prevention using automated systems to monitor (hereinafter AMS) the rock condition detecting geodynamic events with an energy from 1.0 E+02 J, and an accuracy of a few meters (in the area of particular accuracy). AMS is a software and hardware complex for local forecasting of rock-bump hazard based on continuous seismic monitoring of underground mines of Kirovsk branch of JSC "Apatit". Local forecasting of rock-bump hazard based on continuous registration of seismic activity increases mining safety under difficult geodynamic conditions. Telemetric subsystem of seismic data collection (Kirovsk and Rasvumchorr) includes: a network of seismic stations hosted in the mine tunnels underground, cable communication lines and a telemetry controller [1].

2.1 Seismic analysis

When registering a seismic event in the Khibiny Massif, it was determined seismic setting of controlled production area is not steady. In effect, there is a whole spectrum of geodynamic events observed in Kirovsk branch of JSC "Apatit": tectonic rockbumps, rockburst, strainburst, seismic induced rockfall and rock bulking.

According to the results of microseismic monitoring at Kirovsky mine (Km) from 1997 to 2017, AMS had recorded 92 seismic events with an energy of 1.0 E+07J. For the period of January 1, 2007-Janury 31, 2017, AMS had registered 489 seismic events at the Yukspor deposit with an energy of 1.0 E+06J, most of them were recorded in the area of pillar mining and overlying rock base. That is related to stress redistribution and rupture intergrowth into the overlying rocks. It is known that not only geological and tectonic structure, stream conditions, methods of excavation affect the seismic response, but the progress rate and the amount of simultaneously blasted explosive materials as well.

Figure 2 shows the time series of distribution of registered natural events with an energy of 1.0 E+06 J and ore distribution at the Yukspor deposit. With the increasing ore extraction volume, seismic activity at the deposit raises as well.

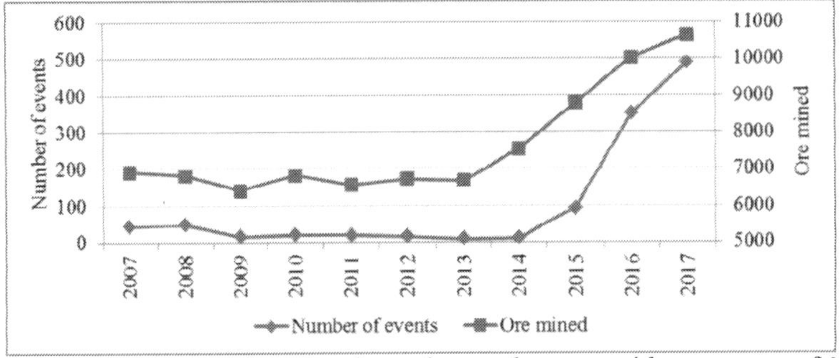

Fig. 2. Time series of distribution of registered natural events with an energy of 1.0 E+06 J and ore distribution at the Yukspor deposit.

Since the beginning of pillar recovery at the level +320m of the Yukspor deposit (Fig.3), the number of seismic events with an energy of 1.0 E+06J in the overlying rock base had sharply increased. This reflects fracture intergrowth and gradual caving of overlying rocks.

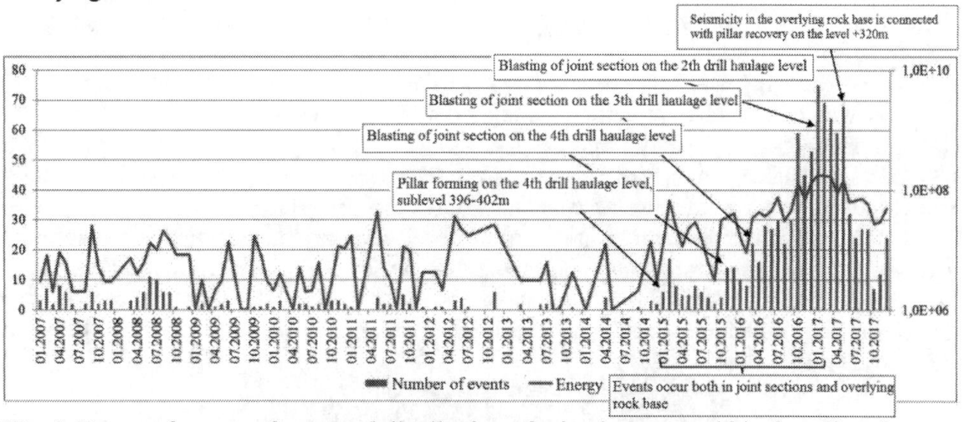

Fig. 3. Energy change and temporal distribution of seismic events within the affected area of the overlying rock base at the Yukspor deposit.

As a result of mining operations, the undermining of hanging wall occurs, which causes extension zones in one part of the rock and increasing compressing strain in another. In case of rockfall, stress stabilization in the affected area occurs [2, 3, 4]. If fracturing and timely recovery of the overlying rocks do not occur during the excavations, then sudden rockfall can be dangerous because of the air strike underground and be a threat to the objects on the surface. As for stabilization of mine tunnels, caving is not a threat, and after caving, due to the reduction of the mounting base, and therefore the decrease of bearing pressure, the stability of hanging walls increases.

As statistical data analysis shows, these processes are accompanied with an increase in the number of seismic events and energy. When registering series of seismic events or events with energy greater than 1.0 E+06 J, according to the approved methods, a zone of local rock-bump hazard is declared, people and equipment are removed from the outlined area.

3 Rock stability and status categories of mine workings during pillar recovery

Kirovsk branch of JSC «Apatit» assessed a stable rock condition and the condition of individual mine tunnels basing on physical properties of the rocks composing the deposit and the magnitude of affective stresses. In accordance with strength assessment and rock fracturing and external features of rock-bump hazard on the outline, status category is determined. There are five status categories A - mine working remains stable, B – rock fracturing, C – rock bulking, slight seismic induced rockfall, D - seismic induced rockfall, E – strainburst and intense seismic induced rockfall, microbumps and rockbursts are possible [5].

At the preparation stage and subsidiary pillars development at the sublevels +320m, rock pressure was observed in the cross drifts and mine tunnels. In the areas of intersections, the rockfalls with geological structural heterogeneities up to 15 meters were noticed.

In case of convergence of mining operations, breaking of joint sections on the overlying sublevels and mine workings made in a hanging wall of the ore deposit fall into the affected area of pillar mining. Compressive stress increases, and, therefore, rock pressure intensifies.

February 5, 2017 and March 27, 2017 at the sublevel +334m, geodynamic events occured. The Commission on rockbursts classified them as microbumps having appeared as a result of a number of negative factors, both natural and self-induced, exacerbating the rock condition in general:

- geological features of rock structure (vertical fractures filled with secondary minerals);
- overlying rock failure border and breaking border of the sublevel +334m are located in the affected area;
- seismic induced fracture intergrowth in the overlying rocks, as the rockfall mounting base is located on the border of stable and non-stable rock exposures.

To forecast further pillar condition at the level +334m, mathematical simulation of stress-strain state in software system SigmaGT was carried out. Classification of rock-bump hazard was made for all mine workings of the pillar. When defining the category as "Dangerous", repeated measures on rock load relief are applied and the rock support system is improved (Fig.4).

Fig. 4. Improved rock support system of haulage entry of the Yukspor deposit.

Since the beginning of mining of joint section at the sublevel +402m in 2015, Kirovsk branch of JSC "Apatit" has significantly revised the approaches to the the types of rock supports. New types of rock supports are implemented - reinforced sprayed concrete, self-attaching anchor bolts, resin-grouted roof bolts and others [6]. Regulations on mining operations on apatite-nepheline deposits were updated: "Guidelines for safety operations on dangerous and burst-prone deposits", "Guidelines for support erection and control of excavations at the mines of JSC "Apatit".

As practice shows, new approaches to the types of supports, definition of their parameters and stabilization of excavations, provides mining safety under the conditions of high compressive stress and seismic impact from blasting.

Conclusions

Automated control systems to detect geodynamic danger zones in on-line mode, effective interaction of production and science, scientific-research and experimental-industrial operations on the development and the introduction of new (and improvement of the existing) means, ways and methods to reduce the risk of adverse situations related to the transition of mining tunnels into unstable condition, allow effectively designing, passing and operating under difficult geodynamic conditions of the Khibiny deposits [7-9].

We note that in order to reliably solve the problem of preventing dangerous manifestations of rock pressure in underground mines, multilevel systems of complex geodynamic monitoring should be created, which may include a number of measurement methods and technical means that ensure effective registration and in-depth analysis of a wide range of spatially and temporally changing geophysical parameters wave fields (geoacoustic, microseismic, deformation, etc.).

Joint interpretation of geoacoustic and microseismic monitoring data, correlated with geological and mining technical models, provides high reliability of forecast estimates [10, 11].

References

1. P.A.Korchak, S.A.Zhukova, P.Y.Menshikov Formation and development of monitoring system for seismic processes in the production area of JSC "Apatit". *Mining journal* **No. 10** (2207). Viva Star. 96 p. (2014).
2. V.S.Kuksenko, I.E.Inzhevatkin, B.A.Manzhikov and others *Physical and methodological basis for rockburst forecasting.* Physical and technical problems of mine tunnels development. **No.1.** p. 9-22. (1987).
3. V.S.Kuksenko, V.A.Petrov. *Physics of solids destruction and earthquake forecasting /* / Physics: issues, history, and people. L.: Science. p. 79-84. (1986).
4. A.V.Lovchikov Conceptual approach to the problem of rockburst forecasting at the mines. *Geomechanics in mining operations in highly stressed areas.* Apatity. p. 95-108. (1998).
5. Guidelines for safety operations on dangerous and burst-prone deposits (the Khibiny apatite-nepheline deposits) (Joint-stock company "Apatit") /Corporate authors//the Mining Institute KSC RAS, JSC "Apatit". Apatity. 112 p. (2016).
6. "Guidelines for support erection and control of excavations at the mines of JSC "Apatit" /St. Petersburg-Kirovsk / Corporate authors. 56 p. (2017).
7. Ortlepp W D. Observation of mining-induced faults in an intact rock mass at depth. International Journal of Rock Mechanics and Mining Sciences. **37(1/2)**, 423–436, (2000).
8. Cai M., Kaiser P. K., Tasaka Y., Maejima T., Morioka H., Minami M. Generalized crack initiation and crack damage stress thresholds of brittle rock masses near underground excavations. International Journal of Rock Mechanics and Mining Sciences. **41(5)** 833–847, (2004).
9. Diederichs M.S., Carter T., Martin D. Practical rock spall prediction in tunnels. Proceedings of the 2010 World Tunnelling Congress, Tunnelling Association of Canada, Richmond, paper 517. (2010).
10. I. Yu. Rasskazov, A. V. Gladyr, P. A. Anikin, V. S. Svyatetskiy, B. A. Prosekin, Mining Journal, **8**, 9 (2013).
11. A. V. Gladyr, Mining Information and Analytical Bulletin, **6**, 220 (2017)

Investigation of regularities of brittle fracture formation around mine workings in overstressed rocks at the mines of Kirovsk branch of JSC «Apatit»

Pavel Korchak[1*]

[1]Kirovsk branch of JSC «Apatit», Kirovsk, Russia

Abstract. The article is devoted to the investigation of brittle fracture formation in underground mines of Kirovsk branch of JSC "Apatit" by applying numerical simulation and subsequent comparison with the results of field data. It includes the results of studies of rock failure forms with different directions, values and stress ratio, as well as the measures to reduce stress by drilling relief holes. Brittle fracturing and dependencies of its formation based on many factors, including mining, geological and geomechanical factors were researched. The results serve as the basis for the next research aimed at searching and grounding the effective ways to stabilize mine tunnels in case of brittle fracturing.

1 Introduction

At the Khibiny apatite mines, all forms of dynamic rock pressure are noted: from external signs of rock-bump hazard (rock bulking, dynamic seismic induced rockfall, strainburst) to the actual rockbursts of both local and regional origin. All these events are to some extent related to the features of stress-strain rock condition, characterized by increased horizontal (almost horizontal) stresses, much higher than the vertical ones.

The main, envisaged by the regulations, way to reduce stress on the outline (turn into unhazardous condition) is drilling the relief holes and boreholes [1]. This method is based on the application of rock pressure energy for interhole pillar rashing. In this case, in the outline adjacent to the line of relief holes, due to the stretching force, sloughing zone is formed, which provokes cutter break forming, resembling in configuration the dynamic rockfall, which also effects further stability, and, therefore, should be taken into account when choosing a method of stabilization.

Currently, one of the main priorities to improve the stabilization in underground mines of Kirovsk branch of JSC "Apatit" is to explore the methods to support and to maintain the excavation in the overstressed rock mass. This search activity is based on the studies presented in this article.

In this regard, the objectives of the study were as follows:

* Corresponding author: Pkorchak@phosagro.ru

1. Study of regularities of brittle fracture formation, including the increased stresses, different directions of their activity, as well as instantaneous and fatigue rock strength;

2. Study of regularities of brittle fracture formation including relief holes' availability.

2 Theory and research procedure

Rock failure next to the mine working at the mines of Kirovsk branch of JSC "Apatit", as well as at the other enterprises, depends on stress condition, rock strength and structural rock faulting. If rock stress is low *σ_1 / σ_(c,i)<0.15,* the rockfall into the developed space is mainly determined by the level of fracturing, that is, the number and the orientation of fractures relative to the mine tunnel position. However, as stress level increases, the rockfall into the developed space is associated with new fracture formation oriented parallel to the outline. At medium depths, the fractures are localized in the peripheral outline area, but as the stress increases, the area of brittle fracture also expands. The rock volume in the brittle fracture zone determines the support load and must be defined to improve the reliability of temporary or permanent support parameters [2].

Brittle rock failure close to the excavations is defined as rock bulking or rock spalling, as a result a V-shaped zone is formed, and the rockfall can occur both as monotonous almost quasi-static failure, or as dynamic failure, local rockburst. The size of these zones varies from a few centimeters to a meter, sometimes even more. In general, regularities of brittle fracture formation at the mines of Kirovsk branch of JSC "Apatit" coincide with widely recognized, stated in various sources, for example, in [3].

At the first stage of the study, it was necessary to calculate stress-strain rock condition close to the excavations and to assess the brittle break zone for the mines of Kirovsk branch of JSC "Apatit".

The studies were carried out on the grounds of numerical simulation including the nonlinear Hoek–Brown failure criterion [4].

On the grounds of plane deformation, brittle fracture formation near the excavations without relief slot along the length of the mine working or incomplete response of relief holes were studied. The calculations were made for the inclination angles of principal stress areas in the global coordinate system, 0, 15 and 30^0, that corresponds to the mining terms at the mines of Kirovsk branch of JSC "Apatit" [5, 6].

Boundary conditions of the models - standard for geomechanical issues. Shifts in the direction perpendicular to the model boundaries are prohibited. The initial field of stress condition was set as a tensor of principal stress. When rotating basic areas relative to the global coordinate system, stress tensor was recalculated as follows

$$\boldsymbol{\sigma}' = \boldsymbol{T}\boldsymbol{\sigma}\mathbf{T}^{\mathbf{T}}, \tag{1}$$

where σ - stress tensor in the global coordinate system; σ' - stress tensor in the global coordinate system after rotation of basic areas by corresponding angle; T - stress transformation matrix [7, 8].

Stress-strain state calculation is performed for the conditions when the ratio of maximum stress on the outline to the strength of undisturbed rock sample corresponds to 0.5, 0.7, and 1.0. Taking the rock strength equal to 150 MPa, and given the magnitude of vertical stresses equal to 15 MPa, we obtain the following initial rock stress (table 1). The ratio between stresses is chosen using actual (full-scale) measurements of their values at the mines of Kirovsk branch of JSC "Apatit".

Table 1. Initial field of stress condition

$\frac{\sigma_{\theta.max}}{\sigma_{c.i}}$	σ_1, MPa	σ_2, MPa	σ_3, MPa
0.5	30	15	15
0.7	40	20	15
1.0	55	27.5	15

When performing the calculations it is accepted that the value of long-term rock strength relative to its initial value, that is, the index of long-term strength, is 0.9 on average.

3 Research data

Research data showed that brittle fracture zone at $\frac{\sigma_{\theta.max}}{\sigma_{c.i}} = 0.5$ is mainly formed in the peripheral outline area, and its size varies from 0.15 to 0.25 m. Depending on the orientation of principal stresses relative to the global coordinate system, the position of brittle fracture zone shifts from the dome to the wall by approximately the same angle. When increasing the initial stress to $\frac{\sigma_{\theta.max}}{\sigma_{c.i}} = 0.7$, the area of brittle fracture expands to 0.8-0.9 m, and it adopts a V-shape. Subsequent stress increasing $(\sigma\_(\theta.\,max)/\sigma\_(c.\,i) = 1.0)$ leads to the sharp growth of brittle fracture zone. Over time, strength range also leads to the growth of brittle fracture zone by 1.3 – 1.8 times. The higher the value is, the smaller the value was at the time of mine development. Absolute sizes of brittle fracture zone obtained for all calculated cases are represented in table (table 2). The drawings predicting brittle fracturing are shown in figures 1 and 2.

During the second stage of the study on the grounds of spatial model, the formation of brittle fracture zone with relief slots both along the length of the excavation and advance relief holes had been studied. Based on spatial calculations, the growth of limit-state area along the length of the excavation (figure 3), including relief holes, was observed.

Size assessment of brittle fracture zones along the length of the mine tunnel is made based on elasticity solution, which does not include stress redistribution due to the rocks reaching their limit state. At the same time, it is seen that the growth of brittle fracture occurs in the bottom-hole zone, and mostly at the distance equal to the width of the excavation, the development of this area is completed. Thus, if significant geomechanical and geodynamic processes caused by the excavations are expected, it is necessary to perform the advanced unloading of the rock mass.

According to the results of this study, important practical conclusions were made, namely:
- relieve slot significantly reduces outline stress and reduces rock area, potentially dangerous for the loss of stability, and therefore is an effective way for safety operations at the mines of Kirovsk branch of JSC «Apatit»;
- after development of interhole pillars (after dynamic rockfall) in the extraction area or around an additional fracture zone connected with a decrease in rock strength may form over the time (fatigue rock strength), it means that supports systems in these areas must be designed considering this factor. It is also worth noting that the results are fully correlated with the results of full-scale measurements carried out at the mines of Kirovsk branch of JSC "Apatit". The examples of cross sections with brittle fracture are shown in figure 4.

Table 2. Dimensions of brittle fracture near the excavations under various mining and geological conditions

Inclination of principal stresses *	Time for brittle fracture formation	Size of brittle fracture		
		$\frac{\sigma_{\theta.max}}{\sigma_{c.i}} = 0.5$	$\frac{\sigma_{\theta.max}}{\sigma_{c.i}} = 0.7$	$\frac{\sigma_{\theta.max}}{\sigma_{c.i}} = 1.0$
0^0	Straight after excavations	0.22	0.89	1.85
	Upon reaching long-term strength	0.39	1.15	-
15^0	Straight after excavations	0.19	0.85	2.03
	Upon reaching long-term strength	0.34	1.18	-
30^0	Straight after excavations	0.26	0.81	1.72
	Upon reaching long-term strength	0.46	1.07	-

Note: * The inclination of the areas of principal stress relative to the global coordinate system

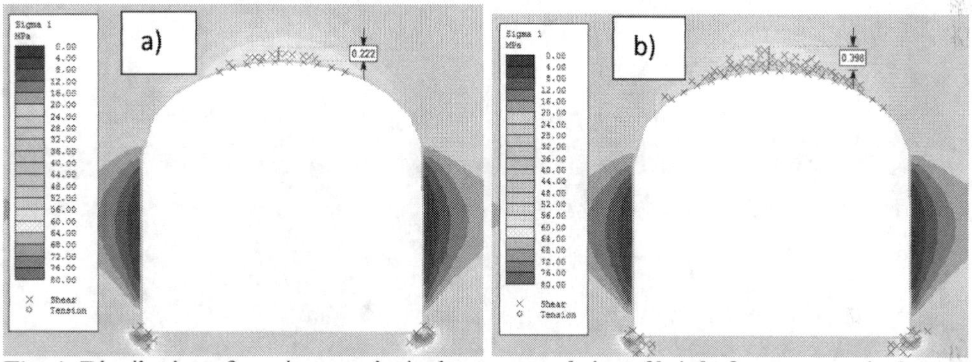

Fig. 1. Distribution of maximum principal stresses and size of brittle fracture zone in the peripheral outline area (ratio of principal stresses 30:15:15): a – in straight and b – long-term strength, the angle of inclination of principal stresses is

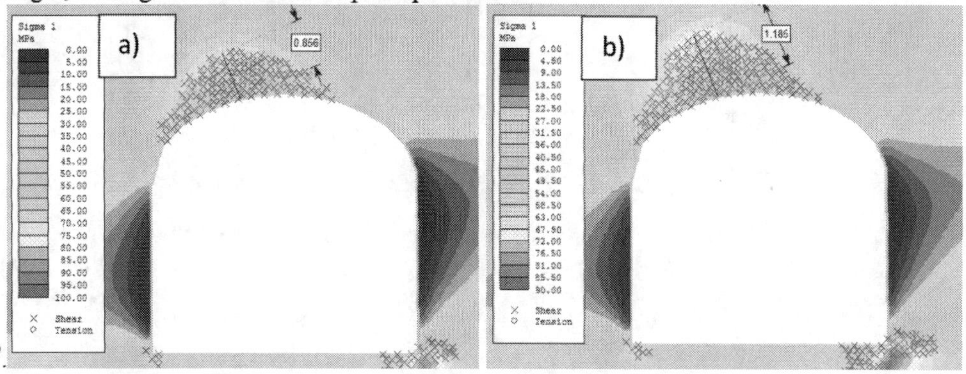

0^0.

Fig. 2. Distribution of maximum principal stresses and size of brittle fracture zone in the peripheral outline area (ratio of principal stresses 40:20:15): a – in straight and b – long-term strength, the angle of inclination of principal stresses is 15^0.

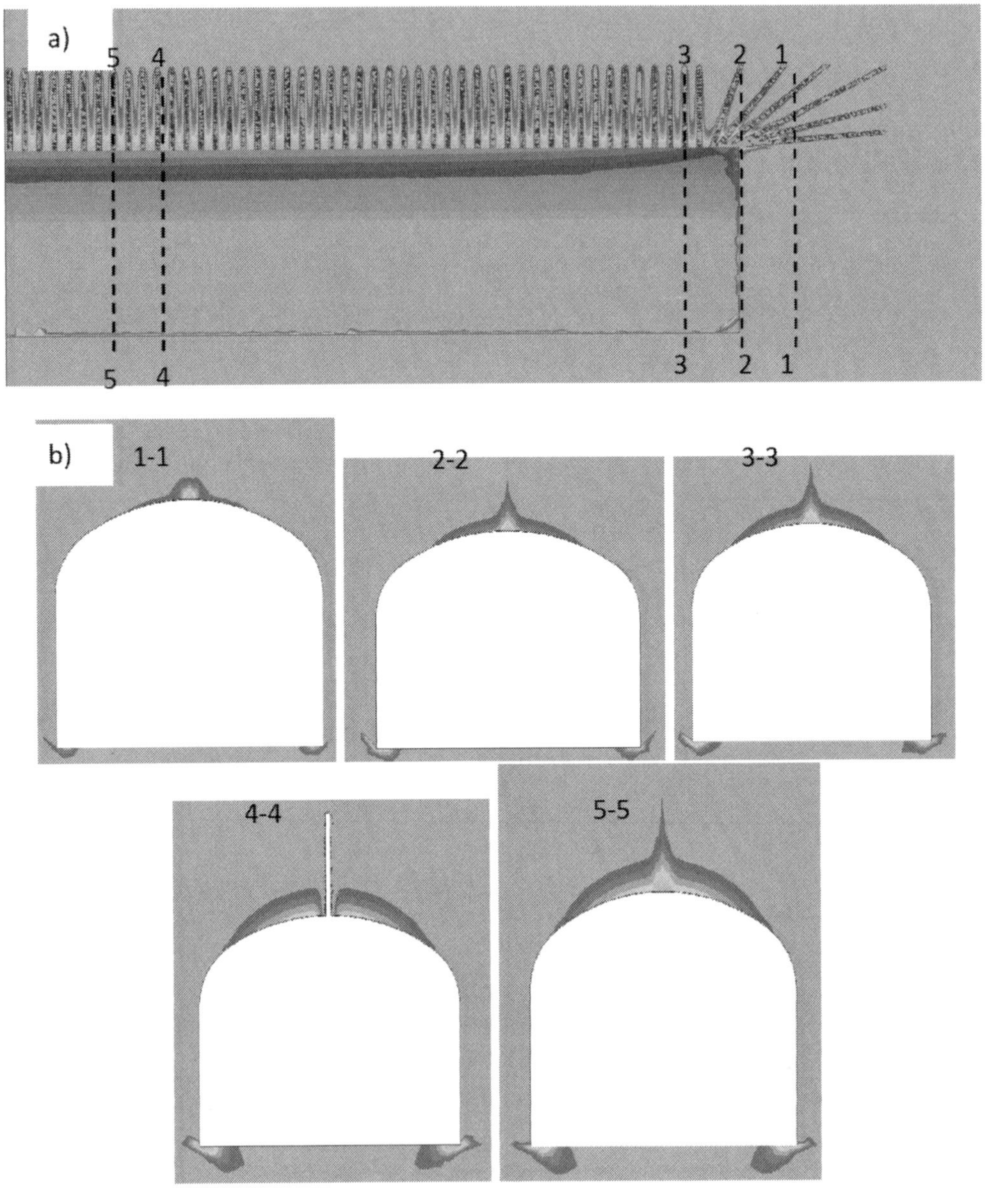

Fig. 3. The size of limit-state area, characterizing brittle rock fracture with vertical position of relieve slots (the results were obtained based on elasticity solutions): a – longitudinal section; b – cross sections along the length of the excavation.

Fig. 4 Brittle outline fracturing at the mines of Kirovsk branch of JSC "Apatit".

4 Conclusions

The development of geomechanical processes in highly stressed rock massifs, as shown in the paper, is a very complex process. The forecast of such processes, using the traditional prerequisites for deformation and destruction of rocks, does not allow to obtain reliable sizes of brittle fracture zones. In this paper, a technique has been adapted in which the rock strength mobilization process is not realized immediately, but as micro and macrocracks form and develop, which allows matching between the observed actual zones of brittle fracture of the rock massif in the vicinity of Kirovsk branch of JSC "Apatit" mines and calculated values.

On the whole, the assessment of the geomechanical state of the rock massif in the vicinity of the production located in a highly stressed rock mass allowed not only to determine the size of the brittle fracture zone presented in Table 2, but also to obtain important regularities for practical purposes.

Analysis and processing of the results showed that in the next stage it is necessary to carry out a geomechanical justification for the use of high-performance and reliable ways of fixing and supporting workings in the conditions under consideration, taking into account the results obtained, which will be the subject of further research. It should also be said about the need for joint use of modern measuring complexes that provide reliable registration and in-depth analysis of a wide range of geophysical wave fields that vary in space and time [9,10].

References

1. "Guidelines for safety operations on dangerous and burst-prone deposits (the Khibiny apatite-nepheline deposits)", Apatity-Kirovsk. (2016).
2. C.D. Martin and N.A. Chandler. The progressive fracture of Lac du Bonnet granite. International Journal Rock Mechanics Mining Science & Geomechanics Abstracts, **31 (6).** 643–659. (1994).
3. C.D. Martin, R.S. Read and J.B. Martino. Observations of brittle failure around a circular test tunnel. International Journal Rock Mechanics And Mining Science. **34 (7)**. 1065–1073. (1997).
4. C.D. Martin, P.K. Kaiser, and D.R. McCreath Hoek–Brown parameters for predicting the depth of brittle failure around tunnels. Can. Geotech. J. **36.** 136–151. (1999).

5. Diederichs M.S., Carter T., Martin D. Practical rock spall prediction in tunnels. Proceedings of the 2010 World Tunnelling Congress, Tunnelling Association of Canada, Richmond, paper **517**. (2010).
6. Cai M., Kaiser P. K., Tasaka Y., Maejima T., Morioka H., Minami M. Generalized crack initiation and crack damage stress thresholds of brittle rock masses near underground excavations. International Journal of Rock Mechanics and Mining Sciences. **41(5).** 833–847, (2004).
7. Ortlepp W D. Observation of mining-induced faults in an intact rock mass at depth. International Journal of Rock Mechanics and Mining Sciences. **37(1/2),** 423–436, (2000).
8. Potvin Y. Surface support in extreme ground conditions—HEA Mesh. In: Safe and Rapid Development in Mining. Perth, Western Australia, 111–119, (2009).
9. I. Yu. Rasskazov, A. V. Gladyr, P. A. Anikin, V. S. Svyatetskiy, B. A. Prosekin, Mining Journal, **8**, 9 (2013).
10. A. V. Gladyr, Mining Information and Analytical Bulletin, **6**, 220 (2017).

Development of multi-channel portable impact control device for local assessment of the state of the edge parts of the rock massif

Alexander Konstantinov[1*]
[1]Mining Institute of Far Easten Branch of Russian Academy of Sciences, Khabarovsk, Russia

Abstract. The improvement directions of geoacoustic methods and software and hardware used for local rock mass impact hazard assessment are proposed in the article, in particular, comparison is made with the successfully proven Prognoz L device. Three main directions for its modernization are proposed, including the improvement of constructive elements, providing for the use of discrete components and wireless technologies advantages, a new approach to software development and the use of a "full" operating system. Thirdly, it is hardware improvement of the local control device, which in the future will provide source-locating possibility of acoustic emission in the massif.

1 Introduction

At the present geoacoustic signal security systems development is becoming more urgent because of mining sites volume increasing and high mining work speed [1, 2]. Such systems belong to non-destructive testing methods and therefore is absence of requirement of the rock mass continuity violation are their main advantage with the ability to large volume control [3]. The entire burden of providing expert assessment of rock mass state falls on smart systems of data processing and interpretation that obtained through passive registration of natural processes caused by mining [4]. This determines the effectiveness of their use in real dynamic production conditions.

The Mining Institute FEB RAS together with the NC Geomechanics St. Petersburg Mining University is developing an approach to shock hazard local control on based of software and hardware systems using geoacoustic methods in connection with task relevance.

1.1 Overview of tools used

Prognoz L is one of the devices of local control of mining pressure, that allows to carry out local control of shock hazard, combining modern technical means and flexible software, and is successfully applied on a number Russian mines.

The following criteria are used to determine the impact hazard degree by analytical software of the local control device: the average acoustic emission signal intensity $N1_{AE}$

* Corresponding author: alex-sdt@yandex.ru

and the amplitude allocation index b. Scientific and practical usage and laboratory tests of the methodological support of the local control device are showed sufficient precisy and valid assessment of the impact hazard category corresponding to the initial destruction moment for strong rocks. From other side, the flexibility of the calculation parameters allows for the adjustment of analytical algorithms taking into account the features of almost any mining enterprise.

It should also be noted that the flexibility of the software provided by device developers allows producing additional methods of analysis and processing of rock mass state monitoring results.

As a result of portable device Prognoz L tests conducted in Kirov mine JSC "PhosAgro" [5]:

– Detailed analysis results of signal wave forms and trends of acoustic emission spectral-energy parameters for a wide range of mountain-geological conditions of rock pressure development are obtained;

– A qualitative model of geoacoustic phenomena in mine areas is formulated;

– The acoustic emission signals classification by spectral composition and their correlator forms is carried out, which determine their belonging to a certain source type;

– The modern algorithm of information flow processing and analysis based on the pattern recognition concept is developed. The algorithm is based on spectral-correlation analysis of random processes [6, 7].

Based on the above, it can be concluded that the developed device has undeniable advantages, including:

– Light weight and compact dimensions, allowing to carry out device transportation and to carry out necessary measurements;

– The presence of a graphical display, allowing with sufficient comfort and acceptable information to adjust built-in software parameters and analyze measurements results, etc.

2 Features of the developed device

More than biennial experience in the Prognoz L operation at various industrial objects allowed us to formulate requirements for the device hardware and software modernization and the development of its extended version, which has a number of advantages.

These requirements can be divided into three groups:

– Construction improvements;

– Software improvements;

– Hardware improvements.

2.1 Constructive features

As a design improvement, it is proposed to abandon the film keyboard as the main method of information input and control of the device. Touch screens are now becoming the most common and convenient to use. The additional keyboard in this case can be only an auxiliary input tool. The keys of this keyboard should be assigned to the most popular control command of the built-in software, as well as the device on and off buttons.

It should also be noted that since the device is supposed to be used in low light conditions, it is proposed to equip an additional keyboard with additional illumination.

Currently, the Prognoz L device uses a built-in battery to provide power supply. To replace the battery, it is necessary to disassemble the device in stationary conditions, which is unacceptable in the conditions of continuous measurement works at industrial objects. It is proposed to develop and implement a power supply system with duplicating of the supply elements with the possibility of operational "hot" replacement of the power elements of the device as a constructive improvement.

It is necessary to provide the device protection from external influences not lower than IP65 because of usage of the device in aggressive climatic environment conditions, in high humidity conditions and dust for most of the time. At the same time, it is should be provided sealed plugs for all external interfaces as a mandatory condition. It should be provided, if possible, wireless technology to increased protection, communicate with external devices and to charge the device's batteries.

A STM32 family microcontroller is the Prognoz 1 device computational core currently. The used microcontroller does not allow full operating system using and therefore, as a built-in software is used specially developed by the C language programming [8].

High performance of the created systems is advantage of this approach, and main disadvantages are need for hardware reprogramming of each device instance when making changes to control program code and control program development and modernization complexity.

The use of Intel Compute Stick [BLKSTK2m364CC] nettop based on x64 architecture is the most optimal approach in this case.

2.2 Improve the software

It is proposed to use Windows 10 IoT Enterprise - a product of Microsoft, made on the Windows 10 core, as an operating system.

The advantages of this operating system are:

– Intending for use on commercial devices (in accordance with the license agreement): kiosks, ATMs, thin clients, POS-systems, medical systems, information boards, etc.;

– Available in the Long Term Servicing Branch form (LTSB - long term maintenance assembly), which means a complete functional updates rejection that can make unplanned changes to the OS running on a specialized device. Using Windows 10 IoT Enterprise on embedded systems ensures stable and predictable device operation over a long period of time;

– High demand among software developers and users;дружелюбный графический интерфейс;

– Embedding capabilities: device lock, write filters, pop-up notification lock, etc.);

– Extensive software development capabilities using the .NET Framework and .NET Core platforms and embedding many other technologies and software solutions with extensive data display and processing capabilities;

– Support for Enhanced Write Filter (controllable write filter), which allows to protect the device software from viruses, trojans, as well as inexperienced users, limiting access to the file space.

The use of the Windows operating system allows to reduce time and cost of developing specialized software registration, visualization and analysis through application of developed tools to quickly develop software using third-party modules and libraries [9, 10], including as common mathematical analysis packages as MATLAB. This will allow more successful implementation and testing of various algorithms [11, 12].

The use of a full operating system with a developed graphical user interface allows you to equip the updated device with additional software.

As an additional software can be recommended:

– Reference database;

– Graphic image plans for underground horizons of the mining enterprise;

– Software tools for reporting on the results of measurements;

– Automatic extensible database of measurement reports.

2.3 Improve the hardware

Is should be paid to the technical capabilities of analog information input when forming requirements for improving the local control device. Currently, it is possible to process one channel of analog data with a frequency of 100 kHz and a resolution of 16 bits. It is not possibility of locating the source of acoustic emission in the array in this case. The need for a linear location of signals was formulated at the initial device development stage with the possibility of expanding in relation to a flat and three-dimensional location as a result of the practical use of the local control device. Note that the existing technical characteristics of existing local control device hardware do not allow to fulfill these needs.

To solve this problem, it is proposed to use analog-to-digital transformation industrial modules developed by the research and production group R-Technology. As the ADC selected RT USB3000, which is a compact universal 8-channel ADC, connected to a computer via USB 2.0 interface.

The device equipment additionally includes a 2-channel DAC and input/output digital lines.

The device can be used as a multi-channel oscilloscope, spectroanalyzer or logic analyzer, as well as a full-fledged recorder with the ability to save data on the computer's hard disk without breaks and recording time restrictions.

Circuit design of the input stages provides:

– High-precision measurements whatever of the signal source output impedance;
– Possibility to measure high voltage signals directly through resistive dividers;
– Low passage between channels and commutation noise lack;
– High input impedance of all channels in any operating mode.

Small dimensions and weight, as well as the ability to power directly from the USB bus to allow the use of the RT USB3000 in a compact mobile measuring complexes on the base of the laptop, the possibility of simultaneous acquisition, processing, visualization and storage of data without gaps for an unlimited time.

The included software for your computer (OS Windows) allows you to:

– Start working with the device immediately after connection, without prior calibration and programming;
– Process, visualize and save data to your computer's hard drive in real time.

Note that the maximum frequency of digitization is 750 kHz per channel if simultaneous use of four analog-to digital conversion channels, and 375 kHz if eight channels. This significantly exceeds the current local control device's capabilities [13].

The use of four or more channels of analog information input from the acoustic emission receivers will allow further to locate of acoustic emission sources on the plane and in the volume after the development of the corresponding software.

Sensors are selected depending on the measurement modes used in the multi-channel local shock hazard control device and differ in the type of built-in electronics. Recommended types of sensors are shown in the table 1.

Table 1. Recommended sensor types.

Characteristic	GT250	GT350
Type	low frequency with amplifier	broadband with amplifier
The electroacoustic conversion coefficient, dB Rel. 1 W / m / s	> 90	> 75
Operating frequency, kHz	50	120
Bandwidth, kHz	40 ... 100	100 ... 800
Gain	10	10

The primary frequency converters have built-in electronics, therefore, it is should be supplied constant voltage to the output lines simultaneously with reading the analog signal from the sensor output, which will be used to ensure the operation of the built-in electronics. This involves the development of a specialized preamplifier, which combines the sensor's power supply functions and the signal amplifier to ensure the coordination of the levels at the sensor output and the input of the analog-digital conversion module. Also, it is should be provide the possibility of software change of the gain to ensure the widest dynamic range of measurements as a necessary condition of the development of a specialized pre-amplifier. The use of digital outputs of the RT USB3000 analog-to-digital conversion module is preferred to control the amplifier and switch the gain.

There is a block diagram of a multichannel local impact hazard monitoring device at the figure 1.

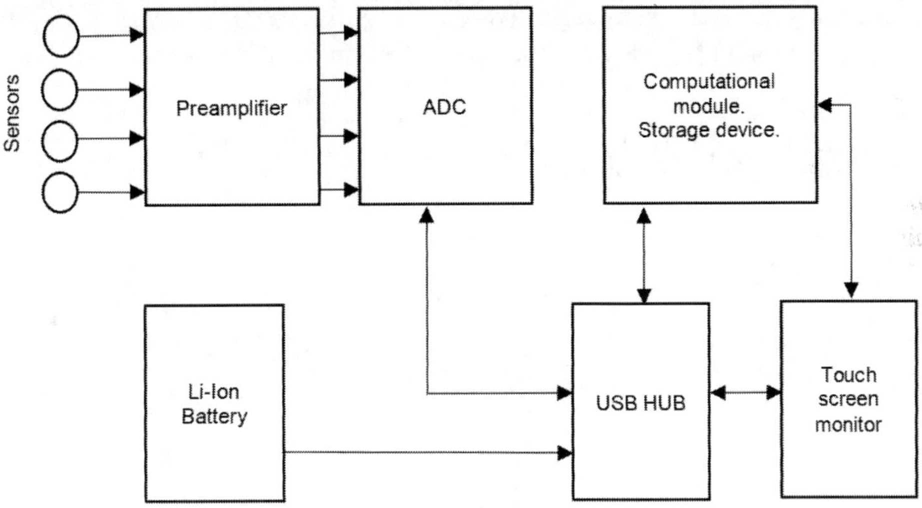

Fig. 1. Block diagram of a multichannel device for local shock hazard control.

3 Conclusion

On the one hand, proposed improving the structural, hardware and software will allow you to obtain a better instrument for local control of impact hazard in the rock massif, giving its users many new functions for use in the field, as well as convenience and comfort during use. On the other hand, such an approach to the development of a portable device will allow its developers to more quickly implement advanced tools aimed at implementing impact hazard local control, correct software errors and improve the technical device characteristics. Thus, the set of proposed tools is an important and necessary step in the geoacoustic tools development of impact hazard local control in the rock mass.

References

1. I. Yu. Rasskazov, *Monitoring and rock pressure management in mines of the Far Eastern region* (Mountain Book, 2008).
2. I. Yu. Rasskazov, B. G. Saksin, V. A. Petrov, B. A. Prosekin, Physical and technical problems of mineral development. **3.** 3 (2012).
3. A. V. Gladyr, Mining Information and Analytical Bulletin. **6**. 220 (2017)
4. I. Yu. Rasskazov, S. V. Cirel, A. O. Rozanov, A. A. Tereshkin, A. V. Gladyr, Physical and technical problems of mineral development. **2**. 29 (2017).
5. A. A. Tereshkin, D. S. Migunov, P. A. Anikin, A. V. Gladyr, M. I. Rasskazov, Problems of subsoil use. **1 (12)**. 72 (2017).
6. Z. Reches, D. A. Lockner, J. Geophys. Res., **99**, B9, 18159 (1994).
7. T. Backers, O. Stephansson, E. Rybacki, Int. J. Rock Mech. Min. Sci., **39**, 755 (2002)
8. I. Yu. Rasskazov, A. V. Gladyr, D. S. Migunov, P. A. Anikin, A. A. Tereshkin, Certificate of registration of the computer program RUS 2014613552 (2014)
9. A. V. Gladyr, Certificate of registration of the computer program RUS 2017660983 (2017).
10. A. V. Gladyr, Certificate of registration of the computer program RUS 2017660981 (2017).
11. T. Backers, S. Stanchits, G. Dresen, International Journal of Rock Mechanics and Mining Sciences, 42, 7-8, 1094 (2005).
12. A. O. Rozanov, 75th EAGE Conference and Exhibition Incorporating Spe Europec (2013).
13. I. Yu. Rasskazov, D. S. Migunov, P. A. Anikin, A. V. Gladyr, A. A. Tereshkin, D. O. Zhelnin, Physical and technical problems of mining. **3.** 169 (2015).

E3S Web of Conferences 56, 02025 (2018) https://doi.org/10.1051/e3sconf/20185602025
VII International Scientific Conference "Problems of Complex Development of Georesources"

THE GEOMECHANICAL STATE OF THE MINE «MNOGOVERSHINNOE» LOWER LEVELS MONITORING

Arkady Avdeev[1*], *Elena* Sosnovskaya[2] and *Roman* Krinitsyn[1]

[1]Institute of mining- the Ural branch of the Russian Academy of Sciences, Ekaterinburg, Russia
[2]Irkutsk National Research Technical University, Irkutsk, Russia

Annotation. In the low level mining of the Mnogovrshonniy mine, there was a harmful rock stress manifestation in the form of displacement and collapse of rock blocks, which adversely affected the safety of mining operations and reduced the mining efficiency. There was an urgent need to conduct special studies of the geomechanical execution process at the mine. In the process of research the basic physical and mechanical rock characteristics were determined, the initial natural ground stress, man-made stress in mines and pillars was measured, the stability degree of underground cavities was determined. Based on the current geological and surveying documentation analysis and visual observations, the areas of mine workings that are dangerous because of the rock stress and displacement manifestations were established. For the identified areas, projects to set the observation stations to monitor the parameters of the geomechanical state of excavations that affect the safety of mining operations have been developed. The most optimal methods and observing techniques for specific conditions of the mine were chosen. Based on the developed projects, a complex of observation stations on the low mine level was set. Several series of observations were made. The research helped to identify the factors that had the most influence on the mine stability. The assessment of the underground cavities stability degree was carried out, measures to ensure the safety of mining operations were proposed, and the directions for the further research on the creation and improvement of the rock stress observation stations system at the mine were proposed.

1 Introduction.

The Mnogovershinnoe deposit has a complex geological structure and faulting. There are several ore bodies: Glubokoye, Promezhutochnoye, Severnoye, which are being mined simultaniously and often combined. The underground working of ore bodies Tihoye and Olenye has started. Severnoyer and Verhneye ore bodies mining is complicated because of a significant amount of unextinguished underground cavities and closely spaced quarry

*Corresponding author: avd8691@yandex.ru

workings. The day surface is substantially damaged by underground operations, especially in the area of Glubokoyer and Severnoye ore bodies.The depth of mining operations at the mine currently reaches 400 meters and more.

The low level Mining of the Mnogovershinniy ore showed harmful rock pressure manifestations - the displacement and collapse of rock blocks, which adversely affects the safety of mining operations and reduces the efficiency of the field mining. For example, in 2012 there was a collapse of the entry site to the underground cavity of the underlying Promezhutochnoye ore body-producing block. In 2014, the balmstone appearance was observed at the lower level on the roof and the sides of the vehicle and ventilation entry 380-1, waste roadway, 327, orts 327-1-2, 327-1-1 and near the block rising 352-327. Intensive water inflow in the form of the water drip was noted [1].

There was an urgent need to conduct special studies of the development of geomechanical processes at the mine to improve the safety and efficiency of mining operations.

The IRGTU (Irkutsk national research technical University) and the Institute of mining, Ural branch of Russian Academy of Sciences conducted a comprehensive research of geomechanical condition of the structural elements of mining systems in 2012-2017 In the course of studies in accordance with current instructions of the Federal Budgetary Institution Scientific and Engineering center "Power Safety" there was an urgent need to build complex rock stress observing stations for the operative monitoring of the current manifestations of rock displacement and rock stress processes.

2 Methods of research

In 2012 on the basis of normative documents of the Irkutsk Research Institute of precious and rare metals and diamonds, Research Institute of mining geomechanics and mine surveying — intersectoral scientific center "VNIMI" and Unipromed institutions, and many years of practical experience in the study of rock stress at the gold-mines, experts of the institutes of the Irkutsk National Research Technical University and Institute of mining, Ural branch of Russian Academy of Sciences developed a package of normative-technical documentation for the mine, including:
guidelines for determining the design parameters of mining systems; instructions for the protection of structures, natural objects and mining from the harmful effects of underground mining; instructions for safe mining; instructions for the organization of comprehensive monitoring of geomechanical condition; conclusion about the potential rock-bump hazard of the mine. In the developed documents for the specific conditions of the mine, the main objects of protection were selected and the basic methods and techniques for monitoring the parameters of geomechanical and geophysical condition of areas potentially dangerous for the manifestations of displacement and rock stress, and protected mine workings were set.

Dangerous and protected areas are:

- Mining, disturbed by steeply dipping tectonic faults, which possibly contain water, causing the collapse processes.
- II category of protection mining (ventilation and transport entries, etc.), in the security pillars which had suffered partial dredging of the mineral
- Mining sites of the III category of protection, which are located in the zone of influence of the stopes and after the expiration of the pillars expected service life and the roof exposures of the underlying operational blocks chambers.
- Mining sites, where the processes of displacement and collapse of individual blocks of rocks by visual and instrumental surveying observations were identified.

During the stoping the protected objects are: temporary over- and under-track pillars, inter-chamber pillars, and the exposure of the walls and roof of the chambers. Long-term effect monitoring stations (slice discharge, pairs of landmarks, the profile lines landmarks) setting is expected to be conducted using transport and ventilation entries, as the main protected workings of the II category [4]. The III category protection excavations (block rising, a subsidiary crosscuts and entries, haulage orts, etc.) are recommended to be performed using the following methods: sound-ranging control, analysis of fracture by core exploration wells, visual observations. In areas potentially dangerous for the manifestations of displacement and rock stress, surveying measurements of deformation and convergence of the working walls were proposed, along with the visual observation and slice discharge. To forecast the category of rock-bump hazard of potentially dangerous areas of the low mine level, the core disking is used as the basic method.

The category of protection of mine workings and their separate sites, dangerous with the displacement and rock stress manifestations are set based on the current geological and surveying documentation analysis and visual observations. The relevant projects of monitoring stations settings to track the geomechanical condition of mine workings that affect the safety of operations were developed later. The type of stations, their spatial position and methods of observations were specified for specific geological and technical conditions of mining.

3 Research result

The main geomechanical factors that complicate the field mining were identified. The physical and mechanical characteristics of rocks and ores were determined, the regularities of the tectonic structures formation were revealed, and the initial stresses of the rock mass were determined by the natural measurements [1].

It was found that rocks and ores of the Mnogovershinniy Deposit have high elastic characteristics and are capable of brittle destruction under load. Prone to brittle destruction are andesites, diorites-porphyry, siltstones, sandstones, ore metasomatites, unconventional – basalts and tuffs.The initial stress of the rock mass was measured using the slice discharge according to the Mining Institute, Ural branch of RAS method (table.1) [2].

The measurements were carried out in August 2012 in Glubokoe ore body in the runaway 253/183 and in the block rising pass 183/2511-2. The depth of mining operations at the site of measurement was 284 m. The stress was measured in slots. In October of the same year, measurements were carried out in the transport and ventilation entry 245/254 and access crosscut № 1 at the depth of 380 m in 21 slots. In may 2017, held in the transport and ventilation entry 395 m (runaway 1 410/395) and in the cutting №30. The depth of mining operations at the site of measurement was 350 m. the Total number of slice discharges - 12

Analysis of the measurements results of natural stresses allows us to note the following.

The values of natural stresses in the rock mass on average are approximated by formulas

$$\sigma_v=-\gamma H \qquad \sigma_L=\sigma_t= - K\gamma H = -1.23..1.24\ \gamma H \qquad (1)$$

With σ_v, σ_L, σ_t - natural vertical, horizontal longitudinal and horizontal transverse stresses of the rock mass, MPa

γ - the density of rocks and ores, MN/m3; H - The depth of mining operations, m; K - The horizontal stress coefficient.

It is obvious that the magnitude of the horizontal stresses measured are close to the values of the stresses according to the hypothesis of gravitational tectonic stress (hypothesis E. I. Shemyakin, M. V. Curleni, N. P.Vloh, etc.) [2,3,5,6].

The spread of this coefficient is explained by the uneven distribution of natural stresses in the rock mass. Taking into account the fact that the difference between the depths of all three stations does not exceed 100m, it is advisable to average the coefficient by the results of all three measurements to predict the natural stress state of the field as a whole. The resulting the horizontal stress coefficient will be 1.23..1.24.

It seems appropriate to carry out in the process of further research the clarification of the stress state of the massif with the reduction of mining operations and the mining of the other ore bodies in the Deposit.

Table 1. Results of field measurements of natural stresses

Orientation of stresses relative to the ore body stretch	Mining depth m	Results of field measurements	
		The number of individual stress measurements	Stress value, MPa
Runaway 253/183 and block rising access 183/2511-2			
Vertical σ_v	284	42	-7.7±1.0
Longitudinal σ_L	284	18	-11.8±2.7
Transverse σ_t	284	18	-11.1±1.2
The transport and ventilation entry 245/254 and access No. 1			
Vertical σ_v	380	54	-10.3±0.7
Longitudinal σ_L	380	24	-10.6±0.8
Transverse σ_t	380	24	-11.3±1.3
The transport and ventilation entry 395 and the cutting №30			
Vertical σ_v	350	16	-9.7±2.3
Longitudinal σ_L	350	16	-11.2±2.7
Transverse σ_t	350	16	-11.1±1.9

In addition to the measurements of natural stresses, long-term rock stress monitoring stations in the form of slots in the pillars and walls of excavations were set in the mine using the method of slice discharge. It should be noted that the service life of such stations mainly depends on the degree of safety of mine workings.

Rock stress measurement station set in 2012 in the transport and ventilation entry 245/254 and access No. 1, in the runaway 253/183 and block rising access 183/2511-2 were destroyed in the mining process (flooding of mine workings). An effective method of observation of processes of rock pressure is the deformations of paired landmarks mine monitoring. It is obvious that the use of this technique is convenient for extended workings of II and III categories of protection [4]. In November 2014 in the area of the stope 380-5 in the transport and ventilation entry 380-1 was studied. The depth of mining operations on the site of measurement was 180-200 m. pair landmarks were laid in the amount of 8 pieces, i.e. 4 pairs landmarks. Landmarks were laid on expanding anchors with the depth of 12-15 cm, with a diameter of 12 mm. Two series of observations were made - in 2014 and 2017. It was found that the total deformation between the pairs of landmarks were 1-14 mm over a three-year period of measurement, i.e. there was a slight compression of the walls of the workings. Given the error of surveying measurements, it can be concluded that the deformation is small and does not exceed the permissible, equal to 100 mm for displacements in the workings of II category protection. Walls of transport-ventilating entry 380-1 are sustainable. It is recommended to continue the measurements at the station.

To control the mine working and pillars walls stability the station was set in 2014 in the form of a number of relieve slots in block C320-15, on the field entry 327-1 and entry orts (Fig.1). The destruction of certain pillars and the reinforced water flow were found on the site. Three series of surveying instrumental observations were made: in November and December 2014, and in May 2017.

The results of the observations allow us to note the following: the nature of the stresses in the slope is compressive, in the field entry - compressive too. The resulting stresses in them are in the horizontal plane (vertical slots) – minus 13.4 MPa, in the vertical plane (horizontal slots) – minus 13.9 MPa on average.

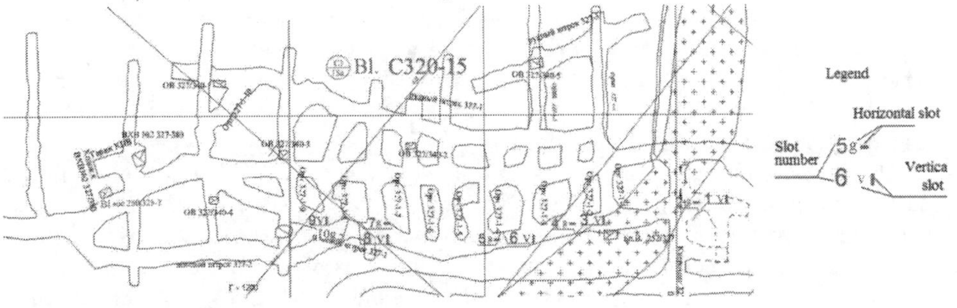

Fig 1. Observation station plan to control of geomechanical state of the mine workings by the relieve slots (327 m level)

The strength limits of rocks in dry and water-saturated state, respectively, equal to 103 and 66 MPa were determined from the samples taken on the studied level. The allowable compression stress is not less than 0.7 compression strength limit. Consequently, the allowable stress in dry workings will be minus 72 MPa, in watered-minus 46 MPa (minus means compression of rocks). It is obvious that the actual stress in mining is much less than acceptable at any degree of water cut. The crosscut 327 m level and the slope junction, the southern wall of the field entry 327-1 are sustainable.

The pillar No.1 (slots 5 horizontal, 6 vertical), there is a stable lateral extension with a small vertical compressive stresses. Apparently, this is due to the gradual destruction of the pillar or the possible opening of the fracture in the area of the slot No. 6. The pillar No. 2 (slots 8,9) has a small tension in the horizontal plane that marks a gradual unloading of the pillar. The resulting horizontal stresses in the pillar are close to the transition from compressive to tensile and are +0.3....-1.7 MPa. With this dynamics, the pillar number 2 will also gradually collapse.

It is recommended to continue to make regular visual observations and to assess the stability of the other pillars in a timely manner. It is necessary to provide the possibility of the collapsing pillars extraction. In May 2017 activities on the development of a rock stress monitoring stations network continued. Observation stations were set in the transport and ventilation entry 395 m (runaway 1 410/395) and ore body Glubokoye cutting №30 .Depth of mining works at the site of -350 m. Relieve slots in the amount of 12 pieces. The results of stress measurements at the station on the mining level of 395 m allow us to note the following. The nature of the stresses in the cutting number 30 and transport and ventilation entry 395 m (runaway 1 410/395) - compressive.

The stress level in the cutting No. 30 averages: in the horizontal plane (vertical slots) or minus 9.4 MPa, in a vertical plane (for horizontal slots) – - 12.5 MPa, in the transport and ventilation entry 395 m stresses in the horizontal plane will be -9.5 MPa, in a vertical -12.3 MPa, in the runaway 1 410/395 -8.7 MPa to 15.6 MPa, respectively.

Allowable stress in compression in the area of observation stations in the dry state of the mine workings is equal to minus 60 MPa, due to flooding minus 52 MPa. It is obvious that the actual stress in the mine workings, constituting minus 9.4 ÷ minus 15.6 MPa is significantly less than acceptable at any degree of water cut. Currently, cutting No. 30, transport and ventilation entry 395 m and the runaway 1 410/395 are sustainable.

The core material of the drilling camera exploration wells № 3 cutting № 30 was selected on the same site to assess the degree of danger of rocks by the method of core disking. The analysis of core material of the disking zone on did not reveal small discs with a thickness of 1-5 mm, the danger of the rock-bumps in the Glubokoye ore body lower level does not exist. It should be noted that the calculated critical depth of rock-bumps is 300-440 m (in the mathematical modeling of the stress-strain state of mine workings, conducted in the zone of stopes bearing pressure) [1]. Therefore, in addition, in order to confirm the rock-bump resistance of the rock mass, during the excavation of new workings, experimental field observations were carried out by the acoustic emission device SB32 №002 [7-10]. Measurements were made on the mining level of 305 m in the area of the Glubokoye ore body after rock blasting.

The results of the measurements showed that the total number of pulses measured during 40 minutes is 39. The average activity of acoustic emission is 0.97, i.e. approximately 1 signal per minute. This distribution of sound pulses corresponds to the steady state of the rock mass. The results of emission measurements confirm the hazard category "not dangerous" at the current depth of mining operations in the development of the Glubokoye ore body.

It should be noted that at the mine by the staff of The Mining Institute, Ural branch, Russian Academy of Sciences (senior researcher Liping J. N., senior researcher, V. V. Bodin) were also carried out geophysical researches of the degree of rock works stability in the the rock stress station setting area in 2012 in the runaway 253/183 and block rising pass 183/2511-2. As experimental methods, seismic transmission of the massif with the construction of tomographic scheme of the site (elastic wave velocities trend) and seismic profiling of the massif on mine workings were used.

The tectonic structure of the rock mass was estimated based on the results of seismic transmission at the station location. Two zones of tectonic disturbance were discovered: one perpendicular to the field entry 250-1 and the junction of the field entry and the orts 250-1-4250-1-6. It is suggested that the position of mineralization zones can also be predicted using the seismic transmission method at the mine. According to the results of seismic profiling of the massif, a correlation between the magnitude of the horizontal natural stresses σ_h and the velocity of the longitudinal seismic wave V was established

$$\sigma_h = -18.87 + 0.00507 \cdot V \text{, m/s} \quad (2)$$

Unfortunately, both methods have not been further applied at the mine due to the high complexity and cost of work.

4 Summary

Mnogovershinnoye Deposit stand the gravitational tectonic stress. On the upper levels, the rock mass is characterized mainly by a stable state. The earth's surface, minings and stopes in the area of mining Olenye, Glubokoye, Tihoye ore bodies are currently stable. Geoinformation researches of the tectonic structures formation regularities, evaluation of exploration wells core disking, measurements of stress did not show the areas of the rock massif with the rockburst hazard [1].

The factors reducing the stability of mine workings at the Mnogovershinniy Deposit are established: steep tectonic disturbances, water saturation of rocks, working off of contiguous ore bodies, the depth of mining operations of 300 m or more, a high degree of earth's surface damage, the influence of nearby open-pittings.

Thus, the mining of ore bodies Severnoye and Verhneye is complicated by a high degree of the earth's surface damage and quarry workings. Mining of the Promezhutochnaya ore body is complicated by the high damage, steeply dipping tectonic faults, water-bearing rocks. It is necessary to make provisions for the setting of the long-term observation stations in the form of profile landmarks on the earth's surface over these ore bodies and paired, profile landmarks or slice discharges - on the lower levels. The calculated critical depth of the rock pressure manifestations in dynamic forms during the stoping is 300-440 m

[1]. At these depths, the mine workings carried out in the zone of bearing pressure of the stopes have the calculated stresses above the critical rock bumps.

At present, the maximum depth of mining operations has not yet reached the deadline. However, preparatory mining works at depths of 300-350 m are carried out (Glubokoye ore body). On the lower levels, according to the results of observations for the period 2014-2017, partial destruction of the pillars was established, especially in the flooded areas. Therefore, it is necessary to continue periodic instrumental and analytical studies of natural and man-made stresses of rock mass on the lower levels of the mine. With the external signs of rock and ore rock-bump hazard appearance, it will be necessary to implement the appropriate measures to forecast and prevent of rock bumps.

The mine is encouraged to continue work on the observations on the old and setting the new observation stations for monitoring of geomechanical condition parameters of the mine workings along with the further accumulation of subsurface voids at the existing levels and the depth of mining operations increase. Methodological guidance on the rock stress assessment and its harmful manifestations should be undertaken by specialized agencies (Institute of mining Ural Branch of RAS, Institute of mining Far Eastern branch of RAS, Irkutsk Research Institute of precious and rare metals and diamonds, etc.)

References

1. The study of geomechanical conditions Mnogovershinnoe gold deposits for the purpose of predicting its potential rockburst hazard. Sosnovskaya E. L. Bulletin of ISTU. 4. 82-88 p. (2015).
2. Geomechanics and Geotechnology. Zubkov A. V., Ekaterinburg: Uro RAS. 335 p. (2001).
3. Technology of the gold deposits mining. Neganov, V. P., Kovalenko V. I., Zaitsev, B. M., Sosnowski, L.I. Under the editorship of V. P. Neganov. M. Nedra. 336 p. (1995).
4. Temporary rules of structures, natural objects and mining protection from the harmful effects of the underground mining on gold deposits- Irkutsk: Irgiredmet. 76 p. (1994).
5. Strength and deformation of rock massives. Shuplecov U.P, Yekaterinburg: Uro RAS. 195 p. (2003).
6. Vloh N.P. Management of rock massifs strains in underground mines. Nedra. 208 p. (1994).
7. Barbezat M., Brunner A.J., Flueler P., Huber C., Kornmann X. Acoustic emission sensor properties of active fibre composite elements compared with commercial acoustic emission sensors. Sensor and Actuators A. Physical. **1**. pp.13-20. (2004).
8. Zandt G., Leidig M., Chmielowski J., Baumont D., Yuan X. Seismic detection and characterization of the altiplano-puna magma body, central Andes. Pure and Applied Geophysics. **3-4.** pp. 789-807. (2003).
9. Zang L., Ma S., Ren M., Jiang S., Wang Z., Wang J. Acoustic emission frequency and b value characteristics in rock failure process under various confining pressures" Yanshilixue Yu Gongcheng Xuebao **10** pp. 2057-2063. (2015).
10. Reuber G., Kaus B.J.P., White R.W., Schmalholz S.M. "Non lithostatic pressure during subduction and collision and the formation of (ultra) high-pressure rocks" Geology. **5**. pp. 343-346. (2016).

E3S Web of Conferences **56**, 02026 (2018) https://doi.org/10.1051/e3sconf/20185602026
VII International Scientific Conference "Problems of Complex Development of Georesources"

The capability of detailed study of coal- and ore-bearing structures using electrical resistivity tomography (ERT) by means of "well bore-surface" observation system

Nikolai Shkabarnia[1,2*], *Grigorii* Shkabarnia[2,3], *Petr* Gorelov[4] and *Svetlana* Korneeva[1]
[1] Mining Institute of Far Easten Branch of Russian Academy of Sciences, Khabarovsk, Russia.
[2]Far Eastern Federal University, 8 Sukhanova Str., Vladivostok city, Russia.
[3]V.I. Ilicheva Pacific Oceanological Institute Far Eastern Division of Russian Academy of Sciences, Vladivostok city, Russia.
[4]Geophysical Service of the Russian Academy of Sciences, seismic monitoring department, Vladivostok city, Russia.

Abstract. Coal open-cuts of Khabarovsk and Primorskii krai are characterized by multiple coal beds of small thickness. Important is the task to evaluate structural architecture of landslide blocks distinguishing insecure beds of small thickness. In detailed study of ore deposits it is also needed to distinguish thin beds and rock-fracture zones. Therefore, in order for detailed study of such structures, it is proposed to develop a ERT technology with the use of "well bore-surface" observation system. On this stage, only theoretical justification of proposed technology is given. In order to ascertain the technology capability, the calculation of apparent resistivities for three-layer structures with various correlations of geometrical and electrical parameters was conducted. Calculations were performed with uniform spacing of 20 electrodes in a well bore and 40 of them on the surface with polling of Wenner and Schlumberger four-electrode arrays, axial array and their three-electrode designs. Regularities of electrical field for three fragments of apparent resistivity section were established based upon the analysis of modeling results: electrodes only on the surface (horizontal profile), electrodes in the well bore (vertical profile), electrodes in the well bore and on the surface (mixed profile). Outstanding interest is represented by anomalous areas of mixed profile which are described by new regularities of the field and criteria of distinguishing beds of small thickness. Theoretically it has been demonstrated that new technology is effective with distinguishing beds of small thickness on different depths of geological section **Key words:** coal beds, insecure beds, electrical resistivity tomography, apparent resistivity, anomalous areas.

* Corresponding author: shkabarnya@mail.ru

1 Introduction

The majority of operating coal open-cuts of the Far Eastern south is characterized by numerous (20–30) coal beds of rather small thickness (meters and seldom tens of meters). Bedding is monoclinal only on separate areas and complicated by a great number of fractures which are required to be distinguished and detected also in order for effective coal mining. During the research, mining and geological conditions of the area of Luchegorsk open-cut (Primorskii krai) using ERT, surface observation systems were distinguishing and detecting coal outbreak under quaternary deposits at the depth from 8 to 50 meters [1, 2]. On the area of the open-cut, the extremely important is the task to evaluate structural and tectonic structure of landslide blocks distinguishing insecure beds of small thickness. In the course of evaluation and forecast of landslide processes, the most dangerous were thin beds (3–6 m) of high plasticity Neogene clays embedding under the complex of soft-firm clays and sand-shale deposits of Quaternary age with the thickness of up to 30 meters. The method of electrical tomography from surface allowed confidently distinguishing and detecting [3] only intercalations and pockets of plastic clays with the thickness of more than 7 meters amidst Quaternary age. The experience in using this method in the study of coal deposits of the Far East Region confirmed the effectiveness of this modification of resistivity method against traditional electric soundings and profiling.

Similar problems appear as well as in case of detailed study of ore deposits structure in Norilskii district [4]. In the Amur Oblast, dense network of well bores with the depth of down to 60 meters were required in order to distinguish thin (2-4 m) bouldery and pebble-bed with pockets of fat clay containing mineable concentrations of gold on Iasno-Polianskoe deposit (Amur Oblast). The use of dense network of vertical electrical soundings (VES) which have the same capabilities as electrical tomography didn't allow detecting this productive bed. Thus, despite the success in the field of geophysical technologies development in order to solve various geological tasks it is necessary to commonly use dense network of well bores.

Therefore, we think that further development of electrical tomography by way of detailed study of coal- and ore-bearing structures extinguishing thin bed and pockets is related to the application of "well bore-surface" or "well bore-surface-well bore" synchronous observation systems. Theoretical justification of proposed electrical tomography method in this article was conducted based upon electrical prospecting basic principle [5, 6] – electric field modeling principle. We consider detailed study of mass rock formations on deposits of minerals in seeking to resolve various tasks as fundamental problem of modern stage of electrical prospecting methods development.

2 Theoretical basis and conditions

The solution of direct problem and the development of software for calculation of apparent resistivities (ρ_r) for horizontally-layered model with unspecified location of sources and receivers on daylight surface and inside structure (along well bore) have been performed and given in the work [7]. In the course of the solution of Laplace's equation, cylindrical coordinate system, interface conditions, conditions in the vicinity of a source and at infinity, Hankel transform and linear filtration were used. The case when the source is located inside the structure comes down to the introduction of dummy boundary inside the bed with the same conductivity and going through the source.

In order to determine the capabilities of new method, mass calculation of electrical field parameters and its analysis for the purpose of establishing the regularities of behaviour of abnormal field areas and criteria of distinguishing and detecting interbedded layer was required. The calculations results were represented as apparent resistivity sections (ρ_r)

where the location of electrodes in a well bore and on surface were plotted against X-axis, and against Y-axis — installation spacing. Sections represent 2D images of electrical parameters of the structure as three fragments: 1 – electrodes only on the surface (horizontal section), 2 – electrodes in the well bore (vertical section), 3 – electrodes in the well bore and on the surface (mixed section).

In the course of analysis, many factors, primarily, the types of resistivity sections and observation systems, various correlations of physical and geoelectric parameters were taken into account. On the first stage, all four types of three-layer structures (H, A, K, Q) with relative parameters were considered: geometrical – to the distance between adjacent electrodes (a), electrical – to specific resistivity of the first or the third bed. The thickness of beds were changing from 1a to 16a and specific resistivity — from 1/16 to 16.

The majority of calculations was conducted with the uniform spacing of up to 20 electrodes in a well bore and 40 — on surface with the polling of four Wenner electrode system (AMNB, AM=MN=NB), Schlumberger array (AMNB, MN<1/3AB), axial array (ABMN, AB=BM=MN) and their three electrode designs. Taking into consideration great information scope, regularities of the field and criteria of distinguishing bends only for Wenner electrode system are considered below. In this case, the section included 19 lines and 570 values of ρ_r with the initial spacing of 1.5a. Fragment 1 includes 57 values, fragment 2 — 247, fragment 3 — 266.

3 Modeling results

Apparent resistivity section for three-bed model of H type (h_1/a=8a, h_2/a=2a, ρ_1= ρ_3=1, ρ_2/ρ_1=1/4) is given in fig. 1a. Special attention during the analysis has been paid to the shape of abnormal areas which were distinguished enveloping isolines with the largest gradients and their connection with the geometry of interbedded layer (superface, subface, extent within a fragment). Note that geometrical sections on the area of development of landslide processes and deposits with interbedding ore bodies may be approximated by such models.

As seen, the least informative for distinguishing interbedded layer with the largest spacing of AB/2=19.5a is the fragment 1. This is quite an expected result since it corresponds to section ρ_r with traditional method of research by surface observation systems. Here isolines ρ_r are located in parallel to x-axis with gradual reduction of values down to 0.8a on spacing of 19.5 and thus specify either horizontal location of boundary at depth or gradient change of structure resistivity.

With electrodes located on vertical profile by the shape of abnormality on small spacings (1.5, 3.0), the thickness of the second bed (2a) is determined steadily, and the value ρ_r=0.34 are close to true bed specific resistivity. On large spacings, the area of low values of ρ_r extends to AB/2=9, where ρ_r=0.72. Zones of increased values of ρ_r related to screening effects are specified around the zone. In fig. 1a, such zone is well detected from fragment 2 section on fragment 3 section where ρ_r reach the value of 1.01 on spacing AB/2=9.

In case when electrodes located synchronously on vertical and horizontal profiles of fragment 3 section, except for the zone of increased values of ρ_r, abnormal area of low values (0.72-0.79) which has wedge-like shape and reflects horizontal horizontal location of the second bed on trial profile is well distinguished. Extreme values of the zone on various spacings are located along the line nearly in parallel to boundary between fragments 1 and 3 and the continuation of this line to the vertical profile defines the situation of the second bed mid.

With the reduction of bed thickness (h_2=1a) and specific resistivity ρ_2/ρ_1 =1/16) for this type of model, noted regularities of the field on fragments 1 and 3 sections change slightly (fig. 1б). In the first fragment of section, ρ_r isolines are also located in parallel to x-axis specifying horizontal horizontal bedding of possible boundary at the depth where value ρ_r on the largest spacing equals 0.72. In the field of section ρ_r of fragment 3, wedge-like abnormal area with narrow range of extreme values which reflects horizontal bedding of the second thin bed on the profile under study is distinguished. The continuation of lines of these values to the vertical profile defines the depth of bedding.

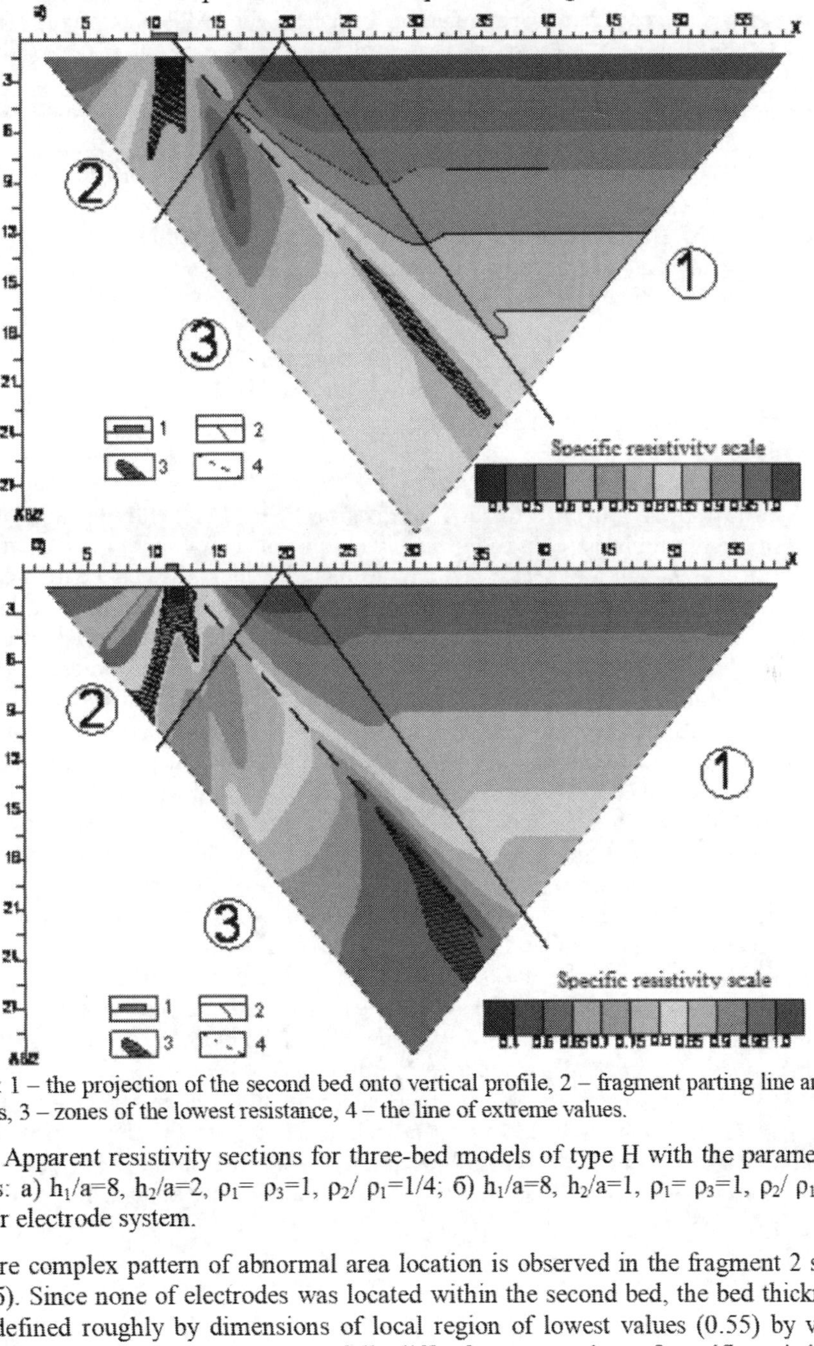

Legend: 1 – the projection of the second bed onto vertical profile, 2 – fragment parting line and their numbers, 3 – zones of the lowest resistance, 4 – the line of extreme values.

Fig. 1. Apparent resistivity sections for three-bed models of type H with the parameters as follows: a) $h_1/a=8$, $h_2/a=2$, $\rho_1=\rho_3=1$, $\rho_2/\rho_1=1/4$; б) $h_1/a=8$, $h_2/a=1$, $\rho_1=\rho_3=1$, $\rho_2/\rho_1=1/16$, Wenner electrode system.

More complex pattern of abnormal area location is observed in the fragment 2 section (fig. 1б). Since none of electrodes was located within the second bed, the bed thickness is to be defined roughly by dimensions of local region of lowest values (0.55) by vertical profile. As seen, these values are essentially differ from true values of specific resistivity of the bed. On different sides from area abnormality, zones of anomalous values related to screening effects are distinguished.

"Well bore–surface" observation system may confidently distinguish and detect beds of increased resistivity and small thickness. Such conditions are met in resistivity sections with beds of brown coal, effusive rocks and sandstones among clays, argillites and aleurolites. Anomalous areas of horizontal profile on apparent resistivity sections for type K model are characterized by known regularities of a field for bedded structures. The field analysis on the fragment of vertical profile demonstrates that anomalous area with increased values defines the depth of bedding and approximate thickness of the second bed. The location of thin bed of increased resistivity and its extent are well defined on mixed profile. Wedge-like area is distinguished on small spacings (up to AB/2=9.5) by values of ρ_r being less than 1.0, and on large spacings by values being more than 1.0, by isolines of which it is possible to establish the depth of bedding and the extent of bed on trial profile. Extreme values (ρ_r=1.19) differ from true bed resistivities. The continuation of these values lines allows establishing approximate bed location. At large for this model, field regularities are nearly analogues to type H model but with the opposite sign.

Capabilities of distinguishing interbedded layer for types A and Q models slightly drop, but as well as in case with types H and K, main field regularities are fulfilled. The most informative are also sections of fragments 2 and 3. Their anomalous zones allow distinguishing bed at the depth and its extent by a profile, and apparent section — closer to true specific resistivities.

Sections for type A with parameters as h_1/a=8, h_2/a=2, ρ_2/ρ_1=4, ρ_3/ρ_1=16 (fig. 2a) and h_1/a=4, h_2/a=1, ρ_2/ρ_1=4, ρ_3/ρ_1=16 (fig. 2Б) are given in fig. 2. The depth of bedding and the thickness of the second bed are distinguished by abrupt concentration of isolines of ρ_r with the values of 2, 4, 6, 8 on fragment 2 sections. At that, the thickness may be defined in the first case in more reliable manner (fig. 2a). This implies that main factor for bed discovery is a ratio between thicknesses and distance between electrodes (h_2/a). Values of ρ_r in the central part of concentration of isolines with the range of 3.9 – 4.2 almost define true specific resistivities of the second bed. Long wedge-like areas which reflect horizontal orientation of a bed are distinguished on fragment 3 sections (fig. 2), and continuation of extreme values isolines to vertical profile defines an approximate depth of bedding. Value of ρ_r on spacing AB/2=28.5 in case when h_2/a=2 equals 2.66, and with h_2/a=1 — 4.20. But conditions for distinguishing the second bed in the first case (fig. 2a), are absolutely better.

On fragment 1 sections, gradual increase of ρ_r means, as well as for type K, presence at the depth of a bed of increased resistivity or gradient change of structure resistivity.

Regularities of the field for type Q models is analogous to fields of type A, but with the opposite sign. At that, interbedded layer with the reduction of resistivities is distinguished in more reliable manner with depth increase. Thicknesses and the extent of a bed nearby a well bore are defined by concentrations of isolines on fragment 2 section, in the center of anomalous area ρ_r values are close to true specific resistivities. In fragment 3 section, anomalous wedge-like areas allow defining the conditions of bedding, and the continuation of extreme values lines to vertical profile shoes its location. On large spacings ρ_r are close to true specific resistivities. Field regularities demonstrate that the increase of bedding depth without sacrificing other parameters of the structure do not influence distinguishing it in the sections of fragments 2 and 3. The least informative in distinguishing bed is also fragment 1 section.

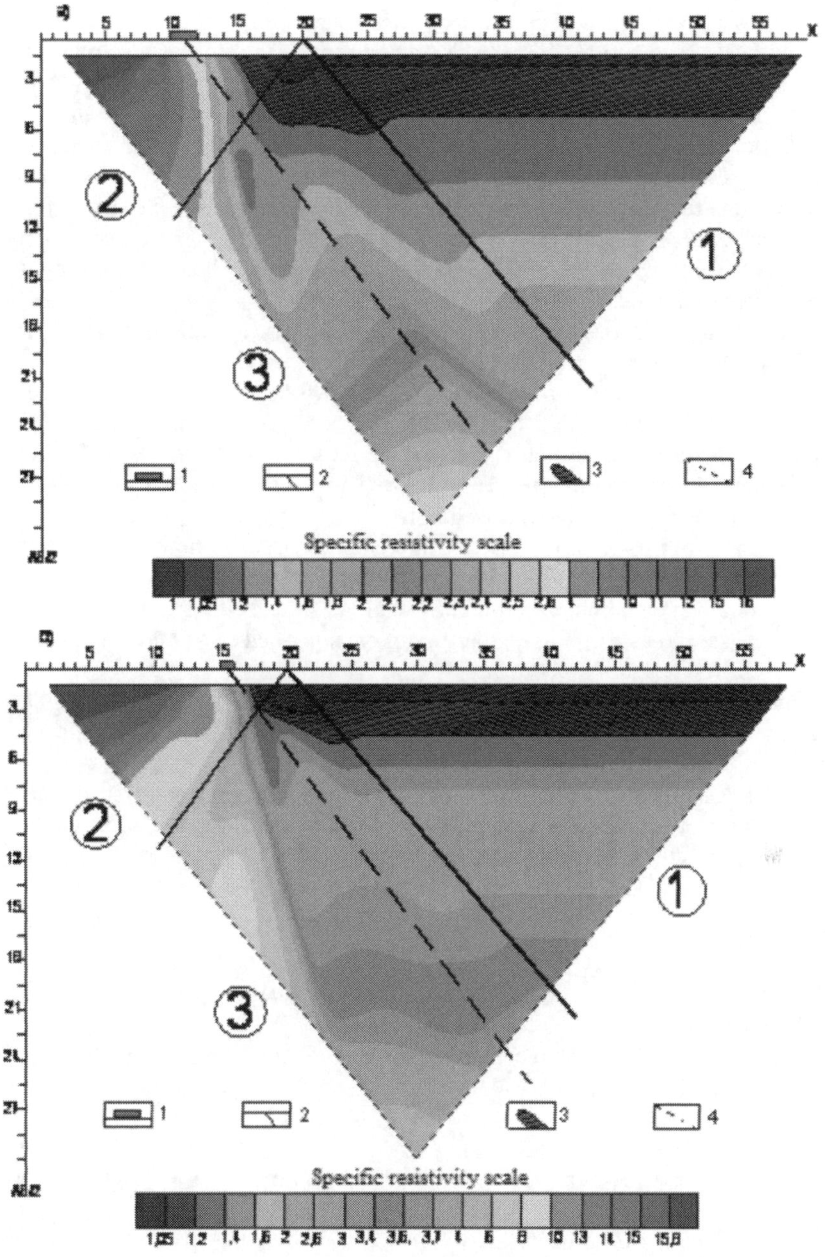

Fig. 2. Apparent resistivity sections for three-bed models of type A with the parameters as follows: a) $h_1/a=8$, $h_2/a=2$, $\rho_2/\rho_1=4$, $\rho_3/\rho_1=16$; б) $h_1/a=4$, $h_2/a=1$, $\rho_2/\rho_1=4$, $\rho_3/\rho_1=1$. Legend in fig. 1.

4 Conclusions

Thus, based upon modeling of electrical field for selected three-layer structures and "well bore-surface" observation system, the regularities of anomalous areas of apparent resistivity sections have been established and criteria for discovery and detection of thin beds at different depths. Regularities sufficiently depend on section fragments which are defined by the location of electrodes: on the surface (horizontal profile), in a well bore (vertical profile), in a well bore and on the surface (mixed profile). As follows from the analysis of sections, the effectiveness of the proposed electrical tomography method for distinguishing bedding thickness, extent of interbedded layers and its electrical properties.

"Well bore-surface" observation system reliably distinguishes thin beds in types H and K resistivity sections and slightly worse for types A and Q. The most informative are the areas of sections of vertical and mixed profiles. Field regularities demonstrate that the increase in depth of bedding of interbedded layer with the retention of other parameters of a model do not influence the conditions of distinguishing it. The least informative in the discovery of a bed is the section of horizontal profile.

The following field peculiarities should be used in analysis and interpretation of practical apparent resistivity sections obtained by new methods:

- anomalous areas of vertical profile section are characterized as known field regularities in geophysical well logging (apparent resistivity method, lateral sounding), electrical tomography observation system in case of dense location of electrodes by vertical profile (in a well bore) allows confidently defining thickness, extent and specific resistivities of thin beds at various depths close by a well bore;
- anomalous area of mixed profile section are characterized by new field regularities and criteria for distinguishing interbedded layers, wedge-like areas on section allow specifying the location of a bed by depth, detect its low-angle or tilted bedding away from a well bore on trial profile as well as defining regularities of change of electrical properties of a bed and host medium;
- anomalous areas of horizontal profile are characterized by known field regularities for bedded structures and surface observation systems, they are the least informative compared to reviewed profiles.

In sum, in order to put the proposed method into practice of geophysical activity, bore hole cables with electrodes are required. The technology of such cables manufacture is known in the procedures of geophysical well logging.

References

1. V.Ch. Miasnik, I.V. Kalinin and others Mining magazine, - M.: Mining book. **12.** pp. 16-20. (2006).
2. I.Yu. Rasskazov, G.N. Shkabarnia, N.G. Shkabarnia *Journal of Mining Science.* **Volume 49, Issue 3**, pp. 388-394. (2013).
3. I.Yu. Rasskazov, G.N. Shkabarnia, N.G. Shkabarnia. *Journal of Mining Science.* Volume 49, Issue 5, pp. 772-778. (2013).
4. V.A. Kulikov, A.A. Bobachev, I.D. Gruzdeva, A.G. Iakovlev. Geophysics. **1**. pp. 27-34. (2013).
5. B.S. Svetov, M.N. Berdichevskii. Geophysics. Publishing house Gers. **2**. pp. 4-11. (1998).
6. N.G. Shkabarnia, G.N. Shkabarnia. Mining information-analytical bulletin. The world of mining book. pp. 377-390. (2007).
7. A.A. Furman. Vadose Zone Journal. **2**, p. 416-423. (2005).

E3S Web of Conferences **56**, 02027 (2018) https://doi.org/10.1051/e3sconf/20185602027
VII International Scientific Conference "Problems of Complex Development of Georesources"

The problems of monitoring the deformation processes in the integrated development of the underground space of metropolitan cities

Dmitriy Neguritsa*[1]

[1]Federal State Autonomous Educational Institution of Higher Education "Peoples 'Friendship University of Russia" (RUDN), Engineering Academy of Peoples' Friendship University of Russia, Moscow, Russia

Abstract. By analyzing the existing regulatory documents regulating the surveying and geodetic monitoring of buildings and structures from the negative impact of mining and construction works, conclusions were drawn about the problems of its implementation in the conditions of modern metropolitan cities and the imperfection of the methodology for observing deformational processes in the construction of underground structures. New approaches to the organization of monitoring of deformation processes in metropolitan cities were proposed.

1 Introduction

Despite the crisis phenomena in the economy, modern metropolitan cities are developing dynamically. One of the main trends is the complexity of development: the increase in the number of storeys, the development of underground space, the development of transport above-ground, ground and underground infrastructure, the subway, the construction of various communications [1-3].

Currently, the territory of Moscow is overloaded with objects influencing change in engineering geological and geotechnical conditions in the city. Due to urbanization and consequently, the acceleration of human impact, the increase in the geomechanical load on the massif and the degradation of its bearing capacity takes place [4].

2 Factors affecting the development of deformation processes in the development of underground space

Underground space of cities is a valuable georesource, the cost, resource intensity and development efficiency of which is determined by its geomechanical, geotechnical properties, geosituational features and the dynamics of their changes in the development.

In the modern city-planning situation, when underground structures and communications are being built, buildings and structures are being erected, their geomechanical processes are being reconstructed, having different directions, the time of

* Corresponding author: neguritsa_dl@rudn.university

passage and the intensity affect a large number of objects of different responsibility levels. This leads to deformation of existing buildings and structures falling into the zone of influence, varying degrees of their damage and even to the possible occurrence of emergencies [5-7].

The study of the causes of accidents and deformations of buildings and structures shows that in most cases they occur not because of errors in engineering calculations, design, installation of above-ground structures, but due to miscalculations made during engineering and geological surveys, design and construction of foundations and casting concrete, incorrect operation of buildings, leading to a deterioration in the properties of soils in the foundation, as well as due to insufficient account of the joint influence of several underground structures.

Therefore, to control the safe flow of geomechanical processes and the forecast of their development, it is necessary to take into account all the factors that influence the development of deformation processes [8].

At the moment, factors that influence the development of deformations are well studied, methods for forecasting and controlling their development and calculation methods have been developed, there exists an instrument base for measuring displacement and deformation, but from the engineering point of view there is a lack of implementation of an unambiguous and effective solution to this problem.

This is due to a number of reasons, namely:

-the heterogeneous structure of the soil-rock massif and the variability of its properties at different degrees of intensity of the impact thereon;

- the complexity of describing deformation processes due to the influence of various sources and the uneven spread of their influence in different parts and in different directions;
- Nonlinearity of development of deformation processes in the soil-rock massif;
- Mutual influence of structures with dynamic loads and passive structures and communications;
- The impossibility to know the true state (degree of wear and safety factor) for a large number of structures and communications;
- The development of an emergency situation so quickly that the use of preventive and protective measures is impossible;
- The discrete development of deformation processes in time and space, which can be the reason for their uncontrolled development.

Emergency situations on objects usually arise not under the influence of one factor, but when combined and added together, including a certain sequence of their influence.

At the same time, there is some inconsistency between the current regulatory documents on monitoring, ensuring industrial safety and the methodology for their application in urban underground and surface construction, since under certain conditions the development of geodynamic processes can be instant and irreversible, which completely excludes the use of protective or preventive measures.

3 Principles of geomonitoring organization

To date, a fairly large number of methods for monitoring deformation processes have been developed, but the most effective ones in terms of *accuracy - reliability - efficiency - cost* are the geodetic and mine surveying methods that provide high accuracy, efficiency and reliability.

Existing methods of forecasting the expected shifts and deformations, as a rule, allow obtaining their greatest values.

To control the development of deformation processes in the process of construction and operation of an underground structure, geomechanical monitoring is performed. Geomechanical monitoring is a complex of measures to control the development of deformation processes performed under a special project whose main task is to identify the signs preceding the creation of emergency situations.

Geomechanical monitoring is carried out under a special project, which is an integral part of design documentation and undergoes industrial safety expertise.

The geomechanical monitoring project consists of the following sections:

- Geological and geomechanical description of construction conditions;
- Technological description of the process of erection of an underground structure;
- Forecast of development of deformation processes;
- Examination and determination of the maximum and permissible deformations of buildings and structures falling into the zone of the construction influence;
- the definition of a set of protective and preventive measures that must be applied to individual buildings and structures for which allowable and limiting values of deformations are less than predicted ones;
- The description of the structures of observation stations, the construction of reference points, the methods of their laying;
- The description of applied measuring devices and the equipment;
- The procedure of measurement and processing of received data;
- The analysis of the obtained measurement data, comparison of measured values of shifts and deformations with permissible and limiting values;
- The preparation of a report on the development of deformation processes;
- Making expert decisions on the protection of objects, buildings and structures.

When performing geomechanical monitoring, the most important stages are the stage of forecasting the development of deformation processes and the stage of monitoring the development of displacement processes.

Determining the zone of influence of construction and the forecast of deformation development is one of the most complex, time-consuming and responsible parts of the project since geomonitoring is organized based on this data.

For today the Russian Federation lacks a uniform forecasting technique. The existing regulatory framework lags behind the development of construction technology and does not take into account the specifics of modern construction and technologies applied. Thus, for example, the Instruction on Observations of Earth Surface Displacements and Objects Located on It in the Construction of Underground Structures in Moscow, M .: 1997 was published more than 20 years ago and does not reflect the current trends in the development of the underground space of the city. The forecast methods used in the construction industry using the finite element method and the adoption of an elastic or elasto-plastic medium do not provide sufficient accuracy since the soil stratum containing the underground structure consists of soils that are very heterogeneous in their geological composition and physico-mechanical properties. When preparing the forecast, it is necessary to take into account a large number of influencing factors that can be conditionally divided into four groups: geological, technological, geomechanical and geosituational.

The additional risk of occurrence of emergency situations is caused by the fact that the impact on other underground facilities, in particular, those that are below the constructed ones - tunnel and station buildings of the subway, etc. are often didn't take into account

The next stage after the forecast of the geomechanical monitoring is the design and laying of observation stations and the performance of observations.

The laying of observation stations and the production of observations in the city are complicated by the following: the surrounding areas are almost completely asphalted, except for small lawns, the existing building is very dense, in the areas of construction there is very dense pedestrian and vehicular traffic, which can significantly complicate the use of modern high-performance equipment, tacheometers with servo drive, working in automatic mode and laser scanners. In the case of particularly important buildings and structures, for example, monuments of culture, permanent monitoring systems are being installed, and it is necessary to install not one but several instruments.

One of the problems with observing the shifts is to ensure the accuracy and reliability of the obtained measured values of the shifts and the actual course of the process, taking into account that in some cases the accuracy of the measurements can be compared with the measurement values [9].

The task of ensuring the stability of the position of the initial reference points in each of the cycles is solved by laying a clump of at least three initial reference points, whose position control is performed with each observation cycle.

In modern conditions, to ensure the effectiveness of measurement work, it is necessary to constantly improve the design of benchmarks to ensure their continuous service during both the construction and also the operation of the underground facility. When laying the observation stations, a large number of benchmarks are used, having a different design, depending on the laying location destination and the service life. Usually three types of signs - basic, intermediate and deformation - are used at observation objects.

The main reference points are the reference points of the city geodetic network and are included in the main trunk way to monitor their mutual position. Some of them are located outside the zone of possible deformations from the production of construction, installation and mining works, so as to avoid the influence of settlings from the structure being built on the measurements made.

Intermediate signs serve to organize the optimal scheme for transferring the mark from the initial benchmark to the deformation one.

Deformational benchmarks are simultaneously points for determining displacements in plan and height and are fixed in such a way that their location and quantity allows revealing possible deformations. As deformation marks, the reinforcing bars laid at the base of the buildings and along the profile lines in the adjacent ground mass are used, at the end of which there is a hemispherical head for the installation of the rail and reflector.

The largest number of benchmarks is laid directly into the walls of those buildings and structures that fall into the zone of influence of construction. Deformational benchmarks are laid along the perimeter of buildings at a distance of 5-20 meters, with mandatory installation on the corners and characteristic protrusions of buildings. Marks are laid at the same height, approximately 0.5 meters from the surface of the earth. The deformation mark can have various designs. To each benchmark, a unique serial number is assigned, applied with paint on the wall of the building or a lodge at the ground bench. It is advisable to make a digital photograph of the benchmarks laid.

Measurements are made of both horizontal and vertical shifts. In the city of Moscow, horizontal shifts sometimes exceed vertical ones.

The accuracy and frequency of measurements are established by the project and is determined depending on the monitoring objectives and the conditions for the construction. Proceeding from this, the instruments and methods of measurement are selected. In complex conditions of the city, a complex method of observations with the use of various measurement technologies is often used. At the same time, settlings and horizontal shifts are measured.

In addition to measuring the shifts of objects falling into the zone of influence, measurements are made of the underground structure shifts. On the "wall in the ground", beacons are installed with a step of 5-10 m and horizontal and vertical movements are measured. In addition, as far as the erection and accessibility to the "wall in the ground" is necessary, laying benchmarks on all floors of the complex is necessary to establish the movement of the wall in the construction process. It is also necessary to observe individual structures and parts of the complex in order to establish the values of their deformation.

In the immediate vicinity of the projected complex, deep benchmarks are laid to observe the deformations of the soil massif.

The processes of shifting and deformation can develop smoothly, discretely, or move from one stage of deformation to another.

With the smooth development of the processes, observations and their processing are not problematic and the maximum reliability of the information obtained is ensured.

With the discrete development of the processes of shifting and deformation, difficulties arise both with the interpretation of the results obtained and with the determination of the origin of the local manifestation of the dynamic process. This makes it difficult to process observation data, as well as to assess the actual condition of a building or structure, which is especially important for monuments of architecture and art in modern metropolitan cities. The appearance of local dynamic manifestations of deformations alters the description of the development of deformation processes using standard techniques for processing measurement data for 3-6 benchmarks.

4 Assessment of the accuracy of observations

The analysis of the processes of shifting and deformation both in space and time, as well as their unevenness in one cycle of observations, is of particular importance in the processing of measurement results.

Obviously, when observing deformations of buildings and structures, it is almost impossible to determine in advance which mark will receive the maximum shifts and which areas will be deformed as much as possible.

Therefore, to calculate the accuracy, the maximum displacement standard σ_{Smax} is determined for the weakest point of the network.

To determine the error of the settling under high-precision geometric leveling, when observing the deformations of buildings and structures in the conditions of close urban development, used in determining the vertical component of the displacement vector:

$$S_{ji} = (H_{I0} - H_{Ii}) + \sum_{j=0}^{j} h_{j0} - \sum_{j=0}^{j} h_{ji}, \quad (1)$$

where: $(H_{I0} - H_{Ii})$ – the difference in the marks of the initial benchmark in the zero and i-th cycles of observations, (the error of the change in the height of the starting point for a period of time t_i between the zero and the current observation cycle), $\sum_{j=0}^{j} h_{j0}$ and $\sum_{j=0}^{j} h_{ji}$ –the sum of all the excesses, respectively, in the zero and i-th cycles.

This error is summed from the sum of the error of settling or jacking of the benchmark and the change in the length of the benchmark structure under the influence of seasonal temperature fluctuations.

To reduce the negative impact of these factors, it is advisable to lay a clump of reference points, the superelevation; between which is determined with the greatest possible accuracy in these conditions to identify the most stable reference point by the criterion of stability. In practice, the Tarnowski stability criterion is used quite often:

$$K = 2\mu_H\sqrt{2n}, \quad (2)$$

where: μ_H –the standard for determining elevations at a station, n –the number of stations between the reference points.

The range of values from + K to -K where it is impossible to detect settlings.

After the measurements, calculations of the slopes, curvatures, and horizontal deformations are performed, allowing to comprehensively evaluating the development of the shift and deformation processes.

In the process of monitoring the shift and deformation of structures of significant height, it is necessary to pay special attention to the observations of the roll, which will allow performing accident-free construction and installation work and monitoring the development of deformation processes.

The use of a modern hardware base for measurements will make it possible to perform measurements with high efficiency, but it will require special algorithms for processing the results of measurements obtained with their help, analyzing them and, on the basis of the received data, adopting technical, technological or managerial decisions [10].

In the time frame established by the project, as well as in the event of abnormal situations, reports on the development of deformation processes and recommendations on the implementation of protective measures are prepared and sent to other interested specialists.

Using the GIS tools to analyze the development of deformation processes, it is possible to establish the parameters and the nature of deformation processes development both in space and time, which will allow not only to monitor their progress and establish the signs preceding the emergence of an emergency situation, but also to clarify the patterns of development of deformation processes in these conditions.

Measurements are made until the displacement process is completed and in the first two years of the operation of the underground complex.

5 Conclusions

Thus, the main goal of geomechanical monitoring is the creation of a system for observing the development of deformation processes in the soil massif and the earth's surface, which ensures the timely detection of signs preceding the emergency situations in such a way that it is possible to take the necessary preventive and protective measures in advance. For this purpose, a forecast of deformations is made, schemes of observation stations are compiled and monitoring of the state of the surface, objects and structures during the construction and installation works is carried out.

References

1. B.A. Kartoziia, A.V. Korchak, A.N. Levchenko, Mining Magazine, **9**, 60-64 (2018)
2. Wout Broere, Tunnelling and Underground Space Technology, **55**, 245-248 (2016).
3. Priscilla P. Nelson, Tunnelling and Underground Space Technology, **55**, 32–39 (2016).
4. V.E. Merkin, D.S. Koniukhov, Underground and tunnels, **1-2**, 18-23 (2017)
5. K. N. Trubetskoi, Mining Magazine, **7**, 19-23 (2015).
6. K. N. Trubetskoi, M.A. Iofis, Markscheider bulletin, **4**, 27-30 (2007)
7. A.Z. Vartanov, I.V. Petrov, A.V. Fedash, Mining information-analytical bulletin (scientific and technical journal), **10**, 160-164 (2015)
8. E.Iu. Kulikova, A.V. Korchak, A.N. Levchenko, *Risk management strategy in urban underground construction* (Moscow State Mining University Publishing House, 2005)
9. D.L. Neguritsa, Problems and prospects of integrated development and conservation of the earth's interior, 175-178 (IPKON RAS, 2018)
10. D.L. Neguritsa, Kh.S. Alafar, Bulletin of Russian University of Peoples' Friendship. Series: Engineering research, **1**, 22-27 (2016).

Factors influencing the characteristics dependence of zonal structure rock failure around lined opening

Lyudmila Ksendzenko[1*]

[1]Far Eastern Federal University, Vladivostok, Russia

Abstract: For the safe mining operations, it is necessary to take into account a large number of negative factors, one of which is zonal failure (disintegration) of rocks that occurs after developing deposits at great depths.Therefore, investigation of the behavior of zonal mesostructures regularities around deep mine openings is relevant.In this paper dependences from influencing factors of the first failure zone radial extent, of the depth of the contour zone and the first failure zone merging, and the depth of the second failure zone appearance are established based on the non-Euclidean model of the zonal failure rock mass around deep lined openings and the Odintsev's criterion of the failure under compression.The influencing factors are the uniaxial compression strength of the rock σ_c (MPa); the opening depth H (m); the ratio of empirical coefficients γ_3 / γ_1 ; the values of the rocks elastic modulus, E, and the Poisson's ratio, ν . The study was carried out both for hard and weak rocks at values resistance of support 0.5 and 1 MPa. Key words: lined opening, rock mass, zonal failure, non-Euclidean model, depth of zones merging

1 Introduction

The disruption of the massif solidity at great depths has a periodic, zonal character, in the literature it is known as "zonal disintegration." Under these conditions, the deformation of the surrounding rocks also acquires an oscillatory periodic character.The rock massif is a hierarchically-block medium [1, 2].

The description of the block defective media behavior within the framework of classical models of continuum mechanics encounters fundamental difficulties: 1) the inability to take into account the conditions for the disruption of the continuity of the medium as a result of large compressive stresses action; 2) the difficulty of representing the rock mass by a medium far from the conditions of thermodynamic equilibrium, and others. The influence lining of the opening on the character of the massif zonal destruction has not been adequately studied at present. To ensure the ability to manage the characteristics of the failure zonal structure, depending on the values of the influencing factors, special evaluation studies are needed.

* Corresponding author: ksendzenko@mail.ru

In the paper, the dependences of some characteristics zonal structure of the rock massif failure around the lined opening from the influencing factors are investigated. A defective field around a lined underground opening of a circular section corresponding to the meso-level of the massif [3] is described by a non-Euclidean model, where the stress components are determined by the relationships [4]:

$$\sigma_{rr} = \sigma_\infty(1 - r_0^2/r^2) - (E/2(1-v^2)\gamma^{1,5}r)\{aJ_1(r\sqrt{\gamma}) + bN_1(r\sqrt{\gamma}) + cK_1(r\sqrt{\gamma}) - \\ -(r_0/r)\left[aJ_1(r_0\sqrt{\gamma}) + bN_1(r_0\sqrt{\gamma}) + cK_1(r_0\sqrt{\gamma})\right]\} + P(r_0^2/r^2), \quad (1)$$

$$\sigma_{\varphi\varphi} = \sigma_\infty(1 + r_0^2/r^2) - (E/2(1-v^2)\gamma)(aJ_0(r\sqrt{\gamma}) + bN_0(r\sqrt{\gamma}) - cK_0(r\sqrt{\gamma})) - P(r_0^2/r^2) + \\ + \frac{E}{2(1-v^2)\gamma^{1,5}r}\left\{aJ_1(r\sqrt{\gamma}) + bN_1(r\sqrt{\gamma}) + cK_1(r\sqrt{\gamma}) - \frac{r_0}{r}\left[aJ_1(r_0\sqrt{\gamma}) + bN_1(r_0\sqrt{\gamma}) + cK_1(r_0\sqrt{\gamma})\right]\right\}. \quad (2)$$

Here P is the resistance of support (MPa); $\sigma_\infty = \gamma_r ЧH$, γ_r is the unit weight of rock (kN/m^3), and H is the opening depth (m); r_0 is the radius of opening (m); r is the distance from the center of opening to the current point of the rock mass (m); E is the modulus of elasticity (MPa); v is the Poisson's ratio; γ, c are the model parameters (m^{-2}); J_0, N_0, K_0, J_1, N_1, K_1 are the Bessel, Neumann, Macdonald functions of zero and first orders, respectively.

At the same time, cracks related to the macro level for the rock massif are considered not to be interacting, and the failure conditions defined in [5] are applicable to them:

$$K_1 = \sqrt{\pi l}\left(\gamma_1\sigma_{\varphi\varphi} - \gamma_3\sigma_{rr}\right) \geq K_{1c}, \quad (3)$$

where l is the half-length of the rock mass fracture faults and is assumed to be equal to the minimum half-length of a tensile macrocrack (m), which his unstable in stress conditions; $\sigma_{\varphi\varphi}$, σ_{rr} are the maximum and minimum principal stresses, respectively (MPa)determined in (1), (2); γ_1 and γ_3 are the empirical factors; K_1 is the coefficient of stress intensity (МПа·м$^{1/2}$); K_{1c} is the fracture toughness of rock material (МПа·м$^{1/2}$). The resistance of the support P is interpreted as the magnitude of the radial stresses at the contact of the support with the rock mass.

2 Hard rocks zonal failure around a lined opening regularities investigation

The following characteristics of the zonal structure of the rock massif failure around deep rock openings are investigated: the radial extent of the first failure zone, the depth of the contour zone and the first failure zone merging, and the depth of the second fracture zone appearance at the values resistance of the support P = 0.5 and P = 1 MPa of rocks at Nikolaevskij mine (Dalnegorsk, Primorsky Krai, Russia) (Table 1).

Table 1. Physical and mechanical characteristics of the studied rocks

Type of rocks	UCS (MPa)	Young's modulus (MPa)	Poisson ratio	Fracture toughness (MPa× m$^{0.5}$)	Volume weight (kN·m^{-3})
The hard rock	100-250	17000-35000	0.15-0.26	1.25-2.5	26-28

2.1 The first failure zone radial extent dependence on the influencing factors for hard rocks

Dependence of the radial extent of the first fracture zone on the depth of the tunneling for different values of uniaxial compressive strength σ_c of the rock and the ratio of the empirical coefficients γ_3 / γ_1 for the resistance of the supports P = 0.5 and P = 1 MPa is reflected in Tables 2, 3, 4.

It is established that with an increase of opening depth, the radial extent of the first failure zone increases and the faster, the less the ratio γ_3 / γ_1 (Table 2, 3, 4). The radial extent of the first failure zone increases with increasing opening depth, the faster, the less the strength of the rock on uniaxial compression was established. The radial extent of the first fracture zone is shown to decrease if resistance of the support rises, but insignificantly (Tables 2, 3, 4). For example, at $\sigma_c = 100$ MPa (Table 2), the divergence in the radial extent of the first failure zone does not exceed 4% at values resistance of support P=0.5 and P=1 MPa.

The influence of the support resistance decreases if the strength of the rock on uniaxial compression rises (Tables 2, 3, 4). It is shown if the ratio γ_3 / γ_1 rises, the radial extent of the first failure zone decreases the faster, the less the strength of the rock on uniaxial compression at a fixed depth. The radial extent of the first failure zone is established to decrease if ratio γ_3 / γ_1 increases and the faster it is, the greater the opening depth (Tables 2, 3, 4).

Table 2. Comparison of the radial extent of the first failure zone for resistance of supports 0.5 and 1 MPa, different opening depths, ratios γ_3 / γ_1 and fixed values σ_c =100 MPa

P, MPa	H, m	γ_3 / γ_1		
		0.7	0.8	0.9
0.5	450	0.349732	0.293720	0.247328
1	450	0.342596	0.288524	0.243385
The divergence, %		-2.04	-1.8	-1.6
0.5	550	0.639650	0.527899	0.440048
1	550	0.623972	0.517409	0.432437
The divergence, %		-2.5	-2.0	-1.7
0.5	650	0.906963	0.715899	0.586888
1	650	0.871841	0.698732	0.575536
The divergence, %		-3.9	-2.4	-1.9

Table 3. Comparison of the radial extent of the first failure zone for resistance of supports 0.5 and 1 MPa, different opening depths, ratios γ_3 / γ_1 and fixed values σ_c =120 MPa

P, MPa	H, m	γ_3 / γ_1		
		0.7	0.8	0.9
0.5	600	0.393720	0.327695	0.271992
1	600	0.388829	0.324128	0.269312
The divergence, %		-1.2	-1.1	-1.0
0.5	800	0.680011	0.555470	0.455874
1	800	0.670064	0.548771	0.451058
The divergence, %		-1.5	-1.2	-1.1
0.5	1000	0.946749	0.742747	0.599215
1	1000	0.926651	0.732255	0.592274
The divergence, %		-2.1	-1.4	-1.2

Table 4. Comparison of the radial extent of the first failure zone for resistance of supports 0.5 and 1 MPa, different opening depths, ratios γ_3 / γ_1 and fixed values σ_c =180 MPa

P, MPa	H, m	γ_3 / γ_1		
		0.7	0.8	0.9
0.5	800	0.300844	0.249089	0.202957
1	800	0.299306	0.247940	0.202091
The divergence, %		-0.5	-0.5	-0.4
0.5	1000	0.576396	0.473812	0.384159
1	1000	0.573267	0.471533	0.382470
The divergence, %		-0.5	-0.5	-0.4
0.5	1500	1.039442	0.830789	0.662640
1	1500	1.032250	0.826183	0.659448
The divergence, %		-0.7	-0.6	-0.5
0.5	1900	1.442766	1.077890	0.841364
1	1900	1.422335	1.070620	0.836876
The divergence, %		-1.4	-0.7	-0.5

2.2 Dependence of the contour merging depth and the first failure zone on the influencing factors for hard rocks

As the opening depth is increasing, the radial extent of the contour zone rises, and the radial extent of the first intermediate zone decreases to zero, which results in the merging of the contour zone and first zone of failure. Fig. 1*a, b* shows the dependence of the depth merging of the first failure zone with the contour zone on the uniaxial compression strength at values of support resistance P = 0.5 and P = 1 MPa, respectively.

It is established that with increasing in the rock uniaxial compression strength, the depth of merging of the contour and first failure zones increases and the faster it is, the larger the ratio γ_3 / γ_1. The influence of the values of support repulse on the depth of merging of the contour and the first failure zones for different values σ_c and ratios γ_3 / γ_1 is shown in Table. 5. It is shown that with a decrease in the uniaxial compression strength σ_c, the influence of the support on the depth of merging of the contour and the first failure zones becomes more significant, but the difference in the depth of merging does not exceed 4%.

Dependences of the depth of merging of the contour and first failure zones on the ratio γ_3 / γ_1 for different values of the rock uniaxial compression strength are shown in Fig. 1*c, d*. It is established that with increasing ratio γ_3 / γ_1, the depth of merging of the contour and the first failure zones increases and the faster it is, the larger the value σ_c. At the same time, the influence of support resistance is insignificant (Table 5).

Table 5. The depth of merging contour and first failure zones (m) at values the resistance of support P = 0.5 и 1 MPa, different values of the uniaxial compression strength of the rock and ratio γ_3 / γ_1

γ_3 / γ_1	σ_c, MPa					
	100		120		180	
	P = 0.5 MPa	P = 1 MPa	P = 0.5 MPa	P = 1 MPa	P = 0.5 MPa	P = 1 MPa
0.7	814.4	841.6	1066.2	1087.5	1967.6	1986.6
0.8	969.6	995.4	1274.1	1299.2	2416.6	2439.4
0.9	1154.0	1184.2	1534.6	1564.2	3013.7	3041.6

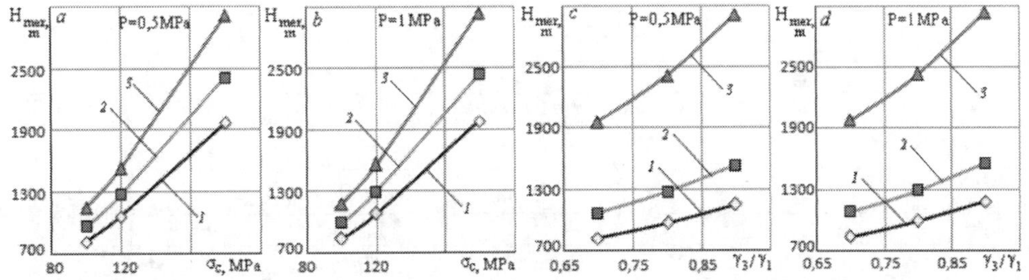

Fig. 1. Dependence of contour merging depth and first failure zones for hard rocks on: *a*) the uniaxial compression strength of the rock σ_c at the values γ_3/γ_1, respectively: *1* – 0.7; *2* – 0.8; *3* – 0.9 (*a, b*); the ratio γ_3/γ_1 at the values uniaxial compression strength of the rock σ_c (MPa): *1* – 120; *2* – 150; *3* – 180 (*c, d*); (*a, c*) – at resistance of support P = 0.5; (*b, d*) – P=1 MPa

2.3 Dependence of the second failure zone depth appearance on the influencing factors for hard rocks

Dependences of the depth at which the second zone of failure appears on the uniaxial compression strength of the rock at resistance of the supports P = 0.5 and P = 1 MPa respectively are shown in Fig. 2 *a, b*. It is established that with increasing in the rock uniaxial compression strength, the depth at which the second zone of failure appears also increases. With an increase in the ratio γ_3/γ_1, the second zone of failure appears at a greater depth (Fig. 2 *c, d*).

The influence of the values of support resistance on the depth at which the second zone of failure appears for different values σ_c and ratios γ_3/γ_1 is shown in Table 6. It can be noted that with the decrease in the uniaxial compression strength of the rock σ_c, the effect resistance of support increases and that the difference in the depth at which the second zone of failure appears for resistance of supports P = 0.5 and P = 1 MPa does not exceed 1.5%.

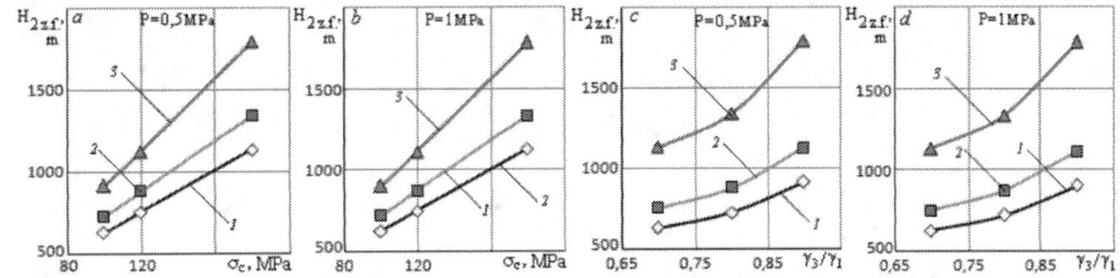

Fig.2. Depth dependence at which the second zone of failure appearsfor hard rocks on: *a*) the uniaxial compression strength of the rock σ_c at the values γ_3/γ_1, respectively: *1* – 0.7; *2* – 0.8; *3* – 0.9 (*a, b*); the ratio γ_3/γ_1 at the values uniaxial compression strength of the rock σ_c (MPa): *1* – 120; *2* – 150; *3* – 180 (*c, d*); (*a, c*) – at resistance of support P = 0.5; (*b, d*) – P=1 MPa

Table 6. The depth at which the second zone of failure appears (m)

γ_3 / γ_1	σ_c, MPa					
	100		120		180	
	P =0.5 MPa	P =1 MPa	P =0.5 MPa	P =1 MPa	P =0.5 MPa	P =1 MPa
0.7	631.6	625.3	758.4	752.6	1134.8	1130.1
0.8	730.9	722.2	884.2	876.0	1345.1	1338.4
0.9	916.2	903.0	1128.7	1115.9	1796.4	1785.1

3 Weak rocks around a lined opening zonal failure regularities Investigation

The study was carried out at coal mine named Artyom (Shkotovo settlement, Primorsky Krai, Russia) (Table 7).

Table 7. Physical and mechanical characteristics of the studied rocks

Type of rocks	UCS (MPa)	Young's modulus (MPa)	Poisson ratio	Fracture toughness (MPa × $m^{0.5}$)
The little hard rock	8-15	500-700	0.14-0.16	0.112-0.168

Fig. 3 shows the dependence the radial extent of the first fracture zone on the mining depth at the resistance of supports P = 0.5 and P = 1 MPa, respectively. It is established that with an increase of mining depth, the radial extent of the first failure zone increases and the faster it is, the less the ratio γ_3 / γ_1 (Fig. 3*a, b*). It can be seen from Figures 3*c, d* that the radial extent of the first failure zone increases with increasing depth, and the faster it is, the less the σ_c.

It is shown that with increasing the resistance of the support, the radial extent of the first failure zone decreases and at a depth of 90 m the difference in the radial extent reaches 30% (Table 8). Hence, there is a significant influence the resistance of support on the radial extent of the first fracture zone for weak rocks.

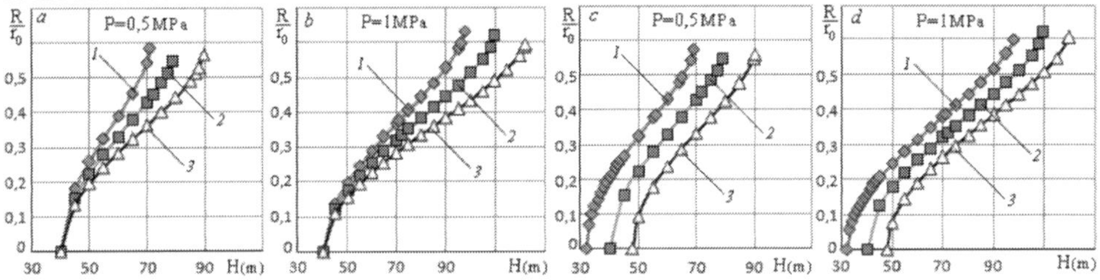

Fig. 3. Dependency graphs of the radial extent of the first failure zone on the opening depth for the weak rocks: (*a, b*) – at ratio γ_3 / γ_1 :*1*– 0.7; *2* – 0.8; *3* – 0.9, σ_c=10MPa and values resistance of support *a* – 0.5; *b* – 1 MPa; (*c* , *d*) – at ratio γ_3 / γ_1 = 0.8 ; σ_c : *1*– 8; *2* – 10; *3* – 12 MPa and values resistance of support *c*– P=0.5; *d* – P=1 MPa

The patterns behavior of the rock massif zonal failure characteristics, established for hard rocks, is also preserved for the weak rocks. The influence of resistance of the support for the weak rocks is significant and reaches 30%. It is found out that the deformation modulus E and the Poisson ratio have practically no effect on the radial extent of the first failure zone, both for hard and weak rocks.

Tablel 8. Dependence of the first failure zone radial extent on opening depth at resistance of the support P=0.5 and 1 MPa, $\gamma_3/\gamma_1 = 0.8$ и $\sigma_c = 12$ MPa

H (m)	The resistance of the support P=0.5 MPa	The resistance of the support P=1 MPa	The divergence, %
50	0.0938	0.0766	-18.4
70	0.3349	0.2647	-21.0
90	0.5485	0.3859	-29.6

The discussion of the results

The notions of contour (near the opening boundary) intermediate zones and fracture zones are related to the coefficient of normal stress intensity K_1, which depends on the gravitational stress.

For a certain value of the gravitational stress, an inequality (3) is satisfied, which indicates the zones appearance of fracture. The maxima of the oscillating function K_1 correspond to the zones of failure of the rock massif, and the minima to the intermediate zones it is assumed. The contour zone of failure is meant the zone enclosed between the contour of the opening and the first intermediate zone.

In the future, it is necessary to investigate the dependence on the influencing factors of the zones number, the position of the last failure zone, and others.

Conclusions

1. The radial extent of the first zone of failure increases if the depth rises, up to the confluence of neighboring zones; the radial extent of the first fracture zone decreases if the σ_c rises; the radial extent of the first fracture zone decreases if the ratio γ_3/γ_1 rises.
2. The merging of the contour zone and the first fracture zone occurs at a greater depth if the σ_c rises; the depth of the merging of the zones increases if the ratio γ_3/γ_1 rises.
3. With increasing in the σ_c, the depth of appearance of the second fracture zone also rises. If the ratio γ_3/γ_1 rises, the second zone of failure appears at a greater depth.
4. The radial extent of the first fracture zone decreases if the resistance of support rises. The insignificant influence of the resistance of support on the characteristics of the zonal fracture structure for hard rocks and a noticeable influence for weak rocks is shown.

The author thank Professor Vladimir Makarov for his important recommendations in the process of writing this paper. The research is supported by a grant from the Ministry of Science and Higher Education of the Russian Federation No. RFMEFI58418X0034.

References

1. M.A. Sadovsky, DAN USSR, **247,** 829. (1979)
2. P.V. Makarov, Physical Mesomechanics, **7,** 25. (2004)
3. V.V. Makarov, M. A. Guzev, V.N. Odintsev, L.S. Ksendzenko, Journal of Rock Mechanics and Geotechnical Engineering, **8**, 164. (2016) .doi.org/10.1016/j.jrmge.2015.09.010
4. L.S. Ksendzenko, GIAB, **8**,364. (2007)
5. V.N. Odintsev, Rupture destruction of a brittle rocks mass (Moscow: IPKON, RAS, 1996)

Research of the geological factors of coal deposits in the south of the Far East by electrical exploration methods

Nikolay Shkabarnya[1,2*], *Grigory* Shkabarnya[2,3], *Evgenia* Nechaeva[2] and *Svetlana* Korneeva[1]

[1] Mining Institute of Far Easten Branch of Russian Academy of Sciences, Khabarovsk, Russia
[2]Far Eastern Federal University, Vladivostok, Russia
[3]V.I. Il'ichev Pacific Oceanological Institute, Far Eastern Branch, Russian Academy of Sciences, Vladivostok, Russia

Abstract Coal deposits in the south of the Far East are characterized by complex geological factors, which is due to the inclined occurrence the enclosing layers and coal seams, high pressure of underground water, the presence of water-bearing sandy-pebble and gravel-pebble sediments, the presence of numerous faults and various inclusions, the presence of island permafrost. In the exploration of deposits, the low efficiency of previously used electrical exploration methods, including electrical sounding and profiling, was noted. In this paper, it is shown that increasing efficiency in the solution of some mining and geological tasks is associated with the use of the electrical tomography method. This method has advantages in the conditions of weak differentiation of rock resistivity due to the high density of obtained parameters. With the use of electrical tomography, the tasks in the allocation of tectonic disturbances and coal seams uplift under quaternary sediments, estimation of the landslide slopes structures on the coal pit edges have been solved. Successful results were obtained in isolating the weathered layer of the bedrock, sandstone outcrops under the soil-vegetative layer in the upper part of the section, the bedrock surface on the deposit's flanks and the covers of effusive rocks on the surface.
Keywords: coal deposit, weakened layer, weathered layer, electrical sounding, electrical tomography.

1 Introduction

For effective exploitation of coal deposits in the Khabarovsk and Primorye Territories, reliable information on geological factors is necessary, which can be ensured by the use of modern electrical exploration methods. The main object of coal mining in the Khabarovsk Territory is the Urgalsky coal deposit, and in the Primorsky Territory - the Bikin brown coal deposit.

The Urgalskoe deposit is characterized by relatively favorable geological factors for mining. Therefore, geophysical methods were used here in limited volumes to identify the

* Corresponding author: shkabarnya@mail.ru

areas of subsidence in the worked-out space. During the studies, electrical exploration using the transient electromagnetic method and seismic exploration using the refracted and reflected wave methods (in the CDP modification) was used. As a result of the work, the structural-tectonic structure of individual sections was studied, the places of subsidence of the worked-out space in the working zones of coal seams were determined, the conditions for the occurrence of the coal-bearing complex in the 50–150 m depth interval were established. Complex geological factors are observed in other coal deposits of the Khabarovsk Territory: Urgal-Soloniy, Ust-Umaltinsk, Olonzhinsky. However, their studies using geophysical methods were not carried out.

In the Primorsky Territory, the study of geological factors within coal deposits was carried out in much larger volumes. This is especially true for the Bikin brown coal deposit. Here, the relatively simple plicative structure of the depression is everywhere complicated by faults and fault-shifts. Numerous coal beds lie among the host rocks of the Paleogene-Neogene, which are represented by sandstones, siltstones, argillites and carbonaceous argillites. Quaternary sediments cover with a continuous cover all ancient formations and are represented by clays, loam and sandy loam. On the local areas of the surface the covers of effusive rocks lie, represented by andesite and andesite-basalts of Paleogene age.

Coal open-casts are characterized by very complex geological and hydrogeological conditions. The water-physical and physical-mechanical properties of the rocks vary sharply both in depth and lateral. There is a large number of synclinal overlapping aquifers with an unequal water availability. The sides of the open-casts are in the water-saturated state due to unloading of the pressure waters of unopened coal seams, as well as the aquifers uncovered by the side. Throughout the entire period of mining, landslides of the sides and ledges were observed, which considerably complicate the mining operations.

Geophysical surveys of the open-casts during the 70-90s of the last century included vertical electrical sounding (VES), symmetrical and dipole electro-profiling combined with parametric boreholes. As a result of the VES curves interpretation, the thickness of Quaternary sediments and coal-bearing strata, the character of the buried surface of the basement, the boundaries of the Cenozoic sediments distribution, the patterns of changes in the physical properties of different layers of the section were established.

In the upper part of the section, sandy-argillaceous sediments and Quaternary clays were confidently distinguished by resistivities in the range of 15–40 Ωm, their thickness reached 30 m. Below on the section, separate boundaries in the thickness of the Cenozoic rocks were noted, on which the approximate thicknesses of coal-bearing strata were determined. The identification and tracing of the faults according to the VES data was carried out, as a rule, from the change in the depths of coal seams occurrence.

However, it should be noted the low efficiency of previously performed electrical exploration in solving some geological tasks related to the study of coal seams, the identification and tracing of weakened layers, weathered layer of the basement and various local inclusions. First of all, this is due to the fact that most of the coal seams and weakened layers have small thickness (5–15 m), and the resistivity of the rocks overlaps. Therefore, in order to reduce the amount of drilling work during the study of geological factors in the open-casts, a decision was made to set the work by the method of electrical resistivity tomography (ERT). This method has advantages in conditions of weak differentiation of rocks by resistivity due to the high density of the obtained electric field parameters [1–3].

2 Methodical research

To effectively solve the mining and geological tasks, a priori geological and geophysical information was preliminarily studied at each section of the Bikin deposit and mathematical modeling of electric fields was carried out in the heterogeneous media conditions [4, 5].

When analyzing the modeling results, the regularities of the anomalous areas of the electric field were determined, and the detection criteria for multidirectional boundaries and local heterogeneities were established, which were used at the key stage of interpretation during the formation of background geoelectrical models. Quantitative interpretation of field data was carried out by the Res2Dinv program [6] using background models. Studies ended in the formation of geoelectric sections, which, after geological interpretation, were presented in the form of geologic-geophysical sections (Figures 1, 2, 3).

Experimental electrical tomography studies at the Bikin deposit were carried out to solve various tasks. In some areas, outcrops of coal seams (with resistivities of 50–65 Ωm) under quaternary sediments among argillites (10–25 Ωm), siltstones (20–40 Ωm) and sandstones (60–110 Ωm) were identified and traced. The outcrops were identified at depths of 8 to 50 meters with an average thickness of 15 meters. In the study of fault zones, it has been established that the zones of crushing rocks filled with sandy–crushed-stone sediments, gravel and pebbles (and with clays and loams near the surface) are more effectively determined. Resistivity of these rocks varies in the range from 10 to 120 Ωm, and the thickness reaches 30 m.

ERT field works in a large volume were carried out on the landslide slopes, where the spatial position of the weakened layers was determined. Weakened layers are represented by soft-plastic clays in Quaternary sediments at depths of up to 30 m and very soft plastic clays of Neogene age at depths of more than 30 m. Moreover, the latter, due to their small thickness (3–6 m), were identified (by low resistivities of 10–14 Ωm) only in conjunction with the underlying water-saturated argillic-like loam of Paleogene age, with a thickness of more than 10 meters.

Recently, for the purpose of a detailed study of the upper part of the geological section and mapping of the basement surface, investigations were carried out on the flanks of the Bikin deposit. The area of work is characterized by a shallow bedding of the basement, represented by fine-grained and very fine-grained sandstones and siliceous-argillaceous shales of the Triassic and Jurassic ages. Above the basement lie Quaternary sediments and overburden rocks.

As a result of quantitative interpretation of field materials, taking into account the geometric and electrical parameters of background models, geoelectrical sections were constructed. They confidently identified the conditions of occurrence and patterns of changes in the electrical properties of quaternary sediments, fill-up soils, and, most importantly, of the low-resistivity complex of the basement weathered layer. The roof of the complex is traced at depths from 7 to 30 meters and more. After the geological interpretation of the geoelectric sections with associating of the geoelectric and lithologic horizons the geologic-geophysical sections were compiled.

Figure 1 shows the geologic-geophysical section on one of the profiles along depression edge. It can be seen that the roof of the weathered layer is traced at depths of 10 to 25 m. Judging by the patterns of variation in resistivity (45–100 Ωm), the weathered layer is the most homogeneous part of the section with highly watered segments (45–55 Ωm). In the central part of the section, at the depth of 45 m with a resistivity of 130 Ωm, a surface of durable sandstones of the Triassic age has been identified. Sand-gravel sediments and fill-up soils in the upper part of the section are distinguished by resistivity from 100 to 700 Ωm. As can be seen, the electrical properties of the upper part change sharply along the profile. In the areas of pickets 130 and 360, alleged tectonic faults are marked.

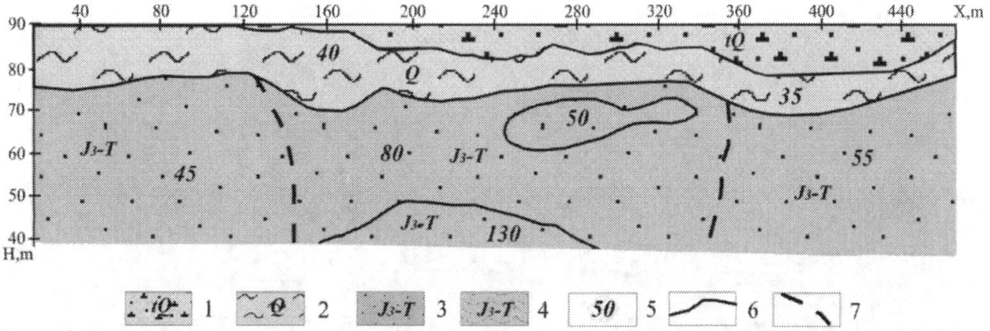

Legend: 1 - technogenic quaternary sediments, fill-up soils, sandy-gravel sediments; 2 - alluvial quaternary sediments, clays, loams; 3 - Triassic-Jurassic sandstones; 4 - Triassic-Jurassic sandstones with lenses of watering; 5 - resistivity in Ωm; 6 - layer boundaries; 7 - alleged tectonic faults.

Fig. 1. Geologic-geophysical section along the basic profile in the area for tracking the weathered layer of the basement rocks.

In the course of research in several areas of the deposit, in the geological section the protrusions of the Triassic sandstones were identified and traced. These protrusions significantly complicate the mining operations. Taking into account the topicality of the such task solution, first, based on the analysis of geological and geophysical information, a background geoelectric model with a sandstone protrusion was developed (Fig. 2, a). The model parameters were used in calculating the theoretical sections of apparent resistivity for wedge-shaped media, on the basis of the analysis of which criteria for isolating such heterogeneities were determined. The main criteria were a special form of isolines of apparent resistivity with different concentrations of them over the top of the object.

Figure 2,b shows a geologic-geophysical section, on which the sandstone protrusion is marked with high resistivities (110 Ωm). Its top is traced among quaternary clays (14–24 Ωm), in the lower part – among low-resistivity (25-30 Ωm) argillites and siltstones of Paleogene-Neogene age. Coal seams at this depth are not found. As can be seen, the boundaries of the layers reflect the detailed structure of the geological section at the site of investigation.

In some parts of the deposit, electric tomography made it possible to identify and trace the covers of effusive rocks represented by andesites and andesite-basalts of the Paleogene age. The thickness of this rocks within the depression sometimes reaches 30 meters, which also complicates the mining operations. Figure 3 shows a geologic-geophysical section, in which under the soil-vegetable layer of low thickness, effusive rocks with resistivity of more than 70 Ωm and a thickness of 17 meters are distinguished. Sandy-argillaceous and slightly plastic clays of Quaternary age are shown below in the section. As can be seen, the resistivity of these sediments decreases with the depth from 50 to 14 Ωm.

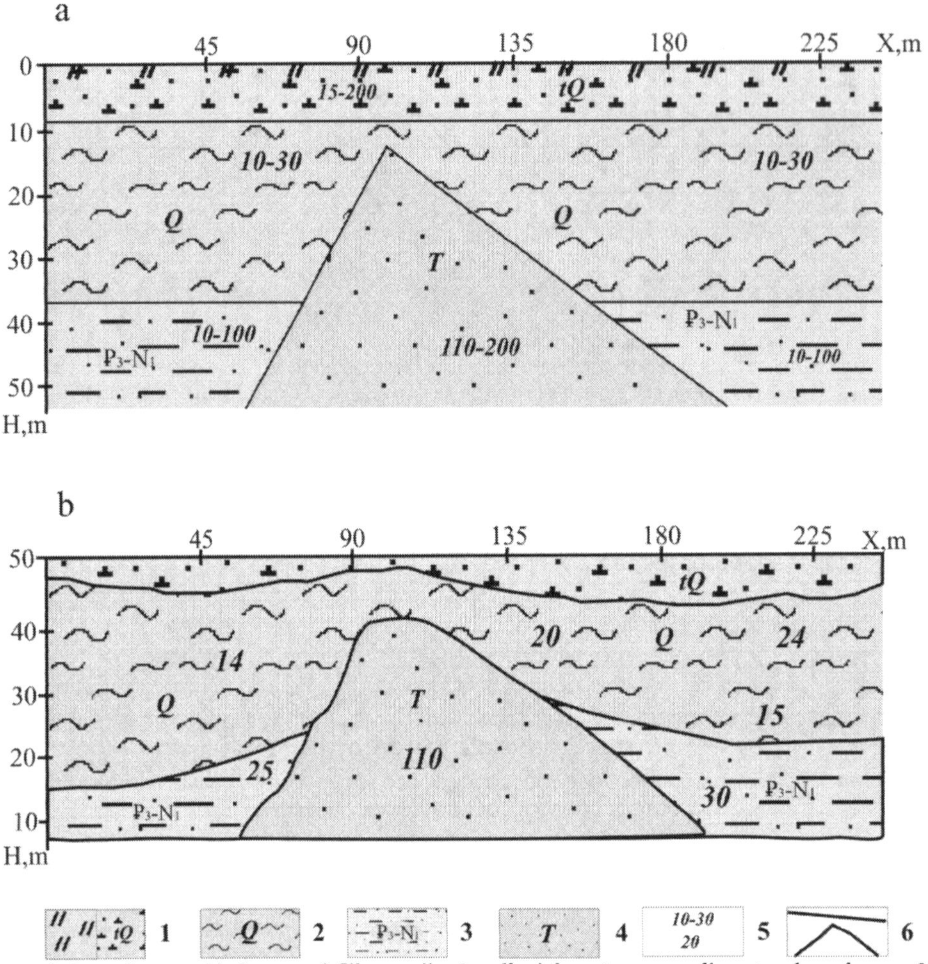

Legend: 1 - soil-vegetable layer and fill-up soils; 2 - alluvial quaternary sediments, clays, loams; 3 - Paleogene-Neogene sediments, mudstones, siltstones, sandstones; 4 - durable sandstones of Triassic age; 5 - resistivity in Ωm; 6 - boundaries of layers and protrusion of sandstone.

Fig. 2. Background geoelectric model (a) and geologic-geophysical section (b) at the site of tracking of the sandstone protrusion.

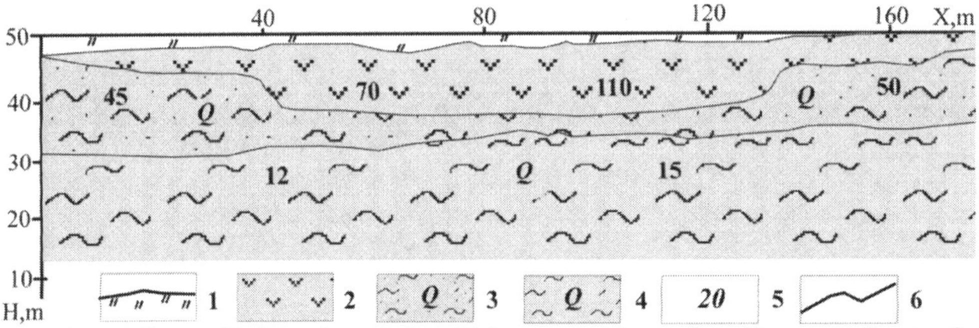

Legend: 1 - soil-vegetable layer; 2 - andesite of Paleogene age; 3 - sandy-argillaceous sediments of Quaternary age; 4 - alluvial clays of Quaternary age; 5 - resistivity in Ωm; 6 - layer boundaries.

Fig. 3. Geologic-geophysical section in the area of the effusive cover.

3 Conclusions

Thus, according to the research of geological factors at Bikin brown coal deposit the following main results have been obtained.

1. Previously conducted work by the method of electrical sounding (VES) and profiling (EP) allowed to determine the occurrence conditions and physical properties of quaternary deposits, and sometimes the approximate parameters of coal-bearing strata and weathering layer. However, these methods do not allow us to reliably study the geological factors of coal open pits.

2. Electrical tomography (ERT) has advantages in comparison with the VEZ and EP methods for solving mining and geological tasks in conditions of weak differentiation of rocks by resistivity, due to the high density of the obtained electric field parameters, as evidenced by the results of recent years.

3. As a result of the ERT exploration, segments of the coal seams uplift under quaternary sediments were established, weakened layers have been determined and traced. In the course of investigations on the flanks of the deposit, the thicknesses and electrical properties of quaternary deposits and fill-up soils were studied in detail, and, most importantly, the bedding conditions and properties of the basement weathered layer have been determined.

4. In several areas, by the ERT method, protrusions of durable sandstones among alluvial quaternary sediments and cover of effusive rocks have been identified and traced.

The introduction of the electrical tomography for studies of geological factors of coal deposits in the south of the Far East will contribute to their more effective mining.

References

1. Dahlin, T. *Computers & Geosciences.* **27**, pp.1019–1029. (2001).
2. Kenma, A., Binley, A., Ramirez, A. and Daily, W. *Chemical Engineering Journal.* **77**, pp.11-18. (2000).
3. Rasskazov I.Yu., Shkabarnya G.N., Shkabarnya N.G. *Journal of Mining Science.* **49 (5)**, pp. 772-778. (2013).
4. Griffiths, D. H.; Barker, R. D. *J. of Appl. Geophysics.* **29 (3),** pp. 211-226. (1993).
5. Rasskazov I.Yu., Shkabarnya G.N., Shkabarnya N.G. *Journal of Mining Science.* **49 (3),** pp. 388-394. (2013).
6. Loke, M.H., Acworth, I. and Dahlin, T. *Exploration Geophysics.* **34**, pp.182–187. (2003).

Selective Desintegration and Concentration: Theory and Practice

Tatyana Aleksandrova[1*], N. Nikolaeva[1], Holger Lieberwirth[2] and Alexander Aleksandrov[3]

[1]Federal State-Funded Educational Institution of Higher Professional Education "Saint Petersburg Mining University", Saint Petersburg, Russia
[2]University of Technology "Freiberg Mining Academy", Freiberg, Germany
[3]Mining Institute of Far eastern branch of Russian Academy of Sciences, Khabarovsk, Russia

Abstract. Investigations on the possibility of selective grinding and benefication of quartz-pecoraite ores of the Ufaleiskoye deposit have been carried out. Methods of computer X-ray microtomography have been used to study the morphometric features and pore volume parameters of samples, and these data was compared with results of quantitative microstructural analysis. The performed research shows the powerful capabilities of X-ray computer microtomography method for solving scientific and practical problems related to the study of physical and mechanical properties of rocks, the determination of ore minerals liberation, the evaluation of their content, the analysis of the granulometric composition and the spatial distribution of their grains. Analysis of mineralogical and geochemical composition of ores showed that these ores meet the criteria for selective disintegration. An analysis of flotation experiments results showed that when material is grinded in a roll mill, the consumption of reagents is 25% lower than when grinding in a ball mill if the same degree of nickel extraction is achieved. **Key words:** selective disintegration, flotation, nickel, separation, microtomography.

1. Introduction

When developing new deposits and designing concentrating factories, they often face problems of processing complex and fine-grained mineral raw materials, which leads to additional costs and an extension of the crushing and grinding scheme for sufficient intergrowth liberation. This trend can lead to an increase in electricity consumption four or more times by 2030 compared to today [1-2]. Thus, a reduction in specific energy consumption in the processes of disintegration and mineral benefication is an obvious task, taking into account the tendency of saving natural resources, as well as reducing the impact on the environment during the processing of mineral raw materials.

One of the ways to solve this problem is the use of selective disintegration [3], which would allow not only to break the rock mass to the required size, but also to carry out pre concentration at the stage of classification to obtain a preliminary concentrate and put the waste rock into the tailings.

*Corresponding author: alexandrovat10@gmail.com

2. Theoretical approaches to selective disintegration

Selective grinding leads to optimal formation and propagation of cracks in one or at once in several components and, consequently, causes differences in the distribution of the valuable component over the size fractions (Table 1).

Table 1. Graphical examples of selective and standard disintegration

Raw material	Selective disintegration		Standard disintegration
	Selective crushing	Selective grinding	
Initial material	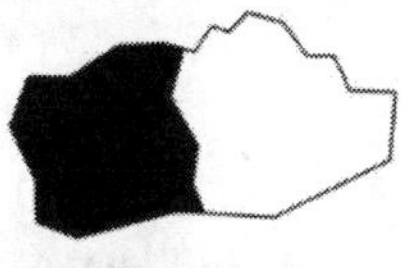	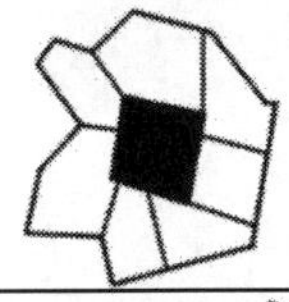	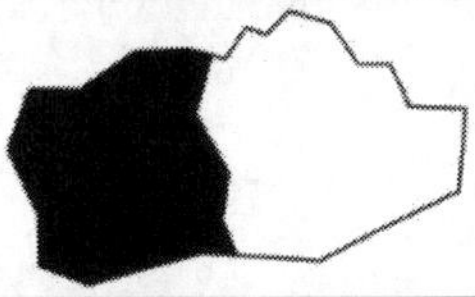
Crushed product	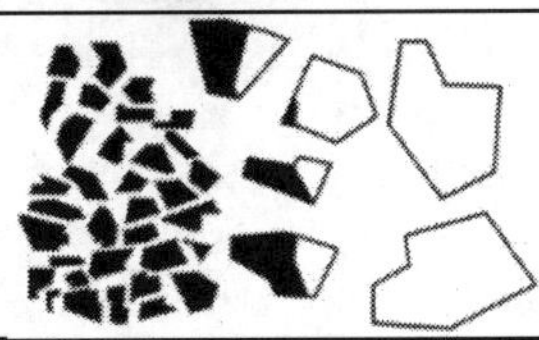	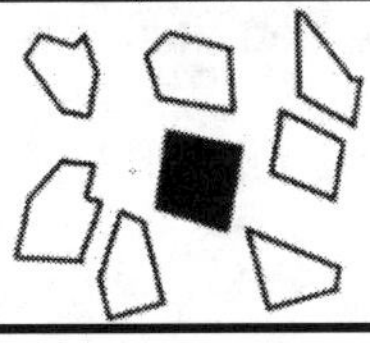	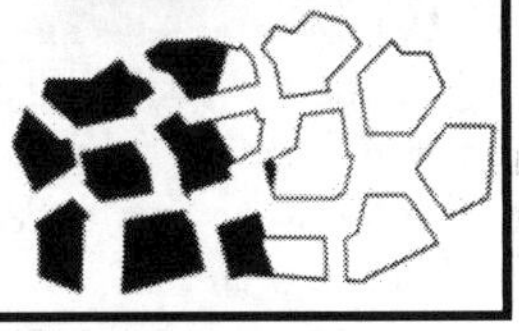

The material's ability for selective disintegration is determined by differences in the physical, chemical, structural, and other properties of the material. Systematization of theoretical data on selective disintegration allowed revealing the system character of selective grinding (Fig. 1) [4-6].

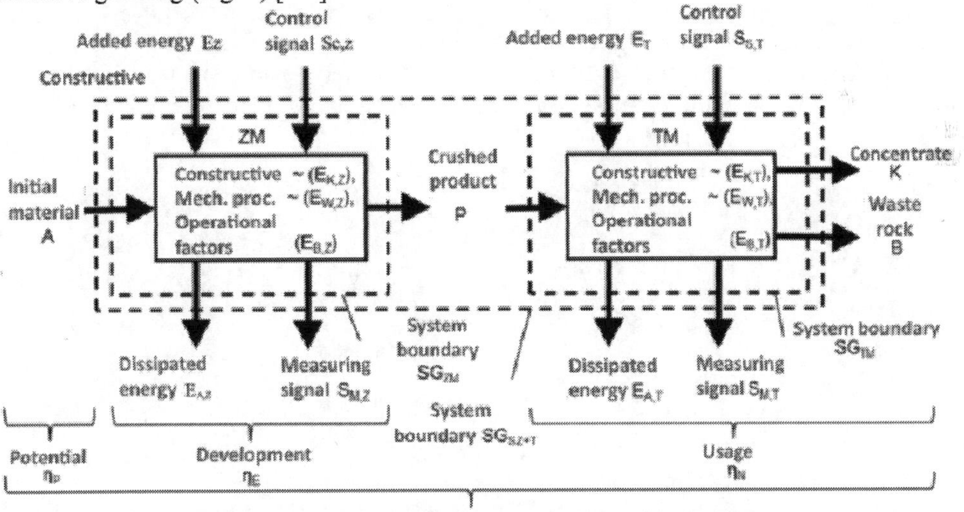

Fig. 1. Black box model for system chain in case of selective disintegration

But in order to qualitatively assess the possibility of selective destruction, it is necessary to quantitatively determine the structural and texture parameters of the ore. For this aim, the quantitative microstructure analysis method (QMA), developed at the Institute of Mineral Processing Machines of Freiberg Mining Academy TU [7-8], is successfully used.

For the implementation of selective disintegration effect, an in-depth study of the mineral components properties of the ore is necessary. Usually, geologists give a description of the texture and structure of ore formations by analyzing microsection or polished sections. This method has limitations, since it is two-dimensional and can only give a qualitative description. Obtaining more detailed information on mineral raw materials will allow to reveal the dependencies between the corresponding parameters of breaking, the characteristics of the product and their inherent mineralogical-technological and structural-texture features, which is also an urgent task in the study of selective disintegration processes and the development of devices for their implementation.

Although selective disintegration is sometimes used in concentrating plants [9-10], systematic studies of the processes of selective disintegration are still lacking, as well as systematic data on the influence and interrelation of mineral-technological, structural-texture and physical-technical properties of minerals, rocks and ores with the processes of selective disintegration and flotation.

3. Results and discussion

Quartz-pecoraite ores of the Ufaleiskoye deposit were investigated as an object of research. The deposit has been well studied by Russian geologists [11-12]. Three zones of weathering are distinguished in the field profile: the oxide-iron zone (goethite ores), the nontronite zone (chamosite ores and the zone of leached serpentinites (chrysotile-lizardite and nepouit-chrysotile-lizardite ores). Major part of nickel is contained in the nontronite (chamosite) zone. In some places, this zone can reach a thickness of about 50 m. Ores are a moist clay-like mass with fragments of leached serpentinites, which requires a rather complex process for their benefication. Despite the fact that the main ore zone of the deposit is the zone of nontronite (chamosite) clays, the underlying leached serpentinites also contain nickel concentrations that are profitable for extraction.

A number of typical samples were analyzed by computer tomography.

Sample 1. The distribution of ore minerals in the entire rock volume is shown in Fig. 2, from which it can be seen that it is possible to distinguish on a tomogram dense ore minerals from the silicate minerals of the matrix as effectively as possible, while the latter minerals are quite difficult to distinguish from each other, due to their close X-ray density.

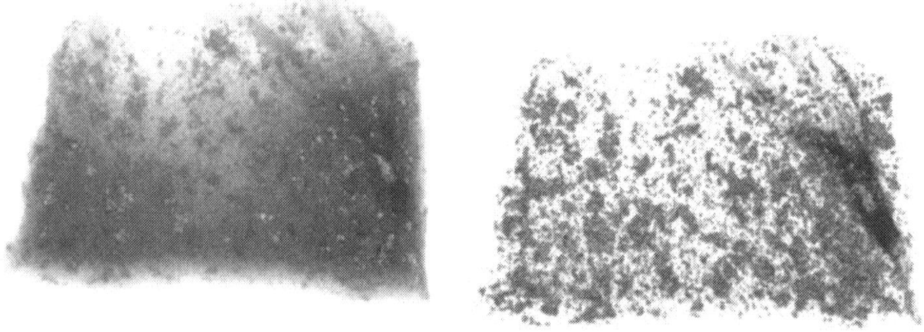

Fig.2. Distribution of ore minerals (green) in a silicate matrix (gray)

Using the CTvox and CTan software (Bruker, Belgium), it is possible to examine in detail and analyze not only the size of the grains composing the rock, but also their mutual relations with each other, which is especially important for technological mineralogy. Ore minerals are rather evenly distributed throughout the rock volume, sometimes forming individual grains with an average grain size of 0.07 and fine aggregates 0.26 mm in size (Fig. 3A, B), sometimes up to 2.0-4.5 mm (Fig. .3, C).

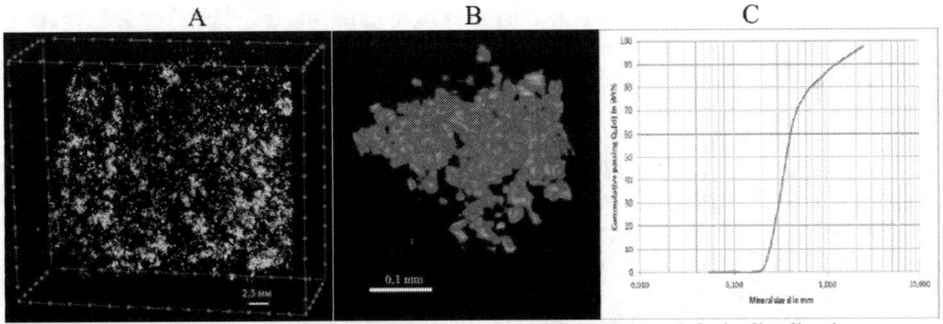

Fig. 3. Aggregates of ore minerals in the rock volume (A, B) and their distribution according to the maximum diameter, mm (C)

Ore minerals have a quite isometric shape, sphericity coefficient calculated as the ratio of the short to the long axis (Bruker, Belgium) is 0.78 (this coefficient reaches 1, when the shape of the grain is almost isometric, Fig. 4A). The specific surface of ore minerals, calculated as the ratio of the total surface area of minerals to their volume and representing the degree of sinuosity of the boundaries, is 31.86 mm^2/mm^3. The orientation of ore grains and aggregates is marked, expressed quantitatively as the deviation of the longest axis of grains from the vertical in degrees, (Fig. 4 B).

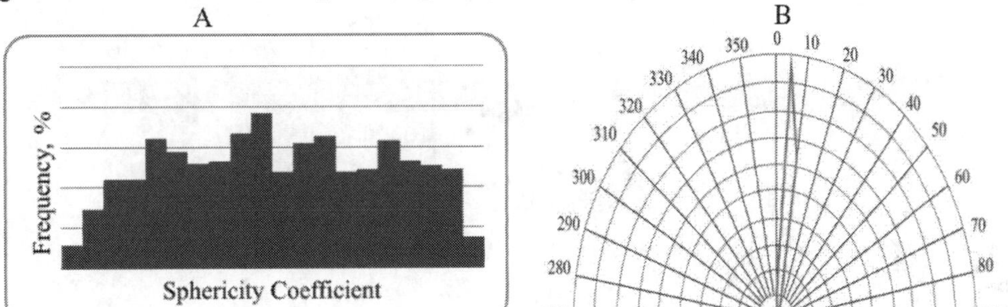

Fig. 4. The distribution of ore minerals aggregates according to sphericity coefficient (A) and orientation of longest axis (B)

One of the most important characteristics for rocks disintegration, linking the strength of the rock with the defectiveness of its structure, is porosity. It is determinant when interpretation of the strength properties and behavior of the rock during disintegration [13-15]. The total porosity of the cupriferous sandstone sample is 1.4%, the opened porosity is 0.68%, and the pore connectivity is low. In general, subcapillary pores size is less than 0.0002 mm, the predominant pore size is 25-50 μm, there are single pores larger than 0.1 mm in diameter. The pore sphericity factor is 0.62. The pore density, calculated as the ratio of the number of pores per unit volume of rock, is 9.10 mm^{-1}.

Similarly, samples No. 2 and 3 were analyzed. Comparative analysis of samples porous volume based on CT results revealed that despite the highest total porosity (3.05%), sample 3 has a minimum pore density (2.03 mm^{-3}), that due to the small pore sizes (15-40 microns) and their isometric shape (Ksf 0.80) leads to increase of rock the strength of the rock. Sample 1 has intermediate parameters of pore volume and medium strength, and sample No. 2 has a minimum strength at minimum parameters of total porosity and maximum values of pore density. By results of researches it is possible to speak that strength of the rocks decreases with reduction of total and closed porosity parameters, as well as pore sphericity coefficient. Rock strength also decreases with increasing of closed pores total quantity and its density.

Thus, the porosity parameters obtained by the CT method correlate with rocks strength differences, specifically with the strength index Is(50) [16, 17], whereas the total porosity, determined from the values of the mineral and bulk densities, allows to indirectly estimate the rock strength and does not provide information on the size, shape of the pores, their connectivity and spatial orientation. In addition, the undoubted advantage of the X-ray tomography method is the possibility of 2D and 3D data visualization with the help of specialized CTvox and CTan software that help analyze not only the size of grains and their aggregates, but also their distribution in the total rock volume and mutual relations with each other .

Since during benefication of serpentinites it is necessary to take into account that the nickel content in the rock may be too low, selective grinding may be required to create a high-quality concentrate in a certain fraction (Figure 5).

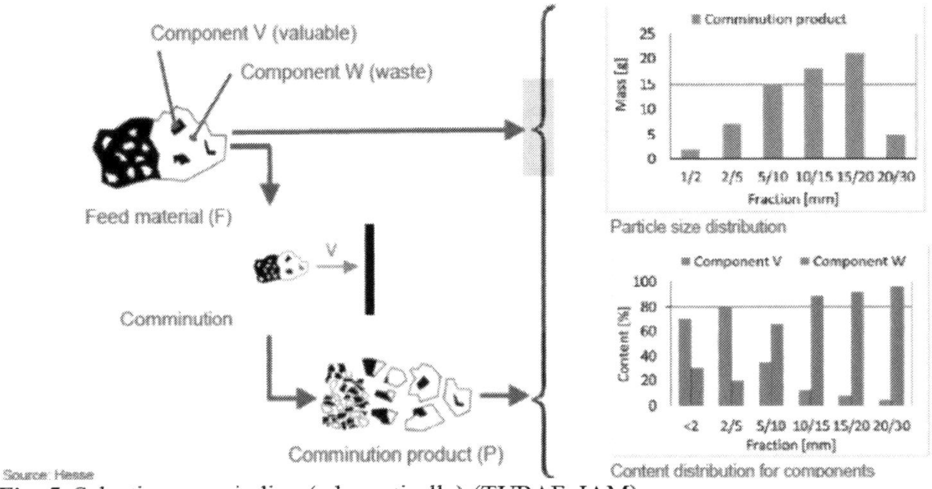

Fig. 5. Selective ore grinding (schematically) (TUBAF, IAM)

In this case, high-grade nickel ores can be separated from the bulk of serpentinites.

If the nickel content in serpentinites is quite high, then to extract it from the grinded product by subsequent flotation, it is possible to propose a roller mill method (Fig. 6).

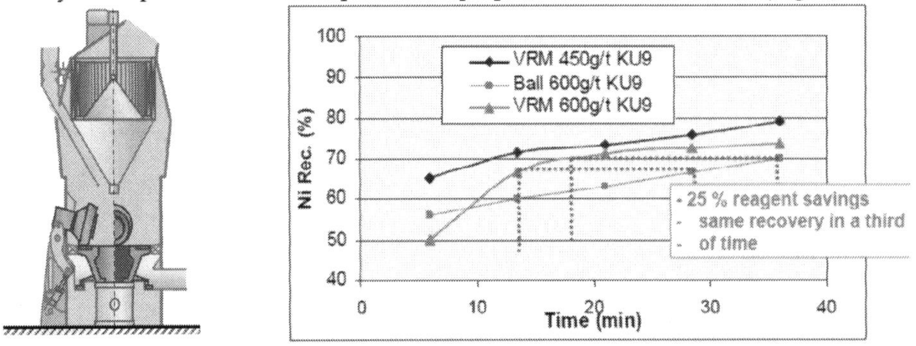

Fig. 6. Results of nickel flotation extraction when grinding in a roller mill (Loesche) and a ball mill

An analysis of flotation experiments results showed that when grinding material in a roll mill, the consumption of reagents is 25% lower than when grinding in a ball mill when the same degree of nickel extraction is achieved.

4. Conclusion

Morphological characteristics of serpentinites were studied using computer microtomography to study their behavior during disintegration. Rocks containing ore in the hypergene deposits of weathering rinds of the Urals are characterized by a different set of properties that must be taken into account when disintegrating and benefication them.

To increase the efficiency of metal extraction, it is necessary to apply selectivity to valuable and gangue minerals, depending on technical characteristics. Wet and sticky laterites should be separated from solid serpentinites at the stage of ore preparation and pre-concentration, which will allow separately processing both types of ores, as well as it is possible to isolate the third grade (loose) for increasing the efficiency of nickel extraction.

Acknowledgment

The work is carried out with financial support of Ministry of Education and Science of Russian Federation according to grant program "Research and development in priority areas of development of Russia's scientific and technical complex for 2014 – 2020"

References

1. Norgate, T., Haque, N. Energy and greenhouse gas impacts of mining and mineral processing operations. Journal of Cleaner Production, **18,** pp. 266-274. (2010)
2. Norgate, T., Jahanshahi, S. Reducing the greenhouse gas footprint of primary metal production: Where should the focus be? Minerals Engineering, **24**, pp. 1563-1570. (2011)
3. Pevnivtsev, et al.: Selektivnoye razrusheniye mineralov. Moskva: Nedra, (1988)
4. Hesse, M.: Selektive Zerkleinerung von Erzen und Industriemineralen bei Prallbean-spruchung. Dissertation, Technische Universität Bergakademie Freiberg, 2017.
5. Hesse, M.; Lieberwirth, H.: Selective comminution for pre-concentration in mineral processing of lead-zinc-ores. ISBN 978-3-940276-64-3.
6. Hesse, M.; Lieberwirth, H.: Selective comminution to increase efficiency in the beneficiation of mineral recourses. ISBN 978-5-94211-724-5.
7. Ohser, J., et al.: Quantitative Gefügeanalyse Theoretische Grundlagen und Anwendung. Freiberger Forschungshefte. doi: 10.1002/maco.19950460411.
8. Popov, O.: Beitrag zur mathematisch-petrographischen Gefügecharakterisierung für die Beurteilung der Festgesteine hinsichtlich ihrer Aufbereitung und ihrer Produkteigenschaften. Dissertation. Technischen Universität Bergakademie Freiberg, (2007).
9. Puffe, E., 1955. Erfahrungen bei der Prallzerkleinerung in Bergbau-, Aufbereitungs- und Hüttenbetrieben. ERZMETALL, VIII, pp. 570-580.
10. Taggart, F., 1956. Handbook of mineral dressing – ores and minerals. John Wiley & Sons, Inc
11. Talovina I.V. Geochemistry of the Ural oxide-silicate nickel deposits. / St. Petersburg Mining University. 270 p. (2012)
12. Gayfutdinova (Duryagina) A.M. Osobennosti geologicheskogo stroyeniya platinonosnykh kor vyvetrivaniya Svetloborskogo i Nizhnetagil'skogo massivov, Sredniy Ural. // Problemy osvoyeniya nedr v XXI veke glazami molodykh: materialy mezhdunarodnoy molodezhnoy nauchnoy shkoly. **2.** P. 37-39. (2012)
13. Vaysberg L.A., Kameneva Ye.Ye. Izmeneniye struktury gornykh porod pri tsiklichnom zamorazhivanii i ottaivanii // Obogashcheniye rud. **2.** p. 28-31. (2015)

14. Talovina, I.V., Aleksandrova, T.N., Popov, O., Lieberwirth, H. Comparative analysis of rocks structural-textural characteristics studies by computer X-ray microtomography and quantitative microstructural analysis methods. Obogashchenie Rud. **3,** P. 56-62. (2017) DOI: 10.17580/or.2017.03.09
15. Talovina, I.V., Lieberwirth, H., Alexandrova, T.N., Heide, G. Supergene oxide-silicate nickel deposits: Mineral-geochemical composition and peculiarities of processing. Eurasian Mining. **2017 (1).** P. 21-24. (2017) DOI: 10.17580/em.2017.01.06.
16. ISRM – International Society for Rock Mechanics (1985): Suggested method for determining point load strength. Commission on Testing Method. Working Group on Revision of the Point Load Test Method. Int. J. Rock Mech. // Min. Sci. & Geomech. Abstr.. **22 (2).** P. 51–60. (1985)
17. Raaz, V. Charakterisierung der Gesteinsfestigkeit mit Hilfe eines modifizierten Punktlastversuches. In: Z. Geol. Wiss. 30. **3**. S. 213–226. (2002)

Promising deposits of complex iron ore raw materials of Urals and its processing new technologies

Sergei Kornilkov[1*], *Andrei* Dmitriev[2] and *Aleksei* Pelevin[3]

[1] Federal State Budgetary Institution of Science Institute of Mining of Ural branch of RAS, Yekaterinburg, Russia
[2] Federal State Budgetary Institution of Science Institute of Metallurgy of Ural branch of RAS, Yekaterinburg, Russia
[3] Federal State Budgetary Educational Institution for Higher Education Ural State Mining University, Yekateriburg, Russia

Annotation. There have been provided general geological and metallurgical characteristics of some complex iron ore raw materials of Urals, which contain not only iron but also titanium, vanadium, chrome, and nickel. There has been provided the description of such deposits. It is shown that the degree of extraction of certain products is determined by the choice of extraction, benefication, pyrometallurgical processing procedures.

1 Introduction

In general iron-bearing ores of the Ural region are the complex ones, and the fuller extraction of associated components from them is an important reserve for increasing production of not only iron but also of non-ferrous, precious metals and rare elements at all stages of their processing. In connection with that the aim of developing new and improving existing benefication methods and processing procedure at metallurgical enterprises appears, because the complex ore processing leads to production prime cost reduction, providing a significant ores saving.

Among Urals complex deposits are titanium-containing ores (Kachkanarskoye, Kopanskoye, Medvedevskoye deposits) and brown chrome ores of Serov deposit, which contain chrome and nickel, besides iron. Titanium-containing sands of Yaregskoye deposit oil-containing sandstones of the Komi Republic also bear mentioning, it contains about the half of titanium dioxide reserves of Russia. The equivalent Athabaska oil sands deposit is situated in Canada. Titanium isn't developed in both these deposits because of the absence of an effective technology.

Having regard to the above, the aim of the research is the mining and technological justification of the mining and metallurgical complex raw materials extension (ilmenite titanium magnetite ores), obtained in virtue of the observation of concentration properties

* Corresponding author: kornilkov@igduran.ru

and technologies of their complete metallurgical processing with pyrometallurgical methods at the complex extraction of iron, titanium and vanadium, including:

- the evaluation of mineral resources base of titanium-containing ores of the Nothern and Southern Urals;
- the evaluation of the ilmenite titanium magnetite ores pyrometallurgical processing availability;
- the development of titanium-containing ores, titanium-containing sands and brown chrome ores processing procedure.

2. General characteristics of complex ores deposits

Russia possesses large balance reserves of titanium dioxide, which exceed 500 million tons and takes the fourth place in the world after Ukraine, China and Australia. In connection with the reunion of Crimea and Russia the loading of Crimea's Titan, pigmental dioxide production plant in Armyansk with the capacity of 120000 tones a year, which works with Ukrainian raw materials, is of interest.

The half of titanium reserves of Russia is enclosed in magmatic type deposits, which contain iron and vanadium, besides titanium. At the present time only the ores of Kachkanar deposits group (16.2% Fe, 0.13% V_2O_5, 1.23% TiO_2) are processed, and only iron and vanadium are extracted by the blast furnace – converter procedure.

Titanium-containing ores of the deposits like Medvedevskoye and Kopanskoye ones are generally considered as a source of iron, and, possibly, of vanadium. Still these ores can be considered as a source of titanium and pigmental titanium dioxide. Moreover, pigmental titanium dioxide hasn't been developed in Russia for 25 years.

It's important to notice, that in Urals and in Eastern regions the iron-ore basis of ferrous metallurgy of Russia is mostly represented by the ores of a similar type: Kopanskoye, Chineyskoye, Kruchininskoye, Kuranakhskoye, Bolshoy Seyim deposits, which contain more than 30% of titanium reserves of Russia.

Brown chrome ores Serov deposit is represented by complex raw materials (Fe_{gen} = 36-38%, Ni – 0.18-0.39%, Cr_2O_3 – 2.5-3.0%), which can be used rationally only if their processing technology foresees the most complete extraction of all alloying components and metallurgical extraction products utilization. In that respect the usage of the virgin charge is preferential for naturally alloyed chromium-nickel steel production (up to 50 grades according to Uralgipromez evaluation), and also for the high quality prefabrication for Ural semi-integrated plants. Serov ores are clean by sulfur and phosphorus, they almost don't contain cooper, and attract attention because of scrap materials, which are the source of unfavorable admixtures in the steel-melting process.

The general metallurgical characteristics of the above-mentioned deposits is provided in the work [1].

3. Technologies of separate excavation and benefication

The results of beneficiation of samples of the Gusevogorsky deposit ore mineral types showed, that ore with low concentration of titanium has a better beneficiation ability, than ore with normal concentration of titanium has. It has been determined that it is rational to divide ore with low concentration of titanium and ore with normal concentration of titanium to different technological types and process them separately because of differences in beneficiation ability and grindability. Herewith the accumulated technological and economical indexes of the separate beneficiation of ore natural types may be higher than bulk mining ore beneficiation indexes [9].

Table 1. Characteristics of titanium raw materials processing technologies

Beneficiation plant, object	Beneficiation procedure	Content in concentration		Processed raw materials
		Fe,%	TiO_2,%	
Ore deposits				
Kusinskaya	Magnetic gravitational flotation	60	45 (up to 48)	bucked ore
Olekminskaya	Magnetic highly magnetic electrical	62	48	bucked ore
Bolshoy Seyim deposit	Magnetic highly magnetic electrical	-	46,5	rough ilmenite concentrate
	Magnetic higly magnetic electrical	-	50,67	rough ilmenite concentrate
Kopanskoye deposit	Magnetic gravitational	24,4	44,3	ilmenite titanium magnetite ore
	Magnetic flotation	13,8	45,2	titanium magnetite ilmenite ore
	Magnetic highly magnetic electrical	19,2	50,6	ilmenite ore
Medvedevskoye deposit	Magnetic flotation	60-63	4-6	ilmenite titanium magnetire ore
	Magnetic gravitational			titanium magnetite ilmenite ore
	Magnetic electrical	-	45-48	ilmenite ore
Alluvial deposits				
Novopokrovskoy e «Steklyanka»	Gravitational electrical magnetic	-	45-47	bare titanium zircon sands
Tarskoye	Gravitational electrical magnetic	-	54,2	bare titanium zircon sands
Butkinskoye	Magnetic electrical	-	48	bare titanium zircon sands
Man-made mineral formations				
Gusevogorskoye	Magnetic gravitational	-	40,1	tailings of the Wet Magnetic Separation-I,II,III
Volkovskoye	Magnetic gravitational	-	41,25	tailings of titanium magnetite flotation
Kusinskoye	Magnetic electrical	36,8	53	mature tailings
Gold deposits barrows	Magnetic	-	44	mature tailings

The following issues have been discussed at bringing grounds for the re-engineering of titanium-containing deposits with separate iron and titanium, iron and vanadium ore types:

-Substantiation of factors, which effect on ore natural types separation;

-Improvement of excavation and ore preparation technological procedures at forming separate flows of titanium-containing iron ore raw materials.

- Development of the geoinformational data processing procedure at geometrization of ilmenite titanium and manganese ores natural types.

The performed classification of titanium-containing raw materials (Table 1), from which titanium concentrates are received for the further processing, showed, that the following may be a raw materials base:

1. ore deposits (ilmenite titanium magnetite, titanium magnetite ilmenite, ilmenite);
2. alluvial deposits (titanium and zircon deposits);
3. man-made mineral formations (titanium magnetite ores processing tailing, gold-containing deposits beneficiation tailings);
4. titanium magnetite concentrates melting slag.

The data from Table 1 certifies that titanium concentrates, produced by different technological procedures, but the technologies of production of titanium from slag have been studied least of all, although they are promising from the point of view of the mineral wealth complex usage.

The evaluation of concentrates magnetic properties, performed at the Institute of Metallurgy of Ural branch of RAS showed that titanium magnetite of the type of ore with low concentration of titanium is less hard-magnetic than titanium magnetite of the type of ore with normal concentration of titanium. Coercitive force is 82 against 93 Oe (0.0083 against 0.0093 T).

It lets us to tell that the efficiency of classification in hydraulic cyclones and the efficiency of beneficiation in drum magnetic separators for wet beneficiation would be higher at the beneficiation of ore with low concentration of titanium.

Lesser magnetic hardness of titanium magnetite in ore with low concentration of titanium of Gusevogorskoye deposit is one of the factors, which encourages the reception of concentrate with larger content of iron of this type and a specific reduction of energy consumption necessary for its receipt.

According to the data of detailed and operational reconnaissance with the help of GIS software it has been performed geometrization of ores quality specifications in terms of excavated open-cuts (Northern, Main, and Western open-cuts). There have been determined zones with different concentration of valuable components (Fe, Ti, V). It has been shown that the distribution of vanadium in ores depends on magnetite quantity in ores. It has been developed a prototype of the Cloud data base with restriction on viewing specific sections and editing.

Preliminary calculations show that the minimum content of TiO_2 in concentrate shall be less than 7% in order that the titanium dioxide content would be acceptable for the further processing. These conditions may be implemented due to the separate development of the Gusevogorskoye deposit part with high content of titanium, and, particularly, Kachkanar titanium magnetite deposit.

Such approach may be used at Medvedevskoye and Kopanskoye deposits operation. The Institute of Metallurgy of Ural branch of RAS has shown that Medvedevskoye and Kopanskoye deposits concentrates may serve as raw materials for high quality pigmented titanium dioxide production [8].

4. New pyrometallurgical technologies

New suggested technologies of excavation and beneficiation of Gusevogorskoye deposit titanium magnetite ores with getting iron-ore concentrates with low and high content of titanium [2, 3] let to offer new pyrometallurgical technologies of these concentrates processing according to the blast furnace – converter procedure and metalling – electric smelting procedure [4].

There have been studied two laboratory samples of acidified iron-ore pellets, obtained from Gusevogorskoye deposit concentrates with high and low content of titanium. The recovery grade (recoverability according to GOST 23581.11-79) is provided in Table 2.

Table 2. Pellets recoverability

Sample	**Pellets recoverability, unit fraction**
High content of titanium	0.75
Low content of titanium	0.98

The result of the research of acidified pellets strength after the low-temperature recovery according to GOST 13930 are provided in Table 3.

Table 3. Pellets strength, %

Index	**Sample with high content of titanium**	**Sample with low content of titanium**
$LTD_{+6,3}$	65.71	69.92
$LTD_{-3,15}$	12.87	11.39
$LTD_{-0,5}$	2.17	1.67

The result of the research of acidified pellets softening temperature interval according to GOST 17212-84 are provided in Table 4.

Table 4. Pellets softening temperature interval

Index	Sample with high content of titanium	Sample with low content of titanium
Temperature of softening starting, °C	1210	1180
Temperature of softening ending, °C	1300	1310

Estimate indexes of blast-furnace smelting of acidified pellets with high content of titanium are provided in Table 5. The calculations have been performed with the help of math model [5]. The content of TiO_2 in slag amounts to 14.82 %, which shall not cause problems at the blast-furnace operation.

In Table 6 estimate indexes of metallized pellets with high content of titanium in ore thermal furnace are provided. The calculations are performed with the help of math model [6]. The examination of rationality of processing of concentrate with high content of titanium according to the metalling – electrical smelting procedure certify, that metallized pellets with the content of general iron – 79.79 %, TiO_2 – 4.21 % and V_2O_5 – 0.86 % can be obtained from the received material. It is shown that their smelting will lead to the formation of slag, containing TiO_2 (22.88 %), insufficient for economical receipt of pigment titanium dioxide or titanium sponge from it.

Table 5. Estimate indexes of pellets blast-furnace smelting

Cast iron content, %	*Slag content, %*
Ti – 0.15 C – 4.67 Fe – 94.38 V – 0.54 Mn – 0.13 Si – 0.07	CaO – 28.9 SiO_2 – 23.7 Al_2O_3 – 17.07 FeO – 0.61 V_2O_5 – 0.33 **TiO_2 – 14.82**
Blast-furnace volume – 2200 m^3 Productive capacity – 7210 t of cast iron a day Coke consumption – 322.5 kg/t of cast iron	

Table 6. Indexes of metallized pellets smelting

Metal composition, %	*Slag composition, %*
Ti – 0.05 C – 2.50 Fe – 96.21 V – 0.484 P – 0.03 Si – 0.7	CaO – 9.64 SiO_2 – 21.20 Al_2O_3 – 22.41 FeO – 7.50 V_2O_5 – 0.90 **TiO_2 – 22.88**
Capacity of thermal furnace – 33 MVA Productive capacity – 563.7 t of metal a day Energy consumption – 1011.5 kW·h/t of metal	

Serov brown chrome ores' particularity is a big amount of alumina contained in it. One of the proposed methods of reducing Al_2O_3/SiO_2 in blast-furnace slag is to melt brown chrome ores in the blast-furnace together with silicate nickel ores or other ores. It will allow to get tradable cast iron, containing chrome and nickel, and subsequently to smelt alloy steel from it [7].

By complex researches of technologies for beneficiation and development with ores natural types separation and of advanced metallurgical processing of mineral and man-made raw materials it has been determined that also in Urals it is being created a new mining and metallurgical complex raw materials base not only for the iron production, but also for titanium raw materials production with iron, titanium and vanadium separation by means of pyrometallurgical methods for the organization of titanium magnetite concentrate processing by metalling – electric smelting procedure with the further utilization of slag, containing high amount of titanium.

Primary deposits direct processing with ilmenite and rutile concentrates obtainment may be recognized as a prospective type of titanium raw materials production technology.

The work has been performed with financial support from the Ural branch of RAS Program, project No. 18-5-2345-56.

References

1. L.I. Leontyev, N.A. Vatolin, S.V. Shavrin, N.S. Shumakov. Pyrometallurgical processing of complex ores. - Moscow: Metallurgy. 432 p. (1997).
2. S.V. Kornilkov, A.N. Dmitriev, A.E. Pelevin, A.M. Yakovlev. Mining magazine. **5**. P 86-90. (2016).

3. A.N. Dmitriev, G.Yu. Vitkina, R.V. Petukhov, S.V. Kornilkov, A.E. Pelevin, A.Y. Fishman, T.V. Sapozhnikova, K. Y. Shunyaev. Advanced Materials Research. **834 - 836**. P. 364-369. (2014).
4. A.N. Dmitriev, R.V. Petukhov, G.Yu. Vitkina, Yu.A. Chesnokov, S.V. Kornilkov, A.E. Pelevin. Defect and Diffusion Forum (Diffusion in Solids and Liquids XI). **369**. P. 6-11. (2016).
5. A.V. Chentsov, Yu.A. Chesnokov, S.V. Shavrin. Logic-statistic balance model of blast furnace smelting process. Ural Branch of Russian Academy of Sciences, Ekaterinburg, Russia (2003).
6. A.N. Dmitriev, G.Yu. Vitkina, Yu.A. Chesnokov. Use of Information System "Interactive Calculations in Ferrous Metallurgy", Proceedings of 2014 – Sustainable Industrial Processing Summit/Shechtman International Symposium. June 29 – July 04, 2014. Non-Ferrous, Iron and Steel. **3**. (2014).
7. Yu.A. Chesnokov, A.N. Dmitriev, Z.N. Pastukhova, V.P. Perepechaev. Collected works of the All-Russian Conference "Researches in the field of man-made formations and wastes processing and utilization" with the elements of the school for young scientists. 24-27.11.2009. Pages 427-432. (2009).
8. Andrey Dmitriev and Leopold Leontiev. Journal of Materials Science and Engineering B 7 (11-12) P. 268-271. (2017).
9. S.V. Kornilkov, A.V. Glebov, A.E. Pelevin, A.N. Dmitriev. Titan. **4**. P 8-11. (2017).

The technology of gold-containing concentrates recovery from the coal combustion products of the Yerkovetskoye brown coal deposit (the Amur Region, Russia)

Anatoly Sorokin[*1,2] and *Oleg* Ageev[1]

[1] Amur Scientific Center, Far Eastern Branch of the Russian Academy of Sciences, Blagoveshchensk, Russia
[2]Institute of Geology and Nature Management Far Eastern Branch of the Russian Academy of Sciences, Blagoveshchensk, Russia

Annotation. The results of experimental research of the native gold distribution in the coal combustion products (CCP) obtained out of the brown coals of the Yerkovetskoye deposit combustion material in the experimental technological complex "Amur" (ETC "Amur") are described in the article. The technology of the formation of slag, fly ash and products of wet cleaning of flue gases in the process of various combustion modes of coal combustion is examined. The possibility of the obtaining of enriched gold-containing concentrate by gravity and magnetic methods from the brown coal combustion products and extraction of native gold out of the concentrate is shown.

1 Object of study

Yerkovetskoye brown coal deposit was chosen as the object of research. It is located in the Amur region, 75 km east of Blagoveshchensk. It is operated on the southern outskirts, in the "Yuzhny" quarry, where a coal layer up to 9 m thick is discovered with interlayers and lenses (up to 10 cm) of clays, aleurites and tuffs. Upper Miocene sands and clays and Quaternary sediments are found in the supra-angular part. Coals of the deposit are brown coals, technological group 2B, are of the medium degree of coalification, dense with interlayers of loose. The quality of the coals (in %) is the following: humidity 36.5, ash content 17.5, volatile yield 44, and sulfur content 0.3. The lowest heat of combustion is 12.81 MJ / kg. The balance reserves of 1077.8 million tonnes. [1]

2 Technology of the CCP obtaining

To study the process of transfer and distribution of gold and other valuable elements in the products of coal combustion (furnace slag, fly ash, smoke and technogenic water) formed during the coal combustion, the Amur Scientific Center (AmurSC) FEB RAS created an

* Corresponding author: amurnc@ascnetl.ru

Experimental-Technological Complex "Amur "(ETC" Amur ") in 2017 [2]. The particular qualities of the complex are the autonomous production of CCP at the coal combustion stage, controlled combustion of fuel, the presence of an afterburner chamber for volatile carbon compounds, and a combined technology for cleaning flue gases and man-made water from harmful and toxic impurities.

Brown coals of the Yerkovetskoye deposit were investigated. The gold content in the initial products of combustion of brown coals (Table 1) is determined by neutron activation analysis (NAA) at the Institute of Chemistry, FEB RAS, the gold content in technogenic water is determined by atomic-adsorption analysis (AAA), at the Institute of Geology and Nature Management (IGNM) FEB RAS [3].

Table 1. Average content of gold in brown coal combustion products.

	Coal combustion products	Au content	NAA	AAA
1.	Furnace slag source	g/t	0,25	-
2.	Products of flue gases (fly ash) coarse (dry) cleaning	g/t	0,30	-
3.	Products of thin (wet) cleaning of the flue gas (sludge)	g/t	1,71	-
4.	Technogenic water	mg/l	-	0,00030

The graph (Fig. 1.) shows the increased gold content in flue gas cleaning products in comparison with furnace slags. The transformation of gold during the combustion process and the form of its transportation with the flue gases has not been investigated.

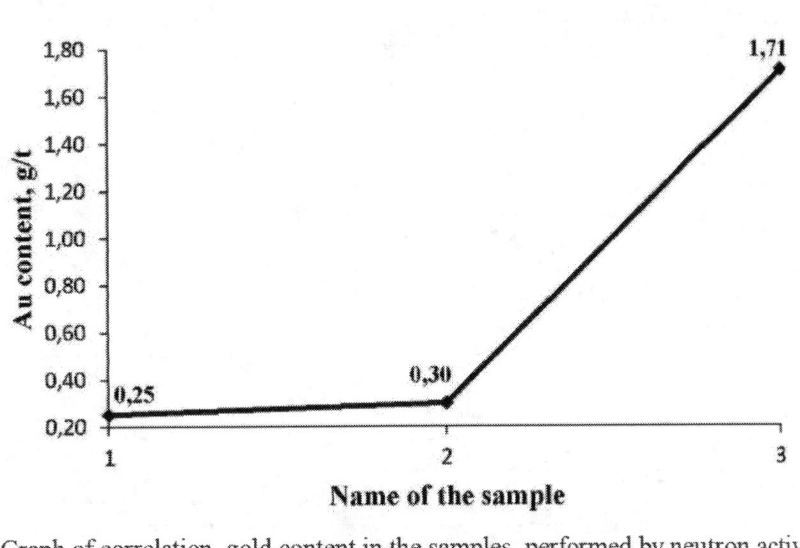

Fig. 1. Graph of correlation, gold content in the samples, performed by neutron activation analysis. Sample name: 1 - slag initial; 2 - products of coarse (dry) cleaning of flue gases; 3 - products of thin (wet) cleaning of flue gases.

3 Preparation of gold-containing concentrate from coal combustion products (CCP).

To discover and study the forms and features of finding native gold in coal combustion products (CCP), the AmurSC FEB RAS developed a technology of gravity and magnetic enrichment of the CCP for obtaining gold-containing concentrate.

3.1 Enrichment of combustion slag

The slag, which was formed from the combustion chamber, formed as a result of coal combustion by layer method, was crushed on a jaw crusher up to 1 mm fraction, then it passed through a wet magnetic separator WMS-0.1. The resulting magnetic fraction (metal oxides) can be a raw material for the metallurgical industry.

The non-magnetic fraction (mixture of underburn, slag, sand, clay, etc.) was enriched by the gravity method (Fig. 2). To separate the non-magnetic fraction of underburn and light particles from the composition, a slightly inclined gateway of rectangular cross-section was used. The drug mats, which allow creating a turbulent flow in the near-bottom layers and retaining the heavy particles that are deposited on the bottom, were laid at the bottom of the gateway. The isolated light fraction (underburn) is the raw material for the production of coal briquettes.

The heavy fraction, after grinding (up to -0.1 mm), was fed to the concentration table SKO-0.5. On the table were received the final products of enrichment - concentrate and table tails. Tails of the table - a light enrichment fraction (purified from underburning, iron oxides and other heavy elements) are ready raw materials for use in road construction and the production of building materials.

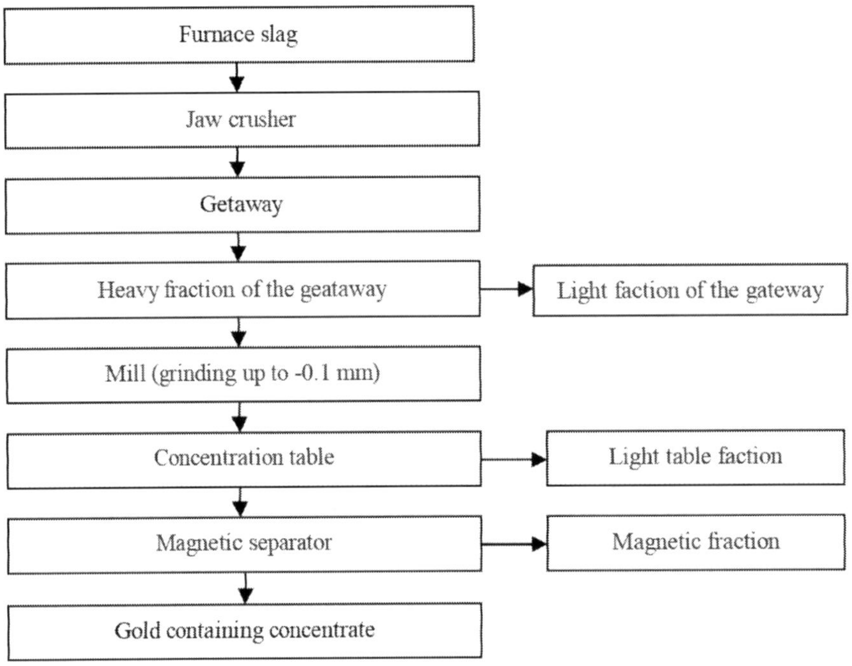

Fig. 2. Technological scheme for gold-containing concentrate obtaining from combustion slag.

Gold content in furnace slag in enrichment products was analyzed by the assay-analytical method in assay laboratory of AmurSC FEB RAS (Table 2).

Table 2. Gold content in furnace slag beneficiation products

Flue slag enrichment products	Au content, g/t
Heavy fraction of table (concentrate)	8,31
Light table faction	0,05
Magnetic fraction	0,88

3.2 Fly ash enrichment

A two-section ash collector is used in the ETC "AMUR" for rough cleaning of smoke from medium size dust and particles (up to 10 microns). The principle of the ash collector is dry gravity enrichment, based on the rate difference of incidence of the particles, having different density or fineness in the gas flow. There is a significant slowing down of the particle flow in the ash collector, which facilitates the precipitation of all particles. To adjust this process, as well as to obtain the desired fraction of fly ash, adjusting flaps are used. Fly ash accumulates in the hoppers of the ash collector and is periodically removed.

The resulting fly ash (Fig. 3) is distributed on a sieve analyzer A-50 by size classes on fractions from 0.5 to 0.04 mm. After that, each faction is enriched individually. The magnetic fraction was extracted on the wet magnetic separator then the non-magnetic fraction is enriched on the concentration table (Fig. 3).

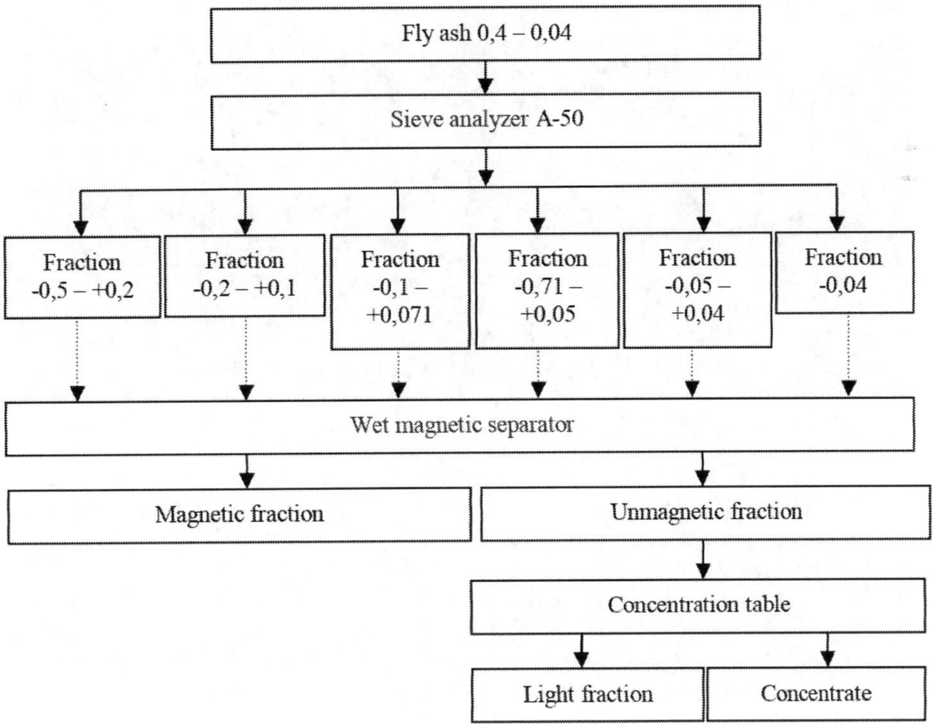

Fig. 3. Chart of fly ash beneficiation for obtaining gold concentrate.

Gold content in furnace fly ash in enrichment products was analyzed by the assay-analytical method in assay laboratory of AmurSC FEB RAS (Table 3).

Table 3. Gold content in the products of fly ash enrichment.

Products of the fly ash enrichment	Au content, g/t
Heavy fraction of the table (concentrate)	2,25
Light fraction of the table	0,93
Technogenic water sediment (filtration)	1,54

The resulting gold-bearing concentrate was studied by the gold panning method at the Institute of Geology and Geochemistry of the Far East Branch of the Russian Academy of Sciences, with the release of particles of native gold (Fig. 4, 5).

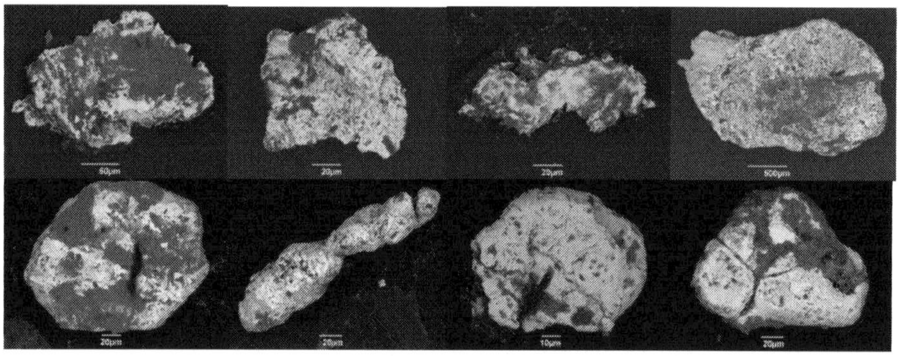

Fig. 4. Native gold extracted from combustion slag.

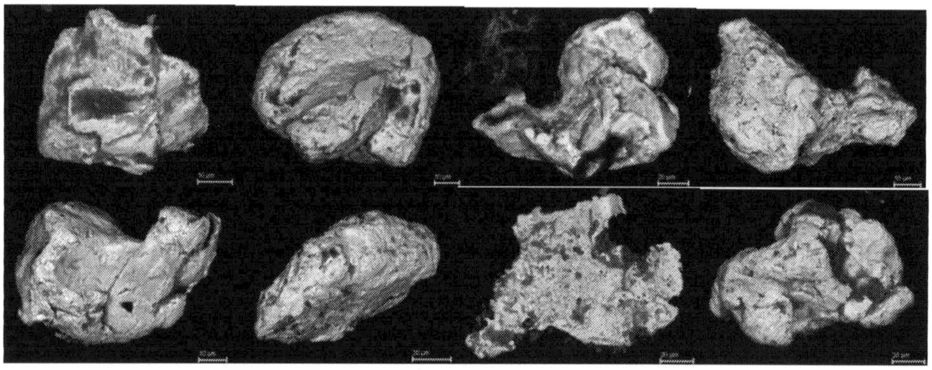

Fig. 5. Native gold in the fly ash.

As a result of magnetic and gravity enrichment we've obtained the following:

1) Magnetic fraction consisting of ferrous compounds in form of magnetite and paramagnetic compounds of heavy metals. The magnetic fraction is the raw material for the metallurgical industry.

2) Light fraction (table tails), which is an aluminosilicate form of silicon, can be used as raw material for cements and besclinker binders, concretes (heavy, light, cellular), porous aggregates, silicate, ceramic, heat-insulating and other materials.

3) The heavy non-metallic fraction (table concentrate) contains a number of elements, including native gold.

The gas stream purified of dust (more than 10 microns in size) and as well as the volatile products of fuel combustion consisting of hot gases containing oxides of carbon, nitrogen, sulfur, and water vapor, is extracted from the ash collector into the scrubber by the smoke exhauster.

Main function of the scrubber is the deep cleaning of the flue gasses from the impurities.

The operating principle of the action of wet cleaning is based on the intensive crushing of the irrigating liquid with a dusty gas stream moving at high speed. As a result of this interaction, dust particles penetrate into the liquid upon impact, and in the form of sludge are continuously discharged to the filter unit for cleaning.

3.3 Enrichment of the products of the wet flue gas cleaning

The flue gases wet cleaning output is the sludge, obtained as a result of purification of technogenic water by a coarse filter, was not subjected to gravity enrichment (due to physical limitations of the method).

The sludge was washed in bromoform, divided into fractions: magnetic, electromagnetic, non-magnetic heavy and non-magnetic light fraction, which were investigated by the gold panning method using microchemical reactions and the immersion method with the release of particles of native gold (IGNM FEB RAS) (Fig. 6).

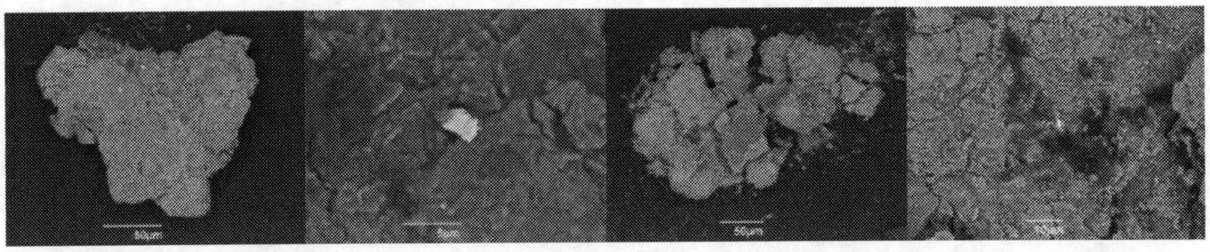

Fig. 6. Gold conglomerates extracted from sludge with the presence of microparticles of gold (white phase).

4 Conclusion

The conducted research of slags and fly ash showed the presence of significant amount of native gold, which can be concentrated by gravity and magnetic methods.

Only by combining physical methods of enrichment with hydrometallurgical operations the effective gold extraction out of the CCP can be achieved. The "chemical concentrate" is obtained as the result of this combination. [4–10].

Finished raw materials for various industries were obtained in the process of gravity and magnetic enrichment of CCP: magnetic fraction - raw materials for the metallurgical industry [11, 12]; light fraction - "underburn" - raw materials for the production of coal briquettes [13, 14]; light fraction of the concentration table - raw materials for road construction and production of building materials [15].

References

1. Mineral and raw materials base of the Amur Region at the turn of the century / I. A. Vasiliev et. al.; Ed. I. A.Vasiliev. - Blagoveshchensk: Committee of Natural Resources Amur region, P. 168 (2000).
2. Sorokin A. P., Ageev O. A. Formation of the coal combustion products in the experimental and technological complex "Amur" // Complex use of the potential of stone and brown coals and creation of combined ecologically safe technologies for their

development: All-Russian Conf. with intern. participation: reports. Blagoveshchensk: AmurSC FEB RAS, P. 100-102 (2017).
3. Ageev O. A., Yudakov A.A., Ivannikov S.I., Zubenko I.A. Distribution of macroelements and gold in coal combustion products (experimental results) // Complex use of the potential of stone and brown coals and creation of combined ecologically safe technologies for their development: All-Russian Conf. with intern. participation: reports. Blagoveshchensk: AmurSC FEB RAS, P. 129-132 (2017).
4. L. Canada, T. Heput, E. Ardelian. Methods for recovering precious metals from industrial waste // IOP Conference Series: Materials Sciences and Engineering. IOP Publishing. **106**(1) P. 12-20 (2016).
5. J. He, A. Kappler. Recovery of precious metals from waste streams //[Microbial Biotechnology] **10**(5) P. 1194-1198 (2017).
6. Prafulla Kumar Sahoo, Kangjoo Kim, M. A. Powell, Sk. Md. Equeenuddin. Recovery of metals and other beneficial products from coal fly ash: a sustainable approach for fly ash management // International Journal of Coal Science & Technology. **3**(3) P. 267–283 (2016).
7. L. Tolhurst. Commercial Recovery of Metals from Coal Ash. Lucid Insight Ltd, Oxford, UK, Conference: (2015) World of Coal Ash [www.worldofcoalash.org]
8. K.V. Ramesh, D. Sandhaya Rani. Coal-Gold Agglomeration: An Alternative Separation Process in Gold Recovery. Conference: Chemical Engineering Letters. At: Visakhapatnam, **1**(1) P. 39-40 (2015).
9. S. Dai, V.V. Seredin, C.R. Ward, J. Jiang, C. Zhao. Composition and modes of occurrence of minerals and elements in coal combustion products derived from high-Ge coals // International Journal of Coal Geology. **121**(1) P. 79-97 (2014).
10. S. Yu. Bratskaya, A.S. Volk, V.V. Ivanov, A. Yu. Ustinov, V.A. Avramenko. A new approach to precious metals recovery from brown coals: Correlation of recovery efficacy with the mechanism of metal–humic interactions // Geochimica et Cosmochimica Acta. **73.** Is. 11(1). P. 3301-3310 (2009).
11. A.S. Shoumkova. Magnetic separation of coal fly ash from Bulgarian power plants // Waste Management and Research. **29**(10) P. 1078-1089 (2009).
12. G.C. Hanl, N. I. Um, U. Kwangsuk, H. C. Cho, J. Ahn. Recovery of ferromagnetic material by wet magnetic separation in coal bottom ash // Geosystem Engineering. **12**(1) P. 26-30. (2009).
13. H. Carrasco, H. Sarper. Developing Engineered Fuel (Briquettes) using Fly Ash from the Aquila Coal-fired Power Plant in Cañon City and locally available Biomass Waste. Final Report. Colorado State University – Pueblo.
14. USOO.5916826A United States Patent (19)11. Patent Number: **5,916,826** White (45). Pelletizing and briquetting of coal fines using binders produced by liquefaction of biomass. Date of Patent: Jun.29, (1999).
15. I. Kula, A. Olgun, V. Sevinc, Y. Erdogan. An investigation on the use of tincal ore waste, fly ash, and coal bottom ash as Portland cement replacement materials // Cement and Concrete Research. **32(2)**. P. 227-232. (2002).

E3S Web of Conferences **56**, 03004 (2018) https://doi.org/10.1051/e3sconf/20185603004
VII International Scientific Conference "Problems of Complex Development of Georesources"

Processing of refractory gold sulphide carbon-bearing ores by means of their pre-oxidation with water-and-gas emulsions prepared in electrochemical and photochemical reactors

Artur Sekisov[1*], *Paul* Korpi[2], *Anna* Rasskazova[1], *Natalia* Litvinova[1] and *Tatiana* Konareva[1]

[1]Mining Institute of Far eastern branch of Russian Academy of Sciences, Khabarovsk, Russia.
[2]Limited Liability Corporation "IG Copper LLC", Russia.

Abstract. In connection with the depletion of mineral resources base of gold, the engagement of refractory ores reserves in the processing becomes a necessary condition to stabilize gold market. Standard methods of processing of refractory sulphide carbon-bearing gold ores and float concentrates extracted from them include pressure oxidation and bacterial oxidation enabling the liberation of dispersive gold and oxidation of carbon having parasitical sorption with further cyanation. The alternative solution of the aforementioned problem is pre-oxidation of refractory ores with water-and-gas emulsions. Theoretical basis for active water-and-gas emulsions formation with the use of electrolytes solutions electrolysis in combination with ultraviolet exposure is stated in the article. Given method of water-and-gas emulsions preparation ensures the formation of active oxidants complex. Gold recovery into pregnant solutions up to 55.7 – 73.5% by chloride scheme was reached in laboratory experiments on samples of sulphide carbon-bearing ore, herewith 18% was recovered using cyanic solution. Based upon experimental findings, process scheme for combined oxidative preparation of sulphide carbon-bearing ore with dispersive deportment of gold was developed.

1 Introduction

Sulphide carbon-bearing gold ores, due to the presence of sorption and active particles of organogenous carbon and residual organic compounds causing re-precipitation of gold diluted in the course of leaching with themselves, are characterized by exceptionally low rate of its recovery (as a rule not exceeding 20%). Gold ores containing carbonaceous matter and sulphidic minerals are generally referred to as double refractory. Extraction of the gold from such ores usually requires an oxidation step to both liberate the ultra-fine gold from the matrix of the sulphides and remove the carbon that preg-robs the dissolved gold. Besides well known pressure oxidation and roasting microwave roasting can be applied to oxidize both the sulphides and the carbonaceous matter [1]. The ability of pyrite

* Corresponding author: sekisovag@mail.ru

and chalcopyrite to adsorb gold is a well-known fact. The degree of adsorption onto the ore or the activated carbon is a function of the kinetics of the adsorption process. Copper, silver, zinc, nickel and iron complexes counteract preg-robbing, and serve to stabilise the aurocyanide complex by precipitating prior to gold precipitation. When cyanide or cyanide-species present in solution, the adsorption onto activated carbon is in preference to the adsorption onto the minerals [2].

Besides, gold may be present in ores of that mineragenic type as bi-element clusters of Au-C type as well which is reluctant to cyanation. Formation of AuSx is indicated at the gold surface of double refractory gold ore (sulphide with carbon coatings) in a cyanide medium [3]. For instance, the presence of carbon with graphite structure matrix consisting of ether and thiocarbamide functional groups is established in ores of deposit of Goldstrike, Carlin Trend. It is assumed that carbon not only sorbed gold during ore genesis and retained bonds with it in ores, direct ones as Au-C as well as through Au-S-C $(NH_2)_2$ thiocarbamide functional groups, but formed bi-element cluster compounds with gold in the course of metamorphous process and carbonation.

In connection with the depletion of mineral resources base of gold, the engagement of reserves of such refractory ores in the processing becomes a necessary condition to stabilize gold market.

Currently two process schemes of processing are primarily used to treat refractory sulphide carbon-bearing gold ores or float concentrates extracted from them: scheme with pressure oxidation and bio-oxidation (bacterial oxidation) [4]. [5]. These schemes are applicable only for ores the gold content of which is at least 2.5 - 3 g/t. Therefore, the development of alternative technologies for raw and poor sulphide carbon-bearing ores not requiring significant energy costs and reagent consumption is an actual problem in gold-mining industry. One of the future solutions for the problem is the use of ozone as oxidant for carbon-bearing ores in gold leaching. This idea was justified as far back as in works of 30-s - 40-s of XX century by an outstanding Russian scientist and metallurgist, a corresponding member of Academy of Sciences of the USSR — I.N. Plaksin. Detailed study of oxidation with sorptive and active carbon-bearing inclusions of gold ores from Carlin Trend were conducted by American researchers (gold recovery with subsequent cyanation reaches about 90%) — D.J. Sheiner, R.E. Lindstrom (US Bureau of Mines) in 60-s - 70-s of XX century.

Recent geological exploration demonstrated great interest to Carlin-style mineralization. Gold zones range from 105 to 400 meters in depth, and steepen from flat (10 degrees) to moderate (45 degrees) dips to the east, as they approach the western strand of the NBFZ. Gold is associated with sooty sulfides, silica, carbon, clay, barite, realgar and orpiment in addition to elevated As, Hg, Sb and Tl. High-grade (> 6 g/t Au) gold occurs in both the upper and lower gold zones. Some of the best gold intercepts include up to 15 ppm gold content [6]. However, despite the high level of dispersive gold recovery reached with the use of ozone, technical difficulties and high energy costs which exceeded such for electro-chlorination method of Carlin-type oxidation prevented this method from being used widely in mining industry. It should be noted that more than 30 years later the research of using ozone as oxidant for carbon-bearing and sulphide ores was successfully carried on in the USA by W.J. Zaleski, W.P. Van Antverp, Ph. A. Lincoln and in the Russian Federation in Moscow Institute of Steel and Alloys (Ph.D. in Engineering Science L.N. Krylova and others) [7]. Nevertheless, the cost of this method of oxidation thus far precludes it from wide use in hydrometallurgy.

2 Theoretical basis for preparing active water-and-gas emulsions containing reactive oxygen intermediates

From the viewpoint of the authors of the article, oxidation of sulphide and carbon-bearing gold ores is practical to conduct not by means of sparging of slurry with ozone-containing gas mixture, but using to this end the solution containing highly active oxidants — the group of peroxide compounds of oxygen with hydrogen (the so-called reactive oxygen intermediates), including, first, hydrated hydroxyl-radical and hydrogen peroxide as well as superoxides of alkali metals. The combination of electrolytes solutions electrolysis with UV exposure ensures the formation of active oxidants complex. In case of UV light exposure within the range of wave lengths of 180 – 250 nm of water-and-gas emulsion being formed during electrolysis, highly active oxidants are generated in the volume of gas bubbles, such as: atomic oxygen, (O^{*-}) superoxide – radical oil, ozone, (OH^*) hydroxyl-radical creating, in case of inter-reaction with film water surrounding gas bubbles, radical ion and radical-containing clusters. In this case the formation of hydration shells happens around centres of ionization — inoculating ions, radicals and radical ions of hydrogen, oxygen and their compounds, particularly H^+, OH^-, OH^*, O^{*-} and others [8].

The most active cluster-forming particle is OH^* hydroxyl-radical the oxidation/reduction potential of which reaches the value of 2.8 V while such for atomic oxygen is 2.42 V and for ozone – 2.07 V [9].

From this fact it emerges that not only oxygen atom unpaired electron, but also, to a greater extent, active hydrogen atom or its ionic form — proton as part of hydroxyl-radical — changes structural and energy parameters of electron shells of molecules (atoms) contacting with them.

Since the system consisting of excited oxygen and hydrogen atoms is metastable, it is possible to suggest that hydrogen atom electron occasionally moves to a free orbit of oxygen atom electron shell. Herewith, inner dissociation of $OH^* = (O^{*-}{*}H^+)$ occurs in such radical. Hydrogen atom as part of hydroxyl-radical, shifting towards conventional border of electron shell from time to time, becomes an unbound particle — proton with rather high energy. Accordingly, there is a possibility of the fact that its location is not limited only by electron shell of oxygen radical ion associated with it. Therefore, the system $(O^{*-}{*}H^+)$ may be considered as metastable radical ion pair of $O^{*-}{*}p^+$ as well which has, on one side, the function of a proton acceptor, on the other side — the function of an electron acceptor. p^+ primary proton separated from OH^* radical penetrates into electron shell of another nearest water molecule. Thus, hydronium ion ($p^+ + H_2O = H_3O^+$) emerges. O^{*-} oxygen radical ion left free polarizes and attracts one of hydrogen atoms from adjoining water molecule with its densed electron shell. Thereat, hydrogen atom being polarized ionizes while moving away from hydrogen radical ion and eventually forced dissociation of water molecule with the formation of OH^- hydroxyl ion and (p^+) proton takes place. Free (p^+) proton inter-reacting with oxygen radical ion reproduces oxygen hydroxyl-radical and, accordingly, $(O^{*-}{*}H^+)$ metastable pair. Counter migration in ordered chain of centres of hydronium ions and hydroxyl ions formation leads to their inter-reaction in the end: when a proton penetrates inside an electron shell of hydroxyl ion (H^+H^-), ion pairs or metastable pairs of hydrogen atoms being joined by ($H^*_{(e+e-)}H^*$) electron-positron (exciton) bindings may originate. As a result, local clustering of adjoining water molecules takes place with the formation of at least its dimers: $H_3O^+ + OH^- = H_4O_2$. The cycle of dissociation induced by hydroxyl-radicals and association (clustering) of water molecules repeats involving other adjoining molecules in it. Radical ion hydrated clusters created in such a manner basically contain collective systems of electrons or protons with sufficient energy potential capable of breaking bindings between collective electrons of clustered carbon, sulphur and dispersive gold associated with them.

In the course of processing water solution of sodium bicarbonate in electrochemical and photochemical reactors superoxide carboxy-containing compounds (H_2CO_4 monopercarbonic and $H_2C_2O_6$ percarbonic acids) may be synthesized. Such superoxide compounds are metastable, that's why inside cluster structure they dissociate producing highly active OH* hydroxyl-radicals (through intermediate phase of hydrogen peroxide extraction): $H_2CO_4 = H_2O_2 + CO_2$, $H_2C_2O_6 = H_2O_2 + 2CO_2$, $H_2O_2 + OH^- = H_2O + O^{*-} + OH^*$ or radical ions of hydroxyl and carboxyl-cations — ($2OH^{*-} + C_2O_2^+$, $2OH^{*-} + C_2O_4^+$). Obtained products of dissociation inter-reacting through hydration shells form associates in reverse manner i.e. hydrated clusters with reduced inoculating part — superoxide compounds. Thus, self-oscillations of hydrated clustering – de-clustering, occur in the solution. The reality of such self-oscillations in hydrocarbonate water solutions is proved experimentally during researches conducted in Moscow State University under the guidance of professor, doctor of biological sciences V.L. Voeikova [10]. Therefore, such solutions, or more precisely water-and-gas emulsions retain their activity for relatively long time (tens of hours), while free hydroxyl-radical itself may exist for not more than 10^{-12} s. Due to this the presence of superoxide compounds in leaching solutions will allow keeping them active for sufficient time.

During pre-processing of ore with small (to 3%) content of sulphide minerals using such activated solutions, clustered dispersive gold with collective electron shells (including hybrid shells, levels 5d-6s) after the oxidation of sulphur bonded chemically with it evolves and may fragmentarily interact with hydroxyl-radical and with atomic hydrogen. Herewith, weakening of bindings between clustered gold atoms may as well occur. It is known that even in natural water there is a stable form of gold – $AuOH(H_2O)^0$ at wide Eh—pH interval. Therefore, under certain conditions, in mineral environment-process solution system, it becomes possible to perform de-clustering of gold using active hydrated complexes and its relocation in atomic (ionic) form or as part of AuOH*, AuH* compounds to the surface of micro-cracks and interstices of concentrating mineral. Subsequently, upon contact with leaching solution, the formation of more stable gold complexes take place with corresponding complexing components directly as well as through the phase of secondary hydrated gold clusters formation.

In the course of preparation of ores containing dispersive gold with higher content of sulphide minerals and arsenopyrite as well as with the presence of carbon black in the amount of 0.1% - 1% it is reasonable to use combined preparation including the use of water-and-gas emulsions containing percarbonic acids and sulphuric and peroxide combinations in various linkings and sequence. The oxidation of carbon in sulphuric and peroxide medium occurs by means of its inter-reaction with metastable $H_2S_2O_8$, H_2SO_5 permonosulphuric and persulphuric acids which may be as well obtained via processing initial solution in electrochemical and photochemical reactors.

In the course of processing refractory ores with obtained water-and-gas emulsions carbon contained in them oxidises to monoxide with metastable permonosulphuric acid by the reaction as follows: $H_2SO_5 + C = H_2SO_4 + CO$

Dispersive gold which retained binding with carbon after the oxidation of the latter forms $Au\text{-}Au(OH,HCO_3)$ complexes by inter-reaction with hydroxyl-radical.

3 Materials and research methods

In order to perform geotechnological assessment of mineral raw materials, experimental studies on gold carbon-bearing argillaceous carbonated ores from one of deposits of Southern Ural were conducted. Prior to leaching, preliminary oxidative preparation of a sample was made by pelletizing ore crushed to fraction of –5 mm +0 mm with the addition of cement and its watering with photoelectroactive solutions with sulphuric and peroxide and carbonate and peroxide base. Further, pelletized sample weights were being mixed and poured in percolation columns of a bench (fig. 1) as solutions also used after combined oxidation (direct activation chlorination

Fig. 1. A bench for geotechnological testing of ores using standard and activation schemes of percolation leaching

1. Watering of pelletized materials with active carbonate and peroxide solution was being performed during 3 days in the first percolator. Three days after it was registered that iron and copper concentration in output solution dropped, pH rises from 4.8 in first portions of drain solution to 6.8 in the last portions of drain from percolator.
2. Ore sample weights in the second percolator after pelletizing was saturated with the 10% solution of sodium chloride in amount of 430 ml in order to form hydrochloride-peroxide medium in it after which the second pause of 2.5 days was maintained in order for gold diffusion leaching to occur and carbon-bearing inclusions sorption activity reduction to continue.
3. The watering of ore in the first (reference) percolator was being performed by feeding active alkali cyanic solution with the concentration of 1.5 g/l with the flow rate of 300 ml/day.
4. Active chloride solution obtained by photoelectrochemical processing was fed into the second column with the flow rate of 300 ml/day.

4 Results and discussion

Oxidative preparation of ore by pelletizing and its watering with photoelectroactive solutions with sulphuric and peroxide and carbonate and peroxide base enabled essentially increasing gold recovery during subsequent percolation leaching with active chloride (73.5%) and cyanic (55.7%) solutions what to a significant degree higher than 18% value of direct cyanation.

5 Conclusions

Thus, activation chloride system allows ensuring dispersive gold recovery, associated mainly with sulphur and carbon, suitable for heap leaching.

Besides, the effect of using chloride complexes is as well achieved by the fact that deeper passivation of carbon black at the second main stage of leaching is ensured and accordingly the amount of gold re-precipitated by it from pregnant solutions significantly drops (preg-robbing effect).

Based on experimental findings, in order to realize combined oxidative preparation of carbon-bearing sulphide ore in production, a process scheme given in fig. 2 was developed.

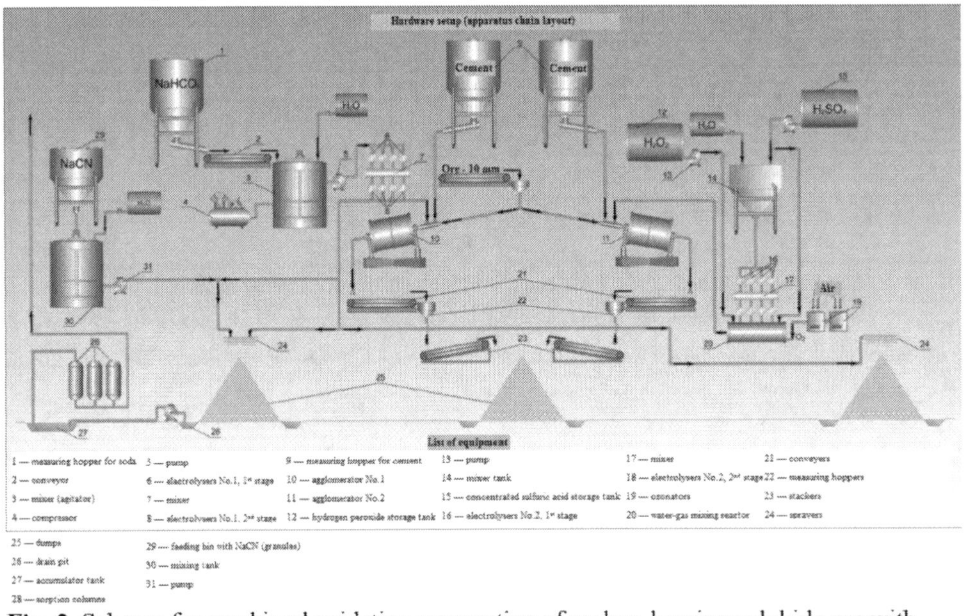

Fig. 2. Scheme for combined oxidation preparation of carbon-bearing sulphide ore with dispersive deportment of gold

The selection of complexing compound option (cyanide or chloride) used after the oxidation of ore from a specific deposit may be determined only based upon feasibility analysis and testing of samples weighing more than 50 tons.

Acknowledgement

The research was carried out with the support of scientific grant "Development of technology of cyanide-free leaching of refractory gold from low-grade copper and porphyritic ores" provided by the Government of Khabarovsk Territory under the contract No. 115/2018D, 28.06.2018.

References

1. R. K. Amankwah, C. A. Pickles, Min. Eng. **22**, pp. 1095-1101 (2009).
2. K.L. Rees, V. Deventer, Hydromet. **58**, pp. 61-80 (2000).
3. H. Tan, D. Feng, G.C. Lukey, V. Deventer, Min. Eng. **18**, pp. 1361-1372 (2005).
4. D. D. Gertenbach, Minerals & Met. Proc. **33**, pp. 178-186 (2016).
5. S. Ubaldini, F. Veglio, L. Toro, C. Abbruzzese, Min. Engineering **13**, pp. 1641-1646 (2000).
6. M. R. Jackson, D. C. Mathewson, S. R. Koehler, W. M. Pennell, L. J. Garside, New Concepts and Discoveries: Geological-Society-of-Nevada's Symposium - New Concepts and Discoveries (2015).
7. M. E. Denisov, B. P. Rudnev, L. N. Krylova, Yu. S. Kuchmina, Nonferrous metallurgy **3**, pp. 25-30 (2014).
8. A. G. Sekisov, Y. I. Rubzov, A. Y. Lavrov, J. of Mining Institute **217**, pp. 96-101 (2016).
9. A. G Sekisov, All-Russian Conference on Challenges for Development in Mining Science and Mining Industry **53** (2017).
10. V. L. Voyeykov http://www.delphis.ru/journal/article/aktivnyi-kislorod-organizovannaya-voda-i-protsessy-zhiznedeelnosti (an electronic source, 2005).

The prospects for secondary processing of tailings remained after the beneficiation of fluorite ores of Primorye Krai using a highly selective combination of reagents

Lidiya Kienko[1*] and *Olga* Voronova[1]

[1]Mining Institute of Far eastern branch of Russian Academy of Sciences, Khabarovsk, Russia.

Abstract. The study of the composition and preparability of two samples of tailings dumped at the Yaroslavl Mining Company revealed the following contents: CaF_2 - 15.2 and 20.7%, $CaCO_3$ - 13.3 and 10.2%, and Zn - 0.49 and 0.38%. The volume of the stockpiled wastes is estimated at 30 million tons. The need for fine material grinding at the stage of preparation for flotation was shown, and the efficient fluorite concentration was possible with technical solutions found. The determined optimal parameters of the flotation acid-base condition provided neutralization of the negative influence of salts and slimes left by the previous processing cycle. It has been established that a mixture of ammonium and lignosulfonates was effective as a modifier. The extraction of sample 1 yielded 52% of fluorite extraction to the concentrates with 92.2-92.8% fluorite content, and more than 69% when beneficiating the sample 2. Upgrading of the concentrate quality is only possible through transferring to flotation tails of the most contaminated fluorite grains impregnated with silicate minerals, but this is followed by a significant extraction reduction. The concentration of sphalerite to a separate product occurs with low effectiveness. The inclusion of sulfide flotation in the circuit head can improve the fluorite cycle efficiency: extraction of CaF_2 to the concentrates of FF-92 brand may reach 70%. *Keywords:* mineral mining and processing wastes, fine dissemination, fluorite, secondary processing, flotation, selectivity, modifiers.

1 Introduction

In the midst of a significant reduction in the ore reserves, which are primary sources for the production of fluorite concentrates and progressive decline in their quality, the prospects for stable supply of raw materials to the aluminum industry and other industries using fluorinated compounds might be associated with the involvement of mining and processing plants in the processing of residue waste. The stocks of accumulated waste processing mineral raw materials of production are estimated at a number of enterprises in tens of millions of tons [1, 2]. Thus, the volume of tailings of fluorite ore concentration of the

* Corresponding author: kienkola@rambler.ru

Yaroslavl Mining Company (YAGRK), based on the processing of the ore of the Voznesensky ore district of Primorsky Krai, is estimated at more than 30 million tons. The content of fluorite in them varies between 13-23%. The average value of the carbonate module ($M_к = \alpha_{CaF_2} / \alpha_{CaCO_3}$), which largely determines the preparability of raw materials, is 1.4-1.6.

The ores of the Voznesenskiy fluorite district are quite complex ones due to a number of reasons [3, 4]. One of the main reasons is very close (up to emulsion) mutual germination of mineral phases, which dictates the need for fine grinding. The availability of two calcium-containing components presupposes problems in the selection of minerals in the flotation beneficiation of raw materials. In addition, flotation mica, topaz is present in ores; in separate massifs, the increased content of carbonaceous and clayey components is observed.

2 Materials and Methods of the Research

In the two samples of residue raw materials selected for research, which are the tailings of the primary processing of ores, fluorite is found mainly in either difficult to open splices or in slimes, which further complicates the concentration thereof in a separate product. The processing flow scheme for such a product should include simultaneous operations aimed at opening fluorite with the removal of films of secondary formations and stabilizing the dispersed pulp composition, taking into account the presence of sludge of various nature. The following main components were determined by chemical analysis: CaF_2 – 15.2-20.7 %, $CaCO_3$, respectively, 13.35-10.2 %, SiO_2 – 30.85-32.2 %, Zn – 0.49-0.38 %. For the successful development of techniques for fluorite separation with obtaining branded fluorite concentrates, it is necessary to estimate the beneficiation of each sample taking into account the main determining factors:

1. The granulometric characteristics of the sample, the peculiarities of the mutual germination of mineral components, the necessary preparation of the material by size.
2. Physico-chemical parameters of the water-mineral suspension providing a minimal effect on the process of residues of primary beneficiation reagents and secondary surface formations [5, 6].
3. The existing practice of flotation separation of finely dispersed mineral phases with similar adsorption characteristics.
4. Identifying the most promising areas of research and searching for new technological solutions that provide the required level of breeding.

The research of the granulometric composition of the samples has shown that the selected tailings material in the original form contains significantly fewer grains of the size controlled by the concentrator of the YAGRK (-0.044 mm) than the current waste of processing of fluorite ores (51-53% against 70-85%), obviously, with the natural processes of segregation occurring in the tailings pond. Therefore, the feed was subjected to grinding in a ball mill for 20 minutes, which allowed obtaining a material close in size to the composition of the milled ore during primary processing thereof.

The preliminary research data on the possibility of extracting fluorite from the raw materials prepared in this way are provided in Table 1. The obtained results indicate some features of flotation beneficiation of secondary raw materials.

Calcite content in dedicated concentrates, particularly in experiments using developed technology based on the use as a modifier of mixes with ammonium fluoride [7], can be estimated as extremely low, indicating the process selectivity. At the same time, the content of fluorite in the concentrates does not reach the high-grade level (not more than 93.28% CaF_2). When beneficiating ore material with ammonium fluoride compositions and dosage of lignosulfonate, such low contents of calcite corresponded to the production of high purity concentrates (95% CaF_2 and more).

Table 1. The results of the experiments to extract fluorite from tailing samples using the standard technology of the YAGRK ore-processing plant.

Experiment	Product	Content, %		Extraction of CaF_2, %	Note
		CaF_2	$CaCO_3$		
1	Concentrate	92.11	0.96	37.65	Sample 1 NaF-1,000 g/t Ic=83.7
	Intermediate feedstock 1-6	20.40	25.23	49.60	
	Tailings	3.34	6.87	12.75	
	Flotation feed	15.02	13.21	100	
2	Concentrate	91.05	1.04	49.92	Sample No.2 NaF-900 g/t Ic=42.67
	Intermediate feedstock 1-5	21.58	19.67	44.29	
	Tailings	2.65	3.59	5.80	
	Flotation feed	20.89	10.19	100	
3	Concentrate	93.28	0.59	49.21	Sample 1 NH_4F+ LS-700+150 g/t Ic=136.7
	Intermediate feedstock 1-6	19.89	23.67	42.65	
	Tailings	2.05	8.89	8.14	
	Ore	15.05	13.00	100	
4	Concentrate	93.27	0.49	55.22	Sample No.2 NH_4F + lignosulphonate-600+100 g/t Ic=92.03
	Intermediate feedstock 1-6	20.67	20.49	39.7	
	Tailings	2.22	3.95	5.08	
	Ore	20.83	10.14	100	

The extraction of fluorite by technology with the use of a new composition of modifiers in open-type experiments amounted to 49.8-55.22%. The use of the traditional regime provides a much lower performance. Selective separation of fluorite and calcite when replacing sodium fluoride with ammonium fluoride mix with lignosulfonates increases due to several factors. The fluorine content in the NH_4F molecule is significantly higher, compared with NaF (51.3 and 45.2%, respectively). Dissociation of the molecule NH_4F is much more active than that of the NaF molecule, because of the highest solubility of ammonium fluoride. In addition, in connection with the use of sodium carbonate in the flotation process as a medium regulator, sodium Na + cations are already present in the pulp at the time of filing of sodium fluoride, somewhat shifting the dissociation process of NaF towards the slowing down. Lignosulfonates have a depressant effect on a wide range of minerals, including calcite and silicates.

Mineralogical viewing revealed that, despite the finish grinding, which provides the content of the -0.044 mm class for about 90%, fluorite particles are marked with minute sprouts and primings of other minerals, such as quartz, tourmaline, garnets, mica, etc. Figure 1 shows photographs of fluorite concentrate made with the JEOL scanning electron microscope, equipped with the JCM-6000 PLUS energy dispersive analyzer.

Even with a low degree of magnification, the images clearly show that pure fluorite grains in the isolated concentrates are practically absent. Insufficient disclosure of mineral components, as well as the presence of various emulsion impregnations in fluorite grains, is the most probable causes, which do not allow obtaining high purity concentrates. During almost the entire period of operation, before stopping for reconstruction, the Yaroslavl Mining Company produced concentrates of FF-90 grade, and of the FF-92 grade in fewer amounts, containing 90 and 92% CaF_2, respectively.

Fig. 1 Fluorite concentrate in varying degrees of magnification.

At the present time, in connection with the increased demand for high-grade concentrates, both on the domestic and world markets, the issue of improving their quality has risen sharply [8, 9]. To derive products that in their composition correspond to the FF-95, FF-97 grades with satisfactory extraction indices, from such low-quality raw materials is almost impossible. However, under the conditions of the planned conversion of cryolite production technologies, the content of the consumer in CaF2 concentrates below 93-94% is not satisfactory. In the cycle of experiments with a further reduction in the size of the boundary grain and the identification of the effect of fineness of grinding on the flotation indices, it has been established that the optimum grinding size corresponds to a content of the -0.044 mm class of not less than 95-97%. Further increase in the fineness of grinding leads to a decrease in technological indices, which is caused by increasing slurry fractions volumes [10-12].

3 Results and Discussion

Exploring the possibilities of stable flotation of residue tailings and obtaining the highest quality of fluorite concentrates from them requires detailed analysis of the selectivity dependencies of mineral separation on the complex of parameters affecting the process [13, 14]. One of the main factors determining the efficiency of interaction of ore components with reagents in the flotation medium is the neutralization of the effect on the process of residues of the primary beneficiation reagents formed during storage of the material in the tailings pond of the films and their slime-like chips that have arisen during the grinding process. Achieving the optimal acid-base composition of the liquid phase of the pulp can significantly reduce the activity of the chemical and physico-chemical interaction of the flotation pulp components. Fig. 2 presents the results of studies of the influence of the dosage of the medium regulator on the flotation results.

With an increase in the consumption of calcined soda, there is a steady increase in the extraction of fluorite into the concentrate during the flotation of both samples. With the increased dosage of Na_2CO_3 from 1 to 2.5-3.5 kg/ton, the quality of concentrates increases: the content of CaF_2 increases from 89.61-92.45% to 93.67-93.96%. The selectivity index of the separation of fluorite and calcite also increases when each of the two samples is dressed. Higher consumption of the alkaline regulator leads to a decrease in the selective characteristics of flotation; the extraction of fluorite stably increases.

Further studies to optimize the performance parameters for the purpose of extracting concentrates of higher grades were carried out in various directions: varying the composition and consumption of the modifiers, increasing the number of recleaning operations and the corresponding concentrations of reagents in them. In addition, the expediency of sulfides inclusion in the flotation head with the utilization of a fine flotation mixture of pyrite, pyrrhotite and sphalerite was investigated. The possibility of selecting a sulfide product and isolating sphalerite into a separate concentrate was also studied. The results of the experiments are shown in Table 2.

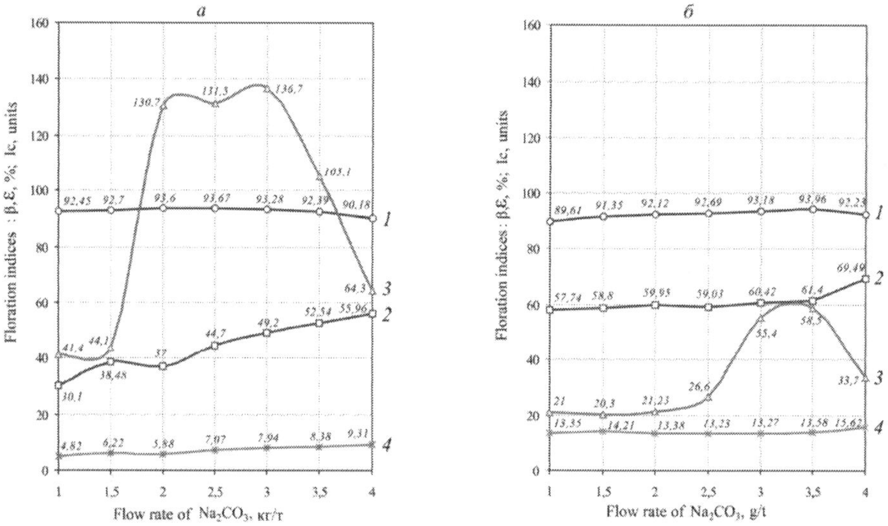

Flotation indices
Consumption of Na_2CO_3, kg/t

Fig. 2. Dependence of fluorite flotation indices from the residue raw material of sample 1 *(a)* and 2 *(b)* on the consumption of calcined soda: 1 – the content of CaF_2 in the concentrate; 2 - extraction of fluorite into concentrate; 3 - index of selectivity of separation of fluorite and calcite ($I_C = \varepsilon_{CaF_2} / \varepsilon_{CaCO_3}$); 4 - concentrate output.

Table 2. The results of research on the selection of the technological regime of fluorite flotation from tails of sample No. 2

Experiment	Product	Output %	Content, %		Extraction CaF_2,%	Consumption NaF+NH_4F +LS (bas.fl./recl), g/t
			CaF_2	$CaCO_3$/ SiO_2		
26	Concentrate	15.36	92.21	1.48/2.74	67.93	0+500+100
27	Concentrate	15.23	92.80	1.18/2.4	67.86	0+500+100
28	Concentrate	14.77	92.82	0.79/2.48	65.28	300+300+100
30	Concentrate	12.33	93.27	0.4/2.339	55.22	0+600+100
31	Concentrate	11:73	93.46	0.39/2.33	52.95	200+400+100
32	Concentrate	13.48	93.76	1.08/2.29	60.88	0+400+100/100+20 (1-2 recl.)
33	Concentrate	8.94	95.07	0.4/1.87	40.71	0+600+100/100+20 (1-3 recl.)
29	Collective sulphide flotation in the head of the process					0+500+100/100+20 (1-2 recl.)
	Concentrate	16.06	92.21	1.68/2.68	70.48	

Note. In experiments 26, 28, 30, 31 - six, in the experiments 27, 29, 32, 33 - seven recleanings of the foam product;

The maximum extraction of fluorite into concentrates with a CaF_2 content of over 93% was 60.88%. The modifier was dosed both in the main flotation and in the recleaning. The increase in the costs of the modifier and the number of its supply points for the purpose of a more intensive concentration of fluorite made it possible to isolate the concentrate containing 95.07% of CaF2, however, its recovery has sharply decreased.

Silicon dioxide is a harmful impurity in fluorite concentrates, which has a negative effect on the process of their chemical processing, limited by GOSTs and technical conditions. Chemical analyzes revealed that the concentrates with a CaF2 content of 92.21-92.80%, contain 2.74-2.40% of SiO_2. As the concentration of CaF2 in the concentrates increases, the content of silica also decreases by 0.3-0.4%. The obtained results testify to the compliance of SiO2 contents with GOST requirements for fluorite concentrates of FF-92A and FF-92B grades. It should also be noted that the minerals carrying silicon dioxide in the concentrates are mainly micas and amphiboles. They lack free silica.

Flotation of fluorite according to the scheme with the preliminary extraction of sulphides (experiment 29) was not accompanied by an increase in the quality of fluorite concentrate, despite a general decrease in flotation sludge. However, the extraction of fluorite into a concentrate containing 92.21% CaF2 was more than 70%, which allows considering this variant of the technology as quite a promising one.

In periods of its stable work, in addition to processing fluorite ores and releasing fluorite concentrate, the YAGRK processing plant periodically processed zinc-fluorite raw materials. Flotation tailings were stored in the common tailings pond [15]. The possibility of concentrating sphalerite from residue tailings was studied according to the scheme in Fig. 3. Initially, collective sulphide flotation was carried out in a slightly alkaline medium created by calcined soda, and later - the concentration of sphalerite in a highly alkaline medium with a dosage of calcium hydroxide.

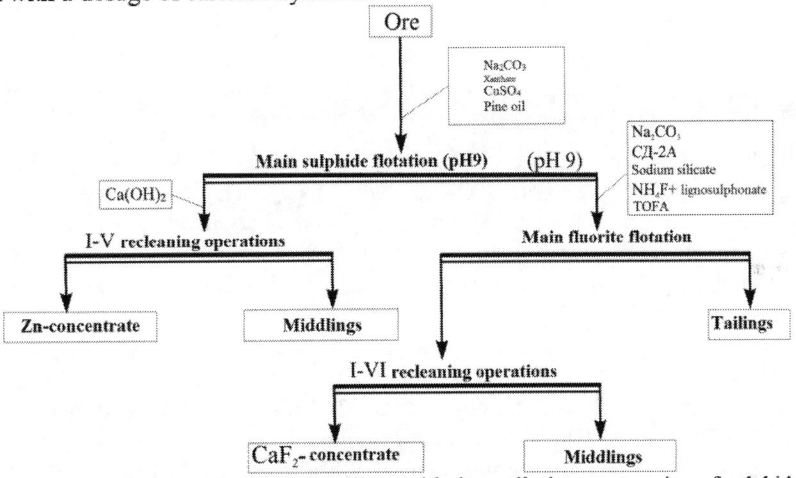

Fig. 3. Scheme for beneficiation of residue tailings with the preliminary extraction of sulphides.

The results of the research showed that beneficiation according to the scheme with the complex extraction of sphalerite and fluorite from the sample of residue tailings No. 1 allows obtaining zinc and fluorite concentrates. At the same time, zinc flotation is inefficient, which is predetermined by the low content of sphalerite in the feedstock. Extraction of zinc into a concentrate with a content of 42.11% was 38.84%, and the extraction of fluorite in a concentrate containing CaF_2 - 92.98% - 50.01%. When beneficiating the sample No. 2, it was possible to isolate only a low-quality zinc product with a zinc content of 33%. The feasibility of beneficiating residue raw materials in accordance with a scheme that includes a sulfide cycle can be finally assessed by economic calculations, as well as further studies on semi-industrial equipment with a continuous flotation cycle on a full scheme.

4 Conclusions

The composition and preparability of two samples of residue tailings of the Yaroslavl Mining Company, selected from the tailing dump of the enterprise, were studied. It was established that the content of fluorite in them is 15.2-20.7%; zinc, represented by sphalerite - 0.49-0.38%; calcite 13.35-10.2%.

The mineralogical analysis found that almost all the fluorite grains are permeated with the finest sprouts and adherents of other minerals. Increasing the fineness of grinding does not allow achieving a significant increase in the degree of fluorite isolation.

The technology of flotation of fluorite from the residue tailings of the presented samples is proved and proposed, using ammonium fluoride-based modifiers. An effective regimen providing a decrease in the activity of the reagents used at the primary processing stage and secondary coatings was proposed.

Fluorite concentrates obtained by direct flotation in the open cycle contain 92.21-93.76% CaF_2. The extraction of fluorite with a content of 92% of CaF_2 may be more than 52.54% for sample No. 1, and up to 67.86-69.49% for sample No. 2. The preparation of concentrates of higher purity is a complex task, which is due to the presence of inclusions in the fluorite grains. When the content in CaF2 concentrates increases to 93.28-93.76%, the extraction in samples 1 and 2 decreases, respectively, by 3-6.5%. The extraction of fluorite into a high-grade concentrate containing more than 95% CaF2, was 40.71%.

Beneficiation according to the scheme with the preliminary flotation of sulphides allows to increase the extraction of fluorite up to 70.17% with the content of CaF_2 in the concentrate of 92.86%. In this case, the concentration of zinc in a separate product passes with low efficiency.

References

1. N. I. Grekhnev, I. Yu. Rasskazov. Tich. geologiya journal, **35**, No. 2, 107 (2016).
2. V. A Chanturiya., L. A Vaysberg., A. P. Kozlov. Obogashch. rud, **2**, 3 (2014).
3. L. A Kiyenko, L. A Samatova. O. V.Voronova. Gorn. zhurnal, **12**, 63 (2013).
4. V.Z. Schestovets, N. V. Egorov, V.E. Pavlov, L.V. Krylova. Gorn. zhurnal, **9**, 26 (2000).
5. O. S. Bogdanov, *Teoriya i tekhnologiya flotatsii rud. [Theory and technology of ore flotation.]* 364 (M.: Nedra, 1990).
6. Gao Zhiyong, Bai Ding, Sun Wei, Cao Xuefeng, Hu Yuehua. Minerals Engineering, **72**, 23 (2015).
7. L.A. Kienko, O. V. Voronova, *Patent No. 2646268 Sposob obogashcheniya karbonatno-flyuoritovykh rud [Method for beneficiation of carbonate-fluorite ores]*
8. Y. J. Li, F. Y. Sun, Y. Zhou, L. Zeng. Journal of Spectroscopy, **2015** (2015).
9. A. V. Fat'yanov, L. G. Nikitina, S. A. Shcheglova, O. L. Gorn. Zhurnal, **3**, 82 (2011).
10. V.I. Klassen, D. I. Nedogorov, I. Kh. Deberdeyev, *Shlamy vo flotatsionnom protsesse [Sludge in the flotation process]* 160 (M.: Nedra, 1969).
11. L.A. Kienko, O. V. Voronova. Fiz. Techn. Probl. Razr. Polezn. Iskopaemykh, **1**, 176 (2014).
12. Y. Yu, L. Ma, M. Cao, Q. Liu. Minerals Engineering, **114**, 26 (2017).
13. F. Hernainz, M. Calero. Chemical Engineering and Processing, **40**, 269 (2001).
14. S.A. Kondratyev, N. P. Moshkin. Phis. Techn. Probl. Razr. Polezn. Iskopaemykh, **4**, 150 (2014).
15. L.A. Kienko, O. V.Voronova. Gorn.. zhurnal **2**, 68 (2015).

Iron manganese ores comprehensive processing

Margarita Gurman[1*]

[1]Mining Institute of Far eastern branch of Russian Academy of Sciences, Khabarovsk city, Russia

Abstract. This article illustrates the results of the study of leaching the resistant industrial products of manganese with the subsequent deposit of Mn from the solution with ammonia and hydrogen peroxide oxidizer. The chemical beneficiation of industrial products under optimal conditions aids in converting 96-98% manganese and its attendant iron into concentrates (in the form of sediments) containing 30.53-40.44% Mn and 10-15 % Fe, sludge cakes containing 80.36-83.81 % SiO_2. It shows that the use of hydrometallurgy in combination with the beneficiation of hematite-braunite ores using gravity and magnetic radiation aids in the 10.39% increase of total manganese recovery in mineral form, (with 76-87.19 % increase in chemical form) and allows for integrated use of valuable components. *Key words:* hematite-braunite ores, resistant industrial products of manganese, acid leaching, deposits, manganese extraction, combined scheme.

1 Introduction

Over ten minor mineral deposits of manganese are known on the territory of the Far East, and the largest of them (with balance stocks of manganese ores 6.53 million tons) is the Poperechnoye deposit, which is part of South Khingan deposit. The studies of the material composition and beneficiation capacity of the manganese ores in the Poperechnoye deposit were held in the 1950's (Chebotarev M.V, Chebotareva N.A., Suvorov F.S., Shchugol S.D., Toropkov S.D.). Previously obtained data on the ores are nowadays complemented with geochemical, petrographic, mineralogical and technological studies involving present-day physical-chemical methods [2-6]. As a result of iron manganese ores analyses, that were held in the Institute of Mining of Far East Division of Russian Academy of Sciences, it was found that ores are resistant to beneficiation. They are characterized by the complex micro- and finely banded texture of layers, besides having uneven alteration of hollow rocks and layers of ore, finely imbedded dispersive granularity of the braunite, hematite and magnetite in jaspilites, their intergrowth with the minerals of the hollow rocks, the similarity of the physical properties of separable minerals, and the low content of metals. The ore was found to have precious metal mineralization, characterized by pure gold and its thin spots, mixed with platinum and silver in the magnetite and hematite jaspilites. Combining the methods of magnetic separation (in low field and high field) and ore

[*]Corresponding author: mgurman@yandex.ru

beneficiation by gravity, based on the concentration of pure gold, illustrates the possibility to obtain manganese concentrates containing 37.85-46.46 % Mn and golden "heads" containing 12.76-37.17 g/ton Au. The process of ore beneficiation by gravity and magnetic radiation has generated the industrial products (with low content of manganese, but having a large amount of quartz), that can be fully developed based on the chosen scheme. Manganese ore decomposition is known to be achieved by means of processing the ore with HCl, HBr, H_2SO_4, HNO_3 or their mixtures depending on the mineral form of the manganese, which will interact with certain reagent [8,9]. This research was held for the purpose of conducting the acid leaching of resistant industrial products of manganese with subsequent extraction of manganese from the solution.

2 Research techniques

The research was focused on the resistant industrial products of gravitational and magnetic beneficiation of the hematite-braunite ores containing 11.7-32.1 & manganese. The industrial products were presented by the braunite, its intergrowth, its parts attached to quartz and hematite, hematite jaspilites with crystal spots and the thin layers of braunite; magnetite jaspilites with spots of thin-scaled hematite. According to the X-Ray diffraction analysis data, braunite is basically present in the form of Mn_2O_3 and $Mn_2O_3{\cdot}MnSiO_3$. Braunite modifications were revealed - $CaMn_{14}SiO_{24}$ и $(MnFeCa)_7(SiFe)O_{12}$, with tephroite - Mn_2SiO_4 and bixbyite – $(Mn,Fe)MnO_3$ as well.

Manganese mineral oxides are dissolved in hydrochloric acid, which is the dissolvent for the attendant mineral oxides of iron, aluminum. The manganese and iron, resultant from this reaction, are turned into a solution, and the insoluble sediment of silicon dioxide remains in the sludge cakes of leaching. The extraction of manganese out of the solution was implemented by means of precipitation in the form of hydrated oxides in the solution of ammonia mixed with hydrogen peroxide [8, 10]. The addition of H_2O_2 accelerates oxidation [11], while the sediment is immediately turned into a brown-black hydrated manganese dioxide. The iron and aluminum, present in the solution, are precipitated together with the manganese. For the separation of alkaline earth metals that can also be attracted to the manganese sediment the manganese was subject to reprecipitation.

3 Results and discussion

According to the obtained data, hydrometallurgical processing of manganese industrial products (sample 1-3) allows for the conversion of 95-98.8% manganese and iron into a concentrate (in the form of sediment) with its 57.43-76.82 % output (pic 2). Test 1, containing 29.8-32.1 % manganese, 7.9-8.4% iron and 18.62-20.77 % SiO_2, resulted in the concentrate containing 39.1-41.8 % Mn и 10.4-10.7 % Fe with its residual content of 0.64-1.45 % SiO_2. Test 2, containing 17.6-18.1 % Mn, 9-9.3 % Fe и 33.19-36.34 % SiO_2, led to the extraction of the concentrates containing 30.1-32,1 % Mn, 15,1-17.5 % Fe with its residual content of SiO_2 – 0.64-1.28 %. Test 3: Mn – 11.7-12.6 %, Fe – 8-9 % iron, SiO_2 – 34.05-35.97 % resulted the concentrates with lower content of Mn – 20-22.3 % и 10.3-10.6 % Fe with its residual content of SiO_2 –1.34-2.78 %.

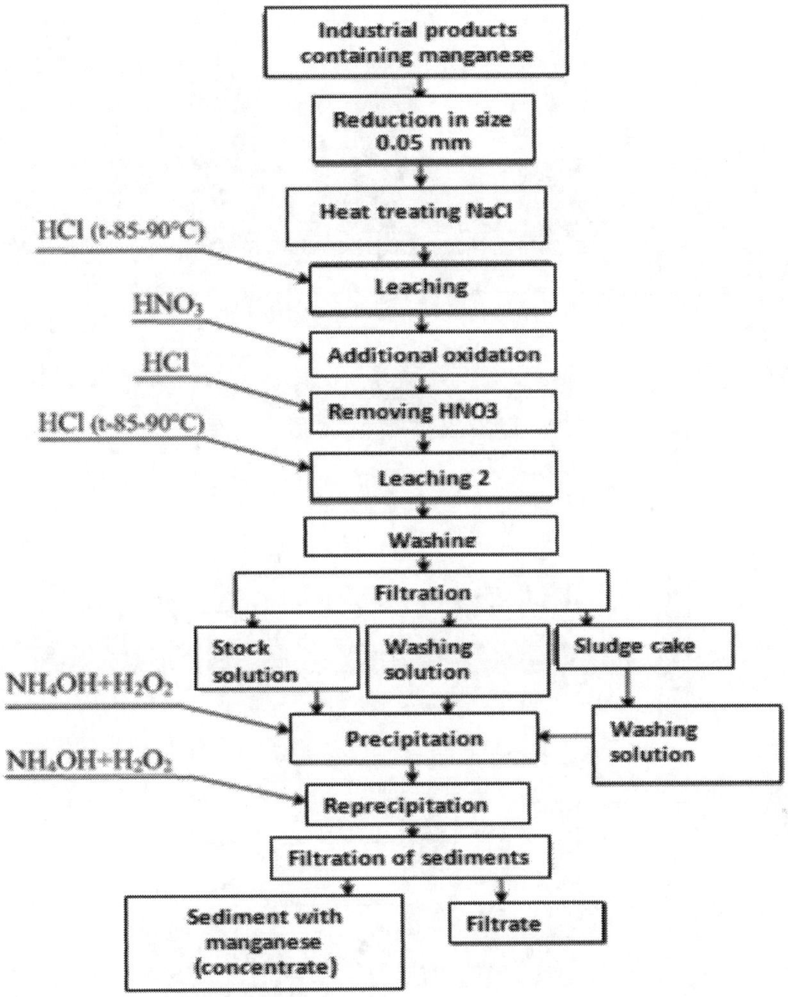

Fig. 1. Hydrometallurgical scheme of processing industrial products of hematite-braunite ore beneficiation.

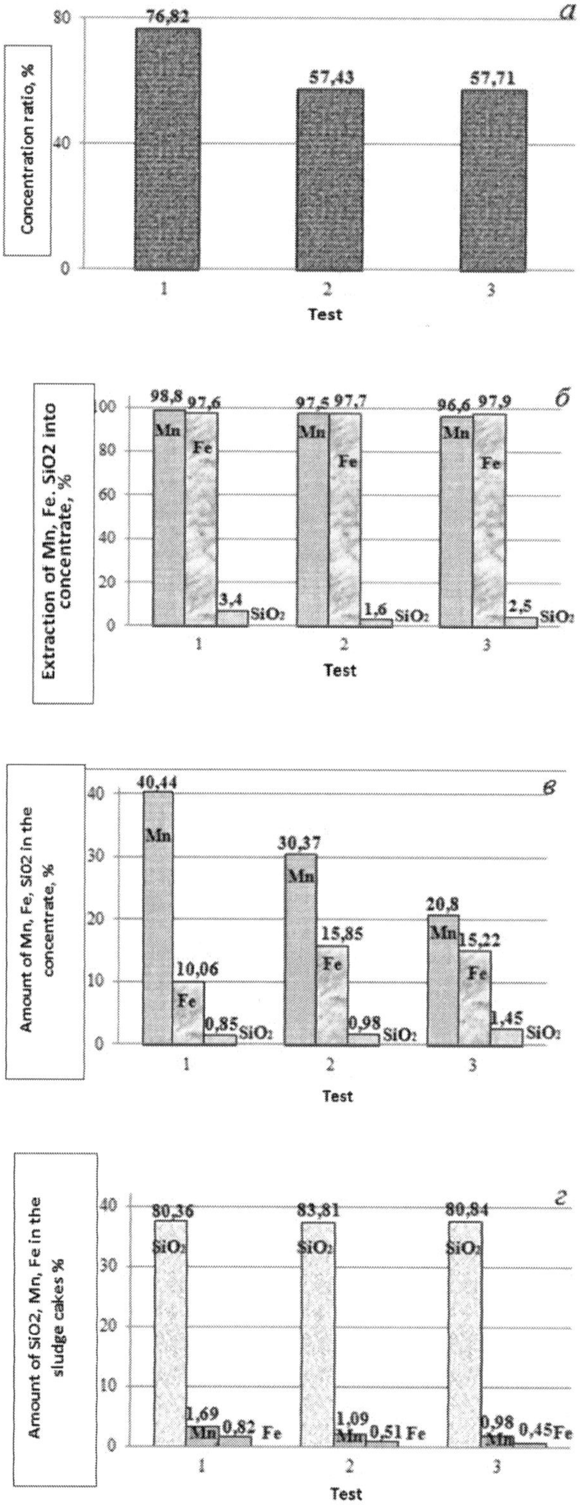

Fig. 2. Technological rates of processing industrial products.

Table 1. Rates of extracting manganese in mineral [7] and chemical form under the combination of beneficiation and hydrometallurgy techniques.

The name of the products	Output, %	Contents, Mn, Fe %		Extraction, Mn, %
Wet magnetic separation				
Concentrate 1	11.43	44.31	6.82	23.69
Concentrate 2	0.41	37.05	5.02	0.71
Total amount of concentrate	11.84	-		24.40
Semiproduct	2.71	16.96	8.35	2.15
Jigging – centrifugal separation Kn – concentration on the table				
Concentrate jigging – table	1.35	37.00	6.47	2.33
Concentrate Kn – table	1.74	38.71	6.77	3.15
Reduction concentrate – Kn – table	4.89	41.23	6.54	9.43
Total gravitation concentrate including the head of the table, Au, g/ton	7.98 0.50	39.95 12.76-37.17	6.52	14.91
Industrial product	14.94	29.19	6.53	20.40
Tailings	62.53	13.04	6.45	38.14
Dry magnetic separation				
Concentrate -1+0.5 мм	2.86	47.10	3.81	6.30
Concentrate -0.5+0.2 мм	8.00	42.07	5.55	15.74
Concentrate -0.2+0.02 мм	1.85	39.58	8.01	3.42
Total concentrate Смс	12.71	-		25.46
Industrial product	10.52	12.05	8.97	5.93
Tailings	33.60	2.72	4.27	4.27
Sludge -0.02мм	5.70	9.30	6.11	2.48
Reduction in size – rewashing of industrial product				
Industrial product concentrate	4.55	46.05	7.20	9.80
Industrial product	10.39	21.81	6.71	10.60
Acid leaching				
Concentrate 1 from solution	5.81	38.53	11.18	10.47
Concentrate 2 from solution	1.53	29.20	14.57	2.09
Industrial product	9.11	19.15	13.75	8.16
Sludge cakes after leaching	12.87	0.76	0.59	0.46
Total amount of products:				
Concentrate (in mineral form)	37.08	43.00	7.92	74.57
Concentrate 1 from solution	5.81	38.53	11.18	10.47
Concentrate 2 from solution	1.53	29.20	14.57	2.09
Industrial product	9,11	19.15	13.75	8.16
Sludge cakes after leaching	12.87	0.76	0.50	0.46
Final tailings	33.60	2.66	4.31	4.27
Total amount of tailings	46.47	2.17	3.25	4.71
Total amount of hematite-braunite ore	100.00	21.38	6.66	100.00

The acid leaching under the optimal parameters of the process resulted in revealing the possibility of forming manganese concentrates containing 38.53-40.44 % manganese (61.27-66.46 % MnO_2) with residual content 0.85-1.45 % SiO_2. The obtained concentrates, depending on the purpose, can be subject to chemical cleaning for the purpose of separating some additives. The quality improvement of manganese concentrates is the subject of further studies.

The sludge cakes of decomposition of the industrial products contain 80.36-83.81 % SiO_2, with calcium and aluminum being their related components, respectively, CaO – 2-8.5 %; Al_2O_3 – 4-5 %; magnesium, sodium and potassium are present in small amounts. According to electronic and optical microscopy data, sludge cakes were found to have separate uncovered grains of braunite and tephroite, whose presence is probably conditioned by the incomplete decomposition, as a result of their thin intergrowth with the minerals of hollow rocks.

As a result of acid leaching, 10.47% manganese containing 38.53 %, with the removal of 12.87 % material n the form of sludge cakes containing 0.76 % Mn (tbl) was extracted from resistant industrial products of gravitational and magnetic beneficiation of the hematite-braunite ores.

4 Conclusions

The performed studies have shown the possibility of acid leaching of resistant industrial products of beneficiation of hematite-braunite ores and extraction of manganese concentrates in the form of sediments containing 38.53-40.44% manganese (61.27-66.46% MnO2) and 10-15% Fe having residual content SiO2 – 0.85-1.45 %.

The use of hydrometallurgical techniques in combination with the operations of beneficiation allows for the 10.47% increase of total extraction of manganese in mineral form, and the 76.8-87.19% increase of it in chemical form.

The suggested combined scheme of processing hematite-braunite ores allows for the complex extraction of valuable components; obtaining manganese concentrates, gold density, and the sludge cakes of acid decomposition (with the content of 80.36-83.81% SiO_2) can also be recommended for glass manufacturing.

References

1. G.I. Arkhipov, Mineral resources of mining industry in the Far East. Strategic evaluation of exploration capabilities, (2017).
2. T.V. Bashlykova, G.A. Pakhomova, B.S. Lagov, A.B. Zhivaeva, T.B. Doroshenko, A.R. Makavetskas, T.O. Shulga, Technological aspects of subsoil use, 241. (2005).
3. N.V. Moiseenko, S.V. Shchipachyov, N.S. Sanilevich, T.B. Makeeva, Geology, mineralogy and geochemistry of precious metals of Eastern Russia: new technologies of processing precious metal raw material, 72. (2005).
4. L.P. Tigunov, E.G. Ozhogina, E.G. Litvintsev, E.S. Bronitskaya, S.I. Anufrieva, E.A. Kalish, Mining journal, **2**, 78. (2007).
5. I.Y. Rasskazov, B.G. Saksin, Physical-technical problems of processing of minerals. **1,** 13 (2014).
6. M.A. Gurman, L.I. Shcherbak, Physical-technical problems of processing of minerals. **3**, 157, (2018).
7. M.A. Gurman, L.I, Shcherbak, Physical-technical problems of processing of minerals, **1**, 144, (2018).
8. A.K. Lavrukhina, L.V. Yukina, analytical chemistry of manganese, (1974).
9. Alafara A. Baba, Lateef Ibrahim, Folahan A. Adekola, Rafiu B. Bale, Malay K. Ghosh, Abdul R. Sheik, Sangita R. Pradhan, Olushola S. Ayanda, Ismail O. Folorunsho, Journal of Minerals and Materials Characterization and Engineering, **2**, 230 (2014).
10. K.A. Baklanova, O.P, Boyarshinova, Mineral raw material analysis, 350 (1956)
11. Z.E. Golbraich, Practical work on inorganic chemistry (1979).

Beneficiation and hydrometallurgical processing of gold-containing sludge

Sergei Ivannikov[1*], *Oleg* Ageev[2], *Svetlana* Bratskaya[1], *Michael* Medkov[1], *Evgeny* Shamrai[1] and *Aleksandr* Yudakov[1]

[1] Institute of Chemistry, Far East Division of Russian Academy of Sciences, Vladivostok, Russia.
[2] Amur Scientific Center, Far East Division of Russian Academy of Sciences, Blagoveshchensk, Russia.

Abstract. The research results of technogenic objects in the Far East of Russia, in particular, the silt of the settling pond "Chalgany" and the waste of the Chalganov kaolin plant are given. The presence of valuable components in the investigated objects is shown, in particular: gold, noble metals, rare earth elements. It is established that gold in the investigated objects is represented mainly by a difficult-to-extract finely dispersed class of fineness (<5 μm). The methods of preliminary concentration and hydrometallurgical processing, which allow extracting gold from the studied technogenic objects, are proposed. The effectiveness of non-cyanide methods of ammonia-thiosulphate and thiocarbamide leaching was investigated. It is shown, that the efficiency of gold recovery from the silt of the settling pond "Chalgany" by the thiocarbamide method reaches 93.5%. It is established that the waste from the Chalganovskiy kaolin plant, after recovering the concentrate of precious metals from them, can serve as a raw material for the production of chemically pure silicon and silica, as well as aluminum, its compounds, and refractory fibrous materials.

1 Introduction

Currently, the situation in the gold mining industry of the Russian Federation is characterized by a critical imbalance between production and buildup of stocks, which is exacerbated by a decrease in the quality of ores, a decrease in the efficiency of gold mining and the accumulation of significant amounts of technogenic waste. With the depletion of ore reserves containing coarse gold, arises the task of involving into processing of placer mines containing finely dispersed gold.

The prospects for processing of gold-containing ores are largely related to the solution of the problem of refractory ore benefication, which is contingent on the cohesion of gold in sulphides (arsenic pyrite, pyrite) and a high proportion of fine-grained gold [1]. In connection with this, various technogenic wastes, including technogenic gold mining wastes, become a competitive source of mineral raw materials for gold mining. These objects are formed due to the imperfection of the applied gold mining technologies, when, depending on the type of the deposit and methods of gold recovery, 10 to 50% of metal can

* Corresponding author: fyajkfqn@mail.ru

remain in the used material [2]. Since gold deposits are recovered by gravitational methods, mainly coarse gold fractions are extracted, it means, that such technogenic objects can contain significant gold reserves, represented by particles ranging in size from tens of microns to micron, densely associated with mineral components of various nature [3, 4], what makes such raw materials close to poor, refractory ores. At the same time, one should pay attention to the fact that technogenic gold-containing objects, as a rule, are located near the infrastructure that has remained since opening of the deposit. Moreover, such objects can be a promising raw material for extraction of other useful components. For example, gravitational methods of gold recovery lead to natural beneficiation of sand. Also, gold placers can contain minerals with stannum, zirconium, tantalum, niobium, tungsten, rare earth elements (REE), etc. Practically for all methods of gold recovery by gravity separation, these minerals can be concentrated together in the place of gold-containing fraction [5].

Since the degree of gold extraction by gravitational technologies is inversely proportional to the size of gold grains[6], the effective use of hydrometallurgy methods turns out to be a promising approach for efficient extraction of fine-grained gold from technogenic waste. At the same time, the most widely spread method of cyanide leaching of gold has its limitations due to the environmental hazard of the reagents and the decrease in efficiency of gold recovery from carbonaceous and organic raw materials as well as from raw materials high in copper, zinc, nickel, antimony, and arsenic. The last fact can be critically important when working with complex technogenic gold-containing raw materials. An alternative to cyanide leaching, capable to work efficiently with cyanide-resistant ores of complex composition, are the methods of thiocarbamide and ammonia-thiosulphate leaching [7, 8].

This article is devoted to the analysis of prospects of gold and other valuable components recovery from technogenic objects in the Far East of Russia, in particular, from the silt of the settling pond "Chalgany" and Chalganovskiy kaolin plant wastes. Hydrometallurgical methods of ammonia-thiosulphate and thiocarbamide leaching are considered as environmentally and economically promising technologies for gold recovery.

2 Materials and methods

The content of noble and rare-earth metals in technogenic gold-containing wastes was determined by atomic absorption spectrophotometry (AAS) and inductively coupled plasma mass spectrometry (ICP-MS) in the Laboratory of Analytical Chemistry and Micro- and Nano-Studies of the Common Use Center of the Far East Geological Institute of the Far East Division of Russian Academy of Sciences (FED RAS) and the laboratory of Khabarovsk Innovation Analytical Center in Institute of Tectonics and Geophysics of the Far East Division of Russian Academy of Sciences.

To estimate the amount of Au, Pt, and Pd by the AAS method, the samples were successively decomposed with a mixture of concentrated acids HF + HNO_3 at a ratio of 2:1, then with HCl + HNO_3 at a ratio of 3:1 and HCl, followed by coprecipitation with Te according to TSNIGRI-2005. Measurement of Au, Pt and Pd was carried out in the electrothermal atomization regime using the atomic absorption spectrophotometer Shimadzu 6800. For the measurement of Ag by the AAS method, the samples were decomposed by a mixture of concentrated acids HCl + HNO_3 (3:1) and HCl according to HCAM-130-C. The measurement of Ag was carried out in flame using the atomic absorption spectrophotometer Shimadzu 6800.

ICP-MS researches were carried out with the use of spectrometers Agilent 7700 (Agilent Technologies, the USA) and ICP-MS Elan DRC II PerkinElmer (the USA), according to ПНД Ф 16.1:2.3:3.11-98.

To estimate the amount of gold in technogenic objects, the method of instrumental neutron activation analysis (INAA) was used at a compact installation with a radionuclide excitation source based on ^{252}Cf developed at the Institute of Chemistry FED RAS [9]. The detection limit of gold according to isotope ^{198}Au for INAA from the samples of 50-100 g was 0.2 g/t.

The control of gold content in the samples from Chalganovskiy kaolin plant was carried out by an assay test in accordance with MA ИАЦ-43-2004, ФР.1.31.2004.01195 [10].

The chemical composition of samples was determined by X-ray fluorescence analysis (XRF) at the Institute of Chemistry FED RAS (the Common Use Center - the Far Eastern Center of Structural Research) and the Institute of Geology and Nature Management, FED RAS with the use of Shimadzu EDX 800 HS and XRF-1800 Shimadzu spectrometers. The sensitivity for elements from Na to U was up to 1 ppm.

The studies were carried out on two types of samples:

As the first object of study, the silts of the settling pond "Chalgany" were used, which are the technogenic wastes of gold-containing raw materials of various deposits "Solovievsky placer", "Oktyabrsky placer", deposit "Beregovoe", etc. The samples were taken from the experimental site of the laboratory of Amur Complex Scientific Research Institute "Chalgany" after gold recovery from gold-containing sands using gravitational methods of benefication.

As the second object of study, the wastes from Chalganovskiy kaolin plant were used. Among them were the tailings dumps of suspended particles of clay minerals transported by process water streams, as well as sand line waste and quartz-feldspar sands concentration tailings. Within the tailings dump, 20 wells with a depth up to 2.5 meters have been traversed with sampling every 0.5 meters. In the second section (in the territory of waste storage and concentration tailings) 16 wells up to 1.5 meters deep have been traversed with sampling every 0.5 meters. Total of 36 wells were drilled, 102 samples were taken. The total amount of waste accumulated at kaolin plant is more than 500,000 tons.

3 Results and discussions

According to the mineralogical analysis, silt samples of the settling pond "Chalgany" represent a complex agglomeration of alluvial-forming minerals, typical for silt-detention basin of gold-processing companies: a large amount of clay, silt, the presence of organic inclusions, the presence of a metal scrub and, as a consequence, a large amount of hydrous ferric oxides, and the entire rock has a brown color due to limonite. The processes of oxidation and decomposition of iron-containing minerals (sulphides, carbonates, silicates) occur due to the access of oxygen and high humidity, which is typical for silt-detention basins of mining enterprises. The sample contains angular crystals of pure white quartz, a small amount of bull quartz, as well as a large amount of K-feldspar brownish and pinkish in color with scales of light and black biotite. Sulfide minerals can be found in the sample in small amounts: pyrite, arsenopyrite, chalcopyrite. Separate grains of galenite, tin stone are noted, there is a hornblende in the sample, a pomegranate of light red color.

Mineralogical analysis of primary raw materials showed the absence of visible gold and other noble metals in the rock under investigation. At the same time, according to AAS and INAA, the sample contains a significant amount of gold (78.6 g/t) and silver (57.5 g/t).

X-ray spectral microscopy studies showed that the maximum particle size of gold is about 50 μm, but most of the particles detected in the study do not exceed 5 μm. With a decrease in size, gold particles become more spongy.

To study the possibility of additional concentration of gold, the material was subjected to magnetic separation. The sample of silt was divided into heavy magnetic fraction (magnetite, ilmenite, metal scrub), heavy non-magnetic fraction (black sand) and light residue. The results obtained indicate a fairly even distribution of gold between the magnetic and non-magnetic fractions and the ineffectiveness of magnetic separation. The conducted sieving analysis also did not allow us to distinguish the fraction with which gold would have been associated.

The silts of the settling pond "Chalgany" are characterized by a high content of gold, but, gold is represented in the form of finely dispersed particles hardly yielding to mechanical beneficiation. To extract gold from this object, a hydrometallurgical methods was developed. The investigations were carried out at the Institute of Chemistry FED RAS in two regimes: laboratory, weighing 100 grams each with pulp agitation, and large scale on a sample of 40 kg without agitation, simulating heap leaching. The gold content in samples taken for testing was 80-100 g/t. The tests worked out the methods of ammonia-thiosulphate and thiocarbamide leaching.

A preliminary assessment of the efficiency of ammonium-thiosulphate leaching of gold from aluminosilicate rocks was carried out in laboratory conditions on model clays (kaolin) with artificially introduced gold. An optimized composition of the leaching solution was obtained - 0.1 mol/l $Na_2S_2O_3$; 0.2 mol/l NH_3; 0.015 mol / l $CuSO_4$; pH 10.4; ratio T: M = 1:2.5. The efficiency of gold recovery from model systems within 2 hours of contact with the ammonia-thiosulfate solution exceeded 90% already at a ratio of T:F 1:1. Application of these leaching conditions to real samples of gold-containing silicate and aluminosilicate samples provided high efficiency of gold recovery (78-97%) [11].

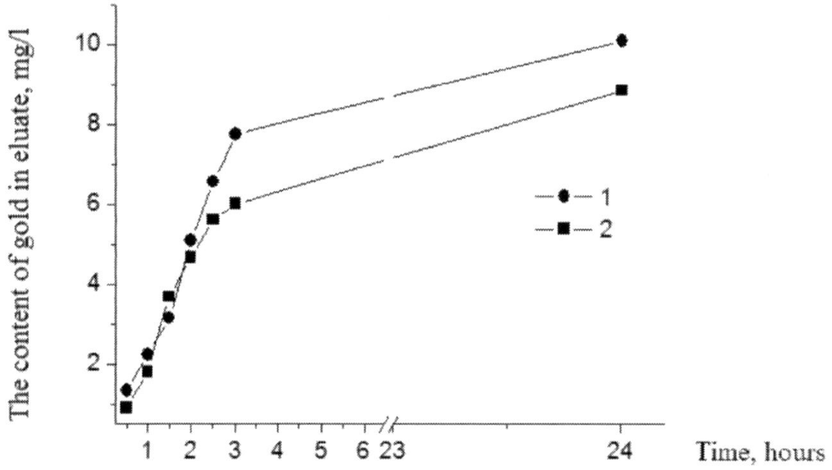

Fig.1. Kinetic of the gold content in the ammonia-thiosulphate leach solution (eluate): 1 - in the laboratory; 2 - in extended tests

When carrying out ammonia-thiosulphate leaching, the reservoirs with a low surface area contacting the atmosphere (0.05m^2) were used, since the arrival of atmospheric oxygen in the presence of iron oxides (the content of which in the sample according to XFA data exceeds 60%) can significantly reduce the rate of ammonia-thiosulphate leaching due to the accelerated oxidation of thiosulfate. For extended tests, a leaching solution was used, optimized at the stage of laboratory tests. The ammonia-thiosulphate leaching tests carried out under these conditions showed an efficiency close to the results obtained in the laboratory (see Figure 1). However, the overall efficiency of gold recovery for the silts of the settling pond "Chalgany", both in laboratory and enlarged conditions, did not exceed 47%. What can be associated with a decrease in the stability of the ammonia-thiosulfate system in the presence of transition metal oxides catalyzing the oxidation decomposition of thiosulfates and the formation of tetrathionates [12, 13]. Additional studies show that in the presence of magnetite, the dissolution rate of gold by the ammonia-thiosulphate solution of the optimized composition drops sharply [11]. Thus, the significant content of magnetite and titanomagnetite in this technogenic raw material is a factor that reduces the effectiveness of the use of ammonium-thiosulfate solutions for gold leaching.

The concentration of gold from the leaching solution was carried out in a laboratory and extended conditions on the ion exchanger AV-17-8 at the solid/liquid ratio - 1:50, the recovery efficiency in both cases was about 80%.

The testing of thiocarbamide leaching was conducted in same conditions as ammonium-thiosulfate: laboratory, weighing 100 grams and large scale on a sample of 40 kg. In both cases, the thiocarbamide leaching was carried out in the sulfuric acid medium in the presence of an oxidizing agent. In the laboratory, optimal conditions for thiocarbamide leaching were determined: 0.65 mol/L $CS(NH_2)_2$; 0.1 mol/L H_2SO_4; 0.09 mol/L $(NH_4)_2S_2O_8$, the solid/liquid ratio – 1:2.

Bulk gold leaching tests were carried out for 5 hours until the concentration of gold in the solution according to AAS data did not stabilize at about 14.4 mg/l. The total percentage of gold recovery by cake was 72.5%, which is lower than the percentage of recovery obtained in laboratory conditions for a sample weighing 100 g, where the percentage of gold recovery was 93.5%. The main reason, which resulted in a decrease in the percentage of gold recovery from the sample, was the absence of pulp mixing.

Gold recovery from the solution is the final stage of processing of almost any type of ores using the technology of thiocarbamide gold leaching and electrolysis is the most perfect method of precipitating gold from a solution of thiourea leaching. Previous studies [14] show that for electrolytic gold recovery the flow design turns out to be optimal, as it supposes the separation of the anode and cathode spaces with an ion-exchange membrane using the lead anodes and stainless steel cathodes. Under such conditions, the efficiency of gold recovery from the leach solution exceeds 95%.

Thus, taking into account a small amount of raw materials for gold recovery from the silt of the settling pond "Chalgany", it is perspective to use a low-capacity heap leaching based on the thiocarbamide process.

Studies of the silt of the settling pond "Chalgany" conducted with regard to ISP-MP method showed that these technogenic wastes contain other valuable components, in addition to gold, including the most part of the rare-earth elements (REE) (see Table 1). Such a high content of REE in the samples makes the development of an integrated approach to processing of this type of raw materials potentially productive.

Table 1. The content of rare and rare-earth elements in the samples of the silt from the settling pond "Chalgany" (according to ISP-MS)

Element	*C, g/t*	Element	*C, g/t*	Element	*C, g/t*	Element	*C, g/t*
Be	0.84	Sr	226.68	**Ce**	819.60	**Tm**	1.11
Sc	13.65	Y	105.57	**Pr**	74.46	**Yb**	8.91
V	283.52	Zr	3406.04	**Nd**	315.08	**Lu**	1.40
Cr	701.40	Nb	120.54	**Sm**	39.56	**Hf**	87.60
Co	42.95	Mo	15.71	**Eu**	1.27	**Ta**	34.62
Ni	161.49	Cd	2.22	**Gd**	37.24	W	472.89
Cu	169.92	Sn	906.23	**Tb**	3.91	Pb	1884.10
Zn	246.04	Cs	3.69	**Dy**	19.40	Th	206.55
Ga	21.95	Ba	831.33	**Ho**	3.18	U	29.72
Rb	63.15	**La**	394.60	**Er**	9.57		

Thus, the silt waste from the settling pond "Chalgany" is a promising source of various valuable components, including gold and REE, but the reserves of this type of technogenic raw materials are very limited (several hundred tons). As an alternative source of gold and REE, the waste of Chalganovskiy kaolin plant was investigated. According to the research conducted by the Amur Scientific Center FED RAS in 2014, the volume of waste generated at this facility was more than 500,000 tons.

X-ray fluorescence analysis was used to determine the chemical composition of the formed waste, including not less than 80% of kaolinite, 14% of quartz, feldspar.

The method of assay test proved the presence of gold samples. According to the assay test, the gold content in 46 samples taken from the tailings dump is from 0.0 to 5.5 g/t (see Table 2). The average content is 0.41 g/t. In 31 samples taken from the waste sands, the gold content was from 0.0 to 0.82 g/t. (see table 3). The average content is 0.20 g/t.

Table 2. The gold content in the waste samples from Chalganovskiy kaolin plant, according to the assay test (Object: Tailings dump)

Sample	C, g/t	Sample	C, g/t	Sample	C, g/t	Sample	C, g/t
Л-I-1-0.5	1.02	Л-II-1-0.5	traces	Л-III-3-0.5	traces	Л-IV-2-1.0	0.29
Л-I-1-1.0	0.18	Л-II-1-1.0	0.10	Л-III-3-1.5	0.18	Л-IV-3-0.5	traces
Л-I-1-1.5	0.16	Л-II-2-1.0	traces	Л-III-4-0.5	0.16	Л-IV-3-1.0	0.10
Л-I-1-2.0	traces	Л-II-4-0.5	4.30	Л-III-4-1.0	0.10	Л-IV-3-1.5	0.10
Л-I-1-2.5	0.14	Л-II-4-1.0	traces	Л-III-4-1.5	0.14	Л-V-1-1.0	0.13
Л-I-2-0.5	0.22	Л-III-1-0.5	0.10	Л-III-5-0.5	0.52	Л-V-2-0.5	5.50
Л-I-1-1.5	0.10	Л-III-2-0.5	0.64	Л-III-5-1.5	traces	Л-V-2-1.0	traces
Л-I-3-0.5	0.20	Л-III-2-1.0	0.10	Л-IV-1-0.5	0.10	Л-V-2-1.5	4.40
Л-I-4-0.5	0.12	Л-III-2-1.5	0.08	Л-IV-2-0.5	traces		

Table 3. The gold content in the waste samples from Chalganovskiy kaolin plant, according to the assay test (Object: sand line)

Sample	C, g/t	Sample	C, g/t	Sample	C, g/t	Sample	C, g/t
X-I-1-0.5	0.10	X-II-1-1.0	0.10	X-II-4-1.0	0.22	X-III-1-1.5	traces
X-I-1-1.0	0.42	X- II-2-0.5	0.10	X-II-4-1.5	0.10	X-IV-2-0.5	0.18
X-I-2-0.5	0.16	X-II-2-1.0	0.37	X-III-1-0.5	0.18	X-IV-2-1.0	traces
X-I-1-2.5	0.10	X-II-2-1.5	0.81	X-III-1-1.0	0.10	X-IV-2-1.5	0.56
X-I-3-0.5	0.14	X-II-3-0.5	0.26	X-III-2-0.5	traces	X-IV-2-2.0	0.10
X-I-3-1.0	0.10	X-II-3-1.0	0.82	X-III-3-0.5	0.20	X-III-3-0.5	0.20
X-I-3-2.5	0.10	X-II-3-1.5	0.50	X-III-3-1.0	0.10	X-III-3-1.0	0.10
X-II-1-0.5	0.10	X-II-4-0.5	0.16	X-III-4-0.5	0.28	X-III-4-0.5	0.28

The content of gold in the waste of Chalganovskiy kaolin plant in relation to the total amount of waste is extremely small, what makes it unprofitable to apply hydrometallurgy methods for metal recovery. To efficiently extract valuable components from this raw material, it is necessary to solve the task of reducing the waste rock that is involved in the process of beneficiation and obtaining the concentrate, which will then go to the beneficiation plant and refining. For this purpose, a team of authors developed a model of a centrifugal concentrating-classifying apparatus [14], which allows, in the course of work, to remove an empty rock, silty and clay fractions and obtain a concentrate of useful components, including noble metals, which will be further processed: a concentration table; low-capacity heap leaching; pyrometallurgy; refining. The application of this approach will significantly increase the gold content in the rough concentrate obtained from the waste of Chalganovskiy kaolin plant, especially for waste from the sand line. And then it is possible to use the method of ammonia-thiosulphate leaching for efficient extraction of gold from this technogenic object.

After the recovery of the valuable components, the remaining part of waste from Chalganovskiy kaolin plant will present the secondary kaolin and may be considered as a raw material for the production of chemically pure silicon and silica, as well as aluminum, its compounds, and refractory fibrous materials.

4 Conclusions

It is established that investigated technogenic wastes are a valuable source of mineral raw materials. The silts of the settling pond "Chalgany" contain a significant amount of gold, precious metals, rare earth elements and other valuable components. A promising approach for gold recovery from the silt of the settling pond "Chalgany" is carrying out of a low-capacity heap leaching based on the thiocarbamide process followed by electrolytic precipitation of gold from the leach solution.

Wastes from Chalganovskiy kaolin plant are also of considerable interest as a source of mineral raw materials. The composition of this raw material includes kaolin, quartz, feldspar, and a small amount of REE and PGM. After the recovery of the valuable components, the remaining part of waste can be regarded as the secondary kaolin - a raw material for the production of chemically pure silicon and silica, as well as aluminum, its compounds, and refractory fibrous materials.

Gratitude

The work was partially (selection, sample processing, and technological research) supported within the framework of the Comprehensive Program of Fundamental Research of the FED RAS "FAR EAST", project 18-3-029 "Research and development of materials and processes for environmentally safe technologies for precious metals recovery from technogenic objects (section 1), and partly (analytical studies of samples and products of technological processing) was supported within the framework of the State task of the Institute of Chemistry FED RAS (topic №0265-2018-0002).

References

1. Yu.G. Safonov, Vestnik OT GTN, **3**, 51 (1998).
2. B.K. Kavchik, V.G. Pyatakov, Irgiredmet .OJSC: Gold mining, **135** (2010).
3. A.P. Van-Van-E, V.S. Litvintsev, G.V. Sekisov, MIAB "Far East-2", 32 (2009).
4. M.A. Medkov, G.F. Krysenko, D.G. Epov, A.A. Yudakov, Zh. "Vestnik FED RAS", **5** 75 (2010).
5. V.P. Molchanov, M.A. Medkov, A.A. Yudakov, V.A. Avramenko, Yu.V. Pruss, "Vestnik NESC FED RAS", **4.** 73 (2011).
6. N.G. Valiiev, *Scientific-methodical and technological foundations of hydro-mechanized preparation of gold-containing rocks for fine gold recovery*, Ural State Mining and Geological Academy, 27 (2002).
7. M.G. Aylmore, D.M. Muir, Minerals Engineering, **14**, 135 (2001).
8. V.I. Radomskaya, O.V. Losev, S.M. Radomsky, "Vestnik FED RAS", **1** 80 (2004)
9. N.A. Shilo, J. Radioanal. Nucl. Chem., **79**, 309 (1983).
10. Method for determination of mass fractions of gold and silver in samples of ores containing gold and products of their processing by assaying and mass fractions of gold by assay-atomic absorption method. Irgiredmet (2004).
11. S.I. Ivannikov, D.G. Epov, G.F. Krysenko, M.A. Medkov, S.Yu. Bratskaya, A.A. Yudakov, "Vestnik ESD FED RAS", **5** 1 (2013).
12. Y. Xu, M.A.A. Schonen Geochim. Cosmochim. Acta, **169** 279, (1995).
13. A.V. Taskin, Journal of Fundamental and Applied Science, **08** 1674 (2017).
14. E.I. Shamray, M.E. Shamray, The patent of the Russian Federation for utility model No.101942 (2010).

Preparability research on metasomatic ores and rocks of "Vostok-2" deposit

Luiza Samatova[1*] and *Elena* Shepeta[1]

[1]Mining Institute of Far eastern branch of Russian Academy of Sciences. Khabarovsk, Russia

Abstract. The research for examination of elemental composition of typical metasomatic rocks and ores have been carried out According to the data of mineral and petrographic analysis of twenty transparent sections in parallel and crossed Nicols, specific associations of minerals in ore of current production at "Vostochnyi" mine have been determined. Flotation testing on beneficiation of tailings of core geological samples has been conducted. Substantial redistribution of valuable mineral from easy ores to diluting rocks has been noted. The percent of greisen ore, with coarse scheelite grains, has become extremely rare and on the average is 2 - 2.5% of gross mass of scheelite-quartz ore. The appearance of skarn minerals relicts and, at times, areas of skarn minerals in scheelite-sulphide ores leads to formation, in the near-vein spaces, of scheelite-biotite, scheelite-amphibole, scheelite-piroxene, scheelite-chlorite and scheelite-mica complex associations. Impregnations of scheelite have been found in quartzous rocks, the composition of mother rocks may be diverse (skarn, skarnified or biotite hornfels, quartzites). The number of dispersed single scheelite grains in rock and anhedral grains in near-vein spaces adjacent to veins with high scheelite concentration has grown. The content of $CaCO_3$ which is conditioned by the presence of Ca-containing minerals in the composition of skarn in the shape of thin (fine) grain crystalline impregnation along the crevices and cleavage faces with formation of thin microveinlets, often together with quartz has increased. The calculation of expected extraction has evaluated the drop of scheelite extraction from feed of Primorskaia ore-processing plant in saleable concentrate of down to 3 %. **Key words**: elemental composition, minerals association, idiomorphic and xenomorphic crystals, the calculation of expected extraction by ore types, flotation

1 Introduction

Scheelite skarn deposit — "Vostok-2" — refers to the main genetic type of tungsten deposits widely mined in Russia. In global balance, these deposits account for about 65% of ore production, whilst reserves amount to not more than 30% (S.G. Solovev, 1997) [1]. Ore bodies of the deposit are represented by steep layer-by-layer pods situated along junctions of marbles and metamorphized diabase porphyrites [2].

* Corresponding author: samatova_luiza@mail.ru

Currently, in the global preparation practice, the expansion of resource base of tungsten takes place towards the searches and mining of deposits which are large by reserves, but with low content of WO_3 0.15 - 0.20% [King Island (Australia), Ferbertal (Austria), Salo (France), Kti-Teberd (North Caucasus), Kayrakty, Koktenkol (Central Kazakhstan), Skrytoe (Russia, Primorskii krai)], the involvement of dradge rock dumps and residue formations [3-9].

With the reduction of the content of valuable minerals in ore, particularly topical is the question of the efficiency of preparation and guarantee of raw material processing feasibility. Feasible mining of new low-grade deposits and residue formations may be reached by way of combined processing flow sheets with preliminary concentration of valuable components in the head ore with the use of radiometric methods of separation [10-13]. However, currently scheelite-containing ores are processed at national plants as per traditional flotation technology where selective extraction of scheelite from finely disseminated ores is a difficult process task [14-17].

In order to define perspective directions with the increase in efficiency of main processes of flotation flow sheet for preparation, detailed study of elemental composition of ores, rocks and preparability evaluation are required.

2 Research materials and methods

In order to solve the task set, samples of ore, rocks and material from tailing of geological mapping cores of lower deposit levels have been selected.

Twenty transparent sections in parallel and crossed Nicols were studied by means of mineral and petrographic method; main associations of mineral, fineness and shape of scheelite grains were figured out.

Flotation benefication of ores and rocks were tested on samples weights with mass of 1 kg. Time of size degradation were defined via distribution of scheelite in flotation grain-size class. The extraction of scheelite in the class 80+0 μm corresponded with 84%, including in sludge (-44 μm) 16%.

Flotation tests have been performed using tap water in open and locked cycles.

Basic flotation scheme included sulphide and scheelite cycles, reagent schemes of bulk cycles were constant in all tests of given research. Main reagents of flotation sulphide cycles: sodium silicate (use rate (C)-5%), dialkyldithiophosphate IMA-I413p (C-1%), pine oil, sodium carbonate (C-10%), sodium oleate of commercial acid B-115 (C-5%).

Recovery (desorption cycle) of rough concentrates was performed by Petrov's method. The solid/liquid ratio was maintained as 1:1 in steaming, steaming duration in sodium silicate medium at the temperature of 80 - 85 °C was 60 minutes. Selective flotation was performed after dilution with fresh water with poor flow of collectors.

3 Experiments results and discussion

Ore types defined, in current stage of deposit mining, have close positional connection, form gradual transitions of ones into others, and it is nearly impossible to outline larger blocks by types of ores and their selective mining.

Intervals of variations of thickness of various types of ores and rocks and contents of main valuable scheelite mineral (see table 1) are represented by core material (selection by 380 well-bore samples).

Materials with the range of content of WO_3 1.93 - 0.03% was studied for preparability in laboratory environment.

The largest values for extraction were reached on quartz metasomatites and scheelite-quartz-sulphide ore, the range of content is 0.08(0.25) - 1.93%. Scheelite was in these ores as local impregnation in quartz veins, quartz areas of hedenbergite skarns, on the boundary of quartz-sulphide. Less often, scheelite may be found in sulphides where grain size is less than 0.03 - 0.2 mm, their shape is more roundish as if with melt-off corners and edges. Sometimes, big crystals of scheelite are crevassed; sulphides, quartz and calcite develop in crevices. Scarcely ever, but limestones lenses may be met in scheelite-quartz-sulphide ore. Then, the content of calcite rises from 5(8)% to 45% which causes the complication of selection conditions and the drop of scheelite extraction into saleable concentrate from 92(90)% to 67%. In current production ores, percentage of rocks (biotite metasomatites, granodiorites, sandstones, quartzites) grew with the increase of scheelite content. In testing for preparability of biotite metasomatites with the content of scheelite of 0.8 - 1.9% (the content of calcite — 1 - 3.5%), high level of extraction into saleable concentrate of 86.5 - 91% is reached. The reduction of contents of down to 0.11 - 0.37% led to the extraction drop of down to 35 - 75%.

Table 1. Distribution of ore and rocks in geological mapping cores

Types and subtypes of ores and rocks	**Layer thickness, m**	**WO_3 content, %**
	Value, min-max	
Scheelite-quartz and scheelite-sulphide ore	0.4 - 9.5	0.2 - 4.5(25*)
Vein-disseminated ore and rocks, including:		
Pyroxene skarn	0.6 - 15	0.10 - 1.5
Skarnified hornfels	1.2 - 14.5	0.02(0.17) - 1.0
Biotite hornfels	0.8 - 18.1	0.01(0.11) - 2.0
Quartz metasomatites	1.0 - 11.8	0.02(0.14) - 0.9
Limestones	0.5 - 7.5	0.01 - 0.08
Granodiorites	1.2 - 2.5	0.01(0.15) - 0.5
Note: * Greisen ore present.		

The increase of scheelite percent in rocks is connected, on one side, with the growth of number of favourable easy scheelite-sulphide veins with idiomorphic crystals. On the flip side, the quantity of dispersed single scheelite grains in the volume of rocks and anhedral grains in near-vein spaces adjacent to veins with high scheelite concentration, where complex associations of scheelite-biotite, scheelite-amphibole, scheelite-piroxene originate, grew up. Ore feed supplied to Primorskaia ore-processing plant for preparation has its valuable mineral percent redistributed by types as follows: from rich easy scheelite-quartz (skarn)-sulphide ore with 80% (the year of 2004) to 60 % (the year of 2017), into rocks accordingly from 1% to 25%.

The results of testing of clean types of ores and rocks for preparability are represented in fig. 1.

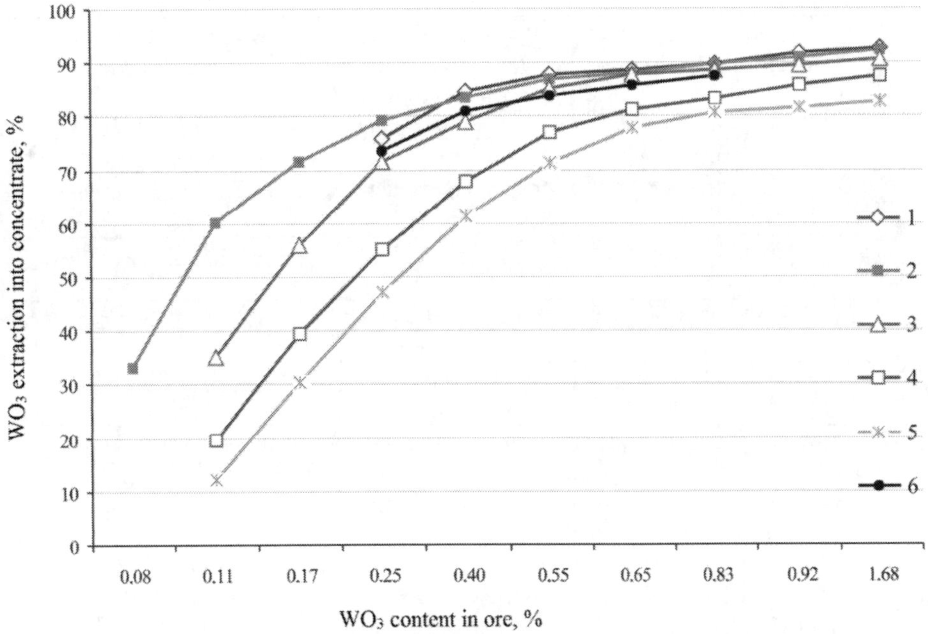

Fig. 1. The dependence of extraction of scheelite into saleable concentrate on the content of WO_3 in various types of ores and rocks: 1 – scheelite-quartz-sulphide ore; 2 – quartz metasomatites; 3 – biotite hornfels; 4 – skarns; 5 – skarnified hornfels; 6 – feed of Primorskaia ore-processing plant

The peculiarity of skarns and skarnified hornfels on lower levels of deposit mining is an increased content of calcites. Limestones lenses are met seldom, at that the degree of their marmorization is grown. The increase of $CaCO_3$ content is conditioned by the presence of Ca-containing minerals in the composition of skarns in the shape of thin (fine) grain crystalline impregnation along the crevices and cleavage faces with formation of thin microveinlets, often together with quartz. The extraction of scheelite and obtainment of saleable concentrate significantly depends on the content of calcite in ore.

Scheelite mineralization in skarns is extremely non-uniform, basically poor, disperse content of WO_3 0.05 - 0.5%, among which areas with thickness of 1 - 3 meters with rich scheelite impregnation and the content of WO_3 1 - 5%, seldom to 7 - 8%, limited to intensively sulphidized and quartz zones, quartz-carbon-bearing veins and their vein walls, occur. Scheelite impregnation is 0.05 - 0.2(1.0) mm, idiomorphic crystals. In altered rocks (skarns, skarnified hornfels, quartzites) along veins and around local areas, scheelite occurs in the form of anhedral grains (the ratio between idiomorphic and xenomorphic scheelite — 10:1) with size of 0.01(0.02) mm to 0.1(0.2) mm. The range of scheelite extraction from skarns and skarnified hornfels of 0(11.5) - 88% is conditioned by the ratio of the contents of two Ca-containing minerals: scheelite and calcite.

The distribution of scheelite by ore types in the feed of Primorskaia ore-processing plant is represented in table 2.

Table 2. The distribution of scheelite and calculation of expected extraction

Ore types	**shl-q, shl-q-s shl-q-($P_x$skarn)-s**	**$P_x$skarn, skarnified (amf) hornfels, q metasomatite**	**bt metasomatite, granodiorite, sandstone**	**Feed**
Period of analysis	Output-content-distribution / expected extraction., %	Output-content-distribution / expected extraction., %	Output-content-distribution / expected extraction., %	Content, / extraction, %
09.2017	15.26 - 3.95 - 74.16 / 93.1	39.56 - 0.33 - 17.54 / 71.6	46.18 - 0.14 - 8.30 / 56.5	0.75 / 86.3
08.2017	17.09 - 2.96 - 75.48 / 92.9	40.54 - 0.26 - 15.50 / 60.6	42.37 - 0.14 - 9.02 / 58.7	0.67 / 84.8
02.2017	16.75 - 2.43 - 61.10 / 92.2	40.12 - 0.25 - 15.18 / 59.1	43.13 - 0.37 - 23.72 / 69.2	0.67 / 81.7

Note: shl-q-s – scheelite-quartz-sulphide ore; shl-q-($P_x$skarn)-s – scheelite-quartz-(skarn-)sulphide ore; $P_x$skarn – pyroxene skarn; skarn (amf) – hornfel-skarnified hornfels; bt metasomatite. – metasomatite-biotite hornfels; q metasomatite – quartz metasomatites.

The increment of scheelite distribution into diluting rock from main types of ores to 14.7% in February in relation to August, 2017 conditioned the drop of the expected extraction from feed with the WO_3 content of 0.67% into finished concentrate by 3.1% (from 84.8% to 81.7%).

4 Conclusions

The drop of scheelite extraction in current production ore from "Vostok-2" deposit is conditioned by several factors:

1. Greisen ore, with coarse scheelite grains, has become extremely rare and on the average is 2 - 2.5% of gross mass of scheelite-quartz ore. The percentage of easy scheelite-quartz and scheelite-quartz-sulphide ores has reduced with growing redistribution of scheelite into diluting rocks.

2. The number of dispersed single scheelite grains in the volume of rock and anhedral grains in near-vein spaces adjacent to veins with high scheelite concentration, where complex associations of scheelite-biotite, scheelite-amphibole, scheelite-piroxene originate, has grown up.

3. The appearance of skarn minerals relicts and at times areas of skarn minerals in scheelite-sulphide ores which leads to formation, in the near-vein spaces, of scheelite-piroxene, scheelite-amphibole, scheelite-chlorite and scheelite-mica complex splices has been noted. Impregnations of scheelite have been found in quartzous rocks, the composition of mother rocks may be diverse (skarn, skarnified or biotite hornfels, quartzites).

4. The content of $CaCO_3$ which is conditioned by the presence of Ca-containing minerals in the composition of skarns in the shape of thin (fine) grain crystalline impregnation along the crevices and cleavage faces with formation of thin microveinlets, often together with quartz has increased.

References

1. V.I. Gvozdev. Irkutsk, Publishing house of Institute of geography of Siberian division of Russian Academy of Sciences, materials of scientific conference, **1**, 33-36 (2005).
2. V.I. Gvozdev. Vladivostok, Dalnauka, 31 (2010).
3. V.I. Gvozdev. Vladivostok, Dalnauka, 630-637 (2006).

4. V.I. Gvozdev, Vladivostok, Dalnauka, 87-97 (2004).
5. L.A. Samatova, V.I. Gvozdev, L.A. Kienko and others. Pacific Geology journal, Khabarovsk city, **6**, 83-96 (2006).
6. G.N. Stepanov, V.I. Gvozdev. New data on mineralogy of Far East, Far Eastern Russian Academy of the USSR, 49-55 (1987).
7. G.N. Stepanov. Mineralogy, petrogeny and genesis skarn-scheelite-sulphide deposits of Far East, M.: Nauka, 178 (1977).
8. G. N. Stepanov. Mineralogy and geochemistry of tungsten deposits, L.: Leningrad State University Publishing House, 234-242 (1971).
9. L. A. Samatova, E. D. Shepeta, V. I. Gvozdev. Physico technical problems of development of commercial minerals, Novosibirsk city, No.3, 152-161 (2012).
10. K. Nienhaus, T. Pretz, H. Wotruba, Sensor Technologies: Impulses for the Raw Materials Indusrty, Shaker Verlag, Aachen (2014).
11. T. Pretz, H. Wotruba, K. Nienhaus, Applications of Sensor-based Sorting in the Raw Material Industry, Shaker Verlag, Aachen (2011).
12. A. Haslam, Developments in the Tungsten Industry - Australia. 21-st ITIA Annual General Meeting, Xiamen, China (2008).
13. Report: "Experience with a sensor based sorting plant separating coarse scheelite ore" ITIA's 23rd Annual General Meeting, Vienna, Austria (2010).
14. L. A. Barskii, O. V. Kononov, L. I. Ratmirova. Selective flotation of calcium minerals, M.: Nedra, 232 (1979).
15. O.V. Konov, D.G. Koshut, I.V. Nesterov, Non-uniformity of minerals and nature of restricted isomorphism of scheelite-powellite, in the book of "Non-uniformity of minerals and crystal growth". M.: Nedra, 128-134 (1980).
16. A.A. Abramov, Kh. Khobert, Non-ferrous metals, **2**, 26-34 (2008).
17. E.D. Shepeta, L.A. Samatova, V.A. Ignatkina. Ore beneficiation., **3**,41-48 (2017).
Research have been completed with the support of Russian Foundation for Basic Research grant No.17-05-00241\18".

Skarn-scheelite ores beneficiation stimulation using ultrasonic treatment

Elena Shepeta[1*] and *Luiza* Samatova[1]

[1]Mining Institute of Far eastern branch of Russian Academy of Sciences Khabarovsk, Russia

Abstract. The impact of acoustic vibrations on physical and chemical properties of circulating water, heterogenic systems of pulps of sulphide and scheelite cycles flotations, collectors (IMA-I413p (potassium butyl xanthate), sodium oleate of commercial acid B-115) have been studied on skarn scheelite-quartz-sulphide ore of Vostok-2 deposit. Ultrasonic treatment has been performed using UZDN-2T apparatus. Working frequency 22 kHz, absorbed acoustic power average for radiator Pak = 46 W, sound intensity average I_{us} = 1 W/cm^2 (rated I_{us}= 2-3 W/cm^2). During all experiments, ultrasonic treatment time was constant — 5 minutes, insonation solutions volume — 250-100 ml, solutions heating temperature in the course of insonation — 25-35 °C. The flotation tests patterns, reagent scheme are approximate to process regulation of Primorskaia ore-processing plant processing ores of given deposit. Utmost increment of extraction in case of collctors ultrasonic treatment has been reached on dialkyldithiophosphate (IMA-I413p), the increment of extraction of chalcopyrite into bulk sulphide concentrate 0.98 %. The best results of testing of ultrasonic treatment impact on scheelite ores beneficiation have been obtained in case of ultrasound impact on circulating water. In consequence of stimulation of precious minerals flotation increment of extraction of chalcopyrite into copper concentrate by 1.4 % and scheelite into bulk rough concentrate by 2.5 % has been obtained.

1 Introduction

Despite the fact that ultrasonic treatment of ores and metals spread already in 60-s of the previous century, up to this day it is considered new and advanced method for future development of mineral deposits beneficiation [1-4]. Such state of things is conditioned by the absence of industrial designs of inexpensive high-performance reliable ultrasonic units. Currently, prevailing are ultrasonic baths for items treatment and the acceleration of physical and chemical processes in liquids. Main manufacturers of laboratory ultrasonic units, in domestic industry, are "Aleksandra-Plius", LLC, units have been supplied to Moscow Institute of Steel and Alloys, "Central research geological-prospecting institute of non-ferrous and noble metals" Federal State Unitary Enterprise, "GVINTSVETMET" Federal State Unitary Enterprise, "Uralmekhanobr" OJSC and etc [5]. Industrial jump in modern technologies and equipment development allows suggesting intensive development

* Corresponding author: Elenashepeta56@mail.ru.

as well as in the field of creation of industrial ultrasonic units what actualizes the researches in the field of ore beneficiation using ultrasonic treatment [6].

2 Experimental technique

Testing of ultrasonic treatment impact on efficiency of beneficiation of scheelite and chalcopyrite on ores of Vostok-2 deposit, in given research, was conducted on sample weights of 125 g. Size degradation was carried out in "Pulverisette-5" planetary mill: solid mineral mass/ball load/liquid substance ratio – 1:3.7:0.8, ball diameter — 20 mm, time of size degradation — 7 minutes, content of flotation class in flotation feed — 73 %, distribution of WO_3 – 80.6 % (including sludge 16.5 %), Cu – 84,7 % which correlates to data of size degradation of one kilogram of sample weights in a ball mill.

Ultrasonic treatment of skarn scheelite-sulphide ore was performed using UZDN-2T apparatus, the radiator of which allows ultrasonic treatment of small quantities of mineral media (to 350 ml). Working frequency 22 kHz. Absorbed acoustic power average for radiator Pak = 46 W, sound intensity average I_{us} = 1 W/cm^2 (data obtained based upon performed researches by mineral deposits beneficiation laboratory). Rated approximate values for the device intensity — I_{us}= 2-3 W/cm^2.

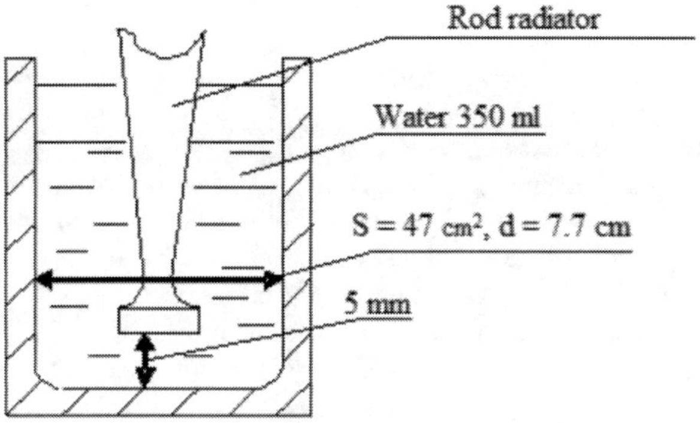

Fig. 1. Exponential rod concentrator (design A)

Exponential rod concentrator with a disc on the end is designed for treatment of media in laboratory breakers with the maximum capacity of 400 ml, ultrasonic vibrations are coupled in the breaker from the top.

During all experiments, ultrasonic treatment time was constant — 5 minutes, insonation solutions volume — 250-100 ml, solutions heating temperature in the course of insonation — 25-35 °C.

Typical pattern of copper-sulphide and scheelite cycles. Main reagent of sulphide cycle: sodium silicate, aeroflot IMA-I413p, pine oil, activated carbon, trisodium phosphate; of scheelite cycle: sodium carbonate, sodium silicate, sodium oleate of commercial acid B-115. The consumption of reagents for bulk cycles were constant in all experiments and corresponding to reagent scheme of Primorskaia ore-processing plant. During finishing cycles, the consumption of reagents was being defined by the quality of rough concentrate.

3 Results and discussion

In case of ore beneficiation, acoustic vibrations obtained most wide spread during treatment of flotation reagents (apolar oils, fatty acids, amines and etc.) which are hardly soluble in water. In production, ultrasound emulsification of oleic acid, tall oil, kerosene, aeroflot, naphthenic acids which allows stimulating the process of reagents dilution in the volume of pulps in case of decrease in reagents consumption by 11-40 % and increasing in extraction of precious minerals (Dzhezkazganskaia, Balkhashskaia, Almalykskaia, Gaiskaia, Akchatauskaia plants) is used [7-10].

Increase in efficiency of flotation of titanium-zirconic sands, gold from ledge ores, coal, cinnabar, antimonite and selection of lead-pyritic and copper-pyritic concentrates under the impact of ultrasound on heterogenic systems of flotation pulps was noted [11-14]. Insonation of bulk sulphide flotations pulps within 5-20 minutes abruptly decreases flotation response of galena, bornite pyrite, arsenopyrite with retention of flotation of sphalerite, chalcopyrite and pyrrhotite. The difference of kinetics and desorption of collectors from mineral surfaces enables increasing coefficient of minerals selectivity process [13].

Based upon research findings received during the beneficiation of carbonate-fluorite raw materials, ultrasonic treatment of water coming for dilution of froths of cleanings allows reducing the quantity of cleaning operations by means of the growth of selectivity of separation process and increasing fluorite extraction by 7-8 % [15-16].

Based upon theoretical research in this field, main factors influencing flotation process activation were outlined: processes of surface cleaning, cracks opening, splices destruction, structure rupture with caving, microflaws, cavity processes going on in liquid and reinforcing on the boundary of liquid and solid phases, alternation of chemical and phase composition in near-surface layer [17].

Stimulation of beneficiation is one of directions in increasing feasibility of mineral raw materials flotation processing patterns, it is especially important while precious minerals content is decreasing in head ore.

For plant finishing field mining, the increase in feasibility of mining is connected with the stimulation of flotation processes by means of ore feed, reagent schemes, process water treatment.

Goal of research performed: determine main promising trends of ultrasonic treatment application with Vostok-2 field tungsten ores beneficiation.

Results of chalcopyrite beneficiation ability in sulphide cycle in case of ultrasound impact on liquid phase of flotation, process (circulating) water and reagents are represented in Table 1

The best result of chalcopyrite flotation response increase has been obtained with circulating water treated with ultrasound.

It was noted in the previous research that ultrasound impact changes physical and chemical properties of water: pH, electrical conductivity, reduction-oxidation potential, xanthate ions, cyanic compounds and other flotation agents break up, sulphide reagents electrolytic potentials change. With the intensity of acoustic vibrations of 10 W/cm^2 and frequency of 22 kHz and the duration of water treatment of 0.5-10 minutes, the extraction of chalcopyrite increases by 12 % [18].

The dependence of pH change and temperature on time of ultrasonic treatment of tap and process water is given in fig. 2.

In our experiments, insonation of circulating water containing 8.7 g/l of suspended solids, with volume of 250 ml with the intensity of acoustic vibrations of 1.0 W/cm^2 and frequency of 22 kHz within 5 minutes (see fig. 2) led to the reduction of water pH by 0.19 (from 8.19 to 8.0) with insignificant increase in temperature from 20 °C to 26 °C. The clarification of insonated water for a day amounted to 28 %, by initial sample — 36 %. Ultrasonic treatment leads to the enhancement of colloidal solution disability. With the reduction of insonated water volume down to 100 ml, as a consequence of the growth of intensity of impact on water, the reduction of pH by 0.59 and the degree of clarification to 19 % was obtained.

Table 1. Results of chalcopyrite beneficiation in sulphide cycle using ultrasonic treatment of water medium and collectors

Item of insonation	Sulphide concentrate			Sulphide middlings			Sulphide flotation tailings	
	Output %	Copper, %		Output %	Copper, %		Copper, %	
		Content	Extraction		Content	Extraction	Content	Extraction
Without ultrasonic treatment								
H_2O tap water.	14.16	1.32	81.15	8.27	0.30	10.77	0.024	8.08
H_2O circulating water	13.19	1.38	79.43	7.16	0.36	11.21	0.027	9.36
Ultrasonic-treated (volume 250 ml, time of insonation - 5 minutes)								
H_2O tap water.	13.29	1.38	79.76	9.78	0.28	11.89	0.025	8.35
H_2O circulating water.	12.82	1.48	80.84	5.65	0.42	10.31	0.025	8.85
Pulp of flotations	14.47	1.31	82.05	5.84	0.30	7.59	0.030	10.36
IMA-I413p	13.16	1.44	82.13	4.97	0.50	10.78	0.020	7.09
IMA-I413p+potassium butyl xanthate	14.09	1.33	81.66	4.25	0.59	10.86	0.021	7.48
Note: content in ore: WO_3 – 0.72 %, Cu – 0.23 %, As – 0.043 %, S – 4.16 %, Au – 1.0 g/t, Ag – 3.3 g/t.								

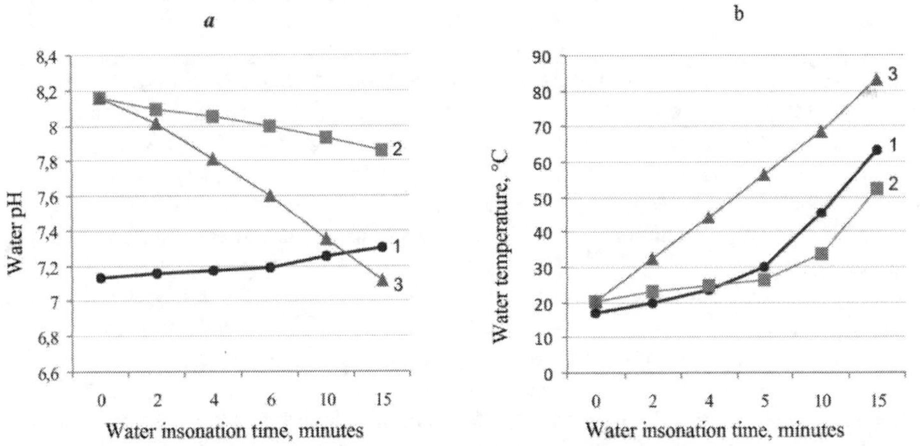

Fig. 2. The dependence of pH (a) and temperature of water (b) on time of ultrasonic treatment: 1 – tap water (250 ml), 2, 3 – circulating water (250 and 100 ml)

In case of using ultrasonic treated circulating water (V – 250 ml, t – 5 minutes) in the process of flotation, the increase in extraction of chalcopyrite into sulphide concentrate by 1.41 % was achieved. The drop of the activity of arsenopyrite flotation in bulk cycle by 4.6 %, pyrrhotite – 3.1 %, was noted.

Saleable concentrate was obtained in copper selection with the increment of extraction by 1.36 %, at that the reduction of losses with copper tailings of selection — 5.96 % which should lead to greater increment of copper extraction in experiments of locked cycle tests.

Insonation of aeroflot IMA-I413p let to reduce losses of chalcopyrite with tailings of bulk cycle by 0.98 %, with treatment of the mixture of aeroflot reagents and butyl xanthate losses reduced by 0.51 % which proves data of partial breaking of xanthate.

Data on ultrasound impact on scheelite flotation are given in Table 2.

Table 2. Values for beneficiation of scheelite from ore using acoustic vibrations

Item of treatment	WO_3 concentrate of basic cycle			Steaming feed			WO_3 flotation tailings	
	Output, %	WO_3, %		Output, %	WO_3, %		WO_3, %	
		Content	Extraction		Content	Extraction	Content	Extraction
Without ultrasonic treatment								
H_2O tap water.	19.59	3.78	94.85	17.95	4.05	93.17	0.050	5.15
H_2O circulating water	27.61	2.62	92.70	16.03	4.35	87.88	0.08	7.30
Ultrasonic-treated								
H_2O tap water.	21.85	3.30	95.01	19.90	3.63	92.61	0.050	5.00
H_2O circulating water	22.35	3.28	95.16	15.45	4.65	93.37	0.048	4.84
Pulp of flotations	20.23	3.65	95.86	10.70	5.32	73.85	0.040	4.14
Sodium oleate	18.57	3.93	94.70	10.45	5.85	79.58	0.050	5.30

Ultrasonic treatment of circulating water led to the stimulation as well as for scheelite flotation. Significant increment of scheelite extraction of 2.46 % was obtained in bulk cycle. Increase in extraction is connected with the reduction of tungsten trioxide content in tailings of scheelite flotation from 0.08 % to 0.048 % with the reduction of rough concentrate output by 5.3 %.

Circulating water insonation let to obtain values of precious minerals extraction close to values on tap water.

Process water ultrasonic treatment is especially important for ore-processing plants during autumn-winter work period when substantial water quality reduction by suspended solids occurs. At Primorskaia ore-processing plant, concentrate is extracted from skarn scheelite-sulphide ore with WO_3 content of 0.56 % in summer (suspended solids in circulating water — 2.9-1.0 g/l) with the extraction of scheelite of 86.9 %, with concentrate quality of 53 % WO_3, with the increase of suspended solids up to 6.0 g/l accordingly 82.6 % and 52.8 % (losses – 4.3 %). With the increase of suspended solids in water up to 12 g/l the losses grow up to 6.5 %.

4 Conclusions

As a result of performed testing of skarn scheelite-sulphide ore from Vostok-2 deposit for beneficiation ability in case of ultrasonic treatment impact (working frequency 22 kHz, absorbed acoustic power average for radiator Pak = 46 W, sound intensity average I_{us} = 1 W/cm^2) it has been established that the greatest increment of precious minerals extraction is obtained with circulating water insonation. Increment of extraction of chalcopyrite into copper concentrate amounted to 1.4 % and scheelite into bulk rough concentrate — 2.5 %.

Best results for collectors ultrasonic treatment have been obtained for dialkyldithiophosphates (IMA-I413p), increase in extraction of chalcopyrite into bulk sulphide concentrate amounts to 0.98 %.

References

1. V.I. Revnivtsev Feldspars and quartz beneficiation. M: Nedra 250 (1970)
2. I.M. Babakov. Theory of oscillations. M., (1966)
3. E. Kikuchi. Ultrasonic sensors. M., (1968)
4. V.A. Glembotskii, M.A. Sokolov, I.A. Iakubovich and others Ultrasound in mineral deposits beneficiation. Almaty (1972)
5. N.M. Lebedev, G.I. Zhirnova, G.V. Sedelnikova and others Technology and equipment for beneficiation ultrasonic stimulation. Source book for IX Congress of CIS countries beneficiating companies, **1,** 338-34, (February 26-28, 2013).
6. V.A. Chanturiiaa. Electrochemestry of sulphides. Theory and practice of flotation. M.: Ore and metals, 272 (2008).
7. Iu.P. Eremin, G.A. Denisov, M.D. Shtern and others Concerning future solutions on using vibration and acoustic oscillations on flotation processes. Ore beneficiation, 3,24-28 (1981).
8. G.I. Eskin, I.I. Gurev, Iu.F. Soluianov. The impact of ultrasonic treatment of melting on properties of zink ingots and plates. Non-ferrous metals, 5, 69-71 (1981)
9. Z. Bonderek, et. al. Prafinaejes ultradźwięki odlewniczego stopu alum. AK 9. *Przeglad Odlewnictwa*, 7, 229-232 (1981).
10. B.A. Agranat. Research of acoustic cavitation erosive activity in organic solvents. Acoustic magazine, **29**, issue 5, 577-579 (1983).
11. A.A. Abramov. The technology of processing and beneficiation of non-ferrous metals ores. M., Publishing house Moscow State Mining University, **1**, 575,(2005).
12. V.A. Tsarkov, V.V. Dobroskokin. Current trends of the development of gold extraction from ledge ores. Mining magazine, **5**, 81-82 (2000).
13. N. Ghaffari. Scavenging Flotation Tailings using a Continuous Centrifugal Gravity Concentrator. The University of British Columbia, Vancouver, Augus. 167 (2004)
14. J.R. Walklate, P.J. Fourie, A history of gravity separation of Richards Bay Minerals. The Journal of the Southern African Institute of Mining and Metallurgy. **106**, 741-748 (November 2006).
15. L.A. Kienko, L.A. Samatova, O.V. Voronova. The impact of pulp ultrasonic treatment on flotation selectivity in carbonate-fluorite ores beneficiation. Promlems of comprehensive development of geological resources. Materials of V National scientific conference in Khabarovsk city, Mining information-analytical bulletin, Instalment 4, 172-179, (October 2-4, 2013).
16. L.A. Kienko, O.V. Voronova, Zh.A. Shagina. Increasing effectiveness of fluorite flotation based on pulp ultrasonic treatment application. Problems of comprehensive development of geological resources: Materials of VI National scientific conference with the participation of foreign scientists. Khabarovsk, **21**, 233-241 (05-07 October 2016).
17. N.M. Lebedev, O.V. Voronin ("Aleksandra-Plius", LLC), S.S. Lozhnikov and others The use of ultrasound for beneficiation stimulation. ("GVINTSVETMET" Federal State Unitary Enterprise) Materials of V Congress of CIS countries beneficiating companies, **3,** 225, (March 23—25, 2005).
18. N.K. Algenbranstova, A.V. Makshanin. Flocculation as a method of gold extraction from tailing resources. Journal of Siberian Federal University. Engineering Technologies.**3**, 283-295 (2011).

Influence of agitation leaching conditions to the behavior of gold in solution

Konstantin Prokhorov[1*]

[1]Mining Institute of Far eastern branch of Russian Academy of Sciences, Khabarovsk, Russia

Abstract. Deposition of gold ions in the process of agitation with a chloride-hypochlorite solution with 8.15 g / l active chlorine concentration in the model material was studied. Fresh chlorite-hypochlorite solution was used in the study. The effect of the initial pH of the medium on the dynamics of gold deposition is considered. It was found that up to 56% of gold precipitates in alkaline medium. Precipitation is not observed at pH<6. It is established that the ore under investigation has buffer properties and stabilizes the pH of the medium to weak-acid (pH = 5.5-6).

1 Introduction

Due to the growing environmental concern of the use of cyanide for gold processing, interest on the use of non-cyanide lixiviants especially with halides has been renewed recently. Despite the fact that cyanide is used commercially for gold extraction, studies are being conducted on alternative non-toxic leaching agents. Based on the works [1, 2, 3], the suitability of chloride systems for gold leaching from ores was determined. In the study [4], the authors found that the leaching of gold from the ore should be properly controlled to achieve a high rate of gold recovery and dissolution. In this case, sodium hypochlorite provides a higher reaction rate with gold compared to the same calcium hypochlorite.

In [5] was noted that hydrochloric acid and NaCl and NaOCl salts solutions significantly increase Cl_2 (Cl_3^-) solubility. In an aqueous HOCl–NaOCl–NaCl–HCl solution, when HOCl is consumed, its concentration is recovered by acidifying (as long as the solution keeps containing NaOCl)

$$NaClO + HCl \leftrightarrow HClO + NaCl \quad (1)$$

By contrast, in alkaline solutions, hypochlorites undergo decomposition, depending on pH, by the equation

$$2OCl^- \rightarrow 2Cl^- + O_2 \quad (2)$$

As a result, the authors note that chloride-hypochlorite solutions can be used to extract gold from gold-bearing industrial raw materials (ores, concentrates). The oxidation of gold by films of solid reaction products can be prevented by conducting the process at a sufficiently low pH.

* Corresponding author: kostyan1986_ne@mail.ru

All studies of the gold dissolution using oxidizers or complexing agents based on halides indicate the importance of controlling the pH of the solution to stabilize the gold complex. Otherwise, gold can be reprecipitated already after its dissolution from ore [6]. And this fact, in our opinion, is the main problem of the gold extraction process by chloride systems leaching.

2 Experimental

This study is an investigation of the behavior of dissolved gold in the process of agitation leaching from a model material based on the base ore of the Malmyzh deposit with a fresh chloride-hypochlorite solution. Previous studies have shown that in the process of gold extraction with a hypochlorite solution in close to neutral medium, the gold is reprecipitated from solution to the ore. Thus, gold is extracted from the ore, but does not remain in solution.

To studying the process of gold reprecipitation in pregnant solutions depending on the pH of the medium a model material was used. A fixed amount of a solution of gold ions of known concentration was added to the base ore of the Malmyzh deposit. Thus, reproducible initial conditions of the initial concentration of gold in the samples were created. For each experiment, a fresh chloride-hypochlorite solution of known composition and pH was taken. The preparation of the chloride-hypochlorite solution was carried out by NaCl electrolysis followed by the addition of hydrochloric acid. The pulp was pre-treated with ultrasound and leached in agitation mode.

At the first stage, the extraction was simulated using a solution of hypochlorite, previously adding gold solutions to the sample. In the process of leaching, the solid to liquid ratio was 1:4. This ratio was chosen to reduce the heating of the pulp during sonication. With a smaller of solid to liquid ratio, during the cavitation process, the pulp is heated up, the liquid is evaporated, chlorite-hypochlorite solution is degassed, and, as noted [6], the decomposition of dissolved gold is more pronounced at higher temperatures. Ultrasonic treatment was performed for 10 minutes. Then the sample has been placed in agitator for 4 hours. After the interaction of the chloride-hypochlorite solution with the model material, an aliquot was taken for analysis. A blank experience was also conducted to verify the absence of gold extraction from the base ore. After conducting a blank experiment, an aliquot of the liquid phase was taken from the pulp for analysis. The remaining volume of the liquid phase after the blank experiment was divided into portions and a solution of gold ions was added for comparison.

The second stage was the study of the behavior of gold ions as a function of the pH of the medium. Equal weighed portions were mixed with a chloride-hypochlorite solution having a different initial pH (3, 5, and 9). In a defined time intervals, 30 ml of the slurry was sampled and filtered. Gold content of the clarified solutions was analyzed by the atomic absorption method on AA-7000 (Shimadzu). According to the data on the amount of gold in solution at a certain time interval, curves were obtained.

3 Material

For the first stage of the study, a solution of gold ions was added to the samples of the test sample (base ore), simulating 0.8 g/t and 2 g/t gold in the ore. In parallel, a blank experiment was conducted without the addition of gold ions, adding it after leaching. Agitation was conducted with a fresh chloride-hypochlorite solution with an active chlorine 8.15 g/l concentration and 9.37 initial pH. The solution was prepared by electrolysis of 20 g/l sodium chloride solution by a Saner 3-50 unit.

For the second stage of the study, a solution of gold ions was added to the samples of the model material, simulating 2 g/t gold in the ore. The extraction was carried out with the same chloride-hypochlorite solution with 8.15 g/l active chlorine concentration and 3, 5 and 9 initial pH.

4 Results

4.1 Investigation of gold deposition in the process of agitation, depending on the initial conditions for the concentration of gold ions.

The prepared pulp with the model material was sonicated for 10 minutes, and then the process was carried out in agitation mode for four hours.

The conditions and results of the experiment to determine the reprecipitation of gold ions during the agitation process with hypochlorite solution are shown in Table 1. 100% gold extraction equals the added amount of gold ions to the base ore.

Table 1. Conditions and results of modeling experiment

№	Composition of pulp	Gold extraction, %	Final pH
1	Base ore+ Au 0,8 g/t	89,59	8,06
2	Base ore+ Au 2 g/t	62,45	7,52
3	Blank experiment	0	8,37
4	Blank experiment + Au 0,8 g/t	100,0	
5	Blank experiment + Au 2 g/t	100,0	

Data given in table 1 demonstrate that gold precipitates (up to 37.55%) in the agitation process with chloride-hypochlorite solution (No. 1, 2). In this case, the amount of precipitated gold is directly depended on its initial amount in solution. Gold added after the process of agitation with blank ore was completely extracted to the pregnant solution (100% extraction, No. 4, 5). The final pH of the medium was neutral or slightly alkaline. Therefore, the next step was to investigate the effect of the pH of the medium on the process of reprecipitation of gold by shortening the sampling intervals for analysis.

4.2 Investigation of the dependence of the gold ions deposition on the pH of the medium.

The prepared pulp with the model material was sonicated for 10 minutes, then the process was carried out in agitational mode for four hours. Aliquots were taken after ultrasonic treatment, and then every hour during the leaching process. The results of gold content in aliquots for different initial pH of the medium are shown in Figure 1.

At the initial stage, the quantity of added gold ions is 100% gold extraction. Further, during the agitation of pulps with an initial pH = 9, more than half of the added gold ions are precipitated and the extraction drops to 44%. In the experiment using a chloride-hypochlorite solution with 5 and 3initial pH, gold deposition is not observed, the extraction is about 100%.

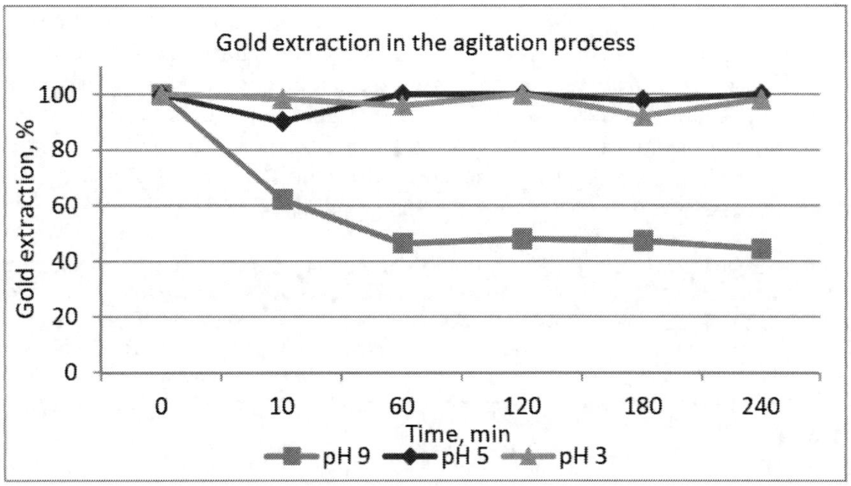

Fig. 1. Dynamics of the reprecipitation of gold ions from the solution as a function of the initial pH of the solution.

During the experiment, pH of the medium was also observed. The pH dynamics is shown in Figure 2. In the experiment with a chloride-hypochlorite solution with the initial pH = 9 during the agitation process, the pH drops to slightly alkaline medium, and at the last hour to neutral one. In experiments with a solution of hypochlorite with an initial pH of 5 and 3 during the agitation, the pH of the pulp has rise to slightly acidic (about 6).

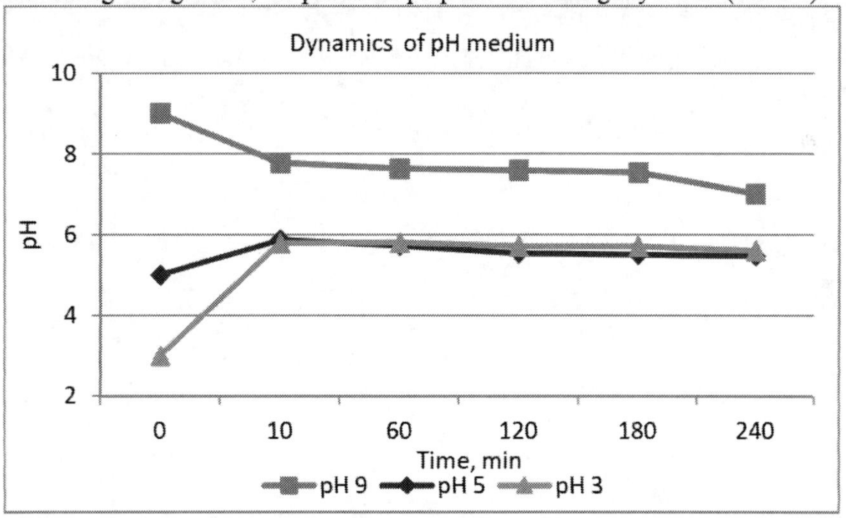

Fig. 2. Dynamics pH of the hypochlorite solution during the pulp agitation process.

The results of this research showed that the ore has buffer properties and tends to stabilize the pH level closer to the weakly acidic (pH = 5.5-6). At pH below 6, gold does not precipitate in the model material, and gold extraction increases. Acidic medium (pH<6) prevents gold precipitation from solution to the solid phase.

5 Conclusions

The process of reprecipitation of gold ions as a function of the pH of the leaching medium has been studied. The results of the study showed that at the initial alkaline pH of the medium, more than half of the gold ions (up to 56%) precipitated from the solution during the first hour of treatment. In this case, the amount of precipitated gold depends on its initial content in solution. When the process is carried out in acid medium, no precipitation of gold from the solution is observed. It is noted that the pulp has buffer properties and stabilizes pH of the medium to weakly acid (pH = 5.5-6). In the process of research of oxidized ore of the Malmyzh deposit it was concluded that it is necessary to carry out the gold leaching processes at 6-5 pH to prevent the precipitation of one.

The author is grateful to the staff of MI FEB RAS, who took direct part in carrying out experimental studies: A.G. Sekisov, T.G. Konareva, A.V. Rasskazova. Research was supported by Far Eastern branch of Russian Academy of Sciences project (18-2-015 "Gold and copper and porphyritic mineralization of the Far East: indicators of geodynamic situations, ore-bearing magmatism and mineralogical-geochemical features of ore and magmatic systems in connection with the solution of problems of the forecast, prospecting, assessment and technologies of development of fields of this type"

References

1. M. G. Aylmore. Developments in Mineral Processing. **15.** 501-539. (2005).
2. G.V.K. Puvvada, D.S.R. Murthy. Hydrometallurgy. **58.** 185-191. (2000).
3. T.A. Harlamova, A.F. Alaferdov, V.M. Bahir. Gornyiy informatsionno-analiticheskiy byulleten (nauchno-tehnicheskiy zhurnal). **3**. 108 – 117. (2015).
4. M. Baghalha. Int. J. Miner. Process., **82.** 1166-1171. (2007)
5. L. F. Kozin, V. A. Prokopenko and A. K. Bogdanova. Protection of metals, **41 (1).** 22-29. (2005).
6. Kwang Soo Nam, Byoung Hi Jung Jeon Woong An Tae Jun Ha. Int. J. Miner. Process. **86.** 131-140. (2008).

SUBSTANTIATION OF FIELD OF APPLICATION OF COMBINED (TRADITIONAL AND PHYSICAL-AND-CHEMICAL) GEOTECHNOLOGIES FOR THE DEVELOPMENT OF GRAVEL AND SECONDARY GOLD DEPOSITS

Artur Sekisov[1*], *Anna* Rasskazova[1], *Alexandr* Lavrov[2] and *Roman* Bogomyakov[1]

[1] Mining Institute of Far eastern branch of Russian Academy of Sciences, Khabarovsk, Russia
[2] Transbaikal State University, Russia.

Abstract. Problems of accumulation of technogenic tailings dumps and as a result of sluicing of gold sands of gravel deposits are considered in article. Additional recovery of fine and dispersed gold from a mineral mass of such technogenic formations can be carried out with use of physical and chemical geotechnologies. The factors influencing on the choice of a way of development of the gold gravel deposit are listed. The mathematical model of a gravel deposit is developed for the solution of the problem of multifactorial assessment of technological options of processing of gold sands. The scheme of the combined development of the gravel gold deposit is offered. Cuvette gold leaching experiments with the photo-electro activated solutions on various reagent basis were executed for the assessment of a developed scheme. Material for the researches was a gravity concentrate (schlich) fraction obtained by tailings gravity concentration (Kia gravel deposit, Transbaikal Territory). The first stage of gravity concentrate (schlich) processing included contacting with photo-electro activated carbonaceous solution ($NaHCO_3$). The first leaching reagent scheme (cyanide solution concentration of 1.5 g/l prepared on the basis of photo-electro activated the peroxide - carbonaceous solution provided 66.8 % gold extraction. The second reagent scheme (active chloride-carbonaceous solution) provided 88.9 % gold extraction.

1 Introduction

The Russian Far East or Far East Russia (FER) is host to a huge gold endowment and has produced more than 6500 t of gold, since the 1860s. Much of this gold has come from several mining districts: Aldan, Upper Amur, Lower Amur, Okhotsk, Allakh-Yun, Yana-Kolyma, Priokhotie, Omolon, and Chukotka. These districts include several gold deposits, mostly of orogenic and epithermal nature, as well as large and very large alluvial placer

* Corresponding author: sekisovag@mail.ru

deposits [1]. The Priamur province is one of the main gold-bearing provinces of Russia with the level of gold ore (about 180 tons) and placer (about 1132 tons) gold [2].

The effectiveness of development of large-scale gravel gold deposits is defined by the establishment of efficient field of use of open, underground and physical and chemical ways of their development. It is caused by the essential variation of mining-and-geological parameters in the plan and for depth of the deposit. Some of large-scale gravel deposits can extend to tens of kilometers. An example of such object is an alluvial in a Kruchina river valley (Transbaikal Territory). Development of this alluvial deposit lasts for more than 100 years (with breaks) with use of dragging, bulldozer ways of open-cast mining and locally underground development is applied. Nowdays one site of this alluvial deposit is planned to development by well (in-situ) leaching. Depth of high-grade layer of gold sands is 18 m. Secondary mineral formations (tailing dumps and tailings of gold sands sluicing) are accumulated in the result of a long-term development of gravel gold deposits. Residual content of precious metal can be comparable with gold content in sections of the deposit near to cutoff grade. In these mineral formations gold is presented by fine-grained, nano-dimensional and encapsulated mode of occurrence. A range of micro-analytical and microbiological techniques is applied for the research of thin gold-rich gravel within bedrock gravity traps, hosting elongated and flattened placer gold particles coated with manganese-, iron-, barium- (Mn-Fe-Ba) oxide crusts. These crusts host abundant nano-particulate and microcrystalline secondary gold, as well as thick biomats [3]. Post-mining placer deposits include tailings, overburden rocks, and non-mined spaces over the mining area (polygon) sides and in its bottom, and all this is a rock mass with its gold grades ranging from 0.12 to 0.26 $g/m(^3)$ to be all reprocessed using modern equipment and advanced technologies. Non-extracted gold differs by its morphology ranging from fine and rather thin platy to coarse native and quartz-aggregated type [4]. Reprocessing of mineral mass of secondary mineral formations, as well as development of deep-laying sites of a gravel deposit and between-pit pillar can be economically, efficiently and ecologically safely carried out with use of physical and chemical geotechnologies.

Sands of gravel deposits are an important source of gold and other valuable heavy minerals, in particular, of an ilmenite, titanomagnetite, zircon, monocyte and rutile [5]. The combination of a gravitational method of mineral processing with amalgamation is traditional for small deposits. Mercury, a dangerous pollutant, is the preferred method employed by small-scale miners for gold recovery. Gold extraction using mercury is comprised of the following four stages: (1) amalgamation, (2) separation of amalgamation, (3) removal of excess mercury, and (4) burning of the remaining amalgam to produce a gold sponge. Mercury can be released into the environment at each stage, which makes the promotion of mercury-free alternatives imperative [6]. It is also possible to reach increase in effectiveness of gravitational processing of gold sands by means of preliminary reagent processing of mineral mass [7].

Therefore, the choice of a way, an order and parameters of development of sites of an alluvial deposit with various conditions of bedding of gold sand layers, their geometrical and structural parameters, has to be solved together with problems of development of secondary mineral formations, and taking into account a time factor.

2 Mathematical modeling of mining operations (variation of the main ways and parameters of development of alluvial deposit sites, gold sands processing and technogenic mineral mass)

Estimation of quantity and quality of reserves of productive gold sands (by unproven and/or operational blocks) is necessary for the choice of a way of development of sites of the gold alluvial deposit and/or field of use of combined development. Multifactorial calculations of parameters of excavation (open pit or cuvettes) can be made taking into account their possible combination with physical and chemical geo-technologies or the subsequent use when involving in development of refractory and hard-cleaning sands. The modern mathematical models of deposits are formed as the numerical matrixes setting space and qualitative parameters of geometrized mineral bodies and bearing strata. Geometrical and qualitative parameters of separate components can be set as initial parameters. Coordinates of wells with the interval indication of the useful and harmful elements concentration can be set as initial parameters too. On the basis of initial data, borders of the homogeneous sites of mineral bodies are automatically defined. Considering not determined character of geological data, it was offered to map divide contours in inter-well space on the basis of expected and statistical determination of coordinates of transitional (between high-grade and low-grade) mineralized zones. This method was designed in Moscow Geological Prospecting Institute (MGPI) in 1983 [8]. The main geostatistical method of outline (drilling out) is Gaussian Process Regression Model. One dividing line between projections of wells with the shift from the center is mapped according this method. Two dividing lines are mapped according MGPI method. The contour of the expected transition zone with the corresponding order of metal content is defined. Values of gold content for the main boundary zones are replaced from concrete values to statistically average for the corresponding category of alluvial sand. Besides the formal selection of zones on value of content of gold (grade) in them characteristics (morphology and the gold particles size, existence of dispersible forms of finding of gold, the sand defining its extraction at enrichment technogenic (technogenic transformed) raw materials (or leaching), existence in it of the accompanying industrially valuable components is considered mineralogical-and-geochemical [9]. Assessment of stocks of secondary deposits is an important point in the choice of technology of development. There are indirect methods of assessment of stocks of technogenic resources, providing essential simplification of calculation of stocks, thus simplifying technology of development of secondary deposits [10].

For the solution of the problem of multifactorial assessment of effectiveness of technological parameters of production and processing of gold sands, the mathematical model of the gravel deposit has to be spontaneous and autocorrected. For this purpose, the initial information which is based on data of detailed geological prospecting, in process of drilling and approbation of wells, has to be renewed periodically. At the same time comparison of the preliminary and earlier specified contours of the industrial mineralization is carried out. Correction of industrial mineralization contours is carried out. The histogram of gold distribution in different fractions is fulfilled. Modes of gold occurrence (free, encapsulated and dispersible gold) are taken into account too.

Setting of rational sequence of mining operations and parameters of pits (cuvettes) can be provided by mathematical models. Mathematical models include economic, and technological criteria, directly or indirectly reflecting the current value and future gold prices. Retrospective dynamics of the prices, an order and parameters of formation and secondary processing of natural and technogenic mineral material have to be considered in the modern computer models.

Duration of development of large-scale and average gravel deposits, variation of morphological-structural and real parameters of productive layers, specification of volumes of overburden rock and low-grade sands should be considered in the process of modeling. As a result, different positions of front of strip mine; variant of productive blocks (with various morphological-structural characteristics and qualitative characteristics of sands) development; ways of gold sand and tailings processing are obtained. Criterion of the choice of efficient option of development of mining operation is the quantity of extracted gold (from primary and secondary sources). The maximal size of gold production ($\boldsymbol{P_{Au}}$)

index will define an optimal variant. This criterion is compound and is expressed by the following formula (1):

$$P_{Au} = \sum_{i=1}^{N} \sum_{j=1}^{M} h \cdot S \cdot \bar{c}_{ij} \left[K_{ore} K_{min} K_{proc} \cdot \frac{P_{yj}}{E \cdot (1 + q_s \cdot n \cdot Z_q)} + (1 - K_{min} K_{ore}) K_{ml}^{\ y} \cdot T \cdot K_{extr} \cdot \frac{P_{yj}}{E'} + \right.$$
$$\left. + (1 - K_{ore} \cdot K_{min} \cdot K_{proc}) \cdot K_{ml}'^{\ y} \cdot T \cdot K_{extr}' \cdot \frac{P_{yj}}{P''} \right] \rightarrow \max,$$

where i, j – the index of the ore blocks and a type of the valuable component in alluvial sand respectively;
N and M – the number of the indexed ore blocks and types of a valuable component respectively;
h, S – height of a bench (terrace), m and the area of the ore block, sq.m;
$\bar{c}_{ij}$ – *j*-component content in *i*–block (g/m^3);
K_{ore}, K_{min}, K_{proc} –ore-bearing coefficient, coefficient of mining extraction and processing extraction (and/or gold leaching from the remained pillar between pit or cuvettes) respectively, as a decimal fraction;
P_{yj} – cost price (expected) of j–component in y–year, rub;
E, E', E" – overall expenses of production for balance reserves, off-balance alluvial sands, tailings of sluicing (and/or gold leaching from the remained pillar between pit/cuvette), rub;
$K_{ml}^{\ y}$, $K_{ml}'^{\ y}$ – coefficients of migration losses from tailing dumps of primary processing and secondary gravity concentrate processing plant tailings in y-year, as a decimal fraction;
T – period of assessment, years;
K_{extr}, K'_{extr} – coefficients of metal extraction for the secondary processing by traditional and geotechnological methods respectively from tailings of sands sluicing, as a decimal fraction;
q_s – gradient of an increase of the area of overburden as a result of increase in peat depth for the definite block, as a decimal fraction;
n – number of overburden benches over i-block;
Z_q – share of mining costs in the common structure of expenses, as a decimal fraction.

The offered criterion (***P_{Au}***) allows to carry out the choice of rational options of development of mining operations and formation and involvement in secondary operation of technogenic formations in the uniform system of mathematical models.

3 Substantiation of the combined flow-sheet of gravel deposit development

When the order of development of a gravel deposit and the field of use of traditional and physical-chemical geotechnologies are established by results of model operation, it is necessary to organize a rational technical communication inside the corresponding flow diagram. Authors propose to combine traditional and geotechnological methods by multipurpose application of mine excavations. This approach can be carried out as follows.

First prospecting and assaying takes place. Delineation of high-grade in gold content sites (blocks) of the gravel deposit is carry out by the results of assay. Then drainage channels are constructed on the borders of blocks. After overburden has been removed, alluvial gold sand is mined by excavator. Pits (small cuts, cuvettes) are formed by bulldozer. The pits (cuvettes) are isolated with the dense waterproof film on the bottom. Tailings (pulp) from gravitational sluice comes to the equipped pits. Rough gravity concentration tailings contain fine fraction with not-extracted fine and lamellar gold. Tailings of processing of rough gravity concentrate are loaded in isolated pits (cuvettes) too. Rough gravity concentrate is processed in cleaner gravity concentration industrial equipment. Excess water is pumped out and returned to washing of gold sand after mineral particles sedimentation in the pit. Inlet wells are additionally drilled inside between-pit pillar. Wells are tubed and filters are installed inside. Leaching solutions containing active oxidizers are injected to productive layer of gold sands. First them iron and manganese are leached from sands. Then gold leaching solutions is injected to the productive layer after a technological pause. Leaching solutions containing active oxidizers and gold leaching solutions drained through productive layer of alluvial deposit. Pregnant solution after the filtration is accumulated in pits (cuvettes) for oxidation of dumped fine fraction of sands washing tails and gravity concentrate (schlich) fraction. Then gold is leached from this mixture. Cuvette leaching of residual gold from tailings of gravity concentration is carried out from the sands extracted by the traditional way.

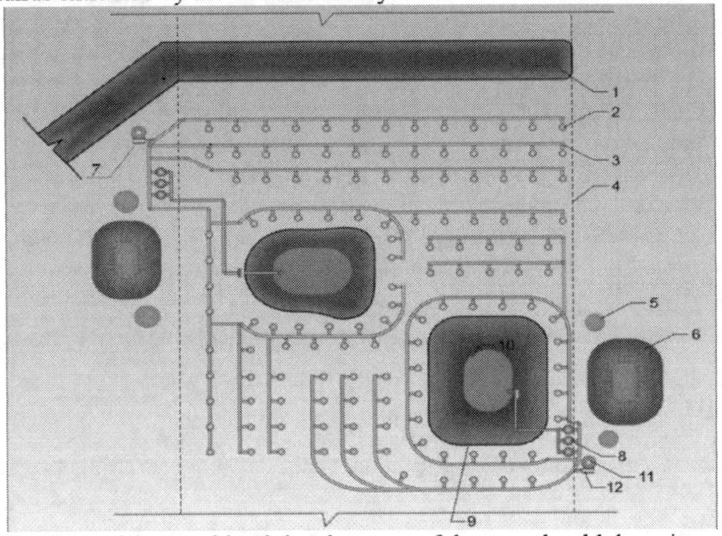

Fig. 1. The flow-sheet of the combined development of the gravel gold deposit.
1. Drainage channel. 2. Inlet wells. 3. Pipeline. 4. Block border. 5. Tails of sand washing. 6. Tailing dumps. 7. Pump. 8. Sorption columns. 9. Pit (cuvettes). 10. Receiver pond 11. Reagent container 12. Pump-measuring bin.

Samples of recirculating oxidizing and pregnant solutions are periodically analyzed for divalent iron, manganese, and gold. Solutions are recirculated and injected to productive layer in case of iron, manganese, and gold content are lower than the threshold. Active solutions provide the required extent of oxidizing preparation of mineral weight and additional gold recovery. Injection of the gold leaching solutions containing the active oxidizers in a layer is begun after the end of circulation through productive layer and cuvette. When threshold concentration of gold is reached, productive solutions accumulated in cuvettes are pumped out to sorption columns. The starved solutions are strengthened and pumped to inlet wells. When gold content is decreased in pregnant solutions down to lower threshold value, mineral mass in cuvettes is pulped by irrigation with this solution. Gold should be adsorbed form the leaching solution first and solution strengthened up to working reagent concentration. The pregnant solutions received in cuvette are pumped to sorption columns. In case of processing of pulped mineral weight, sorption leaching of gold is applied in the cuvette. Application for the Russian Federation patent was applied for this method of combined development of gold-bearing gravel deposit and tailing dumps processing (No. 2018109621, 19.03.2018).

4 Experimental Technique and materials

Cuvette leaching experiments were executed on mineral material of Kia gravel deposit. Magnetite containing gravity concentrate (schlich) fraction of tailings of sands concentration was applied as a research material. Photo-electro activated solutions with various reagent basis were used. The possibility of use of the combined method of development of a gravel deposit with the high gold content in gravity concentrate (schlich) was assessed by results of the experiment. Gold losses reach 55% in mining enterprise developing Kia gravel deposit. More than a half of losses are bound to the compact-grained magnetite of gravity concentrate (schlich) which is washed away with tailings in the gold sands processing.

Gold extraction experiments from magnetite by active leaching solutions were made as follows. The first sample of gravity concentrate (schlich, 2 kg) with 143.5 ppm gold content was placed in a cuvette and contacted with photo-electro activated carbonaceous solution (500 ml). The retention time was 1 month. After iron hydrooxides appeared on magnetite surfaces, carbonaceous solution was deleted. The second stage is characterized by different reagent regime. First is cyanide leaching (1.5 g/l) on the basis of photo-electro activated peroxide-carbonaceous solution. Second is leaching of gravity concentrate (schlich, 2 kg) with 175 ppm the gold content by active chloride and carbonaceous solution. The active chloride and carbonaceous solution prepared as follows. The photo-electro activated solution prepared on the basis of sodium bicarbonate solution with sodium peroxide was mixed with hypochlorite solution and conditioned with hydrochloric acid till pH =6.8-7.1. Also, as well as in the first cuvette, 5 cycles of a leaching were carried out.

5 Results and discussion

Data, demonstrating gold recovery to pregnant solution are given in figure 2.

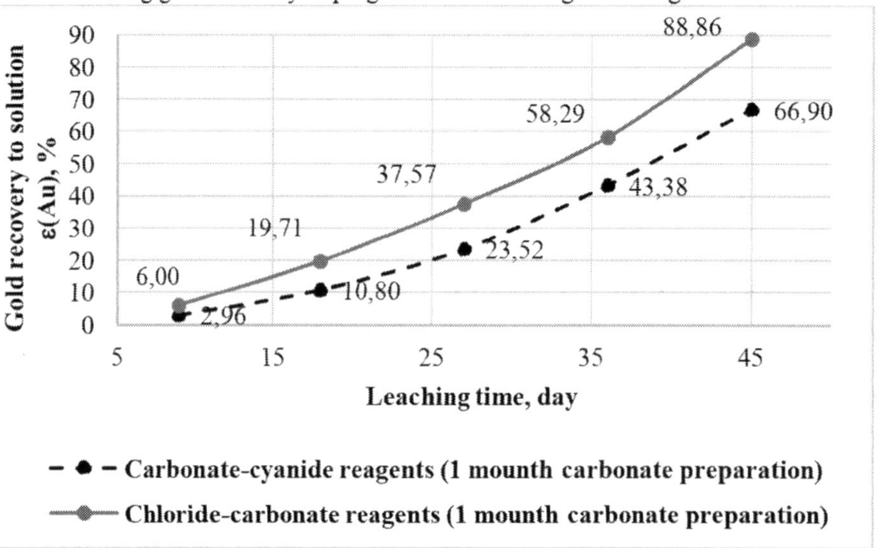

Fig. 2. A kinetics of gold leaching from gravity concentrate (schlich, Kia gravel deposit)

As a result of carbonate-cyanide leaching 66.8 % of gold was extracted for 2.5 months. As a result of chloride-carbonate leaching 88.9% of gold was recovered for the same period. Chloride provide faster leaching kinetics and higher gold recovery to pregnant solution the cyanide-based solution. Preparation of mineral mass by active carbonate solution provided peroxidation and preparation of mineral mass for leaching.

6 Conclusions

Secondary processing of mineral mass of secondary formations and development of deep-laying sites of a gravel deposit and between-pit pillar is a current problem of the modern mining industry. The mathematical model of the gravel deposit is developed for the assessment of the factors influencing the choice of efficient option of development of mining operations and key parameters of development of gravel deposit and processing of productive sands. Authors also offered the combined technology of development of the gravel gold deposit including a well leaching of gold from deep-laying productive layers and a cuvette leaching of gold from washing tailings, provides efficient extraction of fine and dispersible gold. Experiments on a two-stage cuvette leaching of gold from a gravity concentrate (schlich) of Kia gravel deposit are made. The initial gold content in gravity concentrate (schlich) is 143-175 g/t. Duration of the first stage was 1 month. Test was processed by the photo-electro activated carbonaceous solution. Processing of a gravity concentrate (schlich) by carbonaceous peroxide solution provided oxidation of iron to iron hydroxide. The second stage was carried out accoarding two reagent modes. A cyclic leaching (1.5 months) was made by the cyanide solution (1.5 g/l) prepared on the basis of photo-electro activated the peroxide-carbonaceous solution according the first reagent flow-sheet. Gold recovery is 66.8% for cyanide-carbonate solution. The active chloride and carbonaceous solution has provided 88.9 % gold extraction for the same period. It is experimentally proved that processing of gravity concentrate (schlich) material of gravel deposits can be economically efficiently and it is carried ecologically safely out with use of physical and chemical geotechnologies.

Acknowledgement

The research was carried out with the support of scientific grant "Development of technology of cyanide-free leaching of refractory gold from low-grade copper and porphyritic ores" provided by the Government of Khabarovsk Territory under the contract No. 115/2018D, 28.06.2018.

References

1. N. A. Goryachev, F. Pirajno, Ore Geology Rev. **59**, pp. 123-151 (2014).
2. V. A. Stepanov, A. V. Melnikov, J. of Mining Inst. **87**, pp. 5-12 (2015).
3. E. B. Melchiorre, P. M. Orwin, F. Reith **87**, (2018).
4. Y. Pruss, *12h International Multidisciplinary Scientific Geoconference (SGEM)* **I**, pp. 627-629 (2012).
5. J. Wierchowiec, Gospodarka Surowcami Mineralnymi-Mineral Resources Management **87**, pp. 133-147 (2008).
6. R. Vieira, J. of Cleaner Production **87**, pp. 448-454 (2006).
7. T. N. Aleksandrova, A. V. Aleksandrov, N. M. Litvinova, J. of Mining Science **87**, pp. 319-325 (2013).
8. A.G. Sekisov, S.A. Piskunov, B.L. Filatov Patent, Russian Federation, 2044877, Method of excavation of ore intended for leaching, 11.05.95
9. V.N. Oparin, A.G. Sekisov, A.I. Trubachev, B.N. Smolyanitsky, V.S. Salikhov, N.V. Zykov, J. of Mining Science **3**, pp. 70-78 (2017).
10. B. L. Talgamer, International Conference on Innovations and Prospects of Development of Mining Machinery and Electrical Engineering (IPDME) **87** (2017).

Possibility of implementation of dry technology of gold-bearing ores roughing: case study of "Gurbey" deposit

Andrey Matveev[1*], *Ivan* Lebedev[1], *Evgeniy* Lvov[1], *Diulustan* Osipov[1], *Dmitry* Gavriliev[1] and *Ekaterina* Sleptsova[1]

[1] Mining Institute of the North, Siberian Branch, Russian Academy of Sciences, Yakutsk, Russia

Abstract The aim of the research is to identify the ability/feasibility of dry roughing of ore taking as the example "Gurbey" deposit in Irkutsk Region and to develop a rational scheme for processing ores containing free gold. Studies of the dry processing technology of gold-bearing ore at "Gurbey" deposit with the initial weight of 1,123.8 kg were carried out by semi-industrial plants: the DKD-300 combined shock impact crusher, the CMVU-800 centrifugal mill and the POS-2000 air separator developed by CNMI SD RAS. The sample was subjected to two cycles of crushing and four cycles of grinding followed by pneumatic separation. The circulating load is class +12 mm after the DKD-300 crusher, and for the mill, all screening products are +3 mm, as well as the products of classification -3 + 1 mm of the POS-2000 concentrate. As a result of treatment of the concentration products, 23.93 grams of free gold was recovered (concentration on tables: control, recleaning, concentration of the Mosely minerals by analyzer, magnetic separation, heavy medium separation on bromoform). At the same time, it was established that gold from the ore with initial size of 150 mm can be completely stripped and recovered with the POS-2000 air separator with a particle size of up to 100 microns. At the same time gold extraction into the primary concentrate is up to 60%. Based on the results of the research, a basic scheme for dry roughing of "Gurbey" ore containing "free" native gold was developed using the ore treatment devices: RD-MDV-900, DKD-300 crushers, CMVU-800 centrifugal mill and POS-2000 air separator. The technological scheme is fundamental for gold-bearing ores containing "free" gold of fineness over 100 microns.
Keywords: Crushing, grinding, pneumoseparation, concentration, concentrate, magnetic enrichment, gold, stripping, recovery.

1 Introduction

CNMI SD RAS is developing a technology for dry enrichment of gold-bearing ores. For the ores of a number of fields, tests of semi-industrial plants have been carried out: the DKD-300 combined shock impact crusher, the CMVU-800 centrifugal mill and the POS-2000 air separator. Tests have shown the prospects of implementing dry technology of preliminary

* Corresponding author: andrei.mati@yandex.ru

ore dressing with free gold. For a number of deposits, it is possible to create a technological unit for preliminary reduction of the initial ore up to 400 times. At the same time, in most cases, the technology makes it possible to single out a separate enrichment product - primary concentrate. The quantitative and qualitative parameters of this product depend on the fractional composition and nature of the dissemination of gold in the ore. In this paper, studies are done on the feasibility (efficiency) of dry roughing of ore at "Gurbey" deposit in Irkutsk Region, which is a definite stage in the development and implementation of dry technology for the enrichment of gold-bearing ores.

2 Results and Discussion

The aim of the research is to identify the feasibility of dry roughing of ore at "Gurbey" deposit in Irkutsk Region and to develop a rational scheme for processing ores containing free gold.

The choice of ore from "Gurbey" deposit, which contains a wide range of free-gold granulometry by its size and degree of impregnation, is of fundamental importance from the standpoint of investigating the depth of disclosure and extraction of gold.

The ore belongs to the gold-quartz-sulphide low-sulphide type of ores, mainly consisting of lithophilic components (86.0%) with predominance of silicon oxide (59.08%). The ore-forming components are mainly represented by iron and sulfur. The main rock-forming minerals are quartz (22.8%), chlorite (21.7%) and feldspars (20.7%). The sulphide part of the ore sample is represented by pyrrhotine - 2.3% and pyrite - 1.8%. Gold, in general, is very high-grade, almost pure (938 - 964 units). The predominant mass of gold (79.6%) is represented by classes of fineness + 0.071 mm. The fraction of fine, thin and fine-dispersed particles of gold (+0.071 mm) is 20.4% [1].

Studies of the dry processing technology of gold-bearing ore at "Gurbey" deposit with initial weight of 1,123.8 kg were carried out using semi-industrial plants: the DKD-300 combined shock impact crusher [2], the CMVU-800 centrifugal mill [3] and the POS-2000 air separator [4] developed by CNMI SD RAS (Figure 1). The granulometric characteristics of the ore sample given after the primary crushing are presented in Table 1.

а б в

Fig. 1. *a* - the DKD-300 combined shock impact crusher, *b* - the CMVU-800 centrifugal mill, *c* - the POS-2000 air separator

2.1 Methodology of research

In the initial ore contains about 20% of ore fines - a class of size -1 mm. With this in mind, screening was carried out for the original sample by the size classes +12 mm, -12+3 mm and -3 mm. The distribution of the original sample by size classes after classification was as follows: +12 mm - 673.4 kg, -12+3 mm - 197.1 kg, -3 mm -253.3 kg, total 1,123.8 kg. The technological scheme of the studies is shown in Figure 2.

The grain size of +12 mm was crushed by DKD-300, class -12+3 mm by CMVU-800. The -3 mm class was subjected to pneumoseparation by PIC-2000. It should be noted that the nominal size in the feed of the air separator is -1 mm. In this case, an enlarged version of the feed was adopted for maximum concentration into pneumoconcentrate of gold recovered from the ore mass, and also low efficiency of grinding by CMVU-800 was taken into account due to deterioration of its parts.

Table 1. Granulometric composition of the initial sample

Classes of size, mm	Yield	
	kg	%
+100	173.51	15.44
-100+40	240.72	21.42
-40+20	181.61	16.16
-20+10	122.61	10.91
-10+5	57.20	5.09
-5+2	84.62	7.53
-2+1	45.06	4.00
-1+0.0	218.47	19.44
Total	1,123.80	100.00

The concentrate obtained from the air separator was sieved into two classes +1 and -1 mm. Then the class -1 mm was enriched on the concentration table, then concentrate was made by "Henry Moseley" concentrator. For the final separation of free gold, magnetic separation and washing in bromoform were used.

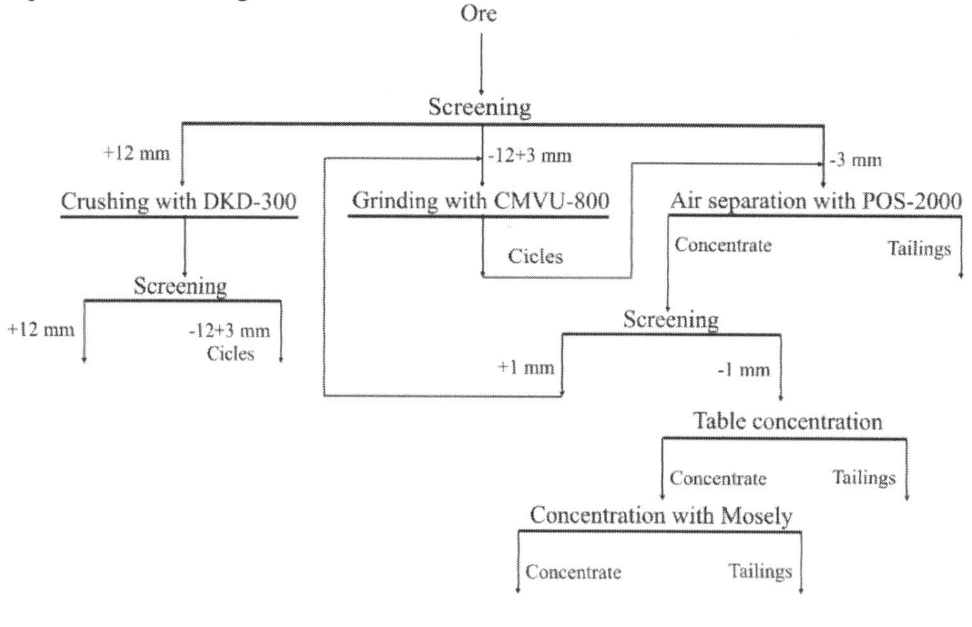

Fig. 2. Technological schemes of samples treatment

The sample was subjected to cyclic crushing and grinding (4 cycles). Class +12 mm after DKD-300 is recyclable for crushing (cycle), for the mill recyclable are all screening products +3 mm, as well as -3+1 mm concentrates after POS-2000. At the same time, processing on CMVU-800 and POS-2000 was carried out strictly simultaneously. For each operation, samples for granulometric and assay analyzes for gold content (tailings of processing) were separately selected and processed.

At the final stage, a balance analysis of the processing products for gold content and calculation of recovery by processing cycles (crushing-grinding-pneumoseparation) was carried out.

The crushing was carried out according to the general technological scheme of processing samples with crushing of +12 mm fractions, including recyclable classes after crushing (the basis for the formation of a circulating load in the technological scheme of crushing).

During the tests, it was found that for a given amount of the sample, two cycles of crushing are sufficient using DKD-300 (Table 2).

Table 2 shows that in the first cycle, the greatest degree of crushing is achieved with a value of 8.4. The second crushing cycle receives only two classes of size -20+10 mm and -40+20 mm (the unbroken product after the first cycle is the basis of the circulation load), which is 19.85% and 10.78% of the total mass of the crushed product. As a result, the degree of crushing for the second cycle falls to 2.28, but the granulometric compositions of the products of crushing of the first and second cycles are quite close, which shows the efficiency of crushing by DKD-300.

Table 2. Granulometric composition of the products of crushing by DKD-300

Classes of size, mm	Yield		
	Initial ore	Cycle 1	Cycle 2
-0.063	5.98	2.93	2.25
-0.1 +0.063	2.64	2.27	2.21
-0.315+0.1	5.76	7.35	6.03
-0.5+0.315	1.88	3.71	2.58
-1+0.5	3.19	7.46	5.46
-2+1	4.00	9.67	8.00
-5+2	7.53	20.73	20.46
-10+5	5.09	15.25	15.47
-20+10	10.91	19.85	29.6
-40+20	16.16	10.78	7.94
-100+40	21.42		
+100	15.44		
Crushing degree		8.40	2.28

Grinding tests were carried out on CMVU-800 according to the general technological scheme of processing samples with grinding the fractions -12+3 mm. During the operation of CMVU-800, in the course of grinding process a fine-dispersed dusty mass of material with a size of -0.063 mm is formed, up to 32% of the total amount of material to be crushed.

2.2 Ore test results by cycles

Initially, the screening products -3 mm after screening of the original ore were tested – 253,300 g (100%). After passing -3 mm material in the air separator, 40% was recovered into the concentrate, while a high loss of mass occurs due to a spill in the feeding scheme of the air separator (conveyor) and in the tail section (fine-dispersed dusty mass). The spill should be attributed to the tailings, since additional blowing of the material by atmospheric wind occurs during transport to the air separator.

As a result of subsequent operations to extract gold from the concentrates, 9.22 g of gold was obtained.

A total of 2 cycles of crushing and 4 cycles of grinding followed by pneumoseparation were completed.

The samples were subjected to grinding and pneumoseparation, the losses in the mass in the form of spillage and fine-dispersed dust amounting to 32%. From the POS-2000 concentrate, 5.74 g of gold were recovered.

Further, a large class of +12 mm of the initial sieving of the original sample was treated. The distribution of the material after crushing by size classes is as follows: +12 mm - 6.28%, - 12+3 mm - 42.9%; -3 mm - 29.8%. Large classes of +12 mm, -12+3 mm form the basis for subsequent cyclic treatment. As a result of the treatment, the following were obtained: class -1 mm - 4.28 g from the products of sieving of the POS-2000 concentrate and class +1 mm - 1.32 g from the products of treatment of sieving of the POS-2000 concentrate. As a result, after processing of the initial sample, only 5.6 g was removed after the re-release by crushing of the large class chippings +12 mm.

Thus, the bulk of gold is in a free state in ore fines, and in the coarsely clastic part, gold is in a subordinate state, which indicates the selectivity of gold recovery at the stage of preliminary destruction.

2.3 Tests of coarsely clastic part of the sample in cycles of crushing and grinding

The initial feed of the crusher is class +12 mm (100%). After the second crushing cycle, the residual quantity of the non-crushed material +12 mm is only 6.28% or 42,300 g. Further processing of these residues is impractical. The bulk of the ore passes into the classes -12+3 mm and is the feed for the mill, the grinding was performed in cycles (return to the grinding of the underground part). Moreover, for the subsequent cycles of grinding the feed of CMVU-800 is the product of sieving +1 mm of POS-2000 concentrators (the mill and air separator work in gear). A total of four cycles of grinding-pneumoseparation were completed. The distribution of the products of grinding-pneumoseparation after grinding cycle 1 is as follows: concentrate - 20.08%, tailings with blowing losses - 17.47%, total losses - 22.82%, i.e., concentrate yield from the operation of grinding-pneumoseparation is 46.8%. After grinding cycle 2, the yield of the POS-2000 concentrate decreases to 6.32%, the loss being 6%, but the concentrate's operating yield is large - 51.21%. After grinding cycle 3, the yield of the concentrate is 2.79%, the loss being 1.62%, the operating yield of the concentrate is increased to 63.26%. In cycle 4, the concentrate yield is 1.27%, the operating yield being 60.48%. In general, there is a high level of the POS-2000 concentrates yield (more than half of the initial material is released to concentrate, this is due to undergrinding and relatively large-size feed of the air separator with -3 mm class.) In the following, it is necessary to adjust the operating mode of the air separator, and apply another variant - treatment by the spiral air separator VPS-500, which has been tried, but the degree of reduction must be linked to the parameters of gold recovery.

2.4 Granulometric characteristic of recovered gold

As a result of treatment of the products of concentration, free gold was recovered by modifying (concentration on the tables: control, re-cleaning, concentration on Mosely minerals analyzer, magnetic separation, heavy separation with bromoform) - 23.93 g (Table

3). Of these, 9.22 g, or 38.56% of the total sample recovered during the processing, was recovered from the -3 mm class of the preliminary sieving of the original sample, i.e., initially free gold in the original sample.

As the gold is released by crushing and grinding in cycles, the amount of free gold recovered decreases noticeably, which indicates a sufficient selective release, depending on the size and nature of the impregnation in the ore (the location of gold in the texture of the ore matrix).

In general, the release gives an additional recovery of 5.45 g of gold. But the recovery parameter is already lower and amounts to 77.56% with the POS-2000 concentrate yield of 45%. The tailings of POS-2000 are characterized by a value already lower - 19 g / t, which is within the limits of reliable identification by the atomic-adsorption method. The high content of gold in the tailings of the concentration table and the "Henry Moseley" analyzer of 17.4 and 34.8 g / t shows approximately equal but low level of enrichment, dry separation, and wet methods of gravitational separation. A possible cause is the thin impregnation of gold in the ore, which has yet to be studied and evaluated technologically.

Table 3. Volume of recovered uncleaned gold

Treatment products	Yield	
	g	%
Pneumoseparation without crushing and grinding -3+1 mm	9.22	38.56
Product of original sample after grinding -12+3 mm	5.74	24.01
Products of crushing -3 mm cycle 1	4.28	17.90
Products of crushing -3+1 mm cycle 1	1.32	5.52
Products of grinding -12+ 3 mm cycle 1	2.56	10.71
Products of grinding -3+1 mm cycle 2	0.51	2.13
Products of grinding -3+1 mm cycle 3	0.25	1.05
Products of grinding -3+1 mm cycle 4	0.05	0.13
Total	23.93	100.00

Figure 3 presents the proposed schematic diagram of dry ore enrichment with a controlled parameter of the degree of crushing and grinding, due to the formation of circulating flows through a classifying device (screen).

It was experimentally established that when circulating loads on DKD-300 in the +12 mm class, the yield is 12.36% and on CMVU-800 in the +1 mm class is 84.41%, which is ultimately characterized by the total yield of the control class of 0.063 mm in size within 35% which allows to completely release and recover with POS-2000 the gold of size up to 100 microns. In this case, gold recovery into the primary concentrate is up to 60%, which corresponds to the initial granulometric composition of gold in the ore, which is confirmed by the enrichment of the products of pneumatic separation with POS-2000 by "wet" gravitational enrichment methods (concentration table, "Henry Moseley" analyzer with recovery and release of free gold). Tailings of dry enrichment are subject to subsequent enrichment using various methods, including flotation or cyanidation.

3 Conclusions

A basic scheme for dry roughing of ores containing "free" native gold has been developed using the ore treatment devices: RD-MDV-900, DKD-300 crushers, centrifugal mill CMVU-800 and air separator POS-2000 with full release and recovery of gold with a size of up to 100 microns into the primary concentrate.

For the developed principle scheme of dry roughing of gold-bearing ores, the limiting condition is the possibility of full release and recovery of gold of a size above 100 microns, according to which all other technological parameters are calculated, first of all recovery, degree of concentration, gold content in tailings, necessary for subsequent enrichment operations.

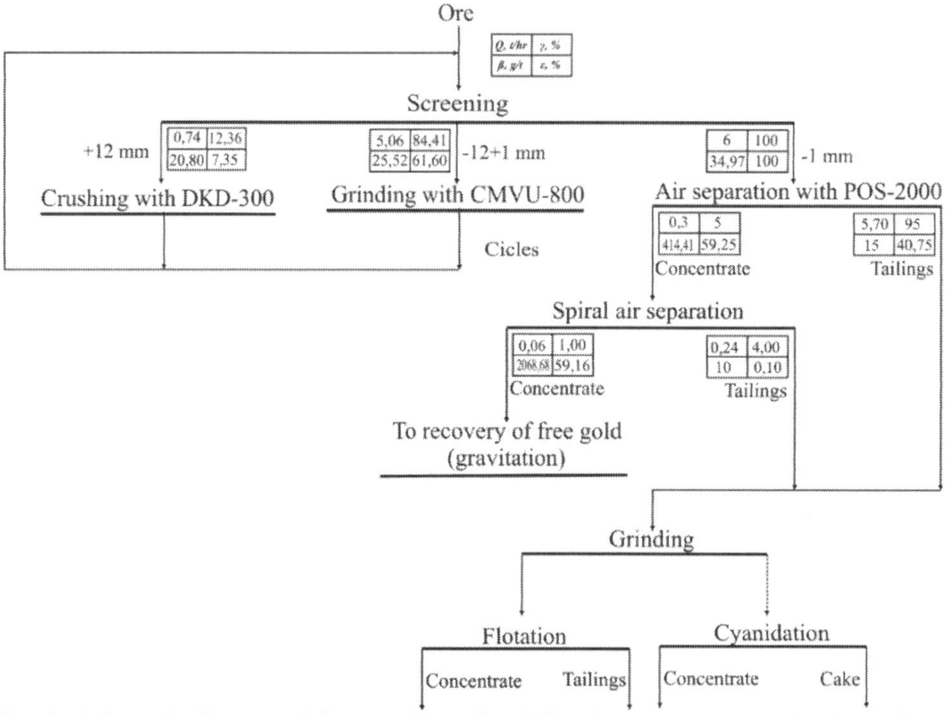

Fig. 3. Schematic diagram of dry roughing of gold-bearing ores: case study of "Gurbey" deposit

References

1. OAO Irgiredmet report on the research work "Conducting research on the technological properties of the TP-1 sample at "Gurbey" deposit" (Final), Irkutsk city, (2017)

2. Patent of the Russian Federation No.2111055. A.I. Matveev, V.P. Vinokurov, A.N. Grigorev, A.M. Monastyrev, CNMI SD RAS of May 20, 1998. *Crusher of combined impact action*. BI No.14.

3. Patent No.2150323 7 B 02 C 13/20. *Centrifugal counter-blow mill* / A.I. Matveev, A.N. Grigorev, V.E. Filippov / CNMI SD RAS. Submitted on October 28, 1997; published on June 10, 2000 // Inventions. Useful models. **16.** Part 2. p. 268. (2000)

4. Patent No.2167005, 7 B 07 B 7/08. *Air separator* / A.I. Matveev, V.E. Filippov, F.M. Fedorov, A.N. Grigorev, V.B. Iakovlev, N.G. Eremeeva, E.S. Sleptsova, A.M. Gladyshev, V.P. Vinokurov / CNMI SD RAS. Submitted on June 11, 1999; published on May 20, 2001. // Inventions. Useful models. **14.** Part 2. p. 346. (2001)

Synthesis scheelite concentrate electroslag remelting for receiving tungsten steel

Evgeny Kuzmichev[1] and *Sergey* Nikolenko[2*]

[1]Far Eastern State Transport University, Khabarovsk, Russia
[2]Institute of Materials Science, Khabarovsk Scientific Center, Far Eastern Branch, Russian Academy of Sciences, Khabarovsk, Russia.

Abstract This article investigates the influence of various reducers of tungsten from a scheelite concentrate on a complex alloying of melted low-carbonaceous steel by tungsten through an oxides slag system. By electroslag remelting of low-carbon steel using ceramic fluxes on the basis of mineral associations are received high-alloyed by tungsten (to 20 wt%).

1 Introduction

Tungsten in steel forms carbides, reduces the tendency of grain growth when heated to high temperatures and increases overmelting, durability and viscosity. In this regard, tungsten is used to receive tool steel and as alloyed steel, which is used for the details operated at high speeds, shock loadings and vibrations.

However, the high cost of tungsten, the limitation of stocks in crust and the technological and organizational difficulties of receiving tungsten steel in the conditions of sustainable development have resulted in a search for new approaches in order to solve these problems.

An effective solution to this problem is the direct processing of raw mineral materials with receiving composite materials, according to the main conclusions of a mineralogical materialogy [1, 2], consisting of reducing the periods between a cycle of receiving raw materials, its processing and a cycle of receiving materials, as well as reducing the periods in each cycle at the expense of using high-level technologies-that is, receiving materials in the region of producing raw materials. We propose using the concentrated streams of energy for receiving materials, including electroslag remelting.

The Far East of the Russian Federation has considerable stocks of tungsten raw materials. In this regard, there was a hypothesis of the possibility of receiving tungsten alloy steel when using electroslag remelting at the application of the fluxes containing a scheelite concentrate in the region of its production. Preliminary research [3, 4] demonstrated such an approach to the problem of receiving tungsten steel. The problem of the complex and rational use of expensive raw mineral materials needs to be solved. In addition, the problem of using these raw materials is connected with the need to reduce the period from the production of raw materials to a cycle of receiving materials and products. The tungsten raw materials go to the western regions of the country and abroad, which

* Corresponding author: nikola1960@mail.ru

considerably increases the entropy of the environment [5, 6]. In this regard, the hypothesis is that in tungsten restoration from a scheelite concentrate the electroslag remelting, the essential role is played by the structure of a reducer, which provides the greatest transition of W in melted steel.

Practical use of this assumption requires research and development of the fluxes providing the maximum transition of tungsten in melted steel-that is, with the greatest possible use of tungsten from a scheelite concentrate ($CaWO_4$). A working hypothesis situation was used that the most active reducers of tungsten provide its greatest transition to melted steel, with elements having the greatest affinity to oxygen in activity with a certain macrostructure.

The purpose of this work is to define the structure of the furnace charge of the fluxes providing the maximum transition of tungsten in melted steel using electroslag remelting. In addition, this work will research the physical, chemical and operational properties of tungsten alloy steel.

2 Materials and Methods

In works [3, 4], the complex use of a scheelite concentrate as a basis of ceramic alloying flux is used because, in addition to alloying components, it contains most of the necessary components such as slag-forming, gas-forming and ionizing.

As tungsten reduces from oxides ($CaWO_2$), it is necessary to use elements that have a greater affinity to oxygen than tungsten. In [7], a number of activities of metals to oxygen in the process of decreasing deoxidizing ability are presented:

$$Al \rightarrow Zr \rightarrow B \rightarrow Ti \rightarrow Si \rightarrow V \rightarrow Mn \rightarrow Cr \rightarrow Mo \rightarrow W \rightarrow Fe \rightarrow Co \rightarrow Ni \rightarrow Cu \quad (1)$$

It is known from [7] that the elements with an affinity to oxygen are more to the left of iron and to some extent are oxidized; therefore, the extent of the restoration of an alloying element from the oxides of subjects is lower. Then more to the left, there is an element from iron. Conversely, to the left of tungsten, there is an element that has a better recovery ability. For tungsten restoration from a scheelite concentrate, elements can be used that are more to the left of it. An experience of the application of carbon for the restoration of elements in metallurgical production is known.

The most as reduces of tungsten active were accepted in relation to Al oxygen and graphite. In addition, as reducers of tungsten and oxides are widely applied in production welding surfacing materials ferroalloys (ferromanganese FeMn and ferrosilicium FeSi). It was supposed that under the influence of high temperatures, ferroalloys would decay on elements and Si, Mn would restore W from oxides in a slag bathtub, according to the activity of the metals.

The furnace charge of flux was formed on the basis of a scheelite concentrate with the introduction of 20 wt% of CaF_2 (fluorite) and 4–16 wt% of the reducer, which increased the contents at the expense of basis reduction. The components of the furnace charge of the flux (scheelite concentrate ($CaWO_4$), fluorite (CaF_2), ferrosilicium (FeSi) and ferromanganese (FeMn)) were crushed to 0.35–0.65 mm. Aluminum was used in the form of powder and shavings with a size of 0.1–0.35 mm, and graphite was used in the form of powder with a size of 0.4–0.56 mm.

Fluxes were made with standard technology. As an additive material, the low-carbon welding wire with a diameter of 3 mm was accepted.

Further experiments were carried out as follows:

—research and choice of the most rational reducer as a part of the flux for ensuring the maximum transition of an alloying element in a melted additive material;

—research of structure and properties of the received alloys.

To conduct the research, both standard and original equipment were used. The electroslag surfacing and remelting were carried out on an installation that was designed and made by authors on modes: I=350 … 1000A, U=20 … 90B. Samples for the analysis were formed in a water-cooled copper crystallizer with a working volume of 40x200 mm^3.

The metalgraphic analysis of alloys was carried out using microscopes of Labomet-2 and modular EU of METAM RV-21 at increases to 1000X. Microhardness was defined on the PMT-3M device. The differential and thermal analysis was carried out on the derivatograf Q-1000. The phase structure of the received materials was studied on the X-ray DRON-7 diffractometer. The research of the element structure of the received alloys was conducted on an X-ray fluorescence scanning spectrometer "MAKS-GV" and a raster electron probe microscope of ISM-35C JEOL with a prefix of the electron probe microanalyzer—an X-ray spectrometer 35-SDS with wave dispersion.

To determine the physic–mechanical and operational properties of alloys, the following devices were used: for hardness measurement—TSh-2M, TK-14, TP-2; for wear-resistance research—the MI-402 car in the conditions of friction without greasing at a load of 50 kg/s and a counter body from the steel 40XH tempered in oil to HRC 60 … 62.

3 Results and Discussion

As shown in fig. 1, the greatest transition of W to steel is observed being used as the reducer of aluminum shaving and ferrosilicium, and the least active reducer in these conditions is ferromanganese. The presented experimental dependences show that it is not dependent on the reducer type. The greatest transition of tungsten to melted steel is observed at the maintenance of a reducer in the flux furnace charge within 8–12 wt%. From the received dependences, a further increase in the maintenance of a reducer does not lead to an increase in the transition of tungsten in built-up metal. In certain cases, it even reduces it.

The electroslag process used in the furnace charge of the flux of Al in the form of shaving and grains graphite proceeds steadily: slag is distributed on all surfaces of the built-up metal evenly, and a well forms on the surface of an ingot to protect the melted metal from air access.

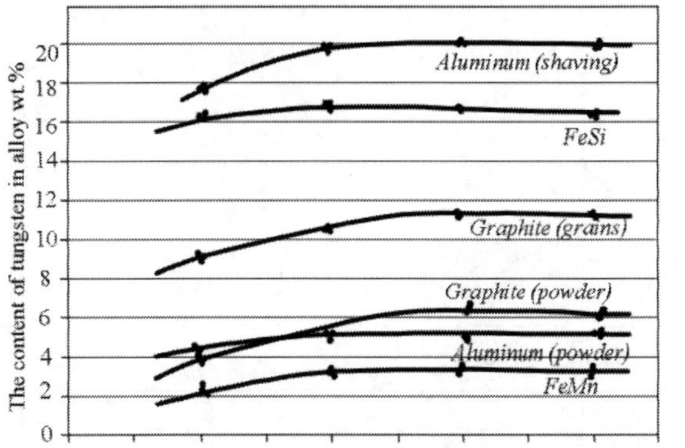

Fig. 1. Dependence of the content of tungsten in an alloy from a quantity of a reducer in the flux furnace charge

As a result of the spectral analysis of the alloys received with using developed fluxes, it was established that all studied reducers allow tungsten to be received from a scheelite concentrate with various degrees of efficiency.

Application as a reducer of ferromanganese allowed 2.87% of tungsten to be received in an alloy that slightly exceeds the results received using flux not contained in the structure of traditional reducers and 6.5 times less than when using flux with ferrosilicium. In our opinion, it is caused by the activity with oxygen manganese, which settles close to tungsten, and it has no sufficient recovery abilities.

The conducted research provided an opportunity to establish that the most effective reducers of tungsten from a scheelite concentrate are aluminum, ferrosilicium and graphite.

A particular interest is represented by aluminum and graphite, which are the most active reducers. However, at their use in the developed flux in the form of powder (C), the powder (Al) transition of tungsten to melted steel appeared to be less than when using ferrosilicium. In this regard, the assumption was made that such a phenomenon was caused by technology factors—reducers were entered into skilled fluxes in the form of very small powders of a small weight. As a result, in the course of remelting, they emerged on the surface of a slag bathtub without being dissolved and partially being oxidized by air.

To check this hypothesis, experimental fluxes in which the preservation of structure the size of fraction of reducers were created. For this purpose, aluminum shavings with a grain size of 0.2–0.4 mm and graphite fractions with a size of 0.4–0.5 mm, FeMn, FeSi with a grain size of 0.35–0.6 mm were used. Table 1 provides the results of the research of a chemical composition and physic-mechanical properties of alloys of the various fluxes obtained with use. As a result of the research of the chemical composition of the received ingots, it was established that at the expense of an increase in the size of the fractions of aluminum, the amount of the restored tungsten increased by 3.5 times and graphite increase by two times, making 20.6 and 11.4 wt% respectively.

The slag that remained after remelting contained: when using aluminum, 1.53% of tungsten, and with a graphite application, 8.38%, which confirmed a solvency of the hypothesis. The greatest amount of tungsten in alloy No. 3.1 was received with flux use as a reducer in which aluminum shaving was applied. According to the spectral analysis, the average content of tungsten in an alloy was 20.6 wt%. Hardness makes 280–300 HB, wear resistance of rather tool steel HS6-5-2-0.5 is 2.73 (Table 1).

Table 1. Physical, mechanical and operational properties of alloys

№ fusion	*Reducer*	*Content of tungsten in alloy wt%*	*Hardness, HB*	*Resistance of steels*		
				W1-11-1/2	*HS6-5-2-0.5*	*HS12-1-1.5-0.5*
1	FeMn	2.5–2.87	158–174	0.192	0.12	0.0187
2	FeSi	17.8–18.6	370–390	2.48	1.73	0.812
3	Al (powder)	5.5–5.6	156–172	1.43	0.647	0.431
3.1	Al (shaving)	19.5–20.6	270–290	2.73	1.87	0.871
4	C (powder)	5.4–5.57	160–170	2.12	1.08	0.442
4.1	C (grains)	10.6–11.4	256–270	3.62	2.08	0.842

The alloy No. 3.1 microstructure (fig. 2.) consists of alloyed by ferrite tungsten with microhardness H_μ= 141–197 HV and pearlite with H_μ= 198–208 HV. In addition, according to the X-ray phase analysis (fig. 3.), intermetallics Fe_2W are also present. Scanning in the reflected electrons (fig. 4) showed the phase heterogeneity of an alloy.

Research of a chemical composition by means of the electron probe microanalysis showed that the matrix of an alloy has a variable structure.

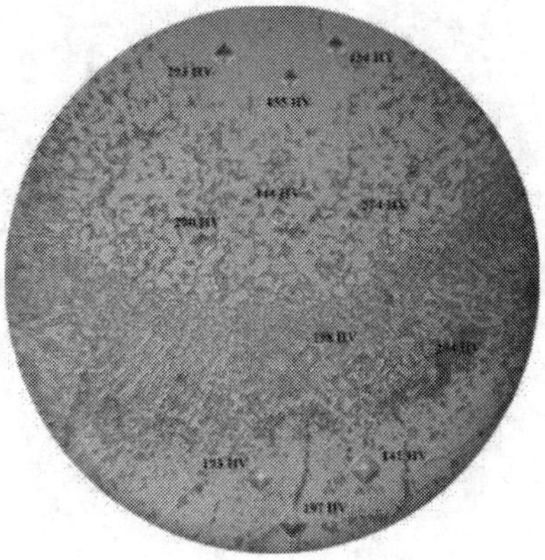

Fig. 2. Alloy No. 3.1 microstructure

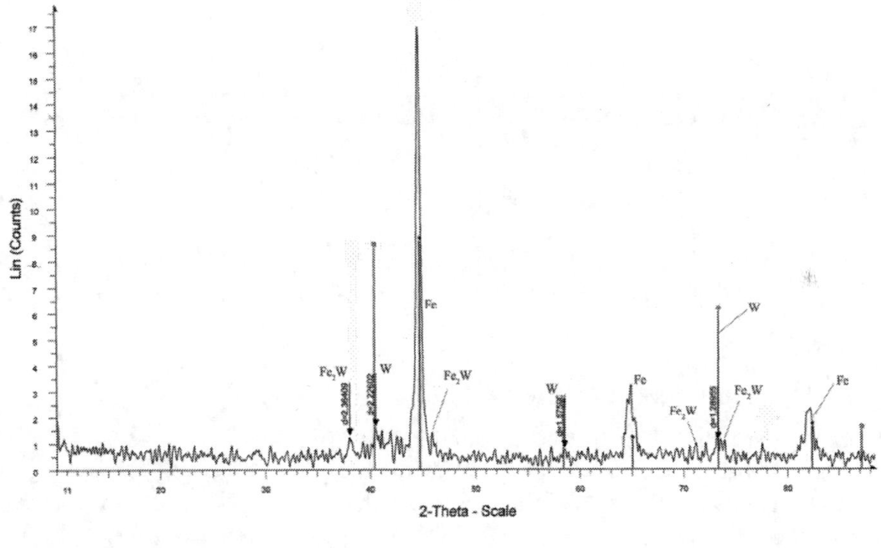

Fig. 3. X-ray diffraction spectrum of alloy No. 3.1.

Point 1 contains 34 wt% of tungsten and 65 wt% iron. In points 2 and 3, tungsten (71–89 wt%) and iron prevail (12–20 wt%), which confirms the data of the X-ray phase analysis on the existence of intermetallics Fe_2W in an alloy and limited firm solution W (Fe). The microhardness of these phases makes H_μ= 293-455 HV (fig. 3).

As shown in fig. 4, grain 4 represents a mechanical mix of light and dark phases. The light phase contains about 70 wt% of tungsten and 30 wt% of iron, which is similar to phases 2 and 3. The dark phase represents a limited firm solution of tungsten in iron.

Thus, according to conducted research, it follows that tungsten in alloy No. 3.1 was distributed unevenly. Partially, the alloy ferrite and cement carbide formed a chemical compound of Fe_2W with iron. Ferrosilicium (alloy No. 2) appeared no less effective a reducer of tungsten from a scheelite concentrate.

As shown, in the X-ray phase analysis (fig. 5) the main phases in an alloy were the ferrite alloyed by tungsten and intermetallics Fe_2W.

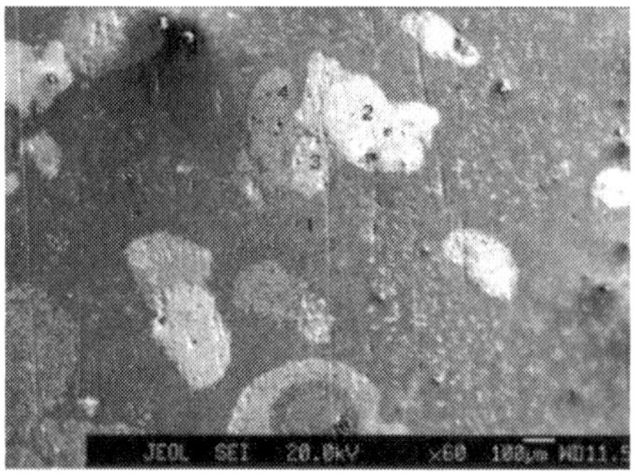

Fig. 4. The image in the reflected electrons of alloy No. 3.1

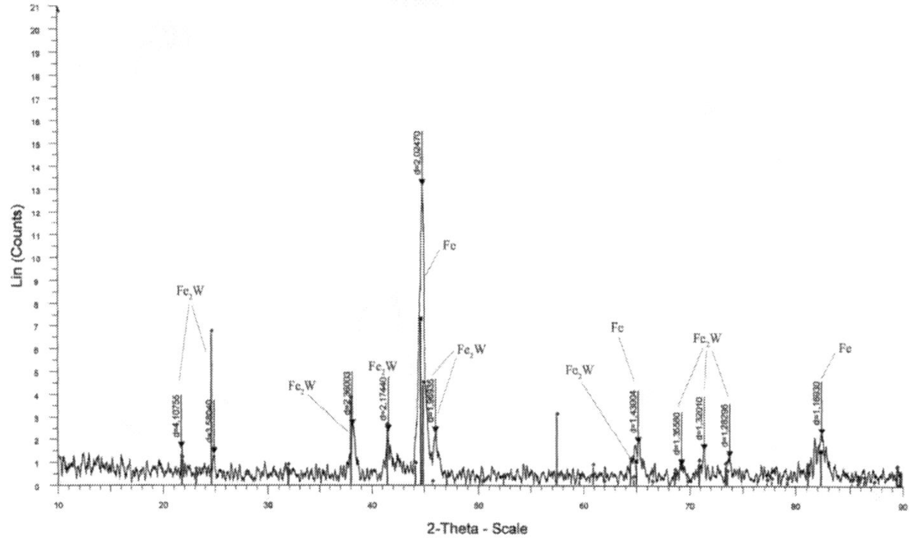

Fig. 5. X-ray diffraction spectrum of alloy No. 2

From the research of a microstructure of alloy No. 2 (fig. 6), it was established that in the matrix from the alloyed ferrite with H_μ= 146–194 HV inclusions intermetallics by Fe_2W (up to 0.5 mm in size) with H_μ= 294–630 HV, which is shown scanning in the reflected electrons, contain about 59 wt% tungsten and 41 wt% iron, which are unevenly distributed. In addition, they are surrounded by the area of the alloyed pearlite, with H_μ= 183–284 HV—the alloyed ferrite divided by a grid with small inclusions intermetallics. It should be noted that the content of tungsten in alloy No. 2 on 8 wt% is lower than in No. 3.1 (Table 2) at a similar phase structure; however, its hardness for 25% exceeds the hardness of alloy No. 3.1. In our opinion, the difference in hardness is caused by the various contents of silicon alloys. In alloy No. 2, the content of silicon made 1.35 wt%, which led to an additional alloying of a low-carbonaceous additive material and an increase in hardness.

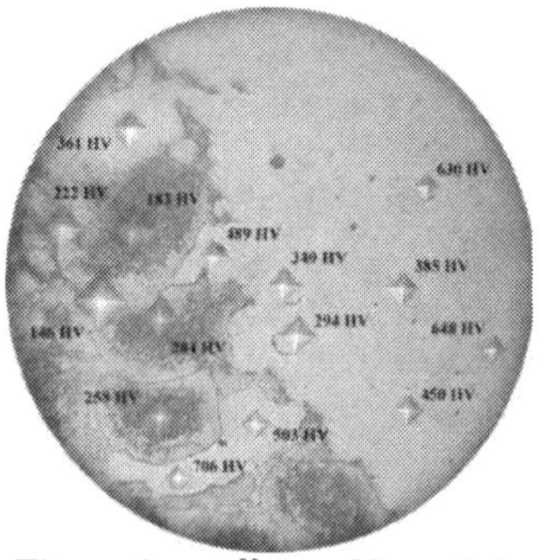

Fig. 6. Alloy No. 3.1 microstructure

Thus, there is the complex alloying of an alloy tungsten (to 20 wt%) and silicon. At the same time, the amount of silicon in alloy No. 2 is 4.5 times higher in relation to alloy No. 3.1, which in turn causes hardness to increase.

Thus, as the main reducer of tungsten from a scheelite concentrate in ceramic fluxes for electroslag remelting, aluminum use in the form of shaving, FeSi and graphite in the form of grains is recommended.

As a result of the conducted research, we recommend receiving the tungsten in complex alloyed steels close to a chemical composition and to fast-cutting tool steels, electroslag remelting with using fiuxes on the basis of scheelite concentrate.

Conclusions

1. The possibility of a complex alloying of low-carbonaceous steel through the slag system is available for the variation of various reducers and their macrostructures. The best results are shown using ferrosilicium and carbon.

2. For the first time, the electroslag remelting of low-carbonaceous steel received alloys with the content of tungsten up to 20 wt%.

3. A number of ceramic fluxes on the basis of the scheelite concentrate were developed and approved, allowing electroslag remelting low-carbonaceous to begin to receive high tungsten.

4. Metallographic and dyurometrichesky analyses and the research of operational properties established a complex alloying by tungsten steel with increased hardness and wear resistance.

References

1. E. Hornbogen. Werkstoffe. (2006).
2. E. Hornbogen, G. Eggeler, E. Werner. Werkstoffe. (2012).
3. E. G. Babenko, A. D. Verkhoturov, V. G. Grigorenko. Basic aspects transport materials technology. (2004).
4. A. D. Verkhoturov Journal of Advanced Materials. **1.** (2003).
5. M. Rühle , H. Dosch, E. Mittemeijer European White Book on Fundamental Research in Materials Science. (2001).
6. V. I. Sergiyenko, Rhee Hosen, Centuries Gostishchev, Yu. I. Mulin, E. Kh. Ri, D. Kh. Ri, S. N. Himukhin Problems and prospects of effective use of mineral raw materials of the Far East for production of metal materials. (2009).
7. A. A. Yerokhin Metallurgiya welding. (1978).
8. D. Dakuort Electroslag melting (1973).

Sorting out of Gusevogorsk deposit titaniferous magnetite ore by technological types by methods of geoinformation modeling and preliminary estimation of their washability

Sergey Kornilkov[1], *Valery* Kantemirov[1], *Andrey* Yakovlev[1*] and *Roman* Titov[1]

Institute of Mining, Ural Branch of the Russian Academy of Sciences, Ekaterinburg, Russia.\

Abstract A more complete and complex extraction of valuable components in the development of mineral resources is achieved through the separation of technological types and grades of ores, their separate extraction and processing, which allows to increase the yield, quality and nomenclature of commercial products and to raise the economic efficiency of the extraction and preparation processes. An important element of the separate processing of types of ores is the technology used for their zoning by grades in the open-pit space. Innovative direction in the development of the forecast of quality and preparation characteristics of mineral resources is the methods of geoinformation modeling with the use of Geological and Mining Information System technologies. The paper describes the technique of modeling and separation of ores into technological grades on the example of the Gusevogorskoye deposit of titanomagnetites, and presents the method of express ore preparability analysis with the estimation of the degree of contrast of qualitative characteristics.

1 Introduction

The most iron ore mining and processing stations (MPS) have adopted gross mining, averaging and ore preparation according to a single technological scheme, which simplifies the organization of extraction and preparation of mineral resources. One of the possible ways to improve the quality of iron concentrate is the separate extraction and processing at the concentrator of various natural types of ores. As a characteristic object for the implementation of these measures, it is possible to consider the deposits of titanomagnetites of the Kachkanar group, for which a number of researchers [1, 2] substantiated the technological possibilities for producing the iron-vanadium and separately ilmenite concentrate and the expediency of additional ore separation into natural grades depending on the content of titanium dioxide.

The Kachkanar titanomagnetite massif is located in the north of the Sverdlovsk region and is represented by two deposits: Gusevogorskoye and Sobstvenno-Kachkanarskoe. The

* Corresponding author:quality@igduran.ru

host rocks are represented by plagioclase porphyrites, effusive diabases, mica and siliceous shales of the Ordovician. Mineralization is confined to pyroxenites, and the structures and textures of ores correspond to the structures and textures of the host rocks. The shape of the ore bodies is the shallow and inclined long deposits of disseminated and banded titanomagnetite ores. The ore at the Gusevogorskoye deposit is extracted by three open-pit mines: Main, Northern and Western.

It is assumed that the titaniferous magnetite ore of the Gusevogorskoye deposit is represented by two major natural types (species) of ore - low-Ti type (the contents of TiO_2 <1.1%) and normal-Ti type (content of TiO_2 ≥1,1%). The main quantity of low-Ti ores is extracted in the Western open-pit mine. The low-Ti grade has higher preparability. At the same time, the smaller the size of grains of titanomagnetite, the greater the difference in iron content in concentrates obtained from low-Ti and normal-Ti ore. On the example of the Gusevogorskoye deposit, a method for geometrization has been developed and tested that allows:

- to carry out the forecast of distribution of qualitative indicators in mineral resources;
- to separate ores into technological grades;
- to assess the possibility of separate extraction and processing;

2 Methods of experimental work

Analysis of literature sources [3] and geological reports has allowed to systematize the current criteria for the separation of technological types of ores of the Gusevogorskoye deposit, depending on the following factors:

1) impregnation of grains of titanomagnetite in ore:

- easy-preparable - large- (>3 mm) and medium-impregnated (1-3 mm);
- medium-preparable - small-impregnated (0.2-1 mm);
- hard-preparable - thin- (0,074-0,2 mm) and disperse-impregated (<0,074 mm);

2) the content of iron in the concentrate, % - an empirical indicator that allows you to predict the preparability of mined ore in industrial plants:

- easy-preparable - the index of preparability being more than 64.4%;
- normal-preparable - the index of preparability being from 62.6% to 64.4%;
- hard-preparable - the index of preparability being from 60, 1% to 62.6%;
- very hard-preparable - the index of preparability being less than 60%.

On the basis of statistical analysis and generalization of operating technological criteria to iron ore, a complex index of preparability has been established, taking into account the mutual influence of the iron content in the concentrate and the impregnated grains of titanomagnetite. The indicated characteristics of ores are used in the current planning of mining operations at the enterprise and are linked to the schedule of ore supply for processing. The evaluation of the ratio of technological grades of ore is carried out in order to adjust the productivity and operating conditions of the equipment in the processing sections of the preparation plant.

Geometrization of the index of preparability has been carried out according to the detailed exploration data of the Gusevogorskoye deposit, and the geological plans, sections, interval borehole columns with data on the content of iron, titanium, vanadium, the impregnations of titanomagnetite and iron content in the concentrate have been used as initial information. The entire array of information was collected in a single geological database, on the basis of which the peg mining models and block deposit models for three open-pit mines have been built: Main, Northern and Western. Also, the correlation-regression analysis has been also performed, which allowed to establish zones with increased and lowered contents of chemical elements, based on the boundary value of iron and the features of the occurrence of ore bodies. The method of geometrization consists in the following sequence of operations:

- data conversion (digitization, data processing, building of numerical distributions characterizing spatial data placement and clustering, compositing);
- building of isosurfaces of qualitative indicators of working horizons by methods of triangulation and return distance with allocation of zones of increased / lowered content of mineral components (Fig. 1);
- building of block models and filling them with interpolated values of qualitative indicators by kriging;
- zoning of low-Ti and normal-Ti ores with the determination of the possibility of their separate extraction and preparation (Fig. 2).
- building of horizon-oriented plans reflecting the size and distribution of the impregnation and complex index of preparability (Fig. 2) of titanomagnetite ores in the working horizons of open-pit mines.

For comparison with the results of block modeling, the developed method for express estimation of the contrast of ores as a technological index characterizing the preparability (using the example of the horizon of +160 m of the Northern open-pit mine) has been used.

The ore contrast is the degree of uneven distribution of the mineral component in the ore determined by its specified volumes. The larger this volume, the less contrast. The greatest contrast is observed in pieces of ore comparable with the volume of ore inclusions, and the smallest contrast being observed in large portions of crushed and mixed ore. [4]

To build a model, the borehole array under investigation is conditionally divided into equal sections-squares (with coordinates along the X, Y axes). A total of 20 arbitrary squares have been selected (see Table 2, Fig. 2) with a step of $Ш_x$ and $Ш_y$:

$$Ш_x = \frac{X_{max}-X_{min}}{5}, \quad (1)$$

$$Ш_y = \frac{Y_{max}-Y_{min}}{5}, \quad (2)$$

Where X_{max}, X_{min} are the extreme marks of the coordinates of the boreholes along the X axis; Y_{max}, Y_{min} are the extreme marks of the coordinates of the boreholes along the Y axis;

A conditional square is assigned a number in accordance with the accepted coordinate system (for example, A3). The following characteristics are calculated in each square: the average content of valuable components in the ore - total iron, titanium and vanadium; coefficients of variation in content; contrast coefficients of content of valuable components. The results of calculations and modeling are presented in Table 2.

3 Results and discussion

Analytical evaluation of the results of the distribution modeling within the Northern open-pit mine (horizon being 160 m) of the impregnation and iron content in the magnetic fraction led to the conclusion that the ores of the north-eastern part of the deposit are more highly prepared.

To assess the possibility of separate mining, the horizon 160 m of the Northern mine has been divided into 18 peg models, each with an ore volume equal to the annual productivity of the medium-duty excavator, the estimation has been performed for the chemical composition and the index of preparability within each model, as well as the statistical characteristics of the distribution of valuable components. On the basis of the analysis it has been found that the individual peg models have a large percentage of off-grade ore in terms of total iron content, while at the same time a part of this ore is characterized by a complex index of preparability as normal-preparable, which makes it possible to use this ore with the organization of in-pit averaging with the aim of increasing the yield of commercial products.

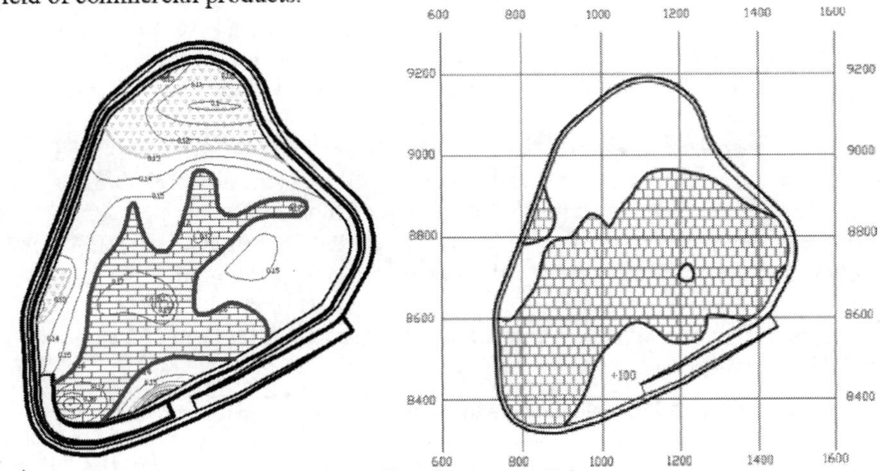

- site with a high content of vanadium (V_2O_5> 0.16%)

- site with a reduced content of vanadium (V_2O_5< 0.13%)

- section of normal-Ti ore (TiO_2 > 1.1%)

Fig. 1. Distribution of vanadium and titanium content at the horizon of +100 m of the Western mine of JSC "EVRAZ KGOK"

A comparative analysis of the distribution in open-pit mines of different technological types of ores according to the above presented estimation methods gives quite similar zoning results. It is established that the northern and western sections of the side are the main concentration of hard-preparable ores at the Northern mine, while these ores belonging to the class of not-contrast ores. The increase in titanium content in the ores of the Northern open-pit mine is oriented to the northeast, and within the Main open-pit mine to the southwest, for the Western open-pit mine from the west to the east. This trend can be traced for all working horizons of open-pit mines (for Northern open-pit mine see Table 1).

Normal-Ti ores have been localized in the Northern open-pit mine in its northeastern part and have been distributed everywhere in the Main open-pit mine except for the north-western section. At the Western open-pit mine, low-Ti ores are located along the northern and southern sides.

Table 1. The ratio of normal-Ti and low-Ti ores at the working horizons of the Northern open-pit mine

Horizon	Normal-Ti, %	Low-Ti, %
+ 205 m	29.8	70.2
+ 190 m	28.2	71.8
+ 175 m	22.6	77.4
+ 160 m	39	61
Total:	25.6	74.4

Within the Northern open-pit mine, easy- and normal-preparable ores, as well as ultra- and high-contrast ores selected by other methods, are located in the central part of the open-pit mine, respectively. The total volumes of low-, normal- and medium-preparable ores established at the horizon +160 (14.6 million m^3) correspond approximately to the volumes of ultra-contrast, high-contrast and contrast ores (10.8 million m^3) selected by the method of assessing the contrast, the difference being ~ 30%. It has been established that practically throughout the studied field, the contrast indices of titanium and vanadium in the ore are higher than in iron. It is assumed that the accuracy of the forecast can be increased by changing the size of the unit cells in the model and increasing their number in the express method for estimating the contrast of ores.

The results of the comparative analysis and the methods of estimating the qualitative characteristics of mineral resources with modeling the placement of technological types of ores in the open-pit space confirm the efficiency of their use in the preliminary stage of studying the geological information. It is established that the contrast between the elements within the horizon +160 m of the Northern open-pit mine differs quite strongly, so on the western side of the open-pit mine there is a higher contrast of iron content than titanium (by 4 times). Table 2 shows the distribution of ore in terms of contrast within the horizon `+160 m in the Northern open-pit mine.

Table 2. Distribution of ore by contrast ratio at the horizon +160 m

Ore type by contrast	**Distribution over the horizon**		
	%	**ths. m^2**	**ths. m^3**
Ultra contrast (M>1.5)	13.6	184.1	2762.2
High-contrast (1.2<M<1.5)	24.7	334.4	5015.3
Contrast (0.4<M<1.2)	33.8	457.1	6855.4
Not-contrast (M<0,4)	26.8	363.1	5446.2
Missing data	1.1	15.1	226.9
Total:	100	1351	20266

It is expected that developing the northern side and the central section makes it possible to minimize fluctuations in the quality characteristics of valuable components. It should be noted that the contrast ratio, characterizing a certain degree of heterogeneity of ore, does not fully reflect the technological results of ore preparation. Its use is recommended for preliminary qualitative analysis of complex ores with various mining methods, various schemes of ore preparation and roughing.

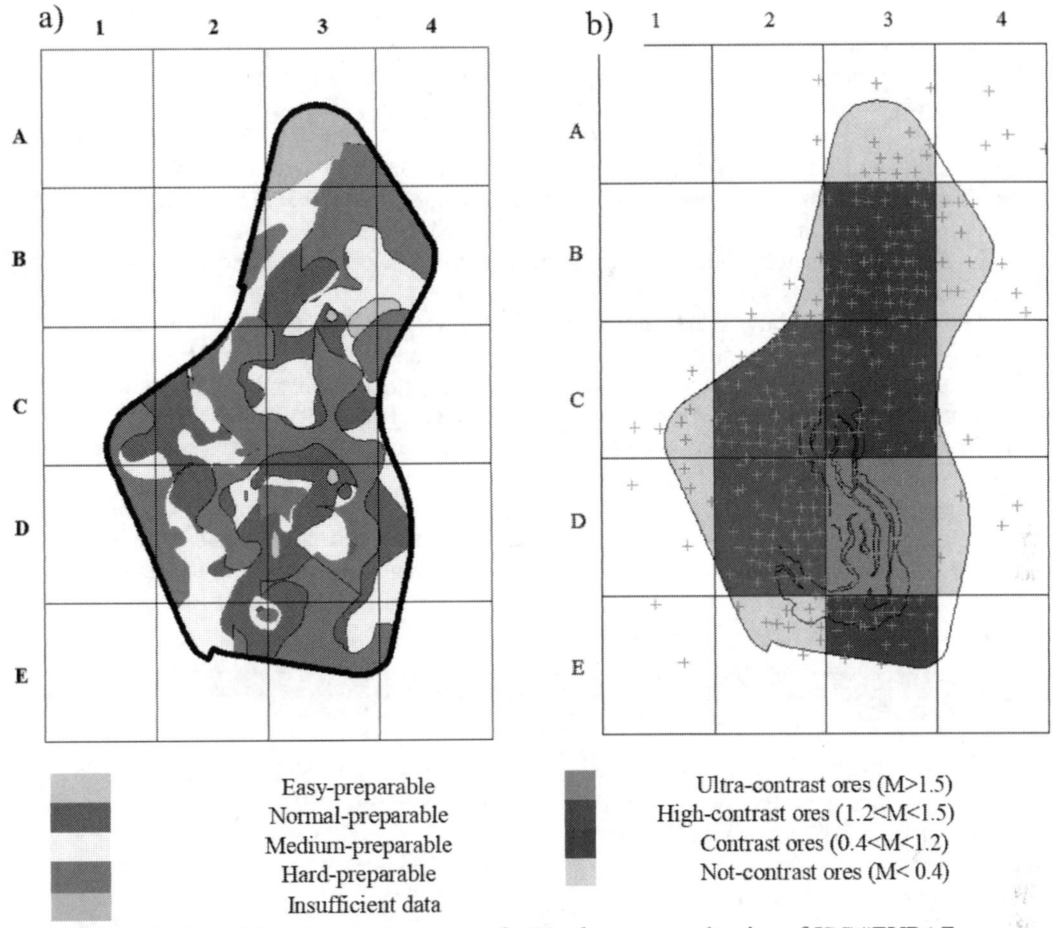

Fig. 3. Zoning of titanomagnetite ores at the Northern open-pit mine of JSC "EVRAZ KGOK": a) on the complex index of preparability (the content of iron in the magnetic fraction and the size of the impregnation); b) on the contrast ratio

4 Conclusions

With a comprehensive assessment of mineral resources, modern standards presume to take into account not only the content of mineral components and deleterious impurities, but also structural and texture features of ores, as well as the requirements of concentrating and metallurgical processing in order to achieve the maximum yield of a quality commercial product.

An estimation of the contrast of chemical and textural-structural properties of mineral resources can be used to justify the choice of technological ore-preparation schemes in the open-pit mine (averaging, grading, selective development, preenriching, etc.). The main advantages of this assessing method are the speed of obtaining the esimation results without using expensive software products. Improving the method requires its further development to improve the accuracy of the forecast and to approach the results of "classical" block modeling using laborious methods of geostatistics.

The studies carried out by IM UrB RAS, IMET UrB RAS, in cooperation with the Mineral Processing Department of the Ural State Mining University, have revealed significant differences in the preparability and grindability of the low-Ti and normal-Ti grades of ore and recommended their separation into various technological types with subsequent processing on separate processing lines. At the same time, the total technological and economic indices of separate preparation of natural types of ores will be higher than those of ore preparation in industrial-scale mining. The results of the research have showed that the low-Ti grade of ore as compared to the normal-Ti ore has higher preparability and grindability and less magnetic rigidity. Therefore, it is recommended to introduce an additional parameter for geometrization, such as the limiting content of titanium dioxide in mined ore, into the current classification of ore preparation, in addition to the indicators taken at the MPS (the grain size of titanomagnetite and the total iron content in the concentrate).

The preliminary analysis has shown that the Gusevogorskoye deposit's concentrates obtained as a result of separate extraction and preparation should be reasonably processed in the near future according to the "blast furnace – converter" scheme.When replacing the ores of the Western open-pit mine with the mineral resources of the Sobstvenno-Kachkanarskoye deposit, the Northern open-pit mine becomes the object that systematically ensures the formation of ore flows of a given quality.

The obtained results allow us to carry out the preliminary horizon-oriented estimation in open-pit mine the degree of variability of mineral components in ore and make a forecast for the distribution of certain types of ores in order to distinguish further the technological types of ores for mining planning tasks and to control the quality of mineral resources and improve the efficiency of ore preparation.

References:

1. D. E. Mansor, "Development of Technologies for Complex Processing of Vanadium Containing Titanomagnetites" / D. E. Mansor and B. S. Tleugabulov // Juvenis scientia. **1**, pp. 13-15. (2016) [In Russian].
2. N. Dmitriev, G. Yu. Vitkina, R. V. Petukhov, S. V. Kornilkov, A. E. Pelevin, A. Y. Fishman, T. V. Sapozhnikova and K. Y. Shunyaev, "The Characteristic of Ores and Concentrates of the Open Society EVRAZ KGOK", Advanced Materials Research. **834 - 836**. Pp. 364 – 369. (2014).
3. M. V. Vasiliev, V. L. Yakovlev et al. Transport of Open-Pit Mines, (Sredne-Uralskoe Knizhnoe Izdatelstvo, Sverdlovsk) [In Russian].
4. L. Ch. Pukhalsky, The Theory of the Contrast of Uranium Ores, (Gosatomizdat, Moscow, 1963) [In Russian].

Study of promises for the use of ultrasound in the technology of secondary benefication of processing wastes

Olga Voronova[1*] and *Lidiya* Kienko[1]

[1]Mining Institute of Far Easten Branch of Russian Academy of Sciences, Khabarovsk, Russia

Abstract. The results of the study, aimed at designing of benefication technology of mine processing wastes of Voznesenskiy ore district, are presented in this paper. By the expert appraisals, the amount of tailings dumps of the Yaroslavskaya Mining Company is estimated to be more than 30 million tons. Sample testing has shown that fluorite content varies from 13% to 23%, calcite – up to 14%, and zinc – in the range of 0.4-0.6%. The research was conducted with tailing samples containing 20.7% of CaF_2 and 10.2% of $CaCO_3$. Mineral grains of the secondary raw material were examined for their surface state and a package of necessary preliminary operations, which provide an efficient material - flotation reagents interaction, was determined. It has been established that extraction of fluorite to concentrates that would match existing requirements is possible in principle. The use of ultrasonic treatment of material in its preparation for flotation is conditioned by an effective desorption of films of different nature from mineral particles. The suggested scheme of fluorite flotation involves preliminary treatment of raw material by ultrasound with dumping of released mud fractions. The optimal operating regime allows obtaining concentrates of higher quality than it is under usual technology without the ultrasonic exposure. In this process, the extraction of fluorite to a concentrate with 94.47% of CaF_2 increases up to 61.11%, which is higher by 2.96% against the original studies.

Key words: mineral mining and processing wastes, fluorite, surface formations, ultrasound, desorption, mud fraction, flotation concentrate.

1 Introduction

The major factor of achieving high efficiency of flotation in the processing of low-grade difficult raw material and processing waste is the creation of conditions for selective adsorbation of reagents on the surface of mineral particles. For the finely disseminated ores, which require fine size degradation, the problem of surface purifying is conditioned by the presence, in prepared material, of a huge amount of mud with the developed surface which surrounds particles of productive size, combining into non-selective aggregate [1, 2]. In the process of mine wastes dressing the problem is aggravated, since mineral grains' surface

* Corresponding author: olya-vo@mail.ru

has residuals of reagents used at the stage of primary processing, various formations conditioned by the long-term storage in the salt-water environment [3]. To ensure necessary contact of mineral particles with reagent and further selection it is essential to supplement the flotation scheme with the operations of surfaces desorptions. Meanwhile, the separation into a different product with further removal from the scheme of desorbed chemical and mechanical surfaces, in order to avoid their negative effect on the technology' further processing, is relevant.

2 Materials and methods of research

Desorption can be performed with the usage of different methods: mechanical, chemical, physical-chemical, and methods based on usage of complicated energetical impulses. One of the most perspective directions is usage of ultrasonic machining. A wide range of different transformations follows the ultrasonic machining of rock slurries. According to the studies of Agranat B. A., Glembotskiy V. A. and others [4-6], the influence of acoustic vibrations on hydro-mineral mixes can be multi-faceted. Thereby, the physical-chemical characteristics of liquid phase, the characteristics of adsorptive and hydrate layers on the surface of mineral particles are changing. Transformation of different physical-chemical characteristics of pulp can be happening under the effect of ultrasound, e.g. conductivity, oxidation-reduction capability, water pH. Moreover, possible destruction of carboxy-containing compounds' colloidal structures, which are used in the flotation process of the majority of calcium minerals, can be followed by increase in dispersibility level of used collectors, intensification of flotation reagents' action, and decrease in their consumption. Traditional methods of mineral surface cleaning are chemical and heat treatment, mechanical rubbing usually doesn't guarantee necessary quality of cleaning, especially cleaning from filming and impurities located in micro fractures and holes. The usage of ultrasound for these purposes seems to be very promising. The removal of natural films and coats together with secondary formations from the surface of mineral particles improves the contact of flotation reagents with a mineral and supports the activations of adsorptive processes [7-9].

Nowadays the usage of processing wastes is the most promising direction of solving the problem of raw material sources shortage for a range of manufacturers in the mining industry. According to the current data, the amount of fluorite ores processing tailings in the tailing dump of Yaroslavskaya Mining Company (YMC), located in the Voznesenskiy ore district (VOD), Primorskiy region, is estimated to be more than 30 million tons. The contains of fluorite is within 13 - 23%, of calcite – not more than 12 - 14%. The average number of carbonate module ($M_к = \alpha_{CaF_2} / \alpha_{CaCO_3}$), which mostly defines the enrichability of carbonate-fluorite ores, amounts in 1.4 -1.6. Moreover, available for mining ores of VOD contains no more than 26 - 29% CaF_2 and less than 20 - 25% $CaCO_3$. All of the ores, being primary raw materials of the company, are considered to be complex ores. The main reasons for this are – extremely thin mutual intergrowth of mineral phases, presence of tiny growths, additional minerals and calcium components, which are close to flotation by characteristics, on the fluorite particles [10]. In technogenic tailings fluorite is presented by the most problematic grains, which were not extracted at the stage of primary benefication. The content and characteristics of processing material determine the necessity of special way of technological evaluation.

The research of possibility of getting fluorite concentrates from processing material was made on the material of tailing sample, taken from the tailing dump of dressing plant of YMC. The chemical analysis discovered the following contents of main components: CaF_2 – 20.7%, $CaCO_3$ – 10.2%, SiO_2 – 32.2%, Zn – 0.38%. During the conduction of research there have been found technological solutions, which can assure the possibility of fluorite concentration together with the achievement of satisfying figures due to rational selection of disruption mode, pH environment in main fluorite flotation and usage of selectively operating compositions of modificators and collectors [11]. During the abovementioned process the concentrates have been extracted, containing more than 93% of CaF_2 with the extraction of 58-60% of fluorite. The decrease of CaF_2 content for 1 - 1.5% gives the possibility to boost the extraction (for 10% and more) due to the redistribution of the most contaminated with additives fluorite grains. That is why the research of influence and additional preparation of flotation feed, aimed at cleaning of mineral grains, desorption from the particles' surface of different chemical, slurry and other coverings, is worth noticing. The additional preparations of the surface can be followed by a reasonable increase of desorption processes selectivity and relevant rise in the quality of concentrates.

3 Results and discussion

The research of optimization of the process of flotation' fluorite extraction from the processing materials with the usage of ultrasonic machining of given material was based on the scheme, given at the fig. 1.

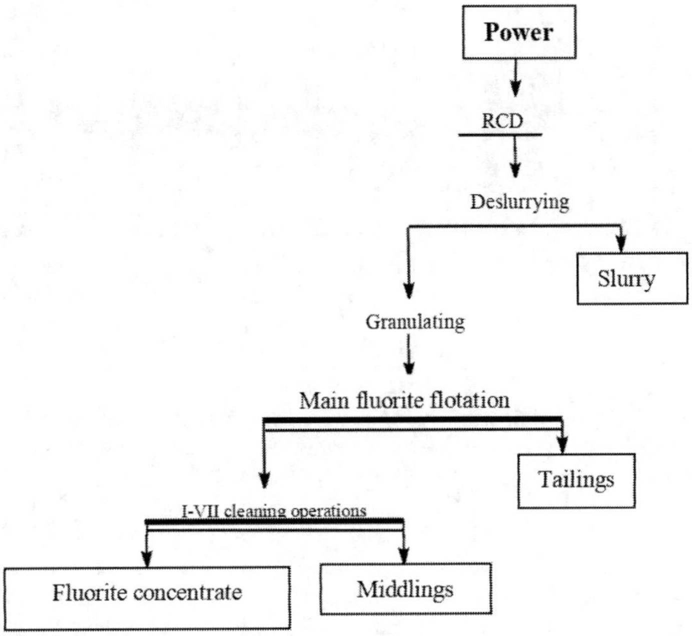

Fig. 1. The scheme of processing tailings flotation with the preliminary ultrasonic desorption of primary cleaning circuit surfaces.

The ultrasonic machining was used on the researched material at its original feed size, before being granulated. The machining was processed with the use of ultrasonic device IL100-6, produced by Saint-Petersburg company "Ultrasonic equipment INLAB". The ultrasound operating frequency – 22 kHz. The operation time – 7 min, which was calculated based on results received earlier, in our ore-based research [12]. Desorbed chemical compounds and slurry particles were extracted into separate product and removed from the scheme, because the possibility of negative impact on the further technology processing is rather high.

The flotation of abovementioned material was done with the usage of TOFA as of collector, modificators – mix of ammonium fluoride salts and lignosulfonates. Depending on the consumption of main reagents (three modes) the concentrates of different quality were extracted. The results of experiments, given at the fig. 2, show major influence of ultrasound on the process. After the treatment we had a chance to get concentrates with the CaF_2 content 93.5 - 94.47%, against 93.46% in the experiment, conducted without preliminary insonation. At the same time we can see the growth of fluorite extraction for 2.27 - 2.96%.

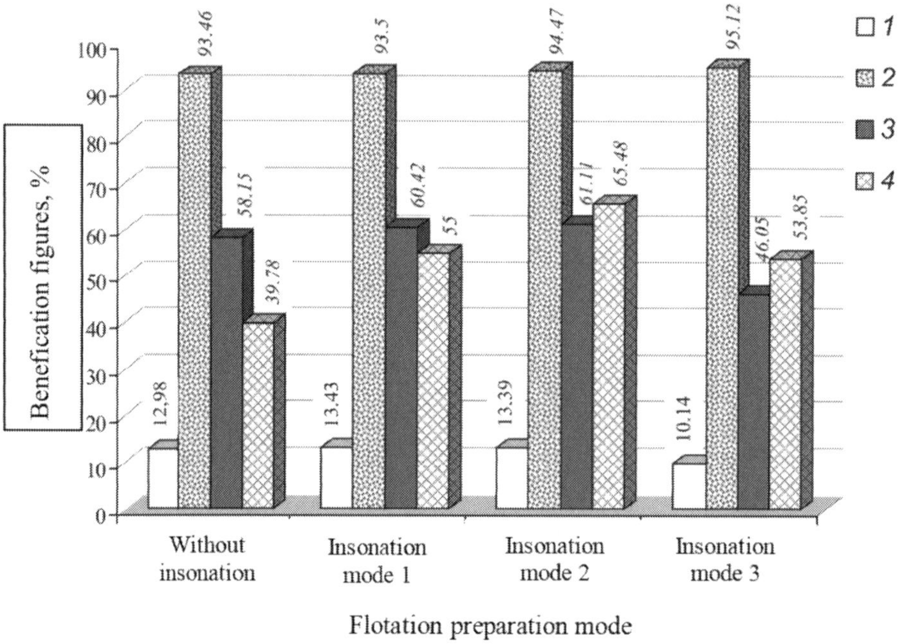

Fig. 2. Comparative figures of fluorite benefication from the processing tailings with and without ultrasonic treatment of given material: 1 – concentration ratio; 2 - CaF_2 content in the concentrate; 3 – fluorite extraction; 4 – selectivity index of fluorite and calcite.

At the experiments with the hardest mode of carbonates drop (mode 3) the concentrate with content of CaF_2 95.12% was extracted. The extraction of fluorite in these conditions decreased till 46.05%. Taking into account that current consumers' market demand for high-quality fluorite concentrate of brand FF-95, FF-97 has notably grown, the result achieved is worth attention. According to the previously made research, during the usual mode flotation, without the usage of ultrasound, the fluorite extraction into high-quality concentrate with the content of CaF_2 being 95.07% accounted for 40.7%.

Acoustic effects let us increase the extraction for more than 5%. Moreover, it is worth mentioning that experiments with the moderate mode of proportion mix of collector and modificators (mode 2) resulted in getting the concentrate with the CaF_2 content of 94.47%, which is quite close to the requirements for the concentrate brand FF-95. The extraction of fluorite in this case is reasonably higher (61.11%).

The content of silica, which is harmful additive and is limited due to the All-Union State Standard (GOST) and technical conditions, is estimated for 1.9 - 2.3%, which is suitable for brands FF-92, FF-92-A.

Conclusions

Therefore, the usage of ultrasound at the stage of preparation for flotation of processing wastes aimed at desorption of primary processing reagents tailings and different coatings with the further separation of slurry fraction ensures effective preparation of mineral particles to the intercommunion with collectors.

In the optimum technological performance with the preliminary processing of materials it is possible to get a concentrate with the content of 94.47% of CaF_2 with the extraction of 61.11% of fluorite. The extraction of fluorite into high-quality concentrate, containing CaF_2 95.12%, decreases down to 46.05%.

In the usual technological performance, without the usage of ultrasound, in all other equal conditions, the extracted concentrate contains CaF_2 93.46% with the extraction of 58.15%.

References

1. V. I. Klassen, D. I. Nedogorov, I. Kh. Deberdeev, Slurries in the flotation process 160 (M.: Mineral resources, 1969).
2. L.A. Kienko, O.V. Voronova. Phys. Tech. Prob. Sol. Minerals, **1**, 176 (2014).
3. L.A. Kienko, O.V. Voronova, S. A. Kondratyev. Phys. Tech. Prob. Sol. Minerals, **1**, 158 (2017).
4. B. A. Agranat, M.N. Dubrovin, N. N. Khavskiy, The basics of physics and ultrasonic engineering, 352 (Moscow, 1987).
5. V.A. Glembotskii, M.A. Sokolov, I.A. Iakubovich and others Ultrasound in mineral deposits beneficiation. 212, Almaty (1972).
6. C. Letmahe, B. Benker, L. Gunther, Intensification of foam flotation using ultrasound, **43,** No. 4, 32, (2002).
7. L.A. Kienko, O.V. Voronova, Zh.A. Shagina. Problems of comprehensive development of geological resources: Materials of VI National scientific conference with the participation of foreign scientists. Khabarovsk, 21, 233-241 (05-07 October 2016)
8. H. Mitome, Action of ultrasound on particles and cavitation bubbles, Proc. World Congress on Ultrasonics, Paris, 2342 (2003).
9. S. G., Ozkan, H. Z. Kuyumcu. Ultrasonics Sonochemistry, **14**, 639 (2007).
10. L.A. Kienko, L.A. Samatova, O.V. Voronova, L.N. Plusnina. Economics and problems of mineral resources exploitation, MIAB, Far East, **9**, 302 (2007).
11. L.A. Kienko, O.V. Voronova. III National scientific-practical conference – Geomechanical and geotechnological problems of mineral resources exploitation of the North, Yakutsk, MIAB, SV **30**, No. 7, 213 (2015).
12. L.A. Kienko, L.A. Samatova, O.V. Voronova. Promlems of comprehensive development of geological resources. Materials of V National scientific conference in Khabarovsk city, Mining information-analytical bulletin, Instalment **4.** 172 (2013).

Efficiency of centrifugal flotation during the refining of gold-bearing enrichment products

Svetlana Salomatova[1*]

[1]Mining Institute of the North, Siberian Branch, Russian Academy of Sciences, Yakutsk, Russia

Abstract. The results of enriching the products of gold-antimony ore processing using centrifugal flotation are presented. A series of experiments was conducted on a laboratory centrifugal flotation machine with peripheral unloading of concentrate (CFM) and on a laboratory mechanical flotation machine (FM). The selectivity of deposition of mineral particles from the flotation mineralized foam flowing on an inclined plane after flotation under different conditions (with a collector and without the use of a collector) has been studied. It has been shown that it is possible to regulate the intensity and selectivity of deposition of material from the foam layer by using a collector. It is established that the combination of the effect of the centrifugal force of fluid rotation and the hydrophobic properties of minerals makes it possible to strengthen the selectivity of mineral separation, flotation in a centrifugal flotation machine can be used for purification operations with the aim of improving the quality of concentrate. **Keywords**: Concentration, centrifugal flotation, mineral particles, hydrophobicity, concentrate, extraction, degree of concentration.

1 Introduction

The solution of the problem of separation of minerals with close technological properties is traditionally carried out by increasing the selectivity of separation in the enrichment processes - using directional reactants, as well as selective changes in the technological properties of minerals with the help of various energy effects [1,2]. One way to influence the efficiency of the flotation process is the use of additional centrifugal forces of fluid rotation - centrifugal flotation [3].

In flotation, sulfide minerals usually pass into the concentrate, as well as rock-forming minerals, mainly in the form of slurry particles. For the subsequent cleaning of concentrates from empty rocks, a combined method of refinement is sometimes applied, which is reduced to an additional classification of the material for sand and slurry fractions and for the production of tailings in the slurry fraction [4]. Decrease in the proportion of sulfide minerals in flotation concentrates in the purification cycle is carried out in the depression mode of some of them. When processing ordinary gold-bearing ores, where the main sulfide minerals are represented by pyrite and arsenopyrite, the process of selective flotation often consists in the separation of these minerals [5].

* Corresponding author: ssalomatova@mail.ru

In the method of flotation developed at Chersky North Mining Institute of the SB of RAS (ChNMI SD RAS), the separation of mineral particles occurs by hydrophobicity in a thin layer on the surface of a rotating flow of water (pulp) [6-9]. The studies were carried out on the products of enrichment of gold-antimony ore. Sentachan gold-antimony deposit is characterized by exceptionally rich ores. [10].

Studies of the selectivity of the precipitation of mineral particles from the flotation mineralized foam flowing along the inclined plane after flotation under different regimes (with a collector and without the use of a collector) have shown that it is possible to regulate the intensity and selectivity of deposition of material from the foam layer by using a collector. The presence of heterogeneity of the flow of water under the foam (comparative results of studies with baffles) on the selectivity of detachment of particles is unimportant when using the collector, but it can have a negative effect on the process of separation of minerals in a "hungry" mode [6].

2 Methodology of experimental research:

Experimental studies were carried out on finely divided ore materials of Sentachan deposit. Preliminary a set of studies was conducted on the research of the substance composition of the materials used. For selection, processing and analysis of samples, existing methods for identifying the granulometric and mineral composition and physical properties of minerals have been used.

For the analysis of the real composition of the separation products, it is envisaged to use assay analysis (for gold), chemical analysis (for antimony and arsenic), spectral analysis of elemental composition on a serial X-ray fluorescent spectrometer Bruker SRS-3400.

Characteristics of the flotation reagent used: Butyl xanthate is the main collector for the flotation of sulfide minerals, it is used in the form of a 2-10% solution at a flow rate of 20-100 g / t [5].

The research uses a centrifugal flotation machine with peripheral discharge of concentrate, designed by the ChNMI SD RAS (protected by the RF patent) [8]. The engineering pecifications of the flotation machine are presented in Table 1.

Table 1. Specifications of the centrifugal flotation machine with peripheral unloading of concentrate

Specifications	Unit of measure	Operation parameters
Consumption of additional water	l / min	2-3
Air consumption	m^3 / min	0.05
Number of body rotations, max.	rpm	400
Motor drive power	kW	under 1.0
Size:		
diameter	mm	230
height	mm	180

Taking into account the natural flotation activity and high concentration in the samples of the minerals under study (pyrite and arsenopyrite), two flotation modes were tested - in a "hungry" mode and in a mode with the addition of potassium xanthate.

It was found that in the flotation variant with the use of a collector, the separation of most of the sulfide minerals under study from the water surface is characterized by reduced dynamics (with the exception of arsenopyrite, all minerals retain retention in the foam layer), and in a series of flotation experiments in a "hungry" mode, the selectivity of minerals separation from the surface water is significantly increased. At the same time, the retention capacity of sulfide minerals on the water surface is significantly higher when a collector is used.

3 Results

A series of experiments on flotation on CFM and on FM carried out with the products of gold-antimony ore enrichment, with the same reagent mode established that the extraction of gold and antimony in the concentrate is higher on CFM (Figure 1), and the extraction of arsenic is lower, which makes it possible to obtain the concentrate with an *As* content of 0.11%.

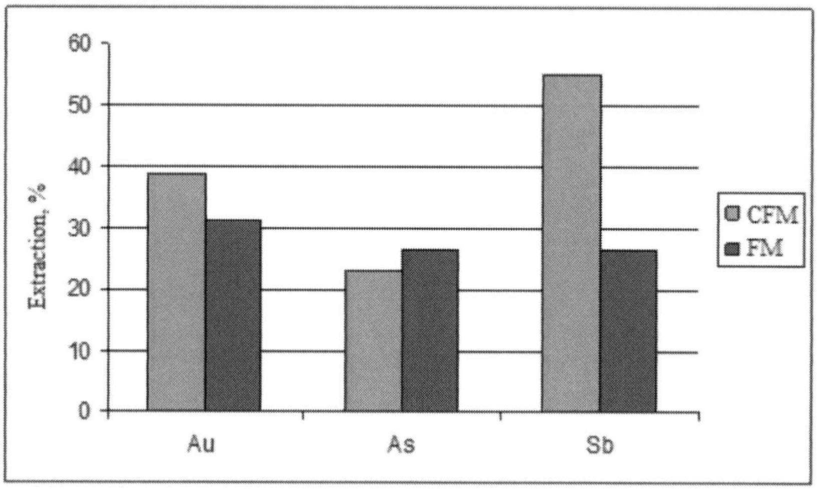

Fig. 1. Extraction of gold, arsenic and antimony into the concentrate during flotation on the centrifugal flotation machine (CFM) and on the laboratory mechanical flotation machine (FM)

Figure 2 shows that by the quality of concentrates the content of arsenic in the concentrate of CFM is lower than in the concentrate of the laboratory FM, while the content of antimony is much higher.

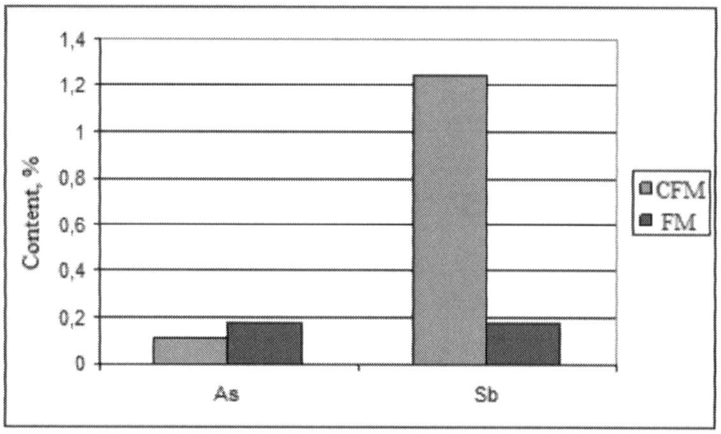

Fig. 2. The content of arsenic and antimony in concentrates of flotation of the centrifugal flotation machine (CFM) and the laboratory flotation machine (FM)

4 Conclusion

It has been experimentally established that the combination of the effect of the centrifugal force of fluid rotation and the hydrophobic properties of minerals makes it possible to strengthen the selectivity of the separation of minerals, flotation in a centrifugal flotation machine can be used for purification operations with the aim of improving the quality of concentrate.

References

1. V.A. Chanturiia, A.A. Lavrinenko, Krasnov, G.D. Mining Magazine, **10**. 48-52 (2006).
2. A.A. Lavrinenko, Krasnov, G.D. Mining Magazine, **2.** 108-117 (2007).
3. Iu.B Rubinshtein., E.Ia. Perelman, I.N. Spivakovskii, Review and information bulletin / Center for R&D and technical and economic studies on heavy and transport machinery (TSNIITEEItyazhmash), - Moscow. **1**. p. 24 (Mining equipment, Series 2) (1988).
4. T.G. Gorbunova, Iu.I. Frolov, The development of technology for the enrichment of ore and placer deposits: Collection of academic papers of VNII-1.Magadan, 3-9 (1985).
5. V.V. Lodeischikov, Non-ferrous metals. **4**. 51–55 (2005)
6. A.I. Matveev, S.I. Salomatova, Flotation of gold on the surface of a rotating liquid; ChNMI SD RAS. - Yakutsk: Publishing House of the Yakutsk Scientific Center of SD RAS, - 141 p. (2008)
7. Patent 2183998 RF, 7B03D 1/02 1 / 24. Flotation process and centrifugal flotation machine / A.I. Matveev, S.I. Salomatova, V.B. Iakovlev, A.M. Monastyrev, N.G. Eremeeva, E.S. Sleptsova; the applicant and the patent holder ChNMI SD RAS; 25.05.2000; published on 07.27.2002 // Inventions. Useful models. **18**. Part 2. p. 170 (2002).
8. Patent 2248849 RF B03D 1/24. Flotation method and centrifugal flotation machine / A.I. Matveev, S.I. Salomatova, A.I. Chikidov, A.M. Monastyrev, V.B. Iakovlev; the applicant and the patent holder ChNMI SD RAS; application 05/08/2002; published on 27.03.05 // Inventions. Useful Models. **9**. Part IV. p. 949 (2005).
9. S.I. Salomatova, Mining information and analytical journal, **11-24**. 240-246 (2017).
10. G.I. Boltukhayev, P.M. Solozhenkin, Non-ferrous metals, **4.** 41-44 (2009).

Strategic approaches of combined electrothermal processing of coal and carbonate mineral raw materials for obtainment of highly efficient synthetic energy carriers and non-fuel products

Anatoly Zhukov[1*]

[1]Far Eastern Federal University, 8 Sukhanov Str., Vladivostok city, Russia

Abstract. Coal gasification and the production of gaseous fuels include three principal directions related to the production of fuel gas: 1) composition and heat capacity of the produced gas; 2) gas generator structures; 3) characteristic properties of the obtained alternative product - low CO content and gas toxicity, which allow making full use of this gas for domestic purposes. In industrial processes of coal conversion, the following combined technologies are used most often: — semi-coking + gasification of fixed ash (low-temperature coke); — semi-coking + hydrogenation of liquid product (tar); — gasification + synthesis of high molecular weight hydrocarbons from the produced SYN gas (CO+H) (Fischer-Tropsch synthesis). The choice of the layout for obtaining SLF (synthetic liquid fuel) can be based on specific conditions, the cost and quality of coal, energy supply, market conditions. The products obtained in the process of gasification and hydrogenation of coals pollute the atmosphere much less than the coal burned in electrical power plants. When implementing the organizational and technological model of innovative production, the first stage includes the following combined approaches for the processing of mineral raw materials and new products:
1. processing of carbonic mineral raw materials: calcium carbide, carbon dioxide (in a gaseous, liquid or solid state);
2. acetylene, plant growth regulators (PGRs), plant protection products (TAKAR).
The second stage includes fuel and non-fuel products:
1. synthetic ethyl alcohol (ethanol), antifreeze, ethylene glycol, dichloroethane, synthetic drying oils, acetone, etc .;
2. carbamide (urea), ammonia, nitrogen in gaseous and liquid states, methanol, gasoline, etc.
Keywords: Strategic approaches, coal, carbonates, mineral raw materials, refining, gasification, synthetic liquid fuel, products, fuel and non-fuel products

* Corresponding author: yul25juk@mail.ru

1 Introduction

Until the middle of the last century, coal resources were the basis of the world and Russian energy. Since the 1950s, due to an increase in the level of oil and gas consumption, there has been a steady decline in coal production and use. At present, the share of coals in the world energy balance constitutes on the average about 30%. According to some calculations, the current level of oil and gas production will be preserved until the middle of the 21st century and already beyond 2050 the demand for coal resources in the world will increase sharply. Today, coal should be considered as energotechnological raw materials and it should be used on a large scale only in a complex way, dividing its potential chemical energy into approximately equal parts between energy (electricity and heat) and chemical products. Organization of energotechnological processing of solid fuel on the basis of large-scale power engineering (thermal and, possibly, nuclear), will allow us to overcome the oil and gas crisis that awaits us in the 21st century. Negative trends in the raw materials base of oil and gas production (the reduction of resources, while it is necessary to maintain high export supplies) require a wide and rapid introduction of new coal technologies targeted at the ecology and energy saving and multiple uses of all types of coal. The coal gasification process is multi-purpose with respect to the composition of the produced gas.

The importance of the issues of technological and economic development of the coal industry, which is a complex production and economic system associated with high costs, the availability of competing energy carriers (oil, gas, nuclear fuel), is increasing due to the need to improve the technologies of mining, combined processing and use of coal, the provision of competitive forecast and industrial reserves.

The immediacy of the problem is growing even more in connection with the strategic importance of hard and brown coal for the energy sector in Russia and all over the world, which by their geological reserves exceed significantly such major energy sources as oil and natural gas, as well as nuclear fuel and potential water resources. Coal, according to the World Coal Institute, accounts for about 90% (according to other estimates - 60-80%) of the energy potential of all organic mineral resources suitable for mining. It is estimated, that comparable geologic reserves of coal in the bowels on the territory of Russia are also several times higher than similar reserves of oil and gas. The Strategy for the Development of Fuel and Energy Resources of the Far Eastern Economic Region until 2020 included the following tasks: "... to explore ways of overcoming the crisis situation by developing Far Eastern coal, oil and gas deposits on the basis of progressive methods of extraction and deep chemical processing of primary raw materials, as well as the introduction of a saving regime..." In this connection, it was necessary: 1) to develop physical and technical and economic and technological models of technologies and technical means for deep, integrated processing of coal and carbonate mineral raw materials; 2) to apply electrothermal and to develop plasma ecologically safe resource and energy saving technologies for processing of coal and carbonate mineral raw materials using several modules (subsystems) to produce competitive fuel and non-fuel products [1,2,3,4].

2 The theory and methods of experimental research

When receiving gaseous fuels, three main approaches are identified, related to the production of fuel gas, a substitute for natural gas and synthesis gas ($CO + H_2$). The composition and heat capacity of the produced gas depends not only on the gasification regimes but also on the structure of the gas generator used. The use of fuel gas allows solving environmental and technological problems in power engineering, metallurgy, and other industries.

The above-mentioned processes are rarely used in pure form during the industrial introduction of brown coal refining processes. The following combined technologies are used most often:

- gasification + gas burning + thermal and electric energy generation;
- semi-coking + gasification of fixed ash (low-temperature coke);
- semi-coking + production of adsorbents;
- semi-coking + hydrogenation of liquid product (tar);
- gasification + synthesis of high molecular weight hydrocarbons from the produced SYN gas ($CO+H_2$) (Fischer-Tropsch synthesis);
- gasification + methanol generation + production of gasoline (Mobil process).

Production of liquid fuels based on Fischer-Tropsch synthesis comprises converting of various carbon compounds (natural gas, hard and brown coal, heavy oil fractions, industrial wood residue) to synthesis gas (a mixture of CO and H_2) with further turning into synthetic "crude oil "- Synthoil. This is a mixture of hydrocarbons, which is divided during the subsequent processing into various types of practically environmentally friendly fuel, free from impurities of sulfur and nitrogen compounds. It is enough to add 10% of synthetic fuel to conventional diesel fuel so that diesel fuel combustion products met the environmental standards.

The composition and heat capacity of the produced gas depends not only on the gasification regimes but also on the structure of the gas generator used. The use of fuel gas allows solving environmental and technological problems in power engineering, metallurgy, and other industries. The main characteristic property of the substitute natural gas is the low content of CO and, consequently, relatively low toxicity, which makes it possible to use this gas for domestic purposes. Synthesis gas is used for chemical refinement into methanol, motor fuels or for hydrogen production. To produce liquid fuels directly from coal, the processes of hydrogenation, pyrolysis, and solvent liquefaction are used. Production of fuel oil (alternative product for residual fuel oil) and motor fuels requires additional application of the processes of hydroprocessing of liquid coal products in order to reduce the sulfur content and other undesirable impurities. "Coal oil", obtained in the process of catalytic hydrogenation of coal, is easy to process [5,9].

The developed and patented technologies provide for a complex waste-free processing of coal and carbonate mineral raw materials with the use of several modules ensuring the production at the first stage: calcium carbide, calcium oxide, carbon dioxide; at the second stage: - calcium carbide derived products: synthetic gaseous and liquid fuels and non-fuel products.

In the new manufacturing unit for the production of carbide (CaC_2), limestone ($CaCO_3$) is preliminarily calcined to produce calcium oxide (CaO) and carbon dioxide (CO_2); batch of calcium oxide (CaO) and hard coal (C) are used to produce calcium carbide (CaC_2). Carbon dioxide (CO_2) escaping during limestone calcination is captured by equipment for trapping of exhaust gases and converted into carbonic acid gas (H_2SO_3). In turn, calcium carbide is processed into plant growth regulators (PGRs) and plant protection products (TAKAR).

Production of acetylene from calcium carbide causes the formation of calcium hydroxide - $Ca(OH)_2$, which is used in construction, in agriculture for production of superphosphate - $Ca(H_2PO_4)$, in water purification systems, etc. Synthesis of carbamide (urea) - $CO(NH_2)_2$ is possible with the participation of carbon dioxide (CO_2) by addition of ammonia (NH_3) or from calcium carbide with the use of nitrogen (N_2). At the same time, for the production of nitrogen (N_2), the separation of the air mixture is used - the rectification of the liquid air or the oxidation of ammonia: $4NH_3 + 3O_2 = 2N_2 + 6H_2O$. Carbamide (urea) is used to increase the efficiency of plant growth regulators (PGRs).

Calcium oxide, calcium carbide, acetylene, methanol, carbamide, PGRs and TAKAR fertilizers are the main derived products of coal and carbonate mineral raw materials; additional products are the following: carbonic acid gas, drinking or calcined water, calcium hydroxide. In turn, the processing of acetylene allows you to obtain: synthetic ethylene and ethyl alcohol, naphthalene, hydrogen; synthetic drying oils, chloroprene rubber. In accordance with the polymerization reaction, the following products are also produced: ethylene, polyethylene (polymer), propylene, polypropylene according to the following methods of acetylene processing (C_2H_2):

1) Acetaldehyde (CH_3CHO) → synthetic ethyl alcohol (C_2H_5OH);
2) Benzene (C_6H_6) → naphthalene ($C_{10}H_5$) → hydrogen (H_2);
3) Vinyl-acetylene (CH_2 : CH·C: CH) → divinyl-acetylene (CH_2: CH·C: C·CH : CH_2) → synthetic drying oils;
4) Vinyl-acetylene (CH_2 : CH·C: CH) → chloroprene (CH_2 : CH·C Cl : CH_2) →chloroprene rubber;
5) In accordance with the polymerization reaction, the following products are also produced: ethylene

$$(nCH_2 = CH_2) \xrightarrow{\text{Temperature, Pressure, accelerant}} (-CH_2 - CH_2 -)n$$ - polyethylene (polymer);

1) propylene $$nCH_2 = CH - CH_3 \xrightarrow{\text{Temperature, Pressure, accelerant}} \left[\begin{array}{l} -CH_2 - CH \\ \qquad\quad\ \ \downarrow \\ \qquad\quad\ CH_3 \end{array} \right]_n$$ - polypropylene.

3 Results and discussion

The results of technological researches and economic and mathematical modeling of processes allow us to consider in more detail the configuration and parameters of the systems of restructuring and diversification of coal-mining energy and industrial enterprises, to agree and determine the operating modes of control subsystems and production organization, to predict the technological and economic performance of newly created and diversified production complexes.

Picture 1 shows the example of physical and technological model of installation for the processing of coal and carbonate mineral raw materials. Similar physical and technological models are developed for the most part of products mentioned above and protected by certificates issued by authors [7,8,9].

The device works the following way.

Limestone is placed in the furnace 1 and burnt. In the process of calcination, the limestone decomposes into lime and carbon dioxide (CO_2). Lime (CaO) is removed by the product pipeline 14 to the dispenser 6, which ensures the selection of the part of lime intended for transfer to the consumer and the part intended for further processing. Part of the lime, intended for further processing, is fed to the second furnace 2, where in the presence of carbon (in the form of coke or hard coal, 20-25 mm in size and less than 1% sulfur content) is processed into calcium carbide. Calcium carbide (CaC_2) is removed by the product pipeline 14 to the dispenser 7, which ensures the selection of the part of calcium carbide intended for transfer to the consumer and the part intended for further processing. A portion of the calcium carbide intended for further processing is fed to the gas generator 4 where it is brought into contact with water and processed into acetylene.

The synthesis of ethanol is carried out according to a known scheme with the production of ethylene at an intermediate stage – C_2H_4:

$C_2H_2 + H_2 = C_2H_4$; $C_2H_4 + H_2O = C_2H_5OH$

Moreover, if the reaction is carried out at a pressure of up to 23 atm and a temperature of 75-80°C, the yield of ethanol is up to 90%, and at a pressure of up to 80 atm and a temperature of up to 280-300°C, the ethanol yield reaches 95%.

In this case, the synthesis of dichloroethane is carried out according to the following scheme:

$C_2H_2 + H_2 = C_2H_4$; $C_2H_4 + Cl_2 = CH_2Cl\text{-}CH_2Cl$.

Ethylene glycol is produced by hydrolysis of acetylene.

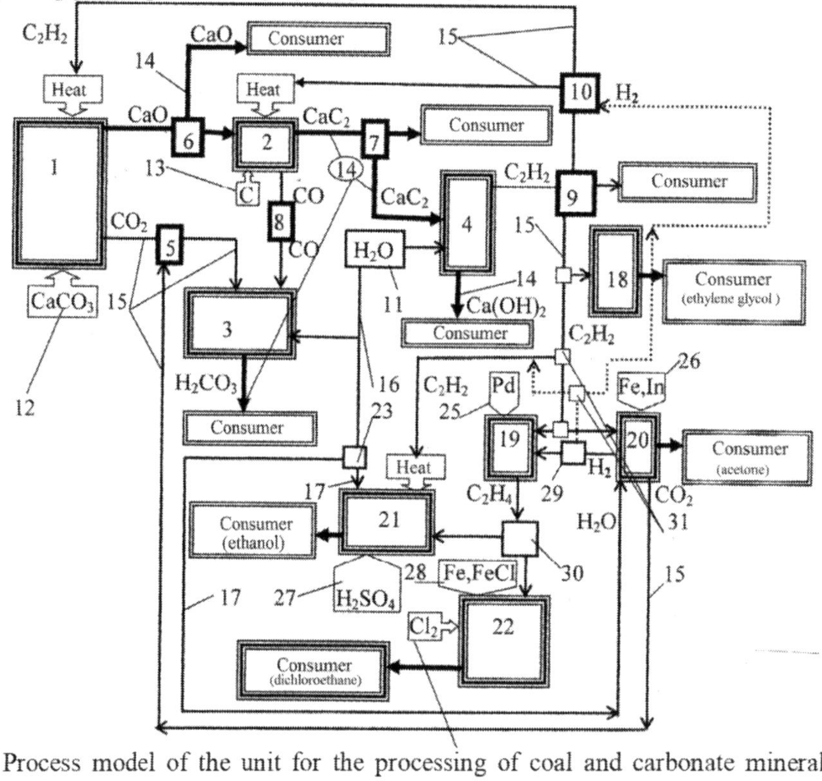

Fig. 1. Process model of the unit for the processing of coal and carbonate mineral raw materials.

The synthesis of acetone is carried out according to a known scheme, at a temperature of about 460°C. Acetone is transferred to the consumer, and gaseous $C_2H_2 + H_2O$ (steam) = $H_3C - C - CH_3 + CO_2 + H_2$ products (process sludge) are recycled locally (CO_2 is used in the production of carbonic acid gas and hydrogen - in the process of ethanol and/or dichloroethane synthesis). Hydrogen combustion is also possible, i.e. that it can be used as an energy carrier in the process of calcination or acetone synthesis, but this is not the most appropriate way to use it (the corresponding bonds are shown in dotted lines).

Thus, calcium oxide, calcium carbide, acetylene, ethanol, dichloroethane, acetone, ethylene glycol are the main products of processing of coal and carbonate mineral raw materials. As additional products, carbonic acid gas, calcium hydroxide, PGRs, and TAKAR are obtained. In case there are consumers and corresponding additional components, the main derived products from the processing of coal and carbonate mineral raw materials can be used for further processing according to the known schemes to obtain, for example, superphosphate, carbamide, ammonia, etc. [9].

The results of technological researches and mathematical modeling of processes allow us to consider in more detail the configuration and parameters of the systems of restructuring and diversification of coal-mining and energy enterprises, to agree and determine the operating modes of control subsystems and production organization, to predict the technological and economic performance of newly created and diversified production complexes.

In this case, optimization of production costs is carried out on the basis of the following criterion:

$$\Pi = \sum_{1}^{m}\sum_{1}^{n}(C_{ji} - S_{ji})V_{ji} \rightarrow max, \quad (1)$$

where Π – is the total profit margin of the coal mining and energy enterprises gained due to the sale of products; m – is the number of coal ranks; n – is the number of methods of processing and use; V_{ji} - is the physical quantity of coal of the rank J, processed and used in the I way; S_{ji} - is the production cost of coal of the rank J, processed and used in the I way; C_{ji} – the price of coal of the rank J and fuel of I type.

Moreover, the following restrictions are imposed:

$C_{ji} \leq C_{ji} \leq C_{ji}$, restrictions are established on the basis of market research;

$0 \leq V_{ji} \leq$ minimum ($v_{ji}^{tech.}$; $v_{ji}^{cons.}$), where $v_{ji}^{tech.}$ – maximum possible volume of production; $v_{ji}^{cons.}$ – maximum possible volume of sales.

The results of calculations of economic parameters of integrated processing of coal and carbonate mineral raw materials are carried out under the conditions of Spasskcement OJSC, Partizanskaya GRES and other industrial enterprises of Primorsky District.

When calculating the technical and economic indicators of the location of competitive production, the following indicators are identified: **return on investment, production cost, sales margins, payback period, profitability index** [4,10,11].

4 Conclusions

I consider it expedient to carry out research and development works together with the Engineering School of FEFU and Institute of mining of the Far East Division of Russian Academy of Sciences for increasing the efficiency of development and use of fuel and energy resources in the following areas:

1) to develop physico-technical and economic-technological model of technologies and technical means for deep chemical stage-by-stage processing of coal and carbonate mineral raw materials;

2) to apply electrothermal and to develop plasma ecologically safe resource and energy saving technologies for processing coal and carbonate mineral raw materials using several modules (subsystems) to produce competitive fuel and non-fuel products;

3) to develop Technical Design Assignment (TDA) for the performance of R&D (research and development), NPD (new project development), TEJ (techno-economic justification) with the solution of the following tasks: consideration of alternative options for locating the enterprises of newly designed mining and chemical complexes in Russia and abroad; to develop innovative technologies and complexes of technical means for industrial processing of coal and carbonate mineral raw materials using plasma reactors with a capacity of 500; 1000 kW.

References

1. A.P. Sorokin, G.P. Avdeyko, A.V. Alekseev, P.Ya. Baklanov, A.V. Zhukov, V.I. Podolian and others. *Strategic development of fuel and energy potential of Far Eastern Economic Region till 2020, p.* 112 (2001)
2. A.V. Zhukov, "Application & Fundamental research" International journal, **12**, pp. 147-150 (2010)
3. E.S. Korneeva, Master's Thesis majoring in Management., Economics and Management Institute of FESTU, p. 45 (2008)
4. E.A. Gnezdilov, A.V. Zhukov, A.D. Iakovlev, Fundamental research journal, **12(2)**, pp. 342-344 (2007)
5. E.A. Gnezdilov, A.V. Zhukov, Fundamental research journal, **9**, pp. 61-64 (2007)
6. A.V. Zhukov, M.I. Zvonarev, Iu.A. Zhukova, *Patent 74912*, Bulletin 20 (2008)
7. A.V. Zhukov, M.I. Zvonarev, Iu.A. Zhukova, *Patent RU 2362735 C1*, Bulletin 21 (2009)
8. A.V. Zhukov, M.I. Zvonarev, Yu.A. Zhukova, *Patent RU 2373178 C2*, Billetin 32 (2009)
9. A.V. Zhukov, D.E. Kusraeva, Ia.A. Zhukova, RJPBBP-RJPBCS – Res. Jour. of Pharm., Biol. and Chem. Sciences, ISSN09758585-India-Scopus, **7(4)**, 2401-2414 (2016)
10. A.V. Zhukov, Mat. on All-Rus. Conf. with Intern. Part. Mult. Use of Poten. of H. and Br. Coal and Dev. of Comb. Ecol. Safe Tech. (Blagoveshchensk, Amur SC of FEB RAS, 2017)
11. A.V. Zhukov, M.I. Zhukova, Yu.A. Zhukova, A.A. Shmelev, Mat. on Intern. Scien.-Pract. Conf. Kn.-int. Tech. Dev. and Use of Min. Res., **4**, 128-133 (2018)

Brown coals of Far East – natural resource for multipurpose industrial processing

Anatoliy Sorokin[*1,2] and *Andrey* Konyushok[1,2]

[1] Amur Scientific Center FEB RAS, Blagoveshchensk, Russia.
[2] Institute of Geology and Nature Management FEB RAS, Blagoveshchensk, Russia.

Annotation. The analysis of the potential economic value of brown coals of the Far East of Russia is carried out when to use them as a complex chemical mineral resource. It was conducted assessment of industrial attractiveness to use coal combustion or coal chemical processing wastes as an additional source of gold, rare metals and rare earth elements. The Sergeevskoye brown coal deposit of the Zeya-Bureya Sedimentary Basin in the Amur Region was proposed as a potential standard facility for the construction of a coal-chemical processing plant to produce resin, montan wax, complex hydrocarbons with associated gold, rare metals and rare earth elements recovery.

1 Introduction

In the economic complex of Russia and a number of other countries, according to experts' forecasts, hard, bituminous and brown coals remain the main sources of thermal and electric power at Thermal Power Plants. In the Amur region, the total predicted resources of brown and hard coal reach almost 70 billion tonnes, of which 43 billion tonnes are concentrated in the Zeya-Bureya sedimentary basin [1]. Balance reserves in the basin in the amount of 3.8 billion tonnes were explored in 7 deposits (Svobodnoye, Sergeevskoye, Tygdinskoye, Yerkovetskoye, Raichikhinskoye, Arkharo-Boguchanskoye, Ogodzhinskoye). They are represented mainly by lignite and brown coal - 99.1 %, the rest - bituminous coal. All deposits are suitable for open pit development. At present, Raichikhinskoye and Yerkovetskoye brown coal deposits of technological group 2B are being excavated, which have reserves of 1.3 billion tonnes. But the main ones are coals of group 1B (2.3 billion tonnes), which are low-quality raw materials that are not suitable without enrichment as energy fuel, but are very promising as a complex chemical and rare metal resource.

At the same time, coal demand as an energy resource will remain relatively stable until 2033, despite the increase in the use of natural gas for power generation. This is due to the forecast of the outstripping demand for electricity, justified by economic factors, including population growth on the planet [2]. Accordingly, along with the development of technologies that increase the efficiency of using brown coal as a fuel, in Russia and elsewhere in the world, research is underway to improve the technologies for coal

* Corresponding author: amurnc@ascnet.ru

combustion, chemical processing, extraction of mineral components and metals from coal combustion products (CCP).

2 Experimental and analytical research results

Due to chemical processing (coal chemistry), brown coal is converted into liquid fuel and combustible gases. In addition, modern technologies make it possible to produce from brown coals more than 130 types of chemical products and semi-finished products, which are further used for the production of more than 5,000 types of products [3], and extract rare metals and rare earth elements from the CCP. One of the most popular chemical products is montan wax, which is extracted from semi-bituminous varieties of brown coal. This is one of the most durable of natural waxes, which finds application in many industries. It is used mainly in the production of plastics, especially in compounds based on engineering plastics such as polyvinyl chloride, polyamide, polycarbonate, thermoplastic polyurethane and styrene-maleic anhydride. Experimental extraction processing of coals of Svobodnoye, Sergeevskoye and Tygdinskoye deposits allowed to obtain from them at least 12% of raw wax in terms of dry coal [4]. In these pilot experiments the technology of liquid phase catalytic alkylation of coal by aliphatic alcohols, developed by the scientists of AmurNC and IGNM FEB RAS, was used. The application of this method contributes not only to a significant deepening of the extraction process and to the growth of extraction of bitumen and waxes, but also to the production of a modified raw wax enriched with high-molecular ester fractions [5].

At the same time, it became possible to exclude from the production process a laborious, but insufficiently effective, step of demineralizing wax. The basic physicochemical characteristics of the modified wax are comparable with those of esterified waxes, for which multistage technologies are used that significantly increase the cost of the final product. As a result, a method has been developed in which the modified coal is enriched with high molecular weight components, and the wax extracted from it, with a resin content near 2.5 wt. %, does not require multistage processing technology.

In addition to wax, other products extracted from brown coal during processing are also of economic value: resins, humates, carboxylic acids, ethers and a solid residue as a raw material for liquefaction processes, production of inexpensive adsorbents, ion-exchange materials and others.

At pilot plants in Primorye, thermal processing of brown coal at the Lipovetskoye deposit was already performed at low temperature parameters (820-920 K). The yield of resin was 8.8% for dry fuel. When a coal had a semi-coked rank, 17 to 30 wt. % of the resin was obtained, based on the combustible mass. Experimental and industrial research was also conducted at the Artyomovskoye coal field in Primorye, where semi-coke (56.2%), resin (4.83%), dry gas (7%), gasoline (0.17%) and organic substances (0.3%) were obtained by thermal decomposition [6].

High porosity characteristics of brown coal make it suitable for the production of carbon-containing materials that can be used as inexpensive adsorbents for purification of drinking water from soluble volatile compounds and organic compounds of low molecular weight. The increased content of functional groups (carboxyl, phenolic hydroxyls) in brown coals stimulates the development of studies in the field of the use of coal as a cheap ion-exchange medium for the recovery of valuable metals or the recovery of heavy metals from aqueous solutions [3].

It should be noted that wide application of brown coal for agricultural and horticultural purposes is possible. Humic acids derived from coals are used as a soil conditioner and as a fertilizer. Brown coal absorbs ammonium on its surface, which turns it into a nitrogen fertilizer. The technology for producing oxidized humates was developed by the scientists of the AmurNC and the Institute of Geology and Nature Management of the Far East Branch of RAS for coals of the Svobodnoye, Sergeevskoye, Sianchik, and Yerkovetskoye deposits [7]. It is established that humates are most effectively extracted with 0.5- 1.0 wt. % solution of NAOH. The yield of humates from coal 1B was obtained in the range of 55- 65 wt. % per organic mass. The best yields were given by low-ash coals. The product was tested in 2006-2007 on soybean fields, wheat, barley sowings in the experimental farming fields of the Institute of Soy of FEB RAS. The dose of 10 g of dry humates per 100 sq. m gives a statistically significant increase in the yield of grain crops of 1.2, 2.0 and 3.8 kg / 100 sq. m, respectively.

Noble metal (NM) mineralization is found in a number of coal-bearing basins of Siberia, Transbaikalia, the Far East, Great Britain, USA, Uzbekistan, China and other countries [8]. In addition to NM, coal is enriched by rare metals and rare earth elements (REE), which are usually associated with Au and Pt in the coal deposits of Siberia (Nazarovo, Borodino), Priamurye (Raichikhinskoye, Yerkovetskoye, Darmakanskoye, etc.), Primorye (Pavlovskoye, Bikinskoye, Khasanskoye, Shkotovskoye) [1, 8, 9, 10]. In recent years, Russia Geological Institute (St. Petersburg) has been working with the aim of making an economic assessment of the metalliferous of coal in coal-bearing deposits. According to their data, the main coal deposits of the Far East contain significant resources of Sc, Ga, Ge, Rb, Sr, Cs, Zr, Y, REE and platinum. At the same time, for most of the metalliferous coal deposits there is no reliable technology for estimating the content of valuable impurity elements, the conditions and mechanisms of concentration, the form of disseminating have not been studied, the technology of their extraction has not been developed, and the prevalence of useful components along the deposit of the coal seam. But with the improvement of technologies for extracting RM and REE from CCP, a significant increase in the profitability of coal processing at coal-chemical enterprises is possible [12].

In brown coals deposits investigated by the authors, besides of decaying organic matter, there is a significant proportion of clay and accessory minerals transported to coal-bearing strata from surrounding mountain massifs during the formation of primary peatlands. According to our data, the abundance of chemical elements in coal and their relationship to a particular mineral form varies depending on the history of the formation of the coal basin as a whole, such as coal-bearing area, even within a single coal seam. The prevalence of gold ore and rare metal deposits, phosphate deposits with rare earth elements in the rocks around the coal basin led to the accumulation a wide range of chemical elements in the coals. They are connected both with the organic matter of coals, and with the composition of minerals (monazite, zircon, cassiterite, apatite, clay minerals and others). At the same time, the total metal content of the Miocene coals (the Sergeevskoye deposit) turned out to be much larger than the Paleocene deposits (Raichikhinskoye and Yerkovetskoye deposits).

Ash content of brown coals in the Zeya-Bureya basin ranges from 13 to 24 wt. %. After coal combustion, the concentrations of some chemical elements in the ash, for example, Si, Al, S, Fe and Ca, can be in the amount of up to 15-30 wt. %. The concentration of metals in slag, ash and adsorbed gas-dust mixture increases several times - for example, for Zn 30 times, Rb - 14 times, Y - 4 times. Among the most valuable RM studied by the authors in brown coals of the Zeya-Burea basin are Li, Be, Zr, Ta, Nb, and from the group of rare-earth elements - Y and Sc. According to [13], for many coal deposits, abnormally high REE contents, with the exception of Ce and Eu, and also in some cases La, Gd and Y, are not typical. However, we have determined the high REE + Y content in the coals of the

Sergeevskoye deposit (Miocene) (up to 200 g/t) against 60-100 g/t in the coals of the Yerkovetskoye deposit (Paleocene). The content of Be, Ta, Ni, Hf, Sn, W, Ti in the coals of the Zeya-Bureya Basin is insignificant. Analysis of data on the distribution of useful components along the Northwestern frame of the basin indicates that Miocene coals in the lower horizons of the coal bed are characterized by industrial concentrations of Li, Sc, Ga and Cs. The exceptions are Y and Zr - their content in the upper part of the coal cuts is higher, as well as the Au content. In the Paleocene deposits of the Raichikhinskoye and Yerkovetskoye deposits, the Au and RM are unevenly distributed. Au is enriched in the middle and lower horizons of the coal seam, mainly in the Yuzhnoye quarry (up to 3.3 g/t). Elevated concentrations of Li, Cs and Ga are also observed in the lower parts of the coal seam, and Y and Zr are predominant in the upper.

The brown coals of the Zeya-Bureya Basin have geodynamic and geochemical affinities with rocky framing complexes and contain Cr, Cs, Cu, Ga, Li, Ni, Rb, Sn, Sr, Ta, Th, Ti, U, V, Y and REE. Our research indicates that these elements, as a rule, are found in clay minerals (in kaolinite, montmorillonite and feldspar) and in tuff tonsteins in coal.

Industrial values of Cs, Ga, Li, Ti, and V are found in coal seams in clay minerals - kaolinite, illite, smectite, and also in chlorite and carbonates. REE and also Y and Sc, are found mainly in phosphates and organic matter of coals [15].

An attempt to determine the profitability of extraction of elements from coal deposits containing Au and Ga in particular is undertaken in China. Approximately, these concentrations were established for Au at a level of 0.2 g / t and 60 g / t for Ga [14]. The average content of Au in fresh coals of different ages is an average of 1.3 g/t. [1].

3 Conclusions

In the Far East region, the issues related to the restructuring and diversification of coal-mining enterprises, the solution of socio-economic problems in some of its districts, the creation of new highly efficient and profitable industrial enterprises and industries are acute. The transition to innovations in the industry should be based on the use of science-intensive technologies for the extraction and complex processing of coal minerals. The explored reserves of bituminous, brown and lignite coals allow to ensure the stable functioning of the enterprises of the fuel and energy complex and newly created enterprises for complex chemical processing of coal for many decades. This is recognized by many scientific, production and research organizations.

At present, the AmurNC and IGNM of FEB RAS continue an economic assessment of noble metal and rare metal mineralization in coal-bearing deposits in the Far East. According to our data, they contain significant resources of Sc, Ga, Ge, Rb, Sr, Cs, Zr, REE + Y and noble metals. At the same time, the conditions and genesis of concentrating gold and metals, chemical and mineral forms of useful metals in coals require further study. With the improvement of technologies for extracting rare metals and REE from the CCP, a significant increase in the profitability of enterprises involved in the extraction, combustion and chemical processing of brown coal is possible.

The effective way of involving coals in complex processing in the Far East should be based on the study of their properties as chemical raw materials and the features of the accumulation and distribution of metals in coal processing products, both thermally and chemically. The prepared project for the development of the Sergeevskoye brown coal deposit (Amur region) can become a pilot object for the construction of a coal-chemical processing plant with the production of resin, montan wax, complex hydrocarbons, fertilizers with associated extraction of noble metals, rare metals and REE.

The work was financially supported by the Far East Program (project No. 18-02-019 "Geological and technological assessment of the resource potential of caustobioliths of the Far East, development of scientific basis for their integrated use with the production of competitive multi-purpose products").

References

1. Young Platforms of the Eastern Edge of Eurasia (Deep Structure, Conditions of Formation and Metallogeny), Ed. corr. member of RAS A.P. Sorokin, Vladivostok: The Dal'nauka, (2013).
2. Lucinda Tolhurst, Lucid Insight Ltd. http://cornerstonemag.net/commercial-recovery-of-metals-from-coal-fly-ash , (2015).
3. J.G. Speight, The chemistry and technology of coal. 3rd ed. Boca Raton: CRC Press. (2013).
4. L.P. Noskova, Solid Fuel Chem., **48**, 4 (2014).
5. L.P. Noskova, A.V. Rokhin, A.P. Sorokin, Solid Fuel Chem., **41**, 3 (2007).
6. Coal Resources of Russia. Coal Basins and Deposits of the Far East (Khabarovsk Territory, Amur Region, Primorsky Territory, Jewish Autonomous Region), M.: Geoinformmark», **5.** Book 1. (1997).
7. A.P. Sorokin, I.F. Savchenko, V.Z. Mezhakov, T.V. Artyomenko, Phys. Techn. Probl. Dev. of Usefull. Resources, 4 (2012).
8. M.V. Golitsyn, V.I. Vyalov, A.Kh. Bogomolov, N.V. Pronina, E.Yu. Makarova, D.V. Mitronov, E.V. Kuzevanov, D.V. Makarov. Georesources. **61,** 2 (2015).
9. V.I. Vyalov, A.I. Larichev, E.V. Kuzevanov, A.Kh. Bogomolov, M.I. Gamow. Region. Geol. and Metallogeny, 51 (2010).
10. V.V. Seredin Geol. Rudn. Mestor., **49**, 1 (2007).
11. A.P. Sorokin, V.I. Rozhdestvina, V.M. Kuzminykh, S.M. Zhmodik, G.N. Anokhin, V.N. Mitkin. Geol. and Geophysics, **54**, 7 (2013).
12. A.P. Sorokin, A.A. Konyushok, O.A. Ageyev. In the collection: Issues of Geology and Integrated Development of Natural Resources of East Asia, Fourth All-Russia. Sci. Conf., Blagoveshchensk: IGNM, Far Eastern Branch of the Russian Academy of Sciences. **2** (2016).
13. S. Dai, I.T. Graham, C.R. Ward. Intern. J. of Coal Geol., **159** (2016).
14. W. Wengfeng, S. Shuxun, H. Weiduo, W. Ran, Z. Jiefang, D. Piaopiao, Q. Yong, X. Shaochun. Fuel, **147** (2015) 15. W.H. Orem, R.B. Finkelman, In: McKenzie F.T., Ed., Sediments, Diagenesis, and Sedimentary Rocks: Treatise on Geochemistry. **7** Elsevier-Pergamon, Oxford, P. 191-222. (2003).

Complex processing of titanium-rare metal raw material

Mikhail Medkov[1], *Galina* Krysenko[1*], *Dantiy* Epov[1], *Pavel* Sitnik[2] and *Valentin* Avramenko[1,2]

[1] Institute of Chemistry of Far East Branch of the Russian Academy of Sciences, Vladivostok, Russia
[2] Far Eastern Federal University, Vladivostok, Russia

Abstract. The paper is devoted to investigation of the complex processing of titanium-rare metal raw materials with ammonium hydrodifluoride. It is stated that fluorination of the main components of the mineral raw materials with ammonium hydrodifluoride proceeds with formation of complex ammonium fluorometallates and simple fluorides. It is showed that in the process of aqueous leaching of the fluorinated mineral raw material niobium and tantalum completely pass into solution together with titanium, iron, and silicon fluoroammonium salts while all the rare-earth elements stay in the insoluble residue as complex fluorosodium salts together with CaF_2. The method of separation of the fluoroammonium salts with obtaining marketable products and isolation of the rare-earth elements from the insoluble residue is offered.

1 Introduction

In most cases, the home-produced rare-metal raw materials contain also titanium though its content is less than that in the widespread kinds – ilmenite and rutile. The typical peculiarity of the unconventional titanium and rare-metal raw materials in Russia from the both exploiting deposits (Lovozerskoye – loparite ores) and perspective for development (Afrikandskoye – perovskite ores) consists in their complex multicomponent composition and low concentrations of the rare metals.

The most studied and widespread way of processing of the titanium and rare-metal raw materials is the sulphuric acid method that is conditioned by simplicity of its equipment and use of cheap and available reagents [1]. At the same time, the sulphuric acid schemes are characterized by large material streams as well as an increased amount of liquid and solid waste. For loparite concentrate, the chlorine technological scheme (named according to the main operation) was realized in the industrial scale [2].

Lately, the fluorination methods of raw materials processing attract great attention. It is especially profitable to use such a way for processing of the polymetallic raw materials because it lets to widen variety and profundity of the valuable components isolation [3]. In this connection, breakdown of a mineral raw material with NH_4HF_2 is particularly interest because of technological attractivity of the way of isolation of the components from the

[*]Corresponding author: krisenko@ich.dvo.ru

natural material via formation of ammonium fluorometallates [4]. Use of ammonium hydrodifluoride lets to achieve breakdown of a raw material at lower temperatures; the fluorination by-products (water and ammonia steam) do not contain fluorine that provides ecological safety of the production and lets to recycle them in condensed state in the processes of ammonia hydrolysis. Moreover, in the hydrodifluoride processing, regeneration of the fluorinating agent is comparatively easily realized. Comparison of the technical and economic indicators of the existing methods of pigment titanium dioxide production showed that use of the fluoride technology let to lower the pigment production cost by half [5].

The present work contains the results of investigation of perspective of the titanium-rare metal raw material complex processing by breakdown of the concentrate with ammonium hydrodifluoride.

2 Materials and methods

Perovskite concentrate of Afrikand deposit and loparite concentrate of Lovozersk deposit were used as the samples. Ammonium hydrodifluoride (analytical grade) served as the fluorinating agent.

Breakdown of the raw material was carried out in a nickel container which was put into a reactor with electric heating. The weighed amounts 50-100 g were used. Fluorination process proceeded at the temperature 160-180 °C and the mass ratios of the concentrate to the fluorinating agent equaled 1:1.7 and 1:1.9.

Leaching of the soluble fluoroammonium salts from the fluorinated concentrate was carried out by triple treatment of the product with water at the ratio solid:liquid = 1:5 and following filtration through a "blue band" filter. The process of pyrohydrolysis proceeds at the temperatures 700 and 800 °C with the constant water steam supply at a speed of 0.7 L/min.

The thermogravimetric investigation was carried at a Q-1000 device in platinum crucibles in air at a heating rate of 2.5 deg min^{-1} using the weighed amounts of 100-200 mg.

The changes of the substance in the processes of heating and leaching were controlled according to the loss of the initial mass as well as by the X-ray diffraction fnd IR spectroscopy methods.

The X-ray diffraction patterns of the samples were recorded at a D8 ADVANCE automatic diffractometer with rotation of a sample in CuKα radiation. The X-ray diffraction analysis was carried out using the EVA search programme and PDF-2 powder database. The IR spectra were registered at a SHIMADZU spectrometer. The samples were prepared as a suspension in Vaseline oil.

The contents of the main components at the various stages were determined by the X-ray fluorescence method at a Shimadzu EDX 800 HS spectrometer (tube with the rhodium anode, vacuum) at the room temperature using the tablets with polytetrafluoroethylene.

The content of thorium and its decay products in the samples was controlled by the radiometric method. The natural long-lived α-radiating ^{232}Th isotope, which initial activity in the sample was determined by γ-radiation of the ^{212}Pb isotope, was used as the tracer.

3 Results and discussion

According to the X-ray diffraction data perovskite concentrate was presented mainly by perovskite $CaTiO_3$ and hematite Fe_2O_3. It also contained silicon, niobium, aluminium, rare-earth and other elements. Loparite concentrate, on the whole, consisted of loparite (Na, Ce,

Ca)(Ti, Nb)O_3 with attendant silicon, iron, aluminium, thorium, and widely distributed rock-forming mineral aegirine which is sodium-iron silicate.

The thermogravimetric study showed that interaction of perovskite concentrate with NH_4HF_2 (Fig. 1, a) started in the dry mixture at 100 °C with heat liberation; above 120 °C one could observe the endothermic effect of the fluorination agent fusion, and further, the reaction proceeded in the melt with the maximum rate at 155 °C and was completed at 200 °C [6]. Interaction of loparite concentrate with NH_4HF_2 also starts in the dry mixture at 100 °C with heat liberation, above 121 °C the endothermic effect of the fluorinating agent melting is added to this process, and further, the reaction proceeds in the melt with the maximum rate at 165 °C and is completed at 205 °C (Fig. 1, b) [7].

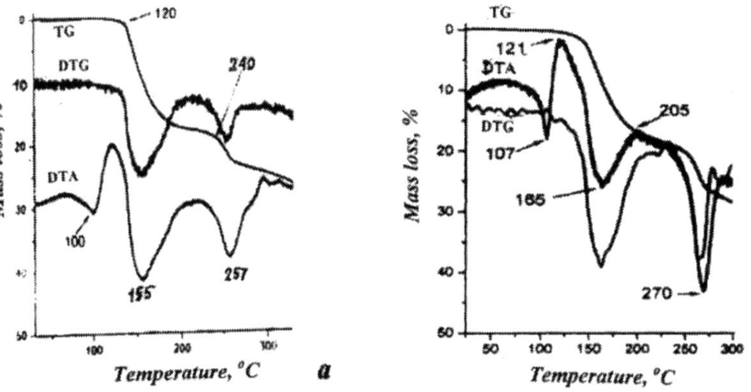

Fig. 1. Thermogravimetric curves of interaction with NH_4HF_2: *a* – perovskite concentrate; *b* – loparite concentrate.

The product isolated at 200 °C is the mixture of fluoroammonium salts $(NH_4)_2TiF_6$, $(NH_4)_3FeF_6$, and $(NH_4)_2SiF_6$ and CaF_2 in the case of perovskite concentrate while for loparite concentrate such mixture consists of $(NH_4)_2TiF_6$, $(NH_4)_3NbOF_6$, and $(NH_4)_2SiF_6$. Thus, fluorination of concentrates can be presented by the following equation:

$CaTiO_3 + 4{,}5\ NH_4HF_2 = CaF_2 + (NH_4)_3TiF_7 + 1{,}5\ NH_3 + 3\ H_2O$; (1)

$(Na,Ce,Ca)(Ti,Nb)O_3 + NH_4HF_2 \rightarrow (NH_4)_2TiF_6 + (NH_4)_3NbOF_6 + NH_4CeF_4 + NaF + CaF_2 + NH_3\uparrow + H_2O$. (2)

Kinetic investigations of the fluorination processes were carried out basing on amounts of liberated ammonia. The results are presented in Fig. 2.

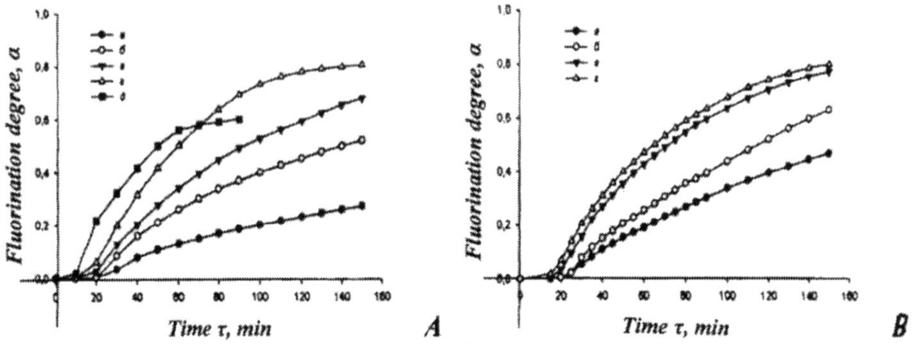

Fig. 2. Dependence of fluorination degree on time at various temperatures: *A* – perovskite concentrate (*a* – 130, *b* – 140, *c* – 150, *d* – 160, *e* – 150°C и 53 % NH_4HF_2); *B* – loparite concentrate (*a* – 140, *b* – 150, *c* – 160, *d* – 165°C).

The kinetic dependence clearly shows that in case of use of the stoichiometric amount of the fluorinating agent the complete breakdown of the concentrates is not achieved that is connected with NH_4HF_2 melting and its following partial evaporation. With a deficiency of the fluorination agent (Fig. 2*A*, curve *e*) interaction proceeds without visible formation of the melt and NH_4HF_2 evaporation that lets to breakdown the mineral raw material without use of an excess of the fluorinating agent provided that the fluorination process is carried out in several stages.

Basing on the experimental data we calculated the values of activation energy, order of the reactions, and constants of action for interaction of the concentrates with NH_4HF_2. For calculation we used the general topochemical Kolmogorov-Erofeev equation [8]. Analysis of variations of the constants of action showed that at the melting temperature of the fluorinating agent interaction of the both concentrates with ammonium hydrodifluoride proceeded at the border of the diffusion and kinetic areas. In the melt of the fluorination agent interaction proceeded in the kinetic area and at a higher rate. The low values of activation energy for interaction with ammonium hydrodifluoride (58.8 and 74.4 kJ/mole for perovskite and loparite concentrate, respectively) show that such concentrates must be easily broken down with this agent.

Basing on the results of the thermogravimetric and kinetic analysis we carried out breakdown of perovskite and loparite concentrates at 160-170 and 170-180 °C and at the mass ratios of the concentrate to the fluorinating agent 1:1.9 and 1:1.7, respectively. The fluorination process was realized by steps, with addition of the necessary amounts of NH_4HF_2 in the two stages. The fluorinated concentrates were friable grey powders. According to the X-ray diffraction data they consisted of the mixture of the complex fluoroammonium salts and CaF_2.

Fluoroammonium salts of titanium, iron, niobium, and silicon are readily soluble in water, therefore it was reasonable to transfer these salts into solution by aqueous leaching at the following stage of the processing. Our study showed that the insoluble residue presented 24-26 % of the initial fluorinated product mass. According to the X-ray diffraction data this residue was CaF_2 in the case of perovskite concentrate and $NaCeF_4$ – for loparite one while the evaporated filtrate consisted of the mixture of fluoroammonium salts titanium, niobium, iron, and silicon. The X-ray fluorescence data on the elemental composition confirm that aqueous leaching of the fluorinated concentrate results in the complete transfer of fluoroammonium salts into the filtrate while the rare-earth elements stay in the insoluble residue.

Niobium, and tantalum can be selectively separated from titanium, silicon, and iron by the following processing of the filtrate using, e.g. extraction. Extraction separation of niobium and tantalum from titanium is carried out with aliphatic spirits when these metals transfer into the organic phase [9]. It is reasonable to separate titanium, iron, and silicon fluoroammonium salts by the sublimation method because the high sublimation temperatures of iron fluorides create conditions for profound separation of volatile ammonium fluorometallates (titanium and silicon) and non-volatile iron salts. The residue after sublimation of fluoroammonium salts titanium and silicon is the slime in which iron exists as FeF_2. Reduction of this slime in hydrogen atmosphere which intensively proceeds at 520-620 °C results in obtaining iron powder while its pyrohydrolysis – in obtaining other valuable by-product – red pigment (ochre). The mixture of ammonium hexafluorotitanate and hexafluorosilicate is easily separated at ≥400 °C under influence of steam in the process of the sublimate pyrohydrolysis on the stage of TiO_2 obtaining [10].

According to the X-ray diffraction data the insoluble residue prepared by leaching the fluorinated loparite concentrate is $NaCeF_4$. The X-ray diffraction pattern of the sample obtained after burning in air at 650°C displays the reflexes of CeO_2, LaOF, and β-$NaNdF_4$ that indicates to the presence of lanthanum and neodymium fluorosodium complex salts in the insoluble residue which are isostructural with $NaCeF_4$ [11]. The elemental composition of the insoluble residue (re-counted for fluorides) is presented in Table 1.

Table 1. X-ray fluorescence results of elemental composition of residue prepared by loparite concentrate processing

Compound	LnF_3	NaF	CaF_2	ThF_4
Concentration (wt.%)	74,4	15,5	9,0	1,0

Thus, the insoluble residue obtained by aqueous leaching of the fluorinated concentrate is the mixture of cerium, lanthanum and neodymium fluorosodium complex salts $NaLnF_4$ and calcium fluoride. It also contains thorium. The main phase in loparite concentrate processing is $NaLnF_4$ while in the case of perovskite concentrate – CaF_2.

In order to isolate the rare-earth elements from the insoluble residue we studied possibility to dissolve this product in nitric acid.

Calcium fluoride is a slightly soluble compound. Solubility of CaF_2 as a function in nitric acid solutions of various concentrations at various temperatures was described in [12]. It was stated that it was maximum in 6 M HNO_3. The cause of decrease of CaF_2 solubility with increase of the acid concentration above 6 M was not investigated but it was suggested it was connected with the salting-out effect.

The investigation carried out on the model mixtures of the rare-earth fluorosodium salts and CaF_2 showed that in the case of acid leaching one could observe interaction of these components with HNO_3 resulting in obtaining respective nitrates as well as their dissolving. HNO_3 amount necessary for dissolving of the samples was determined from the equation

$$NaLnF_4 + 4\ HNO_3 = Ln(NO_3)_3 + NaNO_3 + 4\ HF \qquad (3)$$

and from the solubility isotherm CaF_2 - HNO_3 [12].

The study showed that degree of dissolving of the above-mentioned model mixtures increased with use of more diluted acid solutions, with the temperature rise, and with longer operation time. The elemental analysis indicated that after this treatment, the percentage contents of the rare-earth elements and calcium in the evaporated filtrate was approximately the same as in the initial sample. Thus, acid leaching of the insoluble residue failed in separation of the rare-earth fluorosodium salts from calcium. It is reasonable to transform these complexes into compounds which could more easily interact with HNO_3, e.g. by pyrohydrolysis, into oxides or oxyfluorides.

Pyrohydrolysis of the insoluble residue of loparite concentrate at 800 °C results in obtaining the product consisted, according to the X-ray diffraction data, of the mixture of CaF_2, NaF, CeO_2, NdF_2, $NaLnF_4$, and $Na_3Th_2F_{11}$. Changing the hydrolysis conditions (temperature, rate of steam supply, duration of the process) we could not prepare a product free of $NaLnF_4$ that could be connected with incongruent melting of $NaLnF_4$ within this temperature range with obtaining the rare-earth trifluorides [13].

The rare-earth fluorosodium salts were decomposed by concentrated hydrofluoric acid at slight heating. The residue after dilution of the resulting mixture with water consisted of the rare-earths and calcium fluorides and the evaporated filtrate – NaF. The radiometric data on thorium content shows that after such treatment thorium completely stayed in the insoluble residue. Pyrohydrolysis of this residue at 800 °C resulted in obtaining the product which is the mixture of CeO_2, LnOF, NdOF, and CaF_2.

Investigation of dissolution of the pyrohydrolysis product in nitric acid showed that acid leaching proceeded with interaction of cerium oxide and lanthanum and neodymium oxyfluorides with formation of nitrates. Moreover, one could observe a partial calcium transfer into solution as $Ca(NO_3)_2$, according to the solubility isotherm CaF_2 - HNO_3. The X-ray diffraction data proved that the insoluble residue consisted of CaF_2. The acid volumes necessary for dissolution of the pyrohydrolysis product were calculated by the equations:

$$CeO_2 + 4\ HNO_3 = Ce(NO_3)_4 + 2\ H_2O; \quad (4)$$

$$LnOF + 3\ HNO_3 = Ln(NO_3)_3 + HF + H_2O \text{ and} \quad (5)$$

$$NdOF + 3\ HNO_3 = Nd(NO_3)_3 + HF + H_2O. \quad (6)$$

Acid leaching was carried out with the stoichiometric amount of HNO_3 and with its excess. It was shown that the degree of dissolution of the pyrohydrolysis product achieved ~90 %. The results of the elemental analysis for the rare-earths and calcium in the evaporated to the dry salts filtrates are presented in Table 2.

Table 2. Degree of the rare-earths and calcium transfer into solution in acid leaching of the pyrohydrolysis product (weight of the initial sample 1 *g*)

Conditions of experiment		12 M HNO_3; stoichiometry (V=2.5 *ml*)	12 M HNO_3; 100 % excess (V=5 *ml*)	3 M HNO_3; stoichiometry (V=10 *ml*)
Insoluble residue mass, *g*		0,1097	0,0973	0,0856
Content in evaporated filtrate, *mass %*	Σ REE	43,1	41,9	40,4
	Ca	0,7	1,1	1,4
Degree of extraction, *%*	Σ REE	98,0	96,6	96,0
	Ca	5,4	9,3	13,0

One can see from the Table that in the case of leaching of the pyrohydrolysis product with nitric acid the degree of dissolution of the sample under investigation is influenced by the liquid phase volume: the degree of dissolution increases with increase of the volume of the liquid phase. However, the degree of the rare-earth transfer into the solution decreases that can be connected with increase of the degree of CaF_2 dissolution in HNO_3 under this conditions and respective rise of fluoride-ions content in the solution.

Deactivation of the resulted nitric solution was carried out by co-precipitation of thorium with barium sulfate. Under these conditions, at the same time with deactivation, additional removal of calcium and strontium impurities takes place because of their turning into slightly soluble sulfates. As a result, the solution contains only rare-earth and ammonium nitrates. This nitrate solution may be an initial product for obtaining rare-earths compounds, e.g. carbonates, oxides, etc. Thus, for processing with the purpose of the rare-earths extraction from solutions the stepped neutralization methods are widely used. pH value at the beginning of rare-earth hydroxides precipitation in the process of lanthanide hydroxides precipitation from nitrate solutions varies from 6.95 (Y) to 7.82 (La).

Our investigation showed that addition of ammonia to the nitrate solution to pH≈7 resulted in formation of a gelatinous precipitate which coagulates at slight heating. According to the X-ray diffraction data this precipitate was the mixture of rare-earth oxides. Its composition determined by the X-ray fluorescent method was following (mass %): Ce_2O_3 (59.2), Nd_2O_3 (19.4), La_2O_3 (17.8), Y_2O_3 (0.08).

4 Conclusions

Interaction of loparite and perovskite concentrate s with ammonium hydrodifluoride was studied. The ways of carrying out of their complex processing with isolation of all valuable compounent in such breakdown were determined. The offered scheme of complex processing is universal and may be used not only for loparite and perovskite concentrates but also for other concentrates containing rare-earth elements.

References

1. Rare and scattered elements. Chemistry and technology. In 3 volumes. / Under the editorship of Korovin S.S. Moscow: MISIS. **II**. P. 461. (1996).
2. James B. Hedrick, Shyama P. Sinha, Valery D. Kosynkin. J. of Alloy and Compounds. **250 (1-2)**. P. 467-470. (1997).
3. Karelin V.A., Karelin A.I. Fluoride processing of rare-metals concentrates. Tomsk: NTL (Scientific and technical literature). 221 p. (2004).
4. Rakov E.G. Science and technical resume. Inorganic chemistry. Moscow: VINITI (National institute of scientific and technical information). **15**. 154 p. (1988).
5. Andreev A.A. Fluoride processing of pigment titanium dioxide production. // Theses of reports of all-Russ. theoretical and practical conference. Tomsk 25-26 June 2009. Tomsk: Publisher Tomsk Polytechnic University. P. 27. (2009).
6. Krysenko G.F., Epov D.G., Medkov M.A, et al. Journal "Theoretical Foundations of Chemical Engineering". **50 (4)**, P. 589-593. (2016).
7. Krysenko G.F., Epov D.G., Medkov M.A, et al. Interaction of loparite concentrate with ammonium hydrodifluoride. // Russian Journal of Applied Chemistry. **89 (4).** P. 540-546. (2016).
8. Boldyrev V.V. Methods of investigation of the kinetics of thermal decomposition of solids. Tomsk: Tomsk. Gos. Univ. P. 62. (1958).
9. Rare and scattered elements. Chemistry and technology. In 3 volumes. / Under the editorship of Korovin S.S. Moscow: MISIS. **I**. P. 324-326. (1996).
10. Melnichenko E.I. The fluoride processing of rare-metal ores of Far East. Vladivostok: Dalnauka. 268 p. (2002).
11. J.H. Burns. Crystal Structure of Hexagonal Sodium Neodymium Fluoride and Related Compounds. // J. Inorganic Chemistry. **4 (6).** P. 881-886. (1965).
12. Malcolm W. Wilding and Donald W. Rhodes. Solubility Isotherms for Calcium Fluoride in Nitric Acid Solution. // Journal of Chemical and Engineering Data. **15 (2).** P. 297-298. (1970).
13. Sobolev B.P. The Rare earth trifluorides. Barcelona: Institut d'Estudis Catalans. 520 p. (2000).

INTEGRATION OF FIBER LASERS IN PROCESSES OF MINERAL RAW MATERIAL PROCESSING

Nina Leonenko[1*]

[1]Mining Institute of Far eastern branch of Russian Academy of Sciences, Khabarovsk, Russia.

Annotation. The application of laser technologies (LT) in the economy of the most developed countries. The possibility of using laser technologies for processing technogenic mineral raw materials is considered. The possibilities of integration of fiber-optic lasers into the processes of enrichment and processing of mineral raw materials are explored. The effects of interaction of laser radiation with mineral media - objects of Amur placer technogenic deposits containing submicron gold not extracted by modern gravitational methods are analyzed. The formation of self-organizing gold structures on the surface of a silicate matrix was established, general patterns of agglomeration and concentration of "non-extractable forms" of gold were revealed.

1 Introduction

The wide application and improvement of laser technologies (LT) in the economy of the most developed countries is a global trend of world development. The use of LT is crucial for increasing labor productivity and competitiveness of the national economy, expanding the opportunities for its integration into the world economic system. Laser processing of materials is one of those technologies that determine the current level of production in industrialized countries [1-4]. Distinctive features of the use of lasers in production - high quality of products, high process efficiency, saving of human and material resources, environmental cleanliness. A powerful impetus to the market sector, including laser processing of materials, gave fiber lasers, both small and high power. In 2012, four new companies entered the market - a supplier of such radiation sources, which are trying to compete with the recognized leader in this field - IPG Photonics Corporation. In general, laser technologies (LT) are developed in the EU, USA and Japan, and are reflected in Fig. 1. The countries of South-East Asia (Southeast Asia) are noticeably behind them, but even in these countries the volume of laser sales is about five times higher than in Russia. For example, in China, in Hubei province, which is the concentration of the optical industry in the country, the gross product of laser enterprises in 2005 y. exceeded 75 million $. Laser technologies in the EU countries, in a more general sense - photonics, are viewed only as a locomotive of technological innovations. The photonics market in the world exceeds 150 billion euros, demonstrating a 14% annual growth over the past 10 years.

* Corresponding author: leonenko@igd.khv.ru

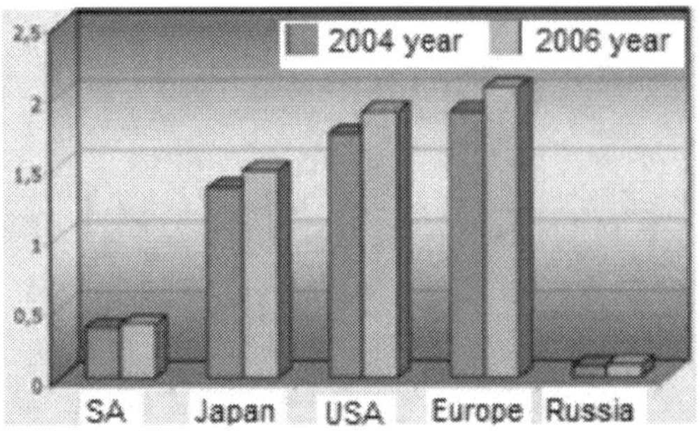

Fig. 1. Sales volume of technological lasers by region the world, bn $ [5]

According to the specialists' research of the general economic conjuncture, the growth rate of world production of laser systems and sources of laser radiation (LI) will remain at the level of 10-15%. Lasers for processing materials sh at present ow the maximum dynamics of growth, both on the international and Russian markets. According to LaserMarketsResearch [5], from 2008 to 2016, the world market for lasers grew from 6.57 to 10.408 billion dollars. The share of lasers was 39%, which used in the processing of materials in the global market in 2016 year: from $ 213 million to $ 1,304.8 million. Lasers for processing materials showed the greatest increase in this time. In 2016, they accounted for 41% of the total market. The table gives a comparison of various technological sources of laser radiation, which are currently used in the processing of materials. At the same time, from the analysis of the problems of processing and extraction of noble metal raw materials, it is known that, a sharp reduction in readily available gold reserves in ores and placers predetermined the search for new low-waste technologies for the extraction of gold and other precious metals from placer technogenic deposits. Involvement in the processing of man-made gold-bearing neo-formations accumulated over a long historical period of gold mining is highly relevant for the Far Eastern region. Such deposits, as a rule, have a high clay content in the sands. The metal is represented in the sludge fractions mainly in dispersed colloidal particles of lamellar or acicular form. The most of particles of gold, of lamellar and dendritic forms, is lost with tails, especially in classes of fineness less than 0.25 mm under gravity enrichment [6, 7]. In mineral media, structural rearrangements occur under the influence of external energy flows on them, so that their state becomes far from thermodynamic equilibrium. Laser treatment of finely dispersed mineral media is one of the promising areas of such studies. This is because laser radiation creates high heat flux densities on both the surface and the volume of the material, sufficient heating, melting and evaporation days. In the work, the processes of interaction of laser radiation with dispersed mineral media have been experimentally investigated with the possible goal of integrating the modern powerful sources of laser radiation into the enrichment and complex processing of mineral raw materials. Among the non-traditional methods of energy impact are: electrochemical, microwave, electroimpulse, electrohydrodynamic, magnetic pulse processing, impact by a stream of accelerated electrons, super-powerful hyper-shock waves and powerful electromagnetic pulses [8]. Laser radiation makes it possible to provide high rates of local changes in temperature in the irradiated medium and temperature gradients, both in narrowly localized sections of surfaces and in the depth of materials due to low temperature conductivity. Such parameters can not be achieved with other methods of exposure. It is known that the properties of micron objects and nanoindivids differ from the properties of their macrohomologists. This is due to the dependence of the specific surface area of the particles on their size and, consequently, to incomparably higher structure perfection and surface effects. Such properties are thermal, electric, magnetic, high aggregative stability. For example, the melting point of the macrograin is 1064 °C gold, and the nanoindivide value of 4 nm is 427 °C.

Table 1. Comparison of different types of lasers. Advantages of fiber lasers*

Parameters	Required for use in industry	CO_2	YAG-Nd with lamp pumping	YAG-Nd with diode pumping	diode lasers	Fiber lasers
Output power,кW	1...30	1...30	1...5	1...4	1...4	1...30
Wavelength, mkm	as less as possible	10,6	1,064	1,064 or 1,03	0,8...0,98	1,07
Beam Parameter Product, BPP, mm x mrad	< 10	3...6	22	22	> 200	1,3...14
Effectiveness, %	> 20	8...10	2...3	4...6	25...30	20...25
Range of radiation delivery by fiber, m	10...300	absent	20...40	20...40	10...50	10..300
Output Power Stability	as high as possible	low	low	low	high	very high
Sensitivity to back reflection	as low as possible	high	high	high	low	low
Footprint, m^2	as less as possible	10...20	11	9	4	0,5
Cost of installation, relative unit.	as less as possible	1	1	0,8	0,2	< 0,05
Cost of operation, relative unit	as less as possible	0,5	1	0,6	0,2	0,13
Cost of service, relative unit	as less as possible	1...1,5	1	4...12	4...10	0,1
Periodicity of replacement of lamps or laser diodes, hour.	as high as possible	-	300...500	2000...5000	2000...5000	> 50 000

*http://www.ntoire-polus.ru/apps_lasers.html

2 Materials and methods

Experimental data were obtained on the ytterbium laser radiation sources LS-06 with a fiber transmission system. The coefficient of efficiency of such an installation is 30%. On the ytterbium fiber-optic installation LS-06, it is possible to control both the radiation power from 0 to 600 W. and the diameter of the local processing zone by placing the optical head over the material being processed. The operating mode of the radiation source is continuous, the modulation frequency of the output power is 5 kHz. The spectral width is 10 nm. The wavelength λ = 1070 nm. Natural gold-bearing dispersed images of mineral raw materials from high-clay technogenic objects of Far Eastern alluvial deposits with the particle size of 71 μm, 40 μm and 20 μm were prepared for research. Gold-containing high-clay aluminosilicate samples in loose form, with a layer thickness of 3 mm, were placed on a special graphite substrate. The optical head was placed above the graphite substrate. It was possible to specify the parameters of defocused radiation. Laser radiation, passing through the ytterbium optical fiber fixed at the input of the optical head, and through the optical head placed vertically and rigidly on the tripod, fell on the samples under study. The substrate was moved at a speed of 1 mm / s. The diameter of the defocused radiation was chosen empirically and was 7 mm.

The material composition of the directed change in the properties of the samples was studied with the help of electron and atomic force microscopy. The images were examined before and after laser treatment. As a result of laser action, the formation of enlarged gold of spherically agglomerated forms on an aluminosilicate matrix was recorded from microscopic analysis data.

Electron microscopic examination of the samples was carried out with a scanning electron microscope LEO EVO 40HV (Karl Zeiss, Germany) equipped with an energy dispersive analyzer INCA-ENERGY. The study of mineralogical objects using a secondary electron detector (SE detector) made it possible to obtain information on the topography of the samples. In addition to the detector of secondary electrons (SE detector), a backscattered electron detector (QBS detector) is used. With the aid of a QBS detector, phases with a higher average atomic number are more pronounced in contrast when the images are obtained than in phases with a lower atomic number. Since the difference in atomic weights of gold and minerals contained in the studied mineralogical objects is large, a more contrasting image is obtained, which makes it possible to visually reveal gold. The relief difference of agglomerated gold surfaces is revealed under the action of continuous energy laser radiation (LI). The sensitivity of the method is ~ 0.1%. The width of the electron beam is ~ 20-30 nm. The penetration depth of the electron beam is ~ 1 μm. The local spectral analysis confirms the agglomeration of gold. When laser processing of gold-bearing mineral raw materials, the character of fast-flowing processes is established, leading to the formation of various structural surfaces of gold, and general patterns of agglomeration and concentration of ultradispersed gold not extracted by gravitational methods are revealed. Figures 2-4 show images of optical, electron, and atomic force microscopy, respectively.

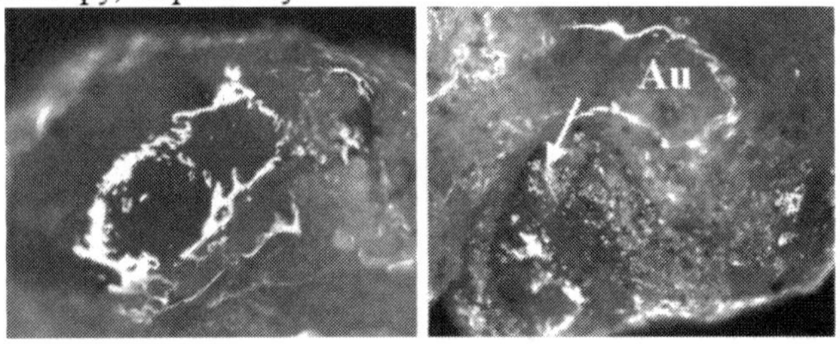

Fig. 2. Images of optical microscopy: micron forms of agglomerated gold, less than 1 mm, on aluminosilicate spectra, samples from technogenic deposits

3 Results and discussion

It was shown in [9-11] that thermocapillary forces make the dominant contribution to the process of coarsening from nanoparticles to sizes of practical importance, up to hundreds of microns. Thus, under the influence of laser radiation, a thermal process of disintegration of the crystal lattice of the mineral occurs, followed by a rapid thermal process of recrystallization, defragmentation and sintering due to a laser exposure of a second duration. Due to this scenario, the chemical homogeneity of dispersed mineral objects increases.

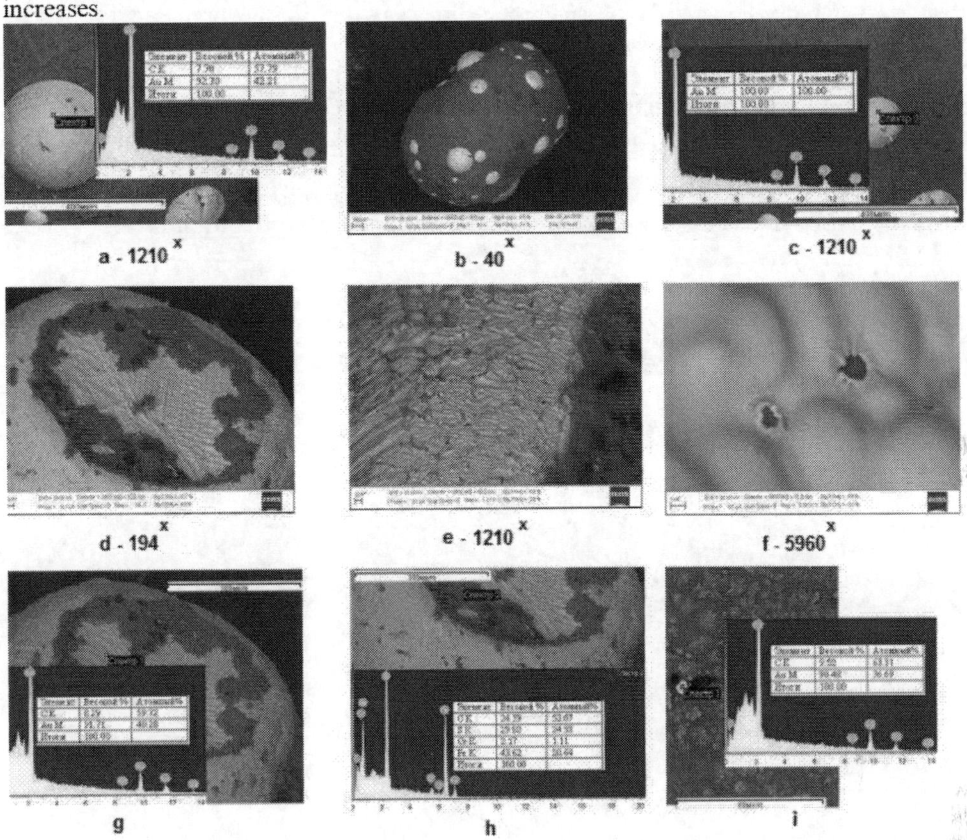

Fig. 3. Images of raster of agglomerated gold of spherical shape and the investigated areas of its surface after exposure to a continuous source of laser radiation LS-06. Risks: 400 mkm – a, c; 300 mkm – g, h; 100 mkm – b; 60 mkm – i; 20 mkm – d; 2 mkm – e, 1 mkm – f.

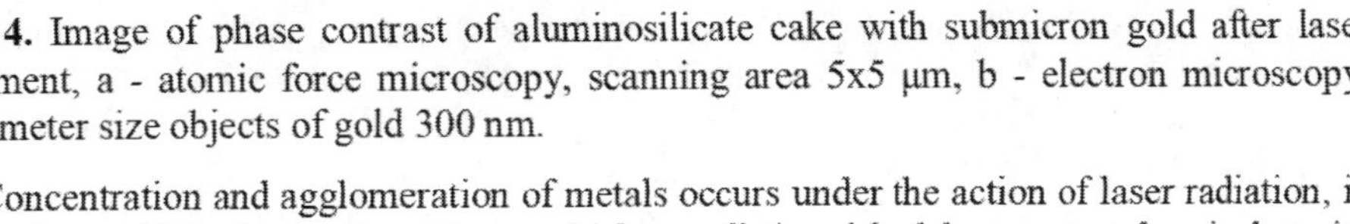

a b

Fig. 4. Image of phase contrast of aluminosilicate cake with submicron gold after laser treatment, a - atomic force microscopy, scanning area 5x5 μm, b - electron microscopy, nanometer size objects of gold 300 nm.

Concentration and agglomeration of metals occurs under the action of laser radiation, in particular, gold in larger formations, which are distinguished by greater chemical purity than the original mineral associations. These studies are the scientific basis for developing new technologies for extracting submicron and nanometric forms of gold and other useful components. On this basis, a sufficiently effective and environmentally safe method of coarsening noble metal particles not extracted by traditional methods based on laser treatment of natural materials of anthropogenic character can be proposed [12]. It is possible that the use of modern laser radiation power sources with fiber-optic ytterbium energy transmission lines will make it possible to solve in the future purely practical tasks related to optimization of industrial technological processes of deep processing of man-made raw materials and extraction of valuable components. This testifies to the practical significance of the method.

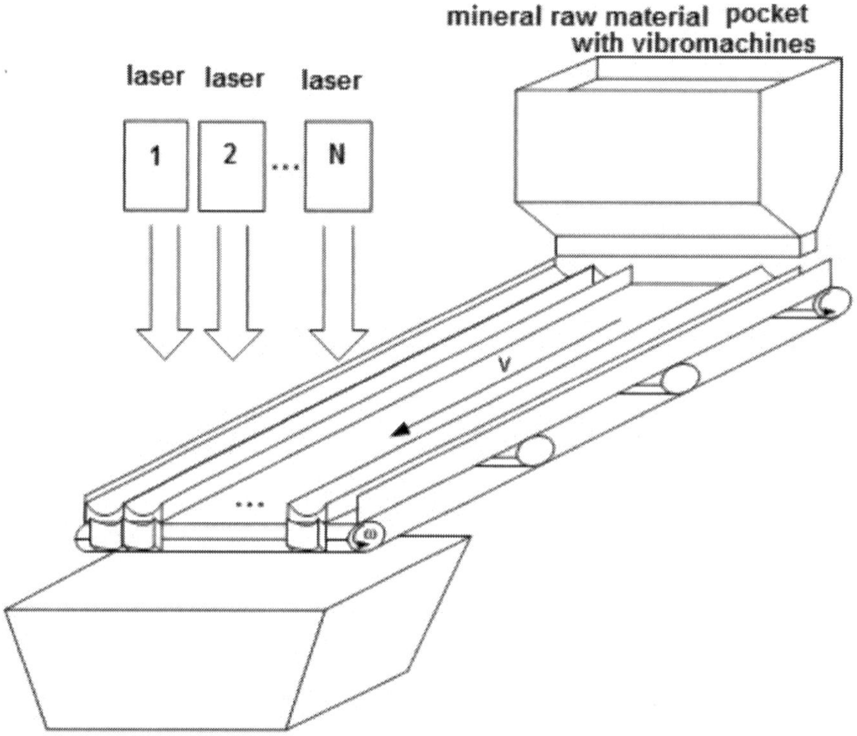

Fig. 5. Scheme of continuous flow technology for the extraction of submicron and ultradispersed gold particles on the basis of power laser fiber technology

4 Conclusion

The given examples of application of non-traditional technologies for extraction of ultradisperse and submicron gold particles allow us to consider that the solution of the problem of hard-to-extract forms of precious metals is possible. The method of laser treatment of gold-bearing high-argillaceous sands is approved. As a result, there is a redistribution of the substance with concentration and agglomeration of gold, which makes it possible to further integrate laser devices in the technology of mineral processing and processing of mineral raw materials. Figure 5 shows a patented scheme of continuous, on-time technology for the extraction of submicron and ultradispersed gold particles based on power laser fiber technology [13].

The article is implemented on a grant from the Ministry of Education and Science of the Khabarovsk Territory, Contract No. 115/2018 D.

References

1. Official site of the innovative Holding "Schwabe" – [Electronic resource] URL: www.shvabe.com (date of the application15.09.2017)
2. Official website of the group «IPG Photonics Corporation» – URL: http://www.ipgphotonics.com/ru (date of the application15.09.2017)
3. V. Kazakevich, S. Yaresko Trends in the development of the market of laser technologies for solving problems of laser processing of materials. Part 1. World Laser Market. "Izvestia of RAS Samara Scientific Center", **16.** 266-275 (2014)/
4. V. Appolonov. Power Optics. Quantum Electronics. **44 (2),** 102-121 (2014)
5. V. Appolonov. Lasers for industrial, scientific and ecological use. Springer Series in Optical Sciences, vol. 201. URL: https://doi.org/10.1007/978-3-319-33359-5_47.
6. Yu. Mamaev, V. Litvinsev, G. Ponomarchuk. *Technogenic placers of precious metals of the Far Eastern region of Russia and their rational development,* 309 (Moscow, 2010).
7. V. Litvinsev, T. Banshchikova, N. Leonenko, L. Shokina. Development of non-conventional technologies for recovery of refractory gold forms from mineral raw materials of production-induced deposits. Mineral processing, **3,** 11-14 (2009).
8. V. Chanturia, I. Bunin. Non-traditional high-energy processes for disintegration and exposure of finely disseminated mineral complexes. Journal of mining science, **3,** 311-330 (2007). DOI: 10.1007/s10913-007-0032-4.
9.A. Kuz'menko, N. Leonenko, V. Kharchenko, N. Kuz'menko, I. Silyutin, I. Khrapov. Termocapillary mechanism of laser-stimulated agglomeration of ultreadisperse and colloidal-ionicdold, Technical Physics Letters, **35(9),** 837-840 (2009).
10. N. Leonenko, E. Vanina, E. Veselova, G. Kapustina. Nonlinear effects of laser radiation and physical model of laser agglomeration process. Russian Journal of General Chemistry, vol. 83, issue 13, 2759-2764 (2013) http://link.springer.com/article/10.1134/S1070363213130240.
11. N. Leonenko, E. Vanina, G. Kapustina, E. Veselova Study of Nonlinear Effect about Laser-induced Processes of Nanodispersed Gold in Mineral Association, Advanced Materials Research, **772,** 355-358 (2013) doi:10.4028/www.scientific.net/AMR.772.355).
12. N. Leonenko, A. Kuz'menko, I. Silyutin et. al. RF Patent № 2413779 (April 7, 2010); 7 (2011).
13. A. Kuz'menko, I. Khrapov, N. Kuz'menko, N. Leonenko. RF Patent № 2541248 (July 17, 2012); 4 (2015).

Obtaining sorbents from brown coals samples of Kharanorskoe coal deposit by way of electromagnetic microwave radiation exposure

Elena Vorsina[1*], *Tatiana* Moskalenko[1] and *Valerii* Mikheev[1]

[1]Mining Institute of the North, Siberian Branch, Russian Academy of Sciences, Yakutsk, Russia

Abstract. This article studies the possibility of intensification of Kharanorskoe deposit brown coal alkaline activation by way of electromagnetic microwave radiation (EMR) exposure in order to obtain sorbents. There is data that presents investigation of qualitative and adsorption properties of sorbents obtained from 0-2 mm brown coal samples impregnated with potassium hydroxide at a KOH / coal of 1 g / g weight ratio and subjected to thermolysis at 800 °C in the thermal shock mode. The authors carried out a detailed analysis of three options for thermal treatment of the coal-alkali mixture ahead of thermolysis: coal-alkali mixture drying to indiscrete mass, EMR treatment and combination of ones. The design of a laboratory microwave oven for heat treatment of raw materials based on the use of EMR at a frequency of 2,450 MHz is presented. It was found out that the effect of electromagnetic microwave radiation with a frequency of 2.450 MHz makes it possible to increase the adsorption activity on iodine by 4-6%, reaching a value of 97%, to significantly shorten the time of preliminary heat treatment of the coal-alkali mixture and to shorten the thermolysis time when sorbents obtain.
Key words: brown coal, alkaline activation, carbon sorbents, adsorption activity on iodine, electromagnetic microwave radiation.

1. Introduction

In the world practice, extensive experience in the production of sorbents from fossil coals has been accumulated. Structural peculiarity of brown coal characterized by increased voids content in comparison with hard coal and anthracites, as well as its significant explored reserves in Russia (103.11 billion tons) and a relatively low price, give priority to brown coal for its use as a precursor for the production of sorbents. Studies have shown that alkaline activation [1-7] is an effective method of obtaining sorbents from brown coal. In addition to the development of the porous structure, alkali provides a high proportion of micropores, a narrow pore size distribution and significantly reduces the ash content of the porous material thorough the reaction with mineral components of mineral raw materials [7].

[*] Corresponding author: e.worsina@yandex.ru

Experimental researches have shown that activation of brown coal from the Kharanorskoe deposit (Zabaikalsky Krai, Russia) with potassium hydroxide followed by thermolysis makes it possible to obtain carbon materials with high values on iodine adsorption activity (more than 85%) corresponding to the level of the best active carbons in the world [8, 9]. In this case, carrying out thermolysis in the thermal shock mode makes it possible to increase the adsorption activity on iodine on an average of 4% [10].

The use of electromagnetic microwave radiation (EMR) is a promising method of influencing on solid fossil fuels that allows to greatly intensify chemical-technological processes and in most cases to receive results that may not be achievable by traditional technologies [11, 12].

A lot of irrefutable advantages and unique possibilities of EMR processing in comparison with traditional heating (high process speed, volumetric heating, high coefficient of efficiency, possibility to carry out the process in the conditions of constant and pulsed energy supply, etc.) creates prerequisites for its using during process of sorbents obtaining.

There are a number of researches on the EMR influence on brown coal, including for the purpose to increase its adsorption properties [13] and technological qualities [14, 15], confirming the effectiveness of microwave radiation use. In this paper, we investigated the possibility of intensification by the EMR of the alkaline activation of brown coals from Kharanorskoe deposit in the production of sorbents.

2. Method of experimental research

As a raw material for the sorbents production, brown coal of the Kharanorskoe deposit (coal grade — B2 grade, size grade — 0-2 mm) with the following qualitative characteristics (%) was used: laboratory moisture of the sample (Wr) - 8.2; moisture of analysis sample (W^a) is 7.6; dry ash (A^d) - 7.9; volatile matter yield on dry ash free basis (V^{daf}) is 45.5; the adsorption activity on iodine (X) - 16.9.

Alkaline activation of Kharanor coal samples was carried out according to the following method. Potassium hydroxide (50% solution) was introduced into the coal by impregnation at a KOH / coal 1 g / g ratio, expressed in grams of KOH per 1 g of dry coal. The coal sample was treated with KOH, mixed manually into indiscrete mass, and held for 2 hours.

Three thermolysis preparation options of impregnated with potassium hydroxide coal (coal-alkaline mixture) were researched:

1 option. Drying to indiscrete mass;

2 option. EMR treatment without pre-drying;

3 option. EMR treatment with pre-drying to indiscrete mass.

The drying of the coal-alkaline mixture to a indiscrete (dry) mass was carried out in a drying chamber at a temperature profile of 105-110 °C.

Preliminary treatment of the coal-alkaline by EMR mixture was carried out in a laboratory microwave oven with a power of 900 W (Fig. 1).

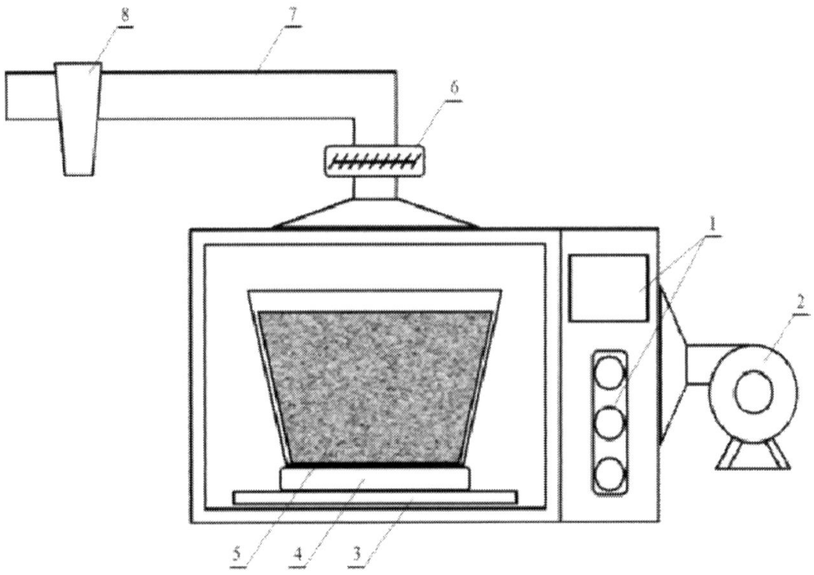

Fig. 1. Laboratory electromagnetic microwave oven scheme: 1 - operating modes indication and control unit; 2 – magnetron centrifugal cooling fan; 3 - rotary bottom; 4 - thermal insulating stand; 5 - heat-resistant glass tank for loading raw materials; 6 - gas heating main axial-flow fan; 7 - gas heating main; 8 - hydro seal

The microwave oven circuit scheme consists of an electromagnetic radiation generator (magnetron), a waveguide, a heating chamber, a magnetron ventilation and cooling system, an excess radiation protection system, a control unit [11]. A number of significant changes have been made to the design of a household in order to provide safe experimental researches using microwave oven. Installing of an additional fan strengthened magnetron ventilation and cooling system. An exhaust system consisting of a centrifugal fan and a gas heating main is installed in order to remove effectively gaseous products from the heating chamber. An additional hydro seal is installed on the gas heating main, that allows to catch the vaporous volatile matters that have produced as a result of raw materials' heat treatment.

The electromagnetic-microwave treatment of the coal-alkaline mixture was processed at a frequency of 2,450 MHz to the maximum possible temperature. In this paper, the maximum possible temperature is understood as the temperature of the coal-alkaline mixture when it reaches to an inflammation of the mixture without open fire, but with following abundant gas liberation. Put in an open laboratory glassware, a sample of the coal-alkaline mixture was placed in a microwave oven. In order to exclude the ignition of the mixture, the EMR treatment process was performed with a pulsed energy supply - 3 min for each of cycles, then the mixture was mixed, its temperature was measured and mixture was again subjected to EMR treatment.

After preliminary preparation of the coal-alkaline mixture in the above-mentioned three variants, the samples were subjected to a thermal shock at 800 °C in a muffle furnace. Isothermal aging time (t) in the investigated variants of obtaining sorbents was 15, 30 and 60 min. After the set time, the samples were taken out, cooled in air, washed from alkali using distilled water filters until a neutral washing water medium was obtained, dried to an air-dry state, and then a laboratory research of the qualitative and adsorption properties was carried out.

3. Results and discussion

The averaged schedule for the temperature set-up under the influence of EMR on the wet coal-alkaline mixture (option 2) and dried to indiscrete mass coal-alkaline mixture (option 3) is shown in Fig. 2.

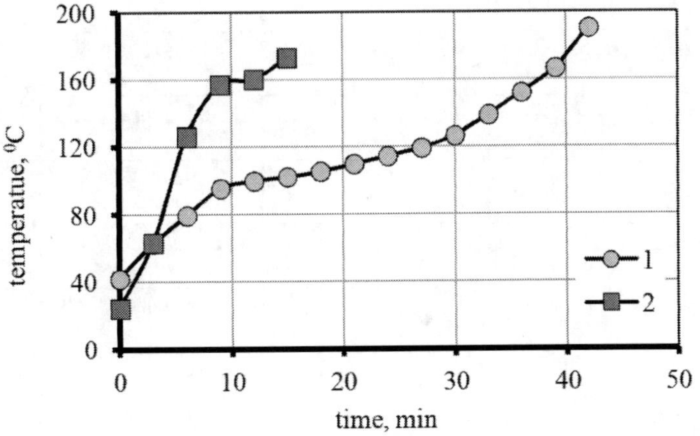

Fig. 2. Averaged schedule of temperature set-up under the influence of EMP on coal-alkaline mixture: 1 - without preliminary drying; 2 - with preliminary drying

According to Fig. 2 that in order to avoid the ignition of a wet coal-alkaline mixture, it is sufficient to apply EMR for 14 cycles (3 min each; total time - 42 min, given the time taken to mix the sample - 47 min) when the maximum possible temperature is 190-191 °C. When EMR is applied to the coal-alkaline mixture dried to indiscrete mass, its ignition begins at a temperature of 172-173 °C in 5 3-min heating cycles (15 min in total, given the mixing time - 17 min). The drying time of the sample (500 g) in the drying chamber at a temperature of 105 °C to indiscrete mass was 16-18 hours. Thus, the maximum time spent for the coal-alkaline mixture treatment was: 1 option - 18 hours; 2 option - 47 minutes (that is 23 times lower than in option 1); 3 option - 18 hours 17 minutes.

The values of the qualitative and adsorption properties, as well as the yield (Y, %) of the obtained sorbent samples are shown in Table. 1.

Analyzing the data of Table 1, it may be asserted that for all the investigated options of thermolysis preparation of samples, carbon samples with a developed porous structure were obtained, i.e. qualitative sorbents. The values of adsorption activity on iodine of the obtained sorbents samples X>80%, which corresponds to the requirements for the quality level of such active coal brands as UAF (X≥70%) and KAD-ground (X≥80%). In all cases the yield of the finished product was Y>50%. At the same time, the application of EMR treatment allowed to increase the sorbent yield by an average of 9.8% for all options with different isothermal aging times periods (in option 2 - by 3-21.7%, in option 3 - by 5.4-13.8%).

In the researched options of the thermolysis preparation of the coal-alkaline mixture depending on isothermal aging time, the adsorption activity by iodine scheme of the obtained sorbents is shown in Fig. 3.

Table 1. Common results of experiment

Option	Coal-alkaline mixture thermolysis preporation		t, min	Technical analysis rates, %			X, %	Y, %
	Drying to indiscrete mass	EMR treatment		W^a	A^d	V^{daf}		
	yes	no	15	11.2	9.6	59.39	82.2	59.4
			30	8.4	8.8	61.30	93.1	61.3
			60	3.6	8.8	50.06	90.1	50.1
	yes	no	15	4.5	7.5	55.78	87.0	61.2
			30	4.0	7.2	64.05	97.1	64.1
			60	3.5	9.1	61.01	95.8	61.0
	yes	no	15	11.0	8.8	65.35	85.0	65.4
			30	9.0	8.4	64.61	91.2	64.6
			60	3.4	9.1	56.97	97.4	57.0

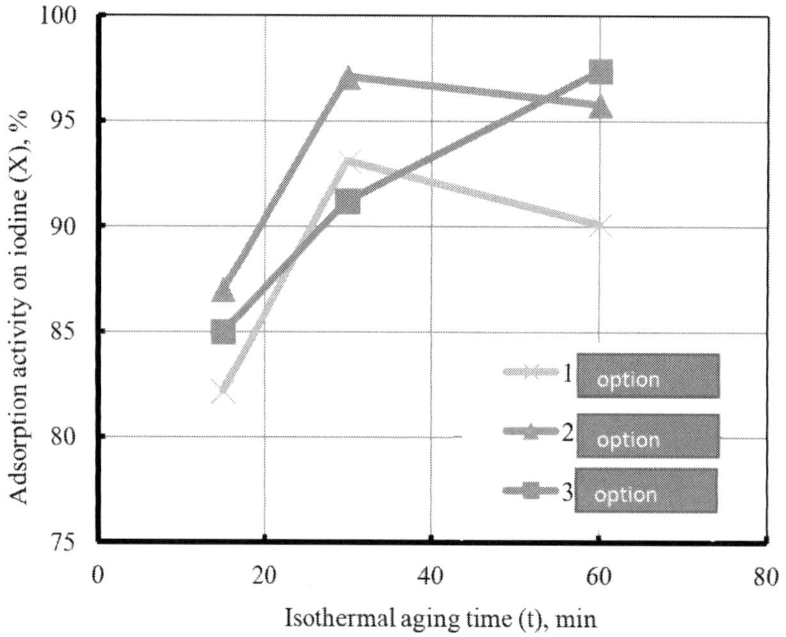

Fig. 3. Adsorption activity on iodine of the obtained sorbents samples in the researched options of thermolysis preparation of coal-alkaline mixture

Scheme demonstrates that the use of EMR treatment in option 2 in comparison with option1 allowed to increase the adsorption activity by iodine, thus making the difference between 4-6%.

At an already short isothermal aging time of 15 min, the adsorption activity values for iodine of sorbents obtained from treated by EMR samples exceed the values of the X samples that don't subjected to such treatment: in option 2 - by 5.8%, in option 3 - by 3.4 %.

With an isothermal aging provided for 60 min, before thermolysis samples' EMR treatment allowed to increase the adsorption activity on iodine to values above 95%, while increasing the results of option 1: without coal-alkaline mixture preliminary drying (option 2) - by 6.3%, with preliminary drying (option 3) - by 8%.

At half-hourly aging of samples in a muffle furnace at 800 °C, value of the adsorption activity on iodine samples, dried to indiscrete mass and treated through EMR before thermolysis (option 3) was obtained slightly lower (by 2%) than samples not subjected to EMR (option 1), which indicates the commensurability of these values. The value of X samples under isothermal aging for 30 min and subjected to EMR without preliminary drying to indiscrete mass in options 1 and 3 for the same duration of thermolysis (option 2) exceeds the values of X and commensurate with the value of X in option 3 at 60 min isothermal aging (with a 0.3%) difference. Therefore, EMR treatment without preliminary drying (option 2) makes it possible to obtain sorbents with a high adsorption capacity on iodine (more than 95%) with significantly lower time spent for preliminary treatment (less than 23 times) and lower time spent for thermolysis (less than 2 times) of the sample directly. Thus, it is possible to exclude the energy and time-consuming drying stage of the coal-alkaline mixture to indiscrete mass from the process of obtaining the sorbents and conduct chemical activation of the brown coal from the Kharanorskoe deposit with potassium hydroxide at a KOH / coal ratio of 1 g / g for 30 minutes.

4. Conclusions

Electromagnetic radiation at a frequency of 2,450 MHz that affects on coal treated with potassium hydroxide (at a KOH / coal ratio of 1 g / g) and followed by thermolysis in the thermal shock mode at 800 °C makes it possible to obtain sorbents samples with a high yield of the final product (more than 50%) and adsorption activity on iodine 85-97%, that corresponds to the industrial-produced active coals ratio level.

Electromagnetic microwave radiation treatment allows to increase the adsorption activity on iodine by 4-6%, to shorten significantly the preliminary heat treatment time of the coal-alkaline mixture (more than 20 times) and to shorten the thermolysis time by 2 times during sorbents obtaining.

References

1. V. A. Kucherenko, T. G. Shendrik, Yu.V. Tamarkina, R. D. Mysyk, Carbon, **48** (15), 4556 (2010).
2. T. S. Manina, N. I. Fedorova, S. A. Semenova, Z. R. Ismagilov, Coke and Chemistry
3. V. M. Mukhin, A. V. Tarasov, V. N. Klushin, *Activated carbons of Russia* (Metallurgy, Moscow, 2000).
4. N. V. Chesnokov, N. M. Mikova, I. P. Ivanov, B. N. Kuznetsov, Journal of Siberian Federal University/ *Khimiya* journal, **7 (1),** 42-53 (2014).

5. T. G. Shendrik, IU. V. Tamarkina, T. V. KHabarova, Solid Fuel Chemistry journal, **5,** 51-55 (2009).
6. M. L. Shchipko, A. O. Eremina, V. V. Golovina, Journal of Siberian Federal University/ *Khimiya*, **1 (2)**, 166-180 (2008).
7. IU. V. Tamarkina, N. N. TSyba, V. A. Kucherenko, T. G. Shendrik, Chemistry and Chemical technology journal, **3**, 132-137 (2013).
8. E. V. Vorsina, T. V. Moskalenko, V. A. Mikheev, Modern Problems of Science and Education journal, 2-3 (2015). http://www.science-education.ru/ru/article/view?id=23990
9. E. V. Vorsina, T. V. Moskalenko, V. A. Mikheev, Problems of Complex Development of Georesources: Mining Informational and Analytical Bulletin, **8 (21),** 214-221 (2016).
10. E. V. Vorsina, T. V. Moskalenko, V. A. Mikheev, Geomechanical and geotechnical problems of Russian North Resources Extraction: Mining Informational and Analytical Bulletin, **11 (24),** 146-154 (2017).
11. D. L. Rakhmankulov, I. Kh. Bikbulatov, N. S. SHulaev, S. IU. Shavshukov, Microwave Radiation and Intensification of Chemical Processes), Moscow: Khimiya, (2003).
12. A. N. Didenko, Microwave Energetics: Theory and Practice (Nauka, Moscow, 2003).
13. O. S. Danilov, T. V. Moskalenko, V. A. Mikheev, Mining Informational and Analytical Bulletin, **3**, 203-208 (2010).
14. A. A. Khaidurova, P. N. Konovalov, N. P. Konovalov, Solid Fuel Chemistry journal, **2**, 67-70 (2008).
15. P. N. Konovalov, N. P. Konovalov, A. A. Khaidurov, Proceedings of Universities. Applied Chemistry and Biotechnology **1(12)**, 74-79 (2015).

Mechanical-chemical activation using solutions containing active oxygen species

Aleksandr Aleksandrov[1], *Natalya* Litvinova[1*], *Tatyana* Konareva[1], *Roman* Bogomyakov[1] and *Natalya* Lavrik[1]

[1]Mining Institute of Far eastern branch of Russian Academy of Sciences, Khabarovsk, Russia

Annotation. The experimental studies of gold leaching from the cyanidation cakes for one of the Far East region deposits is shown in the article. Carbonate-cyanide and hypochlorite-chloride reagents were used for leaching and were electro-photo activated before leaching. Preoxidation of mineral matter by sulfate-peroxide solution were carried out. The mechanical-chemical activation of mineral material was carried out for gold increased extraction. Mechanical activation provided both better cake grinding and formation of micro cracks in the quartz and adular grains. The impact provides the liberation of dispersed nano-sized gold. The change in the granulometric characteristics of initial samples was considered from the upper, middle and lower horizons of the tailing dump as a result of complex physical-chemical effects. It was established that the gold reprecipitated on new activated surfaces of quartz and clay-mica minerals during the leaching process (for the samples of the lower and middle tailing horizons). Carbon in pulp (CIP) technology after the first stage of mechanical activation ensures a loss reduction of the valuable component.

1 Introduction

The main directions of development of research in the field of complex and deep processing of natural and secondary mineral raw materials in the Russian Federation were substantiated on the short time (2014-2020). They include an improvement of physical and physicochemical methods, creation of highly effective energy-saving technologies for ore preparation and selective disintegration of fine-grained ores of complex material composition, increase of the contrast of technological properties of minerals on the basis of application of physical-chemical and energy methods of influence [1].

In1928 P.I. Rebinder discovered the effect of an adsorption decrease in the strength of solids. This discovery was one of the fundamental in physical and chemical mechanics. Since that time the works on using the mechanical-chemical activation process and studying the mechanism of structural transformations of deformed bodies continues at a higher level and captures new areas of application of this process. The classic study of the process of destruction of solids is reflected in the work [2]. A major contribution to the study of the physical-chemical effect of liquid phases on rocks has been made by [3]. It was established as a result of a lot of studies that the maximum decrease in surface energy and

* Corresponding author: nauka22@yandex.ru

strength should be when a solid is in contact with a liquid identical molecular composition. The change in the physical state and chemical properties of minerals during ultrafine grinding is shown in the work [4].
At the present an efficiency of using mechanical-chemical activation has been proven in different directions:

- processing of gold-containing flotation concentrates before subsequent bio oxidation. As a result of a change in the structure of sulphide minerals the duration of bio oxidation is reduced by 3-times and gold recovery is increased by 23% [5];
- mechanical-chemical treatment of gold-bearing sulphides in the presence of oxygen promotes arsenopyrite selective oxidation providing an alternative low-temperature process for extracting gold from ore [6];
- environmental improvement particularly with adsorption purification of waste from the production of alumina – high-energy grinding using acidic solutions [7];
- ultra-centrifugal grinding of zeolites and ash promotes for increase in reactivity [8] and directed change in micro structural formations [9];
- synthesis high-strength composite and materials [10-12]

With the aim of a view to developing of the technological solutions for extracting gold experimental studies of a process of activation leaching of a valuable component basically the mechanical-chemical activation of material were conducted from the cyanide tailings in Far East region in the MI FEB RAS.

2 Materials and methods

The research material is cyanidation cakes of gold extracting plant of one of the mines Far Eastern region. For researches three samples were taken from the upper, middle and lower horizons of the tailing dump with gold content -0,28 ppm, 0,45 ppm and 0,32 ppm.

A modern equipment was used for the research: Fritsch (vibrating screen Analizette 3, divider Laborette 27, ultrasonic bath for cleaning sieves Laborette 17) for preparation of materials for analytically researches; laboratory electrolizer "Saner 5-30-01" for the activation of leaching solutions and the ultraviolet radiation source DRT 400 (UV). Mechanical-chemical activation was carried out in a laboratory roller mill using metal and ceramic balls. The AAS-7000 atomic absorption spectrophotometer was used to measure gold content, silver, iron, copper in the raw materials and products of processing. Electron microscopic study of the initial material was carried out with a scanning electron microscope JEOL (Japan) equipped with an energy dispersive analyzer JCM-6000 PLUS (accelerating voltage 15 kV, probing current 7.475 nA, increase from 20 to 2000).

At the first stage of the works, laboratory samples of the old cyanidation cakes after the sample preparation were leached on the agitation unit accoarding three schemes using different leaching solutions. The sample for the leaching was 300 g. As working leaching solutions used in the first scheme - hydrocarbonate solution, in the second scheme - hypochlorite chloride solution, in the third scheme - sulfate-peroxide solution.
Activation of the leach solution was before to the leaching process (electrolysis, UV irradiation). The pulp was processed in agitation mode for 4 hours. The selection of a liquid sample (filtrate) for analysis made it possible to determine the residual content of CN-ion, Au, Ag, Cu, and Fe. The volume of the solution for the first two schemes was 300 ml, for the third scheme - the diffusion oxidation during the day of 100 ml of the solution, then the addition of 200 ml of active hypochlorite- chloride solution. The researches were in mechanical-chemical leaching scheme at the second stage of work with the initial parameters for intensify the process of gold leaching.

3 Experimental researches results

Before the technological studies a material composition was studied (granulometric characteristic, mineralogical analysis using an electron microscope, chemical analysis of samples).

The granulometric characteristics of the initial samples are presented in Table 1.

Table 1. The sieve characteristic of the initial material.

Size	Sample №1	Sample №2	Sample №3
-0,5+0,2	1,85	0,23	0,20
0,2+0,1	20,18	6,73	5,90
-0,1+0,071	14,23	13,10	14,90
-0,071+0	63,74	79,94	79,10
Total	100,00	100,00	100,00

The size is -0.071 + 0 mm in the second and third samples up to 70% and in the first sample up to 60%. Fine size of particles was determined using a laser diffraction analyzer; the particles are below 10 μm at 50% in the second and third samples, in the first to 42%.

Free and encapsulated gold wasn't find in the main minerals quartz and adular during the mineralogical and electron-microscopic analysis.

Researches of the sample 1 in a scanning electron microscope showed that it consists approximately of an equal proportion of ores and silicates. The sample contains as follows: arsenopyrite, pyrite, feldspar, quartz, pyroxenes, scheelite, apatite, single microspheres of magnetite with an admixture of silicates, titanomagnetite, pyrrhotite, iron scrub, chalcopyrite in the splices. The shapes of the minerals are different, these are individual crystals and fragments of crystals, irregularly shaped angular grains. Up to 30% of irregular grains are fine-grained and splices of minerals in different combinations: silicates with inclusions of sulfides (1-20 microns), sulphides with silicate outgrowths, fine-grained sulphides. Mica, pyrrhotite, titanomagnetite, FeO, Fe, CrFe7-8 intermetallic were also found in aggregate excretions.

The difference of the sample 2 is that it contains fine-grained minerals - sulfides with silicates - up to 50% and the same amount - individual minerals, of which the major part (~ 30%) is composed of silicates. There are from the ore minerals arsenopyrite, pyrite (as well as their splices) titanomagnetite, iron in the form of dendritic formations (scrub?). Scheelite, barite, dense microspheres of magnetite (up to 100 microns) with silicates impurites and in regular intergrowths form a mosaic microsphere are noted rarely. Ore minerals the most often have well-defined crystalline form or clastic. The silicates: K-Na feldspars, Ca-Fe-Mg pyroxenes and single amphiboles, quartz - have an angular clastic shape. Fine-growths minerals (10-50 microns) are lumpy shape.

The sample 3 consist from up to 50% of angular fragments and crystals of individual minerals and as many fine-growth of minerals (10-50 microns) in different combinations. Amount of quartz, dense microspheres of magnetite (some with an admixture of manganese), galena was in the sample 3 more than in samples 1 and 2.

Galena is presented by single crystals and as inclusion in pyrite. In addition to individual grains, there are twin clusters of cubic pyrite crystals, as well as short-prismatic arsenopyrite crystals. Chalcopyrite is present as fine-growth. Other minerals like in the first and the second samples. The results of the chemical analysis of the initial samples for gold, silver, iron and copper are shown in Table 2.

Table 2. The results of chemical analysis of initial samples for gold, silver, iron and copper are shown in Table 2.

Element	**Content, ppm, %**		
	1 sample (upper line)	**2 sample (middle line)**	**3 sample (lower line)**
Au	0,28	0,45	0,32
Ag	2,59	4,45	3,75
Cu	11,95	17,9	10,93
Fe (%)	1,28	1,28	1,15

According to the results of the atomic absorption analysis, in the samples gold is on average 0.35 ppm. The particle size is presumably not higher than 100 nm, because an electron microscopy doesn't define it.

It was extracted only 30% of gold from all three samples by this schemes. The experiments were with saving of the initial parameters by the scheme mechanical-chemical leaching using roller-bearer for intensification of the process of gold leaching. Researches of this materials on re-grinding in the water didn't gave the results using standard cyanide solution 0.5 %. The results are shown in tables 3, 4.

Table 3. The results of an experimental leaching from the first sample with mechanical-chemical activation pulp, which it were prepared of the active carbonate-cyanide solution.

Element	**Content in the first sample(upper line), ppm, %**			
	Au	**Ag**	**Cu**	**Fe (%)**
Initial content	0,28	2,679	11,946	1,28
Mechanical activation in water	0,059	1,421	3,64	27,65
15 min	0,05	1,53	3,66	17,52
30 min	0,07	1,66	3,76	18,137
45 min	0,12	1,63	3,88	18,201
60 min	0,14	1,63	3,90	18,26
Extraction	50,00%	61,96%	32,65%	0,14%

Table 4. The results of an experimental leaching from the second and the third samples with mechanical-chemical activation pulp, which it were prepared of the active carbonate-cyanide solution.

Element	**Content in the second sample (middle line), ppm**				**Content in the third sample (upper line), ppm**			
	Au	**Ag**	**Cu**	**Fe**	**Au**	**Ag**	**Cu**	**Fe**
Initial content	0,45	4,272	17,909	12800	0,32	3,561	11,031	1,1500
Mechanical activation in water	0,096	2,86	4,31	28,29	0,091	1,92	1,80	25,58
30 min	0,27	2,79	4,51	16,64	0,21	1,43	2,66	15,01
60 min	0,27	2,72	4,65	17,41	0,23	1,10	2,76	16,74
90 min	0,25	2,36	4,64	17,81	0,24	1,56	2,88	17,53
120 min	0,32	2,76	4,68	17,82	0,30	2,29	2,90	17,78
Extraction %	71,11	65,31	26,14	0,14	93,75	64,32	26,29	0,15

At the results of leaching from different line of samples the gold content of liquid increase proportionally time of mechanical-chemical activation pulp prepared on the basis of the active carbonate-cyanide solution for all samples.

The gold content increased in three than in mechanical-chemical activation in water.

The change of granulometric characteristics of the material with different mechanical activation duration is demonstrated on figures 1-3.

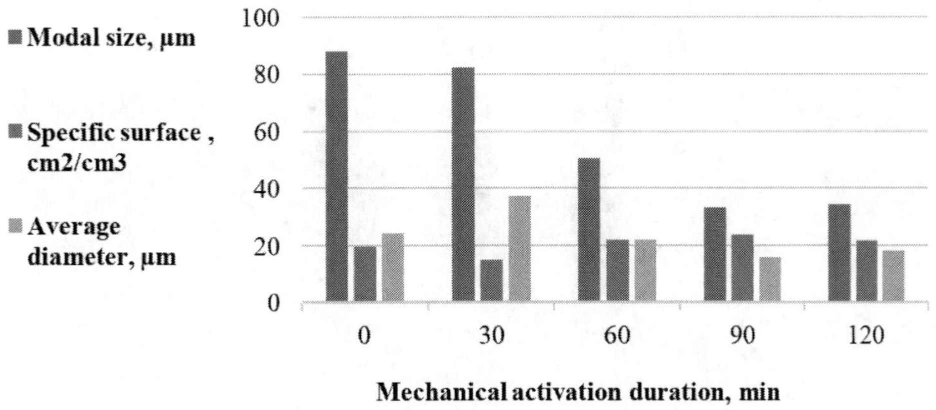

Fig. 1. The dependence of the granulometric characteristics of the cyanidation cakes ZIF from the duration time of mechanical activation of the pulp with the introduction of an activated hydrocarbonate solution (a sample from the upper horizon).

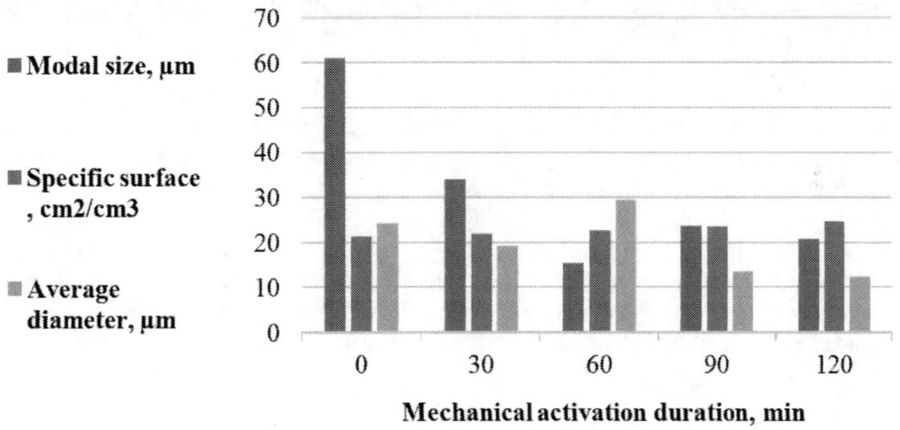

Fig. 2. The dependence of the granulometric characteristics of the cyanidation cakes from the duration time of mechanical activation of the pulp with the introduction of an activated hydrocarbonate solution (a sample from the middle horizon).

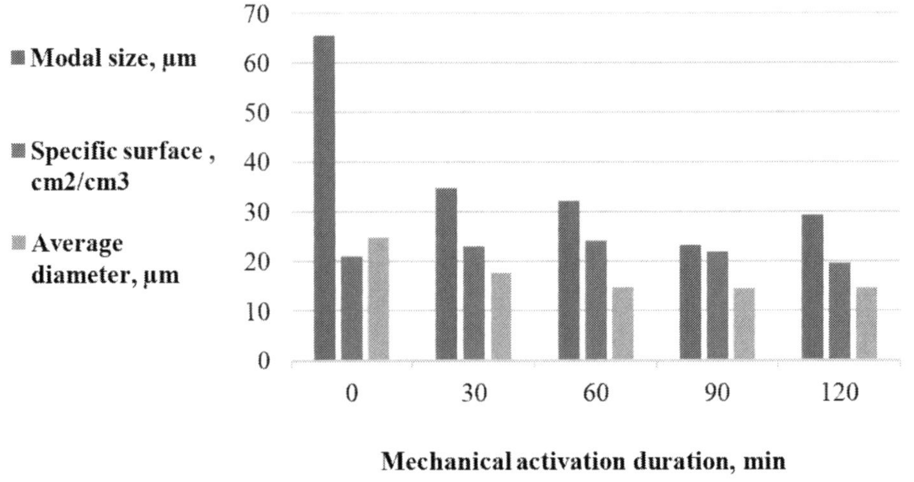

Fig. 3. The dependence of the granulometric characteristics of the cyanidation cakes ZIF from the mechanical activation duration of the pulp with the introduction of an activated hydrocarbonate solution (a sample from the lower horizon).

The granulometric characteristics of materials are the modal size, specific surface and average diameter. Their analyze of change showed that the use of an activated hydrocarbonate solution during mechanical activation of samples promotes to reduce of modal partical size and average diameter and to increase Specific surface of grindable materials. Analezette Nano Tec laser diffraction analyzer was using for this analyze.

The study in the second scheme with including mechanical activation didn't give good results. Probably this was due to the degassing of chlorine and its by-product expenditure on interaction with iron in the metal spheres composition. The time incretion of mechanical activation up to 3 hours and addition active hypochlorite in the pulp led to an iron oxidation and release into solution. Sample agitation with active chlorine with the sodium cyanide 0.35% allowed increasing gold recovery to 50%. The studies by with the third scheme with mechanical activation of active pulp with the chloride preparation made allowed gold 25.71%.

4 Conclusion

It was established that dispersed forms of chemically bound gold is the extracted effectively as a result of mechanical-chemical preparation. In addition to re-grinding of coarse particles the mechanical-chemical activation provides microcrack formation in quartz and adular grains and liberation of dispersed nanoscale gold. According obtained results, gold content in pregnant solution increases proportionally to time of mechanical-chemical activation of pulp prepared on the basis of the active carbonate-cyanide solution for all samples. The gold content increased by three times as a result of mechanical-chemical activation in water medium.

References

1. Chanturiya V.A., Weisberg L.A., Kozlov A.P., Research priority directions in the processing of mineral raw materials. Ore dressing, **2**, 3-9 (2014).

2. Yushchenko V.S., Grivtsov A.G., Shchukin E.D. *Numerical simulation of molecular crystal deformation*. Dokl. Academy of Sciences of the USSR, **215**, 148 (1974).
3. Pertsov N.V., Traskin N.V. *Destruction of solids in solutions*. scientific. tr. International Congress on Surfactants, **3**, 587 - 603 (1978).
4. Pertsov N.V. Physical-chemical mechanics of natural disperse systems, 107 - 117 (1985)
5. Pertsev A.V., Matveenko V.N. Physical-chemical mechanics of natural disperse systems, 125-140 (1985).
6. Pertzov N.V. Kogan BS, Balashov V.N. Model of cracked intrusions of magma under the conditions of manifestation of the effect of adsorption depletion of rocks, **235/6**, 1375 - 1378 (1977).
7. Molchanov V.I. Physical and chemical properties of finely dispersed minerals, 160 (1981).
8. Yusupov T.S. New processes in combined schemes of mineral processing, 202 - 206 (1989).
9. Molchanov V.I. Activation technical means of mineral substances during grinding, **6**, 60 - 75 (1979).
10. Zelikman A.N. Chemistry and metallurgy of rare metals, 30 - 36 (1979).
11. Berestetskaya I.V. Mechanical-chemical activation of the surface of magnesium oxide, **260/2**, 361 - 364 (1981).
12. Avvakumov E.G. Mechanical methods of activation of chemical processes, 305 (1986)
13. Gordon J.K., Asiam, E. K. Influence of mechanical chemical activation on biooxidation of auriferous sulphides. Hydrometallurgy, **15**, 77-83 (2012).
14. Welham, N.J., Minerals engineering, **14/3**, 341-347 (2001).
15. Alighardashi A., Garibi, H. R., Raygan S.H. Water science and technology, **73/4**, 899-908 (2016).
16. Terzic Anja, Piezo Lato, Andrich Lubis. Chemometric analysis of changes in the quality of coal. Science of Sintering, **49/4** (2017).
17. Terzic Anja, Piezo Lato, Andrich Lubis. Chemometric evaluation of mechanical-chemical activated zeolites for use in structural composites. Composites part B-technical, **109**, 30-44 (2017).
18. Wang Weigan, Peng Zhongdong. Nanoscale lithium oxide by mechanical-chemical activation with the assistance of microwave technology as a cathode material for lithium-ion batteries and electrochemical characteristics. Ceramics international, **44/2**, 1425-1431 (2018).
19. Jane Aditya, Omkari Amrish, Saroha Rakesh. Journal of the American ceramic society, **100/11**, 5239-5248 (2017).
20. Zhang Ti, Pan Zhidong, Wang Yanmin. Journal of SOL-gel science and technology, **84/1**, 118-128 (2017).

Experimental research on the formation of mineral-rich areas in residue gold placer deposits

Vladimir Alexeev*[1] and *Tamara* Banshchikova[1]

[1]Mining Institute of Far eastern branch of Russian Academy of Sciences, Khabarovsk, Russia

Abstract. Underestimation of the reserves of raw material base of gold mining (namely residue dump complexes) leads to negative social and economic consequences in the regions engaged in placer gold mining. Nonetheless, their efficient development is possible only through the use of new rational technologies of preparation and deep processing of rock mass for subsequent extraction of valuable minerals. In order to create such a technology, large-scale research was made on the nature and degree of the influence that the basic factors of the technology have on the process of migration and concentration of gold particles in a residue rock mass. **Key words:** residue placer deposits, cryogenic processes, suffusion, filtration flows, gold particles migration, mineral-rich areas formation.

1. Introduction

Exploitation of placer gold deposits of non-ferrous metals, noble and rare metals is of significant importance to Russian Far East, and, despite the discouraging forecast regarding the exhaustibility of natural resources, their development will carry on over many years. Placer deposits include all the basic characteristics of geologic system that reflects the unity and interconnection of the components such as mineral resources, natural processes, technique and technologies, social and economic structures, system organization, production activity, etc. [1].

According to the specialists' estimation [2-6], residue placers possess a great potential. Some gold-mining areas have accumulated significant amounts of auriferous dump complexes – billions of cubic meters of gravel dumps, peats, etc.

The need for a more careful and constructive attitude to the problem of increasing gold extraction out of placer deposits is also connected with ecological state of placer gold-mining areas. The waste products from the primary geogenic deposits have a negative influence on the natural environment, considering that amounts of their rock mass are over 2 billion cubic meters in Far East of Russia alone.

Residue placer deposits are significant reserves of mineral resource base of noble metals, but their development is complicated due to the fact that valuable components in a massif are in chaotic, scattered state, and their extraction is connected with full processing

*Corresponding author: alekseev-vs_83@mail.ru

of all residue formations, which often proves to be unprofitable because of the significant material costs and expenses. Therefore, the problem of finding an efficient method of placer deposits development is acute and is of great financial and social importance.

The research on the residue formations of placer deposits has led to the foundation of in-dump enrichment of a placer deposit, where the particles of noble metals migrate and concentrate into lower layers of the mountain massif, which is chiefly driven by natural factors. These factors include the following processes: hydrodynamic process, including the processes of filtration and suffusion, cryogenic processes and microseismic processes. Significant similarity in genetic technologic nature was found in the analogous processes – natural processes of placer deposit formation, particularly geological differentiation of mineral matter. The term "differentiation" is used to characterize many geological processes, in which primary mineral resources form byproducts that are genetically similar to them but have a different structure or traits.

Unlike the processes of formation of mineral-rich areas of residue placer deposits, the geological differentiation of mineral matter is a lot more time-consuming, although it is necessary to know its characteristics to estimate all the factors that aid in the migration of valuable components of alluvial rock mass into a bedrock-adjacent area of dump complexes. The differentiation of mineral rock mass of residue dump complexes of placer deposits reflects the processes of selective formation of productive layer, i.e. natural division and displacement in rock mass of residue dump complexes of valuable component separate monofractions and enclosing rocks relative to each other depending on their morphological, granulometric, hydraulic and physical characteristics, which is driven by natural and anthropogenic processes [8,9].

One of the characteristics of residue placer deposits is the stochastic nature of gold distribution and other valuable components in a massif, which, in the process of exploration, determines the need for the development and overall enrichment of the rock mass. Therefore, when the amount of valuable components is insufficient, the attempts to process residue placer deposits prove to be futile. For this reason, the formation of residue placer deposits of new kind with selective position of productive layer of sands in lower strata is a major scientific and practical problem. As a rule, a bedrock-adjacent mineral-rich layer is characterized by its homogeneous structure, relatively small amount of productive rock mass, which allows for a more economic and technologic development of residue placer deposits.

The technology itself is the following [10, 11]: a drainage ditch and an accumulating ditch are created in the upper and lower parts of the block of residue sands, which entails the formation of a filtration flow in the block afterwards, and this flow allows the gold particles to move vertically. If such a flow is continuous, it leads to the formation of a mineral-rich layer of sands in the bedrock-adjacent part of the block. It allows for the separation of sands with lower contents of gold and reduces the range of washout.

In order to study the processes, lying in the bases of the technologies, laboratory researches were made in 2008-2009. They were focused on the influence of cryogenic processes on the efficiency of gold particles migration [10]. The results are: the more freeze-thaw actions happen, the deeper is gold distributed based on the capacity of the rock mass; 0,25 mm fraction of gold migrated across the whole depth of the rock mass, where 7.1 % of this fraction reached the bottom; cryogenic processes as an independent factor have little influence on the migration of gold particles.

The further studies were focused on the influence of the processes on gold migration. In the first experiment, gold migration was caused by free flows of water, in the second it was caused by the combination of the processes: free flows of water and freeze-thaw actions. The experiments were held simultaneously on the specially designed unit that consisted of 6 tanks 25x12x10 cm, and each of them was connected to a pipeline with the function of regulation of water flow rate, which amounted to 10 l/h. In the first five tanks, gold with certain thickness stayed on the upper layer; in the sixth tank, designed for freeze-thaw experiments, the upper layer accumulated the gold of all kinds of thickness. In each tank, the migration was caused by the filtration flow during 25 days. One of the tanks got frozen twice at a temperature of -18 °C, and after it thawed at a room temperature, the water flow continued. The results of these experiments are shown in Fig. 1.

Under the complex influence of external factors on the rock mass, gold distribution based on the output of a rock type proceeded more effectively, gold migration into lower layers increased by 20 -64.3%. 0.1 mm fraction reached the bottom of the vessel, and it was 19% more under the complex influence. Content analysis based on the rock length shows that the movement of gold particles in a horizontal plane is caused only by water flow energy.

In 2010, the research on gold migration took place in natural conditions. Holes were filled with gold fractions from placer deposit of the stream "Bolotistyi", which was a permanent source of the water flow. In winter, the rock mass was freezing, and after it thawed, it was excavated layer by layer. As a whole, the experiment lasted 300 days.

Fig. 2 accurately illustrates the change in the content of gold based on the depth relative to the original one: in the upper layer, it reduced by 62%, while in the lower layer it increased by 68.9%. The increase in the content was caused mainly by the migration of 0.25 mm fraction. The horizontal movement of the particles is insignificant, except for the water supply location.

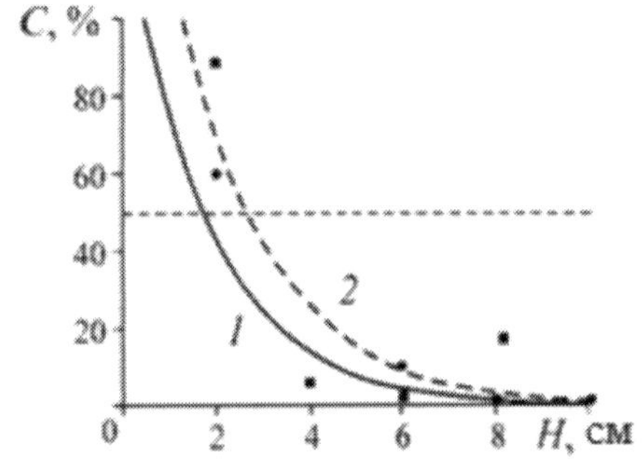

Fig. 1. Dependence of gold content in the areas of concentration on the depth of the rock mass: 1 – under the influence of free water flow; 2 – under 2 additional freeze-thaw actions

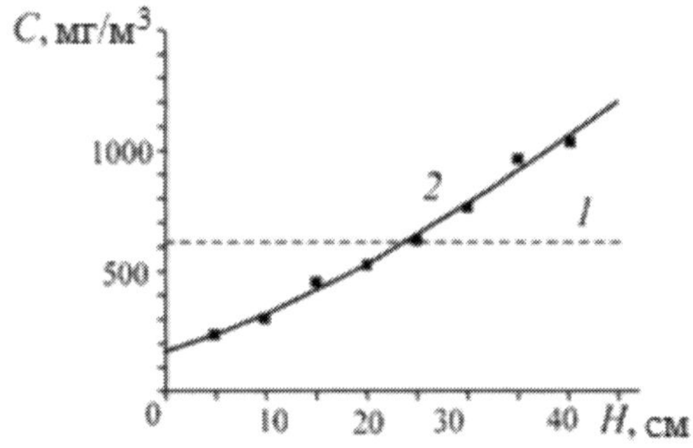

Fig. 2. Change in gold content based on the depth: 1 – the primary gold content in a rock mass; 2 – gold content after the experiment

In 2012, industrial tests began on a test section allotted by "Ros-DV", LLC (Khabarovsk Krai). In the middle part of the deposit from the stream "Bolotistyi", developed in 2006-2007, in close proximity to the silt-detention basin that served as the source of technological water, there was a block of residue sand, 8×15 m in size and 1 m in volume. The materials used were tailings extracted from the processing, that contained small and thin gold. The next stage implies the freezing of the block, but its thawing process generated a filtration flow during the washing process.

Before the next freezing, certain tests were selected. Their results are shown in Fig. 3a. The average content of gold in the four upper samples was 21.6% decreased in comparison to that in hole 1, in hole 2 – 22.9%, in hole 3 the migration is barely visible due to the unprocessed area.

Assuming the limit of the content equal to 125 mg/tons, we will see a cut, where area 1 is the part of the block, where the content of gold reduced (overburden rocks), area 2 is a productive layer the volume of which is equal to the half of that of the sand block. The formation of area 3, presenting a layer of sands with high content of gold (over 125 mg/tons) is explained from the position of this area over the depression curve, which caused this area to not be subject to the influence of water flow, and the migration is caused by the filtration from the surface of the rainwater and "freeze-thaw" process in the rock mass.

In October 2014, (the experiment lasted 2 years), several samples were selected from the test section, and the result of their beneficiation is illustrated in Fig. 3b. The average content of gold in the five upper tests reduced by 21.2% in comparison to that in hole 1, 8.7% in hole 2, and 24.4% in hole 3. Thus, during the 2 washing processes, it was possible to form a mineral-rich layer (with over 125 mg/t), the volume of which is 44% of overall volume of the sands in the experimental block.

The obtained data may be used to expand the theoretical knowledge of the process of gold particles migration, and develop the methods of calculating the basic parameters of the technology that is capable of controlling the factors influencing gold migration.

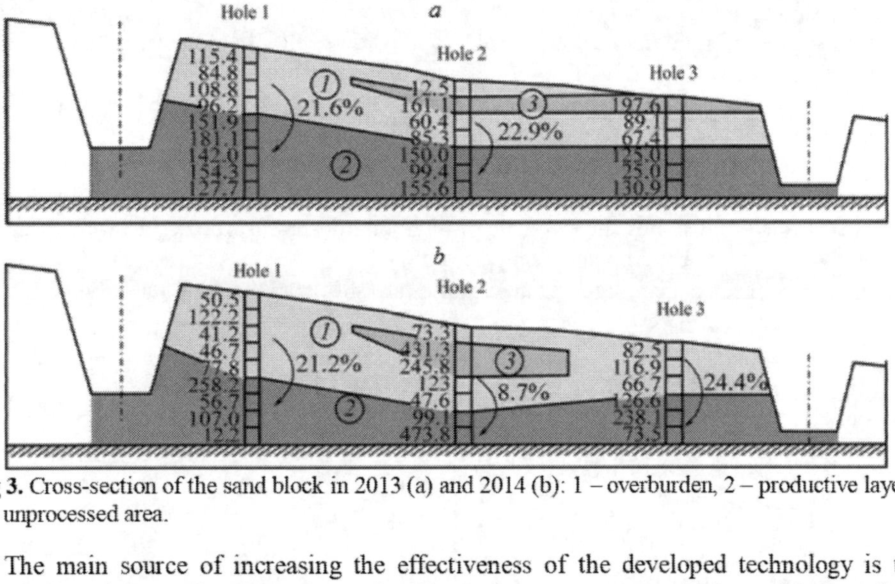

Fig 3. Cross-section of the sand block in 2013 (a) and 2014 (b): 1 – overburden, 2 – productive layer, 3 – unprocessed area.

The main source of increasing the effectiveness of the developed technology is the foundation of the cycle of forming a free flow of process water, which would allow for a significant decrease in its flow rate.

In 2015, for the purpose of studying the influence from the cyclic flood-and-drain process of the rock mass of the residue placer deposit on the migration of gold particles, a laboratory research was held.

2.5-liter tanks were filled with a rock mass with water supply to create a filtration flow. The rock mass was imparted with equal amount of gold fractions: -1+0.5 mm – 150 mg, -0.5+0.2 mm – 50 mg, -0.2 mm – 28.8 mg.

The water supply to tank 1 is constant, to tank 2 – cyclic, and the flood-and-drain process implied 5 days of constant water supply, 2 days of draining; the gold was placed evenly upon the whole volume of the rock mass; process water flow rate in each tank was 12.5 l/ph.

After 25 weeks (25 flood-and-drain cycles), the water supply was stopped, the selection of samples is conducted layer by layer skipping 2 cm. The analysis of the laboratory research emonstrates the following:

- with constant water supply, the content of gold in the 2 upper layers reduced by more than 50% in comparison to the original; the increase of gold content in the third layer was 24.5 %, in the second – 38.0%, in the third – 30.4%; the nature of the change in gold content in different fractions in the same: a little decrease in the upper layer, significant decrease in the second and an increase in the layers below.

- with cyclic water supply, the maximal change in gold content was 4.6 % in comparison to the original; it is possible to observe the migration of only the particles from 0.5+0.2 mm.

In 2016, the experiment was continued in laboratory conditions. Four 20 cm-high tanks were filled with a rock mass, containing evenly spread component of thickness -1.0+0.1 mm. Every tank was fed with water the flow of which was stopped at certain intervals. The first tank was fed with water with no intermission, the second was fed for 2 days with 1-day breaks, the third one was feed day after day. In the fourth tank, the filtration flow was active primarily for 2,5 minutes until the tank was completely full, then it was stopped, but was resumed before the rock mass was drained (approximately in 1.5-2.0 min), i.e., the incomplete draining of the rock mass took place.

After 2 months, the gold was extracted from the samples selected layer by layer; its content is shown in Fig. 4.

The process of gold migration proved to be the most effective with the cycle of incomplete draining. At tank depth of 8 - 20 cm, the content of gold increased by 23.2% in comparison to the original, and process water flow rate reduced by 40 and 50%, respectively.

In 2017, the laboratory researches on the processes of the formation of productive areas of residue placer deposits by means of cyclic influence of filtration flow were continued.Table 1 illustrates the parameters of flood-and-drain process in three blocks of residue rock mass, the experiment lasted 125 days.

The results of the beneficiation of the samples are shown in the diagram in Fig. 5.

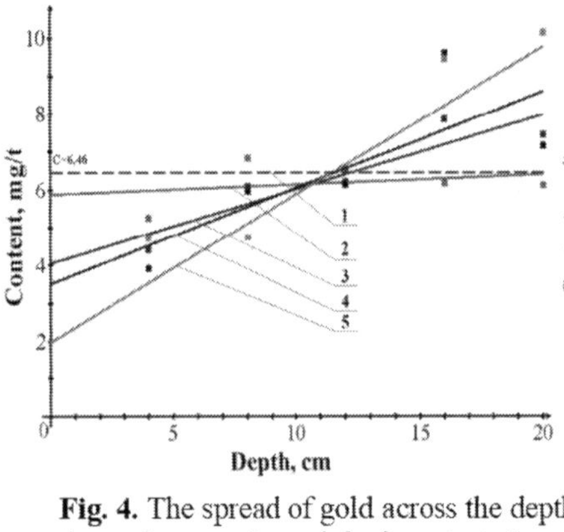

Fig. 4. The spread of gold across the depth of the rock mass: 1 – original content, 2 – with constant water supply, 3 – with cycle 2/1, 4 – with cycle 1/1, 5 – being flooded and incompletely drained.

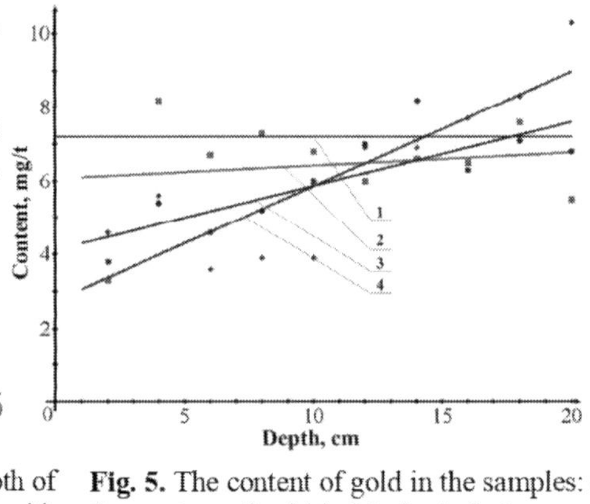

Fig. 5. The content of gold in the samples: ■,♦,● – the content of gold in blocks 1, 2, 3, respectively; 1 – original state, 2 – diagram of content change in block 1, (c_c=0.35), 3 – diagram of content change in block 3 (c_c=0.64), 4 – diagram of content change in block 2 (c_c=0.72).

The analysis of the obtained data shows that the worst migration of gold took place in block 1, the change in the content was mainly caused by fraction -1.0+0.5 mm; the migration of gold particles the thickness of which was below 0.5 mm is insignificant.

The change in the content of gold in tank 3 was also basically caused by fraction - 1.0+0.5 mm, although it is possible to observe the movement of gold particles from fraction 0.5+0.2 mm from the second and third layers being 14 cm deep, and the 73% increase in the content of gold in fraction -0.2+0.1 mm on the bottom of the tank. The effect of gold particles migration is noticeable in tank 2. It is possible to observe the decrease in the content of gold in fractions -0.5+0.2 and -0.2+0.1 mm, and in the 7 upper layers, the content increased by 28% for fraction -0.5+0.2 mm and by 87% for fraction -0.2+0.1 mm.

Table 1. Flood-and-drain cycle parameters

Block number	Water supply, min	Draining, min	Cycle duration, min
1	2	8	10
2	5	5	
3	8	2	

Conclusions

It was established that the migration of gold particles in residue sands is more intensive than with more cycles of "freeze-thaw" actions. Cryogenic processes have a noticeable influence on the migration of gold of small sizes (-0.25 mm). It was experimentally proven that the main factor influencing the movement of the particles is filtration flow.

Under the complex influence on the rock mass, the area of gold concentration is displaced downwards, being increased by 20 - 64% in comparison to the influence of the water flow alone, and it also moves across the length of the dump, with the prevalent influence comes from the volume of water free flow.

On the basis of natural experimental researches it was established that after the influence on the rock mass from the water flow and its freezing in winter, the content of gold in the upper layer reduced by 62% in comparison to the original, in the lower layer it increased by 68.9%. As a result of industrial testing on the test section during the two washing seasons, it was possible to form a mineral-rich layer the size of which is 44% of overall volume of sands in the block.

The migration of gold under cyclic flood-and-drain process of the rock mass is more effective; the content of gold in the laboratory research in the bottom layer increased by 43%, water flow rate reduced by 50%.

References

1. N. P. Laverov, K.N. Trubetskoi. Vesti RAN. **66 (5).** pp. 411-418.
2. A.P. Van-Van-E, Resource base of natural-technogenic gold placer deposits. – M.: Mining book, 268 p. (2010).
3. I.U. Rasskazov, V.S. Litvintsev, U.A. Mamaev. Gold-mining industry. **1**. pp. 14 – 20. (2011).
4. V.S. Litvintsev. PTPDMR. **1**. pp. 118 – 126. (2013).
5. G.S. Mirzekhanov, Z.G. Mirzekhanova, The resource potential of residue gold placer deposits. Max Press. p. 288. (2013).
6. V.S. Litvintsev, R.S. Seryi, T.S. Banshchikova, P.P. Sas. PTPDMR. **2**. pp. 72-79. (2016).
7. R.N. Kopylov. The differentiation of gold in alluvial layer deposits / ed.-in-chief E.D. Izbekov. Yakutsk: Sakhapoligrafizdat. p. 141. (2002).
8. L.V. Pustovalov. Petrography of sediment rocks. Gostopnekhizdat. p. 476. (1940).
9. B.S. Lunev, O.B. Naumova, Atlas of placer deposits geology. **1**: Factors of placer deposit formation. p. 344. (2005).
10. V.S. Alekseev, The foundation of rational technology of forming a productive area under open development of residue placer deposits of Amur River region: synopsis of thesis. candidate of technical sciences - Khabarovsk city. p. 22. (2012).
11. U.A. Mamaev, V.S. Litvintsev, V.S. Alekseev. Pacific geology. **31(4).** 106-112. (2012).

Prospects and possibilities of using ash and slag wastes of power enterprises of Primorsky Krai as technogenic mineral raw materials

Sergei Ivannikov[1*], *Evgeniy* Shamrai[1], *Andrey* Taskin[2] and *Aleksandr* Yudakov[1]

[1] "Institute of Chemistry", Far East Division of RAS, Vladivostok, 690022, Russia
[2] Far East Federal University, Engineering School, Vladivostok, 690091, Russia

Abstract. The results of an investigation of ash and slag wastes (ASW) of enterprises of the energy sector of Primorsky Krai are presented. The averaged contents of the main elements and mineral complexes in Primorsky Krai are given. It is shown that the mineral composition of the ASW data makes it possible to separate the primary raw materials into fractions with different compositions. A scheme is proposed for dividing the initial ash extractors into separate mineral fractions by the particle size and by their physical properties. The predominant concentration of gold, platinum, rare earth elements (REE) and a number of other valuable components in the heavy non-magnetic fraction isolated from the primary ASW was detected. Almost complete absence of gold, noble metals and REE in underburning of coal, magnetic and micro-dispersed fractions of ASW has been demonstrated. A device was offered for complex processing of ash and slag wastes of enterprises of the power industry of Primorsky Krai, which makes it possible to divide the initial ASW into mineral fractions, being raw materials for various industries.

1 Introduction

One of the main problems facing the energy industry, producing electricity and heat using coal, is the problem of waste generated from the combustion of fuel. According to the Federal Agency for Technical Regulation and Metrology, in 2017, the annual increase in ash and slag wastes in Russia is about 21.3-26.7 million tons [1], while the volumes of secondary use of ASW do not exceed 6.3 million tons [1], which led to the situation when the total volume of ASW accumulated at the landfills in Russian Federation is about 1,300 million tons [2], and the area occupied by ash and slag polygons exceeds 22,000 ha [2].

The share of coal in the fuel balance of Russian power plants in 1980 – 2010. did not fall below 25%. At the moment, according to the "Energy Strategy of Russia for the period until 2030", the share of coal in the fuel balance of power plants is forecast to increase to 36-38% with a further increase in coal consumption in physical terms by 64%, while reducing consumption of gas and fuel oil [3]. And this is not surprising, since Russia has huge resources of various quality coals - from brown to anthracite. The total resources are

*Corresponding author: fyajkfqn@mail.ru

estimated at 4,089 billion tons, and the balance reserves are about 272.7 billion tons [4], most of these reserves being energy coals – 3,641.9 billion tons (89%), and only 447.1 billion tons (11%) being coking [4].

Meantime, the volumes of ASW processing in Russian Federation remain one of the lowest in the world today, which leads to further growth of ASW stocks stored in landfills. This fact adversely affects the ecological situation in the ash disposal areas. The main danger is the penetration of effluents from the ash dump into the ground, leading to the entry of polluting substances (heavy metals, radioactive elements, etc.) into groundwater, and then - into rivers and aquatic bodies and, ultimately, human food [5].

At the same time, ASW can serve as a valuable source of various minerals. The ash included in the ASW of energy enterprises contains the following mineral complexes: aluminosilicate microspheres (1-2%), magnetite (5-9%), aluminum oxide (Al_2O_3 20-27%), silicon dioxide (SiO_2 49%), underburning coal (5-15%); oxides of rare earth elements (5-7%), which can find the widest application in industry [6, 7]. In the literature there are numerous reports on the presence of increased contents of precious metals and rare earth metals in ash and slag waste from energy enterprises [8-10]. There have been some attempts to extract from ASW gallium, germanium [11] and scandium [12].

The mineral composition and properties of ASW differ significantly depending on the composition of the coals from which they were obtained, the specifics of combustion technology and other factors [13]. According to the data of Yarmolinskaya [14], for ASW from CHP of the Far East, the following features are characteristic: a relatively small amount of active oxides (CaO + MgO); increased content of sesquioxides (Al_2O_3 + Fe_3O_4); insignificant content of sulfuric and sulfuric compounds: (in terms of SO_3) - 0.36%; acidic surface of particles due to the increased content of amorphous silica (basicity module <1), specific composition of organic waste compounds, increased porosity (41-56%) and bitumen capacity. Existing studies of physical and chemical properties and natural activity of the Far East ASW indicate the possibility of using these materials as raw materials for the production of building materials [15-17].

ASW from energy enterprises can become a promising reserve for replenishing the country's mineral and raw materials base and require comprehensive study with the aim of organizing ASW comprehensive processing that can simultaneously solve both the ecological task of large-scale utilization of ASW by creating building materials, and achieve economic profitability of the process, due to the extraction of valuable components.

2 Materials and Methods

To study the prospects and possibilities of using ASW as technogenic mineral raw materials, samples from the landfill sites of the Primorsky Krai largest cities were taken: CHPP-2 (Vladivostok); CHPP (Artem); CHPP (Bolshoy Kamen); CHPP (Arsenyev); GRES power plant (Partizansk); Primorsky GRES power plant (Luchegorsk). Mechanical sampling was carried out at various depths from the surface (from 0.5 m to 1.5 m) in accessible locations of ash dumps. The distance between the sampling points was from 20 to 100 m, the sample seats were fixed by the navigation system.

To control the gold content in ASW, the instrumental neutron activation analysis (NAA) method was used at the compact facility developed at the Institute of Chemistry of the Far East Division of Russian Academy of Sciences (FED RAS) with a radionuclide excitation source based on ^{252}Cf. Identification of the gold content was carried out according to the NSAM method No. 242-yaf.

To find the chemical composition of ASW, X-ray fluorescence analysis (XFA) was used. The investigations were carried out at the Laboratory of Molecular and Elementary Analysis of the Institute of Chemistry, FED RAS, a member of the Center for Collective

Use, Far East Center for Structural Research (FECSR). An energy dispersive X-ray fluorescence spectrometer EDX-800HS made by Shimadzu, Japan, was used for the measurements. The sensitivity for elements from Na to U was up to 1 ppm.

The identification of rare-earth elements was carried out by inductively coupled plasma mass spectrometry (ICP-MS) using the Agilent 7700 spectrometer (Agilent Technologies, USA) at the analytical center of the Far East Geological Institute of FED RAS.

The content of precious metals in ash and slag technogenic wastes was determined by the method of atomic-adsorption spectrophotometry (AAS) in the laboratory of micro- and nano-investigations of the analytical center of the Far East Geological Institute of FED RAS, on the atomic-absorption spectrophotometer Shimadzu 6800.

In order to identify the chemical and elemental composition of individual fractions of ASW and for the subsequent identification of chemical and technological methods for processing these fractions, the initial ASW sample was divided into individual mineral fractions by particle size and by their physical properties. The general scheme for separating the ASW sample into individual mineral fractions is shown in Figure 1. To separate coal underburning coal, FML-12 unit was used. To separate the magnetic minerals, PBM-32/20 unit was used.

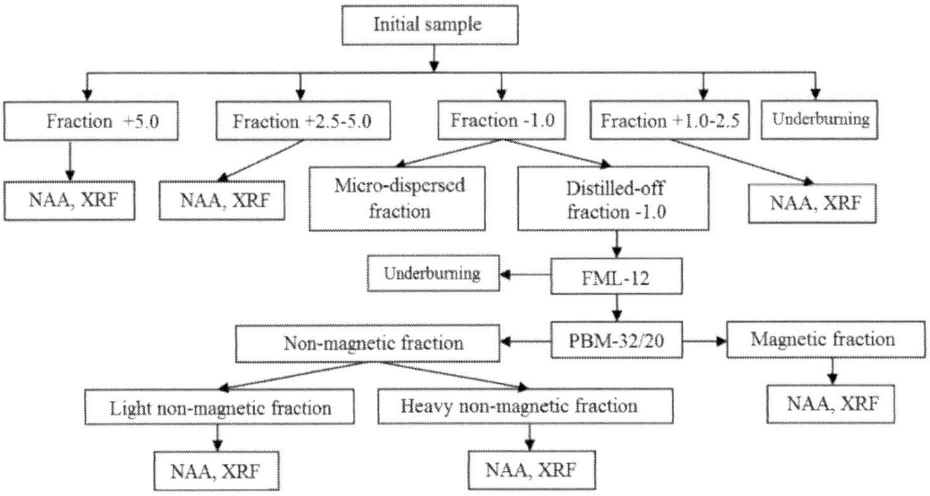

Fig. 1. Scheme of separation of initial ASW sample into individual mineral fractions.

3 Results and Discussion

A survey of ash and slag wastes from the landfill sites of Primorsky Krai showed that 90-95% of the samples consisted of free and bound oxides of silicon, aluminum, iron, calcium, magnesium, potassium, and sodium in chemical compounds. In selected samples, in small amounts (> 1%), the following trace elements are present: boron; fluorine; arsenic; zinc; nickel; molybdenum; germanium; gallium; vanadium; cobalt; mercury; lead; uranium. The chemical composition of ASW is highly dependent on the type of fuel burned from various fields, but for the site of a separate CHP plant, the chemical composition can be considered stable for practical use, being guided by the median value.

The work on the separation of initial ASW samples according to the scheme shown in Figure 1 was carried out and the following fractions of bulk material were obtained: aluminosilicate microspheres; microdispersed fraction; underburning of coal; magnetic fraction; light non-magnetic fraction; heavy non-magnetic fraction.

The distribution of the separated mineral fractions by volume (% of the total mass) was studied, as a result: the magnetic fraction (iron-containing concentrate) accounts for 5-10%;

underburning of coal of 2 types - 8-20%; microspheres of various composition 0,5-1,5%; microdispersed fraction – about 15-20%; light non-magnetic fraction, which is a purified aluminosilicate fraction, accounts for 70-80%.

Conducted by the NAA method, a study of the distribution of gold in the mineral fractions obtained showed a predominant concentration of the metal in heavy non-magnetic fraction (see Table 1), where the gold content increased by 3-4 times compared to the initial samples. At the same time, gold was not found in the underburning of coal, magnetic fraction and microdispersed fraction.

Table 1. The gold content in various mineral fractions ASW, in g / t. (according to NAA)

Sample	Initial sample	Coal under-burning	Micro-dispersed	Light non-magn.	Heavy non-magn.	Magn.
Vladivostok, CHPP-2,	0.10	<0.05	<0.05	0.26	0.39	<0.05
Vladivostok, CHPP-2,	0.05	<0.05	<0.05	<0.05	0.21	<0.05
CHPP Artem, polygon 2	0.12	<0.05	<0.05	0.09	0.28	<0.05
Bolshoi Kamen, CHPP	0.09	<0.05	<0.05	<0.05	0.35	<0.05
Arseniev, CHPP	0.11	<0.05	<0.05	<0.05	0.36	<0.05
Partizansk, GRES, section 3	<0.05	<0.05	<0.05	0.12	0.37	<0.05
Luchegorsk GRES polygon	<0.05	<0.05	<0.05	<0.05	0.15	<0.05

Similar results are observed for other noble metals. Thus, an additional study conducted on samples from the Vladivostok CHPP-2 landfill site using the AAS method showed almost a twofold concentration of platinum in the heavy non-magnetic fraction of ASW compared to the light fraction of ASW (0.006-0.008 to 0.015 ppm). In addition to precious metals, in the heavy non-magnetic fraction ASW, the concentration of rare and rare-earth elements predominates (see Table 2-3).

Table 2. The content of rare metals and rare earth elements in light non-magnetic fraction of ASW of CHPP-2, Vladivostok (according to ICP-MS)

Element	C, mg/kg	Element	C, mg/kg	Element	C, mg/kg
Be	2.01	**Nb**	14.83	**Dy**	4.20
Sc	7.2	**Mo**	3.46	**Ho**	0.93
V	55.3	**Cd**	0.21	**Er**	2.80
Cr	140.7	**Ba**	597.4	**Tm**	0.40
Co	8.8	**La**	29.5	**Yb**	2.77
Ni	60.7	**Ce**	57.4	**Lu**	0.50
Cu	31.3	**Pr**	5.41	**Hf**	6.49
Zn	46.3	**Nd**	23.66	**Ta**	1.35
Ga	10.98	**Sm**	4.98	**W**	3.89
Rb	103.6	**Eu**	0.57	**Pb**	19.04
Y	28.7	**Gd**	4.64	**Th**	11.88
Zr	205.6	**Tb**	0.72	**U**	4.32

Table 3. The content of rare metals and rare earth elements in heavy non-magnetic fraction of ASW of CHPP-2, Vladivostok (according to ICP-MS)

Element	C, mg/kg	Element	C, mg/kg	Element	C, mg/kg
Be	4.86	**Nb**	26.74	**Dy**	8.52
Sc	19.9	**Mo**	4.04	**Ho**	1.86
V	100.4	**Cd**	0.34	**Er**	5.42
Cr	64.96	**Ba**	943.7	**Tm**	0.75
Co	11.64	**La**	60.2	**Yb**	5.97
Ni	30.25	**Ce**	116.5	**Lu**	0.91
Cu	27.82	**Pr**	11.76	**Hf**	7.56
Zn	84.8	**Nd**	50.74	**Ta**	2.24
Ga	20.31	**Sm**	9.51	**W**	4.92
Rb	122.7	**Eu**	1.15	**Pb**	26.37
Y	57.6	**Gd**	9.70	**Th**	22.45
Zr	264.1	**Tb**	1.47	**U**	9.51

Since the studies show the predominant concentration of most valuable components in the heavy non-magnetic fraction of ASW, this fraction can be considered as a consolidated source of valuable elements. The remaining mineral fractions obtained from the separation of the original ASW can be used as a mineral raw material for various industries. In particular, on different equipment, technologies were developed for the making of commercial products with stable liquidity-briquetted high-calorie smokeless fuel based on the coal under-burning (see Figures 2-3), magnetite concentrate, microsphere concentrate, fine-grained sand and magnetite pellets.

Fig. 2. Briquettes of coal underburning, made on a screw press

Fig. 3. Briquettes of coal underburning, made on a roll press

The research conducted by the FEFU and FED RAS in the period from 2008 to 2017 [18-19], allowed developing a technological scheme for a device for comprehensive waste processing at power enterprises of Primorsky Krai. This technological scheme consists of separately functioning blocks connected in a single technological line (see Figure 4-5). In each block a certain intermediate product is extracted: concentrate of precious metals, iron oxides, coal underburning and an aluminosilicate residue.

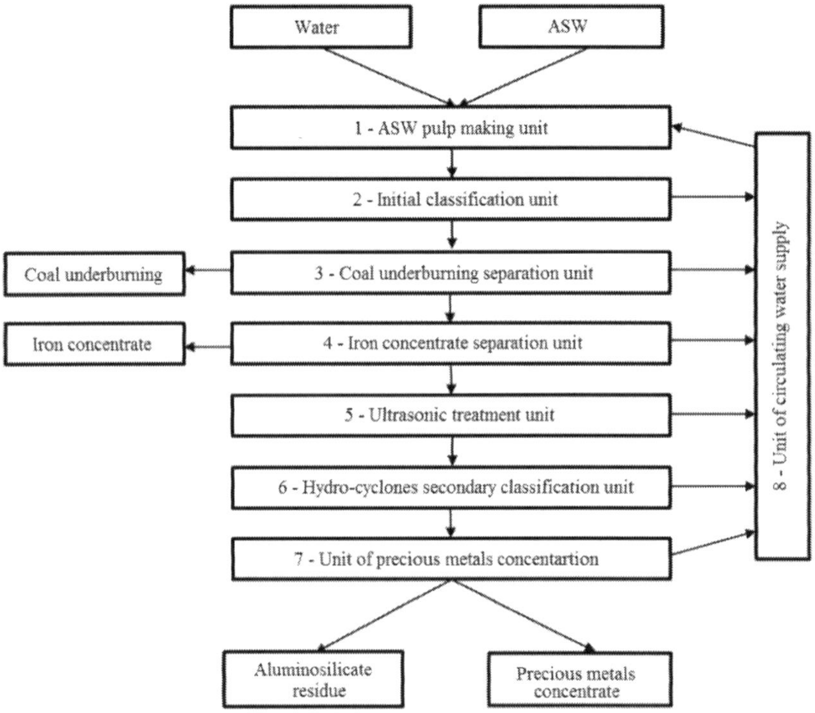

Fig. 4. Block diagram of comprehensive processing plant from ash and slag wastes of enterprises of energy sector of Primorsky Krai

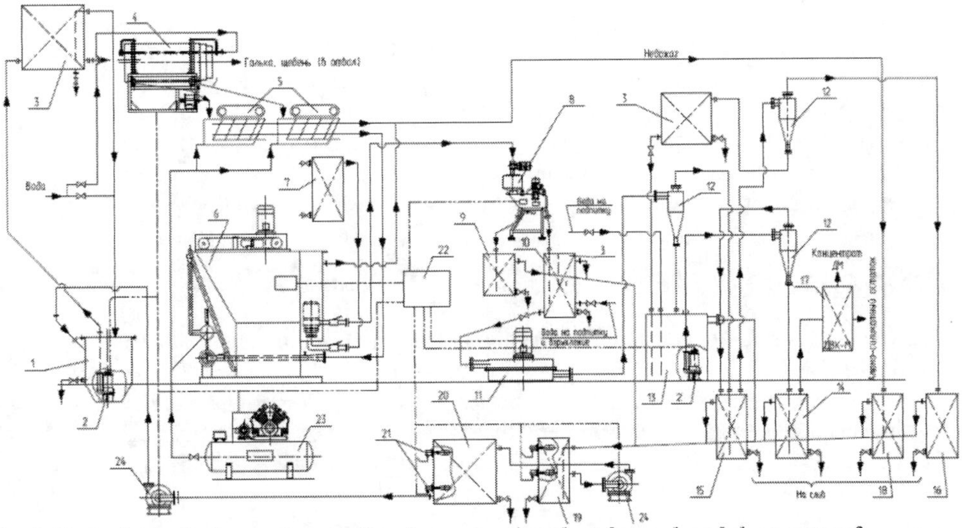

Fig. 5. Technological scheme of comprehensive processing plant from ash and slag wastes of enterprises of energy sector of Primorsky Krai: 1 - mixing container, V = 0.5 m^3; 2 - slurry submersible pump GSZ-4; 3 – recycling container, V = 0.5 m^3; 4 - hydro-grate; 5 – container for coil underburning of the first stage; 6 - flotation unit of the second stage of flotation FML-02; 7 – container of the flotation reagent V = 0.1 m^3; 8 - magnetic separator PBM-P-25-10; 9 - container for magnetic fraction V = 0.3 m^3; 10 - container for non-magnetic fraction V = 0.5 m^3; 11 - dispersant of ultrasonic I100-6 / 9M; 12 - hydrocyclone GPC-75; 13 - container for heavy non-magnetic fraction of stage I, V = 0.5 m^3; 14 - container for heavy non-magnetic fraction of stage II, V = 0.5 m^3; 15 - container for the finely divided fraction of stage I, V = 0.5 m^3; 16 - container for the fine fraction of stage II, V = 0.5 m^3; 17 - modernized centrifugal vibroconcentrator (CVC-M); 18 - container for coal underburning after flotation, V = 1.0 m^3; 19 - a tank-settler of return water, V = 2.0 m^3; 20 - capacity of cleaned recycled water, V = 2.0 m^3; 21 - level sensors; 22 - control panel; 23 - electro compressor B5900V / 270; 24 - centrifugal pump for circulating water.

The main difference of the developed technological scheme from the existing ones, in the comprehensive approach to the processing of ash and slag wastes, the most effective separation of the initial material into separate mineral fractions and the fullness of the recovery of the concentrate of valuable components from ASW. ASW integrated processing technologies provide a good basis for further research. This requires the development of new technological solutions aimed at the extraction of other valuable components besides gold, in particular silver, platinum group metals, REE, silicon oxides, aluminum and iron, and the use of new non-traditional energy methods for the disintegration of mineral complexes, to improve the efficiency of the process extraction of valuable components.

4 Conclusions

The chemical composition of ASW is highly dependent on the type of fuel burned from various fields, but for the site of a separate CHP plant, the chemical composition can be considered stable for practical use, being guided by the median value.

ASW enterprises of the power sector of the Primorsky Krai have a rich trace element composition, which is related to the specific composition of the burned coals and the enrichment of certain trace elements that occur when the organic part of the fuel burns out during combustion. Under the content of coal underburning - up to 20%, ASW of Primorsky Krai refer to raw materials with an average content of combustible substances (5-20%).

The conducted studies show the possibility of efficient separation of ash and slag wastes into separate mineral fractions and separation of the preliminary concentrate of valuable minerals in which most of the gold, BM and REE will be collected. At the same time, the remaining mineral fractions: coal underburning, magnetic fraction containing iron and titanium oxides, microdispersion fraction containing aluminosilicate microspheres and light aluminosilicate fraction can be used as raw stock for the making of commercial products with stable liquidity.

Acknowledgements

The work was partially (selection, sample processing and technological research) supported within the framework of the Comprehensive Program of Fundamental Research of the Far East Division of RAS "FAR EAST", project 18-3-029 "Research and development of materials and processes for environmentally safe technologies for obtaining precious metals from anthropogenic objects" (section 1), and partly (analytical studies of samples and products of technological processing) was supported within the framework of the State task of the Institute of Chemistry of the FED RAS (subject No.0265-2018-0002).

References

1. Information and technical handbook on the best available technologies of ITS 38-2017 "Combustion of fuels in large plants for the production of energy" 280 (Moscow: Federal Agency for Technical Regulation, Bureau of technical and regulatory documentation) (2017)
2. V.V. Slesarenko, A.V. Taskin, Eurasian Scientific Association, **1**, 60 (2015)
3. Energy strategy of Russia for the period until 2030. Approved by the order of the Government of the Russian Federation No.1715-r of November 13, 2009 (2009)
4. Long-term program for the development of Russian coal industry for the period up to 2030. Approved by the Government of the Russian Federation on 24.01.2012 (2012)
5. V.V. Zyrianov, D.V. Zyrianov, Moscow: Information and Research Center
6. V.F. Borbat, L.N. Adeeva, Bulleting of Omsk State University, 2, 141 (2009)
7. P.V. Menshova, Procedia Chemistry, **10**, 184 (2014)
8. A.A. Cherepanov, Geology and Minerals of the World Ocean, **2,** 98 (2009)
9. H. Wu, P. Glarborg, Fuel Processing technology, **105**, 212 (2013)
10. B.S. Ksenofontov, Ecology and Industry of Russia, **9** (2014)
11. N.N. Ezhova, Ecology of industrial production, **2,** 49 (2010)
12. G.L. Pashkov, Journal of Siberian Federal University. Engineering&Technologies, **5**, 520 (2012)
13. L.I. Pugach, Energy and Ecology. Textbook, Novosibirsk city. NSTU Publishing House (2003)
14. N.I. Yarmolinskaya, T.L. Lazareva, The use of waste heat and power industry of the Far East in the technology of building materials. Textbook, Khabarovsk State Technical University (2000)
15. A.A. Cherentsova, Ecology and Industry of Russia, **28** (2013)
16. N.I. Skripnikova, I.Iu. Iuriev, Bulletin of Tomsk State University of Architecture and Construction, **2,** 245 (2013)
17. D.R. Damdinova, Bulletin of East-Siberian State University of technologies and administration, 2, 9 (2016)
18. E.I. Shamrai, Intern. Journal of Energy for a Clean Environment, **17**, 99 (2016)
19. E.I. Shamrai, Modern High Technologies, **3**, 68 (2017)

Simulation modeling of centrifugal concentrator as a basis for increasing its output in enriching industrial waste

Andrey Taskin[1*], *Oleg* Danilov[1], *Viktor* Shalanin[2], *Leonid* Alekseiko[1] and *Oleg* Elkin[1]

[1]Far Eastern Federal University (FEFU), International Center of Mineral Processing Technologies, Vladivostok, Russia
[2] Far Eastern Federal University (FEFU), School of Engineering, Vladivostok, Russia

Abstract. In the research paper a computer model of the operation of a vibrational concentrator used to separate the slurry into individual fractions of a grain size by depositing particles of specified characteristics in the field of centrifugal forces is created and verified using the methods of computational hydrodynamics with the Ansys software complex. Using the model will make it possible to: 1) to study the operation of concentrators of various sizes and designs under different operating conditions, with a change in hydraulic characteristics (speed of treated liquid, speed of rotation of the bowl); 2) in the future study the problems of depositing suspended particles of specified characteristics (density, size) in a centrifugal field and determine the concentration of the sediment in the treatment process. The results of the work will make it possible to increase the efficiency of the enrichment of ash and slag waste due to a more thorough understanding of interaction mechanisms between the centrifugal concentrator and the enriched geomaterial.

1 Introduction

The main issues in today's mining industry include significant depletion of existing mineral resource base, complex material composition and constantly deteriorating quality of ore. This circumstance requires the use of bulk mining extraction at mining plants, which in turn leads to a decrease in the content of valuable component in the ore mass. The alternative source of valuable components – geomaterials of industrial origin (for instance, waste generated by power industry, mining and processing plants, gold mining plants) have a similar set of issues: complex, volatile chemical composition and low content of commercially valuable components, uneven distribution in the material. As a consequence, in order to obtain a unit of commercial product, one is required to process a large amount of raw materials [1-3].

One of the technical challenges in this case is a method of increasing the output of concentrators intended for processing of raw materials. At the same time, selection of optimal operating conditions for mineral-processing equipment at a laboratory or in actual

* Corresponding author: taskin@yandex.ru

industrial conditions is not efficient, as it requires considerable time. An effective solution in the selection of operating modes, according to many experts in the field of enrichment, is the use of computer modeling, in particular simulation [4-7].

This type of modeling makes it possible to quickly assess the effectiveness of certain technological processes in the system under study, the effectiveness of implementing various options for its operation. From the point of view of enrichment, ash and slag waste from power utilities is the most complex object of study. In this connection, the development of a numerical method for calculating hydrodynamic processes in the enrichment of industrial geomaterials in a centrifugal concentrator, as in the case of ash and slag waste, is a priority research and practical problem.

2 Implementation of the concentrator operation model

The modeling of the object consisted of the following stages: 1) creation of geometric model; 2) creation of computational grid; 3) adjusting the solver; 4) solving the problem; 5) verification of the mathematical model.

2.1 Creation of geometric model

A two-dimensional full-scale axisymmetric model of a centrifuge was selected as a geometric model. The dimensions of the computational area of models are made without distortion, on the basis of actually existing apparatuses. The enriched medium is fed into the central part of the bottom of the concentrator bowl and is distributed to the inclined walls of the bowl by the impeller. The water is withdrawn through the upper edge of the concentrator bowl. Geometric models of concentrators No.1 and No.2 are shown in Fig. 1 and 2.

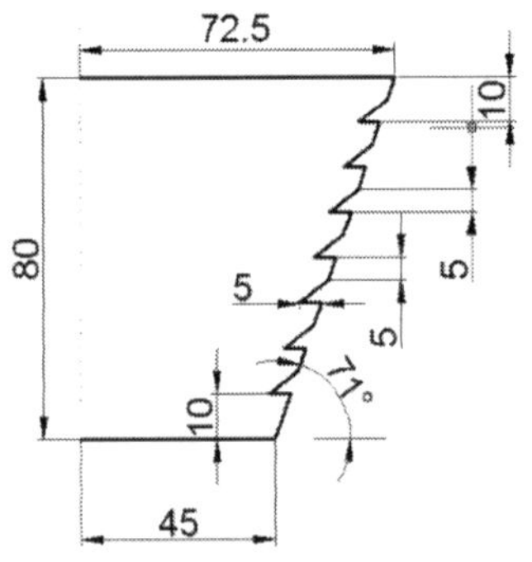

Fig. 1. Geometric characteristics of concentrator model No.1

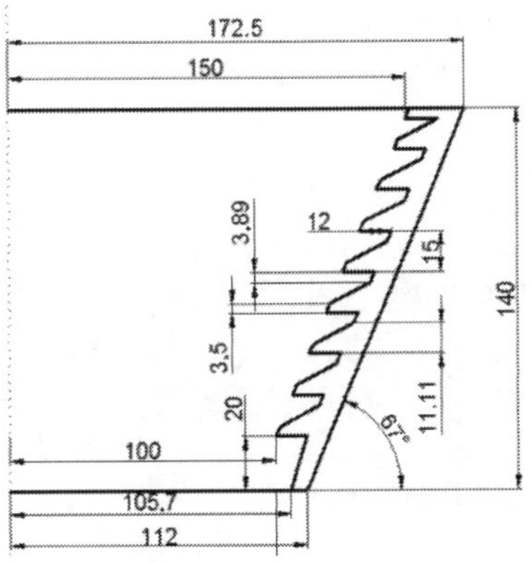

Fig. 2. Geometric characteristics of concentrator model No.2

2.2 Creation of computational grid

The computational grid was created using the Delaunay triangulation method, by constructing a primary computational grid on sides with subsequent construction of a computational grid inside the plane of the model. The total number of cells for concentrator No. 1 was 18,900 elements, for concentrator No. 2 – 17500 (Fig. 3, 4).

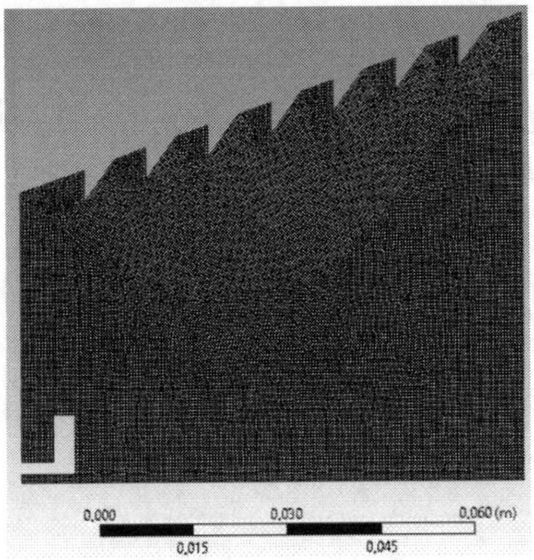

Fig.3. Computational grid for concentrator model No. 1

Fig.4. Computational grid for concentrator model No. 2

2.3 Adjusting the solver

The problem solving was carried out by means of ANSYS FLUENT software. The problem under consideration is nonstationary.

For the mathematical description of turbulence, the k-omega Standard model was used. A mathematical model VOF (Volume of Fluid) with the number of phases = 2 and Implicit Formulation was used to model the tangential water-air discontinuity surface.

The flow characteristics at the "inlet" and "outlet," as well as the direction and the number of revolutions of the concentrator bowl were considered as boundary conditions. In order to implement the joint solution of the velocity-pressure equations, PISO algorithm was used, which is most suitable for solving non-stationary problems.

2.4 Solving the problem

Problem solving was carried out by achieving convergence of the solution at each time interval. The time interval varied from 0.0001 s to 0.001 s. The increase in the time interval was used to reduce the total solution time. The number of iterations at each step did not exceed 50.

3 Results and discussion

3.1 Results of problem solving

As a result of the calculations, liquid flow patterns with a water-air interface and trajectories of flow lines were obtained. The figures present: Fig. 5 – flow pattern of liquid in the concentrator No. 1; Fig. 6 – flow lines in the concentrator No. 1; Fig. 7 – flow pattern in the concentrator No. 2.

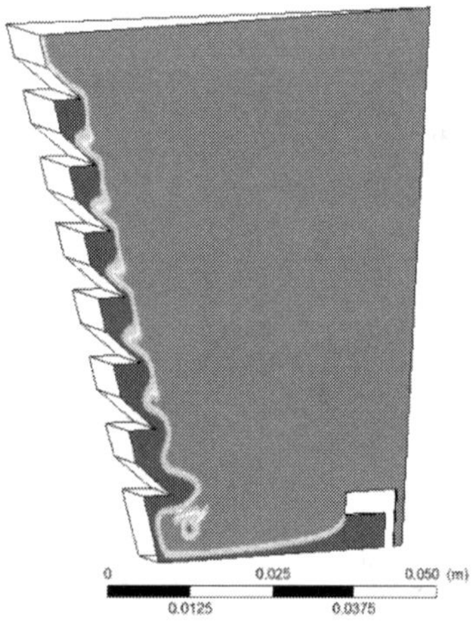

Fig. 5. Flow pattern of liquid in the concentrator No. 1

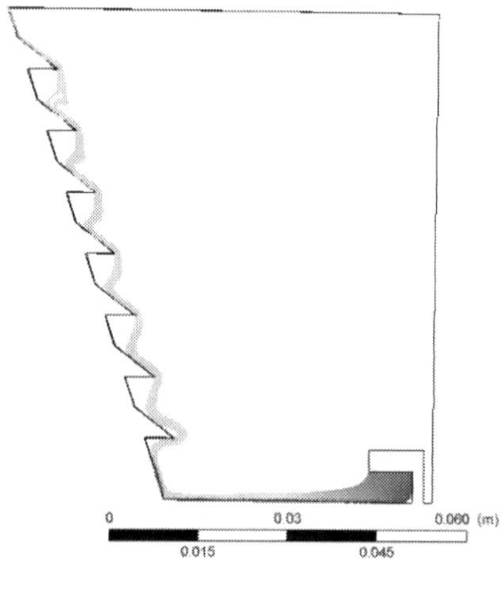

Fig. 6. Flow lines in the concentrator No. 1

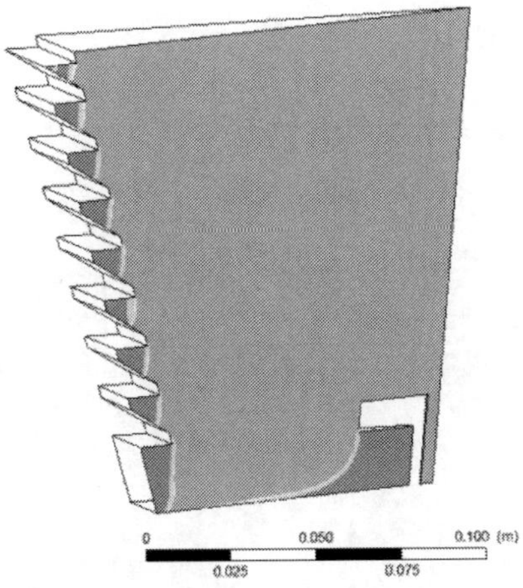

Fig.7. Flow pattern in the concentrator No. 2

3.2 Verification of the model

The absence of similar studies of the hydraulic characteristics of centrifugal concentrators presents a problem for comparative evaluation of the applicability of the chosen mathematical model. In our case, verification was carried out based on the problem of determining the form of free liquid surface in a rotating vessel. Preliminary modeling showed that using the developed model allows obtaining flow patterns adequate to the calculated data. The form of free liquid surface in a rotating vessel is shown in Fig. 8. The results of the comparison of numerical experiment data with the approved theoretical calculation method are presented in Fig. 9 and 10.

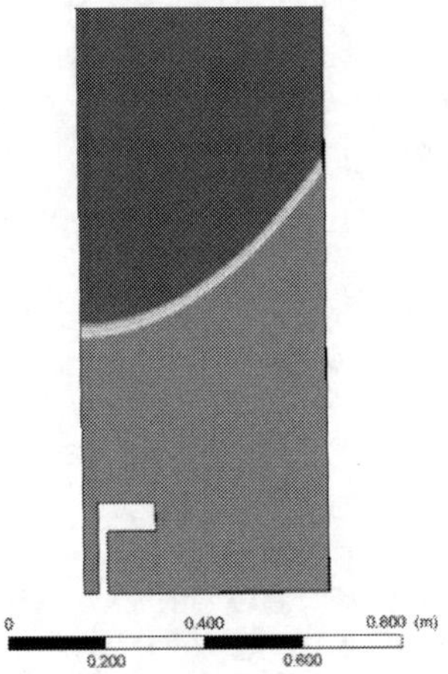

Fig. 8. The form of free liquid surface in a rotating vessel

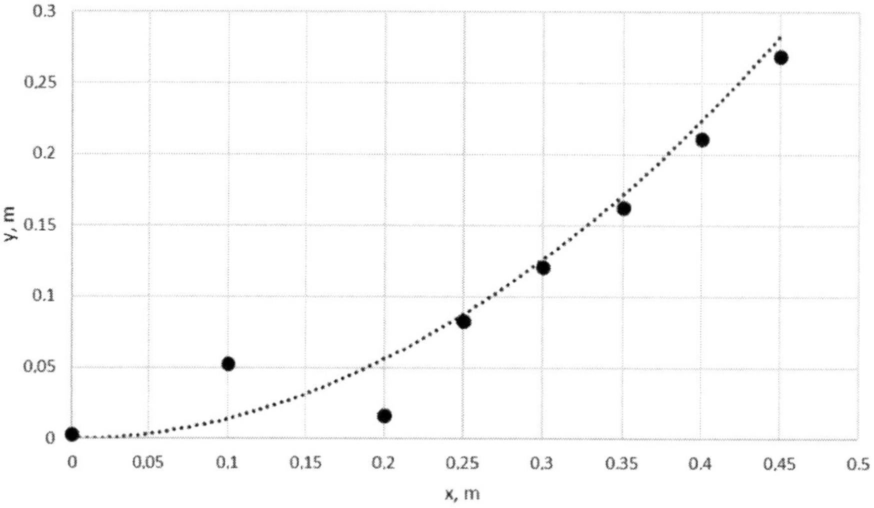

Fig. 9. Comparison of the coordinates of free liquid surface obtained by numerical simulation with calculations data at 50 rpm

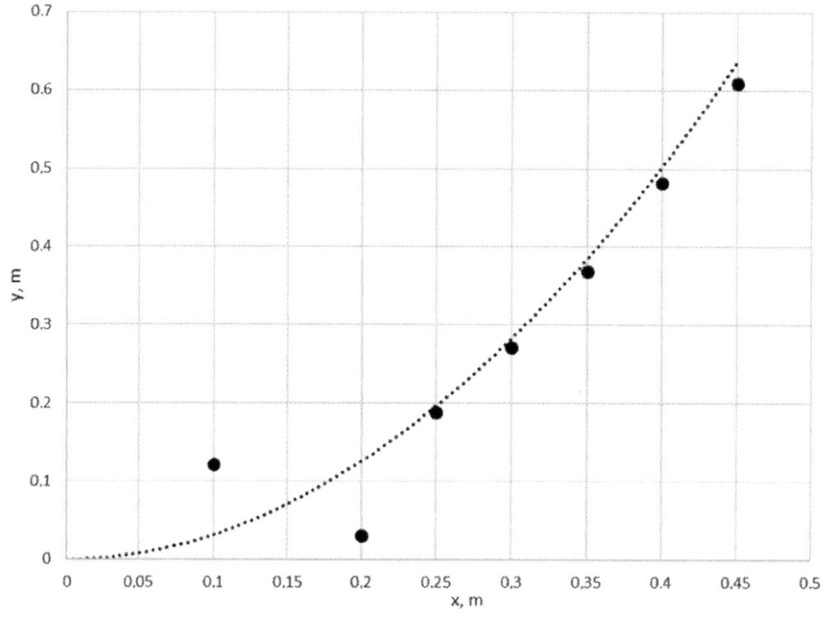

Fig. 10. Comparison of the coordinates of free liquid surface obtained by numerical simulation with calculations data at 75 rpm

4 Conclusion

As a result of preliminary studies, a mathematical model was chosen for computer simulation of rotating flows with a free surface. The numerical model was successfully tested for the problem of determining free liquid surface in a rotating vessel. Using the numerical method will make it possible to:

1. Study the work of concentrators of various sizes and designs under different operating conditions, with a change in hydraulic characteristics (speed of treated liquid, speed of rotation of the bowl).

2. In the future study the problems of depositing suspended particles of specified characteristics (density, size) in a centrifugal field and determine the concentration of the sediment in the treatment process.

It should be noted that the current model needs improvement, as it does not allow to study concentrators with vibration, deformation of the bowl surface (floating bed), with complex movement of the concentrator bowl (rotational, not radial) and the use of magnetic field.

At the same time, its application will increase the output of a centrifugal concentrator and determine optimal regimes for the enrichment of ash and slag waste of various chemical compositions.

5 Acknowledgments

This paper was prepared with the financial support of the Ministry of Education and Science of the Russian Federation, state task in the field of scientific activity № 10.3706.2017/4.6.

References

1. Critical of mineral raw materials and sustainability assessment, Expert Workshop on Security of Supply and Scarcity of Raw Materials (2012).
2. A.S. Kobzev Directions of development and problems of radiometric methods of mineral dressing, Enrichment of ores, **1.** (2013).
3. V.C. Nguyen, A.V. Petrov Technical and economic simulation of technological mineral dressing processes, Proceedings of the 12th International workshop on Computer Science and Information Technologies, **1.** (2010).
4. R. Peter King Modeling & Simulation of Mineral Processing Systems (2011).
5. L. Ergun, Z. Ekmekci, O. Gulsoy and H. Benzer Modeling and Simulation of the grinding circuit in the Madneuli copper concentrator, Physiochemical Problems of Mineral Processing, **38 (231)** (2004).
6. A.Oleynik, V. Skorokhodov Modeling and modeling in the mineral processing industry, Annals of DAAAM for 2006 & Proceedings of the 17th International DAAAM Symposium, **1 (17).** (2006).
7. O.V. Friedman Information technology of the organization of processes of mineral raw materials enrichment, Information resources of Russia, **3.** (2006).

Synthesis of titanium carbide from the rutile concentrate by low temperature plasma

Evgeny Kuz'michev[1*] and *Sergey* Nikolenko[2]

[1]Far Eastern State Transport University, Khabarovsk, Russia.
[2]Institute of Materials Science, Khabarovsk Scientific Center, Far Eastern Branch, Russian Academy of Sciences, Khabarovsk, Russia.

Abstract: As a result of the synthesis of titanium-containing mineral raw materials (rutile concentrate), using high-density energies, the metal titanium and titanium carbide were produced. The article presents the results of experiments on the possibility of using high-density energies for the synthesis of functional materials for industrial use.

1. Introduction

The production and uses of ceramic materials for construction purposes, as well as the production of coatings using ceramic materials as basic components have developed rapidly in recent times. Compounds of metals and non-metals with boron, carbon and nitrogen serve as the basic components for these materials. Carbides, borides and nitrides are extensively used, especially carbides of refractory metals.

The industrial production of such compounds shows considerable promise and is conducted in the direction of its extraction from metal oxides by synthesis or mechanical diffusion. However, the use of a multicomponent mineral raw material that contains titanium for the production of TiC compounds is not possible without additional processing steps. In short, it is necessary to pay considerable attention to the preparation of titanium oxide and to the purification, leaching and precipitation of the resultant solutions, etc. Titanium-containing components are byproducts of the mining of various ores and their extraction is a significant problem. So-called after extraction "tails" (ore waste) from ore mining and processing enterprises are accumulated and then they are partially used in metallurgical production and welding processes or exported to Asia.

Alternative methods of chemical processing include high-temperature and plasma-chemical synthesis.

Methods of titanium carbide production have long been known. For example, from the heating of powdered titanium oxide and soot in coal furnaces in a hydrogen atmosphere at 2250 °C [1], but the productivity of this process is low. The method of mechanical alloying [2 - 4] requires the grinding of the initial charge, consisting of titanium powder and soot, in a planetary mill at room temperature. No free carbon atoms and soot are found in the final product, but the disadvantage of this method is the formation of a non-stoichiometric composition of titanium carbide. The self-propagating high-temperature synthesis (SHS)

*Corresponding author: accord@festu.khv.ru

method concludes with the intimate mixing of powdered reactants and their sintering into briquettes that are used in the process of obtaining carbides at temperatures higher than 1600 °C [5 - 9].

The drawbacks of all these methods are low productivity, the use of rather complex equipment to create high temperatures and vacuum, and the use of "hermetic bomb-reactors", etc. In addition, in the majority of methods it is possible to use only sufficiently pure components, which implies their production from ore concentrates in earlier stages.

The task to be solved is to develop an alternative method for obtaining functional materials, titanium carbide and its compounds, using simple and sufficiently productive equipment that will cover domestic demand in the market, and use dumps in the form of a raw material base and possibly obtain new Ti-based compounds.

Thus, in our work we propose the method of obtaining titanium carbide from a multicomponent mineral oxide-containing raw material by applying high-density energy with the intensity $g>10^3 - 10^4$ W/cm^2.

The initial content of titanium in the investigated material, the rutile concentrate, was about 75%.

2. Methods and materials

In experiments aimed at obtaining a titanium-carbide compound (TiC_x) we used a rutile concentrate and graphite powder.

The preparation of the mixture was carried out by the method of ranking. A portion of the mixture containing 70 wt% rutile concentrate and 30 wt% graphite powder was loaded into the planetary ball mill and ground for 20 minutes in ethanol. The reducing agent (carbidizer) was added to the charge with 10% excess. A charge (P+C) with a particle size of 70-100 mm was obtained as the output.

The charge (P+C) was supplied by the transport gas into the electrode channel through the channels of the plasma torch 5. We used industrial grade argon as the transport gas. The electrode had a tungsten tip, and the diameter of the passageway was about 0.8 - 1 mm. A graphite nozzle was installed coaxially with the electrode. The nozzle was separated from the graphite by a dielectric, ceramic ring 2 (Fig. 1).

In the course of the gas flow through the channels of the plasma torch to the electrode and nozzle region, fine-dispersed particles of charge are transferred from the hopper. An electric discharge is created in the ionization region between the electrode and the nozzle of the plasma torch. In turn, the particles of the charge increase the conductivity of the gas-dust flow, because 30% of the mixture consists of fine graphite powder.

As a result of simulating the gas motion in the chamber of the plasma torch by finite element methods and the nature of the synthesized material deposit, it was determined that the nature of the gas flow is represented by transient processes [10]. This led to the appearance of cyclic characteristics in the motion of particles. The particles move along toroidal trajectories with their subsequent ejection. So a particle can repeatedly cross a hot region of plasma with the gas flow. In contrast to the classical Gaussian character of the distribution of temperatures in the plasma plume over its entire length, the plasma flare is represented by a chaotic motion of the particles, which determines a uniform temperature distribution throughout the reactor volume.

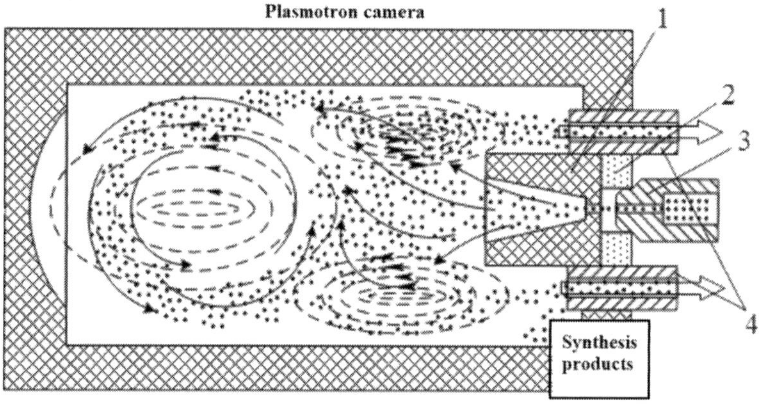

Fig 1. Scheme of the process of plasma-chemical synthesis: 1 – nozzle, 2 – dielectric ring, 3 – electrode, 4 – gas-vapor phase and finely-dispersed discharge tubes.

The obtained samples were studied on an electron scanning microscope "VEGA 3 LMH" (TESCAN) equipped with an energy dispersive spectrometer "X-Max 80" (Oxford Instruments). Photographing samples and searching for micro-inclusions was carried out mainly with the regime of backward scattered electrons (BSE-detector). Using such a regime allows phases with a higher average atomic number to be reflected in sharper contrast compared with phases with a smaller average atomic number in the production of the image.

The phase analysis was also carried out using an electron transmission microscope ZEISS Libra-120 equipped with a HAADF detector and an energy Ω-filter. Investigations were carried out in the modes of lumen, dark field, and electron micro-diffraction. Samples for transmission microscopy were prepared by electrolytic polishing and ionic etching.

The composition of the slag and the synthesized product were studied using a diffractometer "Dron 7" with a tube from Co on a K-line X-ray with a shooting speed of two degrees per minute.

At the end of the experiments we carried out a macro- and microanalysis of the samples obtained during the plasma-chemical synthesis and studied the slag and the precipitate of the vapor-drop phase.

The samples formed during plasma-chemical synthesis were represented by a hard fused material, which was easily ground in an agate mortar.

The results of the analysis showed the presence of the TiC phase as a result of the thermal action on the P+C powder mixture.

In Fig. 2 (a) a photo of the samples of the plasma stream selected from the region of laminar motion with the least impact of turbulent flows is shown. Samples from this region were not affected by a constant magnetic field. The chemical composition of Fig. 2 (a) samples is shown in Table 1.

In Fig. 2 (b) we presented an image of the samples selected from areas that are more prone to turbulent motion of the plasma flow. In addition, on Fig. 2 (b) among the particles represented exclusively by the TiC compound, in the lighter region a sample under the spectrum 10 stands out. Collectively, such samples are often found in the form of condensate in the nozzle area and on the filter surface. In the sample under the spectrum 10 (Fig. 2 (b)) we also detected about 3 wt% of niobium (Nb).

Table 1. Chemical composition of samples obtained as a result of plasma-chemical synthesis presented in Fig. 2 (a).

Specter No.	C	Mg	Al	Si	Fe	Ti	Mn	Other, %	Free C,%
1	15.26	-	0.09	-	-	84.44	-	0.05	0.2
2	19.35	-	0.18	0.01	0.009	80.24	-	0.008	0.203
3	14.18	-	0.12	-	-	85.49	-	0.008	0.202
4	17.22	-	0.11	0.01	0.007	82.44	-	0.008	0.205
5	18.76	-	0.14	0.01	0.009	80.87	-	0.008	0.203
6	11.54	-	0.1	0.03	0.008	88.11	-	0.002	0.21

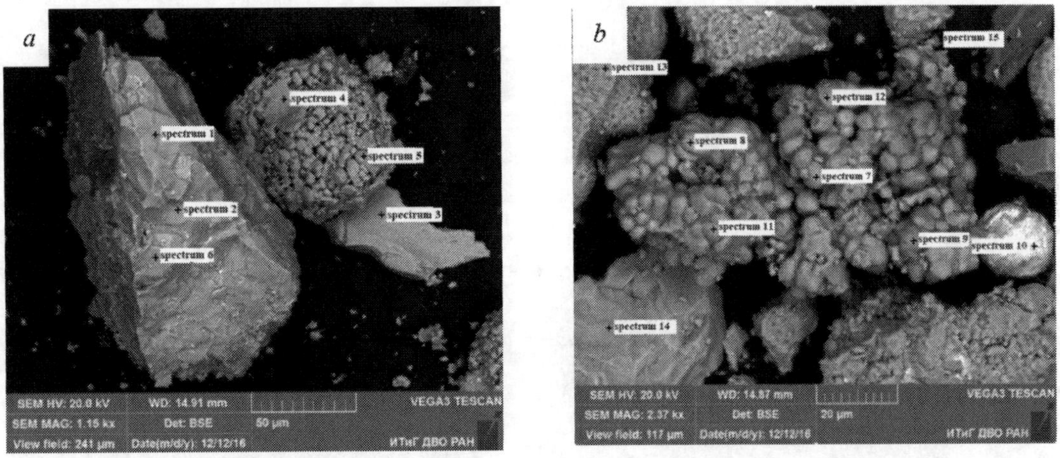

Fig. 2. The image of the sample in the reflected electrons selected from the laminar gas flow region at 50 μm (a), and in the reflected electrons selected from the turbulent region at 20 μm (b).

Chemical composition of Fig. 2 (b) samples is presented in Table 2.

Table 2. Chemical composition of samples obtained as a result of plasma-chemical synthesis presented in Fig. 2 (b).

Spectrum No.	C	Mg	Al	Si	Fe	Ti	Mn	Other, %	Free C,%
7	16.26	-	0.08	-	-	83.47	-	0.004	0.186
8	17.29	-	0.18	0.01	0.009	82.34	-	0.004	0.167
9	13.12	-	0.12	-	-	86.58	-	0.004	0.176
10	16.78	7	0.11	2	14	57.94	2	0.005	0.165
11	17.35	-	0.14	0.01	0.009	82.27	-	0.003	0.218
12	11.76	-	0.1	0.03	0.008	87.89	-	0.002	0.21
13	12.31	-	0.1	0.03	0.008	87.34	-	0.002	0.21
14	14.42	-	0.1	0.03	0.008	85.31	-	0.002	0.13
15	15.24	-	0.08	0.03	0.005	84.51	-	0.002	0.133

In Fig. 3 we present the spectrogram of Fig. 3 (a, b) samples. Insignificant impurities are found in the samples: Al 0.9 – 0.18 wt%, Si – 0.01 wt%, Fe – 0.009 wt%, and free carbon up to 0.22%.

The carbon content in the titanium compound TiC was about 17 wt%, and the Ti content was about 85 wt%. The size of particles presented in Fig. 5 varies from 1 nm to 100 µm. The particles are easily ground and resistant to various types of chemical reagent, which allowed the leaching procedure for separately occurring particles of iron, manganese, aluminum, etc. It also allowed us to evaluate the quality of material which resulted from leaching after washing and drying. The purity of the final product reached 89.5 wt% of the TiC phase and up to 10 wt% of the Ti phase.

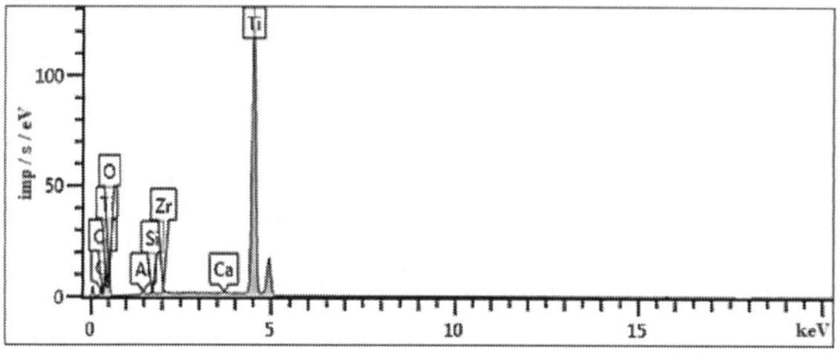

Fig. 3. Spectrogram of a sample of tungsten carbide TiC, obtained in turbulent and laminar flows

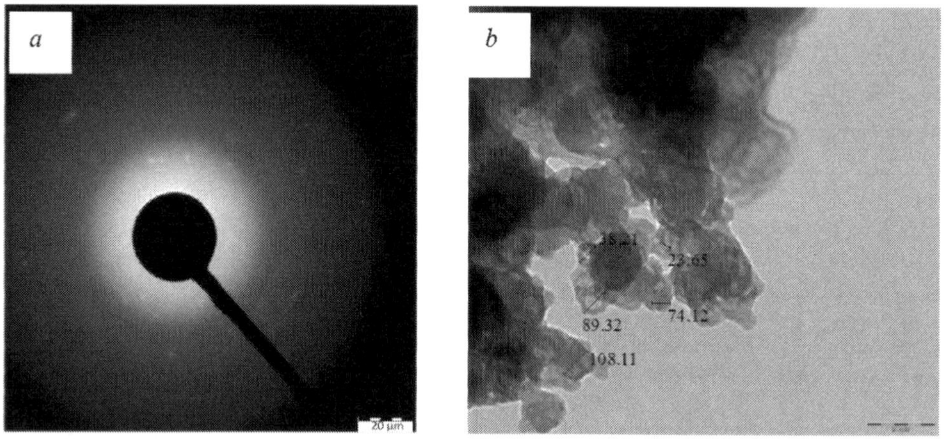

Fig. 4. The image of titanium carbide samples obtained with the transmission electron microscope ZEISS Libra-120, for determining the phase composition: a) the electron diffractogram of the TiC crystal lattice (20µm), b) TiC particles (200 nm)

According to the results of the analysis, the particle sizes in the vapor-drop phase vary from 1 to 150 µm. (Fig. 5). The phase analysis revealed Ti, TiC phases in various allotropic states.

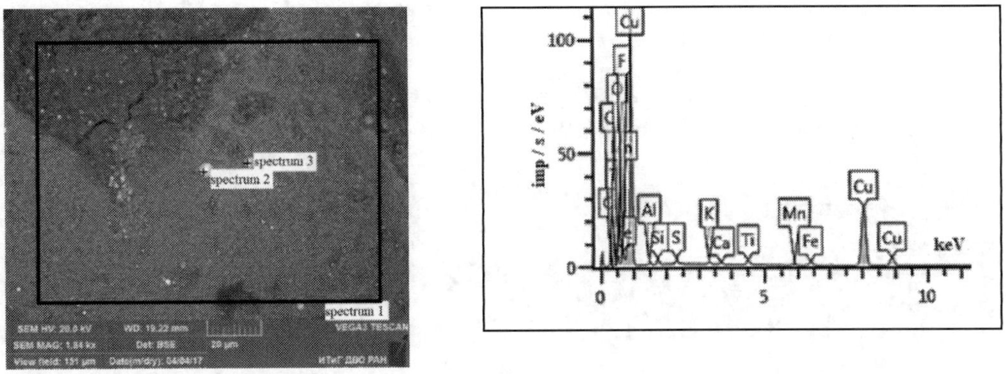

Fig. 5. The image of a vapor-drop phase taken from the chamber walls of the nozzle region and a spectrogram

Chemical analysis also showed the absence of oxygen in the phases, but only under certain conditions for the preparation of samples. In particular, it is necessary to exclude the possibility of prolonged exposure to the atmosphere in the heated state. In the case of atmospheric exposure at 350K to 450K and higher an oxide component (TiO_2) was formed on its surface.

The chemical analysis of the vapor-drop phase samples from the nozzle region of the reactor showed the presence of a multitude of associated elements, including Ti.

Table 3. The chemical composition of the samples (Fig. 2. (b)) of the vapor-drop phase resulting from the plasma-chemical synthesis of the powder mixture

Spectr. No	C	O	F	Al	Si	S	K	Ca	Ti	Mn	Fe	Cu
Spectr. 1	11.1	9.3	7.18	2.7	12.3	6.02	3.77	14.4	8.2	5.63	19.1	0.09
Spectr.2	12.0	12.8	8.24	3.8	11.3	9.11	2.59	13.4	4.5	4.6	16.8	0.49
Spectr.3	13.1	13.7	7.93	2.6	12.1	8.02	3.01	12.6	4.9	3.34	18.0	0.37

The absence of a large number of Ti samples is typical for the samples from the nozzle region. In the process of plasma-chemical synthesis, about 0.09 mass% is deposited from the copper electrode as a result of fracture, and it largely settles on the filter walls, or walls of the reactor in the nozzle region.

3. Conclusion:

It has been established that titanium is reduced in the form of titanium carbide, as well as metallic titanium from multicomponent titanium-containing mineral raw materials by the influence of high-density energy with the intensity $g>10^4 - 10^5$ W/cm^2. The technology of TiC production with the use of plasma-chemical synthesis makes it possible to obtain purity of the final product up to 85.5 wt%.

The use of plasma-chemical synthesis makes it possible to obtain titanium carbide as a high-frequency semi-finished product, with a particle size of 1-150 nm, as a result of a single technological process. With the use of an additional leaching stage, it is possible to obtain 89.5 wt% TiC due to the separation during the high-temperature synthesis of all the constituent elements of the charge when they are condensed.

It is revealed that a high-temperature plasma environment, allows the destruction of the crystal lattice of the mineral, to separate all chemical elements, and to separate Ti from the oxide as well.

The research has shown promising prospects for using high-density energy for the synthesis of refractory compounds from titanium-containing mineral raw materials.

References:

1. Y.N. Tumanov Plasma, high-frequency, microwave and laser technologies in chemical-metallurgical processes (2010).
2. I.A. Zaks, V.G. Kolesnik, S.G. Melamud, M.V. Malygin, E.V. Urusova, Ferrous metals. **11-12** (2001).
3. V.G. Kolesnik, N.N. Mukhtarova, E.V. Urusova, I. Khidirov, B.S. Iuldashev Non-ferrous metals. **12** (2001).
4. T.B. Ershova, N.M. Vlasova, M.A. Teslina, I.A. Astapov Chemical technology. **17 (1)** (2016).
5. B.S. Iuldashev, E.V. Urusova Non-ferrous metals. **Iss 2** (2003).
6. M.E. Deich Technical gas dynamics.(1961).
7. B.A. Kniazev Low-temperature plasma and gas discharge (2003).
8. K. Madhav Reddy, T.N. Rao, J. Joardar Materials chemistry and physics) **128** (2011).
9. D.V. Onishchenko, V.P. Reva, D.V. Moiseenko Inorganic Materials **48 (11).** (2012).
10. U.V. Tsvetkov, A.V. Samokhin, N.V. Alekseev High Energy Chemistry **40 (2).** (2006).

RESEARCH INTO SPECIAL TECHNOLOGICAL EFFECTS FOR FORMATION OF WEAR RESISTANT COATINGS USING MINERAL RAW MATERIALS OF THE FAR EASTERN REGION

Viktor Makienko[1*], *Igor* Romanov[1], *Pavel* Sokolov[1], *Alexander* Atenyaev[1] and *Dmitry* Pervakov[1]
[1]Federal State Budgetary Educational Institution for Higher Education "Far Eastern State Transport University"

Abstract. The results of the study of the possibility of using additional technological influence in the formation of doped coatings are presented. The distribution of thermal fields has been calculated, which makes it possible to estimate and predict the influence of additional influence on the formation of the coating in a simplified manner. Experimental studies have shown that additional technological influence promotes mixing of the melt, leading to an increase in the transition of alloying elements, as well as an increase in the mechanical properties and quality of the weld metal.

1 Introduction

Development and deployment of new materials for coatings and development of the methods for their quality control is an important line of research of materials science. Ferroalloys and pure-metal powders currently used for production of welding-surfacing materials significantly increase the cost of the materials produced. Therefore deployment of mineral raw materials without their deep technological processing when producing materials remains an issue of current importance. Production of welding-surfacing materials from mineral raw materials face problems caused by heterogeneity of the phase and chemical composition of the source mineral materials which in its turn causes irregular distribution of alloying elements in the coatings produced, low efficiency of alloying and, as a consequence – poor service properties [1, 2].

A promising direction of research focused on the improvement of the quality of coatings is the use of special technological methods for affecting the melt that involve simultaneous use of several concentrated sources of energy. In this work we suggest employment of one of the simplest and the most efficient methods of affecting the melt – a special arc of indirect action (SAIA) that serves as an additional source of thermal and electro-magnetic effect. This special arc facilitates redistribution of the temperature gradient and intensifies the molten pool agitation due to the interaction between the magnetic fields of the primary

* Corresponding author: mvm_tm@festu.khv.ru

and indirect arcs which causes the deflection of the primary arc anode spot (of direct action) [3-6].

2 Methodology of experiments

Experimental research was conducted according to the flow-chart below (Fig. 1.) that reflects the sequence of work stages and theoretical prerequisites for use of the special arc of indirect action, as a source of additional impact on molten-pool while forming doped layer. The scheme below presents the major stages of the research from the setting forth of the hypothesis and development of a new technology for special affecting to the deployment of the technology developed to flux cored wires produced using mineral raw materials of the Far Eastern Region.

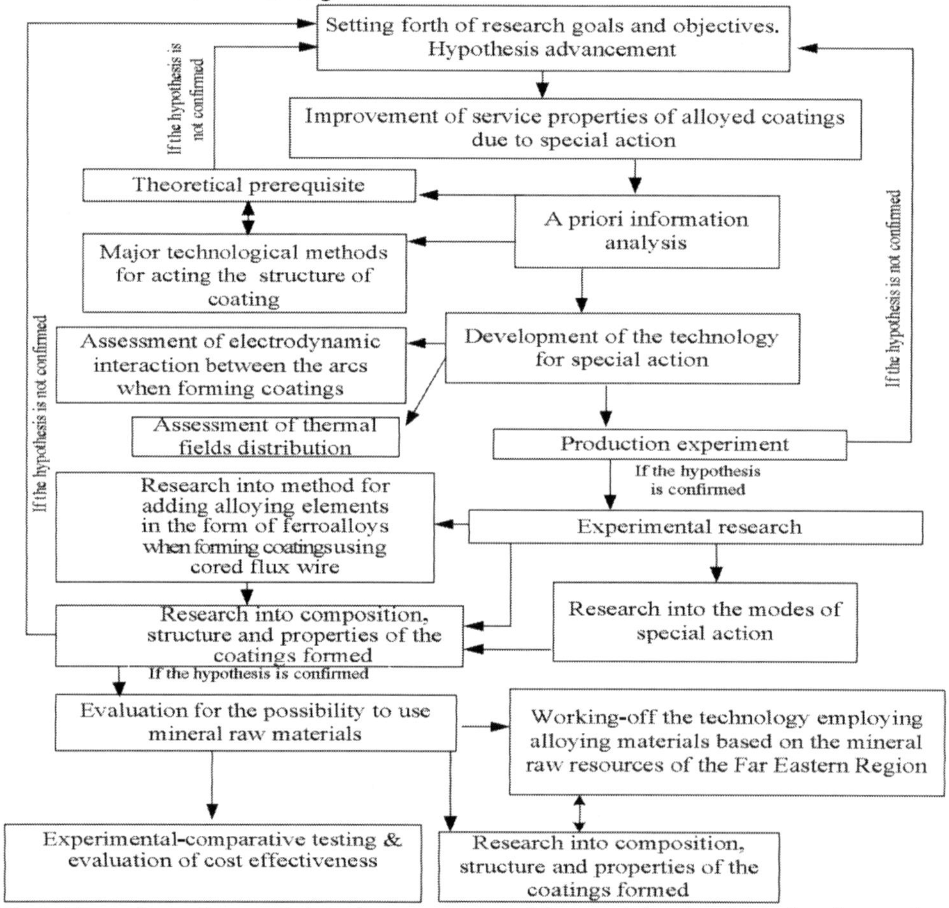

Fig.1. Flow-chart of research into the assessment of the special technological action on the coating formed

While conducting the research we performed theoretical calculation of the thermal fields and the electrodynamics interaction forces involved in the arc processes which allows us to determine the temperature distribution curves and to assess the efficiency of agitation and acceleration of the metallurgical processes under the impact of the special arc of indirect action [7-12].

In addition we used the experimental design methodology based on the experimental statistics methods aimed at determining the dependencies of influence of the special action parameters on the composition of the coatings being formed as well as determining optimal intervals for the variation of the SAIA modes for the maximum efficient alloying of the weld-metal. While planning the experiment we determined input (X_1, X_2, X_i) and output (Y1, Y2, Yi) parameters subject to optimization (Fig. 2). Based on the calculated values we obtained mathematical dependencies that we used to diagram the influence of the special action modes on the composition of the coatings formed.

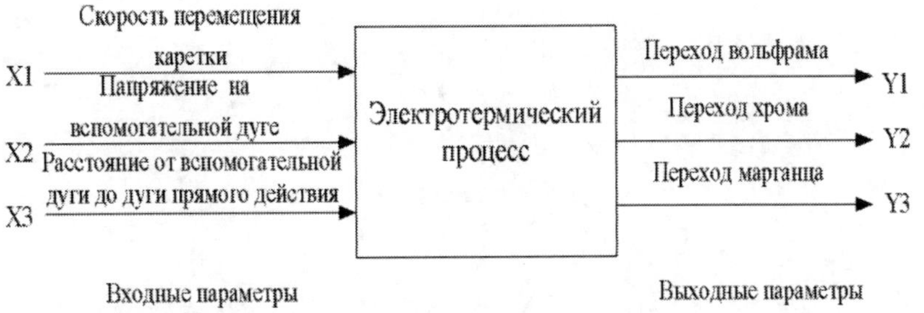

Fig. 2. Calculation model of the research into the influence of the input parameters on the composition of the coating formed.

Frame travel rate	Electro-thermal process	tungsten transition
Special arc voltage		chrome transition
Distance between primary and special arc		manganese transition
Input parameters		Output parameters

3 Results and discussions

Calculation of the distribution of the thermal fields resulted in the temperature gradient distribution diagram presented in Fig. 3.

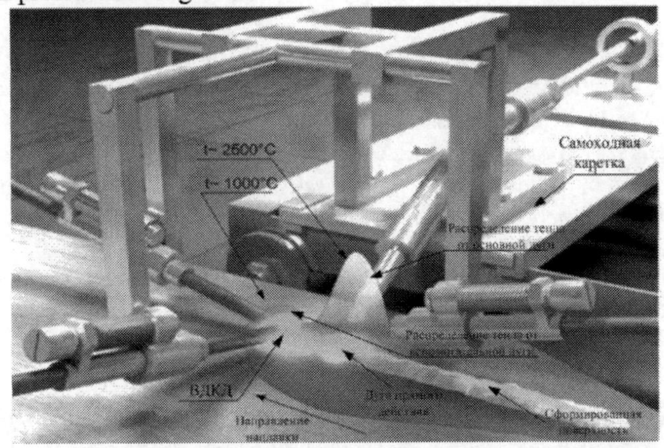

Fig. 3. Diagram of the temperature gradient distribution for primary and special arcs

Calculation of the thermal fields distribution showed that the action of the special arc on the surface being formed facilitates metallurgical processes, improves homogeneity of the structure and increases the transition of the alloying elements to the weld metal. These effects are caused by the redistribution of the heat flows as the heating-up temperature of the surface being formed by the special arc reaches 1000 ^{0}C, and the temperature of the primary welding process is 3000 ^{0}C. Calculation of the electrodynamic interaction force (EDI) between the primary and the special arcs testifies to the possibility to change the spatial position of the primary arc which causes mechanical agitation of the molten metal; if the distance between the arc centers is 4-6 mm, EDI force is 0.10-0.15*10^{-2}H.

According to the methodology of the experimental research the next stage involved experiments aimed at determining the correlation between the input and output parameters of the electric arc process when the coating is formed using flux core wire effected by SAIA. The data obtained were used to plot the match diagram that defines the rational parameters for special action (Fig. 4).

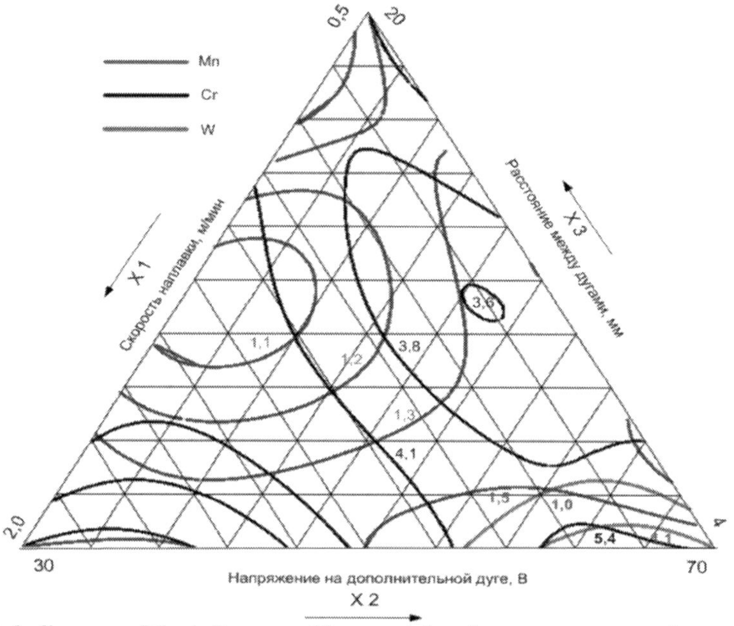

Fig. 4. Match diagram of the influence of the special action parameters on the composition of the coating

Analysis of the diagram shows that the rational modes of special action (the maximum efficiency of the alloying process and the content of the carbide constituent) are observed at the following values of the factors under research: distance between arcs is 4-6 mm; coating overlay rate is up to 0.6 m/min; SAIA voltage is 50-70 V

During the next stage we investigated the structure, the composition and the properties of the coatings formed using SAIA (within the rational range of the modes) and without special arch action; their chemical compositions are presented in Table 1.

Table 1. Chemical compositions of the coatings formed

Chemical element, %	P	Si	Mn	Mo	S	Cr	W	Ti	Zr	C
Without special arc action	0.0217	0.8707	1.3443	1.0251	0.0166	3.6029	0.9140	0.1560	0.0007	0.28
Employing SAIA	0.0199	0.8360	1.7045	1.2526	0.0134	5.3853	1.0818	0.2225	0.0064	0.35

Comparative analysis of the data in the Table 1 shows increased transition of the alloying elements to the metal formed using special arc. The distribution of the alloying elements in the coating has been examined in different points of the structural constituents and is presented in Fig. 5 (points 1-9) and Tables 2, 3.

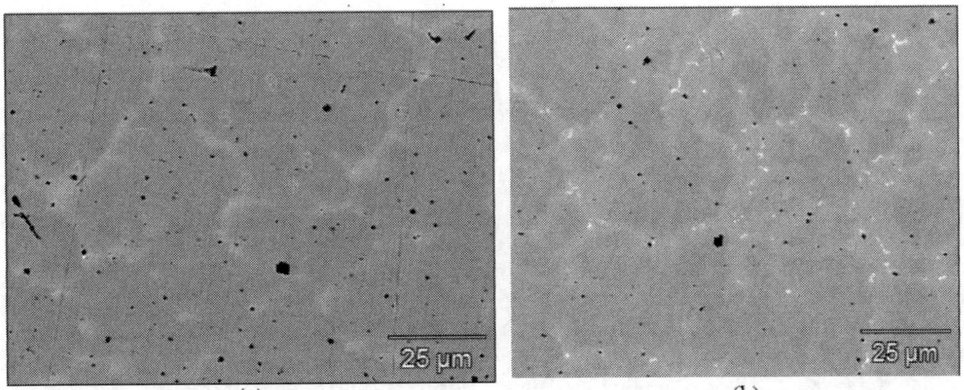

(a) (b)

Fig. 5. Distribution of the elements on the surface of the coating formed:
(a) – without special arc action; (b) – employing SAIA

The distribution of the alloying elements on the surface of the coating was studied by reflected electrons and the data obtained show increased content of alloying elements (Cr, W, and Mn) in the structural constituents of the samples formed using the special arc (Fig. 4 a, b and Tables 2, 3 points 1-3).

Table 2. Distribution of the elements on the surface of the coating formed without special arc action

Point	Element, атом. %							
	C	Si	Ti	Cr	Mn	Fe	Mo	W
1	10.72			12.41	1.81	59.40	12.27	2.86
2	9.07		0.46	11.45	2.04	64.61	9.65	2.33
3	19.95		7.42	7.15	1.48	56.02	6.13	1.85
4	3.45	1.73		6.20	2.26	83.68	2.02	0.66
5	3.61	1.56		5.54	2.04	85.62	1.23	0.40
6	3.87	1.53		5.74	2.13	84.78	1.44	0.50
7	3.10	2.41		4.80	2.15	86.79	0.75	
8	4.25	1.87		3.66	1.48	88.34	0.41	
9	4.10	1.99		3.68	1.26	88.54	0.43	

Table 3. Distribution of the elements on the surface of the coating formed employing SAIA

Point	Element атом. %								
	C	Si	Ti	V	Cr	Mn	Fe	Mo	W
1	22.19			0.38	13.46	2.23	46.67	12.68	2.40
2	24.58		0.20	0.47	13.51	1.95	44.79	11.53	2.98
3	13.09		16.09	0.62	6.22	1.70	51.02	7.57	3.69
4	3.55	2.00			5.89	2.82	84.12	1.09	0.54
5	12.00	2.08	0.19		5.66	2.43	75.94	1.27	0.44
6	3.72	2.03			6.55	2.86	82.49	1.69	0.66
7	2.54	1.72			3.99	2.03	88.81	0.46	0.45
8	2.38	1.71			3.92	1.90	89.33	0.47	0.30
9	3.00	1.50			3.96	1.90	88.99	0.41	0.23

Analysis of Fig. 4 (a, b) also shows growth of the area occupied by the alloying elements of structural constituent (light area points 1-6) with the overall composition of the coating (points 7-9) practically unchanged.

Microstructure of the coating formed without special arc action corresponds to the needled martensite containing retained austenite. In addition, the structure shows presence of carbide phase (1) Fig. 5 (a), hypothetically a complicated carbide of a cementite type which is confirmed by presence of a significant amount (< 5 %) of alloying components in the coating formed. Microstructure of the coating formed employing SAIA is presented in Fig. 5 (b) and it differs from the structure formed by a conventional method (without SAIA). The major difference is the absence of distinct needled martensite structure and increase in the carbide constituent content (1). It should also be noted that the metal of the seam formed by SAIA is of troostite structure and the carbide constituent (light area) of a cementite type (hypothetically [Fe Cr Mo W]$_3$C) is determined by the cut-&-try method applied to the phases and their combining maximally approaching the stoichiometric composition.

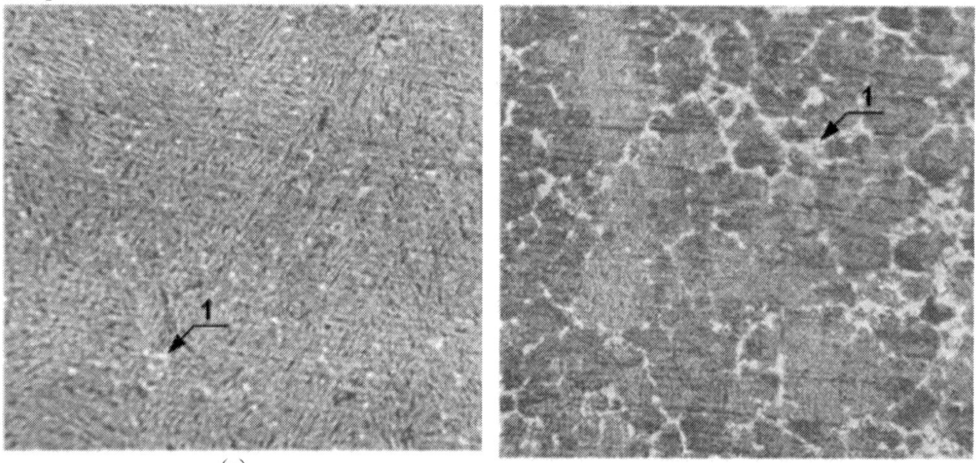

(a) (b)

Fig. 6. Microstructure of the coating formed (x400): (a) - without special arc action; (b) – employing SAIA

Microhardness of the coating formed without special arc action is 456-573 HV. .Microhardness of the coating formed under the SAIA action is 627-742 HV, which is 170 HV higher if compared with the metal formed without special arc action.

The final stage of our research included experimental-comparative testing of the parts of the road-construction machines restored using flux cored wire and SAIA technology. The testing conducted showed that wearability of these parts exceeds this property of the parts restored without using special technological effects of SAIA by 30%.

Conclusions

1. We have scientifically substantiated the possibility for application of special technological effects when forming alloyed coatings and suggested the outline of research aimed at the assessment of the influence of a special technological action by SAIA on the layer formed by the flux cored wire containing alloying elements in the form of ferroalloys and mineral raw materials.

2. We have presented the methodology for experiment planning, theoretical calculation of thermal fields and electro-dynamic interaction of the arcs that allow of the evaluation and prediction of the special arc's influence on the coating formation. Based on the theoretical calculations we have developed experimental equipment for automated deposit welding process.

3. Calculated distribution of thermal fields showed that effect of the special arc on the surface formed facilitates metallurgical processes, results in a more uniform structure and increases the transition of alloying elements to build-up metal. This effect is achieved due to redistribution of heat flows as the heating-up temperature of the surface being formed by the special arc reaches 1000 ^{0}C, and the temperature of the primary welding process is 3000 ^{0}C. Calculation of the electrodynamic interaction (EDI) force between the primary and the special arcs testifies to the possibility to change the spatial position of the primary arc which causes mechanical agitation of the molten metal.

References

1. Verhoturov A. D., Babenko E. G., Makienko V. M. Metodologiya sozdaniya svarochnyh materialov [Methodology for production of weld materials] / Edited by Corresponding Member of the Russian Academy of Science Voronov B.A. – Khabarovsk, FESTU Publ. 128 p. (2009)
2. Lashchenko G.I. Kombinirovannye tehnologii svarki plavleniem [Combined technologies for fusion welding] // Avtomaticheskaya svarka [Automated welding]. **8.** (2012)
3. Saraev Y.N., Lebedev V.A., Novikov S.V. Mashinostroenie [Engineering Industry]. **4 (1)**. pp. 16-2634. (2016)
4. Matsunawa A., Mizutani M., Katayama S., Seto N. Po-rosity formation mechanism and its prevention in laser welding, Welding International. **17(6)**. 431-437. (2003)
5. Rykalin N.N. Raschety teplovyh poley pri svarke [Calculation of thermal fields under welding], Moscow, Mashgiz Publ. 296 p. (1951)
6. Y.G. Gagen, V.D. Taran Svarka magnitoupravlyaemoi dugoi [Magnetically operated arc welding]. Moscow, Mashinostroenie Publ. 160 p. (1970)
7. Goldak J., Chakravarti A., Bibby M. A new finite element model for welding heat source//Metallurgical Transactions B. **15B**. P. 299-305. (1984)
8. Nguyen. N.T., Mai Y.-W., Simpson S. and Ohta A.. Analytical approximate solution for double ellipsoidal heat source in finite thick plate//Welding Journal. **3.** P.82-s-93-s. (2004)
9. Sabapathy P.N., Wahab M.A., Painter M.J.: Numerical methods to predict failure during the in-service welding of gas pipelines//Journal of Strain Analysis. **36 (6).** P. 611-619. (2001)
10. Zhang J. and Dong Y. Method for determining a heat source model for a weld. US Patent No 6,324,491 B1. 2001.
11. Goldak J. [et al.] Computer Modeling of heat flow in welds// Metallurgical Transactions B. **17B**. pp. 587-600. (1986)
12. Ranatowski E., Pocwiardowski A. An analytical-numerical evaluation of the thermal cycle in the HAZ during welding', Mathematical Modelling of Weld Phenomena 4, H. Cerjak (ed.), IOM Communications Ltd, London.pp. 379-395. (1998)

E3S Web of Conferences **56**, 03028 (2018) https://doi.org/10.1051/e3sconf/20185603028
VII International Scientific Conference "Problems of Complex Development of Georesources"

OBTAINING PRESET PROPERTIES IN THE HYDROGENATION PROCESS BY CONTROLLING THE STATE OF PHASE BOUNDARY

Galina Metaksa[1], *Gulnaz* Moldabaeva[1*] and *Zhanat* Alisheva[1]

[1]National technological Satbayev University of the Republic of Kazakhstan, Almaty, Republic of Kazakhstan

Abstract. The introduction of science-intensive technologies that ensure a high level of energy content and restoration of ecological balance in places of extraction and processing of commercial resources is the main condition for the successful development of independent Kazakhstan. This paper concentrates on challenges of studying the mechanisms of chemical effects on oil-reservoir rocks functioning in order to increase oil recovery. Relevance of addressing the challenges is determined by necessity to increase profitability of fossil hydrocarbons mining. The article is prepared based on the results of the works carried out within the framework of the Program of the State Science Academies for 2013-2020 period. "Earth Sciences" geological library Section No.9, "Integrated discovery and conservation of the Earth subsoils" line of fundamental research # 132, innovative processes of mineral deposits development and advenced processing of mineral raw materials", "The Fundamental Basis of Oil and Gas Industry Innovative Technologies" project, No.AAA - 0139-2018-0006. **Key words**: oil, hydrogenation, phase boundary, response, water, influence.

1. Introduction

The oil and gas field development system is formed at the initial stage of the deposit life cycle [11, 15, 19, 22, 25, 27, 31]. Selection of primary, secondary and tertiary methods (modes) for the dedicated operational facilities development is one of the key elements for the creation of the oil field development system [9, 17, 20, 23, 24, 26]. The primary or natural regime for the oil and gas deposits development is named the aggregate of natural forces (energy types) that ensure the movement of oil or gas in the porous layer system to the faces of production wells [10, 12, 14]. Secondary methods of development or methods of reservoir pressuring maintaining (RPM) include methods of injecting working agents as the form of water (flooding) and natural gas [13, 16].

Tertiary methods of development applying in terrigenous and carbonate reservoirs are conditionally divided into two types, first one - the oil recovery enhancement method and

* Corresponding author: gulnazka_1978@mail.ru

the second one – the advanced recovery method, depending on the prevailing influence on the multiplying factors of the equation for the oil recovery coefficient: coverage coefficient and displacement coefficient, respectively [18 , 21, 28-30]. It is necessary to study the behavior of surface and near-surface layers placed in terrigenous and carbonate structures that are oil storage tanks, in order to understand the mechanisms of the oil recovery enhancement, because of any process can occures (melting, dissolution, chemical reaction) at phase boundary. Chemical decomposition of formation water molecules with the help of energy-storage substances (ESS) – is the new research line in the methods of impulse stimulation of formation. In this case, water is a hydrogen and oxygen supplier , which allows to obtain the hydrogenation and synthesis of new compounds. The change in the chemical composition of heavy hydrocarbons through oil hydrogenation under natural occurrence conditions is a pulsed effect that carried out by the help of water decomposition chemical reaction by using of activated aluminum [1-8]. Hydrogen precipitation provides to the hydrogenation reactions of all fractions of heavy oils. The scheme of applicability of the products of this fraction is shown below [1].

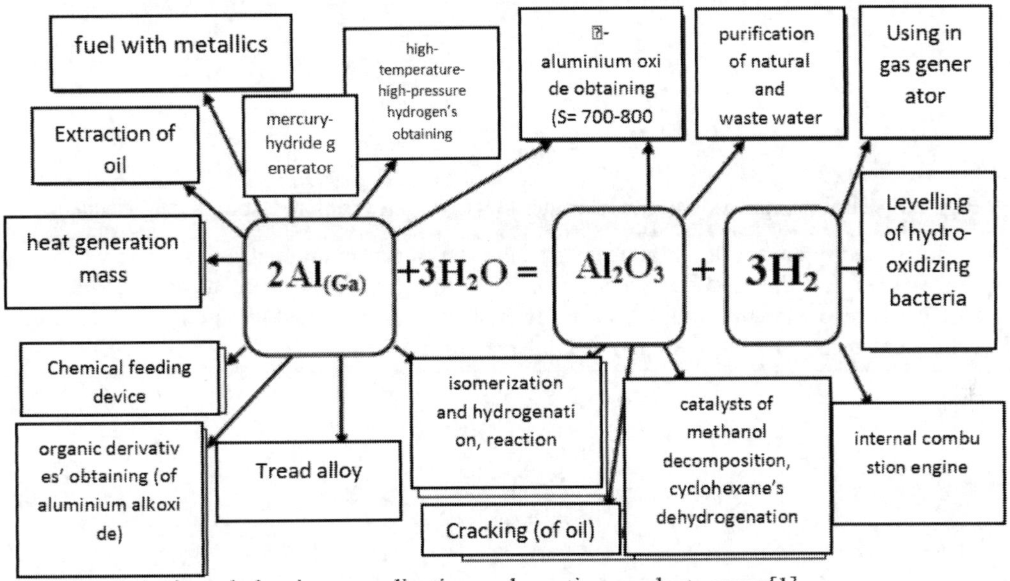

Fig. 1. Activated aluminum application and reaction products areas [1].

2. Method of experimental research

Research group of Kurapov G. G. were carried out industrial tests of the in the oil viscosity change by hydrogenation of its components at the "Embamunaigas" enterprise in 2016. During the first day of the test they received a positive result at enhancing the fluidity and at increasing productivity of the well. Well collation with the precipitated aluminum oxide, which requires a long time to be eliminated (2-4 weeks) is the negative side of the tests.

Thus, the results of these tests demonstrate the necessity to provide additional researches that, with the help of energy-storage substances (ESS), would prevent the negative consequences of the oil hydrogenation processes in natural conditions.

3. Results and discussions

Since any processes start on the reacting components surface, the phase boundary state was studied by measuring the spectral composition of the responses at phase boundary. Here are the results of measurements of the spectral responses of water to an external influence, which was performed from a pulsed generator operating at the water decomposition frequency, i.e. 42.8 kHz (The experimental procedure is described in [9]).

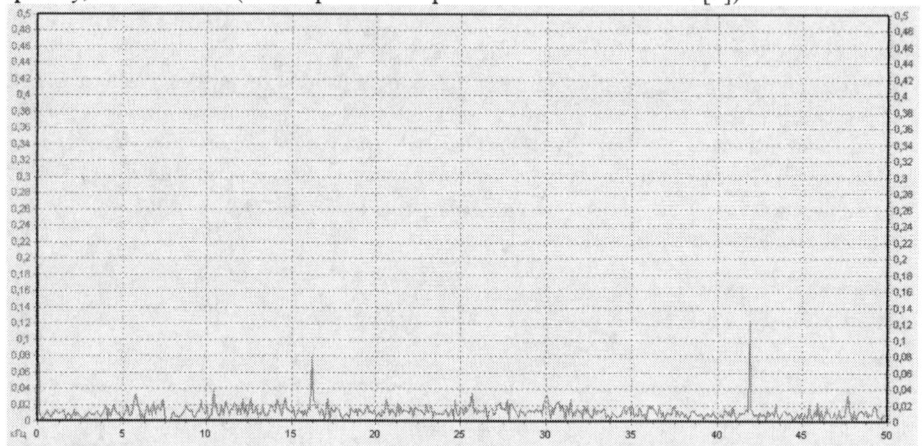

Fig. 2. Spectral composition of faucet mineralization water response at 42.8 kHz frequency

Fig. 2 shows the spectrogram of the initial water of faucet mineralization, from which it can be seen that the response spectrum includes the fundamental frequency (42.8 kHz), which has a maximum amplitude and the first two harmonics, the amplitude values of which are much smaller (2-3 times) than the main one.

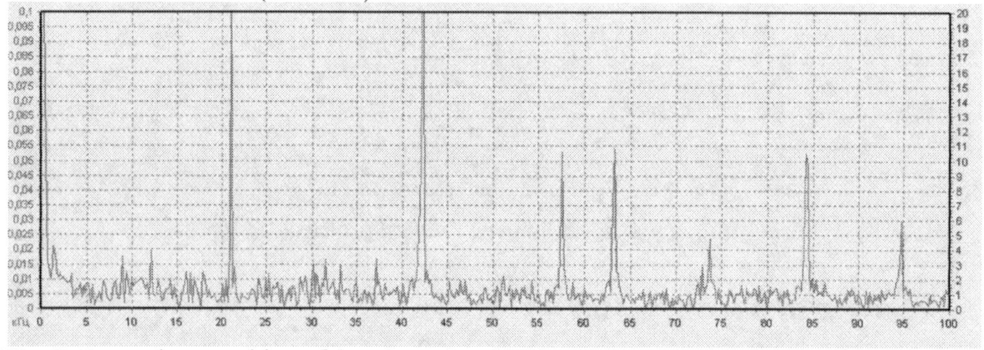

Fig. 3. Spectral composition of the of powdered sulfur response at a frequency of 42.8 kHz

Due to the fact that modern oils contain large amounts of sulfur and its compounds, powder and compressed sulfur pellets have been chosen for the model experiment, their responses are shown in Figures 3 and 4. It can be seen that, at this frequency, powdered sulfur produces responses that do not correspond to the harmonic series of the fundamental frequency of exposure, which is apparently due to the heterogeneity of the fractional composition of the powder (the powder is not sieved). The maximum response is fixed at the frequency of the exposure (42.8 kHz), in addition, responses to 21, 57, 63, 84 and 95 kHz are seen.

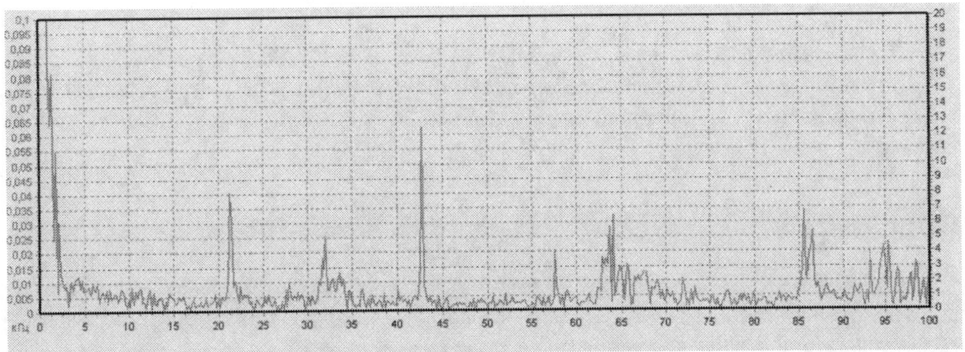

Fig. 4. Spectral composition of the response of pressed, coated with water, sulfur at a 42.8 kHz frequency

Covered with water film, pressed sulfur (Figure 4), gives the maximum response at the frequency of exposure, and the harmonic components of this response have a 2 or 3-watt structure under the same conditions of measurement. Speaking on language of physics this means that the resonant conditions correspond to the other dimensionality of the response structural elements. In other words, another structure of the phase boundary is fixed. For the triplet response, it is possible to represent the layered structure of the phase boundary between phases according to the scheme given below (Fig. 5).

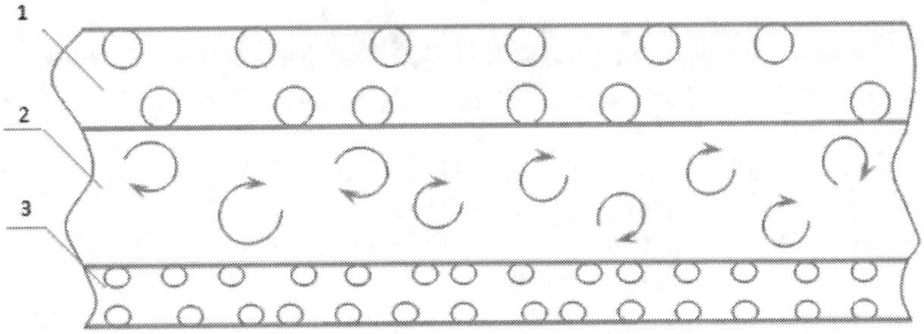

Fig. 5. Hypothetical phase boundary structure
(1-Low-frequency response, 2-Base signal, 3-High-frequency response)

With a shown above phase boundary structure, the conductivity of the base signal, that is, its efficiency, depends on the state of the upper and lower boundaries, and accordingly, in the case of hydrogenation, at least three reaction products can be present at the output, which is confirmed by the results of [9-12]. It is shown here, that this effect can be used to obtain prescribed properties in petroleum compositions via external exposure on the phase boundary by electromagnetic impulse exposure at the frequency of water decomposition. In this paper, additional possibilities for controlling the phase boundary structure were revealed through the use of powders possessing different dispersity. This is a new effect, based on the achievements of other branches of science such as optics (photoconverters), metallurgy [13-14], tribotechnics, etc.

The principal novelty of our development is the use of the water decomposition frequency that experimentally discovered by Puharich A. [15], the physical substantiation of which is still missing in our science. Following the classical logic of calculating the frequency by dividing the propagation velocity of the external exposure by the wavelength (the scale of the water structural components), we can see a complete discrepancy with experiment. The calculation performed according to Kepler's third law [16] gives a logical result:

$R^3/T^3 = K = 3{,}36 \times 10^{-18} = 5.12$ nm (nanometer).

R - interaction range;
T - interaction period.

The aggregate of experimental results show that the state of the phase boundary is the basis for controlling the chemical composition of hydrocarbons in the chemical impulse exposures.

Conclusions

1. To reveal the mechanism for controlling the phase boundary in "oil-water" compositions, an experimental setup has been created to measure the spectral composition of the boundary responses when the water decomposition frequency acts on it.

2. The results of physical modeling on the research of the "water-solid" phase boundary structure have shown that any inhomogeneity, that prevents the transmission of an exposure signal, causes response at the fundamental frequency and at harmonic frequencies, this response may cause a offshift of low or high frequencies.

3. The effect of the appearance of the layered structure of the "water-solid" phase boundary, which allows controlling the mechanism of interaction of the separated phases during the external exposure, is discovered.

References

1. R.G. Samurzina, Production of hydrogen and aluminum oxide in the reaction between aluminum and water. Doctoral dissertation, Almaty city. p. 165. (1984).
2. R.Zh. Abirov, A.G. Mukhamedova, B. Panabekkyzy, Zh.Zh. Abirov, A.A. Nesterkin. Oil and gas field business journal. **5**.15-19. (2016)
3. R.G. Sarmurzina, D.V. Sokolovskii and others. Head of the report of VI Petrochemistry Conference, Atyrau city (former – Gurev), (1984.).
4. Iu.P. Zheltov, N.A. Eremin, E.S. Makarova. Oil Industry. **11.** p. 28-32. (1993).
5. P.F. Bernath. Chem. Soc. Rev. **25.** pp. 111-115. (1996).
6. P. F. Bernath. Annu. Rep. Prog. Chem., Sect. C: Phys. Chem. **96.** pp. 177-224. (2000).
7. Encyclopedia of Spectroscopy and Spectrometry. Lindon J. 2nd Ed. Academic Press. p. 3312. (2010).
8. D.V. Sokolovskii, Hydrogenation in solutions. // Almaty city, "Nauka" Publishing house. p. 364. (1979).
9. USSR Certificate of authorship No.1282462, production process of □-aluminium oxide // G.G. Kurapov and others.
10. Zh.N. Alisheva, G.P. Metaksa. "Sodruzhestvo" Russian - Chinese scientific journal. Monthly scientific journal of research and practice conference. **16 (2)**. pp. 4-9. (2017).
11. V.K. Bekbaeva, G.P. Metaksa. Bulletin of the National Engineering Academy of the Republic of Kazakhstan. **1 (67)**. pp.39-45. 2018.
12. G.P. Metaksa, Theoretical bases of rocks conditions estimation and the forecasting. Academic Publishing. ISSN: 978-3-659-79250-2. (2015).
13. N. Buktukov, G. Metaksa, E. Vikhrachev. Bulletin of national technological Satbayev University of the Republic of Kazakhstan. **1/1 9580.** 198-203. (2007).
14. N.S. Buktukov, G.P. Metaksa, E.P. Vikhrachev, G.Zh. Moldabaeva. Collection of scientific papers of D.A. Kunaev Institute of Mining. **73.** Almaty city. pp. 94-99. (2007).
15. USSR Certificate of authorship. Method of metals and alloys working. // G.P. Metaksa, Kurapov and others., priority as of September 12, (1977).
16. Andria Henry Pucharich // Patent USA. No.9.394.230 dated July 19, (1983).
17. B.M. Iavorskii, A.A. Detlaf. The Handbook of Physics. p. 942. (1974).

The study of transient processes – a new direction in the development of a methodological approach to the solution of problems of complex development of geo-resources.

Victor Yakovlev[1*]

[1]Institute of Mining, Ural Branch of the Russian Academy of Sciences, Ekaterinburg, Russia.

Abstract: The article presents the stages of technology development and methodology of subsoil development in the postwar years to the modern period, taking into account the transition from planned to market economy. The necessity of new methodological approaches to solving the problems of subsoil development based on the principles of consistency, complexity, interdisciplinary orientation. The definition of the strategy for the development of deep-lying complex structural fields with the increase of geological, mining and technical and economic information in the dynamics of mining. The concept of "transient processes" is introduced, the accounting of which laws is the basis for the development of innovative technologies for the evaluation, production and ore preparation of mineral raw materials that meet the requirements of the subsequent enrichment and metallurgical processes of production of commodity products. The features of development of reserves of deposits in the Northern and equated regions, including taking into account the integrated development of territories by creating mineral resource centers. The directions of further research of transient processes, focused on the establishment of the main factors affecting the justification of methods, stages and timing of adaptation of the parameters of the main technological processes in the existing mining enterprises to changing internal and external conditions of their operation and the development of new deep-seated deposits in different climatic and mining conditions. Keywords: Transient processes, complex development, complex field, methodological approach, innovative technology.

For quite a long 300-year period of mining development in Russia the greatest interest for the study of the historical experience of the development of scientific ideas, methodological approaches to the substantiation of the technologies, parameters of mining works and the strategy of exploitation of mineral resources of scientific publications in the

* Corresponding author: yakovlev@igduran.ru

form of articles, monographs of eminent scholars in the field of mining Sciences academicians N. In. Melnikov, V. V. Rzhevsky, M. I. Agoshkova, doctors E. F. Sheshko, M. G. Novozhilov, A. I. Arsentiev, V. S. Khokhryakova and others, as well as the actual parameters and technical and economic indicators of mining enterprises in Russia and CIS countries over the past 50-70 years, i.e. for the period from the beginning of the transition to a complex-mechanized mining enterprises to modern mining and processing plants with a full cycle of technological processes of mining [1].

The continental part of the earth's crust, which is the source of all kinds of mineral raw materials, is represented by almost all the set of chemical elements with different background content, but they are all in a coherent aggregate state. All ferrous and non-ferrous metals are oxidized, so the deposits created by nature are, as a rule, complex and have unique properties and parameters.

The need for research and accounting of transient processes in the technology of their development is due to the factors of natural variability of geological parameters of complex structural fields:

1. Occur in the form of veins, lenses, layers of variable power;
2. Mineral bodies contain inclusions of waste rock or substandard ores of different sizes and shapes, complicated by folded or discontinuous tectonic disturbances;
3. Power and elements of occurrence, stability of rocks, their fracturing, natural blockage, complexity, limits of strength and deformation properties vary widely;
4. The qualitative composition of the mineral changes, there is a natural or random zoning of the total balance of minerals, high unevenness of their content and granulometric composition from solid to fine-grained ores.

All this requires a detailed geometrization of placing a mineral in the bowels and taken into account when choosing mining equipment and technology, adaptive to changes of the geological environment, planning of mining works in the modes of selective dredging and averaging, the creation of innovative technologies of mining and ore preparation, taking into account patterns of transient processes when changing the operation conditions of the mining and processing enterprises. The specifics of the development of deep-lying complex structural deposits of solid minerals is that their development continues for decades, as a rule, begins open Geotechnology, in the conditions of continuous growth of the depth of the working area of the quarry, the growth of geological, mining and technological information, requiring unconditional allocation of stages of formation of career space, changes in the parameters of development systems, the formation of the transport system of the quarry through the use of new modes of transport, etc., that is, practically, revision of most of the design decisions, clarification of the depth of the quarry and, as a rule, the transition to underground or combined development of the field. The branch principle of development of unique fields with large reserves and the content of the main useful component did not demand methodical developments of the system account at carrying out geological prospecting works, justification of standards, the statement of stocks, design, construction and operation of the mining and processing enterprises.

In research and development of the subsoil development strategy should take into account the peculiarities of conditions for the development of mineral resources [2]:

1. Exhaustion and non-renewable reserves of specific mineral deposits.
2. The complexity of deposits and minerals.
3. Variety of natural conditions for the development of the same types of mineral raw materials.
4. The individuality of the technological properties and the conditions of processing of mineral raw materials.
5. High natural variability of geological parameters within a single field.

6. Multiprocessing of mining production, dispersion and unsteadiness of the active part of the main fountains.

7. A variety of alternative ways to ensure a given volume of output.

8. Increase of information about the field as it is developed and making major decisions under uncertainty.

Factors of natural variability of geological parameters of complex structural deposits, substantiating the need for research and accounting of transient processes in the technology of their development:

1. Occur in the form of veins, lenses, layers of variable power;

2. Mineral bodies contain inclusions of waste rock or substandard ores of different sizes and shapes, complicated by folded or discontinuous tectonic disturbances;

3. Power and elements of occurrence, stability of rocks, their fracturing, natural blockage, complexity, limits of strength and deformation properties vary widely;

4. The qualitative composition of the mineral changes, there is a natural or random zoning of the total balance of minerals, high unevenness of their content and granulometric composition from solid to fine-grained ores.

All this requires a detailed geometrization of placing a mineral in the bowels and taken into account when choosing mining equipment and technology, adaptive to changes of the geological environment, planning of mining works in the modes of selective dredging and averaging, the creation of innovative technologies of mining and ore preparation, taking into account patterns of transition processes when changing the conditions of functioning of the mining enterprise. About the necessity of a radical revision of the current by the end of the 80s, ie, almost to the disintegration of the USSR, a methodological approach to solving a range of contemporary technical, technological and economic problems of the mining industry and mining Sciences very clearly set out in an unpublished, and presented in typewritten form in 1987. the report of academician V. V. Rzhevsky "Problems of mining Sciences (Confession of a mining engineer)". The measures taken in recent years in the Russian Academy of Sciences to consolidate the budget theme in the institutions, to concentrate the efforts of scientists on the most relevant scientific areas, to introduce a competitive principle of fundamental research on the programs of the RAS and its departments required the use of fundamentally new methodological approaches [3-5].

An example here is the program of the Presidium of the RAS, The Department of Earth Sciences, integration projects with the institutes of the mountain profile of Ural RAS, Siberia RAS and East region RAS. The institutes of Geology and Geochemistry, Geophysics, metallurgy, industrial ecology, ecology of plants and animals, Botanical garden take part in the implementation of the interdisciplinary project "development of the earth: prospects of expansion and complex development of ore mineral resources of the mining and metallurgical complex of the Urals", coordinated by the Institute of mining. At the present stage, when it is necessary to extract mineral raw materials on the deep horizons of the developed fields or to develop the resources of new regions, it is necessary to take into account the complexity of subsoil development (table.1).

The study of transient processes, the establishment of the causes of their occurrence, the understanding of the nature of the changes and the laws of their development in specific geological and mining conditions is the basis for the creation of a strategy for managing these processes during the entire period of development of deep-seated complex structural deposits – long-term action plan during the exploration, design and development to finished products on the basis of the methodological approach on the principles of consistency, complexity, interdisciplinarity and innovative direction, taking into account the increase in geological data of the Deposit when taking a pre-planned technological and technical solutions in response to changes in the internal and external conditions of functioning of the mining enterprise, including the regularities of their development in the adoption of innovative technologies for assessment production, ore preparation and enrichment of mineral raw materials. In the structural form, the new approaches are reduced to the combined application of program-target methods based on the principles of systematization, complexity and interdisciplinary research and their innovative orientation (table.2).

Table 1. Main difficulties of the current stage of subsoil development and ways of their accounting

Complexities	Ways of accounting and solutions
The difference between the actual geological structure of ore deposits from the laid in the projects on the basis of geological information (size, shape, content, Phys.-fur. properties, associated components, etc.). As a result: low efficiency of technological processes, loss of basic and associated useful components	Formation of database, creation of Deposit model and methods of optimization of parameters of mining enterprise and on their basis: - clarification of the boundaries and inventory; - optimization of mining operations; - modernization of technology and technology; - reconstruction of the enterprise; - complex development of mineral resources
Opening and working out of deep horizons	Creation of specialized loading and transport equipment, technology for safe conduct of open and underground mining
Development of deposits of the North	Development of a step-by-step strategy of geological, technological and economic justification of the parameters of future enterprises

Table 2. The main principles of the methodological approach to the solution of problems of development of mineral resources

Principle	Essence
System	Disclosure of the basic rules of research, operation or development of a mineral object, identification of various types of external and internal relations that affect the construction of a generalized model of the system under study
Complexity	Research of not only the phenomena and processes of development of deposits, but also the possibility of the fullest use of all components of mineral raw materials, development and development of territories and regions of their location
Interdisciplinarity	Combining efforts in related fields of scientific knowledge for a comprehensive assessment of the beneficial effect and the search for innovative approaches to the use of mineral resources
Innovative focus	A mandatory requirement designed to provide a materialized result obtained from investing in research, new technology or technology, new forms of production organization, safety, management, etc.

Among the scientific, technological and organizational and technical issues to be solved in the design and development of complex structural fields of great depth, requiring the study of transient processes, are:

1. Assessment of the volume and quality of reserves, the decision on the expediency of their integrated development.

2. Substantiation of the method of development of a deep-seated field and the possibility (feasibility) of transition from open to underground or combined method.

3. Substantiation of the boundaries of the quarry, its depth, the angles of slope sides, the sequence of development of the working area, the allocation of stages of development.

4. Formation of the transport system of the quarry in connection with the development of the quarry space and its working area, the use of new types, technical means and transport schemes with the replacement of existing, but more often in addition to them.

5. Stage-by-stage development of deposits, formation of temporarily non-working sides of quarries and their subsequent development with the use of special technologies and technical means of blasting, excavation, transport.

6. Opening of new horizons, including the possible change of physical and mechanical properties and material composition of the rock composing them.

7. Development of deposits of complex geological structure of the main mineral deposits in the presence of associated mineral components, significant variability of their content and other properties that require selective development, averaging and periodic changes in the parameters of development systems, ore preparation and enrichment.

Among the most important scientific, technological and organizational and technical issues to be solved in the design and development of complex structural fields of great depth, requiring the study of transients, are:

1. Assessment of the volume and quality of reserves, the decision on the expediency of their integrated development.

2. Substantiation of the method of development of a deep-seated field and the possibility (feasibility) of transition from open to underground or combined method.

3. Substantiation of the boundaries of the quarry, its depth, the angles of slope sides, the sequence of development of the working area, the allocation of stages of development.

4. Formation of the transport system of the quarry in connection with the development of the quarry space and its working area, the use of new types, technical means and transport schemes with the replacement of existing, but more often in addition to them.

5. Stage-by-stage development of deposits, formation of temporarily non-working sides of quarries and their subsequent development with the use of special technologies and technical means of blasting, excavation, transport.

6. Opening of new horizons, including the possible change of physical and mechanical properties and material composition of the rock composing them.

7. Development of deposits of complex geological structure of the main mineral deposits in the presence of associated mineral components, significant variability of their content and other properties that require selective development, averaging and periodic changes in the parameters of development systems, ore preparation and enrichment.

8. Transient processes associated with changes in the physical and mechanical properties of permafrost rocks (phase transitions when the air temperature from +40°C to -60°C), including in the sides of quarries and in the trunks of underground mines.

The necessity of application of innovative technological, technical and organizational decisions in transitional periods of field development it is possible to determine a continuous monitoring of the development of the mining system at all stages of development in connection with changing in time and space of the external and internal conditions of functioning of the mining enterprise.

The results of the research are published in articles and monographs, including collective ones, which reflect the current assessment of the state, innovative solutions and prospects for the development of mining Sciences and subsoil development [6-8].

New approaches to the development of the Northern subsoil are associated with the implementation of the strategy of accelerated gradual development of subsoil and the principle of integrated development of territories:

- accelerated, complex exploration and exploration of deposits;
- express geological, technological and economic assessment of mining and technical capabilities of each ore region;

- creation of a geographical information system to assist with modeling and program complexes for the decision of technological, technical, environmental, economic etc. problems of designing and planning of mining (example: Tarynnahsky mine in Yakutia);
- priority of the goals of reducing the number of employees by increasing productivity and the abandonment of non-core and auxiliary industries;
- one management company (holding) in the territory of administrative education with a single industrial and social infrastructure (example: South Yakutia).

Basic principles of designing technologies for development of Northern fields:

- stages of mining operations;
- high-performance powerful equipment, equipment in the Northern performance;
- step-by-step transfer of mining equipment with diesel drive to electric;
- allocation of technological zones taking into account the cluster structure of ore deposits;
- organization of the operational testing system to optimize the processes of ore preparation and enrichment;
- integrated use of mineral resources by extracting the maximum of useful components and utilization of overburden and enrichment waste;
- an integrated approach to the development of the mining and machine-building complexes;
- advanced or parallel assessments that allow the results of work to be used in advance in the development of decisions, recommendations or the formulation of the main limiting conditions for subsoil use.

The main form of implementation of these principles is the creation of mineral resource centers [9].

Conclusion

The developed methodological approach based on the combined use of the principles of consistency, complexity, interdisciplinary and innovative orientation can be implemented by continuous monitoring of the development of mining and technological system of mining and processing enterprises, timely development of transient processes at all stages of development of the field and implementation of recommendations for improving the parameters and technologies requiring changes in connection with the changing internal and external. The main indicator characterizing the high degree of novelty of research is the further development of the program-target methodological approach to the study of the problems of subsoil development by introducing a fundamentally new concept of "transient processes" and taking into account the laws of their development in the development of innovative technologies for the extraction and ore preparation of mineral raw materials. The essence of novelty of the study of transient processes in the design and development of deep-seated deposits is that if the evaluation of dynamics of development of mining system and indicators of its functioning may need to change the parameters of the equipment or technology of the mining production, it is necessary to establish the need, the totality and consistency of a series of actions (events) to adopt and implement adopted innovative technological solutions. The proposed methodological approach, based on the study of transient processes is universal and can be used in the design of the development of deep-seated deposits, the planning, organization and management of mining and ore dressing of mineral raw materials in existing mines given the growth of geological information, the implementation of the developed innovative events, changes of parameters and indicators of the mining system of the mining enterprise with the development of mining operations.

In general, the proposed methodological approach allows for a more complete account of geological, mining, environmental and organizational features of the assessed fields and their location areas, as well as to provide the prerequisites for the systematic adoption of the most effective energy- and resource-saving technological solutions.

The absence of the state strategic program for the development of the mineral resource base of the Russian Federation, the economy of which is ¾ associated with the use of mineral resources and their processing products, necessitates comprehensive research with the participation of geologists, miners, metallurgists, economists in the framework of the program of the Presidium of RAS "Strategy for the development of the mineral resource base of the Russian Federation as the basis of its economy and national security."

Further studies of transient processes are aimed at establishing the main factors affecting the justification of methods, stages and timing of adaptation of the parameters of the main technological processes in existing mining enterprises to changing internal and external conditions of their operation and the development of new deep-seated deposits in various natural and climatic and mining conditions.

References:

1. Yakovlev V. L. Problems of subsoil use. - № 3 [Electronic resource]: peer-reviewed collection of scientific articles. Ekaterinburg. P. 15 – 26. (2014). //trud.igduran.ru
2. Batugin S. A. Regularities of mining development. IGDS SB RAS. Yakutsk. 116 p. (1992).
3. Trubetskoy K. N. Mountain science. Development and conservation of the earth's interior.Publishing house of Academy of mining Sciences. - 478 p. (1997).
4. Yakovlev V. L. Methodological aspects of mineral resources development strategy. Ekaterinburg: IGD Uro ran. 152 p. (2001).
5. Trubetskoy K. N. Gorny Zhurnal. **1**. P. 15 - 19. (2012).
6. Yakovlev V. L. Herald of Ural branch of RAS. Science. Society. Man. **4**. P. 43 - 49. (2013).
7. V. L. Yakovlev, I. V. Sokolov, G. G. Sakantsev, I. L.Kravchuk. Gorn. **7**. P. 46-50. (2017).
8. Yakovlev V. L. Problems of subsoil use. 2 [Electronic resource]: peer-reviewed collection of scientific articles. - Ekaterinburg: IGD Uro ran. P. 5 - 15. (2017). //trud.igduran.ru
9. V. L. Yakovlev, S. V. Kornilkov, A. P. Van-Van-E, V. S. Litvintsev, G. F. Sklyarov, N. Lavrik, A. Problems of subsoil use. 3 [Electronic resource]: peer-reviewed collection of scientific articles. - Ekaterinburg: IGD Uro ran. P. 38 - 46. (2014).

E3S Web of Conferences **56**, 04002 (2018) https://doi.org/10.1051/e3sconf/20185604002
VII International Scientific Conference "Problems of Complex Development of Georesources"

THE NEW WAVE OF TECHNOLOGICAL INNOVATIONS FOR SUSTAINABLE DEVELOPMENT OF GEOTECHNICAL SYSTEMS

David Kaplunov[1], *Marina* Rylnikova[1*] and *Dmitry* Radchenko[1]

[1]Institute of Complex Exploitation of Mineral Resources of the Russian Academy of Sciences, Moscow, Russia

Abstract. Today the development of information network technologies and the new equipment determine the new strategies in the designing ore deposits exploration with elements of artificial intelligence, software system processing of the geodata, high-precision methods and tools for studying the mineral matter and the structure of mine massifs, renewable energy, and green technologies. In addition, these components are used in the sustainable operations of mine technical systems, for the active expendable of which it is necessary to make a transition to a new technological setup. It is associated with the capabilities of operational synchronous transformation of the taping resources and the adaptation of geotechnologies to specific conditions, to meet the needs of society in georesources.

1 Introduction

In modern world, the trends determine the assigning of a new technological setup of mining production that are notable in the development of information network technologies in the field of development of mineral deposits. The use of machines with artificial intelligence; in systems for obtaining and processing geodata and others are also important.

The development of mining machinery production for the autonomous operation of mining transport means has already prepared a transition to a new technological setup with in unmanned mode, and equipping mining machines with elements of artificial intelligence [1-6]. In the coming years are expected a reducing risk factor of human presence in mining operations dangerous zone. They are the development-cutting heading and breakage faces the zones of intensive massif deformation, including dynamic manifestations of strata pressure, in areas and an increased radioactive background, with unfavorable environmental conditions. High or low temperature of the rock massif, with atmospheric gas pollution, low oxygen content or atmospheric pressure [7, 8] are also important factors. The possibility of mining the mineral products in such zones "with zero input" without reducing of the appropriate in resource development is the key factor in the tap resources in mining. It will characterize the new technological setup of mining in general. In particular, the equipment properties description for composition, structure and condition of the rock massif get to the

* Corresponding author: Rylnikova@mail.ru

new level. Even today, in real time mode at on open-cast the acquisition of collecting and processing information on the stability of open-pit slopes and dumps with the use of georadars are carried out. Therefore, establishment of patterns changes in the state of rock massifs [9, 10 etc.] is being performed also. At the same time, the accuracy of prognosis of landslide phenomena and voids is significantly growing. In the nearest future the appropriate systems will be installed in underground mines. On-line sensors are being introduced to assess the quality of mineral and raw materials stream – ores and rocks, mine waters, filling mixtures and ore processing products. Information of the mine atmosphere the composition is constantly obtained at to the dispatchers' panels. In our paper we describe the software complexes that can capable to estimate accurately the duration of sustainable operation of virtually, technical structure, formed underground chambers to be developed. The development of recommendations to ensure the long-term stability in the dynamics of changes in gas and hydro-mechanical conditions were carried out successfully. In developed countries, renewable energy systems are being introduced in mining industry.

2 Methods

The parameters of the technologies making the new technological setup should be in the form of a single complex adapted to the specific conditions for the development of deposits of solid minerals. Adaptation refers to the operational synchronous transformation of geotechnologies to the properties and condition of a technogenic transformed site of subsurface resources. The changing needs of society regardless to the direct objectives of extracting minerals from the earth interior should be also taken into account. However, the performance of the new technological setup based on the adaptive systems development and technologies is "tailored" to meet the needs of social development. It is a very relevant objective, which was greatly developed in recent years in various fields of human activity. [11, 12]. Thus, at present, there is rapid increase in technological development recorded in the field of mining sciences in implementation of the self-propelled machinery and high-performance mining equipment complexes installed in mines. The possibility of obtaining unlimited volumes of initial data and their processing for adjusting the parameters of geotechnologies, mineral and raw materials flows, formed mine technical designs shows the necessity to shift to a new technological setup. There may become adaptive geotechnologies, when the stages of transition from one type of geotechnology to another in the course of development deposits will be designed including the period after mining operations [13]. Of course it will require the establishment of fundamental regularities and the justification for the stages of the transition to a new technological setup of the sustainable operation of underground mines. It should be based on the establishment of spatio-temporal interrelationships of technological processes in the development and conservation of subsoil.

3 Scientific and methodological basis of research

Creating the scientific basis for the technical re-equipment of mines providing the transition to a new technological setup is related to:

- the exhaustion of the traditional mineral resource base, available to humanity even 10 years ago, due to the long-term implementation of extensive subsoil use concept;
- the transition of the world community to the operation principles of sustainable development of nature and society (the Paris Agreement regulating the measures of reducing carbon dioxide in the atmosphere from 2020 instead of the Kyoto Protocol);

- the large-scale research in the field of intellectualization of the mining complex, renewable energy, the use of equipment and materials of the new generation, the widespread development of combined physical-technical and physical-chemical geotechnologies. Nevertheless mining complex functioning on the principles of sustainable development is associated with unprecedented conditions: the increasing of mining operations depth; the reduction of large-scale deposits of solid minerals with high quality of raw materials; the decrease of valuable components of in the content in ore and increase of harmful impurities; the increase of rebellious ore share; the accumulation of large volumes technogenic formations, comparable to volume of accumulated metals with of promising deposit reserves; the displacement of mineral resource development sites in hard-to-reach areas with undeveloped infrastructure and unfavorable natural and climatic conditions; the leaving of significant volumes of natural and man-made reserves in the depths of the Earth located in complex technogenically altered geomechanical, gas-hydrodynamic and mine technical conditions; the growth of the requirements to the quality of products by the world market; the increase in specific energy costs and others. In this connection sustainable development should be understood not only as a classic definition, implying such a kind of development that allows to ensure the existence of society without threatening to future generations to meet their needs [14-18]. Sustainable development of mining complex should be understood as a set of strategic initiatives to ensure the exploitation of each developing subsoil area for indefinite period time on the basis of the initial design in the stages of transition to a new technological structure of field development. It is performed not only for the development of balance reserves but also for the period after mining of minerals is completed. At the same time the traditional "trilemma" is kept consisting in ensuring the balance of economic, environmental and social components (Fig. 1).

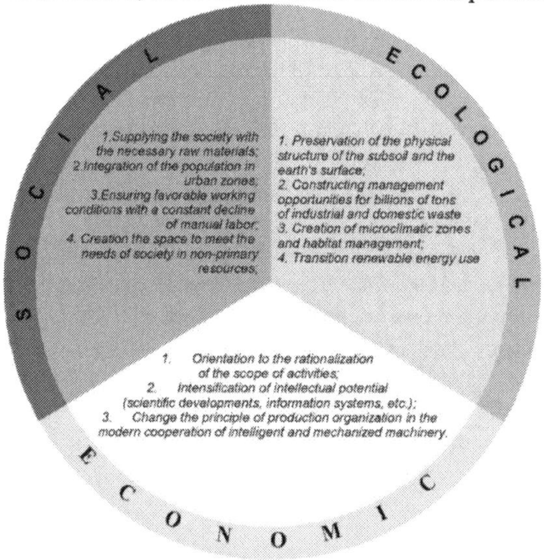

Fig 1. The relationship of sustainable development aspects of mining complex in transition period to the new technological setup

In summary the ecological component of sustainable development comprises:

- the preservation of the physical subsoil structure and the earth surface to maintain the structure, substance and physical fields of the stability in biosphere processes;

- the implementation of management opportunities controls billions of tons industrial and domestic waste, complete ecologically safe utilization of which is possible in the developed mined-out spaces;

- the creation of microclimatic zones and habitat management in the developed underground space of mines depending on its functional use;

- the transition to the use of renewable technogenic energy sources, formed in the implementation of technogenic subsoil transformation processes.

The social component of sustainable development of mine technical systems is:

- supplying the society with the necessary raw materials;
- population integration in urbanization zones;
- ensuring with favorable working conditions by a constant decrease of manual labor share and the influence of the human factor to the implementation of the tap resources processes;
- creation of space to meet the needs of society in non-energy resources - transport communications, location of industrial production, the deep urbanization, etc.

The main design task is the preservation of the production capacity not only for metal products, but also for all natural and technogenic geological resources involved into operation, with an acceptable level of profitability. This requires the economic mechanisms for technical re-equipment. Achieving such goals and complicated tasks may be possible only on establishing regularities and determining the conditions for the transition to the new technological setup of design, comprising the following strategies for the technical re-equipment of the mines (Fig. 2):

- the combining physical-technical and physical-chemical geotechnologies for the most efficient field development [18, 19];

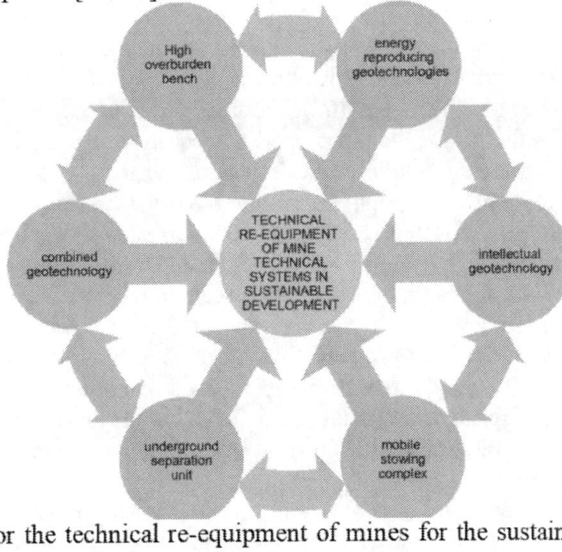

Fig 2 - Strategies for the technical re-equipment of mines for the sustainable operation of mine technical systems

- the change in the part of specifications in open mining operations on the basis of a proper transition to high overburden bench;
- the introduction of energy production technologies using technogenic sources;
- the introduction of equipment with elements of artificial intelligence, capable to operate without human involvement in particularly hazardous areas of quarries and underground mines;
- the use of underground moving or mobile stowing complexes designed for the utilization of rocks from the excavation of mine workings and wastes of internal separation. It is determined for the effective erection of artificial structures with various purposes in the multifunctional use of the mined-out space;
- the use of mobile inter-mineral equipment providing management of the volumes and quality of mineral and raw materials flows. It is possible with the winding to the surface only the amenable ores or even marketable metals, according to the principle "do not get out anything superfluous".

4 Results and discussion

The Institute of Integrated Development Problems of Subsurface Resources of the Russian Academy of Sciences develops technologies designed to provide a new technological setup of mining production in the interests of sustainable development of mining. Some parameters of geotechnologies and the conditions are considered for their implementation in the above mentioned 6 interrelated strategies of transition to a new technological setup.

The combination of physical-technical and physicochemical geotechnologies is the sphere of interests in the field of integrated development of copper-bearing raw materials in the Urals, polymetallic lead-zinc ores of the Caucasus, gold-bearing and uranium ores of the Urals, Siberia, the Far East. Problems of the development of combined physical-technical and physical-chemical geotechnology are currently relevant for the development of diamond pipes by ALROSA, iron ore provinces of the KMA and the Kola Peninsula, and others. Reduction in Cu content is less than 0.5%, Zn is less than 1%, Au is less than 1 g / t and so on. It does not leave an alternative in developing deposits to a combination of open and underground mining operations with various methods of leaching separately prepared ore and the technogenetics solids. A similar situation is abroad [20-22].

Restructuring the reserves to be developed in future, the new trends in the development of open and underground mining should be highlighted. Undoubtedly, the most important strategy of improving open mining is their transition to high overburden bench. It is important to note that the increase in the height of the bench provides the increase in additional extracted reserves, the growth of production capacity in the open-pit increase its final depth. The reduction in the length of the transport routes is also the advantage. The reduction in time for the track side of transport and equipment, the reduction in the number of working decks, reloading points, improvement of the logistic scheme of the mine, the reduction in the amount of equipment in the quarry, the reduction in technological processes may also be performed to advantages. Improvement quality of crushing of the rock is due to increase of operation time explosion of the solid in presence broken-down rock of collapsed rocks. It generally provides the reduction of the negative impact in mining operations to the environment with reducing the number and volume dust and gases and other emissions [23]. It has been proved by the researches that the timely transition to open mining operations by high benches in the period of maximum development of mining operations, provides the increase in the final depth of the quarry and the increment of additional reserves without beating the open pit sides with the current ratio of overburden to economic stripping ratio being equal.

The urgency of researching the conditions and parameters in the application of energy-efficient geotechnologies in the integrated development of ore deposits is due, on the one hand, to the increase in the energy consumption of mines with an increase in the depth of mining (the share of energy costs in the cost of solid minerals is 45-50%), the growing interest in the world to the receiving the and use of renewable energy sources. The share of renewable sources is one of the main indicators of the national economy progressiveness in the overall balance of energy resources. In addition, assessing the energy potential of internal streams of hydraulic mixtures, it is established that they represent an independent group of technogenic energy geo-resources to be realized and usefully used only in the mine technical system. This is a new group of renewable technogenic energy sources (RES), which potential was not taken into account before in the world energy sector. It is determined that, alongside with the increase in mining operations depth, the conditions and opportunities for the beneficial use of renewable energy sources are increasing from the flow of mine water and other hydraulic fluids. This determines a significant renewable energy sources from the mine water streams and filling the mixture at deep levels of ore deposits and provides a significant reduction in the ore extraction energy intensity [24].

It is pertinent to note here that the final extraction of reserves in previously developed large deposits, can now be conducted only on the basis of robotic intellectual geotechnologies the conditions for the development of which are characterized by the increased complexity of mining, So assessing the parameters of intellectual geotechnologies, it was established that the mine technical system previously designed was based on conditions and constraints and associated with human working conditions. Henceforth, with the development of remote and robotic geotechnologies, self-contained means of monitoring the state of the solid and their combinations supports the substantiation of parameters only to ensure the conditions for robotic and autonomously controlled equipment in the zones of direct mining operations. The removal of hygiene requirements for human operation in these zones, as well as the restrictions and requirements of industrial safety, will determine new requirements for the formation of mine technical systems and the justification of the parameters of the habitat in the mine technical system and the areal of its influence. The use of mobile stowing complexes contributes into the rapid utilization of hardening filling mixtures from the stage of excavating of mine workings without winding them to the surface [25]. Small-sized equipment of stowing complexes is mobile and can move along the underground workings after the development of the extraction front. Crushing of rocks is performed in inertial crushers of the new generation with output at least 30% of the class minus 0.074 mm, providing the necessary transport ability of the mixture, as in the case of mill grinding. As a consequence, the length of pipeline transport is reduced. The latter contributes to the involvement in the development of deposits of various sizes and forms of occurrence and increase the efficiency of filling underground voids. Filling mixtures obtained in experimental industrial conditions are characterized by favorable rheologic properties (lowability spread over the cone of StroiNsIL 18-21 cm) and lower moisture consumption. Water consumption is 15-20% lower than in traditional mill grinding at plants. Mixtures are characterized by the acceleration of the hardening time and development of strength. A wide range of equipment size allows to form modular-type stowing complexes in accordance with various production requirements. The operation of the equipment in a direct (open) cycle ensures the minimum of energy consumption. Along with the high energy efficiency, the complex of equipment has low steel intensity and does not require the construction of stationary foundations. Thus with the advent of underground mobile stowing complexes, a new meaning of the technology of internal separation of ores was acquired. The ability to prepare hardening filling mixtures directly in a subterranean mine solves the problem of utilization of ore mining and separation waste that can be effectively used to prepare filling mixtures without being dispensed to the surface. As a result, up to 40% of waste separation can be left in the earth interiors. In modern conditions, when switching to the development of deposits of halving previously classified as non-commercial, the combination of these two separation and bookmarking technologies with the use of mobile complexes provides the very possibility of mastering a number of large mineralization characterized by a low content of valuable components.

The contribution of the mining complex of developed producing countries to sustainable development is represented in the implementation of a technologies set making up the new technological setup of mining. The current state of the mining industry determines the prospects for the development of combined geotechnologies based on the use of equipment with artificial intelligence components operating in "unmanned" zones of deep quarries and underground mines. The introduction of these technologies implies the use of a set of mobile and portable equipment for mine separation, filling mixtures, the production of electricity from the renewable technogenic sources formed in the processes of technogenic subsoil transformation. This study presented interrelated geotechnological processes, the implementation of which provides the formation of the new technological order in mining industry.

5 Conclusions

The study of the conditions for the transition to a new technological setup shows that only on the basis of the interrelationship between intellectual technologies and new equipments, with elements of artificial intelligence, geodatabase software complexes, high-precision methods and tools for exploration subsoil matter, the structure and condition of rock massifs and renewable energy define the new strategies in the design of the development of ore deposits. Content components are of the technical systems of the sustainable operation in mine, for wide practical application of which requires a transition to a new technological setup. It is associated with the possibilities of operative synchronous transformation of geotechnologies to the specifications of the developed site in subsurface resources, taking into account the needs of society in geo-resources.

References

1.Trubetskoy K.N. Gornyi Zhurnal. **5**. Pp. 21-27. (2016).
2. Rylnikova M., Radchenko D., Klebanov D. E3S WEB of conferences. The Second International Innovative Mining Symposium. (2017). DOI: 10.1051/e3sconf/20172101032
3. Rylnikova M.V. JMS. **53(1)**. Pp. 92–101. (2017). DOI: 10.1134/S1062739117011884.
4. Golovina O., Teizer J. Automation in Construction. P. 1–17. (2016). DOI: 10.1016/j.autcon.2016.03.008.
5. Jay Lee, Hung-An Kao, Shanhu Yang. Procedia CIRP 16 Pp. 3-8. (2014).
6. Johansson B. Int. J. Mining and Mineral Engineering. **5 (4)**. Pp. 350-361. (2014).
7. Nikolakopoulos G. A IFAC.. Pp. 66–68. (2015) DOI:10.1016/j.ifacol.2015.10.079.
8. Radchenko D.N. Combined geotechnology: resource saving and energy efficiency. The IX International Mining Conference. MGTU. Pp. 157–160. (2017) https://elibrary.ru/item.asp?id=29304952
9. Osasan K.S. International Journal of Mining Science and Technology. **24 (2).** Pp. 275-280. (2014).
10. Waldir R. Paradella. Engineering Geology, **193**. Pp. 61-78. (2015).
11. Hudec M. Procedia Engineering. **161**. Pp. 1393-1397. (2016).
12. Ferrarini B. Structural Change and Economic Dynamics. **37**. Pp. 52-61. (2016).
13. Kaplunov D.R. Gornyi Zhurnal. **11**. Pp. 52-59. (2017) DOI: 10.17580/gzh.2017.11.10
14. Humphreys D. Resour. Policy. **27 (1)**. P. 1–7. (2001).
15. Pimentel B.S., Gonzalez E.S.S., Barbosa G.N.O.O. J. Clean. Prod. Elsevier Ltd. **112**. P. 2145–2157. (2015).
16. Espinoza R.D. Resour. Policy. **52**, P. 7–18. (2017).
17. Erzurumlu S.S., Erzurumlu Y.O. Resour. Policy. (2014.).
18. Agoshkov M.I., Kaplunov D.R., Shuboderov V.I., Gordin D.V. Combined method for development of ore deposits. P.188. (1992).
19. Kaplunov D.R. Gornyi Zhurnal,. **1**. Pp.14-17. (2018) DOI: 10.17580/gzh.2018.01.01.
20. Epstein R. Oper. Res. **60 (1)**. Pp. 4–17. (2012).
21. Zhao X. J. Cent. South Univ. **19 (11)**. P. 3256–3265. (2012).
22. Marschalko M. Eng. Geol. **147**. Pp. 37–51. (2012)
23. Rylnikova M.V. et all. Gornyi Zhurnal. **1**. Pp. 32-36. (2018)
24. Rylnikova M.V. et all. Gornyi Zhurnal. **11**. P. 71-76. (2017)
25. Kaplunov D.R. Gornyi Zhurnal. **2**. Pp. 101-104. (2013)

Problems of mineral raw material losses of quality and quantity management in developing of large-scale complex-structure deposits.

Sergei Tkach[1*]

[1]Mining Institute of the North, Siberian Branch, Russian Academy of Sciences, Yakutsk, Russia

Abstract. The article deals with the problems of mineral raw material losses of quality and quantity management in developing of large-scale complex-structure and composition deposits of solid minerals. It is shown that a very high degree of mining and geological conditions variability in time and space for the development of mining units is typical for such deposits. This significantly complicates the qualitative and quantitative operating losses setting and accounting of mineral raw materials during its extraction in the framework of existing general and industry regulatory documents. Conceptual principles for face-by-face operational setting of losses and impoverishment of minerals for the conditions of bulk mining of complex-structure deposits with the formation of gross mining flow with economically feasible and specified level of quality (the content of commercial and harmful components) are stated. These principles generally do not contradict effective instructions main provisions and are made to minimize the total operating losses during the processing of several mine sections (faces). **Key words**: mining and geological conditions, mining unit, losses, impoverishment, mining flow, setting, efficiency.

1. Introduction

From the point of view of an economically feasible and environmentally acceptable level of mineral raw material qualitative and quantitative losses, the problems related to planning and organizing mining operations are exacerbated during the exploitation of large-scale complex-structure deposits of solid minerals.

Extremely high diversity and variability of the internal (natural, mining and geological and other operational conditions) and external (market conditions, variability of prices for goods, services, etc.) environment [1, 2, etc.] – main reason for problems exacerbation. It is necessary to simultaneously use the various technologies, development systems and their elements during such deposits exploitation. This significantly complicates the setting and accounting for the solid minerals qualitative and quantitative losses.

Without dwelling on influence of the external environment variability on impoverishment and losses standard ratios technical-economic justification and efficiency of the mining enterprise as a whole, let's look at the effect of mining and geological conditions aggregate of field operation, in particular, the process of normalization of operational qualitative and quantitative losses of minerals on open-cast mines.

* Corresponding author: tkach@igds.ysn.ru

2. Materials and techniques

The whole aggregate of mining and geological conditions (the thickness of productive bodies, dip angles and extent, useful and harmful components content, adjacent-and-ores-formation stability and strength, high-amplitude geological disturbances, etc.) also changes as each extraction ledge (face) moves. The situation is aggravated by the high variability, inadequate texture and detection accuracy in the process of detailed exploration and advanced in-mine sampling of the geological environment main factors [1-5, etc.]. There is no doubt that the authors' assertion ([6], p. 8) that ... "technological standards for quantitative production losses established for each extraction unit (EU) can and should be the basis for assessing the economic efficiency of the extraction completeness of minerals from subsoil". According to subsoil protection rules (Order of the Ministry of Natural Resources of the Russian Federation of March 28, 2012 No. 257), the extraction unit, which accounts for the state and movement of reserves, losses and impoverishment of minerals, is defined as the minimum section of the field having relatively homogeneous geological conditions. Its mining within the framework of the adopted design solutions is carried out by one mining system, a technological scheme of extraction (quarry, ledge, block, lava, chamber, accumulation, deposit, etc.). Within the EU, the reserves are determined with sufficient reliability and a reliable primary accounting for minerals and components extraction (mining) is possible. Due to the extreme variability in time and space of the conditions for large complex-structure deposit mining it would be not possible to justify and identify EU with homogeneous geological conditions for their mining in a certain period of time (year, quarter, and month) even within a single ledge or block.

It is also true [6] that, despite the transition from the planned economy to the market economy of the country in the 90s of the last century, so far they have retained their legitimacy, practical significance and widespread applicability "Standard methodological regulations ..." (SMR) [7]. It should be emphasized that the main SMR and other applicable industry normative documents provisions [8, etc.] do not exhaust the existing opportunities to reduce raw-materials losses in near-contact zones where the formation of the mining flow of large-scale complex-structure deposits is performed by aggregate of faces that different from each other at mining and geological and mine technical conditions of their treatment. Among such conditions: the angle of accumulation dip; ore, coal and wallrock hardness; the content of useful components in the ore or the ash content of coal wallrock; orientation of the near-contact zones on the direction of face movement; gradient of slope and the height of the developed ledge; width of the bench.

It is known that when face flows merge with each other, qualitative ratios averaging takes place in a united mining flow. Along with this, the functions of losses quantities interrelation with impoverishment quantities are also averaged. The nature of these functions is determined by the changing development conditions in faces of the near-contact zones containing mineral resources with the wall rock. This also has a direct effect on the quality of the total mining flow - the content of useful or harmful components. This circumstance makes it very difficult, and in most cases makes it impossible to single out and justify the EU characteristics, which according to the methodological principles of SMR does not allow economically and reasonably to standardize the level of qualitative and quantitative raw materials losses during the development of the near-contact zones.

3. Results and discussion

Face-by-face operational setting of losses and impoverishment of minerals conceptual principles for the conditions of gross development of complex-structure deposits in the formation of a common mining flow of economically justified and regulated quality (useful and harmful component content) are described in [9]. They generally do not contradict to the basic SMR provisions and are aimed to minimize the total operational losses during development of several mine sections (faces). These principles of face-by-face setting are based on a mathematical interpretation of functions of losses quantities interrelation with impoverishment quantities, expressed in terms of the thickness of their layer referred to length unit of the technological surface formed during the development of the appropriate near-contact zone, and their relationship with the values of the bulk mining controlled technological parameters h_l, h_i, Δh , Z, K_l, ΔW (Fig.1). Values of these technological parameters allow to carry out geological survey control and extraction quality level control procedures in the mining process, taking into account the fulfillment of the following conditions:

$$P_o, A_o, a_o = const; L_o \rightarrow min, \quad (1)$$

L_o - total losses of useful mineral resources during working out of extraction blocks (faces), %; P_o - total impoverishment share, %; A_o - harmful components content, %; a_o - useful components content, %.

The optimization calculation method of the total losses minimum in the regulated quality of the mining flow (P_0, A_0 a_0), formed by several extraction units, is based on the use of a well-known method for solving such problems based on the Lagrange optimization function formation.

For the above method, based on the initial data close to the real ones, modeling was performed. Calculations show that the difference of the total losses calculated by the SMR and the present method is very significant in all considered options of the cumulative development of extraction blocks with the regulated impoverishment of the P_o mining flow formed by N faces, with differing mining and geological conditions for their development. In absolute calculus, depending on the of P_o value and the specific conditions of the faces, their difference varies from tenths of a percent to several percent (0.1-5%), in relative terms - from several percent to tens of percent (2-30%). Approximately the same ratios are very likely with the regulated quality of the total mining flow formed by N faces - harmful A_O content (ash content) and a_o useful components.

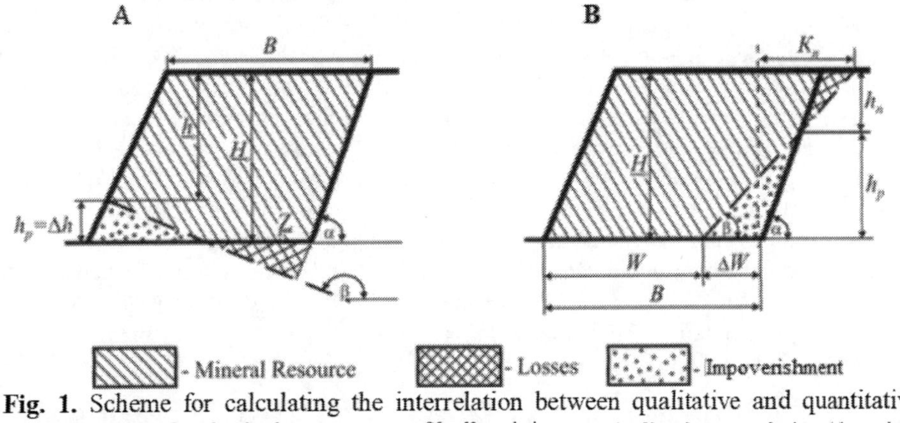

Fig. 1. Scheme for calculating the interrelation between qualitative and quantitative losses including technological parameters of bulk mining: A - inclined accumulation (deposits); B - steep fall accumulations (deposits)

The algorithms of face-by-face optimization of the loss and impoverishment values described in [9] for the regulated quality of the total mining flow in the bulk mining of complex-structure deposits in accordance with the SMR can also be used as an auxiliary tool for setting by the criterion "maximum profit per 1 ton of deplited deposits. Modeling the optimization calculations for the variables P_o, A_o or a_o, we obtain some matrices of the values of the minimum possible total losses and the corresponding impoverishment.

Knowing the ratio of the losses and impoverishment total values, a variant of the minimum possible level of total losses of minerals is determined through the appropriate technical and economic calculations. In addition, it is also possible to optimize (in terms of completeness and quality of extraction) the values of technological parameters that included in the optimization functions: q – bottom-hole (face) productivity, m^3 / unit. time; H - height of the mining ledge (bench), m; B - bench width of the face-to-face sloughing, m; α is the slope angle of the mining ledge, deg.

In [10], the question of determining the optimal height of the mining ledge, taking into account the level of losses for the quarries of "ALROSA", PJSC diamond mining company, was considered. The task of optimizing the height of the mining ledge implies its influence on the damage from loss and impoverishment, with attribution of 1 ton of extracted ore mass. However, in this case, the fact of formation of a common mining flow from several faces and the interdependence of this with the achievement of the goal to minimize the losses of minerals, other things being equal, is not taken into account. The latter approval is also very typical for the requirements (recommendations) of the current regulatory documents on the setting of losses and impoverishment during mining of mineral raw materials.

4 Conclusions

At mining enterprises in modern conditions, in the planning of mining operations and in the management of production processes, during the process of operation digital 3D models of the developed objects are being widely used, constantly replenished and refined, which greatly facilitates and accelerates the very laborious planning and normalization process.

In addition to normalizing the values of losses and impoverishment of mineral resources in the development of near-contact (near-boundary) zones of deposits, from the point of view of management of losses in the quality of mineral raw materials during mining, an important problem is the process of planning the organization of mining operations, taking into account the inhomogeneity of the mineral raw materials harmful and useful components spacing. It is very typical for large complex-structure deposits [1-5, 11, etc.]. Mining and geological conditions high variability and the variability of the content of harmful and useful components of mineral raw materials from large-scale complex-structure deposits presupposes the need to develop for each of them a special system for integrated management of the quality of extracted raw materials. Moreover, in addition to achieving the goal of averaging quality (stabilization), it should be synchronized as much as possible with the process of setting, recording and economic evaluation of losses of solid mineral resources during their extraction. The use of the above-mentioned principles of normalizing operating losses and impoverishment in the aggregate of mining faces in the realization mode of differentiated fields in the field and dynamic operational conditions mode, that operatively take into account the specifics of the conditions for working out specific extraction units will be used in operating quarries and mines to improve the completeness of extracting stocks from the subsoil.

References

1. S. M. Tkach, Methodological and geo technological aspects of increasing of efficiency of mastering of ore and Yakutian placer deposits). Responsible editor: S.A. Batugin; Institute of Mining of the North, the Siberian Division of Russian Academy of Sciences, Yakutsk: Publishing House of Geocryology Institute. p. 284. (2006).

2. S.M. Tkach. Mining Informational and Analytical Bulletin, Problems of integrated development of geo-resources: special issue. **21**. pp. 522-536. (2016).
3. Villaescusa E. Geotechnical Design for Sublevel Open Stoping. CRC Press. 541 p. (2014).
4. Matthews T. Impoverishment and ore loss projections: Strategies and considerations / 2015 SME Annual Conference and Expo and CMA 117th National Western Mining Conference. Mining : Navigating the Global Waters; Denver; United States. pp. 529–532. (2015).
5. Parker H.M. Mining Techn. : Institute of Materials, Minerals and Metallurgy and the ausimm. **121 (3).** pp. 160–176. (2012).
6. K. N. Trubetskoi, E. I. Panfilov. Mine survey and subsoil use. **4**. pp. 3–13. (2015).
7. Typical methodological regulations for the definition, setting, accounting and economic valuation of losses of solid minerals in their extraction (Copy and cartographic enterprises of the All-Union Geological Fund). pp. 224. (1972).
8. Methodological guidelines for the setting, identification and accounting of losses and impoverishment of gold-bearing ore (sands) during mining (Irgiredmet, pp. 265). (1994).
9. S. M. Tkach, Block normalization of losses of minerals with the regulated quality of the miming: Recommendations (Yakutsk Scientific Center, Siberian Division, Russian academy of Sciences). (1992).
10. S. S. Mining, A. N. Akishev, Determination of the optimal height of the mining ledge taking into account the level of losses for deposit quarries "ALROSA", PJSC, Diamond mining company // Mine survey and subsoil use. **6**. pp. 52–55. (2014).
11. E. A. Khoiutanov, V. L. Gavrilov, Management of ash content of complex-structure layer in open mining operations // Mining informational and analytical bulletin, spec. **21.** pp. 163-171 (2016).

Current state and main directions of innovative development of placer gold mining in Far East Federal District

Viktor Litvinsev[1*] and *Petr* Sas[1]

[1]Mining Institute of Far Easten Branch of Russian Academy of Sciences, Khabarovsk, Russia

Abstract Based on the assessment of the resource base of Far East region of Russia, the important role of the mining industry for most regions of Far East Federal District was noted. A decrease in the resource potential of geogenous placers has been established, therefore anthropogenic deposits of placer deposits become an important reserve of placer gold mining. Natural and technological factors that need to be taken into account when selecting and applying those technologies and mining equipment that can reduce their influence are as follows: permafrost rocks, high clayey sand deposits, large content of fine and fine fractions of gold, the presence of high content of heavy fractions of other valuable passing components, etc. A number of developed innovative technologies for development of alluvial deposits, as well as new technological equipment for use on dredges and drags, have been noted. The parameters of the technological processes of the jigging machine on the Alaska-35 washing device (USA) have been analyzed, the matrix of interacting factors (such as the shape of the particles of a mineral, particle size, particle density, height of the bed of the jigging machine, frequency of piston swings, stroke of piston, etc.), the purpose of the study is to obtain graph-analytical dependencies for calculating their rational parameters that ensure effective extraction of metals. **Keywords:** Alluvial deposits, innovative technologies for placer mining, interacting factors, jigging machines.

1 Introduction

Russian Far East has rich and diverse natural resources. In a number of its regions, the mining industry is practically the only sector of the national economy.

The main types of minerals, which are mined in Far East, are gold, platinum, silver, diamonds, tin, lead and zinc, tungsten, coal, boron, fluorite, oil and gas.

Far East takes the second in terms of explored balance reserves of gold, and the first place in Russia for mining from subsoil. In forecasted resources, the leading position currently belongs to core gold (81.3%), with a substantially subordinate value of placer (13.4%) and gold in complex deposits.

---

*Corresponding author: litvinzev@igd.khv.ru

3,305 gold deposits were registered on the territory of the federal district, of which 180 indigenous (including 14 complex) and 3,153 alluvial deposits. The largest number of alluvial deposits is located in Magadan Region (39%) and Republic of Sakha (24.3%), Chukotka Autonomous District - 12.1%, Khabarovsk Region - 10.3% and Amur Region - 9% of the gold deposits [1].

The mining of alluvial gold is a traditional branch of the economy of Russian Far East practically since the development of this territory. For more than a century of history, more than 1,500 tons of precious metal has been extracted from the placers of Far East. The share of gold mining from placers in the total volume of gold mining in Russia was 70-75% at the end of the 20th century, however, at the beginning of the third millennium a definite decline was observed, due both to the development of easily accessible and rich placers, and to a decrease in the volumes of geological exploration. During this period, there was also a significant decrease in the number of prospecting artels, which negatively affected a number of regions of Far East Federal District in the economic and social aspects.

The main resource of placer gold mining in Russia is man-made complexes of geogenous alluvial deposits. Their reserves of rock mass are several billion cubic meters, and the estimated resources in Russia are estimated at over 5,000 tons.

Table 1 presents Far East regions that have a certain number of locations of placer gold, as well as the dynamics of their exploration and development.

The data in Table 1 indicate a very low level of exploration (with the exception of Magadan Region), when the number of recovered placers exceeds the number of explored and registered.

Table 1 - Number of placer gold deposits in the regions of Far East Federal District and the dynamics of their exploration and development

Region	Number of gold placers in Far East Federal District	
	2016 – 2017	2012
Magadan Region	1.242	1.172
Republic of Sakha (Yakutia)	760	771
Chukotka Autonomous District	381	380
Khabarovsk Region	323	342
Amur Region	278	643
Primorsk Region	71	82
Kamchatka Region	49	51
Jewish Autonomous Republic	28	29
Sakhalin Region	21	26
Total	3.153	3.496

2 Current state and analysis of problems in the development of precious metals placer deposits

Given the scarcity of easily accessible geogenous placer deposits with a rich content of large recoverable gold and gold orientation of anthropogenic sets of placers, it is necessary to assess the negative natural factors that should be considered when selecting and applying those technologies and mining equipment that can reduce their impact .

First, the permafrost nature of the rocks of deposits. In the southern part of Russian Far East, placers are almost unaffected by permafrost, although in certain periods some territories fall into year-round cryolithozones, which could not but affect the nature of lithogenesis [2].

Second, the presence of vast fields of argillite, poorly metamorphosed rocks, ore-forming features of the indigenous sources - all this created conditions for the formation of placers containing a significant amount of thin and fine gold and high clay content of host rocks, which is even more so in their anthropogenic objects.

Third, in conditions of a small depth of freezing of rocks (up to 2 meters), the clays of the underlying horizons, both natural and anthropogenic placers, are not affected by frost destruction, retaining their plastic state, which leads to very significant metal losses, not only thin and fine, but also gravitationally enriched, both contained inside the clays and adsorbed from already washed sand by the surfaces of clay "pellets" and carried away to the dump. Only in the Khabarovsk region of such deposits there are up to 30% and only 15% are light-washable.

Fourth, placer deposits with a high content of minerals in the heavy fraction, which in some cases reach 16.4 kg per 1 m3, belong to rebellious placers.

To achieve the optimum development of placer deposits with the use of widely-used washing equipment, it is necessary to identify the most important factors and find the regularities of their comprehensive influence on the choice of rational parameters of technological processes and constructive specifications of the enrichment equipment (Table 2).

Table 2 - Factors influencing the effective development of placer deposits

No.	Factors	Factor description
1	Mining and geological	- Granulometric composition of the sands of placer deposits; - Sieve characteristic of gold; - Morphology of metal; - Presence of associated minerals (hematite, wolframite, magnetite, etc.); - Presence of clay fractions; - Presence of permafrost.
2	Technological	- Possibility of modernization of concentrating equipment parts; - Regulation of parameters of technological processes of concentrating equipment.

At present, only some specific dependencies of the influence of the factors listed in Table 3 on the metal recovery efficiency have been obtained.

At present, in most cases, an open method for the development of alluvial deposits prevails with the use of such domestic washers as PGSh (70%), PKBSh, PBSh (scrubbers - 20%), GGM-5, foreign ones - Alaska-35, etc. The underground method is characteristic mainly for the north of Far East Federal District, it is Chukotka (Riveyem), Yakutia (Kular). Underwater method (dredging technology) is concentrated in Amur region (Solovyovsk mine) and Magadan region (Susumanzoloto). The combined method is mainly open-offshore (Magadan Region) and open-underground (Chukotka) [3, 4].

Machine builders began to focus on increasing the productivity of devices (GGM-3, PPM-5 devices of Magadan Mechanical Plant with a capacity of 100 to 250 m3/h) and complicating their design features in order to increase the extraction of gold, including fine and thin; they started using lapping equipment in the form of centrifugal concentrators, both Russian and foreign. More complete extraction of gold has already become for subsoil users a need for cost-effective development of deposits. However, a relatively high cost of additional equipment, time lost during the rearrangement of complicated devices does not always allow them to be brought into production, and therefore many enterprises have preserved the traditional approaches to gold mining and technological equipment back from the 1960s, which inevitably leads to a high level of metal loss [5].

3 Results

In Institute of Mining of Far East Division of Russian Academy of Sciences, for several years, studies were carried out to create samples of mining equipment that enhance the extraction of gold from placer deposits.

The most effective studies were related to the processing of the efel tails of a dredge on a conical hydro-groove (Solovyovsk mine). A mathematical model was developed for calculating the parameters of a two-phase swirling hydraulic flow of a conical hydro-grate, which is the technological basis for increasing the extraction efficiency of fine and thin gold. The developed sectional design of the hydro-grate is protected by a RF patent.

The prospect of using the dredging technology for the development of anthropogenic placers underwater has been determined, and equipment has been developed to increase the extraction of large fractions of the valuable component from the rock floor of the deposit.

It is possible to give a number of examples of the development of technological equipment in MI FEB RAS, however, the effectiveness of its implementation is weak, since subsoil users are not interested (or do not have the financial and technological resources) to make it in metal, they need, at best, ready samples of mining equipment.

Promising innovative technologies in the laboratory are actively developing those associated with the involvement in operation of anthropogenic sets of placer deposits. This technology is the formation of a productive reservoir of the anthropogenic complex by the impact of non-pressure water flows and the technology of simultaneous development of closely related anthropogenic and geogenic (low-volume or with a small gold content) placer deposits, with the complex extraction of valuable components at a stationary washing device.

In the current period, studies were carried out to assess the complex effect of natural and technological factors on the efficiency of jigging machines. They are an effective technological equipment for extracting various, including small, fractions of gold, however their operation requires high-tech maintenance.

Jigging technology was widely used in 40-60 years of the last century, mainly in the enrichment of coal and placer deposits of tin. In placer gold mining, this technology was occasionally used for finishing operations in the placer gold concentrating site. In the late 1970s, a sharp decline in the development of this field began, one of the main reasons for this being a decrease in the volume of tin mining.

Currently, this technology begins to revive in placer gold mining, but it turned out that, in many ways, the skills and knowledge on the operation of jigging machines were lost. Without knowledge of the operating modes, the conditions for formation of the artificial bed of jigging machine, the rational amount of back water, the magnitude of the piston stroke and other parameters, jigging machines cannot work effectively. Many domestic and foreign manufacturers of jigging machines do not give specific recommendations on their operating modes for particular mineral raw material, because in the enrichment of coal, placer gold and other minerals, in each case there must be special operating modes of the jigging machine.

A great contribution in the field of research of jigging technology was made in the 20th century by Soviet and foreign scientists: Kizevalter, Fomenko, Kuznetsov, Rafales-Lamark, Verkhovsky, Margolin, Fomin, Bochkovsky, Kachan, Samylin, Polkinin, Bert, Taggart, Mayer, Yoshida, Tsimmerman.

As a foreign analogue of jigging machines, it is known to use a jigging machine IHC (Holland), when operating it, about 90% of the sands are released in tailings, but small fractions of gold were also extracted [6, 7].

Questions of the theory and practice of jigging are most fully developed for the separation of large and medium fractions of minerals. The rational parameters of deposition of fine and thin fractions have not been substantiated, either practically or theoretically. Therefore, the purpose of the studies performed was to reveal the patterns of influence of various mineralogical and technological factors on the efficiency of gold recovery using jigging machines of Alaska-35 (US) industrial equipment. Such factors include: the ratio of solid-liquid of super-plate water, the amount of sub-surface water, the frequency and stroke of the piston, the thickness of the bed of the jigging machine, the density of the bed material. The granulometric composition of the mineral mass of sands also has a significant effect.

At the first stage of the study, only a small number of factors that have the greatest impact on the jigging process were examined: this is the mineral density, the amount of subsurface water, the thickness of the jig-bed, the frequency and the stroke of the working piston. The following results were obtained in the study of the parameters of the jigging machine operation using Alaska-35 washing device:

a) Influence of volumes of processed sands on the efficiency of jigging

It has been found that with increasing load, the recovery of the useful component and the yield of the concentrate decreases, since the bed density in the jigging machine increases due to the saturation of its pores with grains of heavy minerals.

b) Influence of volumes of top and back water

With a high density of minerals on the bed of the jigging machine, it is necessary to increase the ratio of solid to liquid (S:L) of the pulp from 1:1 to 1:4. At a higher S:L, a reduction in the recovery of mineral is observed. The supply of back water significantly affects the amount of metal loss. With the help of back water, it is possible to increase the ascending velocity and reduce the down-flow velocity, and to slow the movement of particles from the upper light layers of the bed to the drain threshold of the machine. Thus, the back water promotes an increase in looseness of the bed in an upward flow and a decrease in losses of small classes of gold in a downward flow.

c) Effect of bed height on the jigging process

Bedding is the most important technological factor of the jigging process. Only thanks to the selection of bed material (by density, by particle size), it is possible to achieve selective separation of the mixture and, ultimately, to achieve a high recovery of the useful component.

The results of experimental studies (ten experiments) made it possible to obtain an equation for the dependence of the yield of the undersize on the height of the bed at a constant value of the piston oscillation. The sought equation of quadratic regression has the form:

$$Y = 0.133B^2 - 0.74B + 11.6, \qquad (1)$$

where Y – yield of the undersize, %;

B – bed height, mm.

Only a small part of the interacting factors was analyzed, so in 2017 the research was continued.

In 2017, experimental work was carried out on a laboratory jigging machine produced by TsNIGRI (Central Geological Research Institute for Nonferrous and Precious Metals). The jigging machine consists of 2 chambers, each 100 cm^2 area. The stroke length is adjustable from 1.5 mm to 12 mm, with 8 fixed positions, has 3 vibration frequencies - 180 rpm, 200 rpm and 240 rpm.

4 Description of experimental works

A total sample weighing 18 kg was divided into three parts, then averaged. The average mass of the sample was 2,200 g, the density was 1.7 g / cm^3, the amount of top water was 1 l / min, the amount of back water in chamber 1 was 2 l / min, in chamber 2 - 1 l / min. The time of processing the sample on the jigging machine averaged 25-30 minutes.

Under the influence of the upward flow of water, the mineral mixture is loosened and light-mineral particles with velocity of fall is less than the upward flow rate rise into the upper layers of the material. Particles of the heavy mineral are in this case in a suspended state, and then in the downward flow they occupy the bottom layer.

Tables 3 and 4 show the extraction of gold on the jigging machine when the stroke of the piston is changed at a frequency of 240 rpm and 200 rpm, respectively.

From these tables it is possible to draw a preliminary conclusion that as the piston stroke frequency decreases, the metal losses decrease, the concentration increases in chamber 1, in the oversize and undersize. The stroke of piston in the actual experiments did not have a significant effect, but it is required to conduct repeated studies. It is necessary to continue laboratory work in order to identify the main factors affecting the jigging process.

5 Conclusions

A characteristic of the resource base of Russian Far East is given, an important role of the mining industry for the majority of the regions of Far East Federal District is noted. It is established that an important reserve of placer gold mining is anthropogenic sets of placer deposits.

An assessment of negative natural factors that should be taken into account when selecting and applying those technologies and mining equipment that can reduce their influence has been made – namely, the permafrost of rocks, high clayey sand of placers and a large content of fine and thin fractions of gold, high content of heavy fractions of other valuable passing components, etc.

The availability of developed innovative technologies for the use of placer deposits, as well as new technological equipment for use in dredges and drag was reflected.

Table 3. Extraction of gold on the jigging machine with a change in the stroke of piston at a frequency of 240 rpm

Piston stroke, mm	No. of sample	Mass of sample, g	Yield of Au, %					Total Au, mg
			Top chamber 1	Top chamber 2	Underflow chamber 1	Underflow chamber 2	Jig tailings	
1.5	P-1	2,371.5	55.16	19.66	18.63	4.02	2.53	20,40
3	P-2	2,210	5.82	1.28	82.47	4.07	6.36	12,89
4	P-3	2,200	66.62	6.30	26.56	0.07	0.45	22,22
5	P-4	2,238.5	50.32	10.23	39.45	0.00	0.00	11,73
6	P-5	2,294	75.07	6.19	18.74	0.00	0.00	32,64
8	P-6	2,325	58.73	0.47	40.80	0.00	0.00	21,40
10	P-7	2,380	79.84	0.00	20.16	0.00	0.00	33,65
12	P-8	2,160	56.11	0.00	43.89	0.00	0.00	13,01
Total		**18,179**	–	–	–	–	–	167.92

Table 4. Extraction of gold on the jigging machine with a change in the stroke of piston at a frequency of 200 rpm

Piston stroke, mm	No. of sample	Mass of sample, g	Yield of Au, %					Total Au, mg
			Top chamber 1	Top chamber 2	Underflow chamber 1	Underflow chamber 2	Jig tailings	
1.5	E-1	2,222	73.85	0.00	10.80	0.00	0.00	41,3
3	E-2	2,227.5	64.27	0.00	35.73	0.00	0.00	38,9
4	E-3	2,228	75.59	0.00	24.41	0.00	0.00	51,2
5	E-4	2,223	60.71	0.00	39.29	0.00	0.00	50,9
6	E-5	2,226.5	69.26	0.00	30.74	0.00	0.00	61,15
8	E-6	2,228.5	51.51	0.00	48.59	0.00	0.00	40,85
10	E-7	2,149.5	66.19	0.00	33.81	0.00	0.00	84,3
12	E-8	1,990.5	70.27	0.00	29.73	0.00	0.00	50,45
Total		**17,495.5**	–	– –	– –			419.05

The parameters of the technological processes of the jigging machine with Alaska-35 washing device have been analyzed, the matrix of interacting factors (such as the shape of the mineral particles, the particle size, the particle density, the height of the jig bed, the oscillation frequency of the piston, the piston stroke), was made, studying which will make it possible to obtain graphically-analytical dependences of the calculation of their rational parameters that ensure effective extraction of the metal.

References

1. S.G. Kashuba, V.N. Ivanov. Gold and Technology. **3 (37).** pp. 6 - 9. (2017).
2. V.S. Litvintsev, A.P. Van-Van-Ye, Aleksandrova, TN, Korneeva, SI. A separate article of the Mountain Information-Analytical Bulletin (scientific and technical journal). **3**. 12 p. (2012).
3. Business Directory Far East. Gold of Far East and the Baikal region in 2016. Retrieved at: **minvr.ru**/upload/iblock/40f/Декларация-2017.pdf
4. R.S. Seryi, P/P. Sas, X Congress of Concentrators of the CIS countries. Collection of Materialov. **2**. P. 421-424. (2015).
5. Z.G. Mirzekhanova, G.S. Mirzekhanov, V.S. Litvintsev, Retrospective Analysis of the Formation of Resource-Ecological Problems in the Development of Alluvial Gold Deposits / Pacific Geology. **2**. P. 95-108. (2016).
6. Ackles, D. Gold is where you find it. Authorhouse. Includes an account of the effective use of innovative IHC jigs at Gold Dust Creek in Alaska. 116 p. (2006).
7. Ackels, D., Madonna J. Placer Mining in Alaska with IHC jigs. Alaskan Prospectors and Miners News, Fall Edition. 9 p. (1985).

Integration of production as a way to improve the financial stability of non-ferrous metallurgy enterprises in the Far Eastern region of Russia

Yuliya Arkhipova[1*]

[1]Mining Institute of Far eastern branch of Russian Academy of Sciences, Khabarovsk, Russia.

Abstract. Article is devoted to a question of financial and economic stability of the enterprises. Uncertainty and instability of the external environment can lead to a slowdown in economic growth, to a decline in production and reduction of investment. In recent years, problems of financial instability and bankruptcy have grown into a category of economic security of the country. Annually for these reasons, thousands of businesses cease operations, which has a negative impact on the social and economic situation of the regions. The article examines the theoretical and methodological tools of financial stability of economic entities as an indicator of the effectiveness of the financial management system in commercial organizations. This is ensured by permanent solvency by maintaining a sufficient level of liquidity of assets (finished goods, receivables, short-term financial investments, etc.). The author made attempts to create a methodology for assessing the financial and economic sustainability of mining companies in the Far Eastern region that are engaged in the extraction and processing of non-ferrous metal ores in order to avoid their bankruptcy and as an exit - merger into a vertically integrated company. This will allow to connect the technological cycle at individual enterprises: from the extraction of raw materials and its processing to the production of high-value products and to increase their competitiveness both on the Russian market and on the world market.

1 Introduction

Over the ten-year period in non-ferrous metallurgy there are such processes as concentration, integration and diversification. The active formation of holding structures is continuing, they control a significant part of nonferrous metallurgy and are major producers and exporters of metals [1]. The resulting horizontal and vertical integration has united both a multitude of domestic metallurgical enterprises and their suppliers of raw materials and fuel. Except all pluses and minuses of this phenomenon, it has defined independence of the enterprises of metallurgical branch in relation to the global oligopolistic companies, which control the international market of resources, and respectively own instruments of influence

* Corresponding author: arhipova@igd.knv.ru

on the general price policy, for example, BHB Billiton, CVRD and Rio Tinto which produce 35% of iron ore raw materials in the world [2, 3].

The optimal provision of resources in the holding means satisfying the own needs of the enterprises entering into it, the availability of technological stocks of raw materials and fuel. A particularly significant process is integration in the production of nonferrous metals. According to 2014 - 2015 years, more than 92% of the proceeds of nonferrous metallurgy are generated by four enterprises: Norilsk Nickel, RusAl, SUAL and UGMC. The analysis of international experience has shown that in nonferrous metallurgy integration by the vertical technological principle is most effective. It allows to connect together a production cycle at the separate enterprises: from extraction of raw materials and its processing and to increase their competitiveness to production of advanced processing in the international markets. In a vertically integrated company, efficiency is achieved by reducing the throughput unit costs for the production of final products (usually in the formation of structures that combine the enterprises from the extraction of raw materials to the production of the third and fourth redistribution) and saving costs associated with the use of third-party contractors and intermediaries. There is a decrease in transaction costs (management costs, loss from adaptation, the cost of finding an advantageous partner, etc.) and reducing the indirect taxation of intermediate products, hence the cost of the final product is usually lower than that of independent competitors that are forced to buy raw materials from the supplier always at affordable prices. Therefore, consolidation in a single vertically integrated structure neutralizes the dependence of metallurgists on suppliers of nonferrous metals concentrates.

Vertical integration (inter-regional) in the mineral sector with the participation of the Far East was particularly evident in nonferrous metallurgy, where for many decades only primary redistribution (extraction and enrichment) of raw materials was developed, and the main products were ores and concentrates. This was especially significant in the tin industry: practically all the stocks of Russian tin are concentrated in the Far East (Yakutia, Khabarovsk and Primorsky Krai, the Jewish Autonomous Region), and practically the only refining enterprise in Russia - the Novosibirsk Tin Plant - is located far beyond the region. The Far Eastern tin-mining enterprises and the combine were units of a single technological chain within the framework of a centrally managed unified national economic complex.

2 Materials and methods

Currently, the financial and economic analysis at the metallurgical enterprises is mainly on the study of production and financial reporting, which concludes with an assessment of the risk of bankruptcy. However, the results obtained are often used only informatively in connection with the insufficient elaboration of the algorithms for further management actions. This fact does not allow the maximum use of the information received and deprives many interested persons of the opportunity to manage this type of risk. Risk management will allow the enterprise not only to avoid the state of bankruptcy, but at the same time to increase its competitiveness, investment attractiveness and solvency. The enterprise can effectively develop when it gets net profit which can be reinvested then in development of capacities and the general potential.

Enterprises with the highest risk of bankruptcy can hope only for the help of state authorities, since there is no way for them to fulfill their obligations in any way. Reducing the risk of bankruptcy shows an improvement in the financial condition and the ability to pay off on its obligations with its assets.

Thus, the analysis of the financial and economic situation is the most important aspect and one of the effective ways to assess the current situation. It reflects the instantaneous state of the economic situation and allows you to highlight the most difficult problems of managing available resources and minimize efforts to align the organization's goals and resources with the needs and capabilities of the established market. Analysis of the financial condition is a tool for identifying management problems, it is the choice of the directions of investing capital and forecasting individual indicators, and therefore one can make effective decisions in various areas of the economic activity of the enterprise.

In Russian practice, the analysis of "business activity" is the most popular. This concept in financial activity is defined as the whole range of activities that are aimed at promoting these enterprises in all areas: the sales market, financial activities, the labor market, etc. The increase in business activity is manifested in the expansion of the service sector or the sales market, the increase in the range of goods and services and its successful implementation, the stable development of the staff of the enterprise, the efficiency of the use of the entire resource base (finance, personnel, raw materials). The levels and dynamic of certain "financial coefficients" are analyzed, and are indicators of the achieved results in the activities of the enterprise.

Previous studies [4, 5, etc.] suggest that to date, a variety of methods and techniques have been developed to assess financial sustainability. They are both generally accepted and authoritative, based on the consideration of this economic category as a multidimensional phenomenon. Its take into account different aspects, in particular by assessing the impact on domestic financial stability (liquidity of assets and solvency, capital structure and liabilities, cash flow uniformity, tax costs) and external factors (inflation, exchange rate changes, etc.).

3 Results and discussion

On the example of the selected mining companies of the Far Eastern region, engaged in mining and processing of nonferrous metal ores, an analysis of their stability was carried out. Were studied financial-economic and accounting statements for the three-year period (2012-2014) of the following enterprises: JSC Dalpolymetall MMC, LLC Pravoormyskoe, JSC Tin Ore Company (JSC ORK), PJSC Rusolovo. Were calculated absolute and relative indicators. The most significant coefficients were identified from the set of existing ones (absolute liquidity, fast, current, total solvency, turnover of capital, working capital, turnover of accounts receivable and accounts payable, etc.), which, in the opinion of the author, more fully characterize the financial situation from the point of view of bankruptcy diagnostics. At the same time, the analysis was carried out in such a manner that the solvency indices were first considered, then - "business activity" and then, according to method the A.A. Trifilova [6] determined financial stability. So, for example, if absolute financial stability is determined for PJSC Rusolovo, then individual solvency ratios and "business activity" reflect unfavorable data [7].

Activity of the companies is profitable (net profit) on all analyzed interval except for 2013 of JSC Dalpolimetall MMC when the loss in the sum of 5.5 million rubles has been received. At the same time, it is not typical for the organization to maintain stable growth rates of the main financial indicators, both in general and for other enterprises, for which negative trends are noted and negative indicators are observed.

Due to the ambiguity of results, we will continue researches at the micro level. Each enterprise demands individual approach to carrying out the analysis of a financial state. The determining factor of the specificity of the activity is the branch belonging of the analyzed object. It was suggested that the use of a multilevel (structured) approach would allow not only to obtain an analytical assessment of the financial state of the economic entity's activity in a specific time interval, but also to compare the performance of business activities with the indicators of the mining industry (DPI) and of the Russian Federation as a whole. Deviations in values and competitive advantages with similar leaders in the field were identified. Comparative analysis of financial and economic performance indicators of enterprises taking into account the aspect of hierarchical structuring of the economy will allow to formulate conclusions about the development trends of enterprises as a link of the type of economic activity - DPI in the Russian Federation (RF). Table 1 compares the indicators with the performance of small (including micro) and medium-sized enterprises in general for the Russian Federation, and data on enterprises in the field of mining industry. This will allow performing complex monitoring and control of the financial condition of the enterprise and monitoring negative changes in key economic parameters from the perspective of continuing operations in the future. The analysis was carried out by three companies: LLC Pravoormyskoe, JSC ORK and PJSC Rusolovo. They have the status of small and medium enterprises. JSC Dalpolimetall MMC does not participate in this analysis, since it belongs to the category of large ones.

Investigated enterprises are active participants in the market of marketable products in the sphere of DPI, but economic data indicate a small share of their importance. Let's consider the results of each of them.

LLC Pravoormyskoe. The bulk of the indicators are in an unfavorable range, comparing them with the indicators of the RF and of the DPI. The exception is the profitability of products, which is more than 5 times higher than the indicator of the Russian Federation and 2.3 times of the DPI. Return on capital and reserves is the most important factor in assessing the financial condition of an enterprise and the level of investment attractiveness, and also demonstrates inefficient use of own funds. The coefficient of autonomy shows how the organization is independent of creditors. The smaller the value, the more the organization is dependent on borrowed sources, the less stable the situation. It is 2.5% with a value of 17% for the industry DPI. The enterprise has too little opportunity to extinguish debts at its own expense.

JSC ORK. The only positive thing is the excess of the current liquidity ratio in comparison with industry-wide, other indicators show low values. A similar situation exists in the *PJSC "Rusolovo"*.

Since none of the analyzed enterprises have absolute stability and some of them are in the risk zone, the next step is to evaluate the effectiveness of the creation of a vertically integrated company (VIC) that unites these economic entities. In this case, we include in the calculations of JSC Dalpolimetall MMC due to unfavorable economic data [7]. The effect of the integration results was evaluated from the point of view of maximizing profit and growth in the capitalization of the association, providing an increase in the business value, as well as an increase in the selected VIC financial indicators after the merger of enterprises compared to the sum of these indicators before consolidation (tab. 2). Estimated data showed that the indicators of the possibility (impossibility) of breakeven activity and financial stability are in a favorable range, while the company's share is significantly increased in comparison with the data on the DPI in the Russian Federation.

Calculation and the analysis of relative indicators supplements results of holding analytical procedures of the reporting of VIK (tab. 3).

Usually high growth rates of accounts receivable testify that the enterprise actively uses the strategy of commodity loans for consumers of its products. By crediting them, the enterprise actually shares with them a part of its income. When payments are delayed, the enterprise is forced to take loans to ensure its economic activities, increasing its own accounts payable. In this case, the ratio of receivables shows a decrease, which is a favorable factor. Also decrease in growth rate of accounts payable is observed that reduces risk of insolvency. Coefficient of restoration of solvency more than 1, i.e. the company in the short term will be able to restore solvency, and the coefficient of loss of solvency (more than 1) confirms it.

Table 1. Grouping of indicators of financial and economic activity of enterprises in the system of economic analysis

Index	Values by enterprises for 2014.			RF	Share of enterprises in Russia,%			DPI	Share of the enterprises in DPI, %		
	LLC Pravoormyskoe	JSC ORK	PJSC Rusolovo		LLC Pravoormyskoe	JSC ORK	PJSC Rusolovo		LLC Pravoormyskoe	JSC ORK	PJSC Rusolovo
Basic data, million rubles											
Current assets	4770.2	877.6	6539	4265000	0.112	0.021	0.153	72000	6.63	1.22	9.08
Non-current assets	831.8	247.6	269.3	1833900	0.045	0.014	0.015	65100	1.28	0.38	0.41
Assets	5602	1125.2	6808.3	6098900	0.092	0.018	0.112	137100	4.09	0.82	4.97
Capital and reserves	142.8	80.1	3126.5	1222000	0.012	0.007	0.256	21200	0.67	0.38	14.75
Revenues from sales	240.3	3.1	10	5027800	0.005	0	0	103300	0.23	0	0.01
Short-term liabilities	1557.5	588.6	922	3624100	0.043	0.016	0.025	72000	2.16	0.82	1.28
Financial result (profit / loss)	0.41	0.003	0.34	143135	0	0	0	3652	0.01	0	0.01
Indicators of the possibility (impossibility) of break-even activity,%											
Profitability of production	24.76	-68.4	-	4.6	5.38	-14.9	-	10.6	2.34	-6.46	-
Return On Assets (ROA)	0.01	0	0.005	2.3	0	0	0.002	2.9	0	0	0
Profitability of non-current assets	0.05	0.001	0.126	7.8	0.01	0	0.016	5.6	0.01	0	0.02
Profitability of current assets	0.01	0	0.005	3.4	0	0	0.002	6.1	0	0	0
Return On Capital and Reserves	0.28	0.004	0.011	11.7	0.02	0	0.001	17.3	0.02	0	0
Profitability of long-term liabilities	0.01	0.001	0.012	11.4	0	0	0.001	11.6	0	0	0
Profitability of short-term liabilities	0.03	0.001	0.037	3.9	0.01	0	0.009	5.1	0.01	0	0.01
Indicators of financial stability,%											
Coefficient of current liquidity	303.8	149.2	709.2	117.7	2.58	1.27	6.03	82.5	3.68	1.81	8.6
Coefficient of autonomy	2.5	7.1	45.9	20	0.13	0.36	2.30	17	0.15	0.42	2.7

* Compiled by the author according to Rosstat [8]

Table 2. Indicators of financial and economic activities of a vertically integrated company

Index	The value of the VIC indicator	RF	Share of the VIC, %	Deviation from the RF data	According to the RF DPI	Share of the VIC, %	Deviation DPI from the RF data
Basic data, million rubles							
Current assets	17602.2	4265000	0.413	4247397.8	72000	24.45	54397.8
Non-current assets	3157.1	1833900	0.172	1830742.9	65100	4.85	61942.9
Assets	20759.3	6098900	0.34	6078140.7	137100	15.14	116340.7
Capital and reserves	3435.8	1222000	0.281	1218564.2	21200	16.21	17764.2
Revenues from sales	2075.3	5027800	0.041	5025724.6	103300	2.01	101224.6
Short-term liabilities	5118.9	3624100	0.141	3618981.1	72000	7.11	66881.1
Financial result (profit / loss)	990.5	143135	0.692	142144.5	3652	27.12	2661.5
Indicators of the possibility (impossibility) of break-even activity,%							
Profitability of production	15.28	4.6	3.32	-10.68	10.6	1.44	-4.68
Return On Assets (ROA)	4.77	2.3	2.07	-2.47	2.9	1.65	-1.87
Profitability of non-current assets	31.37	7.8	4.02	-23.57	5.6	5.6	-25.77
Profitability of current assets	5.63	3.4	1.66	-2.23	6.1	0.92	0.47
Return On Capital and Reserves	28.83	11.7	2.46	-17.13	17.3	1.67	-11.53
Profitability of long-term liabilities	8.12	11.4	0.71	3.28	11.6	0.7	3.48
Profitability of short-term liabilities	19.35	3.9	4.96	-15.45	5.1	3.79	-14.25
Indicators of financial stability,%							
Coefficient of current liquidity	343.69	117.7	2.92	-225.99	82.5	4.17	-261.19
Coefficient of autonomy	16.55	20	0.83	3.45	17	0.97	0.45

* Compiled by the author according to Rosstat [8]

Table 3. Analysis of satisfactory (unsatisfactory) structure of balance of VIK

Index	Value indicator			Deviation, +/-		Growth rate, %	
	2012	2013	2014	2013-2012	2014-2013	2013/2012	2014/2013
Ratio of changes in receivables, units.	-	1.9	1.31	-	-0.59	-	0.69
Ratio of changes in accounts payable, units.	-	2.88	0.88	-	-2	-	0.3
Current liquidity ratio at the end of the year	7.97	4.37	3.44	-3.6	-0.93	0.55	0.79
Coefficient of restoration of solvency	-	2.57	2.97	-	0.4	-	1.16
Coefficient of loss of solvency	-	3.47	2.33	-	-1.14	-	0.67

The next step in the study is to assess the risk of bankruptcy. To do this, we use the logit model of the author [9], which was developed specifically for metallurgical enterprises. The logistic model for "large" enterprises includes 6 indicators, which were previously calculated: the coefficient of long-term borrowing (K_2) - 0.78; return on assets (K_5) - 0.15; total profitability (K_4) - 0.57; net profit rate (K_6) - 0.48; the autonomy ratio (K_1) is 0.17; the ratio of debt to equity (K_3) is 5.04.

The total logistic model has the following appearance:

$$Z = -(0.37 + 3.3 * K_1 - 4.7 * K_2 + 0.5 * K_3 + 1.3 * K_4 - 4.23 * K_5 + 0.15 * K_6) \quad (1)$$

where: Z - probability of bankruptcy of the enterprise.

On the scale of bankruptcy risk levels: if Z is in the range from 0 to 0.3 - low risk, from 0.3 to 0.6 - medium level and from 0.6 to 1 - high. In our case, Z = 0.01, which confirms the low probability of a crisis, and as a result, based on the analysis, we can conclude that the effectiveness of combining these enterprises in the VIC.

4 Conclusion

The problem of financial instability and bankruptcy of enterprises in recent years has grown into a category of problems of economic security of the country. Annually on all districts and regions of the Russian Federation for these reasons thousands of commercial structures stop activity that exerts impact on stability of economic and social situation of regions. And therefore strengthening of internal and external control is required. It is necessary to take measures of administrative behavior and control of change of a financial and economic situation in separately taken enterprise. In this paper, attempts are made to create a methodology for assessing the financial and economic sustainability of enterprises in order to avoid their bankruptcy. And as an output - association in vertically-integrated company. The procedure can be used: 1) financial and economic services and internal management system in order to prevent bankruptcy of mining and metallurgical enterprises; 2) executive authorities, Department of Mineral Resources, Federal Tax Service, Federal service of Russia on financial improvement and bankruptcy, departments of the industry; 3) the Russian and foreign investors – for the purpose of assessment of investment attractiveness of the domestic mining and metallurgical enterprises; 4) Russian commercial banks to assess the creditworthiness, solvency and the possibility of bankruptcy of borrowers; 5) rating and consulting agencies - for the preparation of analytical reviews.

References

1. I.A. Kradenykh. Problems and prospects of complex development and preservation of bowels of the earth: Theses of reports of the II International school of sciences of the academician K.N. Trubetskoy, 342, (2016).
2. Lehamn Brothers. The Iron Age. European Metals & Mining. Industry Overview, **64**, (2008).
3. Yu.A. Arkhipova. Formation of vertically integrated production in the mining and metallurgical industry (on the example of the Far Eastern region of the Russian Federation), (2006).
4. T.V. Baronina, S.A. Karpovskaya. Management of economic systems: online scientific magazine, (2011).
5. A.S. Dadayan, A.A. Gurdzhiyan. Economy and business: theory and practice, **9**, 42-44, (2015).
6. A.A. Trifilova. Innovations, **6**, 67 (2003).
7. Yu.A. Arkhipova. Mining Information and Analytical Bulletin, **21**, 374, (2016).
8. *Small and medium-sized business in Russia*, **96**, (2015).
9. V.E. Rygin. Bulletin of the SRSTU (NPI), **5**, 84, (2013).

Method of quantitative assessment of the regularities of natural restoration of biota in zones of technogenic disturbance by extractive enterprises

Jury Galchenko[1*] and *Julia* Ozaryan[2]

[1]Institute of Comprehensive Exploitation of Mineral Resources Russian Academy of Sciences, Moscow, Russia
[2]Mining Institute of Far eastern branch of Russian Academy of Sciences, Khabarovsk, Russia

Abstract. The paper presents the results of field studies of tendencies in plant communities self-regeneration processes in the zone of their technogenic disturbance and on the surfaces of rock spoil heaps. It has been established that in the former case the key factor determining the nature of development of self-regeneration processes is the correspondence of the width of the transitional community to the length of transfer of seeds of primary plant community edificatory and assectator species. In the latter case, the key self-regeneration succession sere passes the bifurcation point at a very early stage of its development.

1 Introduction

The contemporary stage of development of the conflict between humans and the Earth's natural biota is characterised with gradual "greening" of mentality. Both the harmfulness of uncontrolled development of the technocratic civilisation and total futility of biocentrism under the slogan "back to nature" have been acknowledged. The growing contradictions between intellect and nature generated by many centuries-long dominance of the ideology of unlimited human consumption of natural resources inevitably lead to rapid degradation of the Earth's natural biota to the extent of its total destruction. The obvious danger of such processes made this problem a top priority of further development of the entire human civilisation.

The growing understanding of the fact that environmental protection and conservation in the current technocratic civilisation development model is a fight against consequences rather than the root causes of the problem calls for an urgent need for radical revision of the principles of interaction between the technological and biological spheres.

In respect of exploration of mineral resources, especially new mineral deposits situated on territories still unaffected by human impact, the problem of development of principles of design, construction and operation of utility infrastructure facilities of extractive enterprises acquires a no lesser (and perhaps even greater) importance, as it enables to maintain within

* Corresponding author: schtrek33@mail.ru

the biota tolerance boundaries all forms of anthropogenic burden caused by compact human habitation.

At the same time, for mining enterprises whose period of existence is limited by the exhaustibility of its commercial resources, the specifics of regeneration succession in the post-operational period are of particular interest. The research of regeneration of arboreal species in different parts of the territory under survey in the post-operational period provides a general picture of the tendencies in this process.

It is necessary to address two types of tasks:

- development of demutation processes in the zone of technogenic degradation of primary ecosystems of the natural biota;
- development of natural regeneration succession on the surfaces of technology-induced neoformations (primarily rock spoil heaps).

2 Methodology and objects of research

The new approaches were built on principles of sustainable development, which are interpreted in our country as the strategy of sustainable development of the environment and society [1]. The research of plant communities self-regeneration processes was conducted in the zone of technogenic disturbance adjoining the industrial site of an underground mine with the annual capacity of over 1 million tonnes of ore, shut down more than 25 years ago [1]. As well as in a quarry for the extraction of building materials, where there are piles of overburden, where the restoration of vegetation occurs from five to fifty years [2]. The modern phytocoenotic landscapes surveillance methodology consisted in route survey of a mountain river basin and description of phytoecological profiles underlying the valley and adjacent slopes. Geobotanical pictures of the forest association were made and the undergrowth of arboreal species was measured at 2x50 m discount plots. The distances between profiles varied from 2.5 km to 4 km. Secondary profiles were underlying the valleys of some tributary streams [3-5].

3 Results and discussion

According to the findings of this research, the channel pebbled beach and the second bottom of the valley were planted by Populus maximovichzii, Henri, Chosenia macrolepis, Turaz, Fraxinus mandshurica, Rupr, Aser mono, Rupr, Acer mandshuricum Maxim, Suringa amurensis, Rupr, Alnus hirsuta, Turcz, and some other arboreal species. Judging by the nature of regeneration of different species, the natural succession process in this standard natural zone is characterised with an increasing role of Fraxinus mandshurica, Rupr, Acer mandshuricum Maxim, and Aser mono, Rupr, which are gradually superceding Populus maximovichzii, Henri, and Chosenia macrolepis, Turaz, in the edificator synusia.

The integrated effect of anthropogenic objects on the entire territory of the rest of the valley resulted in the formation of a totally different (as compared to the initial) vegetation complex. The forest cover on both slopes of the valley contained only two associations (or types) of planted vegetation deriving from the initial woodland:

- In the bottom section of the slope, it is multi-species deciduous forest consisting of Betula mandshurica Na Kai, Betula costata, Trautv, Populus tremula, L., Tilia amurensis, Kom, Aser mono, Rupr, and Acer mandshuricum Maxim, with insignificant additions of species of the initial forest stand. Abies holophulla, Maxim., is represented by single overmatured trees and rare groups of tall undergrowth. There are also single species of overmatured Pinus koraiensis, Sieb. et Zucc, with pronounced top drying.

- The top section and the ridge are also covered by Kalopānax septemlōbus, Micromeles alnifolia and Betula schmidtii. This means that the initial three-storeyed mixed coniferous-deciduous forest has transformed into a polydominant community of small-leaved deciduous species.

The structure of undergrowth on most of its area fully corresponds to the known specifics of degradation-and-demunation successions caused by disturbances of initial coniferous- deciduous forest stand. Aser mono, Rupr (5,000 species/hectare), Acer mandshuricum Maxim (6,000 species/hectare), Acer pseudosieboldianum, Kom. (3,500 species/hectare), and Tilia amurensis, Kom. (2,800 species/hectare) regenerate better than other species. Pyrogenic species – Betula mandshurica Na Kai, and Populus tremula, L. – hardly regenerate at all. There is practically no coniferous understory of young regeneration either and little hope for its emergence in the future. All this is an indication that after technological and anthropogenic burdens are lifted further succession will most probably be developing towards the formation of multi-species forest stand on this territory with the prevalence of broadleaved species. In other words, the changes in initial biota caused by durable operation of an extractive enterprise have largely acquired an irreversible nature.

The process of regeneration of arboreal species in the zone of transitional community between second-growth forest and part of initial forest remaining in areas with difficult access is totally different. Owing to technological causes of its formation, the border between mature forest stand of these two communities is rather well pronounced and a zone of transition of properties for this part of the plant community is practically non-existent. However, it is quite visible in the case of undergrowth of the main forest forming species. All the remaining primary biota areas in this ecosystem are surrounded by a strip (ecotone) which, in addition to the regenerated second-growth forest community species, also has occasional species of the standard forest-forming community (fig. 1). The intensity of this process for all these species declines as the distance from the mother plants increases depending on the seeds transfer mechanisms. The identified trends in changes of the undergrowth density and its species composition on territories of second-growth forest communities adjacent to the remaining areas of primary plant communities show that the distance of transfer of seeds of the main forest-forming species can be used as one of the criteria of regulation of the sizes of singular land allotments during the formation of infrastructure of an extractive enterprise's surface complex.

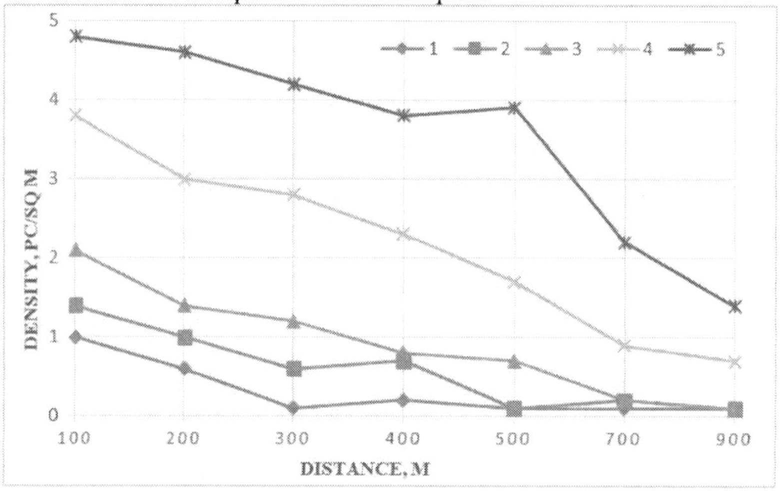

Fig. 1. Dynamics of the main forest-forming species undergrowth density within the zone of the ecotone between the standard and second-growth forest communities (1– Pinus koraiensis, Sieb. et Zucc; 2 –Jurglans manshuriga Maxim.; 3 – Abies holophulla, Maxim.; 4 – Acer; 5 – Betula costata, Trautv.)

The Korfovsky Rock Querry is situated on the borderline of the Bolshoy Khekhtsyr State Nature Reserve, in connection with which the flora and fauna of the adjacent territories are quite diverse. It has a wide representation of various types of deciduous, mixed coniferous-deciduous, and dark coniferous forests. The species composition of the region numbers over 1000 vascular plants.

As the territory of the mineral block allocation used to be covered with forest vegetation, it is obvious that it must be regenerated through environmental measures. This calls for a need to explore the processes of self-regeneration of the territories disturbed in the process of solid waste accumulation of mining production. In addition, the examination of processes of natural regeneration of disturbed lands is important for establishing the levels of tolerance and assessment of resilience of organisms and their self-regeneration potential.

The soils of rock spoil heaps are generally potentially fertile. Moreover, longer periods of overburden rock storage in the spoil heaps lead to improvement of physical and mechanical characteristics and structure of the soils.

It should be mentioned that 1–2 years after the formation of a spoil heap its surface becomes covered with pioneer grassland vegetation represented by 2–3 species, most frequently Artemísia vulgáris and Trifolium hybridium L. [6]

The research of demutation processes was conducted at rock spoil heaps during the survey of their vegetative cover (figure 2, tab.1).

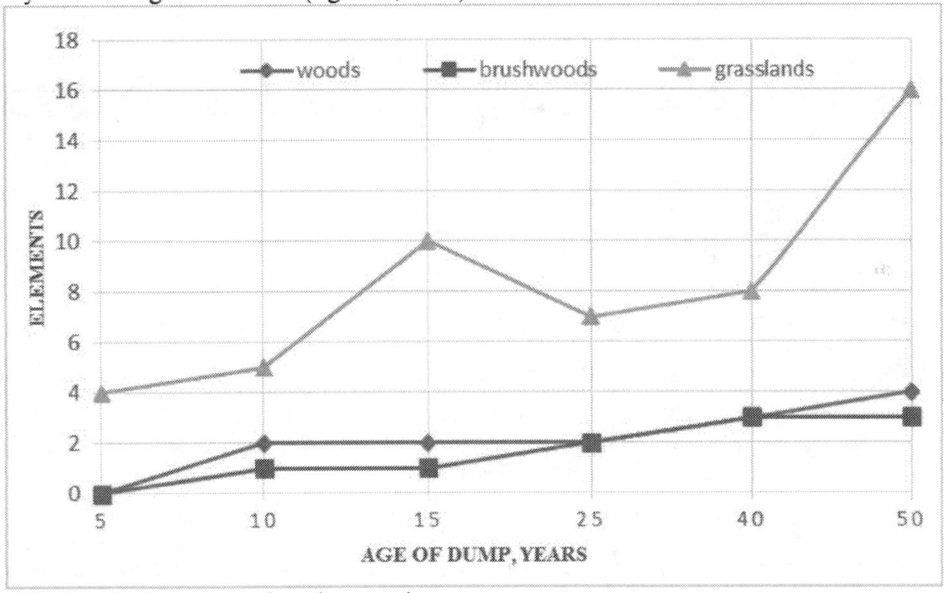

Fig. 2. Change in the number of species over time

The pioneer vegetation of rock spoil heaps is represented by herbs of the following families: legumes: Trifolium hybridium L., Melilotus albus Medik; gramineous species: Alopecurus pratensis, Gramineae, Poa praténsis; sunflower family: Artemísia vulgáris, Tanacetum L., Taraxácum officinále, Crepis tectorum L., and others. Arboreal plants involved in demutation processes include Salix schwerinii, Salix sachalinensis 'Sekka,' Populus maximovichzii, Henri, Betula albosinensis, and Duschekia fruticosa, Betulaceae.

In the first five years regeneration of vegetation is insignificant: occasionally – Artemísia vulgáris, Alopecurus pratensis, Gramineae, Trifolium hybridium L., Taraxácum officinále. The plants are not tall, up to 5 cm high, and their number does not exceed 18 species per 1 square metre. Arboreal plants can be found occasionally at rock spoil heaps

(10 years and more) adjacent to the forest edge: Betula albosinensis, Salix schwerinii, Populus maximovichzii, Henri.

In the ten subsequent years the species composition of the herb stratum is quite varied and is represented by the following species: Oenothera suaveolens, Trifolium lupinaster, Crepis tectorum L., Artemísia vulgáris, Artemísia scopária, Taraxácum officinále, Plantágo média, Pulsatílla pátens, Chamerion angustifolium, and Trifolium hybridium L. Trifolium hybridium L. is known to have a favourable effect on the quality of soils as it improves the nitrogen balance stimulating the accumulation of organic matter and formation of the humus layer. In the later period some species fall out, but new ones appear.

Table 1. Structure of natural regeneration plant community

Areas of self-regeneration (rock spoil heaps, age, years)	Number of species		
	including		
	woods	brushwoods	grasslands
5	0	0	100
10	25	12.5	62.5
15	15.38	7.69	76.92
25	18	18	63
40	21.43	21.43	57.14
50	17.39	13.04	69.56
Background (natural vegetation)	34.78	15.22	49.99

The demutation process is dominated by Salix schwerinii and Populus maximovichzii, Henri, and among herbous plants – by Artemísia vulgáris, Taraxácum officinále, and Trifolium hybridium L.

The variety of herbal species on the ground of the background territory numbers 46 species, which is exactly twice as high as the number of species on the 50-year-old spoil heap. 16 arboreal species, 7 shrub vegetation species, and 23 grass vegetation species have been found here. It should be mentioned that the soil and vegetation cover is more diverse here, and most species are not found in the enterprise impact zone.

Such characteristics of self-organised vegetation processes can be explained by low content of humus in the substance and its unfavourable physical and mechanical features. The soils consist mainly of rocky and sandy mixtures, as a result of which moisture cannot be retained in the substance, and this leads to the absence of the water regime necessary for vegetation.

Key plant communities have been identified on the basis of the conducted analysis, and the key succession seres of the plant community self-regeneration at rock spoil heaps of the Korfovsky Rock Querry was produced (figure 3). The produced succession sere provides a graphic picture of the processes of replacement of some communities with others which is useful for subsequent analysis of the findings.

The first stage of demutation processes is the basis of a succession sere – it includes leguminous forbs and gramineous forbs communities. It has been established that the grass community dominated by leguminous species is being superseded by shrubs with the prevalence of Alnus hirsuta, Turcz., and in the event of presence of gramineous species the dominant species is Salix schwerinii. In addition, in the former case shrubs are superseded by young birch forest, and in the former case – by mixed wood and shrub community.

Apparently, there is no regular pattern in the changing stages of the succession sere. The appearance of some or other species in the plant community depends on the nature of transfer of seeds taking place in a particular season of the year, otherwise it requires other specific conditions of the natural environment. Nevertheless, special mention should be made of total absence of natural regeneration of coniferous species, which indicates that the regeneration succession process under these conditions has passed the point of bifurcation at a very early stage of its development [7].

It is known that depending on their length successions are classified into local, secular, and phylocoenogenetic (evolution of plant associations) successions. All these types of successions are closely connected with each other and overlap. The supersession of coniferous-deciduous forest by mixed deciduous forest is an example of local succession characteristic of the territory under survey. Regeneration of the primary forest plant community will proceed within a different timescale as a result of secular succession.

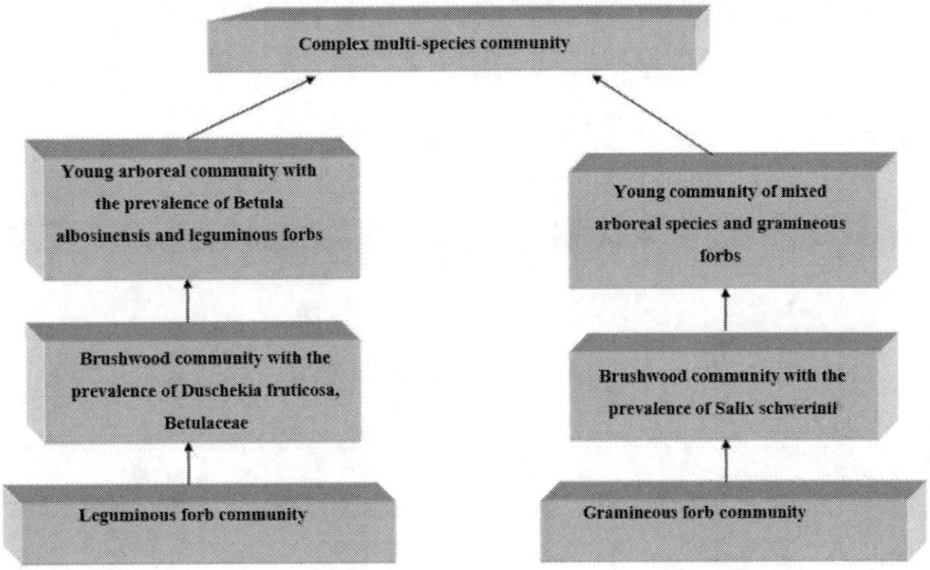

Fig. 3. Succession seres of the plant community self-regeneration at rock spoil heaps

4 Conclusions:

The dominant development of the mineral and raw materials complex prioritises the problem of localising inevitable disturbances of the biota in time and space by creating conditions for development of regenerative successions of plant communities.

The area of technology-induced disturbance of the natural biota caused by operation of an extractive enterprise is considerably larger than the area of the enterprise's land allotment.

The methodology of rehabilitation of territories disturbed by mining works must be based on initiation of processes of natural self-regeneration of plant communities.

The intensity and tendency of demutation processes in the zone of technogenic disturbance of the biota depend on the presence of a source and mechanisms of transfer of seeds of the edificatory group of the plant community.

A complex multi-species community regenerates through development of a key succession sere, but because of bifurcation of the process natural regeneration of the initial biota does not happen at the local succession level.

References

1. K.N. Trubetskoi, Yu.P. Galchenko, L.I Burtsev. Environmental Problems of Exploitation of Mineral Resources under Sustainable Development of the Environment and Society. Moscow: Nauchtekhlitisdat. p. 262 (in Russian). (2003).
2. K.N. Trubetskoi, Yu.P. Galchenko Geoecology of the development of the Earth's interior and ecogeotechnology for the development of deposits Moscow: Nauchtekhlitisdat. p. 360 (in Russian). (2015).
3. Yu.A.Monakov, A.N. Kupriyanov Kuzbass-1: Sat. articles. GIAB. Fascicle 7. (2009).
4. A.V. Bogorodskaya, O.V. Trefilova, A.S. Shishikin. Bulletin of Tomsk State University. **382**. P. 214-220. (2014).
5. Klaas G.J.N., van Lagen B., Buurman P. Geoderma. **100**. P. 1-24. (2001).
6. Yu.P. Galchenko, M.B. Bubnov, Yu.A. Ozaryan. Environmental Systems and Tools. **7**. pp. 25–32 (in Russian). (2013).
7. V.V. Furyaev. Role of Fires in the Process of Forest Formation. Novosibirsk: Nauka. RAS Siberian Publishing Firm. p. 253 (in Russian). (1996).

Revaluation of 'old' ore mining areas of the Amur River region (Russia)

Viktor Kryukov[1*]

[1]Mining Institute of Far eastern branch of Russian Academy of Sciences Khabarovsk, Russia

Abstract. In recent years, highly liquid precious metals have been mined in the Russian Far East. The state fund has no facilities with confirmed reserves. However, there is a real possibility of reviving the 'old' mining areas. Based on the Lower Amur region (Khabarovsk region) the prospects of gold reserves accumulation using 'small' fields (in 50-80-s conception) are considered. It deems new to determine the potential of these facilities from the viewpoint of large-scale mineralization and to involve the territory under consideration in the assessment of the gold-bearing manifestations. 'Reference' deposits were selected and their specific productivity was assessed, which was extended to similar by properties manifestations. Within the Lower Amur region, geological investigations revealed about 300 small deposits and mineral occurrences. 34 objects localized at the clusters of submeridional and sublatitudinal faults intersection, were identified as promising. Basically, the gold-ore blocks were compared with the 'standards': Belogorsky (near-surface), Dyappe (subvolcanic) and Zimovye (hypabyssal). The determining factors were the structural features of mineralization localization, facies conditions, the ores composition and metasomatites, the zonality of ore-metasomatic formations, the erosion of objects and the completeness of hydrothermal processes. The author's assessment is confirmed by the results of exploration works at 15 'small' deposits and ore occurrences, with the transfer of 11 of them to the category of large and medium-sized.

1 Introduction

The aim of this research is to draw attention to potential revaluation of the so-called ore mining areas due to depletion of the deposits national reserves and mineral occurrences. Revision works necessity should be noted in the known mineral areas whose industrial development has reduced significantly due to mineral resources base depletion [1]. The practice of revaluation of 'small' gold deposits in the Far East confirms such works' necessity and effectiveness. So, over the past decade, about ten gold ore and complex gold-bearing deposits have been transferred from the small reserves category to the large [2, 3] in the Amur river Region.

The methodological aspects and results of determining the potential of the 'old' ore mining regions are analyzed on the example of the analysis of the gold content of the Lower

* Corresponding author: kryukov-vg@mail.ru

Amur Region. For prognosis evaluation, the researches usually emphasized the prospects of Mnogovershinnoye, Albazinskoye, Agnie-Afanasievskoye, and Belogorskoye deposits [4, 5, 7-9]. The remaining facilities were classified as small by reserves, and as a rule, non-promising ones. It should be noted that there are no facilities in the exploration practice in the Far East, for which a final evaluation was given during one study cycle. The works on most of the Amur River mineral occurrences in the 20th century ceased after the first cycle, and this served as the basis for assessing the region's as having not high gold content.

2 Analysis technique

Methodical technique for revaluation of the potential of 'old' mining areas were developed at the facilities of the Lower Amur Region. The Lower Amur River gold-bearing area with dimensions of 450x250 km, stretched in the submeridional direction, covering the areas of the Amur and Amgun river basins in their lower reaches, as well as Ulban, Usalgin, Gur, Tumnin and Butu. A considerable number of deposits and mineral occurrences have been established within its boundaries: gold-210 (gold-silver, gold-sulphide and polymetallic-gold formations), copper and molybdenum – more than 40, secondary quartzites – about 30, polymetals – 7, tungsten – 10. Belogorskoye, Mnogovershinnoye, Herpuchinskoye, Pilda-Limuriskoye and Tumninskoye ore-alluvial regions were studied and exploited.

The revaluation includes the following main lines:
- geological-tectonic zoning of the territory;
- selection of standards and justification of their specific productivity;
- comparison of known manifestations and 'small' deposits with the standards;
- the facilities prognosis evaluation.

Geological-tectonic zoning of the territory is based on geological surveys materials Sc. 1: 200 000 and Sc. 1:50 000, the results of prospecting and evaluation works and author's observations (study of individual deposits, mineral occurrences, and decoding results). The main metallogenic elements of the structural framework of the Lower Amur Region, according to author's opinion, are the orthogonal system of faults, tectonomagmatic archings and linear disturbances (fig. 1). The noted structures directly control the location of magmatics, metasomatites and ore mineralization. The standards were selected among the most extensively studied facilities.

The choice of standards and the justification of their specific productivity is deemed to be the most difficult task. With the purpose of revaluating the potential of the territory under consideration, the author, starting from 1991, has analyzed the source materials for a significant part of the occurrences: 172 – gold, 28 – copper and copper-molybdenum facilities, 12 – secondary quartzites, 4 – polymetals and tungsten.

Structural diagrams and layouts for the distribution of circum-ore metasomatites, structural-and-facial columns for circum-ore metasomatites were compiled for the standards, taking into account their vertical extent, the degree of erosion in the categories of weak, moderate, medium and substantially eroded bodies was determined, the completeness and degree of completion of the hydrothermal process was determined, as well as specific productivity. To determine the specific productivity, the ratio of reserves approved by the State Committee for Reserves, of industrial categories or proprietary reserves to the area to which these reserves are distributed was established. The productivity of the ore field of the reference facility served as the basis for determining the resources of the correlated mineral occurrences and deposits.

The unifying signs are the confinement to tectonic elements of fairly large blocks of the earth's crust (the metallogenic zone or region grade), the originality of the genetic series of ore-magmatic systems (pre-, syn- and post-intrusive metasomatites associated with specific intrusive complexes), ore-hosting metasomatites, etc.

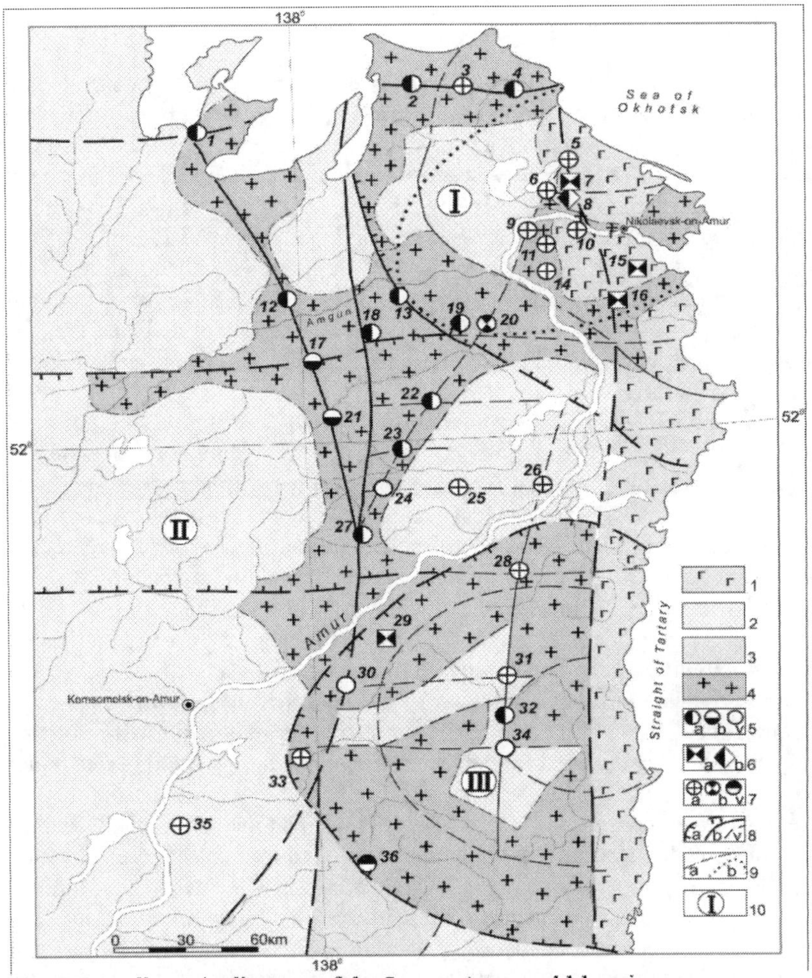

Fig. 1. Metallogenic diagram of the Lower Amur gold-bearing area

1. Basalts, andesites, their tuffs of Paleogene age. 2. Sedimentary complexes of the Early Cretaceous. 3. Paleozoic metamorphic complexes. 4. Areas of Cretaceous and Paleogene magmatics. 5-7. Deposits and mineral occurrences. 5. а) gold-quartz, б) gold-mercurial, в) gold-tungsten. 6. Secondary quartzites а) deposits, б) mineral occurrences. 7. а) porphyritic, б) polymetallic, в) tungsten. 8. Faults. 9. Boundaries of: а) areas, б) Takhtinskoye massif . Ore bodies: 1. Kutynskoye. 2. Manguliyskoye. 3. Mayak. 4. Mnogovershinnoye. 5. Belaya Gora. 6. Blagodatnenskoye. 7. Iskinskoye. 8. Seredochnoye. 9. Kabachinskoye (Sberegatelnoye). 10. Bukhtyanskoye. 11. Tyrskoye. 12. Albazinskoye. 13. Chulbatkanskoye. 14. Dylmenskoye. 15. Krugliy Kamen'. 16. Gryada Kamenistaya. 17. Imskoye. 18. Angochikanskoye. 19. Oktyabrskoye. 20. Chayatynskoye. 21. Bichinskoye-1. 22. Pokrovsko-Troitskoye. 23. Agnie-Afanasievskoye. 24. Uchaminskoye. 25. Dyappe. 26. Kholanskoye. 27. Delkenskoye. 28. Limonitovoye. 29. Shelekhovskoye. 30. Ezhovoye. 31. Zimovye 32. Levodzhegdagskoye. 33. Poniyskoye. 34. Oemkuyskoye. 35. Malmyzhskoye. 36. Dzhaurskoye.

On these grounds the deposits of the near-surface, subvolcanic and hypabyssal depth facies were separated. The first group is represented by the standard – the Belaya Gora deposit, which is compared with the manifestations, and small deposits of Blagodatnenskoye, Bukhtyanskoye, Dylmenskoye, Kabachinskoe, Limonitovoye, Mayak, Polyanka, Tarakanovskoye. The group of subvolcanic deposits includes the standard of Dyappe, as well as Agnia-Afanasyevskoye, Albazinskoye, Pokrovsko-Troitskoe. The Zimovye reference standard refers to the hypobyssal one, which compares Ledvozhegdagsky, Mnogovershinnoye, Oemka, Uchaminskoye.

3 Research results

Within the area the author distinguishes metallogenic elements (from the north to the south): the Lower Amur metallogenic region, the Evoron-Udylskaya metallogenic zone and the Tumninskaya metallogenic region. The metallogenic regions represent tectono-magmatic archings, while the metallogenic zone – a system of Gorstian- and trough-shaped structures (fig.1). The distribution of intrusive rocks, metasomatites and mineralization within the boundaries of the influence of certain structures, emphasizes the nature of the latter. Thus, the zones of sublatitudinal strike serve as ore-supplying and controlling, while submeridional ones serve as the controlling structures. In this case, the 'step' in the spatial position of the deposits makes 30, 60, less than 15 kilometers. The role of diagonal faults is in ore distribution (ore-distributing structures). Disturbances of orthogonal and diagonal systems, are noted as ore-hosting.

Within the Lower Amur Region, Mnogovershinnoye deposit is exploited (4 – here and further in the parentheses are the objects numbers in fig.1). From Mnogovershinnoye ore cluster to the west, within the sublatitudinal structure, the promising mineral occurrences of Mayakskoye (3), Manguliyskoye (2) and the Kutynskoe deposits (1) are noted. A favorable situation is characteristic of deposits controlled by the ore-distributing structure of the northeastern orientation. Belaya Gora (5), Blagodatnenskoye (6), Kabachinskoe (9), Tyrskoye (11) and Chayatynskoe (20) deposits are in the range of its influence.

As a standard of near-surface deposits, ***Belaya Gora*** was chosen, the ore mineralization of which is localized within the lateral apparatus of a large stratovolcano of the Cainozoic age. Mineralization is basically a tubular stockwork and a blanket-like deposit. In the author's opinion, mineralization, confined to ring structures in the headstream of the Kolchanka spring, as well as the bodies of the submeridional and northwestern strike under the alluvial deposits of Belogorskoye and Verkhnyaya, can be of large importance.

The gold reserves of the deposit equal to 24.5 tons (deposit and stockwork). Resources of P_1 category – 3 tons, and are localized in ore bodies controlled by a sub-latitude structure (the Zayachiy, Kolchanskoye, Levoberezhnoye, Ogorodnoye-1, -2, Pavlovskoye areas). The total area of the proved area is 2.2 square km. The specific productivity, taking into account the transfer of resources into the category of reserves, is 11 t/km2. The forecasted area includes the proved area, marked sites with resources of category P_1, and also the area of the Kolchanka spring basin area (in its headstream), which totally equals to 8.6 square km. Taking into account the parameters of the mentioned structures, the prognosis evaluation of the ore field may equal to more than 100 tons of gold.

In the Evoron-Udylskaya metallogenic zone, the fields are being prepared for development (Dyappe, Polyanka, Chulbatkanskoye, etc.). Pokrovsko-Troitskoye (22), Agniye-Afanasyevskoye (23), Uchaminskoye (24), Delkenskoye (27) deposits are confined to the ore-distributing structure of the northeastern strike are referred to the promising facilities. Besides, three latitudinal-oriented chains of promising mineral occurrences (from north to south) are noted: 1) Imskoye (17), Angochikanskoye (18), Oktyabrskoye (19), Chayatynskoe (20); 2) Bichinskoe-1 (21), Elovoye, Pokrovsko-Troitskoe (22); 3) Uchaminskoe (24), Dyappe (25), Holanskoye (26). In the spatial position of the deposits, a clear regularity is established: the 'step' between them, both in latitude and meridian, is 30 and 60 kilometers.

The Dyeppe deposit is a standard for subvolcanic type. The ore field has an oval shape in plane and dimensions of 4x3.5 kilometers. The oval is extended in the submeridional direction. Within its limits, 39 ore bodies - zones of mineralization, which have the form of a linear stockwork, have been identified and studied, lens-and rod-shaped, and vein, in the form of relatively isometric blocks of stringer-porphyry mineralization. They are separated into 4 areas: Central, North-West, South-West and Eastern with a total area of about 6.8

square meters. The most extensively studied is the Central area, drilled by spacing of 20x20 m. Authorized reserves of the Central area, limited to an area of 0.7 square km, are 7.6 tons, which determines the specific productivity of 11 tons per unit area. The predictable mineralization exceeds 100 tons, localized on an area of 6.8 square km.

For some of the analogs, significant reserves of gold have already been proven: Albazinskoye - 112 tons, Delkenskoye - 42.5 tons, Kutynskoye - 18 tons (table 1).

Tumninskoye metallogenic region is characterized by a rather specific metallogenic profile. An important part is played by alluvial and ore gold ones. Gold alluvial deposits were developed in the first half of the previous century, and since the 1950s, hardrock gold deposits (Oemku) began to be developed. Among the leading commercially valuable minerals of the territory, secondary quartzites should be referred to as a source of aluminum and other components, including gold. The Shelekhovskoye deposit (29) is the most prepared for development. Tungsten mineralization, detected in the central and intermediate zones of the taxon is deemed to be a promising one. Copper-molybdenum-porphyritic bodies and insignificant polymetals occurrences are known within the region. The ore controlling is the fracture of the submeridional strike, traced from the city of Kholan in the basin of the Yai and Jegdag rivers and further to the south into the basins of the Zimovye, Tumnin rivers, and the headwaters of the Gur River. It determines the position of different metals deposits: gold - the Kholanskoe, Levodzhegdagskoye (32), Oemku (33), the porphyritic formations - Limonitovoye (28) and Zimovye (31), polymetals. The role of ring disturbances affects the location of secondary quartzites: Shelekhovskoye (29), Salasinskoye, Elovoye, Karginskoye, Pulsinskoye, Bystrinskoye and others. Within the outer ring of the arching, there are also molybdenum-tungsten with gold, Ezhovoye (30), and gold-silver mineral occurrences.

Zimovyinskoye deposit (ore field) is identified as a standard for bodies formed in hypabyssal conditions within the intrusive-dome structure. Within a single framework of sub-meridional zone, 4 areas are differentiated for 5 km (from the north to the south): Pestroye, Tsokolnoye, Taimen and Zolotonosnoye. Authorized reserves for the Pestroye area of 0.8 square km (essentially tungsten with gold mineralization) were determined by A.M. Ezheley (2001) as 19.6 thousand tons of tungsten oxide and 2 tons of gold. The results of the Tsokolnoye assessment, having an area of 0.5 square km (gold-tungsten with sulphides mineralization) are distinguished by a higher gold grade, the reserves being equal to 10 tons of gold. Specific productivity for these areas will make 24,000 tons of tungsten oxide and 10 tons of gold. The predictable gold resources for the entire body of about 10 square km. exceed 50 tons, and tungsten oxide - 100 tons. The predictable specific productivity of gold in this type of deposits is 5 tons per 1 sq. km.

High prospects of the Lower Amur region bodies, apart from their favorable structural position, the author assigns to a number of factors:

- the scale and completeness of the processes of circum-ore metasomatism. Thus, at the Pokrovsko-Troitskoye deposit, metasomatic bodies containing mineralization are traced along a strike for 6 km, with outbreak width of about 2.5 km. Similar parameters are characteristic of Agnia-Afanasyevskoye, Dylmenskoye, Diappe, Holanskoye, etc.;
- by wide manifestation of large-scale mineralization. A well-mapped stockwork on the Dyappe field did not attract geologists' attention to the morphology of ore bodies at that time. At the same time, a variety of morphological types is characteristic of other objects. Hardly identifiable columnar or rod-shaped ore bodies deserve a special attention;
- vein systems with a moderate degree of deposits erosion cut are replaced by vein-disseminated mineralization or stockwork at the depth of a number of bodies;
- by sufficiently high gold content in metasomatites. For example, at the Dyappe deposit, in a stockwork, tapped by boreholes, it amounted to 3 and 5 g/t for a thickness of 7 and 24 m, in metasomatites of Pokrovsko-Troitskoye by channel samples – up to 7 g/t, in an industrial

sample weighing more than 21,000 tons from 'unchanged' green and black-colored sedimentary rocks of the Agniye-Afanasyevskoye deposit – 3.17 g/t;
- underestimation, in 2-2.5 times, of gold content in samples analyzed in the laboratory of the Lower Amur expedition for the period from 1985 to 1989. In the southwestern part of the area, porphyritic-and-gold mineral occurrences are recorded. At one of them, Malmyzhskoye, the works of 2005-2013 not only confirmed the porphyritic type of the body, but also established authorized reserves in the amount of 298 tons of gold and 5.6 million tons of copper. Researchers emphasize the identification of a new, unique deposit [3].

Table 1. The potential of gold ore facilities of the Lower Amur River Region (tons)

No	Deposits and mineral manifestations	Potential				Author's assess-ment, resources
		Data of 1960-1990		Data of 2004-2017		
		Reserves	Resources	Reserves	Resources	
1	Albazinskoye	12		112	30	101-500
2	Agnie-Afanasievskoye	1.817	11.7[1]	0	70[3]	51-100
3	Angochikanskoye			0	0	
4	Belaya Gora	4[3]	17	24[7]	35	51-100
5	Blagodatnenskoye		5[1]	10	15[7]	
6	Delkenskoye		5[4]	42.5[5]	10	51-100
8	Dyappe	1.948	14.86[2]	7.6	40	101-500
9	Zimovye		10		12[3]	101-500
10	Zolotaya Gora		27		74[7]	51-100
11	Kabachinskoye		24.7		63[7]	51-100
12	Kutynskoye		10[3]	6.6	24	51-100
13	Limonitovoye		27[3]	0	0	
14	Polyanka		5[3]	17[7]	3	51-100
15	Tarakanovskoye		5[1]	0	0	
Total for the area		15.765	162.26	219.7	376	> 1500

Note: data – in italics Yu. I. Bakulin, 2004; bars [1]- The Program on gold, 1996; [2]- M.K. Dyachkov, 1989; [3]- V.Ya. Bespalov, 2006; [4]- V.A. Kaidalov, 2008; [5]- V.A. Filippov, 2012; [6]- V.G. Vologin, 2014; [7]- V.N. Priputnevich, 2014

4 The results discussion

As a result of the comparative analysis, 13 gold ore, 4 secondary quartzite, 13 porphyritic, 4 gold-tungsten, 1 polymetallic and 1 zeolitic bodies were identified as promising. The forecasts were tested in recommendations to the Khabarovsk Krai Government and subsoil users, as well as at various specialized conferences in the period 2007-2017.

The author's forecast formulations are confirmed by the geological study data of recent years. As a result of geological exploration work conducted from 2004 to 2014, eleven bodies (a table) of fifteen have been positively estimated within the Lower Amur area, studies continue on one of them, while on three – they have been stopped. The results of the geological study of recent years disprove the idea of low prospects for the objects, since very favorable data have been obtained on the territory high potential with reference to large-volume gold mineralization with moderate contents of the metal.

The table gives author's estimate for a number of areas. In this case, the facilities category by groups is designated subject to possible reserves (resources): 11-50; 51-100; 101-500. Referring to a particular group is determined by zones' mineralization parameters. When a significant number of structural, morphological, metasomatic, geochemical factors coincided with the standards, the productivity of the standards was accepted, but in case of mismatch, the reduction coefficients were used with an interval of 25%. A part of the mineral occurrences and deposits remains unrated since there are no reliable grounds to determine their ores' industrial type. Prospects of such mineral manifestations are unconditionally higher in comparison with the areas located outside the ore controlling structures.

5 Conclusion

Thus, the problem of the depletion of the mineral resources base of the 'old' ore regions in the Russian Far East can be resolved by revaluation their potential. On the example of the Lower Amur Region within which gold mining from ledge ores began as far back as the 19th century, opportunities for reassessing the potential of the territory are considered. Currently, the operation is continued at the Mnogovershinnoye deposit, while the Albazinskoye and Belogorskoye fields development has been started.

More than thirty deposits and gold mineralization and other metals, whose spatial position is subject to certain structural patterns, are distinguished as promising, within the Lower Amur area of gold mineralization. The orthogonal system of faults determines the position of the most promising bodies. A 'step' in the location of deposits is identified, equal to 15, 30 and 60 kilometers. The standards of the near-surface (Belaya Gora), subvolcanic (Dyeppe) and hypabyssal (Zimovye) levels of mineralization are defined.

Among the selected facilities, Agnia-Afanasyevskoe, Blagodatnenskoye, Dylmenskoye, Dyappe, Kutynskoye, Uchaminskoye deposits are in the distributed fund. Within their limits, exploration work is currently under way. As a result of the geological study, the gold reserves of this territory increased from 15.7 tons to 219.7 tons (apart from reserves of the Mnogovershinnoye deposit). The rest of the deposits and mineral occurrences represent the non-distributed fund. The Lower Amur Region possesses real prospects for accumulation of the mineral resources base. It will allow planning the functioning of a mining and smelting enterprise in this ore mining area for a long period.

References

1. B.K. Mikhailov, B.I. Benevolskiy, S.S. Vartanyan, A.G. Volchkov, I.F. Migachev Subsoil exploration and protection. **9**. p. 76-83. (2012).
2. A.V. Boyko, V.G. Vologin. Subsoil exploration and protection. **9**, p. 39-43. (2012).
3. A.F. Chitalin, A.A. Efimov, K. I. Voskresenskiy, E. K. Ignatyev, A. G. Kolesnikov. Mineral resources of Russia. Economy and management. **3**. p.65-69 (2013)
4. Bakulin Yu.I. Systematization of tin-bearing and gold-bearing ore systems for forecasting purposes/MG of the USSR, DVIMS (Far East Mineral Resources Institute). 'Nedra',. 192 p. (1991)
5. V.D. Melnikov
6. V.G. Moiseenko and L.V. Eirish. Gold-ore deposits of Russia's East. Vladivostok, 352 p. (1996).
7. N.P. Romanovskiy, Yu.F. Malyshev, Duan Zhuyan', Chzhu Tsun' and M. V. Goroshko. Pacific geology, **25 (6)** P. 3-17. (2006).
8. V.I. Sukhov. Pacific geology, **19 (5)**. p. 5-20. (2006).
9. Khomich V.G. Metallogeny of volcano-plutonic belts of the northern link of Asian-Pacific interaction zone Vladivostok: Dalnauka, 1995. 343 p.
10. L.V. Eirish Pacific geology. **2.** p. 67-80. 1991
11. V.L. Shevkalenko, collected papers Tectonics, the depth structure and geodynamics of Asia's east. III Kosygin readings. Khabarovsk, ITiG, p. 266-278. (2001).

Migrational activity of elements and chemical soil pollution during deposit exploitation in Amur River region and Okhotsk Sea region

Aleksandra Makhinova[1], *Aleksey* Makhinov[1*] and *Viktoriya* Kuptsova[1]

[1]Institute for Water and Environmental Problems, Siberian Branch of the Russian Academy of Sciences, Khabarovsk, Russia

Abstract. We studied mechanisms of geochemical soil systems formation. This work provides geochemical characteristics of soils under the influence of mining enterprises. It also shows the mechanisms of sewage water enrichment by heavy metals during sulphidic deposit exploitation and their interaction with soil substance. We also describe migrational activity of elements and the conditions of their concentration in the soil volume. It is shown that migrational activity of elements depends on both their chemical nature and the level of aggressive fractions of humus acids and background-forming elements (Fe and Mn) content. The level of heavy metal concentration in soils formed in the area of influence of deposits showed to be exceeded. We suggest new approach to complex evaluation of element reallocation mechanisms in soils. **Keywords:** migrational activity of elements, migration conditions, element concentration mechanisms.

1. Introduction

Increased demand on precious metals in the world leads to intense exploration of gold, silver and platinoid deposits. Russia is the third in the world on the known reserves of gold and platinum [4, 15, 16]. More than 60% of deposits exploited in Russia are located in the Far East. Extraction technology leads to soil devastation [8, 10]. In the mountain regions near Amur River and Okhotsk Sea the most danger comes from overburden mining, which contributes to chemical pollution of the area. During the evaluation of ecological state of the soil heavy metals become an important point of interest, and all of them are xenobiotics, with lead and cobalt being the most poisonous for human and animals [7].

Large amounts of groundmass are extracted during overburden mining, which leads to multiple increase of the area of their surface of contact to water and oxygen from atmosphere. This problem can be seen not only on overburden grounds, but also in the deforested areas. Sewage water also deals a vast amount of damage to the soil.

The ecological problem of soil devastation and their chemical pollution becomes more and more pressing in deposit exploitation regions [10, 16].

*Corresponding author: amakhinov@mail.ru

Evaluation of migrational activity of elements in the soil volume and the influence of mining industry on the chemical soil pollution is the main objective of this research.

2. Subject and methods of research

We have studied natural soils and soils under the influence of ore mining production. Soils within 500 m from the site of overburden removal have profile deformation (faults and thickening of horizons), they are conventionally called industrial.

The research was conducted from 2000 to 2017 in the following deposit regions: Tas-Yuryakh, Ryabinovyi, Komsomolskaya zalezh, Konder, Mnogovershinnyi.

We analyzed soil samples selected by standard five-spot method in the horizon of 0-30 cm at 14 sites 10 by 10 m [23]. All 5 samples from each site were mixed together to form a joint sample.

Physical and chemical properties of the soils were examined using conventional methods [1]; pH of the soil water suspension was determined by potentiometry using glass electrode. Iron and manganese total form content was determined by acetate method. Total carbon content was determined by dichromatic oxidation method [2]. Total micro-element content was determined on the spectroscope No. 835 (Lif200). Overall, 85 specimen have been analyzed. Most of the samples were analyzed in the certified lab by FGBI Agrochemical Activity Center "Khabarovsky". Part of the samples were re-analyzed in IVEP FEB RAS. Audit samples were examined in the shared analytic center by PGI FEB RAS.

Dynamic models were used to describe material migration processes. For evaluation of probable areas of material migration, geographical interpolation methods were used. Majority of numeric values for elements concentration were processed by conventional statistical methods.

3. Findings

The area of research is located in the temperate latitudes of East Asia. It is characterized by humid temperate continental climate with monsoonal features. Annual precipitation ranges between 650 and 800 mm. Areas where deposits are exploited by surface mining have mountain landscape. Sulphide-bearing ores with high levels of Fe, Mn, Cu, Zn, Pb, Ni, Cd are stocked at the surface in large volumes. Alternate wetting and drying of the loosened rock under violent fluctuations of temperature gradient (within 24 hours, within a year) contribute to sulphide oxidation producing complex cations $[FeHSO_4]^+$, $[FeHSO_4]^{2+}$, $[CuHSO_4]^+$ [11, 19]. In case of rapid snow melting and atmospheric precipitation, surface and subsurface flows get formed enriched with fine fractions and heavy metal compounds. These enriched sewage waters contribute greatly to soil pollution.

3.1 Geochemical characteristics of background soils.

In the harsh environment near Amur River and Okhotsk Sea topsoil is formed low-contrast. It is characterized by low level of soil variation. The main constituent parts of these background soils are channery brown taiga soils with signs of humic-illuvial leaching. Their position in the landscape defines the quality of organic matter and intensity of humic-illuvial process. All variations of soils have these features: a) low thickness of soil profile (up to 40 cm, in the lower parts of the slope up to 70 cm); b) low level of finely dispersed grained soil content (4-6%) [9]; c) weak bacterial decay of forest litter and moss, which leads to accumulation of organic matter and formation of humus horizon on the soil surface; d) presence of lasting seasonal frost.

Upper humic-cumulose horizons, 7-12 cm thick, produce humus with high concentration of water soluble violent separates. Structural groups of brown ulminic and fulvic acid, representing 58-88% of the total composition, react even at temperatures lower than 15°C [14, 19]. Soils have acidic reaction of the environment (pH_w 3.9-5.8), their profile is enriched with iron and manganese oxides, and their soluble forms represent 35-45% of the total composition. Distribution of oxalate-soluble forms of amorphous iron correlates to the level of fulvic acid structural groups content [10, 13].

Geochemical features of soil volume determine physical and chemical state of solid, liquid and biogenic soil phases [6, 12, 20]. Processes of molecular interaction between soil phases and gravity moisture provide conditions for solution and dissolved matter efflux by subsurface flow. Geochemical element activity at the phase boundary in the soil system is defined by the level of iron and manganese content. Environment reaction and humus quality support and control the state of migration flow [21, 22].

Exploitation of ore deposits has influence on geochemical properties of the soils. Increased level of polyvalent metals composition (Fe, Cu, Pb, Zn) in the lower parts of slopes indicates the processes of their migration by subsurface flow.

3.2 Migrational activity of elements and migration flow.

Element compounds in soil are defined by their geochemical activity, which is related to the ionic radius and ionic element coordination (table 1) [5].

By the degree of cation geochemical activity and ability to create compounds with humic acids of brown taiga soils they form the following chain: $Fe^{3+}>Al^{3+}>Fe^{2+}>Zn^{2+}>Cu^{2+}>Pb^{2+}>Co^{2+}$. Relocation of elements in the soil volume by their migrational activity depends on the compound composition, their location in the activity chain, and proportions of different forms of Fe^{2-3}, Mn^{2-5-7} in the mineral composition of soils [5]. Migrational activity of cations is defined by the concentration of Fe-Mn, humic substances and soil solution.

Table 1. Geochemical and migrational activity of elements in the hypergenesis area [5]

Ion coordination	A°, mean radius	Geochemical activity of ions in minerals	Migrational activity of elements in soils		
			3< pH 5.5 marsh zone	5.5< pH <6.5 mixing zone	6.5< pH <7.5 accumulation zone
6	0.70	Fe^{2+}, Fe^{3+}, Mn^{2+}, Ni, Co, Zn, Cu^{2+}, Cd,	$Fe^{2+}>Fe^{3+}>Mn^{2+}>Cu>Pb>Co>Ni$ $Cd>Cr>Zn>Sn^{2+}$	$Fe^{3+}>Fe^{2+}>Mn^{2+}>Cu>Pb>Co>Ni>Zn>Sn^{2+}>Cd$	$K>Na>Fe^{(2-3)+}>Mn^{2+}>Pb>Cu>Zn>Co>Ni>Cd$
8	1.00	Mn^{2+}, Cu^{2+}, Zn^{2+}, Pb^{2+}	Elements migrate within soil solutions, organo-mineral and organo-disperse colloids, balance in the solution correlates with their degree of litophility and multivalent cations oxidation in various molecular compounds.		
8–12	1.40	Cu, Zn, As, <u>Cd</u>, <u>Sn</u>, <u>Pb</u>			
Chalcophilic		Fe, Co, Ni, Mn, P, Sn, Cu			
Siderophilic		Zn, Co, Fe, Cu, Pb, Zn,			

Notes: ions in parentheses have transient properties

Migrational activity is the ability of elements to form soluble compounds, an integral quantity indicating the state of migration flow.

Migration flows in soils determine structural groups of soluble humic acids (HA) that bind Fe and Mn in heteropolar salts and define chemical properties of polymetals.

Heteropolar salts are water-soluble. Metals in complex salts can engage in chemical exchange and replacement reactions [20,21].

Background elements (Fe, Mn) and structural groups of humic acids in oxygen-rich soil solutions determine the direction of geochemical processes in soils and formation of migration flows (table 2). Solutions rich in organo-mineral compounds of iron and manganese form buffer zones and contribute to reduction of Cu, Zn and Pb mobility in soil and their settlement [18].

Table 2. Factors determining migrational flows in soils

HA structural groups	Geochemical processes in soil	$pH_{wat.}$	Fe/Mn solubility	Predominate migrational flows in soils
COOH	Ion exchange	<4.2	S/S	**L** – lateral (Ls – subsurface, Lg – overground)
CAr--OH	Complex formation	3.2–4.0	S/S	**R** – radial (Rv – descending and Rm – ascending)
>C=O	Oxidation-reduction	>4.5	S/P	**W** – stagnant regime (Wv – accumulative-exudational and Wz – stagnant-cryosolic)
C_6H_6	Sorption, coagulation	>5.5	P/N	
–CHn	Hydrophobic interactions	–	N/N	**T** – indusrial Tr – natural-industrial

Notes: S – soluble, P – poorly soluble, N – non-soluble

Low pH values (<4.5) indicate high levels of H^+ ion concentration in soils.

Consistent patterns of migration flow occurrences and their zones of migration are shown on the map chart of geochemical-landscape zoning of the Amur River basin (see fig.1). New principles and approaches to geochemical-landscape zoning allowed us to reproduce regional background and the connection between migrational flows and conditions of element distribution in the soil volume in the most informative manner.

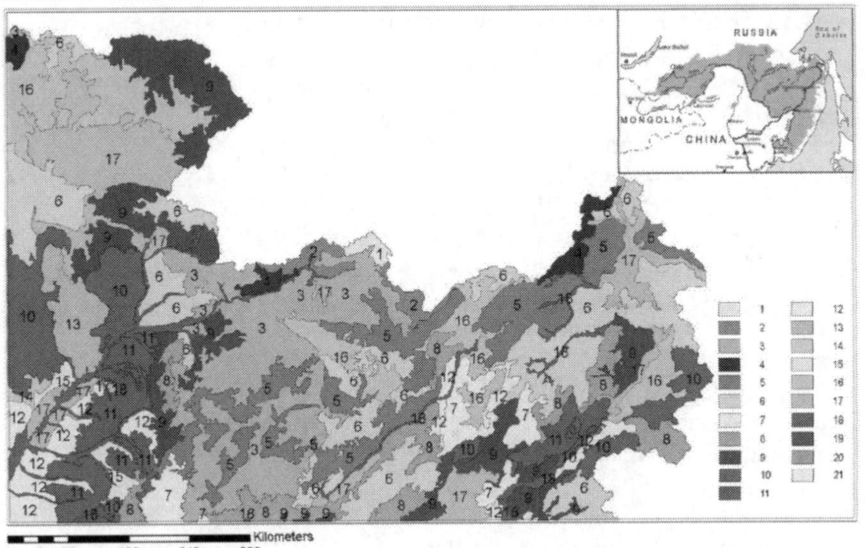

Fig. 1 - Landscape-geochemical zoning of the Amur Basin (fragment of the map)

1. Automorphic cryoturbation zone, 2. Automorphic zone of cryoturbation and mixing, 3. Lateral trans-eluvial zone, 4. Lateral trans-accumulational zone, 5. Lateral-eluvial zone, 6. Lateral-radial zone

of humatogenesis, 7. Lateral-radial zone of humatofulvogenesis, 8. Radial-lateral for Fe^{+3}, 9. Radial-lateral adsorptive zone, 10. Radial-eluvial for Fe^{+3}, 11. Radial zone of intensive trans-accumulation, 12. Radial zone of weak transit, 13. Radial zone of intensive accumulation, 14. Radial zone of exudative accumulation, 15. Radial zone of highly disperse weathering, 16. Zone of radial-aquatic biogenic accumulation, 17. Aquatic zone of inner-accumulation, 18. Super-aquatic trans-eluvial zone, 19. Agrozone, 20. Technozone, 21. Technozone of weathering products.

3.3 Mechanisms of chemical soil pollution in ore mining production areas.

We evaluated chemical soil pollution by element concentration values in industrial soils relative to background soils [6, 12, 22]. We looked only at the elements that had concentration values above spectral analysis susceptibility level. Analysis of element concentration in industrial soils indicated their high levels of concentration. Average values are shown in table 3.

With the absence of lateral flow, reallocation of element concentration in soils separated by distance (***r***) is described by the following continuity equation:

$$DUj(r)\sum^{n}{}_{j}Cj_{r}\leq[U_{j}(x+r)-U_{j}(x)]^{\wedge}t = \beta W(Cj-C)(grad\alpha_{i}C_{ji})_{r}{}^{\wedge}t, \quad (1)$$

where $DU_j(r)\sum^{n}{}_{j}Cj_r$ is element concentration in the specific soil volume; β is interphase exchange coefficient; W is volume water content; Cj is interphase balance concentration at the point r in the time $^{\wedge}t$, Cj depends on the level of organic compounds and content of fine fractions of silt in horizons. In multicomponent systems the total value of element concentration is connected with their migrational activity gradients ($grad\alpha_r$).

Heavy metal distribution in soils presents certain consistent patterns. Element content in background soils does not exceed medium values for the lithosphere. High levels of iron and manganese concentration in the profile are due to the content of soil material and biogenic accumulation [11, 20].

Levels of toxic elements (Cu, Pb, Zn, Mn, Co) content in industrial soils are higher than those in background, reaching maximum concentration in upper organic horizons. This can be clearly seen within 200 m from the pollution source. In the range of 500 m down the slope the level of element content decreases slightly.

Migrational activity of Fe^{3+} and $Mn^{(4-7)+}$ determines behavior of low charge cations $Cu^{(1-2)+}$, Zn^{2+} and Pb^{2+}. For example, specific affinity of {Fe-Mn to copper hydrate ions $Cu(H_2 0)_6]^{2+}$ contributes to coagulation processes that result in formation of copper residue, which is indicated by blue coating on the soil ped surface.

In lateral flow, element migration in the content of geochemical-soil flows with low sorption follows gravity force laws [17]:

$$F_g=P_b gh=[P_b(W-W_m)(W_p-W_m)_z(g/L)](\sin\alpha)=[\beta Wg/L](\sin\alpha), \quad (2)$$

where F_g is flow gravity forces; P_b is soil density; g is gravity acceleration; h is wet soil height; W is mobile moisture volume; W_p and W_m are total and maximum soil water capacity; β is interphase exchange coefficient; L is active porosity; z is coordinate; $\sin\alpha$ is surface slope.

These equations were used to determine cation accumulation levels (table 4).

We evaluated the degree of chemical pollution by concentration accumulation factors (Kc) relative to the average content in the lithosphere (percentage abundance) and the regional background according to the following formula:

$$Kc = Ci/Cb(r), \quad (3)$$

where Ci is level of concentration of the element in the measuring point, Cb(r) is the background (regional) level of its content for the given soils. Average values of chemical components anomaly levels in soils are shown in table 4.

The total accumulation levels of chemical elements were high on all sites of industrial soils. These data indicate the presence of chemical pollution of topsoil and formation of industrial soils.

Table 3. Level of element content in soils and their migrational activity

Research region	Distance to ore storage location, m	pHw	C_{org}, %	Levels and rows of migrational activity of elements	Average values of total element content							
					%		in mg/kg					
					Fe	Mn	Zn	Pb	Cu	Co	Ni	Cd
Mnogovershinnyi[1]	200	4.9	8.2	**a**{Fe,Mn,Cu,Pb,Zn, **f**{Fe,Mn,Pb,Cu,Ni,	4.80	0.87	39.90	28.30	4.70	19.3	10.0	32
	500	5.2	4.1		3.41	0.84	29.90	32.50	Tr.	16.4	Tr.	1.4
Mnogovershinnyi[2]	2000	4.9	5.3	**d**{Fe,Mn,Pb,Cd,Zn,	4.10	0.47	29.90	8.30	1.10	10.3	Tr.	0.9
Tas-Yuryakh[1]	200	5.6	6.1	**a**{Fe,Pb,Mn, Zn,Ni, **f**{Fe,Mn, Ni,Pb,Co	4.66	0.77	66.80	28.20	15.10	21.6	12.0	39
	500	5.1	2.9		3.60	0.86	24.50	27.20	1.50	15.5	9.0	0.6
Tas-Yuryakh[2]	2800	5.7	3.9	**a**{Fe,Pb,Mn, Cu,Co	3.10	0.36	31.90	6.30	1.10	11.2	Tr.	0.2
Ryabinovyi[1]	200	4.5	3.8	**a**{Fe,Mn,Pb,Zn,Cu **f**{Pb,Fe,Mn, Cu,Zn	5.07	0.68	94.30	31.5	18.45	15.1	17.18	17
	500	5.0	2.9		4.31	0.64	83.00	26.0	15.80	9.0	11.00	1.3
Ryabinovyi[2]	2000	4.9	6.9	**d**{Fe,Mn,Zn,Cu,Pb	3.96	0.51	31.00	5.6	8.02	10.2	11.36	0.1
Average content of elements in lithosphere (percent abundance of elements) [3]					4.65	0.85	50	10	20	80	58	340

Notes: soils in the influence zone[1]; background characteristics[2]

Table4. Average values of accumulation levels of some chemical elements in natural-industrial soils

Sampling sites	pH	CA%	Levels of element accumulation (Kc/Kc1)							
			$Fe_{tot.}$	Mn	Zn	Pb	Cu	Co	Ni	Cd
Mnogovershinnyi deposit										
200 m from HRP	4.9	8.2	1.0/1.1	1.0/1.9	0.8/1.3	2.8/3.1	0.2/4.0	0.3/2.0	0.2/-	0.1/32
500 m HRP, 300 m TD	5.2	4.1	0.7/1.1	1.0/1.9	0.6/1.0	3.2/3.8	Tr.	0.2/1.6	Tr.	tr/1.5
Tas-Yuryakh deposit										
200 m from HRP	5.6	6.1	1.0/1.5	0.9/2.0	1.3/2.1	3.0/4.8	1.7/3.0	0.3/--	0.2/-	0.1/180
500 m HRP, 400 m TD	5.1	2.9	1.1/1.2	1.0/2.1	0.5/0.8	2.9/4.7	0.7/14	0.3/--	0.2/-	tr/3.0
Ryabinovyi deposit										
200 m from HRP	4.5	3.8	1.1/1.2	0,8/1,2	2.0/3.0	3.1/6.0	0.9/2.1	0.2/1.5	0.3/1.5	Tr/170
500 m HRP, 300 m TD	5.0	2.9	1.0/1.1	0.7/1.1	1.8/2.4	2.6/4.9	0.1/1.9	0.1/0.9	0.2/0.9	Tr/13

Notes: Kc is microelement accumulation relative to percentage abundance: Kc1 is relative to region background; region background is the content of dispersed indicator elements in background soils.

4. Conclusions

Distribution of element concentration in the soil volume happens due to flow filtration by descending flow and lateral subsurface flow or by slope wash.

Interaction between sewage waters and soil substance reveals the following: 1) adsorption of fine fractions less than 3 μm in size by organic substances; 2) adsorption of dissolved compounds by fine fractions of silt; 3) molecule breaking of complex soils in soil solution; 4) synthesis of organo-mineral complexes in the presence of microorganisms.

Distribution of elements in background and industrial soils has certain consistent patterns: 1) background soils contain predominantly iron-group elements that have high solubility and migrational activity in organo-mineral complexes; 2) polyvalent cations in industrial soils (Fe^{3+}, $Mn^{(4-7)+}$) get bound by aggressive fractions of organic acids, which contributes to the decrease of Cu, Zn, Pb mobility and their accumulation.

This research was conducted with the financial support of the FEB RAS project No. 09294-2018-0001.

We thank I.V. Perminova, Doctor of Chemistry and professor of the MSU Faculty of Chemistry, for valuable advice in preparing this article.

References

1. Ye. A. Arinushkina, Chemical soil analysis guidelines (1970).
2. N. P. Belchikova, Agrochemical methods of soil exploration, 56 (1975).
3. A. R. Vinogradov., Geochemistry, **7**, 555. (1962).
4. D. V. Voronin, E. A. Gavelya, S. V. Karpov, Ore enrichment, **3,** 38 (1994).
5. Yu. Ye. Saet, B. A. Revich, Ye. P. Yanin, Environment geochemistry (1990).
6. M. A. Glazovskaya, Geochemistry of Natural and Technogenic landscapes (2007).
7. V. V. Ivanov, Ecological Geochemistry of Elements (1994).
8. I. A. Kradenykh, Pacific technology, **6:25,** 98 (2006).
9. A. F. Makhinova, Topsoil of lower reaches of Amur river region. (1989).
10. A. F. Makhinova, A. N. Makhinov, M. N. Shevtsov, Mountain journal, **4**, 83 (2006).
11. A. F. Makhinova, A. N. Makhinov, V. A. Kuptsova, V. V. Yermoshin, J. of Geochem. Exploration, **45 (6).** 89, (2013).
12. A. F. Makhinova, A. N. Makhinov, V. A. Kuptsova, Liu Shuguang, V. V. Yermoshin, Pacific geology, **33:2**, 76 (2014).
13. D. S. Orlov, N. I. Biryukova, D. S. Sukhanova, Organic composition of the Russian Federation soils (1996).
14. I. V. Perminova, Chemistry and life, **1**, 50 (2008).
15. Ye. A. Radkevich, A. I. Khanchuk, Pacific ore belt: materials of new research (2008).
16. B. G. Saksin, Predictive evaluation of regional geochemical influence on natural environment of extracting companies in nonferrous industry in the East of Russia (2012).
17. A. V. Smagin, Soil science, **3**, 328 (2003).
18. A. I. Trufanov, Topics of Far Eastern geography, **15**, 147 (1975).
19. S. B. Shustov, L. V. Shustova, Chemical basics of ecology (1995).
20. N. B. Uryev, High concentration disperse systems (1980).
21. T. Caprial, H. Beck, P. Harter, Soil Sci.Soc. Amer. J. **54**, 415 (1990).
22. N. Senesi, E. Loffredo, Chemical Processes in Soils, 563 (2005).
23. GOST 28168-89. Soils. Sampling. (2008).

On the strategy of utilization of resources of nonferrous metals in the far eastern region

Gennady Arkhipov[1*]

[1]Mining Institute of Far eastern branch of Russian Academy of Sciences, Khabarovsk, Russia.

Annotation. Analysis of current status and use of lead and zinc resources in the Far Eastern region, on the territory of Primorsky region for many decades and such analysis is of practical importance for evaluation and selection of areas in this field of activity. Russia safeguarded stocks and lead and zinc resources in the long term. On the territory of the Far East region there are significant deposits of these metals, but for single resources most of them substantially less large deposits in other regions of Russia – Siberia, Altai, Ural. Significantly to support lead-zinc industry in the region is possible in case of exploration of new deposits of ores with high quality. Lead and zinc minerals industrial scale in the region undoubtedly has and can be represented by a leading geological-industrial types of fields, but there is currently little studied. You must not reforming the present state of the industry and its innovative development: need to develop intra-regional use of raw materials in the form of the final product, that will give the region's resources development of new incentives. Non-ferrous metals resource potential of the region, with an effective strategy to use them can be a significant factor in changing the geopolitical significance of the far eastern part of Russia. **Key words**: fields, resources, lead, zinc, production, economic growth, Far East region, «Dalpolimetall»

On the territory of the Far Eastern region, the number of ore deposits identified only reaches 1200. Analysis of the level and completeness of the use of the identified natural mineral resources, taking into account their potential, naturally belongs to one of the most important places in the study of the region's economy. The analysis of the current state and use of lead and zinc resources is of practical importance for the assessment and selection of activities in this field.

Resources and production of the Far East region. Proven reserves and estimated resources of lead and zinc deposits are most often part of the same complex polymetallic deposits in association with other useful components – tin, molybdenum, tungsten, gold, silver, rare metals. Deposits containing lead and zinc resources are shown in figure and table 1 [1].

Reserves of zinc and lead on the territory of the Republic of Sakha (Yakutia) are counted in 5 deposits and make up: zinc C_1 category - 48 thousand tons, C_2 - 2075 thousand tons; lead categories A+B+C_1 - 826 thousand tons, C_2 - 1466 thousand tons. There are no developed deposits. Prepared to absorb deposits Prognoz (owns the company «Prognoz-

* Corresponding author: arhipov@igd.khv.ru

Silver», owned by Canadian company Silver Bear Resources), Verhne Menkeche (company GeoProMining), Sardana (company «Siberian nonferrous metals» of the holding «Summa», then «East-Siberian company»). Deposits of Kutinskoe and Zarnica are on standby. Prognostic resources of lead of the Republic are estimated at 800 thousand tons, zinc – 2550 thousand tons.

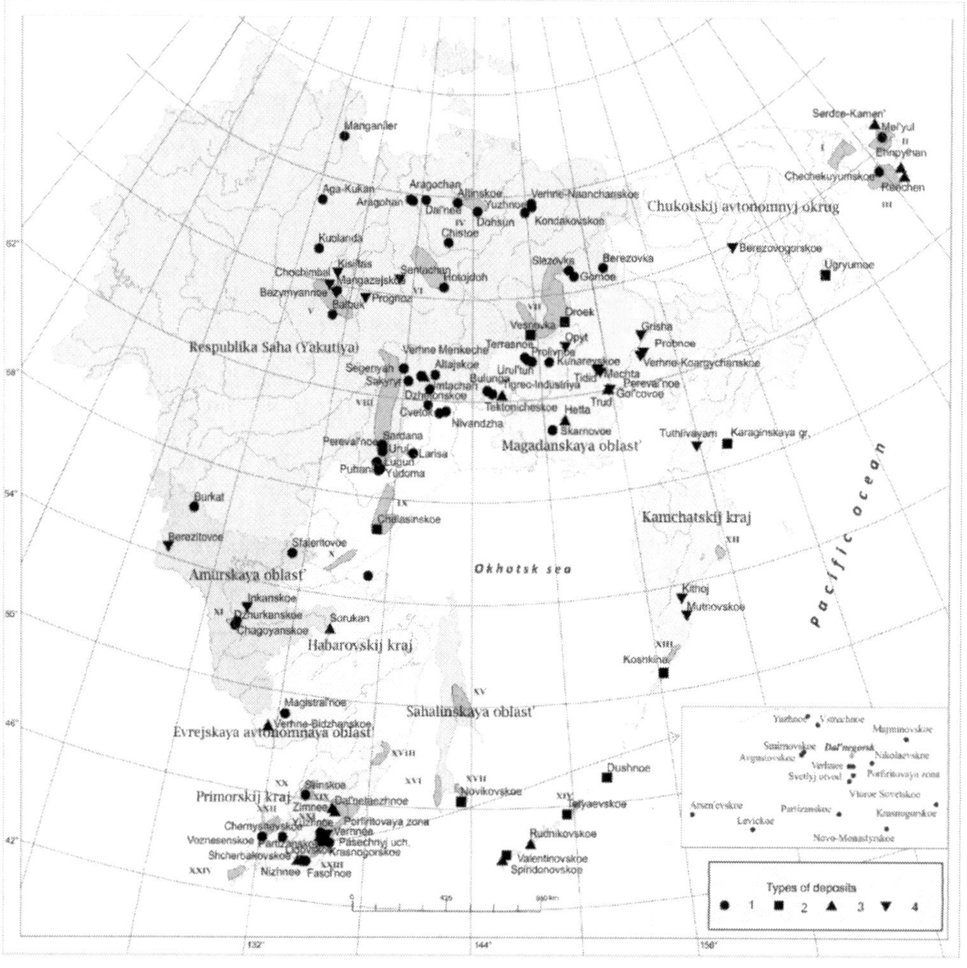

Fig. 1. Deposits, ore districts and promising to lead and zinc zones in the Far Eastern region. Sidebar - Dalnegorskij ore district. Types of deposits: 1 - lead-zinc (with copper, silver etc.), 2 - copper-lead-zinc, 3 - tin-polymetallic, 4 - gold-silver (with lead and zinc). Areas and zones are marked by numerals [2]: I – Krestovaya, II – Beringovskaya i Chegutunskaya, III – Chaplinskaya, IV – Berelekhskij, V – Sedyan-Kuel'skaya, VI – Tas-Hayahtahskaya, VII – Taskano-Omulevskaya and Prikolymskaya, VIII – Yudomo-Majskaya, IX – Chelasino-Maltanskaya, X – Ayano-Shevlinskaya, XI – Chagoyanskij, XII – Kozlovo-Kronockij, XIII – Asacha-Loptkinskaya, XIV – Kuril'skaya, XV – Srednesahalinskaya, XVI – Shebuninskaya, XVII – Anivskaya, XVIII – Samarginskaya, XIX – Kolumbe-Malinovskaya, XX – Ussurijskaya, XXI – Verhneussurijskij, XXII – Hankajskaya, XXIII – Dal'negorskaya, XXIV – Pos'etskaya.

There are more than 20 explored deposits of lead and zinc on the territory of Primorsky krai. Deposits Nikolaevskoe, Partizanskoe, Porfiritovaya zona, Verhnee, Yuzhnoe, Svetlyj otvod, Majminovskoe belong to «Gorno-metallurgicheskaya kompaniya «Dal'polimetall» («GMK «Dal'polimetall», «ГМК «Дальполиметалл») and are developed. Voznesenskoe deposit belongs to the company "RUSAL" as fluorite spar, development, and zinc ore are stored without processing (since 2013 development temporarily stopped). Most of the unallocated deposits (Arsen'evskoe, Verhnee, Silinskoe, Smirnovskoe, Shcherbakovskoe, etc.) were previously developed for the complex of components, including tin, lead, zinc, copper, etc.

The reserves of lead in the region are more than 1.5 million tons (including reserves of categories A+B+C_1 – almost 1 million tons), zinc – 2.2 million tons (categories A+B+C_1 – about 1.5 million tons). Lead resources are estimated at 1.6 million tons, zinc - 2.5 million tons.

Table 1. The main deposits of lead and zinc in the Far East region (data from VSEGEI, IAC Mineral, etc. sources)

Deposit	Reserves, thousand tons		Deposit	Reserves, thousand tons	
	lead	zinc		lead	zinc
Republic of Sakha (Yakutia)			Porfiritovaya zona	14	24
Arsen'evskoe	592	1926	Vtoroe Sovetskoe	7	13
Verhnee	145	35	Dal'netaezhnoe	7	11
Silinskoe	143	174	Avgustovskoe	6	11
Smirnovskoe	135	132	Arsen'evskoe	2	5
Shcherbakovskoe	65	47	Verhnij rudnik	1	5
Primorsky krai			Novo-Monastyrskoe	5	11
Nikolaevskoe	608	684	Pridorozhnoe	1	0,5
Partizanskoe	125	344	Voznesenskoe	-	210
Zimnee	140 (115)	206 (169)	Khabarovsky krai		
Krasnogorskoe	124	182	Festival'noe	4,4	1,9
Shcherbakovskoe	121 (117)	89 (82)	Pereval'noe	21	-
Smirnovskoe	93 (85)	134 (123)	Pridorozhnoe	0,7	0,5
Fasol'noe	92	111	Dzhatonskoe	56	93
Yuzhnoe	88	87	Nivandzhinskoe	31	29
Vstrechnoe	38	42	Magadanskaya oblast		
Svetlyj Otvod	37	37	Dukatskoe	83	80
Silinskoe	26 (19)	36 (26)	Gol'covoe	66	14
Levickoe	26	35	Mechta	32	34
Majminovskoe	61	72	Tidid	19	28

Note. In brackets – the remaining reserves at the beginning of 2013.

In the Khabarovsky Krai, lead and zinc contain deposits of the Komsomolsk tin district, the main useful component of which is tin. Ore deposits of lead and zinc – the ordinary and the poor, the deposits of reserves of these metals is small, in the processing of complex ores at the processing plant of the company "Dal'olovo" lead and zinc concentrates were allocated in the 1990s has accumulated substantial volumes of tailings, enrichment, containing over 100 to 115 thousand tons of lead and zinc.

In different geological-structural zones of the region revealed numerous small deposits and occurrences of lead and zinc: pre-explored deposits Chayatynskoe (inventory of lead – 109 thousand tons and zinc – 204 thousand tons), Majskoe (5.6 thousands ton and 21 thousands ton, respectively), Ampaarynd'inskoe (96 thousand tons and 101 thousand tons), Dzhatonskoe, Detan'dzha (71 thousand tons and 73 thousand tons), Nivandzhinskoe. Within Udoma–May trough a few dozen lead-zinc stratiform type ore manifestations form Lugun-Puhanil promising area inferred resources where estimated on lead 250 thousand tons and zinc also 250 thousand tons.

In the Magadan oblast, 4 deposits are considered, the ores of which contain lead and zinc. Until the 1990s, polymetallic concentrates from these deposits were sent to the country's plants for processing.

According to Rosgeolfond, VSEGEI, IAC Mineral, lead reserves in the Far East are accounted for in 34 deposits and amount to about 2.3 million tons (by categories $A+B+C_1+C_2$), zinc reserves are more than 4 million tons in 28 deposits. More than 90% of the region's reserves are concentrated in the Primorsky krai. Inferred resources of lead in the region is about 2.6 million tons, zinc – approximately 5.2 million tonnes resources of lead are 9 deposits (scale of reserves of up to 1 million tons each) and in more than two dozen small objects. For zinc, one deposit is large (more than 1 million tons), eight are medium (0.5 – 1 million tons), the rest are small.

Mining in the district DFO in recent years (2008-2017) was on the lead concentrates 10-16 thousand tonns annually in concentrates zinc – 14-32 thousand tons. Primary production provides the company «GMK «Dal'polimetall».

Mineral resource base of Russia. According to organizations of the Ministry of natural resources and ecology of the Russian Federation – Rosgeolfond, Iinformation and analytical center "Mineral", VSEGEI, in Russia there are more than 100 mostly lead and about 150 mostly zinc deposits (often these are the same deposits). Deposits containing more than 1 million tons of reserves of lead – 3, more than 1 million tons of zinc reserves – 13, they make up more than 75% of all reserves of lead and zinc of Russia, who lead amount to 19.5 million tons (categories $A+B+C_1+C_2$), inferred resources of about 40 million tons, for zinc – 60 million tonns, inferred resources of about 110 million tonns.

A significant number of large deposits with a high content of lead and zinc is located in the Altai region – Korbalihinskoe, Zarechenskoe, Stepnoe, Zaharovskoe, Talovskoe, Rubcovskoe. Reserves of deposit Korbalihinskoe are estimated at 26 million tonns of ore, containing nearly 470 thousand tonns of lead, and 2.2 million tonns of zinc, and 330 thousand tons of copper. The mine capacity of 1.5 million tons of ore per year with the production of about 120 thousand tons of zinc concentrate and 16 thousand tons of copper is planned to be brought to full capacity by 2021.

Deposits of Buryatia (Ozernoe, Holodninskoe and others), as well as Altai, have a high content of useful components. Since 2009, the corporation «Metals of Eastern Siberia» («Metropol» group) has been developing the Ozernoe deposit, whose reserves of $B+C_1$ categories are 1,31 million tons of lead, 6.9 million tons of zinc, and 3.9 thousand tons of silver. In 2010, the GOK with a capacity of 1 million tons of ore per year was commissioned, and it was planned to reach its design capacity in 2016, but the full – scale commissioning of the field did not take place. Delayed development and other large zinc and lead deposits (Holodninskoe, Nazarovskoe, Kyzyl-Tashtygskoe, Nojon-Tologojskoe).

Production of lead ore in Russia is about 180-240 thousand tons per year, production of lead concentrate - up to 320 thousand tons, production of refined lead - up to 110 thousand tons per year. More than 75% of Russian lead is extracted at the deposit Gorevskoe (Krasnoyarsk krai). Russia accounts for 2-3% of world production of lead, ores and lead concentrates are exported abroad: 90% of all exports go to China, the rest – to Kazakhstan and the Republic of Korea. In Russia there are no factories for the processing of primary lead. The first domestic plant for the production of lead from the concentrates of the deposit Gorevskoe is planned to be built in Sorsk (Republic of Khakassia). The consumption of refined lead in Russia is about 16 thousand tons per year. Russia's relatively small domestic demand for lead is fully satisfied by its production of secondary raw materials. The strategy of development of non-ferrous metallurgy of Russia for 2014-2020 and until 2030 assumes growth of consumption of the refined lead in Russia no more than by 5% a year.

Production of zinc concentrates in Russian fields is of the order of 350-360 thousand tons per year. About 75% of production is in the Urals. The main production is provided by the enterprises of the Urals Mining and Metallurgical company (UMMC). The production of refined zinc is up to 250 thousand tons per year (about 3% of world production). The consumption of refined zinc in Russia is about 225 thousand tons per year. Export of zinc ores and concentrates from Russia is about 70-140 thousand tons. Up to 50 thousand tons of refined zinc is exported from Russia annually to the Netherlands, Turkey, Belarus, etc. Annually about 30 thousand tons of refined zinc are imported to Russia.

The global mineral resource base. According to [3], the world reserves (probable and proven reserves) of lead in 2018 are 88 million tons, production in recent years – at the level of 4.7 million tons, zinc – 230 million tons and 12.6-13.4 million tons, respectively (table 2). The world identified resources (approximately reserves of category C_2 of the Russian classification) of lead are estimated at 2 billion tons, zinc-about 1.9 billion tons.

Table 2. Reserves at the beginning of 2018 and production in 2016 and 2017 of lead and zinc (million tons) by country [3]

Lead				Zinc			
Countries	Reserves	Mine production		Countries	Reserves	Mine production	
		2016	2017			2016	2017
Australia	35	0.453	0.450	Australia	64	0.965	1
China	17	2.340	2.400	China	41	4.800	5.100
Russia	6.4	0.250	0.250	Peru	28	1.330	1.400
Peru	6	0.314	0.300	Mexico	20	0.670	0.680
Mexico	5.6	0.232	0.230	India	11	0.682	1.300
USA	5	0.346	0.313	USA	9.7	0.805	0.730
India	2.2	0.147	0.150	Kazakhstan	13	0.340	0.360
Bolivia	1.6	0.075	0.079	Canada	5.4	0.322	0.340
Sweden	1.1	0.079	0.080	Bolivia	4.8	0.490	0.500
Turkey	0.86	0.076	0.070	Sweden	3.8	0.257	0.260
Other countries	7	0.400	0.390	Other countries	33	1.890	1.520

China is the world's leading producer of lead and zinc. Then follow Australia, USA, Peru, Mexico. They jointly provide 70% of the world's lead production and 75% of zinc in concentrates. The Republic of Korea is a major producer of lead – about 500 thousand tons/year. Over the past decade, zinc production in the world has increased by 40%, the most significant growth occurred in Asia. There are about 15 companies in the world producing lead in concentrates in the amount of 10-260 thousand tons per year and more than 20 companies producing zinc concentrates in the amount of 40-740 thousand tons per year.

Conjuncture. The market of lead and zinc, as well as the entire mineral market, is subject to constant fluctuations. According to the International Lead and Zinc Study Group, world periodically there is a scarcity and a surplus of concentrates of lead and zinc on the level of several tens of thousand tons. Starting in 2012, the production volume of zinc in the world has increased steadily, on average by 2-3% annually, which was stimulated by the growing steel production in the world.

In the world market of lead and zinc there are multidirectional and quite significant price fluctuations. In the period 2001-2017, the price of lead was in the range of 476-3235 USD/t (peaked in 2007), in zinc – 886-3235 USD/t (peak 4400 dollars/t was in December 2006). In 2016, 1 ton of lead cost $1819, in 2017 - $2309. Zinc in 2016, was the fastest growing in the price of industrial non-ferrous metals: during January-October 2016 the price has increased by 60% (on the London stock exchange zinc was worth 2320 USD/t.) The average annual price of zinc in 2016 was 2093 USD/t, in 2017 – 2896 USD/t (the forecast of some analysts to 2018 – 3258 USD/t, 2019 – 3057 USD/t.)

Experts estimate that by 2020 there are 30 possible projects, the cost of which is about 8.4 billion dollars. and what if all these projects are initiated on time, they will produce about 2.2 million tons of zinc per year by 2020. The World is secured for many decades the mining and production of concentrates of lead and zinc.

Analysis of the use of resources in the region. Primorsky krai until the 1990s. was one of the main actors in the USSR for the extraction and processing of lead-zinc ores, despite its high cost and transport costs for deliveries to customers in western part of the country. The modern state lead-zinc industry in the region is such that it is not to be considered as a significant economic factor for the far Eastern Federal district and Primorsky krai: the gross regional product which mining is 2-2,5% (most of this income accounts for coal), in the structure of gross value added of the activity «extraction of minerals» – 1-1, 2%, investments in fixed capital in the industry «mining» – 1.4-1.8 billion rubles (of the total investment of the edge of 123-140 billion rubles.).

The leading enterprise for the extraction and processing of lead-zinc ores in the region is «GMK «Dal'polimetall», which has in addition to the mines of the enterprise «Central processing factory» and «Lead plant». The company's annual ore production in previous years reached 955 thousand tons, including at the Nikolayevsky mine up to 500 thousand tons. The processing factory has an actual annual capacity of more than 700 thousand tons of ore (project 1200 thousand tons) and produces lead (more than 12 thousand tons) and zinc (more than 19 thousand tons) concentrates, in passing in lead concentrate are extracted silver, bismuth, copper, in zinc concentrate – silver and cadmium. The capacity of the processing factory allows to produce at least 20 thousand tons of lead and 32 thousand tons of zinc in concentrate. The Lead plant produced lead from concentrates in the late 1990s; zinc was not produced by the plant. In the future, the reconstruction of the plant with an increase in the production capacity of lead to 78.8 thousand tons per year was expected. However, for economic reasons, the plan was not implemented and the production of lead from primary ores decreased to 1.4 thousand tons. The plant switched to the production of lead from battery scrap. In 2016, «GMK «Dal'polimetall» produced 13,765 tons of lead concentrate and 25,512 tons of zinc concentrate. Income from sales amounted to RUB 2.5 billion. In 2017, production was approximately at the same level. The company exports the entire volume of lead and zinc concentrates to the Republic of Korea, Japan and China.

The mineral resource base of «GMK «Dal'polimetall» is concentrated in the nearby territory (within 20-40 km from the processing factory). According to the company, the reserves of ore of categories $B+C_1+C_2$ are about 25 million tons with an average content of 2.73% lead, zinc – 3.58%, silver – 59.5 g/t. There is a possibility of increase in reserves in all fields operated by the company. Cheremshansky ore cluster with forecast resources of 26 million tons of lead-zinc-silver ores is promising.

«GMK «Dal'polimetall» characterizes its position in the industry as "an average power mining company producing up to 15% of lead and zinc in the same concentrates from their total production in Russia." The position of the company is unstable despite the not only negative, but also positive changes in the situation. This shows a comparison of production and sales of the main products in 2013 and 2014 (according to the reports of the company published on its website). Revenues for 2014 amounted to 1822 million rubles. Due to the fall of the ruble against the dollar, the increase in revenue from the sale of commodity concentrates in 2014 compared to 2013 amounted to 216,061 million rubles, but the sales price was lower than the world average. In recent years, company has been experiencing difficulties in marketing due to its high cost, as the work is carried out on worn-out and obsolete equipment, and low competitiveness of its products.

Meanwhile, according to the company's own assessment, its position and prospects in the region are the most favorable in comparison with other mining companies of the region and the region. In the area of its activity there are deposits of lead, zinc, silver, gold, tin, tungsten, boron, coal, building materials and other mineral resources. Diversification of production for the extraction and processing of these resources can provide the company with a stable and constant growth of gross product for the long term. The geographical location, natural conditions and environment of the territory in general are the most favorable among the subjects of the far Eastern Federal district: it is located on the coast of the sea of Japan, in the North it borders with the Khabarovsky krai – the industrial center of the Far East, in the West – with China and in the South with the DPRK, has stable relations with the Republic of Korea. The company has access to the ice-free harbor. Neighboring countries (Japan, Republic of Korea) do not have their own mineral resources base, are a stable market. China, although it has more resources of the same types of minerals, but cannot fully load the capacity of its processing enterprises, and therefore is also a market for the products of the company.

The main and decisive negative factor is the fact that «GMK «Dal'polimetall» does not have its own funds, as a result, the company cannot count on an annual sufficient profit, is not able to pay dividends to its shareholders.

Conclusions. Russia is reliably provided with reserves and resources of lead and zinc for the long term without the Far Eastern region, as production is steadily carried out in the regions with the existing infrastructure, with existing mining enterprises that claim to create Russia's leading centers for the extraction of lead and zinc. The Far Eastern region is currently in the position of a territory whose mineral raw materials (with the exception of gold and silver) do not need a region or country and are sold in nearby countries. The degree of diversification and innovation of the regional economy is not very high, having a pronounced focus on the extraction of precious minerals with a low degree of processing of other small amounts of ore resources.

In the Far Eastern region there are significant deposits of lead and zinc. Substantially support the lead-zinc sector of the region is possible in the case of exploration of new deposits with high-grade ores. The industry needs cardinal solutions for its new organization, management of a specialized and diversified holding that would take in its sphere of interests all known and promising deposits and objects of lead and zinc, including their valuation, exploration, the creation of a modern processing base (possibly based on assets «GMK «Dal'polimetall» or a new high-tech base in the Komsomolsky ore district of the Khabarovsky krai in conjunction with other industries (tin, copper, nickel, rare metal), as well as production of the final. The construction of a chemical and metallurgical production in the area of Solnechny GOK in the Khabarovsky krai to produce a wide range of products based on polymetallic ores from the Far East and Transbaikalia was reported back in 1982 [4].

In the region, due to its spatial and economic isolation, it is necessary to create new industrial sectors for it and to develop intra-regional use of raw materials, the construction of own metallurgical enterprises, and the production of final products.

References

1. G .I. Arkhipov. *Mineral'niye resursy gornorudnoy promyshlennosti Dal'nego Vostoka. Strategicheskaya otsenka vozmozhnostey osvoyeniya.* Khabarovsk: Institut gornogo dela DVO RAN. 820 p. (2017).

2. V. I. Sukhov., Yu. I. Bakulin., N. P. Loshak i dr. *Metallogeniya Dal'nego Vostoka Rossii.* Khabarovsk: Dal'nevostochnyy NII mineral'nogo syr'ya. 217 p. (2000).

3. Mineral commodity summaries 2018: U.S. Geological Survey, 200 p., https://doi.org/10.3133/70194932

4. M. E. Uskov, A. S. Mardoyan, Yu. G. Melik-Stepanov. Issues of development of ferrous and non-ferrous metallurgy in the Khabarovsk Territory. Scientific-practical conference "Problems and main directions of integrated development and placement of the productive forces of the Khabarovsk Territory until 2000". Section 6. Khabarovsk. P. 15-17. (1982).

Forming long-term competitiveness of a mineral sector as part of economy innovative development strategy

Natalia Lomakina[1*]

[1]Economic research institute of Far East division of Russian Academy of Sciences, Khabarovsk, Russia

Abstract. The research was focused on the processes of structural diversification in a Far Eastern mineral sector and aided in obtaining the characteristics of new market position and space specializations in the mineral-raw complex of the region. It illustrates its influence on the dynamics of investments in the region. It highlights the high value of the mineral sector in Far Eastern federal district, its stable competitive positions in the regional and national economy. It shows the conditions of long-term competitiveness of the mineral raw complex of Far East in the national and foreign economies (these conditions were formed as part of foresight-forecast for the development of Pacific Russia up to 2050). The research also gives predictive estimates of possible ways of the structure and dynamics of long-term development of mineral-raw complex of Russian Far East. **Key words:** Mineral sector, foresight-forecast, synthesis of economic and technologic forecasts, Far East

1 Introduction

The national model of development through innovation that is being designed in Russia nowadays implies the transition from commodity-based economy to knowledge-based economy. In addition, the announced desire of the country to move to development through innovation allows for the continuation of the intensive development of natural resources. In fact, the exploitation of natural resources (in particular mineral-raw resources) in foreseeable future remains the source of means for the innovative transformation of the national economy.

This situation actualizes the issues of revealing the opportunities and characteristics of the perspective development of resources regions. An example of a large macro-region of resource specialization is Far East of Russia that is characterized by the abundance of natural resources and the development of resource complexes (particularly mineral-raw complex and fuel-power complex) that are the branches of national specialization.

The opportunities of perspective development for the region are alternative: either constructing a proving ground in the southern part of the Russian Far East, or moving to the concept of industrial diversification on the basis of exploiting the brand-new technologies,

* Corresponding author: lomakina@ecrin.ru

connected with specific resources and conditions of Far East (a variant of "new industrialization") [1]. These kinds of activity also include the sector of national specialization of the region – extraction of mineral resources. But is Far Eastern mineral sector ready for such challenges? And what are the prospects of its long-term competitive development? What are the chances to maintaining traditional raw-material base and forming innovative-technological chains "extraction of new mineral-raw material – inventing new products – forming new branches of mineral-raw complex"? These issues are studied in the article.

2 Theory of experimental research

The research on the theoretical and methodic issues of the contribution from the natural resources (particularly mineral resources) into socioeconomic development [2-7], quantitative evaluation for economic systems of various levels [8-12] not only remains acute, but also moves to a newer theoretical-methodic level [13—18 et al].

The issues of forming and elaborating a conceptual apparatus, applying new methods to estimate the interconnection "resource sector – resources region" are discussed in the works [14, 15, 17, 18 et al]. This method aids in revealing and estimating new factors of forming "return" trajectory of the development of mineral resources in the resources regions. "Instead of fading and exhausting the economic potential … resources region (as a result of exhaustion … of "conventional" reserves of mineral resources) discussion can be focused on "return" trajectory – development and use of previously "overlooked" mineral-raw resources and "unnoticed" sources of "unconventional" kinds of raw material. In addition, the factors leading to the transition to the "return" trajectory are not only the technologies, but also the new quality of institutional environment" [14, p. 12].

The issues of the change in the canonical scheme of the development of mineral resources of the region – transition from fading to constant development through innovation – are acute for raw-material-rich territories of the Russian Far East, examined in the article. At present, they are also characterized by the growth of technological availability of mineral resources, development of "old" deposits under new technological parameters [19].

One of the main aspects of the problem is the consideration of the effective of resources projects and their contribution to the development of the region. These issues are the subjects of such branches of research as resource economy, economy of natural resource use, regional economy, spatial economy; effectiveness of investments, and particularly the investments to public sphere [20-22], natural resource rent, estimation of multipliers and system effects of natural resource projects [23-25 et al] and their regional localization [e.g. 11, 14, 15].

The scientific value of such researches lies in the adaption of various theoretical approaches and practical implementation of the methods of estimating the interconnection "resource sector – resources region" for certain territories in the conditions of forming innovative strategies of economic growth.

3 Results and discussion

The determining role of the natural resources in the economic development has been the distinctive feature of the Russian Far East for all the time of its development [11]. Far Eastern Federal District is the center of all nationwide reserves and almost 100% extraction of diamonds and tin, over 1/3 reserves of gold and silver (45 and 65% extraction, respectively), almost 80% production of tungsten concentrate and 100% extraction of antimony [26]. The research of structural diversification (industrial and territorial) in the mineral sector of Far Eastern District has shown that the most important changes were the following:

- significant decrease (naturally expressed) in the extraction of such metals as tungsten (43.3 % in 2015, compared to 1990), lead (37.3 %) zinc (21.4 %), tin (9%), and copper (7%);
- beginning of forming new industrial specializations of mineral-raw complex of Far East: extraction of raw material for the production of nickel and cobalt (Kamchatka Krai), titanomagnetite concentrate (Amur Oblast), and iron ore concentrate (Jewish autonomous oblast);
- confirmed value of precious metals as "structural leader" of the mineral sector
- change in the internal structure of gold mining in the region: in the 1990's, gold mining from placer deposits prevailed, and after 2010, over 70% of gold in the region is extracted from primary deposits;
- growth in the technological availability of the resource base of gold mining in Far Eastern district that was driven by major companies entering the region and developing new technologies of estimating, extracting, and processing the mineral resources (e.g. overestimation of the deposit "Natalka", autoclave technologies, etc.)

However, with the presence of a number of positive changes, the main result of the changes in the industrial structure of Far Eastern mineral-raw complex is its significant narrowing and intensification of mono raw material nature of this sector.

As far as the territorial structure is concerned, the role of northern regions has remained important, (such resgions as Yakutia, Magadan Oblast and Chukotka Autonomous Okrug). They still produce 4/5 products of the complex. In the southern part, with saving its share in the structure, was subject to significant interzonal changes in 2016, compared to 1990: sudden decrease of the share in Primorsky Krai (6,4 times) with the increased contribution from Khabarovsk Krai (1,3 times) and Amur Oblast (1,2 times). The share of mineral-raw complex in the industrial structure has been increased as a whole in Far Easter District – from 14-15% in the 90's to almost 30% by 2016. The role of the mineral sector in Far Easter District keeps its strategic importance in the national economy.

One of the conditions of economic growth is investment activity. One of the indicators of investments quality can be their structural proportion between investments into raw-material and non-raw-material spheres of regional economy. We have calculated such structural proportions that were formed in the economy of Far Easter district in general and in separate subjects of the Russian Federation in the period from 2005 to 2006 years (table 1). The analysis shows that the share of investments to the extraction of mineral resources in the Russian Federation in the overall structure of investments into the basic capital has been 1,44 times increased (reaching almost 20% of all investments). In Far Eastern District, this growth was 1,87 times, and the share of investments to Far Eastern District reached 40% in the basic capital.

In the group of far eastern regions with conventional mineral-raw specialization, the share of investments into the raw-material sector in the examined period has been growing, with aided in the growth of competition of this activity in the overall structure of the industry (table 1). Significant growth of the share of investments to mineral-raw complex is noticeable in Kamchatka Krai, Jewish autonomous oblast, which leads these subjects of the Russian federation to the regions with mineral-raw specialization.

The dynamics show that for the majority of the far-eastern subjects of the Russian Federation, investments into mineral-raw complex are the basis of investment activity in the region. In confirms the high competitiveness of the mineral sector, although such a structure of investments forms the basis for the future "raw-material" image of Far Easter economy.

Table 1. Investment activity of the mineral-raw complex of in Far Eastern district

	Share of «mineral resources extraction without fuel-power resources» in industry		Share of investments to the extraction of mineral resources in the overall structure of investments into the basic capital, %	
	2005	2016	2005	2016
Russian Federation	2,8	3,0	13,4	19,3
Far Eastern district	31,2	29,6	21,0	39,3
Republic of Sakha	63,4	51,7	45,5	51,2
Magadan Oblast	67,5	83,6	14,1	69,5
Chukotka autonomous okrug	31,7	88,4	0,3	36,4
Kamchatka krai	11,8	16,4	4,4	21,9
Jewish autonomous oblast	5,0	13,3	0,3	36,4

Source: calculations from the official data of Rosstat.

The issue of the conditions of long-term competitiveness of the material-raw complex in Far Eastern District in the national and foreign economies (particularly southwestern parts of Asia) becomes legitimate. Estimation of the opportunities of the long-term development of the mineral-raw complex, its perspective role in the regional and national economy was among the most pressing issues in the formation of foresight forecast of the development of Pacific Russia til 2050 [1]. "Inventory cut" of fundamental and applied results of the researches in the institutes Far East division of the Russian Academy of Sciences in the field of mineral resources has shown that the opportunities of forming innovation-technological chains "extraction of new kinds of minerals - invention of new products – formation of new branches in the mineral-raw complex" in Far East (until 2050) will be based on:

- development of new unconventional kinds and sources of mineral raw material;
- widespread introduction of complex processing of multicomponent and refractory ores;
- activation of industrial development of ore-placer waste of mining industry
- formation of new mining regions within Far East.

In the perspective spatial structure of the mineral raw complex, leading positions are still held by Republic of Sakha, Magadan and Amur oblasts, Chukotka autonomous okrug. The role of Khabarovsk Krai, Kamchakta Krai and Jewish autonomous Oblast will be significantly increased in the structure of the regional mineral sector. One of the leading branches of Far Eastern economy will be the extraction of precious metals and diamonds. The development of mining and extraction of base metals in Far Eastern District is connected with the development of new deposits of copper and polymetallic ores (including those containing precious metals), antimony, niobium, rare earth metals, as well as vanadium-containing, titanomagnetite deposits. With the relevant level of production of the mineral raw-material base, there are reasons to count not only on the reservation and growth of the complex in conventional sub-branches (table 2), but also on the formation of new productions in its structure.

Table 2. Predictive estimates of production output of the mineral-raw complex of Far East

Types of production	2015	2025	2050

Gold, tons	117,8	76-150	45-240
Silver, tons	1418,2	1000-1500	230-1700
Lead, thousand tons	12,3	24	24
Zinc, thousand tons	22,2	35	35
Copper, thousand tons	1,6	50-100	Н.д.
Tungsten, thousand tons	2,6	3,5-4,5	Н.д.
Nickel, thousand tons	4,6	20-25	Н.д.
Titan (dioxide), tons	-	10-15	152,0
Berrylium (dioxide). tons	-	100,0	150,0
Ferriferous ores , concentrate million tons	2,9	30-45 / 10-20	65 / 20
Manganese ores, thousand tons	5,0	400,0	500,0

Source: [26], [1, p. 341].

Thus, it was possible to obtain the estimates of scientific-technologic opportunities of the mineral sector of Far East in the form of results of fundamental and applied researches. To determine the opportunities of long-term development of the sector and its competitiveness, these results need to be "converted" into economic parameters, scenarios and and institutional-economic conditions for such development need to be formed and estimated.

4 Conclusions

As part of the implementation of these goals, the article was focused on the processes and achievement of the estimate of structural diversification (industrial and territorial) in the mineral sector of Far East. It also characterizes the investment activity in the mineral-raw sector, and illustrates its high value, stable competitive positions in the regional and national economy. It shows the conditions for long-term competitiveness of the mineral-raw sector in the national and neighboring economies, formed as part of the foresight-forecast of the development of Pacific Russia until 2050. The most important conditions, along with the productive and technological conditions, providing the stable and competitive development of the mineral sector of Far East in the long-term prospect, are institutional conditions for forming its "new image"

References

1. Synthesis of scientifically technical and economic forecasts: Pacific Russia – 2050 / edited by P.A. Minakira, V.I. Sergienko, Vladivostok: Dalnauka, pp. 912. (2011).
2. Auty R. Sustaining Development in Mineral Economies. The resource curse thesis. N.Y., Routledge. p. 272. (1993).
3. Auty, R. Natural Resources, the State and Development Strategy // Journal of International Development. **9**. pp. 651–663. (1997).
4. Gylfason, T., Herbertsson T., Zoega G. Macroeconomic Dynamics, **3**. pp. 204–225. (1999).
5. Sachs, J., Warner A. Natural Resource Abundance and Economic Growth / NBER Working Paper 5398, Cambridge, Massachusetts. p. 54. (1995).
6. Barbier E.B. Natural Resources and Economic Development. Cambridge University Press. p. 410. (2005).

7. Harmonious Development Through Innovation 2008. CCICED Annual Policy Report. Peking: China Council for International Cooperation on Environment and Development. p. 317. (2009).
8. E.V. Ryumina, A.M. Anikina, Problems of prognostication. **5**. P.106-125. (2007). http://www.ecfor.ru/fp/index.php?pid=archive.
9. N.N. Mikheeva, Evaluating Russian Far East's Resource Sector Based on "Input – Output" Tables // Spatial Economy. **1**. pp. 72-86. (2006).
10. M.N. Uzyakov. Problems of prognostication. **2.** pp. 3-18. (2011).
11. N.V. Lomakina, Mineral Sector of Russian Far East: Potential for Development. Edited by P.A. Minakir. Russian Academy of Sciences, Far Eastern Division, Economic Research Institute. Khabarovsk city. p. 240 (In Russian). (2009).
12. D.S. Kuznetsov. Problems of development of the territory. **6 (92).** pp. 68-82. (2017).
13. O.S. Anashkin, V.A. Kryukov. Mineral resources of Russia. Economy and management. **1**. pp. 24-33. (2015).
14. Resources Regions in a "New Reality". Edited by V.V. Kuleshov. Novosibirsk: Institute of Economics and Industrial Engineering SB RAS. p. 308 (In Russian). (2017).
15. Mineral Resource Sector of the Asian Part of Russia: How to Ensure the Social and Economic Impact. Edited by V.V. Kuleshov. Novosibirsk: Institute of Economics and Industrial Engineering SB RAS. p. 352 (In Russian). (2015).
16. S.V. Doroshenko, A.G. Shelomentsev. Modern technologies of management.. **3 (51).** (2015). Article number: 5106. Date issued: 2015-03-08. Access mode: http://sovman.ru/articl/5106/
17. V.A. Kryukov, E.O. Pavlov, Integral estimation of the efficiency of subsoil use systems (comparative evaluation) // Mineral resources of Russia
18. A.K. Belan, V.V. Schmat, The analysis of the influence from resource and non-resource factors on the growth of economy in Tomsk Oblast with the help of cognitive approach // Vestnik NGU. Series: socioeconomic sciences. **15 (1)**. (2015).
19. N.V. Lomakina. Spatial Economy. **1**. pp. 59-82. (2018). DOI:10.14530/se.2018.1.059-082.
20. O.S. Nagaeva. Regional economy and management: electronic scientific journal. **4 (48).** Art. No.4804. Date issued: 2016-11-05. Available at: http://eee-region.ru/article/4804/
21. E. N. Sindyashkina. Problems of prognostication. **1**. pp. 140-147. (2010).
22. T. S. Novikova, The analysis of social efficiency of investment projects. – Novosibirsk: Institute of Economics and Industrial Engineering of the Siberian Division of RAS. pp. 221. (2005).
23. N.N. Mikheeva, T.S. Novikova, V.I. Suslov. Problems of prognostication. **4**. pp. 78-90. (In Russian). (2011).
24. A.A. Shirov, A.A. Yantovsky. EKO. **2**. pp. 40-58. (In Russian). (2011).
25. Gunton T. Regional Studies. **37**. P.505-519. (2003).
26. Vologin V.G., Lazarev A.V.. The current status and prospects of the development of the mineral-raw complex in Far Eastern Federal District. Materials from the forum Minex-DV. Magadan, 2016. http://minexforum.com/wp-content/uploads/2016/07/1.1.1-Vologin-V.G.-Dalnedra_Sostoyanie-i-perespektivy-razvitiya-min-syr-bazy-DFO.pdf. Date of access: 03.11.2017
27. A.I. Khanchuk, N.V. Lomakina, V.V. Ivanov. Vestnik DVO RAN. **5**. pp. 113-121. (2012).

Methodological problems of natural and natural-technical systems based on the example of Primorsky Region

Lubov Lipina[1*], *Vitalii* Usikov[1] and *Nikolai* Grehnev[1]

[1]Mining Institute of Far eastern branch of Russian Academy of Sciences, Khabarovsk, Russia

Annotation. The paper shows that in the development of indigenous ore deposits, four main types of solid waste are formed, which affect natural systems and form the main processes of mining technogenesis. In general, the technogenic impact of mining processes on ecosystem components is diverse, so it becomes necessary to separate the stages of production. The technological and geological factors for the development of specific mine objects are of significant importance for the ecological state of natural-technical systems (NTS), the results of which should be taken into account in the ecological mapping of the NTS, taking into account the strongest component - the lithogenic base, which determines the geochemical background (the medium in which the chemical reactions take place).

1 Introduction

Natural-technical or natural-technogenic systems (NTS) significantly changed the natural environment. At present, on the Earth, equivalent systems are considered to be natural systems (receiving control from nature), technogenic (receiving control signals from society), both of which form natural-technogenic systems. The growing technological pressure on natural ecosystems leads to their quick and often irreversible destruction, which gradually assumes a global character. At the same time, the paradox of the situation is that the progressive degradation of nature takes place against the backdrop of the rapidly growing expenditures of mankind for its protection; and the energy resources necessary to preserve nature at the present level can be obtained only by technogenic destruction of this nature foundation - the Earth's lithosphere.

In his writings Academician VI Vernadsky noted that mankind has become a powerful geological and, possibly, a cosmic force capable of transforming nature on a large scale [1]. The life activity of mankind is not an external, accidental phenomenon on the earth's surface, it is closely connected with the structure of the earth's crust, enters into its mechanism and in this mechanism performs important functions without which man could not exist. [2] Interpreting Vernadsky's words we can say that no matter what extent the bowels of the earth are exhausted, society cannot exist when their use is no longer possible. Therefore, it is not possible to preserve the natural environment in its natural state during the development of subsurface resources. In this connection, along with the development of

* Corresponding author: geo-lipina@rambler.ru

the concept of state sustainable development, strategic development becomes increasingly important in relation to specific producing regions. They are based are based on knowledge of the current state of the environment in the areas of former, existing and projected mining enterprises, as well as on predictive estimates of the trend of this state change for the future. The patterns of formation and development of regional natural-technical systems can not be studied without cartographic materials.

2 Depth of the problem

Many scientists have noted the interaction of the natural environment and technical objects in the areas of industrial development in their works, such as Buzmakov SA, Van AV, Zinyukov Yu.M., Revzon A.L. [3-5] The scientific works of a number of authors aimed at increasing the geoecological safety of the PTS have been considered: in the works on methodological provision, V.I. Osipova, I.I. Mazur, N.F. Reimers, N.N. Moiseeva, V.I. Danilov-Danilyan, and others; on the assessment of geoecological safety in the interaction of technical systems and the natural environment - V.A. Gracheva, V.M. Kotlyakova, A.L. Revzon, Boris G. Saxin, A.M. Ivleva, TN. Alexandrova, L.T. Krupskaya and others; on environmental design and expertise - K.N. Dyakonova, A.V. Donchevoy and others. It is established that the natural and technical systems that are formed during the mining of mineral deposits include natural, ore-dressing combines that function as a whole, have not been studied well enough. Therefore, the solution of the main issues related to the problems of research of NTS with a developed mining industry based on the example of Primorsky Krai, as well as the creation of a basis for constructing survey forecast ecological maps, is an urgent task.

3 Methods of work and results

A set of scientific methods was used, including: analysis and generalization of the literature data; mathematical and statistical processing of the obtained data, cartographic analysis.

In environmental terms, the main sources of environmental impact are mining and processing plants, the negative impact of which has recently acquired a large scale, including in the Primorsky Territory of the Far East.

Natural and man-made systems are increasingly replacing the natural environment. Analyzing the environmental component of the NTS, which can be said that any production process is necessarily interconnected with the natural environment, the components of which usually act as labor items. At the same time, as a result of the work, substances, energy and information that are as part of natural components that did not previously participate in the technological cycle are formed[6,7].

Considering the development of Primorsky Krai (the Southern margin of the Far Eastern economic region) as a region with a highly developed mining industry, it can be noted that deposits of coals, lead, zinc, tin, gold, silver, tungsten, fluorspar and boron ore are widely represented here.The mining industry supplies raw materials to non-ferrous metallurgy, as it develops polymetallic ores and fluorspar in the territory of the region. There is also a number of developed coal deposits.In the province, tungsten and lead concentrates, tin, boron products are produced. [8] In the volume of industrial production in the Primorsky Territory, the share of non-ferrous metallurgy is 3.2%, but its significance is determined not so much by the total volume of production as by the city-forming nature of its leading enterprises and by the presence in edge of a unique raw materials base for fluorspar, tungsten, polymetals. The leading enterprises of the non-ferrous metallurgy industry in the Primorsky Territory include OJSC MMC Dalpolimetall, OJSC Primorskiy

GOK, OJSC Yaroslavsky GOK, OJSC Lermontovskaya GRK. Primorye produces 80% of the Russian volumes of tungsten concentrate and fluorspar, 30% of lead, 15% of zinc. As far as mining chemistry is concerned, Bor's open-cast mines contain raw materials of boron-containing raw materials, unique for the country.

The extraction of gold in the province was conducted from the middle of the nineteenth century on small but numerous placers, which were quickly worked out. At present, the Primorsky Region is significantly inferior to other subjects of the Far Eastern Federal District with a developed gold mining specialization in the mineral and raw materials sector of the economy. Due to the active geological study of the gold content of the region in recent years, a number of previously known and newly identified ore-placer sites have been reassessed and new deposits have been discovered. Containing gold were polymetallic, tin, tungsten ores, more recently known deposits of the region:: Primorskoye 11.49 g / t gold, Vostok-2 - 1.4 g / t, Silinskoe - 0.98 g / t, Taiga - 0.63 g / t, etc. [9]. According to the official data of the gold deposit on the territory of the region, registered in the State Reserves Committee, according to VSEGEI, 2007 with additions from various sources are: Glukhoye and Morskoye (reconnaissance), Progress and Silan (undistributed), Vostok-2 and Taiga (being developed).

The extraction of ore gold (tailing production) in recent years is mainly carried out at the Vostok-2 field, about 200-250 kg per year. The tin deposits of the Primorsky Territory are of three types: tin-quartz, tin-silicate and tin-sulfide. The Tigrinoye deposit, the largest in the region, is of the first type,the ore of which is of poor quality. Among the small deposits of tin-silicate and tin-sulfide types, there are objects with rich ores. The extraction of tin, at present, is conducted in small volumes.

Nikolaevskoe deposit belongs to the skarn-polymetallic type (Dalnegorsky ore district), which is leading in the extraction of non-ferrous metals in the Dalnegorsky ore district of Primorye. In addition to the basic elements (lead and zinc), industrially significant concentrations in the ores are: Ag, Bi, Cd, In, Se, Te, Cu, which can be extracted and partially recovered during ore processing.

The main mining enterprises of the Primorsky Territory are shown in Fig. 1.

Anthropogenic activity inevitably leads to a disturbance of the ecological balance in nature.

The concept of ecological balance has a profound meaning, since it relies on an extensive system of scientific knowledge and ideas about the status and properties of biogeocoenoses, as a set of abiotic environmental factors within a single territory, interconnected by the circulation of substances and the flow of energy (natural ecosystem). Natural balance due to anthropogenic impact tends to shift. In the development of indigenous ore deposits, four main types of solid waste are formed, which affect natural systems and form the main processes of mining technogenesis.

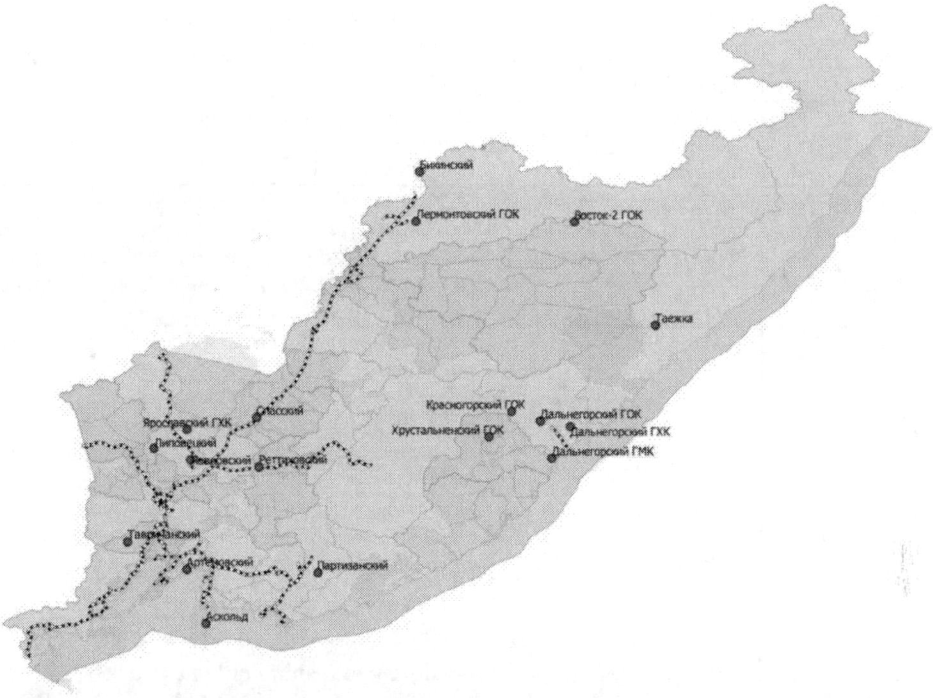

Fig.1. Mining industrial complexes of Primorsky Krai

These include: dumps of overburden, ore-bearing rocks and sub-standard ores; solid phase of processing waste, effluents of liquid saturated solutions; gas emissions. One of the features of the mining industry, as a result of which, the greatest amount of waste is generated, whose volumes are 60-95%, and for rare metals and more from the volume of processed rock mass. Waste, in addition to occupying large areas, often very valuable land (forest fund) lead to significant changes in the environment. Not only the original landscapes change, but the natural geochemical equilibrium also changes. There is an intensive pollution of soils and reservoirs in the adjacent areas, a decrease in their biological productivity. At the same time, the negative processes associated with the violation of the main components of the environment are a long-term impact after the cessation of the production activities of enterprises.

An insignificant amount of solid wastes are losses of ores on transport communications within the mining areas (from mines to concentrating mills etc.), as well as dust emissions from crushing and filtering and drying shops of concentrating mills. In general, the technogenic impact of mining processes on ecosystem components is quite diverse, especially in the full technological cycle of production, so it becomes need to consider the various stages of production (Table 1).

If the stages of development and the stages of individual mining work are to be considered, it is necessary to assess the impact of mining on the environment in the aspect of two issues having important methodological significance: what and in what volumes is introduced and how can the ecosystem respond to this or that intervention.

At the first and second stages of mining, it must be taken into account that ore deposits and their primary haloes are natural foci of chemical contamination of the components of the natural environment by heavy metals, and the products of technogenesis that arise during their industrial development form ecologically dangerous halos and scattering streams that significantly affect the state and metamorphization ecosystems.

Table 1. Stages of development of mining operations and sources of impact on ecosystems

Stages	Stage of development and stages of mining	Sources of impact on ecosystems
I	Exploration and development of the deposit	Mechanical dumps of enclosing rocks and sub-standard ores; mine drainage of mineralized waters of the oxidation zone
II	The stage of extraction and transfer to the surface of the ore mass	Mechanical dumps of enclosing rocks and sub-standard ores; mine drainage of mineralized waters of the oxidation zone; gas-dust extraction during drilling and blasting operations; dusting during transportation, deflation of dumps and ore stores, etc.
III	The stage of ore dressing	Construction and formation of tailing ponds and sludge pits; emergency emissions from slurry pipelines and tailing dumps, ventilation systems of the concentrating mills and transportation of ore; outputs of technological preparations and suspensions into open reservoirs
IV	Metallurgical redistribution of ore concentrates	Technological liquid and gas emissions; metallurgical ash dumps and slag-dumps; emissions of auxiliary industries

When considering the III and IV stages of work, depending on what concentrations and types of toxicants are inherent in this geoecological type of ores, appropriate technological regulations for ore processing should be developed and implemented. In essence, the indicator of toxicity of an ore deposit is, first of all, the level of sulphide content (concentrations) of toxic elements, as well as the most important chemical mineralizers that accompany them, which form migration-capable forms of toxicants in the hypergenesis zone. The latter are formed, first of all, due to the minerals easily soluble in the oxidation zone: sulfates, carbonates, arsenates, phosphates, etc. To determine the value of the toxicity indicator of the deposit, it is proposed to use the total indices of the content of toxicants (ΣKc) in the ores of the deposit.

Based on the research conducted at the Mining Institute of the Far East RAS, it has been established that sulfide-sulfate technogenesis processes are actively manifested in production wastes even with insignificant (from 1 to 5%) sulfur contents, processes occurring in mineral wastes (oxidation of sulfides to sulfates) have a negative impact on the environment. Therefore, technological and geological factors in the development of specific mining objects, the results of which should be taken into account in ecological mapping, are of great importance for the environmental status of mapping of natural and man-made systems. But, it is also necessary to take into account the most "strong" natural component, the lithogenic basis, that determines the geochemical background (the environment in which chemical reactions take place) in ecological mapping of natural and man-made systems. The most illustrative example of the carbonate base: in waters with carbonate hardness, positive ions of heavy metals bind to the carbonate ion and precipitate, the acids are quickly neutralized. The violation of the lithogenic base often leads to a complete transformation of the ecosystem (technogenic complexes). The influence of the lithogenic base is traced at all stages of field exploitation.

Based on the geological map of Primorsky Region [10], a map of the lithogenic substrate was created using geographic information systems (GIS) for Primorsky Region. At the first stage, a legend was developed (Figure 2), which reflects the main classes of the lithogenic substrate, then drawing up the map layout. Using the QGIS program, which is distributed under the GNU GPL license [11], the compiled map is digitized in layers, according to the designations. As the geographical basis of the map was adopted: Open

Street Map (OSM), Russified [12], in the geographical projection lat / long, on the ellipsoid WGS84, freely distributed via the Internet.

The creation of maps of the lithogenic substrate can be considered as the initial stage of landscape-geochemical mapping of the territory.

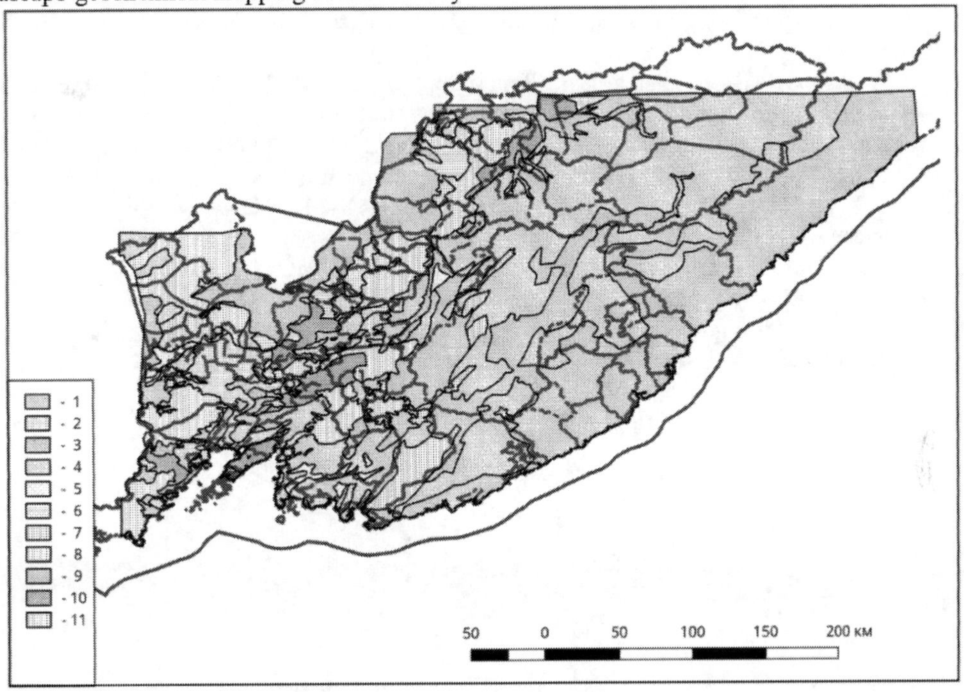

1 - Alluvial quaternary formations (Q_Al); 2 - Alluvial-lacustrine, lacustrine formations of N-Q age (N-Q_Al-lake); 3 - Cenozoic carboniferous deposits (KZ_Coal); 4 - Terrigenous Pz3 - Mz sediments (PZ3-MZ_terr); 5 - Terrigenous coal-bearing Pz3 - Mz sediments (PZ3-MZ_coal); 6 - Carbonate and carbonate-containing rocks (limestones, marbles, dolomites) (Carbonate); 7 - Basic and ultrabasic magmatic complexes, including plateausal basalts (Base_Magm); 8 - Extrusive diabase-basalt complexes and covers of Neogene plateaus of basalts (N db-basalt); 9 - Magmatic (volcanic, intrusive-volcanic) complexes of medium and acid composition (Mz_magm); 10 - Volcanogenic and intrusive-volcanogenic complexes of the Late Mesozoic (Mz-late_Volcan); 11 - Precambrian-Paleozoic granitoids of the Khanka massif, gneisses, crystalline schists (Pre_Cm).

Fig.2. Map of the lithogenic substrate of Primorsky Region

The obtained results can serve as a basis both for preparing the initial information for the development of ecological maps of a new generation, and for creating a database for various classes and subclasses of elementary landscapes.

References

1. Trubetskoi K.N. State and perspective directions of development of mining in Russia // Geoecology. Engineering geology. Hydrogeology. Geocryology. **6**. P. 3-7. (2017)
2. Vernadsky VI Biosphere. M. Thought, (1967).
3. Buzmakov, S.A. Ecology and nature management in conditions of anthropogenic transformation of the natural environment / Geography and region. Materials of the International. sci. Pract. Conf.: in 6 volumes. - PGNIU, P. 22-27. (2015).
4. Wang, A.V. Methodological problems of research of natural ecological and natural-technical systems / INTEREXPO GEO-SIBERIA. **3 (2)**. P. 33-38. (2012).

5. International mining strategy of mining investment-geological science-information integration. Ding Qun-an, Wang Lin, Huang Xing-wen, Jiang Ya. / Ziyuan yu chanue = Resour / and Ind. **4**, p 1-12. (2016).

6. Saksin, B.G. Forecasting the regional geochemical impact on the environment of mining enterprises of non-ferrous metallurgy in the East of Russia. Khabarovsk: IGD FEB RAS. 190 p. (2012).

7. Sekisov GV, Grekhnev NI, Romanov VG, Zykov NV Ecological efficiency of technologies for the development of deposits of solid minerals and methodological principles for its evaluation // Mining Journal. **3**. P. 112-119. (2007).

8. http://www.wiki-prom.ru/region/primorskiy_kray.html.

9. Arkhipov G.I. Khabarovsk: Institute of Mining, Far East Division, Russian Academy of Sciences, 820 p. (2017).

10. "Geological map of the Primorsky Krai", M 1: 500 000 and 1: 1000 000. They composed: V.A. Baksanov; L.F. Navarenko, Yu.N. Oleynik (1988).

11. http://www.gnu.org/licenses/gpl.html

12. http://gis-lab.info/projects/osm-export.html

Consolidation of small and medium mining business in the gold mining industry of the Russian Far East

Irina Kradenykh[1*]

[1]Mining Institute of Far eastern branch of Russian Academy of Sciences, Khabarovsk, Russia

Abstract. The gold mining industry of the Russian Far East has a great development potential, whose economic growth largely depends on the efficiency of developing a unique mineral resource base. To do this, it is necessary to solve a number of tasks, including the management and organization of economic activities of enterprises engaged in the extraction of placer gold in medium and small volumes. Gold mining companies themselves positively assess the prospects for the development of the industry, despite the preservation of external restrictions imposed by Western countries. At the same time, the current position of the modern gold mining industry is determined not only by the state of the Russian economy, but also by a number of branch features related to the management of gold mining at the enterprise level. At the present time, there are increasing questions about the feasibility of consolidation of small and medium mining enterprises, which will allow them to overcome current trends in the economic development of the industry, by combining their own resources. At the same time, in practice, reverse phenomena are observed, indicating that the industry is segmented and the number of mining companies developing placer deposits is gradually increasing. However, consolidation processes, in the form of integration associations, represent one of the directions of development and effective functioning of Russian gold mining enterprises, therefore they require comprehensive research and remain relevant.

1 Introduction

The position of the domestic gold mining industry is determined not only by the state of the Russian economy, but a number of industry features in the management of gold mining at the enterprise level. Currently, in the placer sector, small and medium-sized enterprises have accumulated a lot of problems associated with insufficient resources, both for conducting their activities and for lobbying their own interests at the state level.

There are various ways of development for this group of companies. For example, an increase in the share of small and medium-sized gold mining enterprises, taking into account the provision of comprehensive state support. The next area is the consolidation of

* Corresponding author: kradenyh_igd@mail.ru

small mining companies, which will allow them to continue to be able to influence the strategic development of the industry.

2 Theoretical and practical aspects of the problem

Currently, the processes that form the domestic gold mining, correspond to the trends inherent in the global mining industry. Analysis of the development of the Russian gold mining industry shows that the restructuring carried out in it is based on the consolidation and creation of new gold mining enterprises in the integration processes of vertical and horizontal orientation (table 1).

Table 1. Gold production by the largest Russian enterprises and companies that are part of them

Enterprise	Subsidiaries	Gold production, thous. troy ounces				
		2013	2014	2015	2016	2017
Polyus Gold	CJSC "ZDK "Polyus" CJSC "Pervenets" OJSC "Lenzoloto" LLC "GRK Aldanzoloto"	1652	1696	1763	1968	2160
Polymetal	CJSC "Serebro Magadana" OJSC "Okhotskaya GGC" CJSC "Gold of North Ural" OJSC "Omolonskaya ZRK" OOO "Albazino Resources" Amursky MMC	1282	1431	861	1269	1433
Petropavlovsk	OJSC "Pokrovskiy rudnik" OOO "Malomyrkiy rudnik" OOO "Albynsky Rudnik" OJSC ZDP "Koboldo"	741.2	624.5	504.1	416	439,6

Vertical and horizontal integration processes are expressed in close production and economic relations, both between mining and metallurgical companies. Currently, almost all foreign mining companies are integrated into large companies that have the necessary capacity for the production, enrichment and processing of metals [1]. At present, the stage of initial consolidation and active formation of vertically and horizontally integrated companies at the national level is completed. However, in the gold mining industry, apart from holding structures, independent companies, including mining artels, continue to function, and they still have a significant role [2].

The experience of functioning of companies belonging to the mineral and raw materials sector confirms the interest and significance of economic integration. From the point of view of economic theory, the signs of integration are manifested in the strengthening of production and technical relations, the consolidation of capital, the joint use of resources, in creating mutually supportive conditions with the aim of increasing the aggregate economic productivity of economic entities. From the point of view of corporate management, the integration process involves the reorganization of business and its enterprises. Significant changes in this affect not only the organization and management of production, but also the economic behavior of the company in the market. In the economic literature, basically two forms of integration are considered: mergers and acquisitions – **M&A** (mergers and acquisitions) [3, 4].

The prerequisites for conducting mergers and acquisitions continue to be studied both by representatives of theoretical and economic schools, and experts in the field of strategic management [5]. At the present stage, the most famous theories explaining the reasons for the mergers or acquisitions are the theory of synergy, which was put forward by the researchers M. Bradley, A. Desai, E.N. Kim in 1983. "Synergy" - the word of Greek origin, in translation denoting assistance, commonwealth, complicity. The essence of the theory is that as a result, M&A companies acquire a wide range of different advantages by combining the resources of the companies participating in the transaction [4]. At present, the theory of synergy is considered the most confirmed by both theorists and practitioners in the field of mergers and acquisitions. It should be noted that for successful management of synergistic effects it is necessary to know the sources of their origin.

The impact of synergy is primarily aimed at a qualitative change in the system, with the transformation of the conditions, structure and mechanism of interaction of its elements, resulting in a change in the output characteristics of the system.

Considering as an object of study the organization of the activities of mining artels, it is necessary to take into account a number of differences from the generally accepted operating regimes in other industries. For example, because of climatic conditions development of deposits by open way is only possible in the warmer months. At the same time, development is carried out in hard-to-reach areas located far from developed infrastructure and settlements. Delivery of the necessary goods is carried out mainly in the winter on the so-called winter road (temporary roads). The creation of social infrastructure in the development of the field with a mining period of less than 7 years is considered economically impractical. Due to the fact that the current conditions of production are characterized by the exhaustion of the richest and most favorable for the development of placer gold reserves, mining enterprises are forced to involve poor or very poor gold-bearing ores, which becomes appropriate in the period of high prices in the market of precious metals.

Currently, about 500 enterprises are engaged in the extraction of alluvial gold on the territory of the Russian Federation. In the Far Eastern Federal District, the leaders in the extraction of precious metals are the Chukotka Autonomous District, the Republic of Sakha (Yakutia), the Magadan and Amur Regions and the Khabarovsk Region. The number of enterprises extracting placer gold in the Far Eastern Federal District is shown in Fig. 1

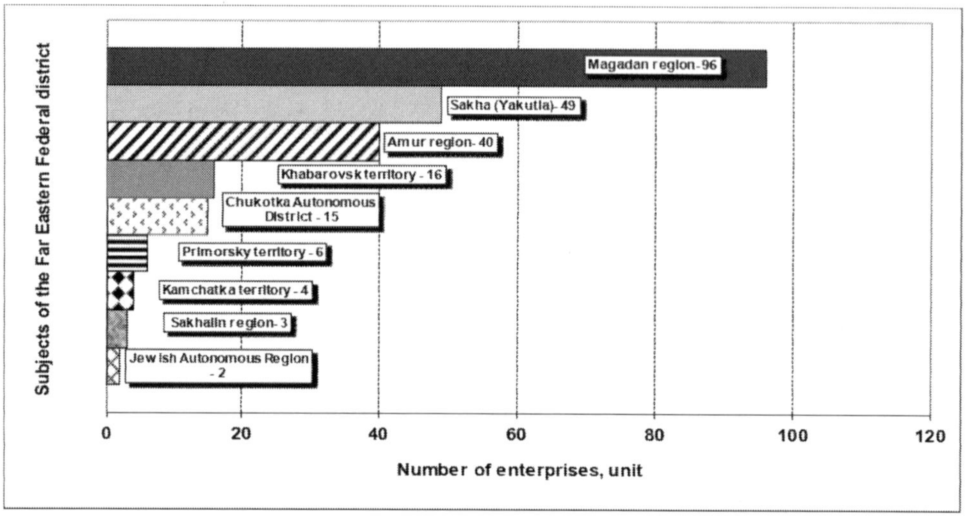

Fig. 1. The number of enterprises extracting placer gold in the Far East

Trends in the development of the gold mining industry of the Far Eastern region can be considered by the example of the Khabarovsk region, the gold content of which is mainly represented by small alluvial deposits.

Some of them, it would be more correct to call "objects of account", since they often represent fragments of previously worked out large-scale placers. The balance reserves of small and medium deposits in the province, as a rule, do not exceed 120 kg [2].

At the present time, enterprises that have been formed on the basis of mining artels, for example, the Amur miners' artel, which produces several tons of gold and platinum annually, continue to work successfully here. Also in the province there are also some production cooperatives, which, according to their organizational structure, are artels. For example, the team of prospectors "Vostok" continues to support the principles of joint work based on the common interest of members of the cooperative. Currently, 19 companies are mining in the region, five of which are developing ore deposits and 14 companies are steadily developing placer deposits.

The gradual depletion of the mineral and raw materials base of alluvial gold and the weak supply of active reserves due to long-term lag in the rate of growth inhibit the development of gold mining enterprises. The objective need to engage in the development of a large number of small deposits and increase on this basis the efficiency of the functioning of small and medium-gold mining enterprises can have a positive impact on the preservation and development of regional infrastructure in areas of traditional alluvial gold mining [7, 8]. In connection with these requires an in-depth study of the management and organization of economic activities of artel prospectors.

In our opinion, in the current conditions, the stable operation of medium-sized and small gold mining enterprises is possible in the case of consolidation either with each other or with larger companies. Horizontal integration is likely to be an objective process of the further stage of development of small and medium Russian gold mining business. The implementation of horizontal integration will allow the reorganized enterprises to expand their own raw materials, provide a higher level of production and become more attractive to investors.

3 Results and discussion

As already mentioned, the goal of companies that have decided to implement the integration process through M&A is to obtain synergistic effects that simultaneously affect the satisfaction of the requests of all stakeholders (shareholders, staff, customers, authorities). Within the framework of gold mining enterprises, who decided to combine their activities through horizontal integration, the following potential synergetic effects should be considered:

- pooling of company funds to purchase more efficient, expensive mining equipment;
- centralization of such types of work as supply, accounting, records management, work related to staff development, etc. In this case, both managerial and operational synergies are achieved through the creation of a new management system and a reduction in operating expenses by combining various services of an integrated company;
- improving the efficiency of work with suppliers, which will allow the combined companies to centralize the necessary purchases and to be able to use any discounts that apply to purchase prices. This aspect is relevant for gold mining companies, because the seasonal nature of the work requires advanced purchase of materials, fuel, spare parts, various mining equipment, etc. In this case, an operational synergetic effect is possible, which occurs due to savings in the implementation of costs aimed at maintaining the current activities of the gold mining enterprise;
- expanding the range of extracted raw materials, by involving not only gold sands, but also other natural components used in various spheres of the national economy, for example, as materials for road construction (such as gravel, etc.). With further processing, the dumps of alluvial deposits are suitable for composites, it is also possible to obtain products for the production of paints or iron ore raw materials, etc.;
- improving creditworthiness is an important for the integrated gold mining company motive based on its size, which allows a larger enterprise to have preferred positions in obtaining bank loans;
- expansion of the commodity portfolio by combining existing licenses or pooling financial resources to acquire additional licenses, which will strengthen the position of the company and ensure its long-term functioning. In this case, a financial synergetic effect is expected that contributes to an increase in cash flows and a reduction in business risks;
- development of poorly developed types of deposits, which will allow the integrated enterprise to strengthen competitive advantages in the long term. Since the richest placer deposits have been developed in the present period, the problem of gold reserves replenishment becomes very urgent. Horizontal integration involves the integration of technical, technological, human and financial resources. As a result, the company has the opportunity to develop previously poorly absorbed non-traditional types of placers (deep-seated, technogenic), the development of which was previously carried out in limited quantities due to complex geological and technical conditions. In this case, not only organizational, but also managerial synergy is possible because of the economy of funds and the increase in cash flows due to the growth in gold production.

Undoubtedly, for the emergence of synergistic effects, it is not enough to unite companies and create a common technological chain of material and technical resources and equipment. The success of the integration strategy depends on the fulfillment of a number of conditions. Of great importance is the social basis, expressed in the methods of management and the creation of a favorable social climate. The synergetic effect is especially pronounced in the presence of joint goals and agreements in the actions between business owners and managers who are moving towards their achievement. In the absence of a system of motivations for the staff of employees of the merged companies, the effectiveness of the overall work is reduced. A negative synergetic effect is possible with a sharp change in the organizational structures of existing companies, which can lead to the loss of individual managerial experience and knowledge. Therefore, the high importance of adapting the organizational structure to the new working conditions, which will save not only a significant part of management developments, but also to introduce new effective elements of the structure in the parent company.

4 Conclusions

As a result of the study, prospects for the development of gold mining enterprises are outlined. A strategy for consolidating existing small and medium-sized enterprises should be identified through their horizontal integration. Its essence is expressed in the intensity of interaction and deepening of production and technical ties, connecting the potentials of enterprises, complementing each other, in the joint use of resources, as well as in ensuring mutually favorable conditions for the implementation of economic activities. After all, the depletion of the mineral resource base of alluvial gold and the weak provision of active reserves due to the long-term lag in the growth rate, constrain the development of gold mining enterprises.

In our opinion, the support of small and medium-sized mining business should be attributed to the priority areas of the industry, as available on the state balance of more than 5 thousand small placer deposits with reserves of about 1132 tons of gold, the most suitable for the development of small gold mining companies This fact is an integral part of the overall strategy of optimization of alluvial gold mining in the region. The objective need to involve a large number of small deposits in the development and increase the efficiency of small and medium-sized gold mining enterprises on this basis can have a positive impact on the preservation and development of regional infrastructure in the areas of traditional alluvial gold mining.

References

1. YU. A. Arhipova. Mountain information and analytical bulletin. **S21**. 374-388 (2016).
2. I.A. Kradenykh, V.S.Litvintsev. News of higher educational institutions. Mining journal. 7. 34-41. (2016)
3. A.E. Molotnikov. Mergers and acquisitions: Russian experience. 344 p. (2006).
4. O N. Misko. International economic integration. 174 p. (2015).
5. V V. Obuhovskij. The economy of the XX century. **1**. 27-46. (2007).
6. S G.Kashuba, V. N.Ivanov, N. V. Dudkin. Gold and technology. **3(25)**. 6-15. (2014).
7. I.A. Kradenyh. Mountain information and analytical bulletin (scientific and technical journal). **S24**. 390-401. (2017).
8. S.B. Avdasheva, V.V.Golikova, T. G.Dolgopyatova Integration processes, corporate governance and management in Russian companies. 197 p. (2006).

Prospects of multipurpose and complex use of the most important types of non-metals in the economy of the Far East region

Galina Sklyarova[1*]

[1]Mining Institute of Far eastern branch of Russian Academy of Sciences, Khabarovsk, Russia

Abstract. The strategic importance of non-metallic minerals is determined by their large-scale and multipurpose use in the main basic complexes of the economy. The Long-Term National Program of the Russian Federation planned the study and assessment of subsoil in the field of non-metallic minerals funding by the federal budget for four major groups of raw materials: agrochemical, non-metallic for metallurgy, environmentally friendly and construction materials. At the territory of the Far East there are geological and economic potential to create raw material bases for development of important economic complexes such as agrochemical (for fertilizer production: phosphoric - apatites, phosphates, potassium - alunites, organic - peat, sapropels; using as structure formers and soil-aggregate stabilizers: carbonate rocks, natural adsorbents, organic raw materials.), metallurgical (refractory clays, alumina slates, alunits), ecological (natural adsorbents-zeolites, bentonites, perlites, diatomites) as well as others - chemical-forestry, mineral-construction, high technologies. Positive results of geological and economic assessments of deposit prognostic models of non-metallic raw materials allow to recommend them as the objects for industrial use. It is recommended to create new mineral non-metal bases not far from the main consumers, which is the most important for remote regions of the Far East. Key words: non-metallic minerals, the Far Eastern region, industrial complexes: agrochemical, metallurgical, ecological, mineral-building.

Non-metallic mineral group is one of the most abundant natural groups, including more than 150 types of mineral raw materials. They include rock and mineral formations, from which it is technically possible and economically expedient to obtain intermediate products. In the Far East region the most important of them are: alunites, apatites, phosphates, high-alumina types of materials (andalusite, dysthene-sillimanite slates), carbonate rocks, zeolites, mineral-construction and other common rocks, their practical and strategic importance is determined by large-scale and complex application.

The main text content is presented mainly by the characteristics of specific non-metals of the Far East region considering the prospects of their traditional and non-traditional use in the development of the most important industrial complexes of the economy: agrochemical, metallurgical (refractory), ecological, etc.

* Corresponding author: adm@igd.khv.ru

Zoning for the main types of non-metallic raw materials was carried out for the purpose of an innovative approach to the subsoil use and planning at the stages of geological study and development of subsoil in the Far East region, it was based on the principles of rational and integrated subsoil use in view of the modern technologies of complex mineral processing.

For territories ranking with allocation of prospects and ore areas, the following factors were used: peculiarities of geographical and economic situation, the scope of productive formations development, the size of reserves and forecast resources, a degree of their development and study, industrial significance and their prospects for development in the near future. A brief description of the ore districts is given as an example.

Specific non-metals of the Russian Far East Alunites. The Far East is the only region of unique alunite mineralization in Russia. In the region, more than 100 large alunite-bearing deposits and occurrences (more than 60 million tons - 5 billion tons) have been discovered. Alunite occurrences are mostly developed in the Khabarovsky Krai within the volcanic zones of the Lower- and Middle-Amur part of the Sikhote-Alin volcanic belt, in the Okhotsk district within the Ulya metallogenic zone. Alunites are hydrothermal-metasomatic formations, containing mineral alunite – $K_3NaAl_3[(SO_4)_2(OH)_6]$, the theoretical composition of it (%) is: Al_2O_3 –36.93; SO_3–38.66; K_2O – 11.37; H_2O – 13.04. The rocks containing more than 30% of alunite can be attributed to commercial (ore with the alunite content more than 34% can be processed without concentration). Presence of considerable amounts of aluminium, potassium and sulfur oxides in chemical composition of alunite define this type of raw material as complex, multipurpose, perspective for metallurgical, agro-industrial, ecological, mineral-construction manufacturing complexes.

In the metallurgical complex alunites can be used as refractory raw materials for alumina production. The alunites of Zagliksky deposit at Kirovobad aluminum plant (at present Gandja Alumina Plant, the Azerbaijan SSR) were used for the first time in the world practice as natural mineral raw material in the production of alumina (aluminum oxide). The alunite content in ore processing concentrates should be not less than 50 – 70%. Concentration of alunite ores, alumina production, requiring a high degree of ore grinding, application of thermal dehydration is very expensive, power-intensive production requiring the complex infrastructure development. Regarding geological and raw materials aspect the aluminium production is expedient on the basis of alunite deposits in the Lower-Amur region. However, due to the lack of power base, expensive technical equipment, the alumina production in the Far East is not planned for the near future. Industrial production of aluminium from alunite ores is technologically complicated, energy-intensive production, which requires higher costs for the industrial base development. This problem for the Far East can be resolved only at the federal level in the long run.

Agro-industrial and ecological complexes: alunites - as raw material for production of potassium sulfate, aluminium sulphate. Potassium is one of the main microelements for agricultural plants. Potassium salts are the main raw material source for potassium fertilizers, their significant reserves are concentrated in the Western regions of Russia. The Far East is the most important producer of soybeans, which requires potassium sulphate fertilizer meal, complex fertilizers, therefore, the assessment and development of alternative raw material sources for potassium production is of vital economic value. The VAMI's (the Russian Aluminium Magnesium Institute) technological researches of alunite ores of the Far East developed the simplified schemes of their processing without preliminary very intensive process of ore fine grinding and obtaining after their concentration very scarce aluminium sulphate and alum, in which the whole alumina as well as sulfate potassium will be binding. Sulfate aluminum is widely used as a coagulant for drinking and sewage treatment, in pulp and paper, textile and other industries. Potassium sulfate is a valuable potassium chlorine-free fertilizer, and alum is also widely used.

Mineral-construction complex. The complex raw materials for manufacturing many types of industrial products can be not only alunite ore, but also in general, the rocks enclosing them: monoquartzites – glass, ceramic and decorative-facing; rutile quartzites – titanium, ceramic and decorative-facing; sulfuric quartzites – sulfuric and ceramic; alunite quartzites – aluminous, sulphuric-acid, potassium, ceramic, cement, brick, and sometimes recovering gold as byproduct, silver-diaspore quartzites – alumina and quartz raw materials; kaolin quartzites – ceramic, recovering gold as byproduct, silver; sericite quartzites – alumina potash and quartz raw materials; propilites – a possible source of non-ferrous and precious metals. Besides, in alunite concentrates gold is defined in amounts of 0.4-0.7 g/t, silver – 2.8-3.1 g/t, in tails, 0.9-1.1 g/t, 3.8-4.4 g/t, respectively. To extract precious metals, the crushed ore prepared for flotation was subjected to cyanidation. 98% of gold and about 63-83% of silver were extracted when the cyanide of potassium 0.64 kg/t, lime 3.7 kg/t were used during the cyanidation time of 8 hours. Heap method of cyanidation allows to process several million tons of ore that allows to extract these metals from ores also.

In general, geological and economic prerequisites of economic complex development of alunite ore deposits are prospective for further development of the Russian Far East economy. The innovative technological processes of alunite raw materials development in the Far East include research-practical works (VAMI, DVIMS (Far Eastern Institute of Mineral Resources), etc.), innovations in technology of alunite ore processing with alumina, sulphate aluminium (coagulant), potassium sulphate (potassium fertilizer) and other mid-products.

Anortozites are complex raw materials for apatite-ilmenite-titanomagnetite and aluminous mineralization. One of the largest in the world the East Asian anorthosite belt was identified at the territory of the Far East.

Apatite-titanium raw materials. In the BAM (Baikal-Amur Mainline) region within the Stanovaya fold-block system, the Dzhugdzhurskaya, Chogarskaya and Baladekskaya structural-mineragenic zones (the Khabarovsky Krai), as well as in the Amur region Kalaro-Haninskiy ore district with deposits of complex apatite-ilmenite-titanomagnetite and apatite-ilmenite ores, are differentiated, their estimated resources are very significant (TiO_2/P_2O_5 mln t): Gayumskoe-23.3/7.6, Bogide-79.8/68.4, Yaninskoe - 70.8/46.4. The technological tests show that complex apatite-titanium ores are most effectively processed by magnetic-flotation scheme, which includes the allocation of titanium-magnetite concentrate by wet magnetic and electromagnetic separation and subsequent flotation of apatite and ilmenite from non-magnetic fraction. Positive results of geological - economic estimation of the predicative models of these deposits allow to consider them as the priority objects for detailed geological and technological studies for assessment of prospects of raw material base for production of phosphoric fertilizers and iron-titanium concentrates for ferrous and non-ferrous metallurgy on the basis of the most prospective deposits of complex apatite-titano-magnetite mineralization in the composition of anortoxite formations.

In the Amur Oblast (District) OOO Uralmining was granted the right to use subsoil (until 2030) for the purpose of exploration and production of titanium-magnetite ores at the Bolshoy Seyim deposit, located at the Tyndinsky district territory, 15 km northeast of the BAM railway line. The P1 prognostic resources of the deposit were tested as 36 million tons of TiO2. According to the preliminary project of this deposit development (1998), the construction of mining-processing complex (GOK) and plant for pigmented titanium dioxide production was envisaged.

Alumina raw materials. The anorthosites are characterized by high alumina content up to 26-30% and bases (predominantly calcium monoxide) – up to 10.8%, silica – up to 53%, iron oxide – less than 4%, sodium and potassium oxides – 5.2%. The requirements for anorthosites as alumina raw material were not developed. In the 1980s, VAMI (the All-Union Aluminum-Magnesium Institute) performed technological testing for the complex processing of Canadian anorthosites with alumina content of about 27%. As a result it has been established that they can be processed for alumina by baking anorthosites with limestone and soda. The forecast resources of anorthosite raw materials of the Far East are determined as 800 bln t (up to 30 m depth). However, almost all of them are located in remote regions, far from the BAM. Only Kavaktinskiy (30 km from BAM), Koloktikanskiy (35 km from the Far East Railroad) and Selemdzhinskiy (60 km from BAM) with total reserves of about 1.5 bln t can be offered for practical estimation. Technological possibilities of anorthosite and leucite raw materials use were studied only at a trial and experimental-factory levels.

High-alumina refractory raw materials. Alternative types of non-traditional refractory raw materials are represented by such minerals as high-alumina rocks containing minerals of sillimanite, disthen, andalusite group; alunites, anorthozites and also zirconium-containing rocks, by preliminary qualitative and quantitative characteristics satisfying the requirements for using as refractory raw materials. [6-7, 9].

In the south of the Far East the ore-bearing formations, prospective for high-alumina raw materials (andalusite-disthen-sillimanite with average content of 20-40%) in the composition of ancient metamorphic masses, are well represented. About 100 deposits and occurrences were registered, for 23 of them total forecast resources in the amount of about 8,958 million tons were estimated, this confirms the huge potential of the territory for these types of high-alumina raw materials. At present, in metallurgy the open-hearth process is replaced by steel making in converters and electric furnaces that will involve usage of more high-resistant types of refractory raw materials. In foreign countries, non-traditional types of refractory raw materials containing such alumina minerals as kyanite, andalusite, sillimanite, as well as zirconium-containing minerals, are extracted and used for modern metallurgy, growth tendency of their demand is marked. The Algaminskoe deposit of zirconium-containing ores with reserves of about 215 thousand tons was prospected, its development can be commercial. For estimation of economic significance of the areas of high-alumina and zircoium-containig formations development, it is necessary to carry out exploration and estimation.

Adsorbent Materials

The interest to use natural mineral adsorbents as raw materials for protection and rehabilitation of environment increases due to growing importance of the environmental problems.

The Far East of Russia has a significant raw material base of various mineral-industrial types of natural adsorbents. Traditionally they are: zeolitolites, opal-kristabolite rocks (opokas, diatomites, tripolites), bentonite and palygorskite clay, vermiculite, perlite. The rocks with active constituents content more than 70% are attributed to the high-quality adsorbents for the most types of natural adsorbents. In addition, experimental technological research has established the possibility to obtain commercial products with adsorption properties from alunite (sulfate aluminum), brucite (brucite-zeolite product). In modern conditions the study and development of significant raw material potential of natural adsorbents of the Far East is very important for consumption and production growth of high-active adsorption-filter materials for industrial and environmental aims.

Fluorite

Fluorite (CaF_2) is used as fluxes in metallurgy, as well as in hydrofluoric acid, freon for laser equipment manufacture. The unique properties of fluorite and its wide use in the industry are associated with a low melting point (1270°) and high chemical activity of fluorine, with the highest electro-negativity. In Russia, fluorite is one of the traditionally scarce material. To reduce import and improve the structure of fluorite reserves, it is necessary to accelerate the development of the identified sites with the balance reserves of Voznesensky and Pogranichny deposits in the Primorsky Krai, Preobrazhenovsky deposit in the Jewish Autonomous Oblast (JAO).

Borates

In Primorsky Krai ZFO GKHK Bor Enterprise with full technological cycle, is carrying out development (since 1959) of Dalnegorsky deposit with extraction of boron ores (reserves of them are up to 86.6% from stocks of B_2O_3 of the Russian Federation), also datolite ores and limestone, datolite ores processing, chemical processing of datolite concentrate. The enterprise consists of a mine (two quarries), a concentrating factory of boric and sulfuric acids manufacturing, a boron shop and other auxiliary shops. Two pit open mining with blasting was used. The reserves life is over 200 years. Treatment at the concentrating mill is carried out by flotation with pre-crushing up to-0.15 mm. The main products are: datolite concentrate, boric acid, calcium borate, boric anhydride, sulfuric acid.

Brucite

In the Far East in the JAO (ZAO Kuldursky Brucite Mine), the only in Russia brucite deposit of the same name, is developed in the open way, its average MgO content being 62.0-65. 0%. Before entering the crushing and sorting station brucite is exposed to ore-picking and large chunks cleaning of the enclosing rocks. The quality of the commodity brucite meets the requirements of TS (technical specifications) 14-8-392-82 with the allocation of 3 grades of Mg (OH) 2 – 60-62%; 4 and 5 – 60 and 55% (off-balance). The content of the grades should be respectively, %: CaO – 1.5-4.0, SiO_2 – 1.5-8.0, Fe_2O_3 – 0.15-1.0. The commercial product is used by OAO Ogneupory (Refractories) to obtain electrical periclase and spinel powders. OAO Ogneupory Trade Company (Yekaterinburg) is the official trade representative of the mine.

Only in Yakutia, the Olekminskoe deposit of *gypsum* is being developed (gypsum stone which is supplied for cement production) and *common chloride*. 5 deposits are explored, four of them represented by rock salts and one – by salt brine. The deposits are known as a part of the halogen Cambrian strata, with thickness of 8 – 47 m with salt saturation from 82 to 100%.

Agrochemical Raw Materials

The analysis of regularities of mineral base formation for agrochemical complex of the Far East Federal District is made for estimation of the prospects of their application in agriculture [1-5]. Within the region there are found practically implemented or estimated minerals as a result of agronomic studies, they are raw materials for fertilizer production (phosphoric fertilizers – apatites, phosphorites; organic – peat, sapropels; potassium – alunites), soil aggregate stabilizers and chemicals for soil composition improvement, mineral fertilizers, materials for pest control (carbonate rocks, zeolites, perlites, diatomites, bentonites, vermiculites, peat, sapropels;).

In practical and experimental conditions the effective influence of natural adsorbents (minerals, rocks or their commodity products) in the processes of rehabilitation of ecological soil and water contaminated with radionuclides, hazardous and poisoned substances was established. Adsorbents are high-dispersed artificial or natural substances characterized by specific chemical properties [8].

It is considered to refer to natural adsorbents the rocks, the structural minerals of which have high adsorption, ion, catalytic filtering and structural properties of selective absorption of some components from the adjacent gas or solution environments. They are traditionally: zeolitolites, opal-kristabolite rocks (opokas, diatomites, tripolites), bentonite and palygorskite clays, vermiculite, perlite [10-11]. For most types of natural adsorbents, the rocks with active components content being more than 70% are attributed to high-quality rocks. Furthermore, experimental technological research showed the possibility to obtain commercial products with adsorption properties from alunite (sulfate aluminum), brucites (brucite-zeolite product). The adsorbent properties of natural minerals are not uniform due to the differences of crystalline structures and chemical composition.

Thus, at the territory of the Far East there are prerequisites for creation and development of raw material bases for agrochemical (fertilizers production: phosphoric - apatites, phosphorites, potassium - alunites, organic - peat, sapropels, soil aggregate stabilizers and chemicals for amendment of soils composition - carbonate rocks, natural adsorbents), metallurgical (refractory), ecological (adsorbent) and other regional complexes of economy. The positive results of the geological and economic assessment of the prognostic models of these mineral deposits allow to consider them as the priority objects for the development of detailed geological and technological researches [12]. Due to remoteness from processing agrochemical companies of Russia, climatic and natural conditions of "risky agriculture", the strategy of agriculture development in the Far East should aimed at development of local agrochemical material sources, the prospects of which are very significant.

References

1. E.M. Aksenov, N.G. Vasiliev, Rudy i Metally. **1.** 5-6. (2009) (In Russ.)
2. G. F. Sklyarova, Mineralno-syrjevoy sektor nemetallov Dalnego Vostoka i ego rol dlya ekonomiki regiona Khabarovsk. 287. (2013) (In Russ.)
3. R.K. Sadykov, R.M. Minulin, Razv. i Okhrana Nedr. **3.** 3-5. (2009) (In Russ.)
4. I. Yu. Rasskazov, G. F. Sklyarova, A.P. Van-Van-E, Problems of Compl. Developm. of Geores., Mater. of V All-Russ. Scienc. Conf. Thirt. Anniv. of Min.Inst., FEB RAS, 428-441, (2013) (In Russ.)
5. A. F. Karpuzov, A. V. Tarasov, Min. Res. of Russ. Econ. and Manag. **6.** 2-12, (2015) (In Russ.)
6. G. F. Sklyarova, Novye Ogneupory (New Refractories), **1**, 6-13. (2014). (In Russ.)
7. G. F. Sklyarova, Otechestv. Geolog. **3.** 3-7, (2012)
8. V.A. Buryak, V.Ya. Bespalov. et al, Novy Geologo-Promyshl. Tip Zirkon Oredenenie (New Geol.- Industr. Type of Zircon Mineraliz.) 216, (1999) (in Russ.)
9. G. F. Sklyarova, V. E. Kislyakov, Mod. Techn. of Min. Res. Devel., Deutschland, Saarbrucken: LAP LAMBERT Acad. Publ. ,120-123, (2012)
10. I. Dwyer. Zeolite structure composition and catalysis, Chem. and Ind., **7**, 258-269, (1984)
11. T.S. Lovering, Epigenetic, diplogenetic, singenetic and lithogene deposits, Econ.Geol. **58**, 5 (1963)
12. A. Alietti, Polimorfism and cristalchemistry of heularites and clinoptilolites, Amer. Miner., **57 (9-10).** 1448-1462, (1972).

Process chains in the development of remote deposits of the North

Vladimir Gavrilov[1*] and *Vladislav* Fyodorov[1]

[1]Mining Institute of the North, Siberian Branch, Russian Academy of Sciences, Yakutsk, Russia

Abstract. The low level of development of remote northern regions of Russia predetermines a very high dependence of the mining and processing enterprises of that area on the import of material and technical resources and the export of products obtained after mining. Infrastructural restrictions and a continuous increase in the costs of goods and services, often outstripping the growth of prices for mineral wealth, increase the risks in the implementation of investment projects and worsen their economic performance. The results of an analysis of long, complex and multi-link process chains formed by mining enterprises of the polar regions of Yakutia are presented. The procedure of supply, extraction, logistics and primary processing of mineral wealth requires improvement. It is shown that mining enterprises by volume and type of output can be divided into four groups. On the basis of a comprehensive review of the factors of external and internal environment, proposals are formulated to improve the management of the delivery of cargos and delivery of finished products within supply chains. The prerequisites for replacing part of the delivered fuel and energy resources with local coal are considered. **Keywords.** Remote area, supply chain, deposit, logistics, export, coal, quality

1 INTRODUCTION

In Russia, the issues of social and economic development of the northern regions of the Far East and Siberia are given great attention [1-4]. The basic elements for these territories are often operating or new mining enterprises, supplemented by logistics, energy and other infrastructure facilities. Effective exploitation of the deposits provides for the expansion of the range, increase in the competitiveness and quality of mineral products, introduction of energy-saving technologies, integrated use of minerals [4].

For mining in the North, the main cost drivers (in addition to extreme natural and climatic conditions) are: remoteness of development facilities from consumption centers; a low level of development of transport-logistical and fuel-energy sectors in the region; significant cost of any types of energy consumed; almost full dependence on the delivery of material and technical resources and outflow of products; limited in time, complex and multi-tiered "Northern delivery". An analysis in [5-6, etc.] shows that the application of traditional approaches, principles and methods of doing business does not compensate for the influence of these factors. And the proposed measures aimed at increasing the

* Corresponding author: gvlugorsk@mail.ru

efficiency of development of deposits in remote areas practically do not provide additional competitive opportunities.

The polar regions of Yakutia, being a case study for analyzing the supply chains of mining and enrichment enterprises, mining, primary processing and logistics of mineral resources, are outsiders among the northern regions of Russia in terms of their level of social and economic development. In general, they can be classified as the territories of perspective development, based on the deposits being surveyed and the developed transport routes, or wildlife areas [1]. In such a situation, it is necessary to develop and implement measures aimed, if not to complete overcoming the existing difficulties, then at least partially neutralizing them. An important direction of increasing the level of management in the development of subsurface resources is various aspects of improving the multi-tiered supply chains of the material and technical resources generated by the regional mining and technical systems for the needs of those systems and the outflow of the products they make.

2 MATERIALS AND METHODS

The basis for studying in the polar zone of Yakutia of existing and promising process chains, generated by deposits of mineral resources of various types and values, was a number of sources of information. Their collection, processing, interpretation and generalization, carried out using the methods of analogies, comparisons, SWOT and PEST analysis, made it possible to assess the current situation with the delivery of goods and outflow of products to/from mining enterprises, and identify ways of improving logistics. The paper considered information on the experience accumulated in that field or the solutions planned for adoption in companies of various capacities (Alrosa, Almazy Anabara, Geoprommaining, Gold Indigir, Tarynskaya zolotorudnaya kompaniya, Prognoz, NordGold, Poisk-Zoloto, Zyryansky razrez, etc.).

In addition, two main points were taken into account. 1. The most effective development and use of mineral resources in the region is achieved by considering all the links of process chains as complex dynamically changing systems for the development of geo-resources, functioning according to criteria agreed between participants (links) [7]. With this approach, process chains oriented to domestic and foreign highly competitive markets can turn into chains of creating real value added. Their functioning can and should be carried out in accordance with the principles and conditions set forth, for example, in [8, 9]. 2. Existing reserves of minerals in the region should be regarded as a probabilistic, to a certain extent, unreliable geological resource. It, in the face of changing, difficult to forecast in dynamics market value of different types and varieties of mineral raw materials, requires additional study at different stages of exploration, design, operational exploration and development of deposits [10]. The purpose of this work is the maximum adaptation of the quality of extracted minerals and the products obtained from them to the existing and / or expected needs of particular end-users.

3 RESULTS AND DISCUSSION

The existing facilities in the polar zone of Yakutia to extract diamonds and gold from placer deposits and coal for local needs are planned to be supplemented with the development of new deposits. According to the strategy of development of Yakutia for the future up to 2030, with the definition of the target vision until 2050, it is planned to gradually develop the deposits of diamonds, gold, silver, rare earths, polymetals, hydrocarbons, coal and other minerals (Fig.).

The specifics of the region in question predetermine an increased attention to the functioning of transport and logistics links in chains related to the inflow of goods (mining machines, general special equipment, spare parts, fuel and energy resources, lubricants, most types of products, overalls, etc.) and outflow of finished products. These elements in a significant, often decisive, degree form the level of accessibility of any field. Accessibility here refers to the property of the system "society - mineral resources", which characterizes the possibility of efficient and safe consumption of resources depending on their condition, demand and the process level reached [11]. In assessing accessibility, one should take into account: 1) geographic and mining-geological features of the region; 2) attractiveness of reserves, depending on their qualitative and quantitative parameters, the idea of which changes at different stages of development of deposits; 3) variation in demand for raw materials with changes in demand and price adjustments; 4) influence of process and organizational levels of production in the mining and processing of raw materials on the efficiency of extraction of useful components at each point in time; 5) attractiveness of subsoil development projects for investors in comparison with alternative options in other regions and industries.

3.1 Supply of fuel, energy and material resources

Enterprises that work out or plan to develop deposits in the area under consideration have a very high dependence on the inflow of virtually all types of cargo. This is due to the low or pioneering level of socio-economic development of the region, the lack of any manufacturers, and often even suppliers of material and technical products, limited opportunities for using local fuel and energy resources, despite their availability.

Companies in the practice of their operation use, with a variety of efficiency, a number of transport and logistics schemes to achieve normal maintenance of their production activities (Fig.1). The following types of transport are used: railway (delivery of cargo to the entrance ports); water (river and sea); automobile (year-round and seasonal for "winter roads"); air (airplanes and helicopters). Due to the multilevel routes of delivery, the mismatch between the timing of river and sea navigation, the work of "winter roads", the shallow water of rivers, their estuaries and ocean parts of the route, the cargo is forced to undergo a series of stages of long intermediate and basic storage. There is an urgent need to maintain large reserves, which leads to the freezing of significant working capital, the attraction of credit money. Long-term (up to 2.5 years) delivery times, change of seasons, precipitation and numerous transshipments generate significant losses in the quantity and quality of both inflowing and outflowing cargos. These losses often exceed normative standards. Considering this, the enterprises pay some attention to the organization of storage and transshipment of goods from one type of transport to another. Warehouses of open and closed storage, warm and cold, own and leased, have found application in logistics.

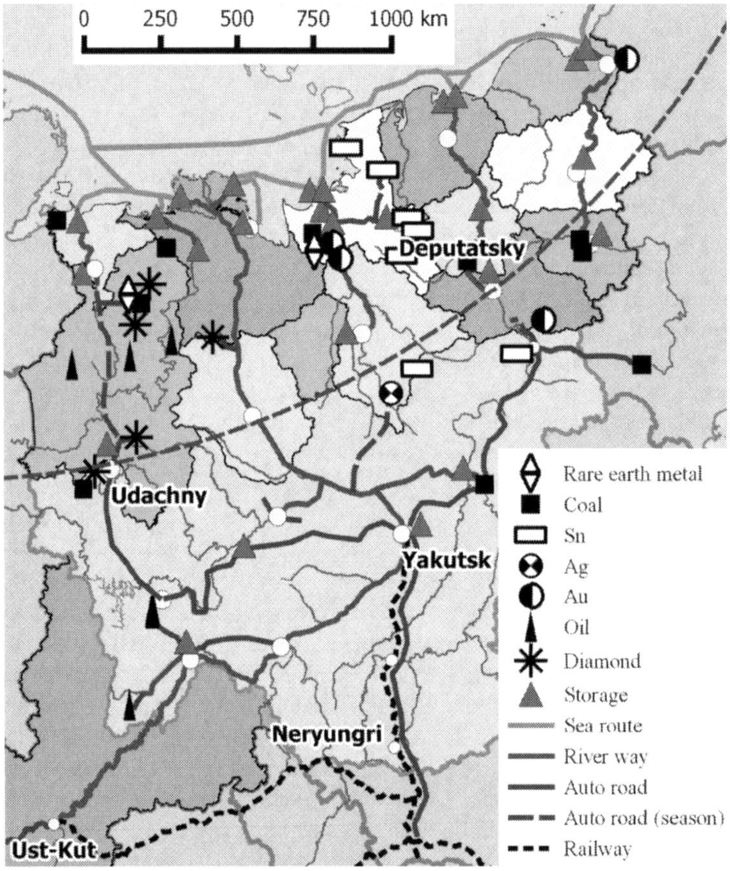

Fig. 1. Deposits of the polar Yakutia area and logistics in the course of their development

In the structure of the cost of almost any enterprise in the North, fuel and energy account for 30 to 60%. In the conditions of a continuous increase in costs for them, often outstripping the rise in prices for finished products, the urgency of replacing expensive fuel with other sources of energy is constantly increasing. Aa analysis of projects for developing deposits in the region shows that medium and small enterprises do not sufficiently consider alternative schemes for covering their needs with local types of resources.

The imposition of a fuel and energy resource map in the region under consideration on a map of the location of deposits to be developed or planned for development shows that under certain circumstances it would be possible to replace some of very expensive hydrocarbons and coal delivered with local resources to reduce total costs.

Practice of operation of *AK Alrosa* shows an example of such a substitution. At present, the use of local natural gas is started as a fuel for highway and public transport. Even conversion of mine machinery to gas fuel is being considered. When Deputatsky MPP (mining of ore and placer tin) worked, the question of developing a local brown coal pit and construction of a coal-fired CHP plant was seriously raised. CHP as a result was built many years later, but it was not fed by local coal, but expensive coal delivered from Dzhebariki Khaya pit (the center of Yakutia) or from even more remote Kuzbass region.

Low demand for local coal, despite its wide distribution, including in the locations of promising deposits (see the figure), is associated with a number of reasons, namely: low quality of supplied fuel, often not satisfying the consumer; traditional opinion of the low efficiency of its use, in comparison with hydrocarbons. Both reasons can be partially neutralized. The first is solved through the integrated management of coal quality in exploration, production and supply to consumers, with appropriate adjustments to planning, production and additional fuel treatment for supplies, including remote consumers [12]. The second solution, similar to the regions and countries leading in this area, is achieved by applying modern small-scale coal processing technologies to produce briquettes, semi-coke, gasification, pyrolysis, use of cogeneration for generating electric and heat energy. The study of the reasons preventing this and the development of proposals for their elimination are a separate topic not considered further.

3.2 Outflow of mineral wealth

From the point of view of transportation of the products obtained, by the degree of influence of the logistic component on the efficiency of development of the area's deposits, georeources can be divided into four conditional groups. 1. Having at the output of a small amount of useful minerals with high consumer value and the degree of readiness (diamonds, precious metals). 2. Forming volumes in the first tens of thousands tons (concentrates of tin, rare earths, antimony, other strategic metals). 3. Creating volumes of finished products, measured in hundreds of thousands and millions of tons per year (hydrocarbons, coal, ferrous ores) and requiring high-performance transport schemes for year-round use. 4. Used for extremely limited domestic consumption in the region (local fuel and energy resources, building materials).

In the structure of the cost price, the share of costs associated with the delivery of mineral and raw materials from the production sites to the places of consumption, processing or transfer of ownership varies from the first percent to the level making seemingly accessible deposits economically unacceptable.

The most simple and effective in the region in terms of logistics are diamonds and precious metals. They are shipped for sorting, faceting, refining mainly by air. Shipping of ore concentrates of less valuable or more voluminous types of minerals by air is often not available. Relatively positive experience of several years of shipping tin concentrate from deposits near Deputatsky township for processing in Novosibirsk can be explained by a favorable combination of circumstances: the presence of particularly heavy aircraft and an airport capable of receiving them, the ability to load aircrafts round-trip, relatively high prices on tin in the market.

Long distances, pronounced seasonality and high cost of services of other types of transport available in the region dramatically reduce the investment attractiveness of deposits with less valuable types of mineral resources. A study of the experience of mining enterprises in the north of Yakutia shows that the solution of tasks for shipping products even for group 2 is already associated with certain difficulties, not to mention group 3. There are cases when, even at distances of 15-20 km, the delivery of ore to the concentrating plant in the conditions of the North-East of Russia becomes more expensive than mining and comparable to its processing. And the decision to deliver mined ore from perspective diamond pipes located 170 km from the concentrator in Udachny led to the need for new trunk road trains, construction of special haul roads.

When considering the enterprises associated with the third group, the complexity of organizing at least relatively normal logistics should be well represented. Earlier it was shown [13] that the implementation of the project on the supply of coal for export from Zyryansky Basin on the Kolyma is economically inexpedient. The cost of ordinary coal already at the mouth of the river was 40-60% higher than the price of steam coal concentrate with higher quality on CIF Japan terms. When organizing large-scale coal and boghead mining at the Taimylyrsk field, located near the sea coast, the issue will arise of integrating the facility into the logistics of the Northern Sea Route with the construction of a deep-sea port and expensive dredging operations. This project can be implemented only under very favorable conditions, for example, if oil and gas production starts in the region. But this is a very distant prospect.

For enterprises in the fourth group, it is also very important what distance they are from consumers. In existing schemes of coal delivery to the polar regions of Yakutia, transportation and storage account for up to 50-65% of total costs at the place of consumption, and in the most inaccessible reaching 80-85% [14]. In such a situation, the question arises about the construction of pits (coal) or quarries (building materials) of low capacity to meet local, very limited needs.

4 CONCLUSIONS

One of the mechanisms for improving the performance of process chains formed in remote polar regions by mining enterprises is to improve the processes of supplying them with the necessary resources and shipping the extracted mineral wealth. The use of results of the proposed differentiation of enterprises into four groups, taking into account the features of the chains, the volume and type of products, makes it possible to reduce the risks of making irrational decisions when assessing and implementing projects.

Almost non-alternative use of expensive hydrocarbon fuel at mining enterprises of the North for the development of mechanical, electrical and thermal energy leads to an increase in the cost of all process operations. The use of local ordinary and graded coal, which can be mined from deposits located in close proximity to mining enterprises that extract other types of minerals, makes it possible to reduce the development costs of both the first and latter. To generate thermal and electrical energy from coal, it is advisable to use cogeneration schemes. Deeper coal processing should be carried out using low-tonnage technologies to produce liquid and gaseous fuels, semi-coke and briquettes. Deep processing will promote the growth of coal consumption in the region and improve the performance of extractive companies.

In the existing institutional conditions in the country, enterprises do not have sufficient incentives and opportunities to implement such activities without governmental support, which requires separate consideration.

The reported research was funded by Russian Foundation for Basic Research and the government of the Yakutia region of the Russian Federation, grant No.18-45-140048 p_a.

References

1. State program of the Russian Federation "Social and economic development of the Arctic zone of the Russian Federation for the period until 2020". Approved by the Government of the Russian Federation 21.04.2014. No.366-p

2. D.A. Dodin, D.V. Kaminskii, O.I. Suprunenko, V.I. Pavlenko, Arctic: Ecology and Economy, **4** (2011)

3. A.E. Sevastianova, Region: Economics and Sociology, **3** (2013)

4. G.V. Baturova, A.M. Konovalov, Region: Economics and Sociology, **4** (2015)

5. S.V. Kornilkov, V.L. Iakovlev, Iu.A. Mamaev, A.P. Van-Van-E, News of the Higher Institutions. Mining Magazine, **6** (2012)

6. M.Iu. Kharitonova, N.A. Macko, Geology and Mineral Resources of Siberia, **3** (2013)

7. S.A. Batugin, S.M. Tkach, N.S. Batugina, V.L. Gavrilov, Mining information-analytical bulletin, **9** (2012)

8. R. Kaplinski, Spreading the Gains of Globalisation: What Can Be Learned from Value Chains Analysis (Translation from English, Moscow, GU VShE, 2002)

9. M.E. Porter, Competitive Advantage: Creating and Sustaining Superior Performance (Translation from English, Moscow, Al'pina Biznes Buks, 2005)

10. Geotechnologies of Open-pit Development of Deposits with Complex Mining and Geological Conditions; Ed. S.M. Tkach (Novosibirsk, Geo, 2013)

11. A.A. Peshkov, N.A. Macko, Availability of mineral resources (Moscow, Nauka, 2004)

12. V.I. Fedorov, V.L. Gavrilov, E.A. Khoiutanov, Advances in Current Natural Sciences, **9** (2017)

13. V.L. Gavrilov, All-Russian scientific seminar "Actual problems, directions and mechanisms of development, productive forces of the North-2014": in 2 parts, **1** (2014)

14. N.S. Batugina, V.L. Gavrilov, E.G. Shepeleva, ECO, **2** (2017)

Experience in the use of "metaknowledge" of geoecology in the working out of the National Concept Sustainable Development of Russia

Sergey Turkov[1*]

[1]FSBMS Computing center of Far-Eastern Branch of RAS, Khabarovsk, Russia

Abstract. Some of the scientific questions and basic methodological regulations, which follows from the practice of using "metaknowledge" of geoecology obtained in the process of research and working out of the National Concept of Sustainable Development of Russia (NCSD), as the official doctrine of its development (1996), presented for discussion. The aim of the study is determination of a special (physically adequate to Nature and Society), information, logical and algorithmically organized subdomain of natural scientific knowledge necessary for a comprehensive idea of the original object of the study: active complicated systems (ACS) of "nature-society" class. The main trend of application of the results obtained is elaboration of automated ("human-machine") systems for planning (support) of optimal management decisions for organizational structures of environmental management at the global, regional and local levels of the planet. A new – "geosystem" – approach to solving the problems of sustainable development ("Sustainable Development", 1995), including the possibility of both external ("technogenic") and internal (or "soft", i.e. restoring the ecological balance) management of the ACS "nature-society" is proposed. In terms of "a new theory of entropy", A. N. Panchenkov (1999) worked out scientific formulations of the following concepts: the system of "nature-society" class, the conflict (in the conditions of "uncertainty"), "metaknowledge" of geoecology, "balance" and "sustainable development", as well as diagrams and formulas for calculation of such states. For practical realization of the NCSD in the country a new and worthy of the XXI century "Industry of Knowledge" should be created, which at the national level should proceed from the formula: "Digital Russia" = ("digital Earth" + "digital economy").

1 Introduction

From social practice it is known, that at present the main problem for our civilization is the solution of the problem of sustainable economic and social development, both of individual regions, and the planet as a whole (G.X. Brundtland commission, 1967). In 1993 it was first formulated and approved by the world community as "the Concept of Sustainable

* Corresponding author: slturkov@gmail.com

Development" ("Sustainable Development"; the CSD in the abbreviation of the United Nations, 1995) and adopted by several countries (United States, 1993; Russia three years later) as the official doctrines of their state development [1]; (FZ of the Russian Federation No. 440 of 01.04.1996).

Today, this "Concept" is perceived by all as some common or "hypothetically" acceptable and science-based criterion function of future directions of development of Nature and Society. However, despite the considerable efforts of fundamental science, it is still not clear how to realize it in the current practice of management in the conditions of modern development of territories and communities of different (global, regional, local) levels of the planet organization, how to achieve the goal formulated in it, or at least, to objectively assess the degree of approximation in it? Adopted and mediated by the community practice of using the main propositions of this theory are absent in the world society. All this leads only to descriptive, but not constructive and normative conceptions of the science concerning its practical realization (in the form of GIS and GRID systems, neural networks, "Blockchain", Conflict Resolution Network and all other AI-technologies), and also extremely limits the development of specific modules, models, methods and algorithms to make optimal control decisions.

The main reasons for the lack of methodically and algorithmically unified, complete and specific solution of scientific and applied problems of CSD at the global and national levels are the following:

- general underestimation of the scientific complexity and uncertainty of this "Concept" as such;
- the fact of its extraordinary broad interdisciplinary, interdepartmental and social manifestation;
- functional unity of global, regional and local levels of its impact on complicated natural and social systems;
- absence of methodological (model and algorithmic) generality and completeness of methodology, theory and conceptual apparatus of "noosphere" thinking and ecological consciousness;
- lack of technological (physical and mathematical) ways to explain and formalize the world around us, as well as technical means and methods of their practical implementation.

The system analysis shows that the main limiting factor in the development of CSD is a wrong choice of the main direction of the scientific research in this field of knowledge, including the original paradigm, or "ratio" (from: [2, p.25; 306]). Today it is becoming more and more obvious that such thinking and such position of an official science seriously block the further development of its basic foundations, and the Concept itself obviously demands fundamental theoretical and technological reconsideration. To fulfil the requirements of its scientific decision (objective function: "min" loss of the original natural matter, with "max" socially necessary material benefits), a suggestion is made to use the theory of "noosphere" by V.I. Vernadsky (1933) as a basic element in the present report. Thus, the following well-known classical definition was used: "... on a qualitatively new form of organization that arises from the interaction of nature and society, which is characterized by the relationship of the laws of nature with the laws of thinking and socio-economic development laws of a society".

2 Methodology and theory of research

It is known that after the scientific publications of V.I. Vernadsky on the theory of the biosphere (1926) and further "noosphere" (in the definition of E. Le Roy and P.T. de Chardin, 1927), more than 80 years have passed. But if the first of them was theoretically and experimentally confirmed by the works of J. Lovelock and L. Margulis ("Geya hypothesis", the 70-ies of the XX century), the theory of the "noosphere" still remains, although very constructive, but a scientific hypothesis. From the point of view of fundamental science, the current situation (except for the modeling of possible consequences of the "Nuclear winter" in Russia, as well as its assessment in the United States [3]) is quite understandable. Thus, in the field of CSD methodology – these are the attempts of the researchers to use the old paradigm (according to I. Newton – "mechanistic" thinking) in the process of its solution.

NCSD of Russia was being developed in the CC, FEB RAS from 2003 to 2017 [4-9]. The initial materials include the term "noosphere" (according to E. Le Roy and P.T. de Chardin, 1927) and its system (functional) interpretation in the view of the last author: *" ... noosphere seeks to become a single closed system, where each element individually sees, feels, wishes, suffers as also are the others, and at the same time"* [10, p.199] and the theory of V.I. Vernadsky (1933) of the same name. The methodology is based on the modern physical ideas about the world around us (Nature and Society) as a materially closed (according to [11]) system. The latter is logically proved by the fact that *" ... to think the world is not only to register it, but to give it a form of unity, which it would be deprived of if it were thought"* [10, pp. 197-198].

The main research method is a "system" approach, but in its qualitatively new – evolutionary – form, or in the direction of its "geosystem" (cybernetic) component. Methodically – this is a way of simultaneous and successive study of the "internal" and "external" management by the processes of functioning and development of active complicated systems (ACS) of the class "nature-society" (state "balance" and "sustainable development"). The general technology of cognition also provides for the implementation of the following commonly accepted in the fundamental science logic scheme: "object" <=> "subject" => "methods of research"; the proposed in "Introduction" criterion function was considered as a complete result of the study for practice. Technologically, this process is organized in the form of a successive transition of logic research from descriptive to constructive and then to normative aspects of the interaction of Nature and Society at all required (from preconditions of the task) levels of global, regional and local environmental management.

The main methods of the research are mathematical apparatus of game theories (GT); game-theoretical modeling; "games with nature" ("Game against Nature"), antagonistic, coalition, cooperative, strategic, etc., required under the terms of setting the objectives of the game, along with the methods of strategic pattern recognition (PR). In modeling, this provides full integration of all available today natural science knowledge, as well as the possibility of computing operations of situational control of ACS "nature-society". As special theories and methods of research, we also used "General theory of systems" by L. Bertalanfi [11] and "infinite nesting of matter" by B. Mandelbrot [12], cybernetics [13], synergetics [14] and the theory of stochastic systems [15], as well as experience and practice in the study of complex branches of knowledge [3].

3 Results and discussion

While studying the initial methodological problems of CSD, it was noted that its practical solution is impossible without transition from the process of differentiation of all known currently natural science knowledge about the world around us ("mechanistic" thinking; "object-oriented" approach) to their full integration ("noospheric" thinking; "normative" approach). In other words, fundamental science needs its original scientific paradigm to be changed. Otherwise, the CSD will in principle not have its full theoretical and experimental confirmation. The areas of knowledge that are directly related to solving the problems of the CSD are shown in Fig. 1. See also: Fig. 1 in [6]. Variants of three possible phase (information) states of scientific paradigms or scientific theories; [16] and the principle of "correspondence" by N. Bor, 1913.

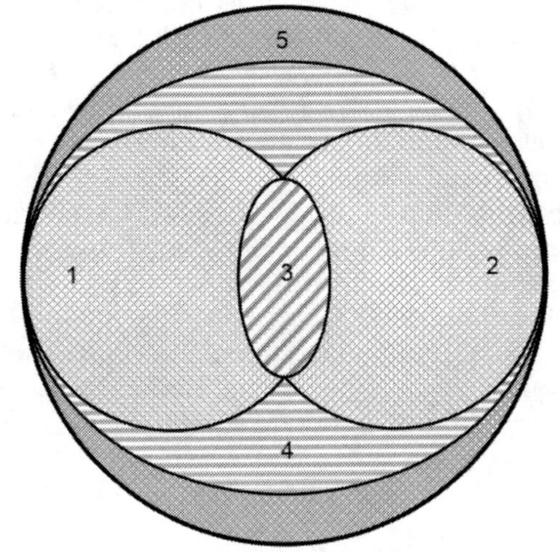

1 – phase space of natural processes; sphere of geographic knowledge
2 – phase space of social (socio-economic) processes; sphere of economic knowledge
3 – natural resources using as interrelated and intersecting processes of nature and society; sphere of knowledge of economic geography and regional economics
4 – physical, chemical and biologic processes of interaction of nature and society; sphere of ecological knowledge
5 – "noosphere" (biosphere) processes of interaction and evolution of the planet geospheres; sphere of knowledge for geoecology

Fig. 1. System conception of the "phase" (information) space of nature management

In the fundamental science, the "thermodynamic" paradigm (according to Newton this is a physical basis of "mechanical" thinking) is known from the second half of the XIX century. But a new – "noosphere" – paradigm being theoretically and experimentally confirmed was developed late in the XX century. It was based on the scientific achievements of the quantum physics (Unified Field Theory, UFT [17], [18]). According to the initial methodology (or ratio, "reasonable beginning, principle, meaning") [2, p. 306]), UFT introduces into natural science in addition to the previously known 4 fundamental interactions the following: gravitational (XYIII century), electromagnetic (XIX century), nuclear (strong and weak, early XX century), and information interactions (late XX century). Thus, it reduces everything exiting in nature to the model: mass {M}, energy {E}, information {I} + 5 elementary interactions of their fields. The model was theoretically proved by A. N. Panchenkov in 1999 ("New theory of entropy" [19]). On the basis of the UFT and this theory, in 2017 we developed "Basic physical and information characteristics of modern paradigms of the system conception of the world" (see: [6]; Table. 1, p. 9).

From gnosiology it is known that our thinking about the phenomenon under study a priori should change under changes of the original paradigm, which in its turn should be reduced to a logically and theoretically related, single unit of knowledge, or "metaknowledge" of geoecology. They are logically (theoretically, methodologically, and technologically) higher-level associated knowledge of the object, subject, and problem (target) sub-fields and research methods of each of the complex of sciences on the Earth, Society, and natural science in general, the objective function of which is aimed at the adoption of effective management decisions in the course of their mutual "co-evolution" (N.V. Timofeev-Ressovsky, 1968). At the same time, and in accordance with the logics of our "geosystem" approach as an initial object of study, it is necessary to proceed to the study of active complicated systems (ACS) of the class "nature <=> society". Physically, such systems are mutually integrated systems and objects integrity, the characteristics of which can not be reduced to the properties of the constituent subsystems and are considered as living systems, where the main driving operation and development mechanism is a conflict (in the conditions of uncertainty). "Conflict" in such systems is understood as " ...

nature's desire to preserve its physical (material-matter), energy and information state as opposed to the human desire to violate it"[4].

Hence (in terms of the "new theory of entropy" of A.N. Panchenkov [19]) the following definitions of the basic concepts of V.I. Vernadsky's "noosphere" theory can be proposed. "Equilibrium " is the state of the ACS "nature <=> society", when in the process of functioning of complicated systems as a whole, the initial position and composition of the structural entropy (H_q) are preserved, which include its thermodynamic (or "frozen", H_f) part: ($H_q \approx$ const); in this case the "structure-attractor "of the system is relatively stable in "space and time". "Sustainable development" is a state when there are qualitative changes in the structural entropy (H_q; it occurs due to "the frozen state", H_f, or thermodynamic entropy: ($H_q \neq$ const); it takes place due to the formation of its new core – the replacement of the old "structure-attractor" by the new one, which will further determine a fundamentally different system, remaining in the future stable in "space and time".

The studies revealed that from the position of the person directly making control decisions (PMD), and in relation to the ACS "nature-society" the general methodology and principal scheme of its actions should have the following form (Fig. 2; using the known principle of "dependence" ("system" + "device") of N. Bor, V. Heisenberg, 1927); development of the control scheme of "teleological" systems from N. Wiener (1950) [13] in the direction of the theory of "noosphere" E. Le Roy, P. T. de Chardin, and V.I. Vernadsky [10].

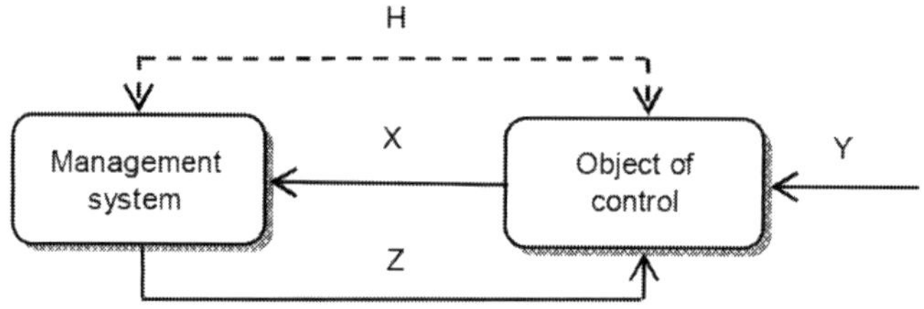

Fig. 2. A general scheme of decision-making under situation management by ACS class "nature-society"

Here the new connection ("H") physically reflects the simultaneous entry of man into the object of control ("object of nature") and the control system (or "subject of nature" – a rational being and a person who has a good chance of making decisions; PMD). This state reveals the "dual" ("biosocial") nature of man (society) and emphasizes its exceptional importance among all other organisms of the biosphere. From here, we can introduce the following general model of ACS management of "nature-society» class:

$$< \{M_x\}, \{M_y\}, \{M_z\}, F\,(f), \varphi, \Psi >, \qquad (1)$$

that is, management is *a process of a rational and purposeful making decision, which is determined by a set of states of the control object and the environment* (*{M_x}, {M_y}*), *a set of functions of the decision-maker* (*{M_z}*), *acting towards the implementation of the full management function* (*Ψ*), *to achieve global and local criteria for optimizing management* (*F* (*f*)), *with due regard to the response of the control object to the control impact and environmental influences* (*φ*).

Technologically, the optimal control resolution can be achieved by the following methodological procedures. The full matrix of the initial data on the region should be analyzed to special information processing for their compliance to the mathematical apparatus of Games Theory. In case some data (uncertainty) are missing, they are restored by using special methods. Further, their processing by statistical methods allows identification of the data (on the principle of proximity of the "spot" of the regional environmental management process, REMP), situation, which are further to be analyzed more closely by the person making decisions.

The "balance" of any complex system (if this process is chosen as a final point of the study), is to achieve its so-called "saddle" point (or such its position when the minimum value in the column coincides with the maximum one in the matrix row) [20]. At the same time, the state of "Sustainable development" of such a system in its original sense refers to the field of long-term planning and forecasting problems; for this purpose the concepts of "vogue" and "structure-attractor" are used in the natural sciences and synergetics (all the necessary schemes and formulas for solving such problems are given in [6]).

For science and management (or "control") structures of particular interest are the following computing technologies. These are graphical methods for solving GT problems (Fig. 3), when in real time it is possible to direct visual observation of changes in the schedules of the solution of consistently studied problems (interpretation, diagnosis and monitoring, planning and reconstruction, forecast, situation management).

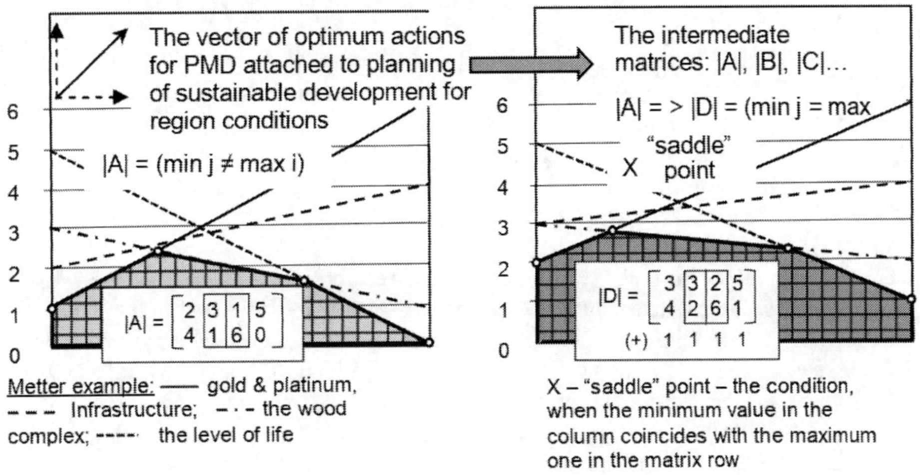

Fig. 3. Graphic scheme of the initial [A] and output [D] matrices

Hence it follows that all the situations can be represented as 2- and 3D - dimensional graphs, and their flat reflections (an analogue of the well-known from geography and ecology concept of "Ecological framework of the territory", EFT). Their combination further allows "mixed" GIS-technologies, as well as to Virtual Information Models (VIM) of decision-making: as a result, it is a simultaneous "vector" + "raster" representation of data and knowledge by regions. In addition, according to the well-known from the "Theory of Games" theorem (see: formulas 2.5.22-2.5.24 [20, p. 50-51]) the decision maker can always calculate the values of single optimal strategy and gains from different games and matrices. Then a complete – strategically optimal – solution implies that all data on the "mixed" strategies of all the players are to be obtained (i.e., throughout probability distributions, or shares, their "clean" strategies); this is what all regional management structures should make use of without exception.

4 Summary

1. Without methodological transition of contemporary natural science from traditional ("mechanistic") to a new – "noosphere" – thinking, theoretical, technological and methodically correct solution of CSD problems is impossible by definition; this fact is clearly confirmed by many years of negative (scientific and practical) experience of their solution. The suggestion we made of the solution of the problems of the CSD can be based

on a new subfield of natural scientific knowledge: "Metaknowledge of Earth and Society" (geo-ecology; specialty 25.00.36).

2. When changing the original paradigm a priori should be changed our general scientific thinking and fundamental ideas about the phenomenon studied itself (the principle of "dependence": "system" + "device" by V. Heisenberg, 1927).

3. On the base of the new – "noosphere" – paradigm it is possible to fully fulfill the computational operations of the situational control of the ACS "nature – society" class (deterministic, probabilistic, uncertain and game-theoretical formulation of the control problems).

References

1. Environmental aspects of the activities of transnational corporations: A surv. N.Y.: UN (1985).
2. Xyojzinga J. Homo ludens. SPb: "Azbuka-klassika" (2007).
3. Moiseev N.N., Aleksandrov V.V., Tarko A.M. A human and biosphere. The knowledge of systems analysis and experiments with models. M.: Nauka (1985).
4. Turkov S.L. The base of control for regional natural resources use. Vladivostok: (2003).
5. Turkov S.L. Decision-making methods in geoecology. MIAB, **4** (2010).
6. Turkov S.L. "Metaknowledge" for geoecology. Prepr. No 225. Khabarovsk: CC FEB RAS (2017).
7. Turkov S.L. Modelling the processes of control for complexity organized "nature-society" systems // InterCarto/ InterGIS-23. M.: **1** (2017).
8. Turkov S.L. Modelling the processes of control for complexity organized "nature-society" systems. Europ. Journ. Techn & Natur. Scien. Vienna-Prague, Austria, **5** (2017).
9. Turkov S.L. The National conception of Sustainable Development for Russia // InterCarto/ InterGIS-24. M.: **1** (2018).
10. Shardin P.T. Fenomen of Human. M: Nauka (1987).
11. The researches by common theory of systems. M.: (1969).
12. Mandel'brot B. Fraktal geometry of nature. ICI (2002).
13. Viner N. Cybernetics or control and connection in animal and machine. M.: Sov. Radio, (1968).
14. Xaken G. Synergetics. M.: Mir (1980).
15. Pugachev V.S., Sinitsy'n I.N. The theory of stochastic systems. M.: Logos (2004).
16. Kuhn T. Structure of Scientific Revolutions. Univ. of Chicago, **2** ed. (1970).
17. Arntz U., Chasse B., Visente M. What the bleep do we know? M.: (2013).
18. Tixoplav T.S., Tixoplav V.Yu. The physics of belief. SPb.: ID "Ves" (2002).
19. Panchenkov A.N. Entropy. N. Novgorod (1999).
20. Ouehn G. Game theory. Ed. LKI (2008).

E3S Web of Conferences **56**, 04016 (2018) https://doi.org/10.1051/e3sconf/20185604016
VII International Scientific Conference "Problems of Complex Development of Georesources"

The application of remote sensing data for the assessment of man-made environments in the coal regions of the south of the Far East

Marina Bubnova[1*]

[1]Mining Institute of Far eastern branch of Russian Academy of Sciences, Khabarovsk, Russia.

Abstract: The paper is concerned with analysis state of surfaces of waste dumps coal deposits according to Earth remote sensing. It is noted, that formed because of open-pit operations at Bikinsky and Raichikhinsky coal mining enterprise areas denudation accumulative violations were determinate by using programs for satellite images interpretation, as well as certain large forms of man-made relief (such as quarries, dumps, sedimentation tanks) were identified, followed by the calculation of their areas and the assessment of the value of environmental and economic damage to the environment, i.e. it was possible to give not only a qualitative description of the pollution process, but also its quantitative assessment. The approach to biomass determination based on vegetation NDVI index is presented. The conclusion is based on possibility of self-restoration of the plant ecosystem on the surface of mining landscapes without special work on the reclamation of disturbed lands. **Key words:** Dumps of coal enterprise, remote sensing of the Earth, satellite images, vegetation indices, self-growth

1 Introduction

The use of earth remote sensing data is currently almost the only opportunity to obtain objective and reliable information about the state of mining landscapes in large areas.

In the mining industry, land degradation as a result mining of deposit is one of the biggest problem. For example, on south of the Far East, a significant amount of disturbed land is the result of the use of open methods of mining, in particular brown coal. At the same time, stripping and containing rocks are transferred to the surface. Non-selective (gross) spallation leads to a chaotic mixing of rocks that differ in granulometric, petrographic, mineralogical compositions. Most of the dumps remain under self-growth and are not reclaimed [1].

Dumps are accumulative forms of techno-relief of different heights and occupied area. Their dimensions often do not go beyond the mesoforms of the relief with an aligned flat top, in the profile having a trapezoidal shape. Physically, the miming dump is a complex structure: the top, from where the small particles and nutrients flush away, the transit zone is the slopes through which the small particles of the rock are transported to the foot of the

*Corresponding author: 138_marina@mail.ru

dump, called the accumulative zone. Here the substances washed away from the top accumulate, so that more favorable conditions are created for the beginning of the process of soil cover restoration [2].

The study of the processes of overgrowing dumps of coal deposits in the southern Far East and their reclamation are devoted to the work of a number of authors: Gusachenko A. Yu., 1992; Krupskaya L. T., 1992; Kurachev V. M. and E. R. Kondrashin, 1994; Radaev V. V., 2004; Kostenkova N. M. and Sabihin V. I., 2007 etc. In these works the main attention was paid to the issues of restoring economic functions of the disturbed mining areas [3-5]..

When solving the problems of assessing the territories of dumps of coal enterprises, a number of issues often arise that are associated with the definition of the boundaries of objects, the allocation of the class of pollutants, the definition of the impact of pollution on vegetation/soil, the detection of zones of deformation of the earth's surface. Earth remote sensing (ERS) technologies provide unique tools for solving such problems. The use of space images in the monitoring of the environment in the areas of coal mining allows you to monitor the pollution of the water basin, to analyze the state of vegetation, to identify the degradation of soil, thus solving the following tasks: a) determining the impact on the environment; b) assessment of pollution; c) determining the boundaries and deformation zones of anthropogenic objects. This article describes some examples of the use of remote sensing data for the assessment of man-made situations in the coal regions of the South of the Far East.

2 Materials and methods researches

Modern technical means of remote sensing of the Earth (ERS), allow obtaining digital images of areas of the earth's surface with high spatial resolution and in a wide range of electromagnetic waves. In addition, mathematical methods for processing digital images (spatial-spectral analysis, coverage of the territory, registration of a sufficiently wide range of electromagnetic spectrum, regular observation of the territory, free access and convenient interface for downloading archives), which can be applied with appropriate adaptation of the algorithmic base processing of digital images of the soil cover of coal-mining areas [6].

The study assessed the spread of soil cover contamination and its self-repair in various mining regions via Landsat satellite images (NASA USA), which is freely available on the USGS website. During the work, the images were downloaded, their selection, the choice of informative ranges of the electromagnetic spectrum [6].

The main working tool for solving the problems of this study was the computer application QGIS (Quantum GIS), which is distributed over the Internet under the license Open GPL [7]. Criteria for its selection in the composition with GRASS GIS were multitasking, multiplatform, extensibility and relatively easy learning opportunities. As a topographic basis, was used the Russian edition vector map Open Street Map, which is freely accessible on the Internet [8]. Separate areas were detailed and refined according to the Cadastral map of the Russian Federation.

From the Internet, digital elevation models (DEM) were downloaded: regular matrices with a sampling rate of 1000 m (SRTM30) and 90 m (SRTM03) and inserting them into a GIS [9]. Combining satellite images with DEM allows you to get a visual representation of mining facilities on the ground, i.e. to build a digital model of the area under investigation [6]. Due to this, it becomes possible to assess the features and character of the distribution of pollutants in mining areas and the factors that influence this distribution.

3 Results and discussion

On the example of Bikinsky and Raichikhinsky brown coal deposits, an assessment was made of the degree of disturbance of the surface of the mining areas by mine workings. The studies were carried out by methods of remote satellite monitoring of the Earth's surface on the basis of an analysis of spatio-temporal variability of the formed pioneer plant groups.

Bikinskoye brown-coal deposit is located in the north-western part of Primorsky Territory. Disposal external: of the total volume of overburden approximately 1/3 is placed in the internal dumps at the horizon of the lower coal seam to be mined and the remainder is transported to external dumps (fig. 1).

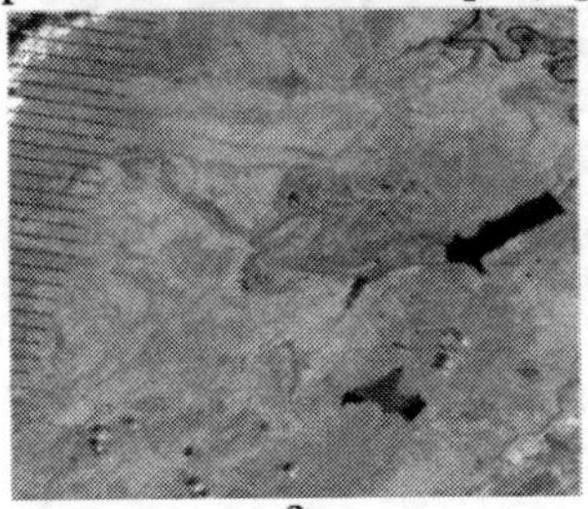

a

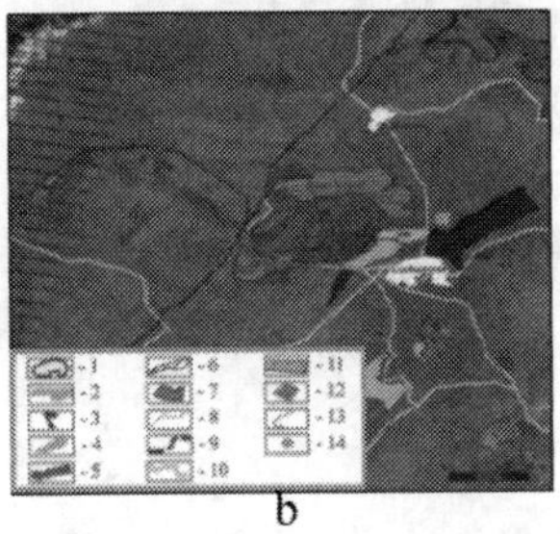

b

Fig. 1. Bikinskoye brown-coal field: a-Multichannel satellite image obtained from the us earth remote sensing satellite Landsat 7 in a combination of spectral channels 4-5-1 from 07.08.2009; b-Maps of the spatial structure of the district of Bikinskoye brown-coal field, with the allocation of man-made objects, based on the decoding of satellite images: 1-settlements; 2-technical complex of the Primorskaya power plant; 3-boilers, boiler emissions into the atmosphere; 4-ash dumps; 5-pond-cooler GRES; 6 - coal pits; 7 - rock dumps; 8 - river network; 9 - Railways; 10 - roads; 11 - auxiliary production; 12 - LLC "Ecomed"; 13-roads; 14-dump sampling points

In QGIS in a rectangular projection EPSG — WGS 84 were measured area of the dumps shown in the picture (fig. 1 b). The area occupied by the rock dumps at the Bikin brown-coal Deposit was 14.54 km^2, 12.34 km^2 was occupied by coal mines, and 8.44 km^2 by ash dumps, the mileage of coal roads was 88.87 km. Total area of disturbed lands in 2009 was 35.32 km^2, which is 65% of the allotted land. The calculation of ecological and economic damage to the environment by disturbed lands was made, which was carried out according to the formula:

$$\mathrm{Ddl} = \sum_{i-1}^{n} B_i \times S_i \text{ , rub.,} \quad (1)$$

where: n - the number of types of disturbed land; i - type of disturbed land; Si - the area of disturbed lands of the i-th type; Bi - the magnitude of the annual economic damage caused to the environment by i-type disturbed land. Value of the damage, inflicted to the environment disturbed lands is determined from Techniques of determining the economic efficiency of recultivation of disturbed lands [10].

Ddl = (270×1454) + (65×1234) + (185×844) = 628930 rub.

The magnitude of the ecological and economic damage from the area of disturbed lands of 3,532 hectares at the Bikinsky brown coal deposit amounted to 628930 rub.

Raychikhinskoe brown-coal field, is located in the southern part of the Amur region, near the town of Raichikhinsk. Area of the field is 50 km^2. The city Raichikhinsk and settlements is almost entirely surrounded by or located next to the dumps, havings a height of 10 to 55 m (fig. 2). Due to this, a peculiar in the form of trench microrelief is created in the city. The use of remote sensing has made it possible to determine the areas of man-caused disruptions at the Raichikhinsky brown-coal field, from them: the area of waste heaps amounted to 11609 hectares, coal sections of 1895 ha and sumps an area of 184 hectares. The total area of disturbed land is 13688 hectares.

According to formula 1, the calculation of environmental and economic damage to the environment by disturbed lands was carried out for Raichikhinsky brown-coal field:

Ddl = (65×1895) + (185×11609) + (21×184) = 2274704 rub.

The magnitude of the ecological and economic damage amounted to 2274704 rubles.

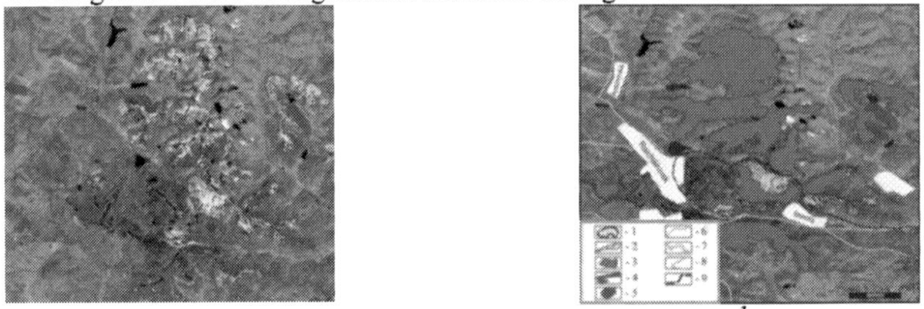

a b

Fig. 2. Raychikhinskoe brown-coal field: a-Multichannel satellite image obtained from the us earth remote sensing satellite Landsat 7 in a combination of spectral channels 5-6-2 from 07.09.2013; b - Maps of the spatial structure of the district of Raychikhinskoe brown-coal field, with the allocation of man-made objects, based on the decoding of satellite images: 1 - settlements; 2 – coal mines, 3 – overburden dump; 4 - settlers% 5 – pit; 6 lake; 7 – river, 8 – highway; 9 – coal road 10 railway road

Thus, with the use of materials for interpretation of space images, was recognized the area of denudation and accumulation of violations of, educated in open coal mining in Bikinsky and Raychikhinsk coal mines, and also identified some large forms of man-made terrain (such as quarries, dumps, sedimentation tanks), followed by the calculation of their areas and the assessment of the environmental and economic damage caused to the environment, i.e. it is possible to give not only a qualitative description of the pollution process, but also its quantitative assessment.

For most problems associated with the study and monitoring of both natural and man-made objects, it is convenient to use their physical characteristics, reconstructed from remote observation data. The greatest interest in our research is represented by vegetative and soil indices. Knowledge of the relationship between the structure and the state of vegetation with its reflective abilities makes it possible to use space images to identify the species composition of vegetation and its quality. A characteristic feature of vegetation and its state is the spectral reflectivity, characterized by large differences in the reflection of radiation of different wavelengths. Various normalized vegetative indices allow to calculate and estimate the volume of biomass in plant systems. The most sensitive spectral channel for counting this type of index is the "extreme red" red-edge range (705-745 nm) [11]. Soil indices characterize the degree of coverage of the observed surfaces by vegetation, and react well to the minimal presence of biomass on the surface of the object. Vegetation is very sensitive to environmental disturbances and most clearly reflects the change in the ecological situation as a result of technogenic impact. The state of vegetation can be considered as an indicator of the level of man-caused stress on the natural environment of the study area. Restoration of vegetation characteristics based on remote sensing data allows obtaining some numerical characteristics of the biomass volume, which can serve as an indicator of the restoration of disturbed lands.

The most popular and often used vegetation index — NDVI (Normalized Difference Vegetation Index). The high chlorophyll content associated with a large phytomass of vegetation leads to lower values of reflection coefficients in the red zone of the spectrum and greater values in the near infrared [12]. The value of the index directly depends on the photosynthetic activity of the vegetation and corresponds to its density at a certain point in the image, which is equal to the difference in the intensities of the reflected light in the red and infrared range divided by their sum. Calculated by the following formula:

NDVI = (NIR-RED)/(NIR+RED), (2)

где NIR – the reflection in the near infrared region of the spectrum, RED – reflection in the red region of the spectrum.

NIKE and RED's attitude to each other allows us to clearly separate vegetation from other natural objects in the pictures. Figure 3 shows an example of the result of calculating NDVI for the territory of Luchegorsky district.

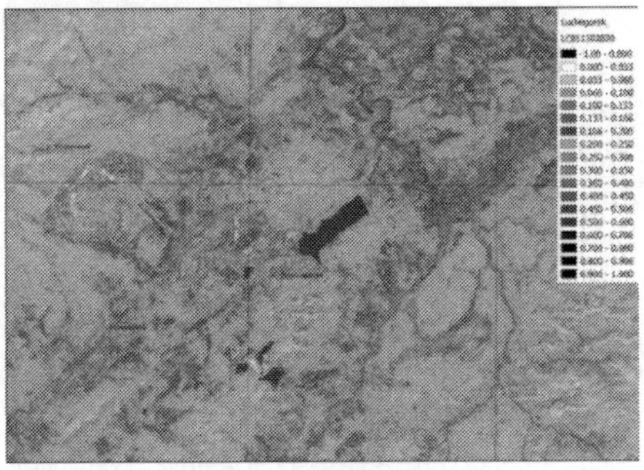

Fig. 3. Combining the NDVI with the digital model of the relief of the location area of the Luchegorsky coal mine

The average value of the index on this territory varies from 0.16 to 0.25, which indicates the appearance of biomass on the surface. The amount of green biomass was influenced not only by anthropogenic factors (violation of the hydrogeological regime, in particular, the formation of a depression funnel, powerful emissions of coal combustion products from GRES), but also natural (flat terrain characterized by meadow, more often grassy vegetation) [13]. The reason for the low values is that at the first stages of settling the territories they have obvious advantages (67,9 %) herbaceous species (taproot), which allow to gain a foothold in the ground and form above-ground shoots, but most of the species are annual grasses and die at the end of vegetation [14].

It was established that the area of bare dumps in 2001 at the Bikinsky deposit of brown coal (coal cuts Luchegorsky 1 and Luchegorsky 2) was 4595.92 hectares, and in 2015 - 1082.64 hectares. Revealed that over the past 15 years have seen overgrowing of about 30 % of the surface of the dump. Especially intensively, these processes occur on the dumps of the Southern and Northern.

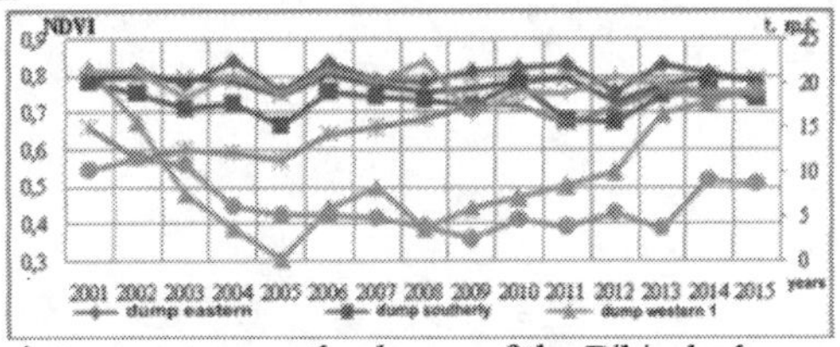

Fig. 4. Indication NDVI and air temperature on the dumps of the Bikinsky brown coal deposit

One of the reasons for the intensive overgrowth of these dumps is a certain temperature regime (fig. 4).

According to the results of the analysis of the space images of the region of studies taken at different times of the year: in winter, in summer, in spring, in autumn (the photographs should be cloudless, but there may be different series and describe different territories), it is possible to construct the dependence of the characteristic values of NDVI on temperatures (t) (fig. 5) and to establish a correlation between the degree of soil degradation and the vegetation response index on air temperature, but this is also how the temperature data vary depending on the region (or are constant values) - a task that is subject to further investigation.

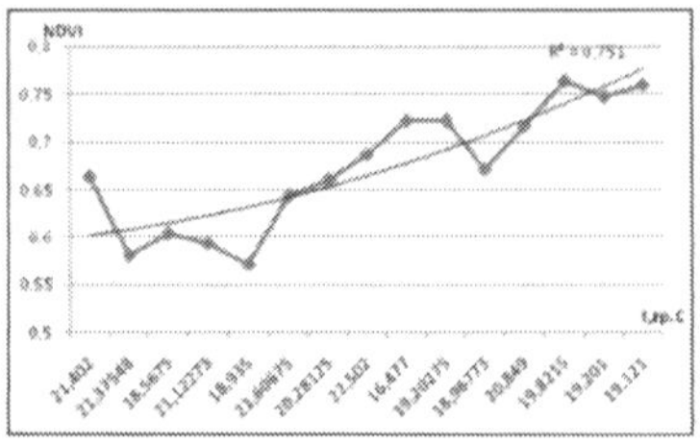

Fig. 5. Dependence of NDVI on temperature

The self-growth of the dumps of the Luchegorsky coal mine is very intensive, therefore there is no need for remediation here. In the stage of stable humus accumulation in soils formed on the dumping rocks, a clearer differentiation of the waste rock to the genetic horizons occurs. This indicates that the emerging ecosystems in the phase of their post-technogenic development have high regenerative capabilities, there is a formation of soils on loose dumps in a relatively short period of time (up to 30 years).

On self-growth dumps of overburden and enclosing rocks of Raichikhinsky brown coal deposit in the period from 2001-2015. on the basis of the vegetative index NDVI, studies were carried out to analyze the dynamics of the state of vegetation over a 15-year period (fig. 6).

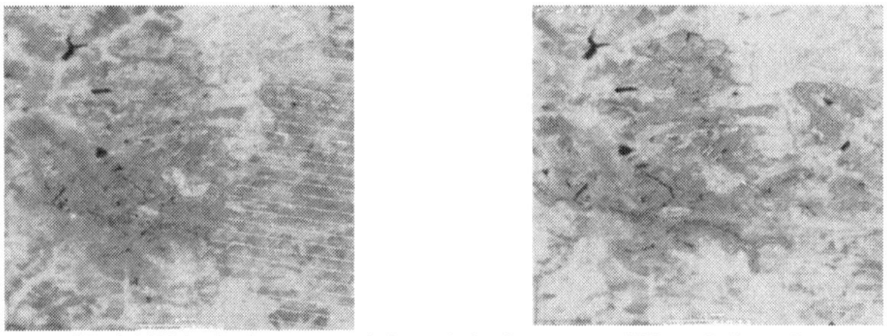

Fig. 6. Dumps of Raichikhinsky brown coal deposit in the system "Vega-Science"

The total open surface of the dumps in 2001 and 2015. made 5565.08 hectares and 1356.58 hectares respectively (calculated in the program QGIS). In this way, the area of the restored territory for 15 years was not less than 4208.5 hectares, which indicates an intensive overgrowth of waste dumps. A stable sustainable growth of biological phytomass and an increase in the vegetative index (fig. 7).

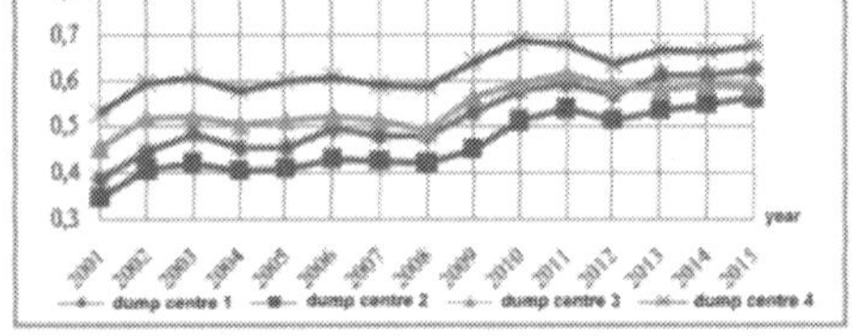

Fig. 7. Dynamics of NDVI secondary phytocenoses of the dumps of Raichikhinsky brown coal deposit

According to our data, the fastest process of overgrowing dumps by vegetation occurs on the western (w-1) and southern dumps (s-1). Perhaps this is due to the predominant South East – Northwest (and vice versa) direction of air flows. The frequency of these winds in Raichikhinsk is about 20% greater than in neighboring areas.

The nature and intensity of the processes of self-restoration depend entirely on the degree and characteristics of the technogenic changes in the biota during the construction and operation of the extracting enterprise. Based on the results of the study of the dynamics of the species composition of phytocoenoses in rock heaps, it is possible to rationally select the species composition of vegetation that is optimal for local natural and climatic conditions for carrying out remediation measures. This will speed up the processes of restoring the natural environment, as well as optimize the costs of environmental protection measures. Thus, it has been established that the properties of technogenic-surface formations differ little from those of self-growing spaces, but the process of natural regeneration proceeds slowly, therefore on disturbed lands it is necessary to combine self-healing of pioneer vegetation and creation of artificial phytocenoses with the help of remediation measures. All this will contribute to improving the environmental situation and positive dynamics of key environmental indicators within the city of Raichikhinsk.

4 Conclusion

The use of remote sensing data in the study of mining dumps provides the opportunity to obtain the following results: to determine the boundaries of dump lands, Identify the areas of denudation and accumulative violations, formed during open coal mining in coal-mining enterprises, and to identify certain large forms of man-made terrain (such as quarries, dumps , sedimentation tanks), with the subsequent calculation of their areas and an estimate of the magnitude of the environmental and economic damage to the environment, i.e. to get an opportunity to give not only a qualitative characteristic of the process of pollution, but also its quantitative evaluation. In the study of dumps of mining enterprises, the zones of anthropogenic transformations of the vegetation cover based on the NDVI spectral index are most reliably identified from remote sensing data, so that the vegetation state can be considered as an indicator of the level of technogenic stress on the natural environment of the study area. This allowed us to determine the zones of self-healing and the causes of intensive overgrowing of the dumps.

References

1. O. V. Polokhin, Fundamental research, **12**, 327 (2014).
2. K. N. Trubetskoi, Ecological problems of subsoil development in the sustainable development of nature and society, 260 (Nauchtekhlitizdat, 2003).
3. S. I. Im, V. I. Kharuk, , Geofizicheskie protsessy i biosfera, **14:1**, 53 (2015).
4. E. V. Pavlova, M. L. Makhrova, G. Yu. Yamskikh, Zhurnal Sibirskogo federalnogo univesriteta. Seriya: Tekhnika i tekhnologii, **8:6**, 706 (2015).
5. L P. Shoo, P. Scarth, S. Schmidt, K. A., Restoration Ecology, **21:4**, 481 (2013).
6. K. V. Myachin, Geoinformatics, **2**, 2 (2016).
7. Quantum GIS [Electronic resource]. – Access mode: http://freesoft.ru/quantum_gis
8. L. Lovgren, T. Hedlung, M. Lindegren, Ehkologicheskaya strategiya razvitiya gornodobyvayushchej otrasli – formirovanie novogo mirovozzreniya v osvoenii prirodnyh resursov, **2**, 72 (2015).
9. Site with archive of data on radar probing of the Earth's surface SRTM3 [Electronic resource]. – Access mode: http://dds.cr.usgs.gov/srtm/version2_1/SRTM3/
10. H. J. Bowen, Environmental chemistry of the elements, 333 (N.Y.: Acad. Press, 1979)
11. V. N. Oparin, Physical and technical problems of mining, **6**, 190 (2014).
12. J. W. Rouse, Monitoring the vernal advancement and retrogradation (greenwave effect) of natural vegetation, 371 (NASA/GSFC Type III Final Report, Greenbelt, 1974).
13. N. M. Kostenkov, V. I. Oznobikhin, Stimulation of the processes of self-growth of dumps of brown coal cuttings in the south of the Far East, 131 (Izd-vo Okarina, 2013).
14. L. N. Purtova, L. N. Shchapova, I. V. Komachkova, Vestnik of the Far East Branch of the Russian Academy of Sciences, **4,** 62 (2010).

Platinum mineralization comparative characteristics of the some Far East deposits

Natalya Lavrik [1*], *Natalya* Litvinova [1], *Tatyana* Aleksandrova [2], *Valentina* Stepanova[1] and *Alexandra* Lavrik[1]

[1]Mining Institute of Far eastern branch of Russian Academy of Sciences, Khabarovsk, Russia.
[2]Federal State Budgetary Educational Institution of Higher Education "St. Petersburg Mining University", St. Petersburg, Russia.

Annotation. In this article shown platinum mineralization comparative characteristics for three deposits: Kondoer-traditional unique deposit and other two probably alternative source of platinum: the Poperechnoe iron-manganese deposit and the Malmyzh copper-porphyry deposit. Platinoids of the Kondoer deposit are the chain Pt>>Ir>Os>Ru=Rh>Pd. The presences of platinum crystals are characteristic, there are over 50 rare and new platinoids minerals in different combinations. There are gold and silver. Platinoids from the iron-manganese ore of Poperechnoe are as Pt>>Rh≈Ir>Ru≈Os>Pd. A scattered dissemination of arsenide sulfate and sulfides of Rh, Ir, Ru, Os are noted in the platinum. Palladium is present as impurities in gold and platinum. The gold content is different-with admixtures Ag, Pb, Cd, Fe. At this stage the platinoids content in oxidized ores of the Malmyzh gold-copper porphyry deposit is Pt ≈ Pd. The gold is present as electrum. There is native silver cadmium.

1 Introduction

There is in the modern world economy a high demand for industrial metals including platinum group metals (PGE), - platinum, palladium, rhodium, iridium, osmium, and ruthenium. A meet growing needs problem in the platinum-metal raw materials became more seriously each year [1, 2].

The Russian Federation balance reserves are 73.8% palladium, 21.7% platinum and other platinoids slightly more than 4%. Native deposits have a leading role in the reserves and production of PGE. The search for rich placer deposits of PGE is currently an actual problem [2]. Therefore, more attention is paid to the study of new ore types and alternative sources of noble metals. These include low-sulfide proper platinum-metal ores of exfoliated ultramafic-mafic intrusions, black-shale polymetallic platinum-containing ores of sedimentary complexes, nickeliferous weathering crusts, man-caused ores (waste from mining, dressing and metallurgical production) [2], iron ore formations [3]. The platinum content is studied poorly in the: fossil coals, asphaltites, porphyry ores, saline sediments, ocean formations and other.

* Corresponding author: lavrik@igd.khv.ru

Although, one genetic type deposits have a differences, the article authors compared the platinum mineralization of three deposits in the Far East: the unique most studied, placer and ore deposit of the concentrically-zoned Kondoer deposit; the Poperechnoe iron-manganese deposit with noble metal mineralization; the Malmyzh a large gold-copper-porphyry deposit, in whose oxidized ores the visible platinum was first isolated.

2 Materials and research methods

Platinum mineralization characteristics of alkali-ultrabasic Kondoer deposit (the most studied platinum source from Far East) took mainly from the basis of literary data and some of the authors' own research.

Mineralogical and technological analysis of samples for noble metals from ore-bearing and enclosing rocks of the Poperechnoe iron-manganese deposit and Malmyzh gold-copper porphyry deposits was carried out using the patented technique for the detection of noble metal mineralization, developed at MI FEB RAS.

At first, samples were crushed to 1 mm. Then the gravitational enrichment without the use of magnetic separation, short mineralogical analysis and electron microscopy was used. Modern Fritsch equipment was using for samples preparation: jaw crusher Pulverizete 1, disk mill Pulverizette 13, ultrasonic bath for cleaning sieves Laborette 17, vibrating screen Analyzette 3. A Stemi 2000 C microscope (Germany, ZEISS) was used for short mineralogical analysis of the monofraction. The electron microscopic study of platinum and gold composition from different ores and altered rocks was carried out using a JEOL microscope (Japan) equipped with an energy dispersive X-ray analyzer JCM-6000 PLUS and a VEGA 3 LMH microscope with X-Max 80 microanalysis.

3 Results research

3.1 Kondoer

Kondoer deposit is located in the north of the Khabarovsk region in the Maya river basin and in the interfluve of her left tributaries Omni and Maimakan.

The geological structure of the Kondoer concentric-zone ultrabasic deposit (one of the largest on the Aldan Shield) has been described and known many times, although many aspects continue to be debatable [4, 5 and ect.]. The region is formed the Archean foundation, Proterozoic platform cover and intrusive rocks of Paleozoic and Mesozoic age [6]. The central part of stock is formed by platinum-bearing dunites and dunite-pegmatites with segregations of chromospinelides. After dunite the olivine-diopside metasomatites zone followed and then the ring of clinopyroxenites, which include by intrusions (J3-K1) - monzonitoids, syenites, gabbroids, kosvits, alkaline diorites [7]. Platinoids ore manifestations are concentrated in ultramafites as accessory submicroscopic sizes in chromospinelides and chromites. The main platinum is in placers with the size from fractions of mm to nuggets weighing 1-3 kg. Stocks of placer platinum are over 90 tons. The average content of platinoids in placers is up 0.2 to 2 g/m3, and to 60-80 g/m3 in the richest areas [4]. The main mineral of deposit is iridic isoferroplatinum containing Pb, Cu and platinoids impurities.

There were found over 50 minerals of platinoids, gold and silver in the placer. Some from them are rare and new, for example, bordnikovit Pd4Cu3Zn, konderite Cu3Pb (Rh, Pt, Ir) 8S16, native Ru, RuOs [8, 9]. Isoferroplatinum forms are diverse: well-cut crystals of regular cubic and octahedral form, twins; teardrops, xenomorphic formations and "porous" dendritic. Heavy concentrated platinum differs in color (tin-white, lead-gray, dark graphite, brown-red) and it's characterized by a magnetic susceptibility (Fig. 1).

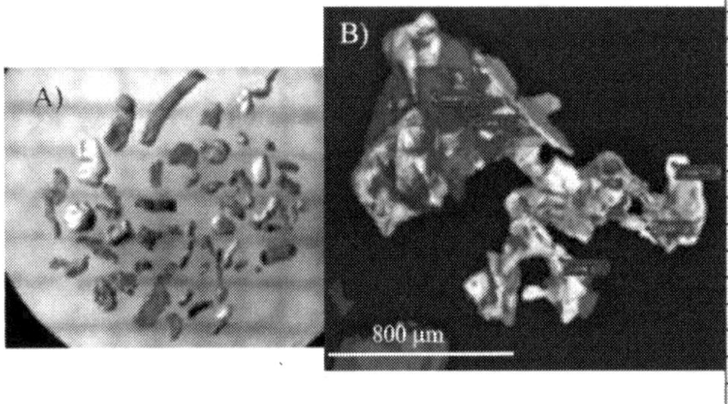

Element	Contents (mass, %)				
	1	2	3	4	5
C	10.52	7.61	-	17.7	16.21
O	3.15	42.52	62.46	3.13	18.95
Mg	-	4.97	2.12	-	-
Al	-	2.67	3.91	0.59	1.29
Si	-	-	23.55	-	1.67
Ti	-	0.32	-	-	-
Cl	1.49	-	4.83	-	-
K	-	-	1.67	-	-
Cr	-	26.35	-	-	-
Fe	7.52	15.18	0.79	-	-
Zn	-	-	0.67	-	37.5
Y	-	-	-	29.44	24.38
Os	-	-	-	39.64	-
Ir	-	-	-	9.5	-
Pt	77.32	-	-	-	-

Fig. 1. Kondoer heavy concentrated platinum : A) platinum particles (0, n-2.0 mm) of different morphology and properties; B) platinum grains of unusual shape and complex composition in intergrowth with chromites and other minerals.

"Crystalline nuggets" of platinum often is covered "Golden shirt". On the mineralogical research it's tin (0.05-1.0 mm) membrane compounds of gold-copper and of platinum-palladium. Membranes of gold compounds on the majority of non-rolled crystals are endogenous [10]. Xenomorphic and dendritic grains of platinum may contain sprouts of magnetite, limonite, chromite, olivine and other minerals (Figure 1).

The distribution of platinoids in zonal massifs is characterized by a series: Pt>>Ir>Os>Ru=Rh>Pd [11].

There is characteristic presence the gold and silver along with platinum and palladium, which have a second role. The gold is bright yellow in plate-shaped and flattened forms. The part of grains is brownish in limonite shirt.

There are found native Cu, Sn, Pb, Ni, Fe, Sb, Bi [10]. The main minerals of heavy fraction are presented as follows: pyroxene, magnetite, titanomagnetite, limonite, chromite, olivine; are found: apatite, almandine and pyrope, amphibole, tourmaline, zircon, pyrite, marcasite, galena, chalcopyrite, rutile, ilmenite, sphene, chrome diopside and spinel.

3.2 Poperechnoe

It is well known that iron ore deposits are one of the most important potential unconventional sources of precious metals [12]. The Poperechnoe iron-manganese deposit (South Khingan ore cluster) locating in Jewish autonomous region in additional to the main ores contains precious metals: gold and silver, platinum and palladium. A lot of specialists have been carrying out in the past years the study of the noble metal mineralization of the South Khingan ore cluster [13-17].

The ore-bearing structure extends in the submeridional direction for 54 km and a width of 6-8 km. The ore cluster and adjoining territory are composed of limestone, carbonate rocks of dolomite composition and shales of the Upper Proterozoic Khingan sequence, as well as Lower Cambrian limestones of the londokovo suite and a variegated complex of rocks of the Kimkan layer. There are fluidites (fluidno-explosive formations) within the Poperechnoe deposit that contain the ore deposit and they are structural and materially complex rocks [16, 17]. Steeply sloping ore formations have been studied to a depth of up to 400 m. Manganese content is 20.4 %. Its stocks are estimated at 5116 thousand tons in categories B + C1. Iron stocks were not counted [17].

Platinum is in all ore types and in the enclosing carbonate breccias. The main part of platinum grains is (up to 0.4-0.5 mm) light gray and angular-lumpy forms, there are flattened, somewhat deformed, cubic crystals, platelets; have a high magnetic susceptibility. The platinum content is variable. In general, it is isoferroplatinum (Pt3Fe) with a variable content of impurities Cu, Pd, Sn, Ni, Co, Cr, V. There are inclusion of well-formed crystals (2-20 μm) in some isoferroplatinum grains as follows: sulphoarsenides of platinumides of complex composition such as hollingworthite (Rh, Ir, Pt, Ru, Fe) AsS-platarsite PtAsS-irisite (Ir, Rh, Os) AsS; Erlichmanite, rutheniridosmin (Os, Ir, Ru), tetraferroplatinum with enstatite are noted (Fig. 2). Platinoids contents of the Poperechnoe iron-manganese deposit can be as by a chain: Pt>Rh≈Ir>Ru≈Os>Pd.

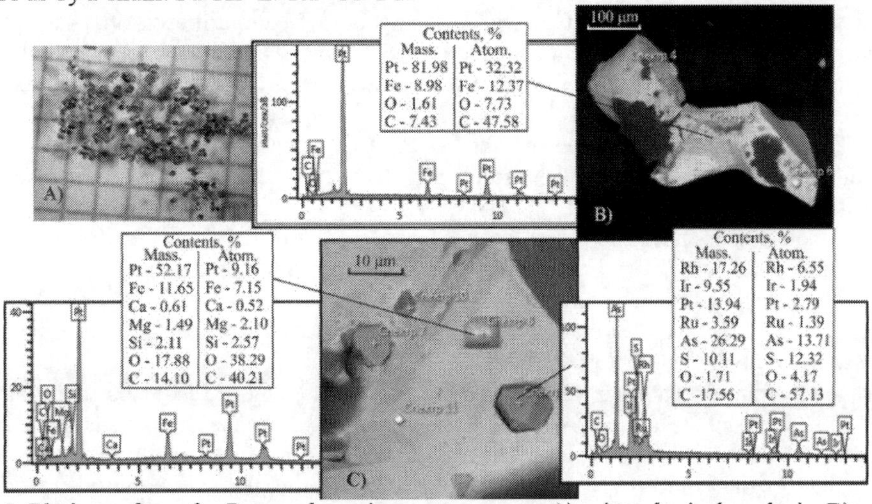

Fig. 2. Platinum from the Poperechnoe iron-manganese: A) mineralogical analysis; B) cranked grain of isoferroplatinum with inclusions of minerals of platinoids; C) crystals of cholinhuortite, tetraferroplatinum, iridosmin and other minerals of platinoids in the grain of isoferroplatinum.

There are weight grains of gold (0.025-0.5 mm) from ores and host rocks in additional to platinum. The gold of light yellow with strong shine has lumpy form, sometimes with elements of a cut of a cube. Its composition is variable from pure gold to electrum and with impurities - Cu, Pd, Fe, Pb, Cd. There are in the paragenetic association as follows: quartz, pyroxene, actinolite, siderite, gypsum, almandine, iron and manganese oxides and carbonates, and feldspars. At the preliminary re-calculation iron-manganese ores of Poperechnoe deposit may contain platinum from 0.22 to 3.6 g/t, gold from 0.45 to 0.79 g/t.

3.3 Malmyzh

Malmyzh copper-porphyry deposits are the main sources of Cu, Mo, Au, Ag, Sn, Re, W. They also contain associated: In, Pt, Pd and Se. Deposits have a low content but evenly distributed copper and a presence of associated impurities [18]. The problems solution of the genesis this deposits and metals source in ores is remain rather contradictory [19, 20].

The large Malmyzh gold-copper-porphyry deposit is located in the Khabarovsk region on the border of the Nanaisk and Amur administrative districts. There are some places of different degrees of study. The geological structure of the Malmyzh is sedimentary terrigenous deposits of the Lower and Upper Cretaceous age with Upper Cretaceous intrusions and diorite-granodiorite dikes [21].

The extensive fields of secondary quartzites and quartz-sericite metasomatites developed on terrigenous and Late Cretaceous intrusive rocks of moderately acidic and average composition are shown on maps. There are 7 long linear quartz stockworks.

At the last geological data of EMX Royalty Corporation [22] in the northwestern part of the Svoboda area of the Malmyzh gold-copper-porphyry deposit was discovered diatreme (depth 650-850 m) composed predominantly by magmatogen-hydrothermal breccias [22].

The total source potential is no less than 10 million tons of copper and 500 tons of gold. Metals average content is Cu – 0,34 %; Au – 0,19 g/t. The primary ore mineralization is presented by nests, dissemination and veins of magnetite, pyrite, and chalcopyrite. Bornite, chalcosine, sphalerite, pyrrhotite, galena, very rarely molybdenite, friable ores are in a much smaller amount. The native gold with Ag and Cu impurities is in chalcopyrite, magnetite and pyrite as very small formations (3-5 μm). The silver is established as an impurity in bornite, chalcocite and with gold, tellurium, selenium and sulfur compounds There are palladium minerals [21, 23].

In 2017 MI FEB RAS together the Far Eastern Geological Institute FEB RAS and LLC "Amur Minerals" began work on the study of gold-copper-porphyry ores of the Malmyzh deposit (Svoboda district).

As a result of mineralogical and technological studies of oxidized ores from Svoboda district, the visible weight grains of platinum, gold, silver cadmium and diamonds were first identified along with characteristic ore minerals. Their origin in the samples is not unambiguous [24]. Oxidized ores are intensively limonitized and clay-diorite porphyrites. The estimated copper content in the sample is 0.09%, gold - 0.52 g/t.

There are 2 grains of platinum (0.06 mg), their form is an elongated flattened shape, similar to smoothed crystals (Figure 3, A), light gray with a matte surface. It contains an impurity of Fe up to 5%. Pyrrhotite and baddeleyite $ZrO2$ (1-3 μm) are singly in the nests.

One grain (0.1 mm) of iron-platinum-zirconium intermetallic (flattened cuboctahedron) is a light, almost white with metallic sheen (Fig. 3, B). The approximate composition is FePt3Zr6. At this stage of studies platinoids content can be characterized as of Pt ≈ Pd.

The weighted light yellow gold has a dimension in the range of 0.1-0.7 mm. The cold content is electrum with silver mass up to 17-26 %. Occasionally there are argentite and galena sprouts (up to 20-50 microns) and impurity of lead up to 6%. Native silver cadmium occurs.

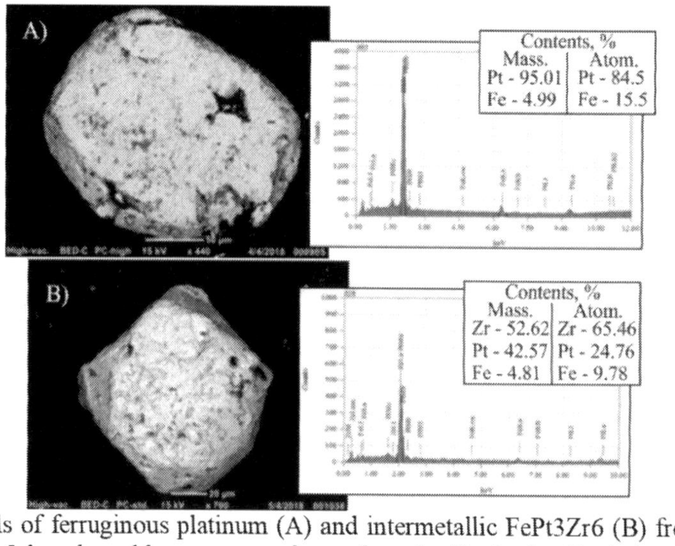

Fig. 3. Crystals of ferruginous platinum (A) and intermetallic FePt3Zr6 (B) from oxidized ores of the Malmyzh gold-copper porphyry deposit (Svoboda district) with electron microscopy.

Crystals of greenish diamonds of 25 μm to 0.4 mm in various form were also found in the oxidized ores. Some from them have a crust of tenardite and they contain N and Al. The origin of diamonds needs in additional research

As well the native copper, zinc, aluminum, zircon, arsenopyrite, jarosite, hematite, covellite, bornite, chalcocite, digenite, azurite and malachite are in the paragenetic association with platinum.

4 Conclusions

The Kondoer traditional unique deposit of alluvial platinum is formed within the alkaline-ultramafic massif of the zonal type. Alluvial platinum mineralization is fairly large, down to nuggets. Indigenous platinum mineralization is dispersed fine-impregnation. It is characteristically the presence of all elements of the platinum group as well as gold and silver. Kondoer platinum is iridic isoferroplatinum with Cu and Pb impurities, which can contain light and heavy platinum minerals inclusions, gold and silver minerals (over 50 rare minerals of noble metals in various combinations). The platinoids content is characterized by series: Pt>>Ir>Os>Ru=Rh>Pd.

The Poperechnoe iron-manganese deposit is an alternative source of precious metals. The platinum of this deposit is isoferroplatinum with Cu, Pd, Sn, Ni Co, C rand V impurities. There are scattered inclusions of sulfides, sulfoarsenides Rh, Ir, Ru and Os in the platinum. Palladium is found as the gold and platinum impurities. The platinoids series from ores of the Poperechnoe is: Pt>>Rh≈Ir>Ru≈Os>Pd. Rhodium predominates here unlike Kondoer.

The Malmyzh gold-copper porphyry deposit also is a potential alternative platinum source. The platinum content of the deposit is poorly studied. Palladium minerals, as well as iron platinum and intermetallic $FePt_3Zr_6$ from the oxidized ores of the Svoboda district have been identified. At the moment platinoids content can be characterized as Pt ≈ Pd.

Thus, by the noble metal mineralization content and characteristic paragenetic associations developed deposits of different genesis we can say that the platinum sources are high-temperature deep.

The Poperechnoe and Malmyzh deposits have good prospects as alternative platinum sources in complex processing.

The work was carried out through financial support of the project of the Presidium of the FEB RAS No. 18-2-015.

References

1. N.P. Laverov, V.V. Distler, Ore Deposits geology, **45/4**, 291-304 (2003).
2. L.I. Gurskaya, D.A. Dodin, Regional Geology and Metallogeny, **64**, 84-93 (2015)
3. N.M. Chernyshov, S.P. Molotkov, O.G. Reznikova, *Bulletin of Voronezh State University. University. Series: Geology*, 2, 137-162 (2003).
4. D.A. Dodin, N.M. Chernyshov, O.I. Cherednikov, Metallogeniya platinoids of large regions of Russia, 302 (2001).
5. L.B. Sushkin, Pacific Geology, **14/5**, 97-102 (1995).
6. G.V. Andreev, Kondoer deposit of ultrabasic and alkaline rocks, 75 (1987)
7. L.V. Eirish, V.A. Stepanov, Pacific Geology. **21/3**, 27-39 (2002).
8. A.S. Mekhonshin, T.B. Kolotilina, News of the Siberian Branch of Earth Sciences of the Russian Academy of Natural Sciences, **2 (41)**, 97-106 (2012).
9. A.V. Lennikov, V.V. Ivanov, RA Octyaborskiy, Deep-seated magmatism, its sources und plumes, 191-227 (2008)

10. L.J. Cabry, J.H.G. Laflamme, The mineralogical Record, 28, 97-106 (1997).
11. V.G. Lazarenkov, A.G. Marchenko, I.V. Talovina, *Geochemistry of platinum elements* (1996).
12. D.A. Dodin, N.M. Chernyshov, V.A. Yatskevich, *Platinum-metal deposits of Russia* (2000).
13. A.M. Zhirnov, GIAB, **5/12**, 115-120, (2009).
14. VG Nevstruev, N.V. Berdnikov, B.G. Saksin, V.I. Usikov, Pacific Geology, **6/6**, 102-111 (2015).
15. A.I. Khanchuk, I.Yu. Rasskazov, V.G. Kryukov, N.M. Litvinova, B.G. Saksin, DAN, **470/6**, 1-3 (2016).
16. N.V. Berdnikov, V.G. Nevstruev, B.G. Saksin, Pacific Geology, **36/6**, 43-57. (2017).
17. V.G. Kryukov, N.M. Litvinova, N.A. Lavrik, V.F. Stepanova, Enrichment of ores, **4/370**, 42-48. (2017).
18. D.S. Bukhanova, P.Yu. Plechov, Vestnik KRAUNTS. Earth sciences, **№ 2/34**, 61-70. (2017)
19. M. Tarkian, D.G. Eliopoulos, M. Economoueliopoulos, Neues jahrbuch fur mineralogie-monatshefte, **12**, 529-537. (1991).
20. He Xiaohu, Zhong Hong, Zhu Weiguang, Acta geologica sinica-english edition, **88/4**, 1288-1309. (2014).
21. D.S. Bukhanova, Geological processes in the conditions of subduction, collision and slip of lithospheric plates, 396. (2016).
22. Electronic resource: www.emxroyalty.com. News Release 2018.
23. 24. V.V. Ivanov, V.V. Kononov, E.K. Ignatiev, *Tectonics, Deep Structure and Mineralization of the East of Asia*, 258-261. (2013).
24. V.A. Pakhomova, D.G. Fedoseev, V.B. Tishkina, S.Yu. Buravleva, Gemmology, 113-122. (2015).

E3S Web of Conferences **56**, 04018 (2018) https://doi.org/10.1051/e3sconf/20185604018
VII International Scientific Conference "Problems of Complex Development of Georesources"

Quality of mineral wealth as a factor affecting the formation of refuse of ore mining and processing enterprises

Svetlana Chrykhalova[1*]

[1]The Federal State Educational Institution of the Higher Education National University of Science and Technology MISiS, Moscow, Russia.

Abstract. Studies are carried out to estimate the amount of refuse generated in mining operations in conditions of deteriorating ore quality. The introduction of static material efficiency is suggested, which shows how much of the extracted rock mass is the finished product (concentrate) and the refuse production factor at mining and processing enterprises.

1. Introduction

The level of product quality and its prime cost largely determine the survival and success of the enterprise in a market environment. Quantitative characteristics of the properties of products that make up its quality are called indicators of product quality.

In mining, the quality of mined ore (mining complex: mines, quarries) has a significant influence on the products of the following processing industries (crushing, sorting, concentrating), responsible for the output of finished products (concentrates, etc.). This impact is made on the technological, economic and environmental performance of enrichment units, determining the material composition of the concentrate, the degree of extraction of minerals from raw ore, the yield of concentrates (commercial ore), their cost, and the number of tailings, their toxicity and residual content of useful components in them [1].

The problem of changing the indicators of product quality in the mining industry has its own peculiarities. By now, people have already used a significant part of minerals, mostly of the best quality.

The most intensive deterioration of the mineral resource base is noted for iron, tungsten-molybdenum, copper, aluminum, etc. ores. A similar situation is observed for mining chemical raw materials: with a very intensive growth in production, there is a decrease in the quality indicators of mined chemical raw materials, which was noted back in the 1970s [2, 3].

Mining is a natural-technical system, which more complicates its study. The main technological process is connected with the extraction of minerals from the bowels, i.e. impact on the bowels of the Earth with the help of various engineering facilities and

* Corresponding author: tchmy@mail.ru

technologies. For the sustainable development of enterprises and society, further improvement of the technological process of extracting minerals is necessary [4, 5].

2. Modeling an object. Choice and justification of the model. Study of the model

By modeling the author means the method of cognition, which involves building and studying a model, and then transferring the obtained data to a modeled object - the original, in this case - a mining processing enterprise [6]. From this definition it follows that the models of the object can be quite a lot, because each of these models can be proposed to identify certain properties of the object.

Inputs receive raw materials - the main material resource that is the subject of processing, additional material and energy resources consumed during the process. Outputs are used for production and refuse disposal.

In this paper, the author uses a schematic macromodel of the mining enrichment enterprise, shown in Fig. 1.

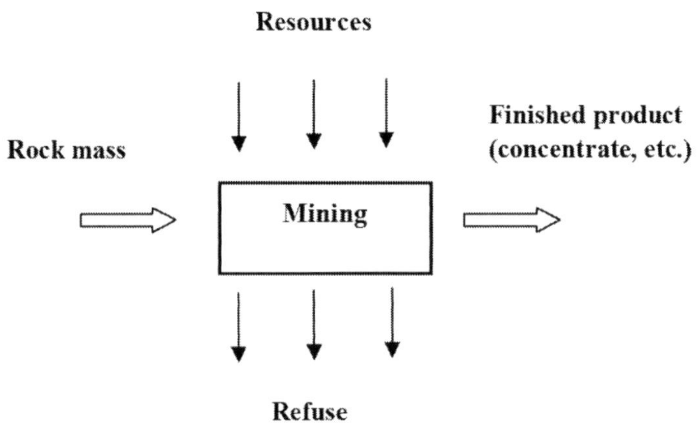

Fig. 1. Macromodel of mining

The main input is specially marked at the scheme - the subject of treatment (raw rock mass) and the main output is the product of processing (concentrate of a mineral). The scheme also reflects the refuse of the main transformation process, which includes overburden and enclosing rocks.

Additional inputs include material - natural and man-made resources, the consumption of which is caused by the technology of production chosen and the engineering facilities used, and additional outputs include the refuse of the production process, resulting from the transformation of additional resources.

The content of the useful component in the ore and the content of ore in the rock mass should be such that it is technologically and economically expedient to extract ore and then convert it to the plant's products. These tasks are solved at the mining and processing plant by mining and processing facilities.

The macromodel of the rock mass is shown in Fig. 2, it reflects the material balance: the distribution of mined rock into mineral resources and waste rock. Minerals can also be represented as a concentrate (finished products) and enclosing rocks, which are an integral part of the mineral. In this study, under enclosing rocks, the author means the part of the mineral resource that does not contain a useful component.

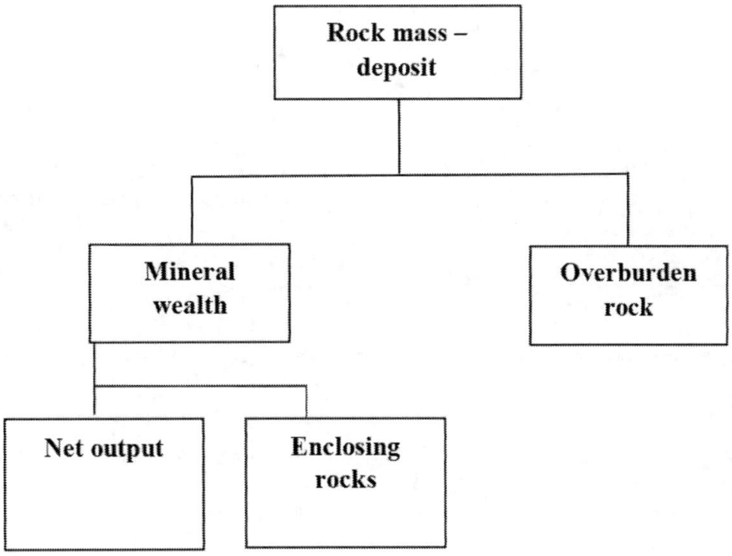

Fig. 2. Macromodel of rock mass

This model is described in some detail in [7].

The formula for calculating the weight coefficient of overburden K_{ovb} is:

$$K_{ovb} = M_{ovb} / M_{min}, \quad (1)$$

where M_{ovb} and M_{min} – the amount (mass) of overburden and minerals recovered from the deposit.

By analogy with the coefficient of overburden, introducing a new indicator - the coefficient of enclosing rocks. Under the coefficient of enclosing rocks, K_{enc} is meant the ratio of the number of enclosing rocks in tonnes or cubic meters to the amount of minerals in tonnes or cubic meters. The coefficient of enclosing rocks, calculated in tons of such rocks per ton of mineral, will be called weight.

The weight coefficient of enclosing rocks K_{enc} is calculated by the formula:

$$K_{enc} = M_{enc} / M_{min}, \quad (2)$$

where M_{enc} and M_{min} – the amount (mass) of enclosing rocks and minerals recovered from the deposit.

The content of new components in the initial rock mass and in the initial useful mineral can be expressed in fractions or percentages.

Based on the structural model of a complex of rocks that form the rock mass of a deposit of ore minerals (Figure 1), writing the following equations:

$$M_{rm} = M_{min} + M_{ovb} \text{ and } M_{min} = M_{conc} + M_{enc}, \quad (3)$$

where M_{conc} – the amount (mass) of concentrate of the useful component.

Dividing the right and left parts of the equations by M_{rm} and M_{min}, obtaining the expressions for the fractional coefficients characterizing the composition of the rocks in question:

$$\mu_{min} = M_{min} / M_{rm} = M_{min} /(M_{min} + M_{ovb}) = 1 / (1+ K_{ovb}); \quad (4)$$

$$\mu_{ovb} = M_{ovb} / M_{rm} = M_{ovb} /(M_{min} + M_{ovb}) = 1 / (1+ 1/K_{ovb}); \quad (5)$$

$$\mu_{conc} = M_{conc} / M_{min} = 1 - K_{enc}; \quad (6)$$

$$\mu_{enc} = M_{enc} / M_{min} = K_{enc}. \quad (7)$$

It is obvious that

$$\mu_{min} + \mu_{ovb} = 1 \text{ and } \mu_{conc} + \mu_{enc} = 1. \quad (8)$$

Introducing the indicator of mining production, characterizing the quality of rock mass - static material efficiency $\zeta_{st.m.}$, calculated by the formula:

$$\zeta_{st.m.} = \frac{M_{conc}}{M_{rm}} = \frac{M_{conc}}{M_{min}+M_{ovb}} = \frac{M_{conc}}{M_{conc}+M_{enc}+M_{ovb}}. \quad (9)$$

This indicator characterizes the quality of the mineral deposit and provides an estimate of the amount of the finished product that can be obtained from the rock mass being developed.

To identify the amount of refuse, introducing the coefficient $\eta_{st.m.}$ – the static coefficient of refuse production, characterizing the amount of generated refuse, depending on the quality of the deposit.

$$\eta_{st.m.} = \frac{M_{enc}+M_{ovb}}{M_{conc}+M_{enc}+M_{ovb}}. \quad (10)$$

3. Implementation of the model in the mining processing industry

Assuming that the macro-indicators characterizing the quality of the mineral are as follows:

α_d - mineral content in the deposit;

α_{om} - useful component content in the extracted ore mass;

β – useful component content in finished product;

θ – useful component content in tailings;

R – dilution coefficient, %;

P – ore loss coefficient, %;

M_{rm} – extracted rock mass, t (thousand tons, million tons);

M_{ovb} – weight of overburden, t (thousand tons, million tons);

M_{ore} – mass of extracted ore, t (thousand tons, million tons);

M_{pr} – weight of processed ore, t (thousand tons, million tons);

M_{enc} – mass of enclosing rocks, t (thousand tons, million tons);

M_{fp} – weight of finished products, t (thousand tons, million tons).

Considering the values of these coefficients for real mining operations according to the data of OAO "APATIT", given in the table 1[8]:

Table 1. Dynamics of changes in the quality of balance and mined ores

Indicator	Year									
	1999	2000	2001	2002	2003	2004	2005	2006	2007	2008
P_2O_5 content in balance ores (at cutoff grade 4%)	15.18	15.1	14.95	14.86	14.68	14.55	14.65	14.34	14.32	14.36
Extraction of ore, million t	28.4	27.9	26.1	27.0	28.2	29.3	29.5	28.6	28.6	26..1
P_2O_5 content in extracted ore, %	14.14	13.96	13.63	13.77	13.63	13.32	13.12	12.91	12.87	12.93
Ore consumption per ton of concentrate, t	3.13	3.16	3.25	3.22	3.25	3.32	3.38	3.44	3.45	3.43
Operational loss, %	8.5	8.0	8.8	9.2	10.5	9.7	10.8	10.1	11.0	11.2
Dilution, %	6.9	6.7	7.0	7.5	7.3	8.3	9.1	10.0	9.9	9.8

Calculating the average indicators for 1999 to 2008. For this period:

- operational losses are on average per year $P = 9.78$ %;
- dilution on average per year $R = 8.25$ %;
- $\alpha_d = 14.70$ %;
- $\alpha_{rm} = 13.43$ %.

Content of the useful component P_2O_5 in extracted ore according to formula [9]

If the content of the component in the volume of balance reserves of ore to be extracted is α_d, the ore loss is P, and the dilution (contamination) with enclosing rocks is R, then the metal content in the extracted ore mass can be found from the relationship [4]:

$$\alpha_{rm} = \alpha_d \frac{1-P}{1+R}. \tag{11}$$

$$\alpha_{rm} = 14.7\frac{1-0.0978}{1+0.0825} = 14.7 \cdot \frac{0.9}{1.08} = 12.23\%. \tag{12}$$

This result is somewhat lower than that presented in the table, which is explained by the variable composition of the ore mass deposit.

Part of this expression, $\frac{1-P}{1+R} = K_\alpha$, determines the degree of change (decrease) in the content of the useful component in the ore mass in the process of ore extraction under the influence of losses and dilution. Therefore, it can be assumed that K_α is an indicator of the quality of mining in terms of ensuring the absolute level of the content of the useful component in extracted ore [9].

4. Discussion

Over the 80-year period of production, by 2009, the P_2O_5 cutoff grade decreased from 25% to 4%, and the content in the processed ore from 29% to 13.6% [3].

In the period from 2014 to 2017, the P_2O_5 content of the flotation concentrate is actually in the range of 12-13%, however, its variability is considerable.

From this study, it can be seen that the extraction of minerals in terms of its natural conditions implies the formation of a huge amount of refuse that can significantly exceed the extracted useful component. Therefore, the management of refuse in mining is an integral part of the technological mining process.

It can be seen from the table that an average of 3.3 tons of ore is consumed per 1 ton of concentrate. This means that 2.3 tons of ore are refuse (tailings of enrichment).

In 2005, the rock mass produced was 101,624 thousand tons, and the production of apatite concentrate was 8,756 thousand tons. Static material efficiency of the enterprise $\zeta_{st.m.}$ = 8.6% for the apatite concentrate. The refuse production coefficient $\eta_{st.m.}$ =91.4%.

5. Conclusions

Thus, it can be concluded from the above study that the extraction of minerals has a low static material efficiency and a large refuse production coefficient. Increasing the efficiency of mining operations is very difficult. The management of the quality of mining should begin in the extraction complex. However, established traditions, significant variability in the properties of the rock mass [10, 11], little information, complexity of production conditions, significant volumes of mining, transportation and loading operations greatly complicate quality management in mining. Currently, there are some recommendations for improving the mining process, but their introduction into production requires significant financial costs [12].

Taking into account what has been said at the first stage, it is necessary to revise the existing Technical Conditions for extracted ore and improve the technological regulations, which leads to the modernization of the extracting complex of mining enterprises.

References

1. Kh.Kh. Kozhiev, , G.G. Lomonosov, Mine management systems for the quality of mineral raw materials. – 2nd ed., stereotype. – Moscow: Moscow state mining university. p. 292. (2008).
2.G.G. Lomonosov, Formation of the quality of ore in open mining. Moscow, Nedra. p. 224. (1975).
3. Iu.E. Bryliakov, A.Sh. Gershenkop, V.N. Lygach. Mining Journal. **9**, 32-36. (2009).
4. Lian-Rong ZHAO, Wei CHEN. Establishment of Assessment Indicator System of Sustainable Development in Mining Industry and Evaluation of Pilot Project. 2014 ICSS 2014. Atlantis Press. pp. 193 – 200. (2014).
5. Pisano, Umberto. Resilience and Sustainable Development: Theory of resilience, systems thinking and adaptive governance. ESDN Quarterly Report N°26, ESDN. (2012).
6. L.G. Raskin, Analysis of complex systems and elements of the theory of optimal control. Moscow: "Sov. radio". p. 344. (1976).
7. S.V. Chmykhalova. Mining Journal **6**, pp. 51-59. (2006).
8. V.V. Belousov. Mining Journal **9**, pp. 58-61. (2009).
9. G.G. Lomonosov, Mining qualimetry: Textbook. - 2nd ed., stereotype. - Moscow: Publishing house "Mining book", Moscow state mining university. p. 201. (2007).
10. A.Iu. Pedchik, S.V. Chmykhalova, S.V. Shevchuk, S.V. Shentseva. Mining Journal **12**, pp. 37-39. (2016).
11. S.V. Chmykhalova. Mining Information and Analytical Bulletin **6**, special issue No. 32 "Safety and ecology of mining production – 2018", Moscow, publishing house "Mining book", pp. 80-85. (2018).
12. Murphy, B., van Zyl, J. and Domingo, G. Underground Preconcentration by Ore Sorting and Coarse Gravity Separation. NARROW VEIN MINING CONFERENCE. PERTH, WA, 26 - 27 MARCH 2012.

Local seismicity alterations in the South Yakutia mining region due to the technogenic impact on its geological environment

Nikolay Grib[1,3*], *Valery* Imaev[1,2], *Galina* Grib[1], *Lyudmila* Imaeva[2] and *Igor* Kolodeznikov[3]

[1]Technical Institute (Branch), Amosov North-Eastern Federal University, Neryungri, Russia.
[2]Institute of the Earth's Crust, Siberian Branch, Russian Academy of Sciences, Irkutsk, Russia
[3]Academy of Sciences of the Republic of Sakha (Yakutia), 677007, 33 Lenina, Yakutsk, Russia

Abstract. Impulse loads, arising due to the high natural seismicity of the South Yakutia region, exercise both direct and indirect effects on the upper part of the Earth's crust during industrial explosions. The direct effects result from nonlinear displacements caused by the blast wave and the subsequent formation of new disturbances. The indirect effects arise due to the activation of structural elements along geological contacts, leading to the emergence of technogenic seismicity foci. The foci of induced seismicity are either confined to the blast points, or located along the tectonic structures crossing quarry fields. The technogenic impact on the geological environment transforms the independent local seismic process, since explosions trigger a chain of local seismic events. The near-surface layers of the Earth's crust become activated in the area of dynamic influence of active faults. Under the influence of explosions, both the number of seismic events and the average level of released energy alter. Impulse loads on the geological environment lead to a spatial redistribution of the foci of low-energy (K <7) earthquakes. The main form of the geodynamic development of seismogenic faults is the movement of their sides in the form of mutual "slippage". Seismic events are manifested only when the aforementioned form of deformation is impossible or difficult to develop, in other words, when the stress-state areas of the Earth's crust develop. Therefore, the shaking impact of blasts can be considered as a factor contributing to the predominance of aseismic forms of fault motion in the form of smooth slippage of their sides. In conclusion, the impact of industrial blasts can not only activate faults around the mining area, but also have an unloading effect on the foci of seismic hazard forming in the interior, i.e. the redistribution of earthquake energy in terms of reducing earthquake energy class.

1 Introduction

Long-term seismotectonic studies conducted in the south of Siberia, on the border of the Eurasian and Amur lithospheric plates, have established the formation of an independent

*Corresponding author: grib@nfygu.ru

Olekmo-Stanovaya zone (OSZ), within which over 40,000 seismic events with a magnitude from M≥2.0 to M = 6.5-7.0 were registered over the 60-year-long period of instrumental observations [1]. In the present paper, we aim to analyse a site located in the central part of the OSZ in the vicinity of the Neryungri city, where various minerals are actively being mined.

The increasing scale and intensification of the Earth's interior exploration have revealed two major issues associated with the interaction between man and nature, i.e., the need to address problems involved with prevention of catastrophic emergency situations caused by technogenic seismicity and research into the possibility of active technogenic intervention into natural seismic processes.

The possibility of initiating earthquakes comes from a hypothesis that the Earth's interior is characterized by an unstable energy state. On its basis, a number of studies [2 - 5] have concluded that atomic explosions lead to a change in the seismic regime even at significantly large distances (thousands of kilometres) from the initiating impact.

An increase in the scale of human impact on the environment typically results in an increase in the load not only on such rapidly-responding ecosystem components as animals, plants, soil, water and air, but also on those its elements that are commonly considered to be unaffected by civilisation [6]. The latter, the most stable ecosystem components, were long believed to include the upper layers of the Earth's crust, as well as the deformation processes occurring therein [7]. It was conventional to refer diverse deformation processes of various space-time scales occurring in the Earth's crust, including earthquakes, anomalous deformations in fault zones, subsidence, landslides, mudflows, etc., exclusively to orogenic and adjacent areas. However, in recent years, the basic perceptions of the Earth's geodynamic state have radically changed. Long-term research carried out on geodynamic polygons has proven deformation processes to occur everywhere, even within ancient platforms and seismically quiet regions. These processes are manifested, e.g., as intense quasiperiodic vertical and horizontal movements along faults with amplitudes of up to 20-50 mm/year [8]. Published data indicates that earthquakes with a magnitude of M>7 can occur even across most ancient platform areas [9, 10].

An increase in the scale of anthropogenic impact is known to intensify technogenic seismicity [11]. As a rule, the centres of technogenic earthquakes exhibit small depths, which fact amplifies their risk to the built environment. The small magnitudes of earthquakes are compensated by the proximity of their foci to the surface; as a result, a relatively weak technogenic earthquake is capable of causing local catastrophes [12]. In recent years, induced seismicity has become one of the problems associated with the geo-environmental consequences of seismic effects from quarry blasts performed in mining regions [13 - 17].

In South Yakutia, the following types of human impact on nature capable of provoking earthquakes can be distinguished: the construction of a water reservoir, open mining operations in the Neryungrinsky coal mine and underground mining operations in other coal mines. Due to the high natural seismicity and proximity of the industrial zone to the city of Neryungri, these forms of anthropogenic impact can significantly affect the level of seismic and geo-environmental risks. The practical significance of solving these problems is extremely high for the economic development of the South Yakutia region [18].

2 Experimental study

The seismological monitoring data obtained for the period from January 1, 2002 to July 31, 2012 was analysed with the purpose of assessing the level of technogenic seismicity in South Yakutia. The comparison of the seismological data with the volume of the rock mass blasted during this period produced the following regularity: an increase in the volume of blasted rock mass leads to an increase in the number of nearby earthquakes. The analysis was carried out applying a summarization approach, i.e., the total volume of the rock mass blasted during a month was compared with the number of nearby earthquakes having occurred during this month. The sample included nearby earthquakes exhibiting a difference in the arrival of longitudinal (P) and transverse (S) elastic waves of up to 6 seconds (within a radius of 50 km from the blast area) and having occurred no later than 10 hours after the corresponding rock blast, as well as the volume of the rock mass blasted during the specified period.

The impact of mining operations carried out in the Neryungrinsky coal mine on the induced seismicity in the nearby area (1^0x1^0) was estimated by calculating the released energy on an annual basis (Fig. 1). The average energy value for the 1979-2005 period is shown to range from K = 9 to K = 10 {J}. The 1989-1990 period is characterized by a peak in the released energy of higher than K = 11 {J}. It should be noted that it is during this period that a great number of uncovering and mining operations were being performed in the Neryungrinsky coal mine. It can therefore be concluded that drilling and blasting operations do have an impact on the overall energy balance of the seismotectonic process [14].

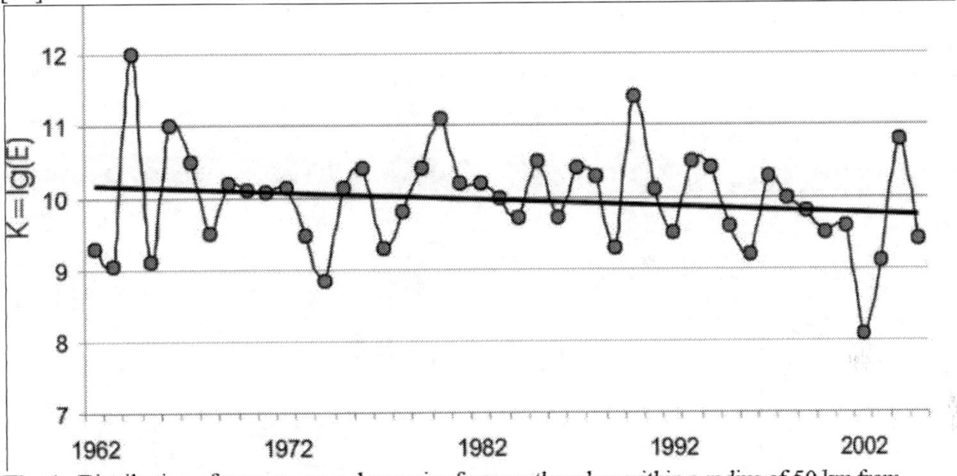

Fig. 1. Distribution of average annual energies from earthquakes within a radius of 50 km from blasting operations

The average energy value for the 1962-2005 period ranges from $E = 10^{10.2}$ to $E = 10^{9.8}$ {J} following a linear function. However, the total energy released yearly was higher for the periods preceding mining operations, which indicates a downward redistribution of earthquake energy (Figure 1). In general, this graph has the form of a linear function with a constant slope.

Therefore, active anthropogenic impact on the geological environment causes a seismic response manifested in the form of a change in the level of seismic activity. Under the influence of explosions, both the number of seismic events and the average level of released energy change. It is established that the centres of technogenic earthquakes are either confined to the points of explosion, or located north-west along the tectonic structures crossing the quarry field.

The patterns of induced and natural seismicity in the area of open mining were established by analysing data obtained from 5-year monitoring observations carried out at the Neryungri seismic station (NSS). Over the 2002-2007 period, its specialists registered over 500 explosions manifested in the form of seismic events, on the average, of the 6th energy class.

3 Results and discussion

According to the monitoring data obtained for the 2002-2007 period, 103 out of 135 (the total number of 211) earthquake cases registered in the nearby area (165 in the degree square) occurred within one day after blasts, with 46 events being registered in the vicinity of the technological platform. The technogenic impact on the environment was transformed into an independent local seismic process. In other words, under the action of rock mass blasts, a local stream of seismic events not correlating with the moments of the blasts was initiated. The spatial distribution of the explosions and earthquakes is shown in Fig. 2.

This means that either a new active tectonic structure appeared, to which 165 out of 211 earthquakes (78%) having occurred within a radius of 50 km from the Neryungri coal mine can be referred, or a local section of the Ammunakta-Muastakhsky and Nizhne-Neryungrinsky active faults became activated. In other words, the seismic process initiated by blasting works, became stationary on a local level. Thus, low-energy impulse loads on the geological environment led to the spatial redistribution of the foci of low-energy (K <7) earthquakes (Fig.2).

The obtained results show that active impact on the environment causes a seismic response manifested in the form of a correlated change in seismic activity. Under the influence of energy loads on the geological environment, the near-surface layers of the Earth's crust were activated in a zone experiencing the dynamic influence of Ammunakta-Muastakhsky and Nizhne-Neryungrinsky active faults.

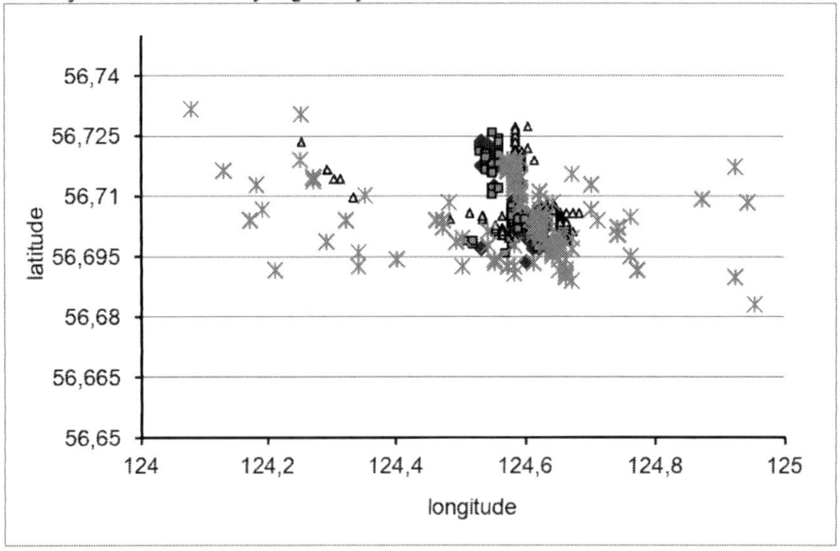

Fig. 2. Spatial distribution of explosion points and weak (K <7) earthquake epicentres near the technological platform of the Neryungri coal mine: ⋇ - earthquakes 2002 – 2007;
◆ 2004 ■ 2005 ▲ 2006 - explosions

The geodynamic development of seismogenic faults is known to result from the movement of their sides [18] in the form of mutual "slippage". Seismic events are manifested provided the aforementioned form of deformation is impossible or difficult to develop, in other words, when stress-state zones in the Earth's crust arise. Therefore, in our opinion, the shaking impact of blasting can be considered as a factor contributing to the predominance of aseismic forms of fault motion in the form of smooth slippage of their sides. Consequently, the impact of blasting can not only activate faults around the mining area, but also have an unloading effect on the foci of seismic hazard forming in the interior, i.e. the redistribution of earthquake energy towards reduction of their energy class.

Since 2003, open coal mining works have been actively performed in small quarries. By 2007, the total energy impact of these enterprises on the geological environment exceeded that produced by the Neryungri coal mine by 10% [14]. However, the launch of the Vostochnaya Denisovka mine in 2013 led to a sharp increase in the number of technogenic earthquakes, with this number reaching 30 cases during 10 hours after a blast. Instrumental observations have shown that the epicentres of the earthquakes initiated by the explosions produced in the Vostochnaya Denisovka mine do not geographically coincide with the epicentres of the earthquakes initiated by the explosions in the Neryungrinsky coal mine. Using the results of explosion and local earthquake monitoring, we have identified the focus of a new earthquake (a square highlighted in grey) related to the explosions in the Vostochnaya Denisovka coal mine (Fig. 3). This site was analysed in detail taking into account its tectonic structure. It is established that the induced seismicity in this site was associated with the tectonic structures stretching from the North to the South and crossing the quarry field of the Vostochnaya Denisovka coal mine.

Therefore, the impact of rock blasts on the geological environment of the-newly launched coal mine caused a seismic response in the form of changed seismic activity, activating tectonic structures in the area of blasting operations in the Vostochnaya Denisovka coal mine that had not manifested themselves before. Under the influence of rock blasts, the number of seismic events changes. The spatial distribution of the epicentres of induced seismicity initiated by the explosions produced in the Vostochnaya Denisovka coal mine is presented in Fig. 4.

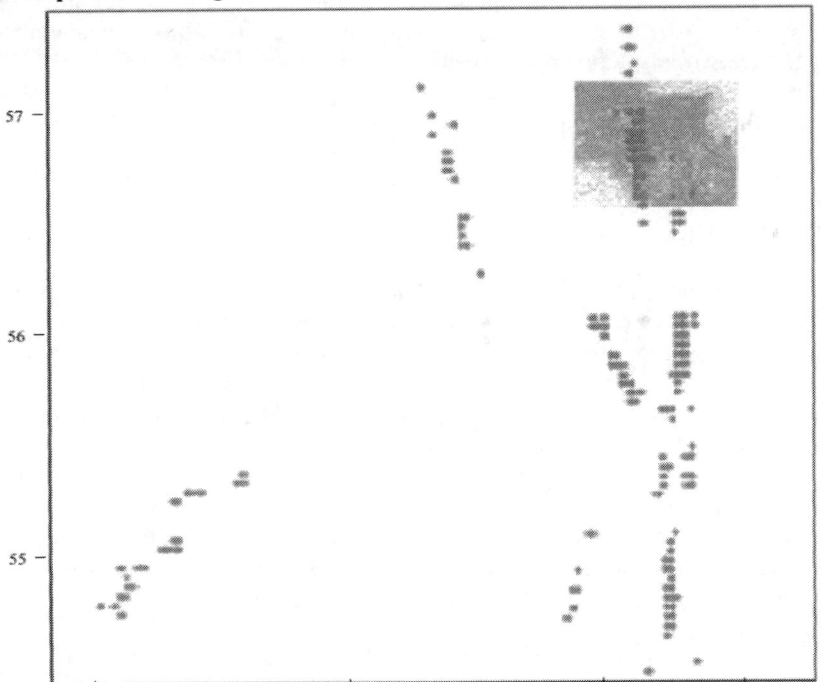

Fig. 3. Epicentres of local earthquakes registered in 2012-2013: • - seismic events

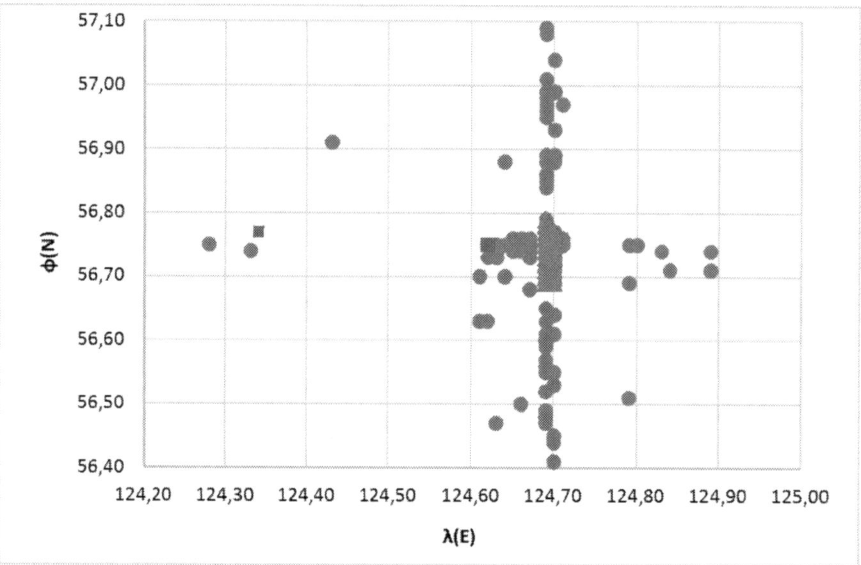

Fig. 4. Spatial distribution of the epicentres of blasts in the Vostochnaya Denisovka coal mine and the epicentres of technogenic earthquakes: ▲ - explosions; ● - earthquakes

The results of seismological monitoring carried out in 2013 have shown that the epicentres of technogenic earthquakes were confined to the rock blocks being exploded or to the tectonic structures crossing the quarry field from the North to the South. It should be noted that the maximal activation of the induced seismicity was related to the explosions on the eastern flank of the mining front, where the tectonic structure had been uncovered by mining operations.

4 Conclusions

It can be concluded that an increase in the technogenic load on the geological and geophysical environment causes irreversible phenomena in the seismic process. Under the influence of rock blasts, the number of seismic events along with the average level of released energy alter. Active impact on the geological environment causes a seismic response manifested in the form of technogenic earthquakes, activating seismic generating structures located in the area of industrial blasts.

The seismic process initiated by blasting works acquires a stationary character at the local level; low-energy impulse loads on the geological environment lead to a spatial redistribution of the foci of low-energy ($K < 7$) earthquakes. Blasting works can not only activate faults around the mining area, but also have an unloading effect on the foci of seismic hazard forming in the interior, i.e. the redistribution of earthquake energy towards reduction in their energy class.

References

1. L. Imaeva, B. Koz'min, V. Imaev, N Grib, S Ashurkov. Journal of Seismology, **21(2)**, 385 (2017).
2. A. Gamburtsev, N. Gamburtseva, I. Galkin, R. Nikolaeva, O. Oleinik. The induced processes in the lithosphere: induced seismicity, 16 (1973).
3. Nikolaev A. Bulletin of the Russian Academy of Sciences, **63(2)**, 113 (1993).
4. A. Nikolaev, I. Galkin. Induced seismicity (1973).

5. A. Nikolaev, G. Vereshchagina. DAN SSSR, **319(2)**, 333, (1991).
6. Wong I.G. Informing Risk Management and Mitigation (2015).
7. N. Grib, A. Syasko, V. Imaev, G. Grib, A. Kachaev. Seismic Instruments, **54(1)**, 101 (2018).
8. Yu. Kuzmin, V. Zhukov. Gornaya kniga (2012).
9. I. Ananyin. Problems of engineering seismology, **29**, 119 (1988).
10. A. Antsiferov, A. Mayboroda, A. Savchenko, V. Tumanov. Transactions of UkrNDMI NAN Ukraine, **11**, 251 (2012).
11. W. Ellsworth. Injection-Induced Earthquakes. Science, **341**(6142), 142 (2013).
12. N. Melnikov. Technogenic seismicity in mining operations: models of foci, prognosis, prophylaxis. Proceedings of International Meeting: in 2 parts, **2** (2004).
13. A. Emanov, A. Emanov, E. Leskov, A. Fateev, A. Semin. Physical mesomechanics, **12(1)**, 37 (2009).
14. G. Grib, N. Grib. Izvestiya of the Samara Russian Academy of Sciences scientific center, **16(1(3))**, 636 (2014).
15. A. Emanov, A. Emanov, A. Fateev V. et al. Earthquakes in Russia in 2012, 104 (2014).
16. V. Sychev. Published summary of Candidate of Physics-Mathematical Sciences (2008).
17. K. Woodward, J. Wesseloo. Journal of the Southern African Institute of Mining and Metallurgy, **115(11)**, 1044 (2015).
18. N. Grib, G. Grib, V. Imaev, M. Tereshchenko. Mining Journal, **10**, 31 (2016).

Geodynamic and structural factors of porphyritic objects localization in Sikhote-Alin

Viktor Nevstruyev[1*] and *Olga* Kozlova[1]

[1]Institute of Tectonics and Geophysics, Far Eastern Branch of the Russian Academy of Sciences, Khabarovsk, Russia.

Abstract Ore bearing porphyritic systems of Sikhote-Alin form linear zones in Cretaceous volcanic belt. They are limited to zones of tectonic disturbances at Moho depths of 19-25 mi (30-40 km). Pacific slab lies at around 340 miles (548 km) below the volcanic belt, which matches the slab depth of porphyritic deposits formation belts in the Andes and Indonesia-Tonga region. Formation of porphyry copper systems is linked to the processes of metalliferous fluids intrusion at slab destruction areas near asthenosphere. **Keywords:** Sikhote-Alin, porphyritic intrusions, gold-copper mineralization, fluids, slab.

Porphyritic deposits are one of the most important world sources of copper. They possess significant and unique reserves of recoverable elements, including molybdenum, rhenium, precious and some other metals. Many issues concerning petrochemistry and deposit formation geodynamic control are still open to debate. Variety of tectonic and structural positions, petrochemical characteristics of primary magma, formation depth and scope of mineralization, as well as other parameters, result in a wide diversity of deposit models. At the same time, many of the characteristics are typical of porphyritic systems.
Majority of researchers agrees that porphyritic deposits are located in distribution areas of igneous rocks influenced by the processes in subduction zones. I and S-type porphyritic intrusions are considered ore bearing, but only some of them actually bear ore. Objects limited to continental margin volcanic belts containing magnetite igneous rocks are the most prospective. Some of them are accompanied by secondary quartzite distribution fields, distinguished by the fullness of manifestation of their facies [1], or spatial cositing of porphyry-copper and acid-sulphate mineralization, as it is at one of the largest gold deposits, Ladolam [2]. Detailed examination of S-type granitoids [3] showed that they form various groups under various conditions. As the result, we suggest distinguishing two subtypes, S1 and S2, which, while having similar chemical composition, form in different time and under different geodynamical conditions, and often contain intrusive contacts. The ore bearing subtype is S2, which is characterized by having higher partial pressure and fugacity of H_2O, CO_2, and HF in fluids.

According to [4], porphyritic deposit formation is heavily influenced by magma enrichment by water and metals in the lower crust, sulphur oxidation in the molten mass, effective metal transfer by hydrothermal fluid released from magma, and ore localization

* Corresponding author: nevstruevvg@mail.ru

processes in the crust. It is important to mention that discovering these factors requires considerable amount of research, which is not always possible to give unique interpretation to.

The analysis of spaciotemporal location of porphyritic deposits in the Andes was conducted based on the plate tectonics model [5]. According to the data, the most important factors for porphyritic deposit formation in the Andes are the age (25 mln years), insignificant incidence angle of slab subduction (10-15°), relatively fast plate convergence (around 10 cm/year), distance from subduction trench (around 2000 km), and aseismicity of subduction volcanic belts.

Reconstructing plate tectonics of the Andes based on the data [5], S. Timmermans [6] came to the conclusion that slab melting point and, therefore, belt aseismicity depends on the distance between the slab and the mantle, rather than speed and angle of slab immersion. The location of slab in relation to the mantle is a determining factor of ore bearing intrusion formation.

Based on the data [5] for the Andes belt, with the incidence angle of 10-15° of subducting oceanic plate, slab immersion depth in the distribution area of porphyritic mineralization at 2000 km from the trench is 535 km. At this depth the subducting oceanic plate begins to flatten out and disintegrate. We assume that with different slab incidence angle this number would change. The distance from the Western Pacific subduction area to the porphyritic intrusion belt of Sikhote-Alin is around 900 km. With slab incidence angle of around 40°, according to the tomography profile [7], slab immersion depth there is 548 km. For Marian type, with the average slab incidence angle of 80°, we see this depth in 150-200 km from the trench, which completely corresponds to empirical data [8, 9]. According to the analysis of ore deposits and slab depth in mineralization areas [8, table 1], there are no porphyritic objects below 600 and above 500 km deep, except for the Porgera deposit, where the older slab, supposedly, is topped off by the newer.

There are 3 porphyritic gold-copper deposits in Sikhote-Alin (Malmyzh and Gurskoye deposits in Khabarovsk Krai and Lazurnoye in Primorsky Krai, all undergoing exploration works), also, there are 91 occurrences and 34 secondary quartzite fields with alumstone facies, which are often connected to porphyry gold-copper mineralization (see fig.). All are limited to magnetite volcanic rocks [1].

In the southern part of the region they form a line in north-south direction, leaning towards the faults on eastern and western borders of Zhuravlevka terrane. In the northern part of the region, potential ore bearing objects form a line in east-west direction, along with a set of objects leaning towards the selvedge of the asthenospheric elevation. Deep structure models lead us to a conclusion that ore bearing structures are connected to mantle diapirism [10].

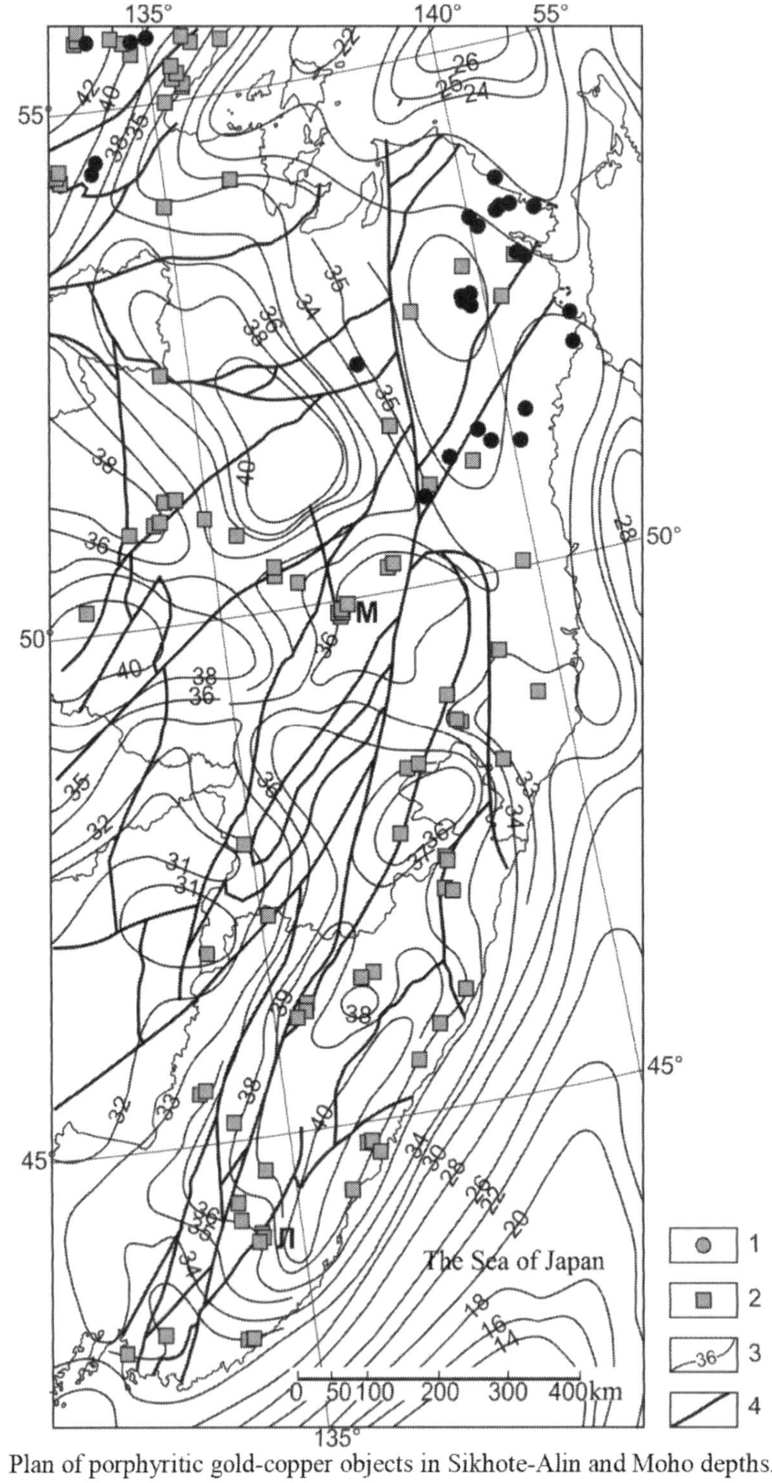

Plan of porphyritic gold-copper objects in Sikhote-Alin and Moho depths.
1 - secondary quartzite fields with alumstone facies; 2 - porphyritic gold-copper objects (deposits: Л - Lazurnoye, М - Malmyzh); 3 - contour lines of Moho depths; 4 - deep faults

From petrological characteristics of igneous complexes in the eastern Sikhote-Alin volcanic belt [11] we can see that intrusive complexes of I and S types are limited to distribution fields in early Cretaceous island arc system (Kema terrane). Metal bearing objects are located close to regional fault area, forming northeastern trending linear zones. In the latitude of 48-51° we can see a transversal strip of contiguous porphyritic gold-copper northeastern trending occurrences, which is probably connected to igneous rocks genesis in slab window area.

Results of geophysical data analysis show that metal bearing objects are limited to Moho depths of 35-40 km. At this depth the upper crust begins to melt, accompanied by slab dehydration and asthenosphere fluids upwelling.

Malmyzh porphyritic gold-copper system is located in the western part of Zhuravlevka terrane near the central Sikhote-Alin fault [12]. High levels of lithosphere fragmentation and decompaction in mineralization distribution regions probably lead to dislocation of numerous metal bearing thermal springs and formation of porphyritic bodies. Rocks containing the deposits belong to ultra-aluminous intrusions of the late Cretaceous period and diorite and granodiorite containing minor intrusions. Majority of ore bodies are controlled by northeastern trending faults.

Lazurnoye deposit, which is the most petrochemically studied, is located near the southern part of Zhuravlevka terrane, between Sikhote-Alin and Furmanov deep faults [13]. Distribution of porphyritic intrusions, including the ore bearing ones, is connected to the Berezovsk fault area, which contains a voluminous northeastern trending minor intrusion belt. The intrusion has two phases characterized by their porphyritic outlook. Rocks of the second phase have deep mantle roots and are related to protolithic melting under conditions of high fluid level [13].

Authors of the summary on Sikhote-Alin intrusions [11] distinguish two stages of **S**-type granitoid formation: Hauterivian-Aptian orogenic stage and Late Albian, during which high volumes of calci-alkalic magnesian granitoids **(S** and **I** types**)** intruded fault systems. Dislocation of Isanagi plate in west-north-west direction during Cenomanian-Maastrichtian age resulted in frontal subduction of the Paleo-Pacific plate, which also influenced the formation of large amounts of plateau ingimbrinites and their intrusive counterparts from I-type acid magmatic melting. Slab disturbance and asthenosphere influence on sub-continental lithosphere resulted in lamproite and fayalite rhyolite invasion, gabbro and A-type alkaline granite intrusions.

To this day, forecasting of new ore regions was based on studying direct evidence of mineralization. Deep structure was seldom factored in it. The consistent patterns of ore bearing regions location in relation to irregularities in lithosphere are shown for gold deposits [14]. As for porphyritic gold-copper objects in the south of Far East, these factors are not yet studied well enough.

The listed sources tell us about the distinct connection between porphyritic gold-copper mineralization and subduction processes. First of all, **S** and **I**-type porphyritic intrusive complexes of Sikhote-Alin passed several stages of formation, but only some of them bear gold-copper mineralization, which is also noted for other regions as well [3 et al.]. The most important factor, in our opinion, is the nature of deep fluids that rise to higher levels of crust and lead to substance melting and, as the result, the formation of porphyritic intrusions.

Potential mineralization of slab magma retinue is based on the composition of its source: rocks from the basaltic layer of oceanic lithosphere, which passes this specialization to new liquids while melting (e.g. slab magma), capable of transferring "ore burden" to large distances and manifest its potential mineralization within the continental crust [15, 16].

The early stage of formation of these types of intrusion begins with influx of fluids to dehydrated slab. Pyritic and gold-silver mineralization happens at this stage, often in the same space. This results in large and superlarge deposits like Ladolam [2].

The next stage is connected to the influx of deep fluids from asthenosphere and decaying slab in the phase of its initial destruction at the depth of 500-600 km. These fluids have drastically high levels of heavy element contents. They are transferred through the same channels (deep faults) that were the leaders of dehydration fluids [4]. As the result, we can see spatial cositing of intrusions of different stages. It is worth mentioning that not all fluids have high element burden that leads to deposit formation. This may result in interference of deep ore bearing intrusions with blank intrusions of the first stage and vice versa. Combining ore bearing fluids of two stages of porphyritic objects may lead to formation of large and unique polyformational deposits.

Predictive evaluation and localization of search areas, given their deep structure based on geophysical data, as well as geochemical and petrochemical research in order to discover sources of ore bearing fluids, may significantly reduce expenses during the early stages of searching.

This research is conducted as a part of Fundamental research program of Far East Division, RAS. Project 18-2-015 "Porphyritic gold-copper mineralization of Far East: indicators of geodynamical framework, ore bearing magmatism and mineralogical and geochemical features of ore-magmatic systems in relation to the issues of forecasting, search, valuation and management plans for deposits of this type."

References

1. L.F. Mishin, N.V. Berdniko, Ores and metals. **3**, 14-24 (2010).
2. Müller D., Groves D. I. *Potassic Igneous Rocks and Associated Gold-Copper Mineralization*. (Berlin : Springer, 2016).
3. A.I. Gusev, Mod. High Technologies **3**, pp 72-76 (2014).
4. Wilkinson, J. J. Nature science. **6(11)**, 917–925 (2013). http://doi.org/10.1038/ngeo1940
5. Butterworth, N., Steinberg, D., Müller, R. D., Williams, S., Merdith, A. S., & Hardy, S. Tectonics. **35 (12)**, 2847-2862 (2016).
6. Timmermans S. Master thesis Faculty of Geosciences Utrecht University.- 36 p. (2017) https://dspace.library.uu.nl/handle/1874/352926
7. Bourova E., Yoshizawa K.,. Yomogida K. Physics of the Earth and Planetary Interiors. 183 (1-**2**), 20–32 (2010). DOI: 10.1016/j.pepi.2010.06.007
8. Garwin S., Hall R. and Watanabe Y. Society of Economic Geologists, Inc. Ec. Geol. 100th Anniversary Volume. 891–930 (2005).
9. Hall, R. and Spakman, W. Earth Planet. Sci. Lett. 201, 321-336 (2002). DOI: 10.1016/S0012-821X(02)00705-7
10. *Deep structure and metallogeny of East Asia* // Vladivostok. Dalnauka (2010).
11. A. V. Grebennikov, A. I. Khanchuk, V.G. Gonevchuk, S.V. Kovalenko, Lithos. 261, 250–261 (2016).
12. A.F. Chitalin, A.A. Yefimov, K.I. Voskresenky, E.K. Ignatyev, A.G. Kolesnikov, Mineral resources of Russia. **3**, 65-69 (2013).
13. V.G. Sakhno, B.I. Gvozdev, P.P. Safronov, F.I. Rostovsky, L.S. Tsurikova. Minerageny problems of Russia GS RAS, Moscow: 301-324 (2012).
14. E.G. Ivolga. Ores and metals. **1**, 14-25 (2013).
15. C.V. Efremov, A.M. Spiridonov, N.A. Goryachev. Mod. Dir. of Geoch. Dev. Ottisk, 87-88 (2017).
16. Huber C.O., Bachmann, J.-L., Dufek J. and Parmigiani A. Geochem. Geophys. Geosyst. **13, 8**, 1-18 (2012). doi:10.1029/2012GC004042.

E3S Web of Conferences **56**, 04021 (2018) https://doi.org/10.1051/e3sconf/20185604021
VII International Scientific Conference "Problems of Complex Development of Georesources"

Mining technological homogeneity and its composition

Gennady Sekisov[1*] and *Alexey* Yakimov[2]

[1]Mining Institute of the Far Eastern Branch of Russian Academy of Sciences, Khabarovsk, Russia
[2]Transbaikal State University, Chita, Russia

Annotation. It is proposed with the initial justification, as a new scientific and production category in the field of development of deposits of solid minerals, it is noted - "Mining technological homogeneity". At the same time, its high relevance, as well as the important role in the solution of the current problem of scientific and practical substantiation and in achieving a significant increase in the efficiency of the development of deposits of solid minerals, in the creation and rational use of innovative mining technologies. The system-complex composition of mining-technological mineral homogeneity as a subject conceptual-terminological category and, mainly, as a scientific and production activity associated with exploration, deposits of solid minerals is presented. The content of this category, as a scientific and industrial activity

1 Introduction

The trend of increasing with years of mining, obtaining and using various mineral products [1,2] under modern conditions is accompanied in most cases and in many countries, including Russia with its Far Eastern region, by a significant decrease in the quality of the extracted and used both natural and man-made mineral raw materials [2,3]. In particular, over the last 25-30 years, the iron content in the ore has decreased almost 1.3 times [1.2], in the ores of precious metals, mainly gold - about 1.3-1.8 times [2].

This is due to a number of objective and subobject factors.

The decline in the quality of ores and some other minerals in demand is due to the following main reasons:

1. The development of so-called rich deposits and the involvement (in increasing quantities) in the development of mineral objects with a relatively low content of useful components. For example, man-made placers with a gold content of another 80 mgr, natural placers - with a gold content of up to 1-1.3 g / t are often worked out in a dredging way.

2. Forced development of deposits of difficult-enriching, so-called resistant ores.

3. High heterogeneity of the material composition of natural and man-made both macro- and mini mesomineral areas involved in development.

* Corresponding author: adm@igd.khv.ru

4. Imperfection of widely used traditional technologies of mining operations, using of which a high dilution of the extracted mineral raw materials occurs, which often reaches 15-18% or more [4].

In these conditions, first of all, rational formation and allocation within the processed areas of the mineral deposit relatively homogeneous elements (mini and mesoexploitation blocks) as well as their subsequent separate (selective) excavation. Mineral homogeneity of such elements is proposed to be expressed terminologically by the category 'technological mineral uniformity. With reference to the mining sphere it is expedient to use the terminological category "Mining technological homogeneity".

2 Scientific and practical justification

It should be noted that the category "Technological mineralogy" [5], which, as, will be shown below, represents (in its own designation) as a very important component of the category "Mining technological homogeneity", If broadly understood as its specific scientific and production activity (Figure 1).

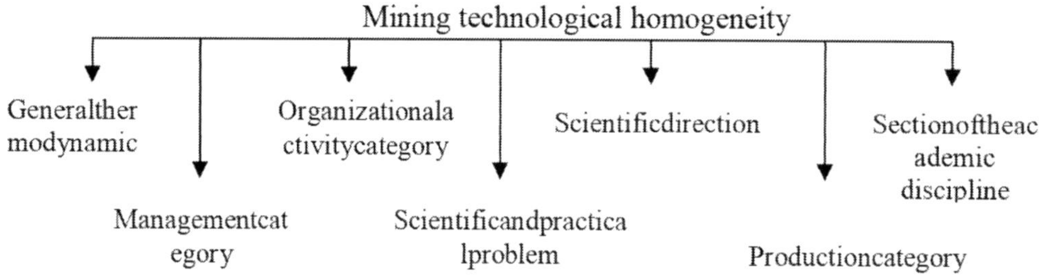

Fig. 1. The composition of the general category - "mining technological homogeneity" in the aspect of the main functionality of scientific-production and educational activities

In terms of overall aggregation, we propose the following categories of mineral uniformity of mining and geological objects: natural-gas-mineral; petromineral; self-mineral; likvamineralnaya, to which it is advisable to classify the mineral-mineral homogeneity and the brine-mineralogical homogeneity; building and pedigree; hydromineral and peat-mineral.

The categories of real-sectoral mineral uniformity are represented schematically in Fig. 2

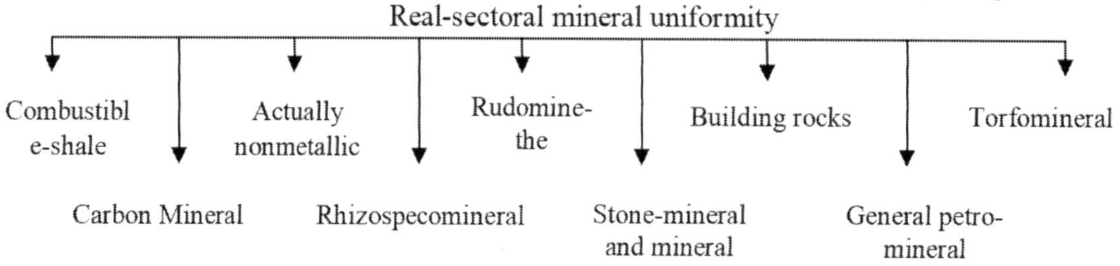

Fig. 2. Categories of mineral uniformity, allocated in the general-sectoral aspect

By origin of the original mineral objects of mineral uniformity, we distinguish the following general categories:

I. Mineral homogeneity of natural mineral objects.

II. Mineral homogeneity of technogenic mineral objects.

III. Mineral homogeneity of natural-technogenic mineral objects, to which, first of all, the previously abandoned springs, mineral panels, mineral bodies and their elements in the side parts of quarries, coal and slate sections belong, in particular.

The distinguished categories of mineral homogeneity for some other important features are represented by the schemes (Figure 3-5) and in textual reflection.

Nowadays, the production-stage mineral uniformity acquires great importance, the initial composition of which is schematically represented in Fig. 3.

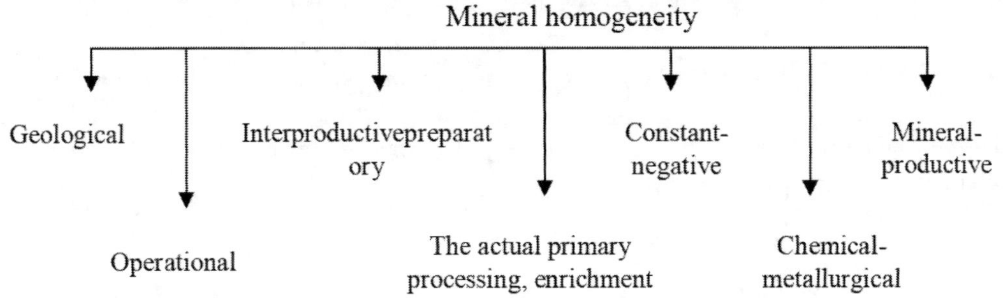

Fig. 3. The initial composition of mineral uniformity in the aspect of the main categories of industrial mineral objects (MO)

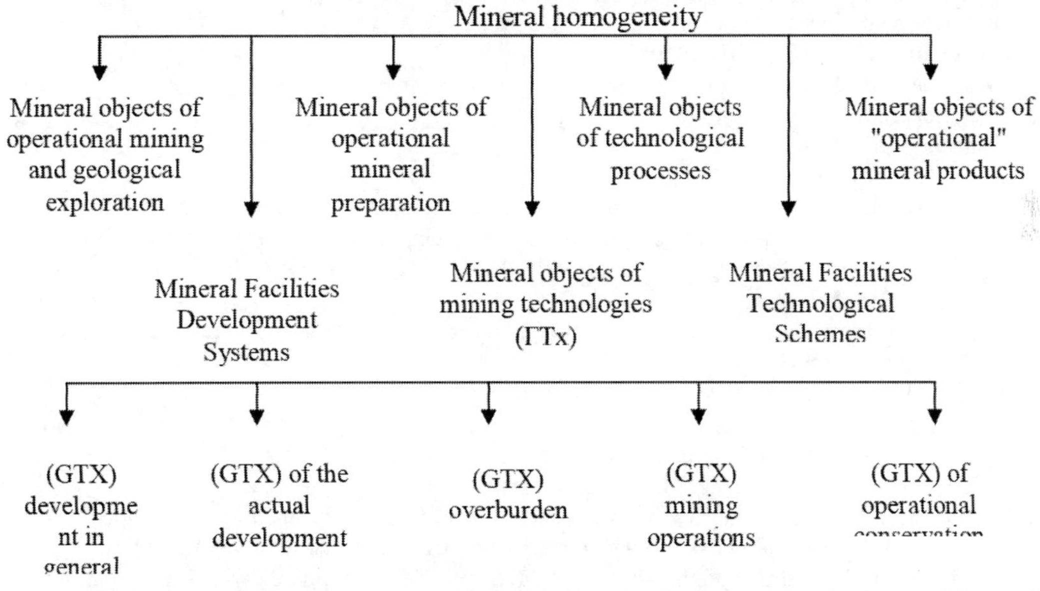

Fig.4. The initial composition of mineral uniformity in the aspect of mineral objects of production operational-technological staging

The distinguished main subcategories of mineral homogeneity in the aspect of spatial localization of mineral objects have a physico-geographical, strictly geological, geophysical, geochemical very complex spatial localization from the point of effective excavation of their MO; quantitative mineralogyogeneity; quantitative chemical-element homogeneity; quantitative mineral uniformity; quantitative complex-material mineral uniformity.

Quantitative-qualitative mineral uniformity is a homogeneity characterized by modern qualities and quantitative features, i.e. manifestations of these features simultaneously in the relationship.

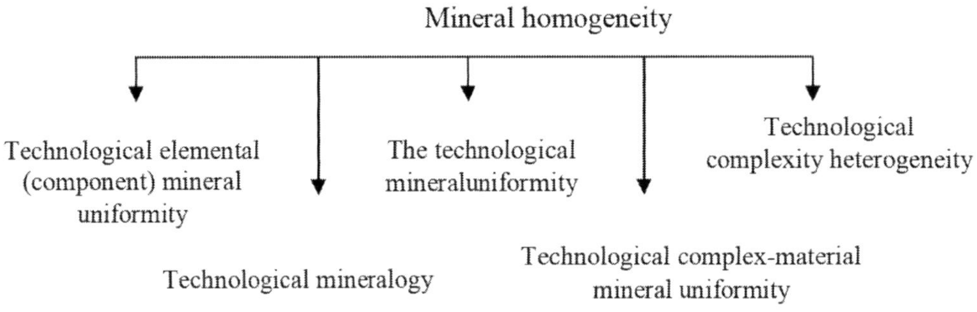

Fig.5. The composition of mineral uniformity in terms of the basic properties of mineral-geochemical materiality

[Under the sub-term "miner-" is meant various possible minerals, especially hard minerals, in particular - ores, coal, etc.]

The uniformity of mineral objects of technological processes of development and development of deposits of solid minerals (especially ore minerals), including:
1. Operational geological exploration.
2. Destruction and loosening of rocks.
3. Reclamation and loading operations.
4. Explosive crushing of rock mass and oversize.
5. Operational mineral preparation.
6. Transportation of rock mass.
7. Warehousing and heaping.
8. Primary processing (mainly enrichment).

In terms of complex quantitative and qualitative features of mineral objects, the composition of mineral uniformity (in a brief reflection) is presented in the form of the following initial subcategories:

I. Qualitative mineral uniformity, in which the quality of gold-bearing ores is manifested in the following indicators [6,7]:
- level of gold content in mineral objects and formations;
- uneven distribution of gold in them;
- the fineness of gold particles and the difference in the shapes of gold;
- Mineral purity of gold and other quality indicators.

II. Quantitative mineral homogeneity, which is not so diverse in its manifestation and is reduced, mainly, to the following sub-categories of homogeneity: quantitative mineralogy - ore of the main object-applied purpose: quantitative geochemical, qualitative geophysical. The composition from the standpoint of the general materiality of mineral uniformity in terms of its basic properties is as follows.

III. Quantitative-qualitative mineral uniformity is homogeneity. Characterized by combined qualities and quantitative features, i.e. manifestations of these features simultaneously in the relationship.

IV. General real-technological mineral uniformity.

V. Technological physico-mineral homogeneity, to which, mainly, homogeneity of the physical properties of mineral objects and formations, and in general - the geophysical properties of mineral objects and formations are relevant.

VI.Technological chemical-mineral homogeneity, which includes geochemical homogeneity.
VII.Technologicalphysico-chemical-mineral homogeneity;
VIII.Technological mineral and technical homogeneity, i.e. uniformity of technical properties of mineral objects and formations.

3 Results

Mineral homogeneity is first introduced and initially justified as a new and highly relevant scientific and production category in the field of mining and the development of mineral objects in particular. The significant mineral heterogeneity of natural and natural-technical mineral objects, especially gold ore and noble metals in general, and, in particular, the relative mineral uniformity of the mineralized and excavated mineral objects of development and excavation within natural and technogenic mineral formations are noted and disclosed.

On the basis of the system complex of important aspects (features), the composition of mineral homogeneity is established as the basic subject aggregate of categories.

In particular, technological mineral uniformity, material, economic, energy and a number of other categories are represented.

At the same time, mineral uniformity in a wide range is presented as a scientific and production activity.

4 Conclusions

The main conclusions include the following:

Mineral homogeneity is one of the most important categories of scientific and industrial activity in the field of mining-geological and economically efficient development of mineral formations and objects and the use of mineral resources in general.

The results of activities in the field of achieving mineral uniformity and its realization, especially - mineralogical - homogeneity, acquire great importance in modern conditions, primarily from the standpoint of providing a number of positive consequences. Particularly important is its role in implementation: a significant reduction in the energy intensity of technological processes and the development of deposits in general; reduction of capital and operating costs; increase in labor productivity; reduction of socio-economic costs of mineral industries; increase the completeness of excavation and use of minerals and mineral resources and subsoil as a whole; improving the quality of mineral raw materials and mineral products derived from it; raising the culture of work and various mineral industries.

The composition and structure of mineral uniformity of mining technology objects and technological schemes are established in a hierarchical sequence and volumetric (using the systemic complex of basic features).Subject to the use of systemic mineral uniformity of mineral objects, it is possible to rationalize traditional and create innovative mining technologies for developing deposits and mining and metallurgical and chemical-metallurgical technologies for the production of mineral products and metals.

References

1. Grovson P.C.F. Miberals handbook 2000-01 Stanstic analyses of the woreds minerals indystry / Endinbridge. Mining jornal books. (2001).

2. Arkhipov GI Mineral Resources of the Mining Industry of the Far East. Overview of the state and development opportunities. "The Mountain Book.".830p. (Mining book - Far East). (2011).

3. Van-Van-E A.P. Resource base of natural-technogenic gold deposits. "The Mountain Book." Publishing house MGGU. 268 pp. (Mining book - Far East). (2010).

4. Sekisov GV, Zykov NV Mastering mineral objects and evaluation methodology. - Moscow: Publishing House "Mining Book". P.295-311. (2012).

5. Mountain Science. Development and reduction of the Earth's interior. / Ed. K.N. Trubetskoy. Publishing House of the Academy of Mining Sciences. P. 392-394. (1997).

6. Bushmin S. A., Belyavsky B. V., Glehovskiy V. A., Borko A. K. ISOCHRON RF-OS AGE of gold from the Magskoe gold-Quartz vein dein deposit (Northern Karelia, Baltic Sheld). DokladyEarthSeiences. **448 (1).** P. 54-57. (2013).

7. Sekisov GV, Sobolev AA, Nigay EV. Mining information-analytical bulletin. **2**. P. 192-201. (2011).

Ecology and mining industry development in high-mountain areas (the Inner Tien Shan)

Yurii Aleshin[1*], *Ilgiz* Aitmatov[1], *Isakbek* Torgoev[1], and *Bektur* Chukin[1]

[1] National Academy of Sciences, Institute of Geomechanics and subsoil development, 720055 Bishkek, Kyrgyz Republic

Abstract. The genesis and transformation in time of risks in the areas of high altitude tailings storage facilities with respect to changing climate are considered. Based on the example of the Kumtor mine in the center of the Tien Shan at a height of 3600–4300 m, three groups of risk factors were allocated concerning a large, ecologically hazardous facility – the tailings dam containing cyanide waste. For one of the factors, calculations of an accidental dam burst were made using the Fault Tree Analysis (FTA) method. It was concluded that the construction of large-scaled and geoenvironmentally hazardous facilities in less researched high-mountain areas has to be be drawn upon detailed and permanent monitoring of the facility itself and its environment. It is crucial to be prepared for operational adjustment of projects and an immediate rehabilitation of constructed facilities.

1 Introduction

Due to the displacement of the mineral resource base of the Kyrgyzstan's mining industry from the foothill and mountainous areas to the highlands in recent years, the conditions for the development of deposits have become significantly more complicated and new problems concerning the ensurance of industrial and environmental safety have arisen. This applies in particular to the gold deposits of Kumtor and Jeruy, located in remote, high-mountain areas of the Tien Shan (the absolute height of mining facilities is 3600–4300 m). Distinctive features of the alpine subsurface development are a severe, sharply continental climate, the permafrost state of rocks and a wide development of complex, modern geocryological and glacial processes, the presence of glaciers, snowfields, and icing [1].

Current trends in climate change, combined with high mining pressure on the weakly stable natural environment of the highlands, lead to the degradation of permafrost areas, the retreat of glaciers, the development of destructive processes in soils and the destabilization of natural and man-made objects, including ecologically hazardous tailings storage facilities (TSF). Up to the present day, sufficient experience and adequate factual data, making it possible to form a reliable forecast for long-term interaction of hydraulic structures with the natural geological environment of the highlands and describe the dynamics of risks at the forecast horizon in decades, have not been accumulated. Therefore, the generalization of operation results of such facilities during the first 15 years and the resulting geo-ecological

[1] Correspondent author: geofizika@list.ru

problems makes it possible to completely identify the complexes of natural and man-made hazards and risks that must be taken into account when choosing the location of their construction.

2 Overview of the facility and risk factors

The tailing pond of the Kumtor operating mine is the largest hydrotechnical structure of the Inner Tien Shan permafrost zone: in the next two years the cumulative volume of solid waste storage will reach 65 million m^3, the height of the retaining dike is appx. 47 m and the ridge is more than 4 km long. It lies in the valley of the Arabel River upstream to its confluence with the Kumtor River, which is the source of the Naryn River (the principal tributary to the Syr Darya River), in close proximity (about 5 km) from the proglacial Petrov Lake (Figure 1) within the transit zone of fluvioglacial and mud-flow masses in the Holocene epoch, which cut through the Middle Quaternary moraines. When choosing the location of the tailings pond in 1995, economic considerations, related to the small distance between the gold extraction factory and tailings pond (6–8 km) and the appropriate morphology of the basin, undoubtedly prevailed.

The practice of subsequent operation of the facility, detailed monitoring of the dam's condition, engineering-geological and geophysical research at the site revealed a number of factors (circumstances and processes) that could lead to the destruction of the facility with catastrophic environmental consequences. Conventionally, there are three groups of risk factors: (1) – internal physical and geological factors in the base and body of the dam,

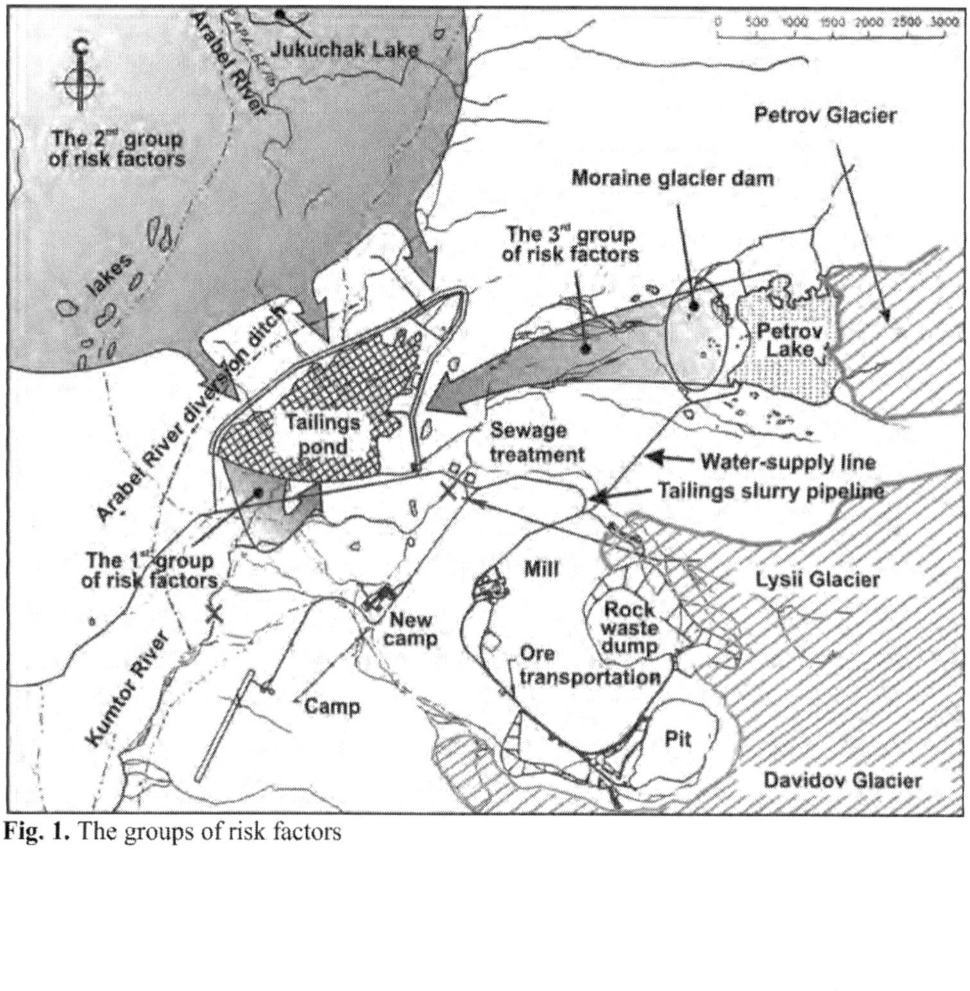

Fig. 1. The groups of risk factors

leading to a loss of its stability; (2) – hydrogeological and hydrological factors within the catchment area; (3) – glaciological and thermodynamic processes in the structures of the moraine and glacial complex of the Petrov Glacier.

The genesis of the hazards from the first group is determined by the peculiarities of the composition, structure, properties, and thermal regime of the alluvial terrace soils on the border with lacustrine loamy silt deposits of low thickness (up to 2 meters) with a gentle slope (1 to 2 degrees) towards downstream at a depth of approximately 4 meters. These deposits with thin lenses and layers of ice have very low strength characteristics (according to the data of [2], their friction angle is between 1.5 and 2.5 degrees). They had been frozen at the beginning of development, but melted during the facility construction and became the cause of the evolution of large deformations and displacements on the dam-base interface at a velocity of up to 6 mm/month. The situation is aggravated by the fact that the dam base lays on the recent and ancient meanders of the Arabel River. Here the underflow, recharged by infiltration of the tailings water, is preserved, in spite of the diversion of the stream flow to the artificial channel beyond the TSF. Geophysical investigations revealed a system of tabetisol areas (Figure 2a) formed by these meanders at the dam site. In the course of its construction and operation, facility rehabilitation required high effort and additional financial costs to stabilize the dam by the construction of a shear key. At the same time new questions arose, concerning the redistribution of seepage flows in the base and body of the dam, as the soil of the shear key freezes.

Thus, insufficiently detailed engineering and geological study of the dam soil foundation, especially changes in the physical and mechanical properties of all the lithological varieties in the section at the boundary cryogenic condition, did not allow geologists, engineers and project leaders to precisely calculate and predict the stability of the object, even in the short term. This is an illustrative example of how important it is to study the internal fine structure of the rock mass in high mountain conditions at the base of engineering structure foundation and the properties of each structural element despite its insignificant thickness compared to the size of the constructed projects.

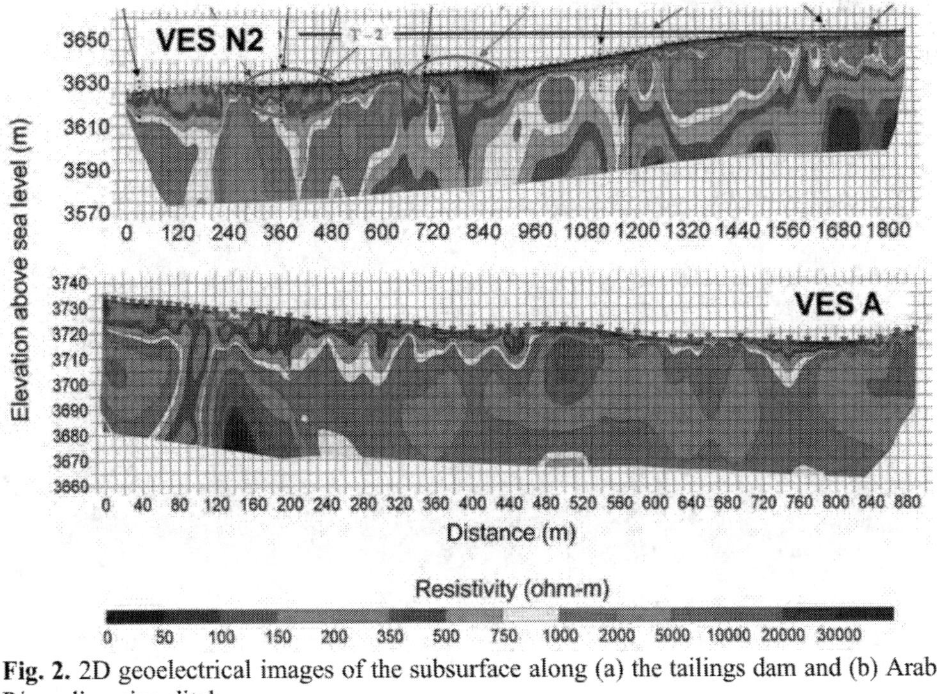

Fig. 2. 2D geoelectrical images of the subsurface along (a) the tailings dam and (b) Arabel River diversion ditch.

The activation of the second group of risk factors is associated with climate change and man-made impact on the large catchment area, which exceeds the tailings area by 55 times. Reserves of moisture in the grounds of this area, including buried ice, may exceed 1 billion m^3, and the water temperature in small lakes reaches 13 °C in summer. Permafrost degradation and development of thermokarst lead to deformations of the diversion ditches that divert the Arabel River, creating conditions for uncontrolled sudden runoff of surface and ground water into the tailings pond. The roofing of frozen rocks along the banks of the diversion ditches is already largely cut up (Figure 2b). Here, deep watered taliks are revealed, acting as reservoirs, collecting and transporting the suprapermafrost water toward the tailings pond. For long-term perspectives after the completion of mining operations, it is now necessary to assess the risks associated with watering the burial material and flooding the entire area abandoned tailings.

Factors of the third group form the danger of Petrov Lake moraine dam breaching (Figure 3), followed by the formation of a heavy debris flood, of which the solid phase will contain large blocks of buried glacier ice [3]. In [4] we discussed various reasons and scenarios for the development of such catastrophic process. During the spreading of debris flow along the outwash plain and the Kumtor River, a stream blockage in a narrow part of the valley and the overflow through a low lateral moraine into the tailings pond, followed by the destruction of the tailings dam and the release of cyanide-containing waste substances into the environment is not unlikely. It should be noted that the small difference in altitude (approximately 10 m) between the Kumtor valley bottom and the lateral moraine that surrounds the tailing pond can be eliminated when removing debris from the Lysii Glacier or in time by the degradation of the moraine itself through buried glacier ice melting. Judging by the special "pockmarked" relief of its surface, these processes are already actively occurring.

The impact of these groups of risk factors is distributed in time as follows:

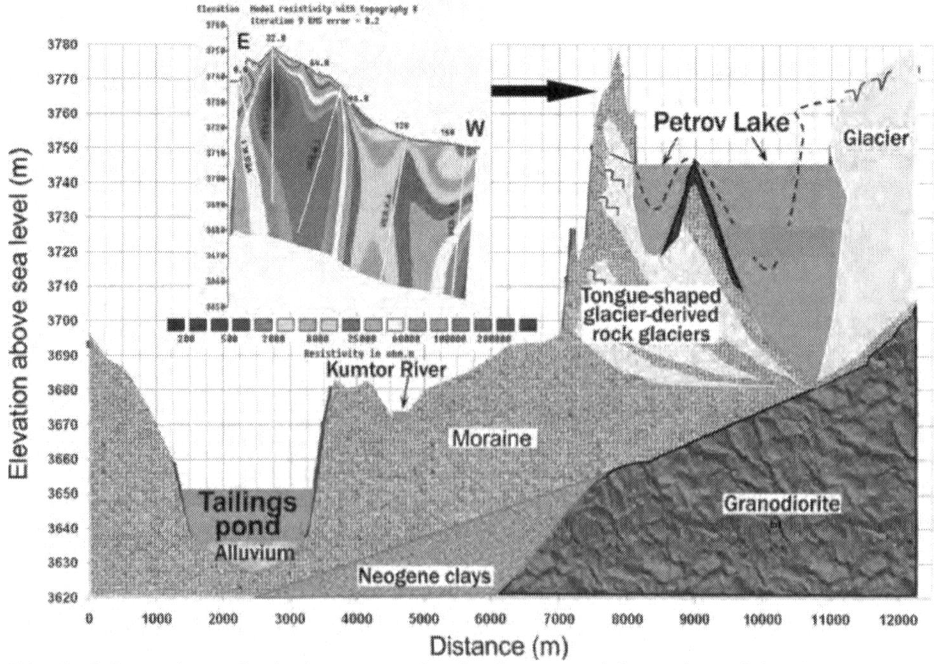

Fig. 3. Schematic geological cross-section in the area of formation of the debris flow and flood hazard in case of Lake Petrov dam breaching. The inset shows the geoelectrical image of the dam cross section.

• The first group – from the beginning of the construction and operation of the TSF with an increased risk of destruction of the facility during the use of the initial project and its gradual reduction as monitoring data are received, engineering-geological knowledge, rehabilitation and adjustment of the project;

• The second group – from the moment of completion of mining operations, conservation of the facility and mine abandonment with increasing risk in time, consistent with changing climatic conditions;

• The third group – from the beginning of construction and operation of the facility with a constant increase in the risk of destruction of the object in time, in accordance to the change of climatic conditions.

At the moment, the knowledge of the second and third groups of risk factors is clearly insufficient to obtain specific numerical estimates of the destruction risk for tailings. It is necessary to carry out detailed geological investigations while the mine is still operating and it is possible to take adequate measures to ensure the safety of the TSF. These investigations must be accompanied by necessary calculations confirming the regulatory safety of TSF and, on the other hand, requiring special and additional technological measures to ensure the implementation of such standards, with the aim to obtain new information.

3 Example of risk assessment of factors in the first group

An example of a rapid and effective response of the mining company to the impact of the first group of risk factors is a set of measures aiming to stabilize the tailings dam, including: careful analysis of the monitoring data; installation of additional equipment to fix and localize the unstable sections of the dam and their displacement planes; further engineering, geological and geophysical surveys in dam body and downstream; the swift development of a stabilization project that provides for the removal of unstable clayey interbed behind its downstream slope to a depth of 10 meters, replacing it with rudaceous rocks, and finally, intensive implementation of the project. We took part in the implementation of some activities.

Immediately after the completion of the rehabilitation project, a risk analysis of the tailings dam failure (in fact, its updated version) was performed. Hazard analysis using monitoring data from 2011 made it possible to identify three main scenarios of tailings failure due to dam collapse:

B1. Dam collapse as a result of the development of seepage deformations in the soils of dam body and foundation.

B2. Dam erosion due to overtopping in case of additional water inflow into the tailings pond.

B3. Loss of stability of the downstream slope and/or entire dam as a result of a shear along the remaining loam interbed areas.

The most dangerous scenario is a hydrodynamic accident in the form of tailings release at the entire dam height in case of stability loss. The modeling of this type of failure was carried out using the "Tailings Flow Slide" code. The main results of the calculation are: the tailings run-out distance is 660 meters for a total time of 70 seconds with maximum flow rate 18.4 m/s. An accident of this scenario is classified as a local emergency situation related to the threat to the lives of up to 10 people. The quantitative risk analysis and development of safety measures are recommended for such cases [5].

The Fault Tree Analysis (FTA) method was used to quantify the risk of failure in case of dam stability loss. Figure 4 shows the "fault tree" for scenario B3 with its elements.

Taking into account the durability of the dam, the average annual probability of a dam failure under static conditions is $R_s = 1.76\times10^{-4}$ year^{-1}; under seismic load $R_c = 3.83\times10^{-4}$

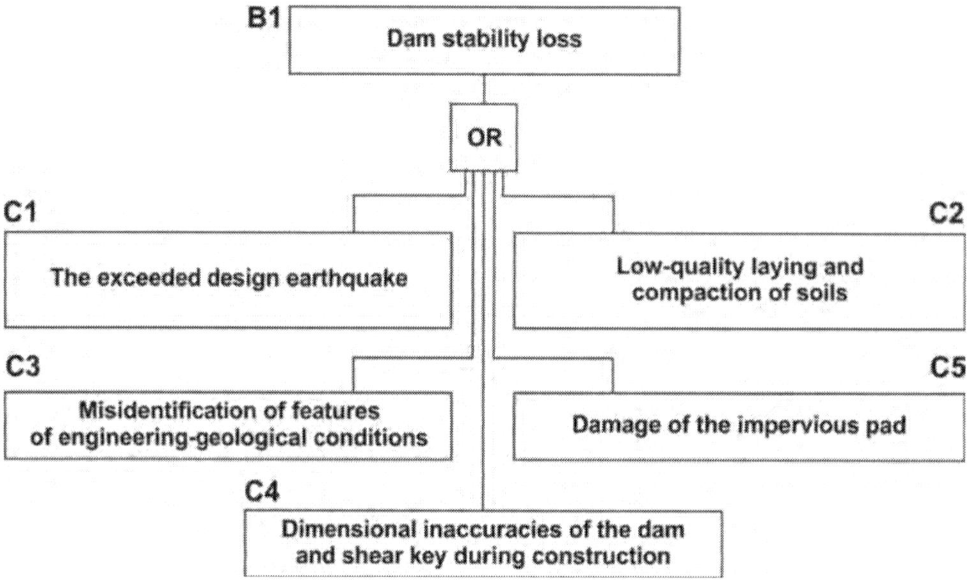

Fig. 4. Fault tree for scenario B3

year^{-1}. The regulatory risk for earth dam of Class 2 is 5×10^{-4} year^{-1} during the operation period.

The comparison of the calculated values with the normative value of risk makes it possible to consider the risk of an accident at the Kumtor TSF during operation in 2012 under static and seismic conditions as acceptable.

4 Conclusion

Taking into account the extremely high geomorphological, seismotectonic and geocryological mobility of the Tien Shan mountain-folded terrain, it is possible to support the thesis [5, 6] on the inadmissibility of using the same data obtained in the early stages or stages of the design or construction of the facility and/or on a nearby site, to assess its current safety, especially for a forecast with any time horizon.

The construction of large-scale and environmentally hazardous facilities in poorly studied high-altitude areas should be based on detailed monitoring of the facility itself and the environment. If necessary, prompt adjustment of the projects and urgent rehabilitation of the newly constructed facility should be established. It is necessary to be prepared for its early close-out, the search for new sites and the relatively high additional costs associated with all these activities. The design and maintenance of objects in a safe way under such conditions have, in fact, a cyclical or permanent character, based on newly obtained engineering, geological and geo-ecological data, as well as on the analysis of the manifested "defects", weakened zones, unstable sites in the interaction of objects with geological medium.

References

1. Yu.G. Aleshin, I.A. Torgoev, *Sergeev's Readings. Engineering geology: state-of-the-art and future prospects*. **6**. 322–326 (Moscow, 2004)

2. G.N. Falaleev, *Stress state of the rock massif and induced geodynamics of subsoil*. 278–284 (Bishkek, 2006)
3. Yu.G. Aleshin, I.A. Torgoev, G.E. Ashirov, K.A. Abirov, *Reducing the risk of natural disasters in mountains*. 19–22 (Bishkek, 2009)
4. I.A. Torgoev, Yu.G. Aleshin, *Geoecology and Wastes of Mining Complex in Kyrgyzstan*, 239 (Bishkek, 2009)
5. I.T. Aitmatov, Yu.G. Aleshin, I.A. Torgoev, Izvestiya NAS KR. **1**. 53–63 (Bishkek, 2011)
6. N.P. Laverov, V.I. Velichkin, B.T. Kochkin, V.I. Malkovsky, V.A. Petrov, A.A. Pek, Geoecology. **3**. 195–206 (2010)

Assessment of technological preparedness coal deposits to a rational industrial development

Marina Pisarenko[1*]

[1]Federal Research Center for Coal and Coal Chemistry Of the Siberian Branch, Russian Academy of Sciences, Kemerovo, Russia.

Abstract. In order to ensure the rationality, saving of mineral resources and to encourage subsoil users to improve production and processing technologies, it is proposed to involve deposits estimated as technologically prepared in the industrial development. This assessment, which is understood to be the compliance of industrial production technologies with mining and geological conditions of the deposits, is proposed to be included in the geological materials for the development of the deposits. As an indicator of determining the technological readiness of the site, use the coefficient of technological readiness-the ratio of the volume of outside-balance resource to the total geological reserves.

1 Introduction

The Kuznetsk coal basin is one of the largest basins in the world, with resources up to a depth of 600 m estimated at 218 billion tons, and balance reserves as of 01.01.2014 are 68 billion tons against the background of the favorable global coal market, the volume of production in the basin is growing. Thus, the results of 2017 production in Kuzbass amounted to 241.2 million tons - this is 59% of the total Russian production, and increased by 1.67 times compared to 1990 (the maximum achieved volume before the restructuring of the coal industry).

The mineral resource base of the Kuznetsk coal basin is represented by coal deposits with a wide variety of mining and geological development conditions, which includes the entire range of geological and technological features, which makes it possible to consider the basin as a testing ground for testing technologies various mining and processing. However, despite a wide range of mining and geological conditions, the existing technological policy of basin development is oriented to an extensive way and consists in using a limited set of coal mining technologies. Thus, the transport system mining with the use of motor transport for the transportation of coal and overburden (about 93% of open-pit coal mining), which is acceptable for a wide range of mining and geological conditions, but is the most costly.

* Corresponding author: iu.kemsc@mai.ru

A longwall mining with smooth roof lowering carries out coal mining by underground method. The economic efficiency of such technological structures is determined by the productivity of the mining face and is achieved with an annual mining of more than 1.0 million tons, which is only possible in favorable mining and geological conditions and at optimum sizes of mine fields: with of length cross-pitch mining 3.0 km, lava from 230 m, of thickness of coal seams is more than 2 m and of angles of incidence of layers to 20 °, disjunctive disturbance up to 50 m / ha, gas content of coal seams up to 10 m3 / t [], etc.

The implemented extensive way of development of mineral resources, aimed at search and development of reserves of sites with a given favorable mining and geological conditions, significantly reduces the mineral potential of Kuzbass (in the extraction of 1 t of coal about 3 t of proven reserves are written off the balance sheet) and does not fully meet the requirements of safety and rationality of subsoil use []. At the same time, the technological platform of coal geotechnologies [] contains a number of promising technologies that, when adapted to specific mining and geological conditions of the sites and improved, will allow to work out powerful and thin layers of inclined and steep deposits, converting outside-balance resource into balance reserves. However, the transition to an intensive path of development of the mineral and raw materials base requires the development of a set of measures, including legislative ones, aimed at encouraging subsoil users to more fully extraction reserves of deposits.

2 Theory

The legal basis for the beginning of the development of the subsoil area by the mining business is the obtaining of a license for the subsoil use. In fact, the license holder receives a legally formalized right to physical destruction of the non – renewable natural resource.

The right of the state as the owner of the subsoil is realized through the state expertise of geological information and project documentation the development of the deposit, their coordination and approval. Based on the results of the state expertise of the feasibility study of the parameters for the calculation of reserves, all the geological reserves of the field are divided according to their industrial importance into balance reserves and outside-balance resource, based on the current capabilities of mining technologies and the economy of mineral raw materials. After the calculation of reserves and their approval, the subsoil user proceeds to the development of the subsoil development project, which focuses on the rational development of only balance reserves, since their development ensures return on investment and profit. The task of preserving outside-balance resource before the subsoil user is not set, although on many of deposits the share of outside-balance resource that do not meet the requirements of the production technologies used is very significant. So, for example, share of outside-balance resource it exceeded 30% in the boundaries of the Ulyanovskaya mine, around 50% on Kolmogorovsky-3 mine, etc. At the same time, share of outside-balance resource reserves are a raw material potential that can be transformed into a real raw material base as a result of improved mining and processing technologies in the foreseeable future. Since in accordance with article 35 Of the law of the Russian Federation "on subsoil" the main task of state regulation of subsoil use relations is the rational use of mineral resources in the interests of current and future generations of the peoples of the Russian Federation, the protection of share of outside-balance resource is a function of the state.

The implementation of this function involves the assessment of the importance of outside-balance resource, the number of which ultimately determines the compliance of technologies of mining the geologicals conditions of deposit, i.e. the technological readiness of the field (site) for mining coal. As an indicator, the value of which determines the technological readiness, the coefficient of technological readiness is proposed. The value of the coefficient is determined share of outside-balance resource in of geological resources. If the value of the coefficient is more than acceptable, the deposit is not technologically ready for mining coal.

The boundary value of the coefficient of technological readiness in addition to mining and geological and mining conditions, depends of economics of mineral raw materials, socio-economic, infrastructure features of the site and requires additional research. As a first approximation for coal deposits, it is proposed to adopt a standard of 20% as a basis. This standard is adopted by analogy with the regulated current document [], according to which the recalculation and re-approval of reserves are made if the share of reserves not subject to development in the total reserves exceeds 20 %. In addition, as shown by the analysis of project documents for the last 5 years, a decrease in the volume of recoverable reserves by 20 % reduces net discounted income (NPV) by an average of 70-80% and such a decrease in reserves is significant for the mining investment project.

Reliable calculation of outside-balance resource involves the correct determination of the parameters of their conditions. The procedure for determining the boundary that separates the balance reserves from outside-balance resource is generally recognized and implemented in the current methodology for the feasibility study of conditions. However, their other (bottom) boundary cannot be established in the same way, therefore are accepted the parameters established by the decision of the Commission of the USSR state planning Committee of July 13, 1960 adopted more than half a century ago. Since their adoption, there have been significant changes in the technology of mining and the economy of mineral raw materials.

The system analysis of 12 and 10 non-traditional technologies of coal mining, respectively, by underground and open-pit methods, which have passed industrial and semi-industrial testing and have the signs of promising technologies, allowed to establish the areas of their use. In accordance with this, it is recommended to take seam thickness the of more than 0.4 m as the "lower" boundary for calculating outside-balance resource of energy coals and coking coals.

The system analysis of coal processing technologies showed that there are no technological restrictions on the ash content of the raw materials supplied for enrichment, and they are determined only by the economy. A series of simplified economic calculations for the complex " the mining and processing" on conventionally grouped by price technological brands of coal allowed to clarify the maximum values of ash content of coal for the calculation of outside-balance resource (fig. 1).

According to studies, the recommended ash content of the coal seam for the calculation of outside-balance resource of technological brands coal: 1- «К, КЖ, Ж, ГЖ» is 59-67 %; 2-«ТС»-53-63 %;3- «КСН, КС, КО» – 43– 55%; 4-«А, Т, СС, Г»– 42-55%; 5 - «Д, ДГ» – 30– 45%. The upper value in this range refers to the coals of light degrees of washability, the lower value in this range refers of very difficult degree of washability. It should be noted that these ash values are obtained by at the accepted price of coal, according to the price list № 003-01-01 and the average costs for the dressing of the rock mass which using in the mining projects.

According to the adopted provision, the technologically prepared deposit is the most of the reserves of which will be extracted as a result of the implementation of design decisions. Since coal deposits are includes from 3 to 20 seams of coal, the capacity and quality characteristics of which vary in the subsoil space, it is necessary distinguish technologically prepared areas within a large field. To do this, build a mining-geometric model in which the simulated indicator is the coefficient of technological readiness. This model to display the spatial variability of the share of reserves left in the subsoil as the toposurface.

As an information basis for the construction of models of technological readiness it is proposed to use of calculation the seams plans of reserves. The method of construction of such models is described in detail in []. In Fig.2 the constructed model is presented of the real mining area (the name of which is not given for the reasons of preserving the commercial secrets of the subsoil user). A feature of this mining area is that the productive thickness before horizon -100 m (abs.) includes 10 coal seams of technological brands coal "Д, ДГ", the average power of which varies from 0.75 to 7.07 m.

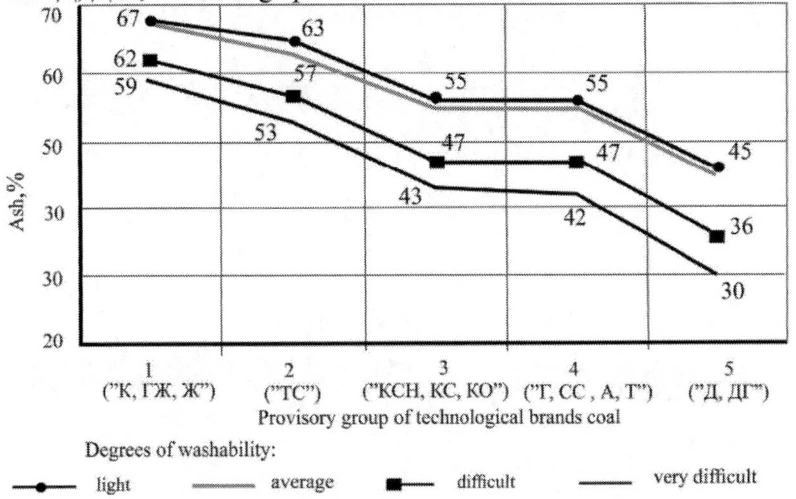

Fig. 1. Maximum ash content of coal for off-balance reserves calculation.

The calculation of balance reserves was carried out according to the parameters of the conditions that meet the requirements of the adopted production technology for the site, namely the longwall (long-pillar) mining with the equipment of the treatment face with a mechanized complex (minimum seam coal of 1.8 m and maximum ash content of coal 35 %).

When calculating outside-balance resource, the following parameters are used - the minimum seam coal of 0.4 m and the maximum ash content of coal is 45 %.

Since within the boundaries of the analyzed area only four of the ten coal seams have a capacity of more than 1.8 m, and the remaining six layers do not meet the requirements of the adopted technology, the coefficient of technological readiness of the deposit is about 0.45-0.6, which is higher than the boundary value, i.e. the deposit is technologically not prepared. This conclusion is confirmed by the practice of mining in this deposit.

During the operation of the coal mining enterprise, mining was carried out on two upper seams. The seam thickness more than 4.8 m was mininged with loss of layer of coal and was about 3.8 m. In addition, after the mining of the upper seams, the subsoil user decided not to work out the underlying seams with thickness of about 1.8 m for economic reasons. As a result, only about 17% of the reserves was mining.

The company is currently closed. The coal reserves contained in the a seam thickness of less than 1.8 m, in the amount of 50 million tons, was transferred to the state balance, and their mining in the near future is unlikely.

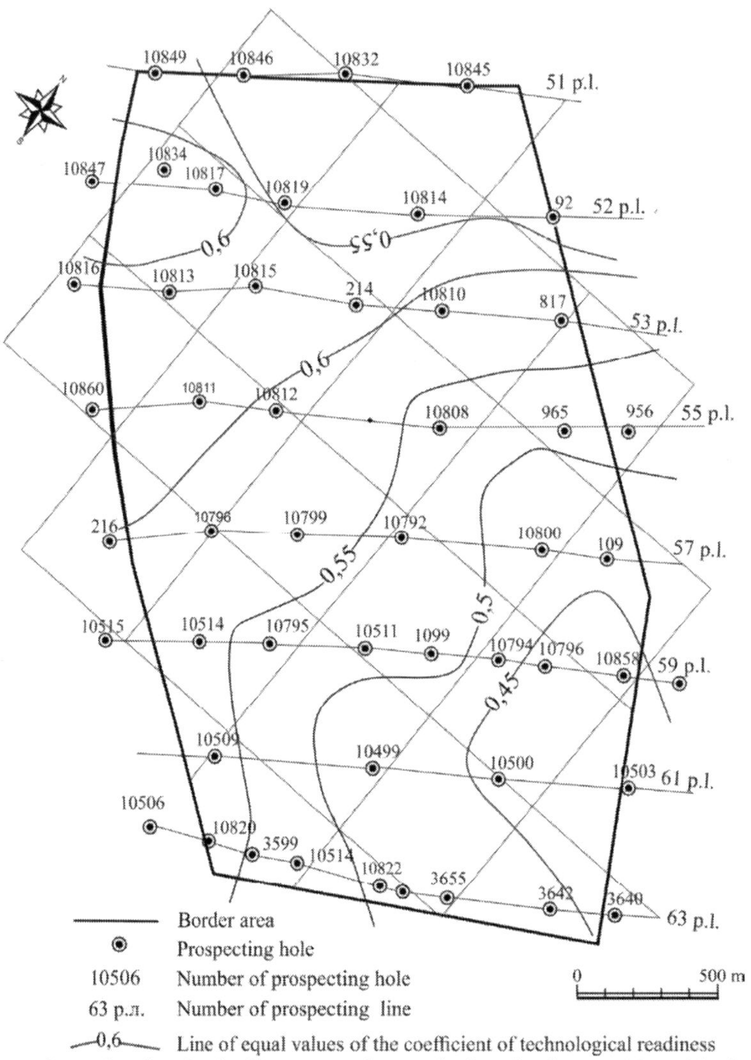

Fig. 2. Toposurface of technologically prepared areas before horizon -100 m (abs.).

4 Conclusion

For to ensure rational development of mineral resources and to stimulate the subsoil to the development and improvement of technologies of mining and processing the current assessment of the preparedness of deposits for industrial mining added by the assessment of technological readiness of the site. It is proposed to carry out this assessment during the state examination of geological materials, and the presence of a positive assessment is the legal basis for the development of technological projects for the mining of deposits.

References

1. S. Shaklein , M. Pisarenko. Mineral resource or Russia. Economics & Management. **5**, P. 25-29. (2010).
2. S. Shaklein , M. Pisarenko. JMS. **50 (3).** P. 527-532 (2014)
3. M. Pisarenko.. Subsoil use of the XXI century **5**, P. 106-110 (2015)
4. V.Yakovlev., G. Karmaev., V.Bersenev , A. Glebov, A. Semenkin, I.Sumina. JMS. **52 (1).** P. 7 (2016).
5. A. Netsvetaev, A. Grigoryan , D. Prugina. UGOL **1**, P. 25-29. (2015).
6. E. Kreinin. UGOL. **7.** P. 58-59. (2006).
7. I. Mohnachuk,, B. Myshlyaev, S. Titov. UGOL **9**, P. 15-17. (2013).
8. H. Foss., M. Bittner. Gluckauf. **6**, p. 139. (2003).
9. V. Klishin., S. .Klishin. JMS. **46 (2).** P. 149-159. (2010).
10. S. Shemyakin, D. Matveev, A. Cheban. Gornyi Zhurnal. **2.** P. 15-19. (2015).
11. A. Makarov, I. Livinsky, I. Rasskazov., B. Saksin, M. Potapchuk. JMS. **52 (3).** P. 438-437. (2016).
11. A.Rasskazova, T. Aleksandrova, N. Lavrik. Eurasian mining. **1**, P. 25-27. (2014).
12. L. A. Samatova, E. D. Shepeta. Eurasian mining. **1.** P. 33-37. (2014).

Geological structure and coal content of the Ytymdzhinskaya cavity of the Gonamsk coal-bearing region in the South Yakut coal basin

Aleksandr Rukovich[1*]

[1]Technical institute (branch) of NEFU in Neryungri, Republic of Sakha (Yakutia), Neryungri, Kravchenko, Russia

Abstract. Southern Yakut coal basin is a huge and reliable base of the high-quality coked and steam coals in the east of the country. The Southern Yakut TPC was begun in the 70-s in the south of Yakutia forming, what predetermined considerable strengthening in this region of exploration works on coal. Now the fields of Neryungri, Elga and certain sites on Denisovsk and Chulmakan fields are developed by an open method. Absence near the operating Neryungri coal mine of the coalfields suitable for open-cast mining, and also finite useful lives of the Neryungri field (till 2018-2020) caused expansion of search works on coal on flanks of the basin, directed at identification of the powerful coal layers suitable for development by an open method.

1 Introduction

The Ytymdzhinskaya cavity is located in the northeastern part of the Gonamsk coal-bearing region in the basin of the river Ytymdzha, the left feeder of the Gonam. The Preytymginskaya depression is located in the cavity's area and is separated from the Gonamsk depression by the ridge Kaltan – Djur on the Gonamsk – Ytymdzhinskaya waterparting [5]. Coal-bearing deposits are developed on the territory of 845,2 sq. km, (fig. 1).

The geological structure of the cavity is presented by deeply – metamorphosed Archean rock mass, partly deposits of the Upper Proterozoic, Mesozoic deposits and residual quaternary deposits. Intrusive formations of different age, breaking Mesozoic sedimentary cover are also discovered in the region [6].

2 Materials and methods

Stratigraphy. The Archean formations, included in the Aldan shield, compose the lower structural level – crystal basement of the exploration area. The formations are presented

*Corresponding author: raul1975@mail.ru

mainly by intensively interlaid dislocated gneiss and crystal shales, quartzrock; and include calcifer interlayers and lenses, granular limestones, rarely diopside rocks, influenced by the process of ultrametamorphism.

Exposure of the Archean rocks is bad, primary rocks are met rarely in the exposures. As a rule, the Archean contacts with rocks, forming the cavities on faults. The exception is the northern margins of the Ytymdzhinskaya cavity, where exposed thickness of the Archean rocks within the order of 35m is the base for the sedimentary complex of the rocks.

Deposits of the Upper Proterozoic have limited spreading, and are developed in the cavity borders. Two strata represent them: terrigene rocks of the Upper Proterozoi and carbon deposits of the Vendian – yudomian suite, overlapping transgressively with angular unconformity on various Archean rocks. In the basin of the river Ytymdzha they build up low limbs of the large-sized synclinal and border mezozoic rock mass deposits mainly from north. These are carbonate shallow sediments of the open sea basin: magnesian limestones, cherty, clay and sandy limestones, marlstones, dolomite limestones.

Mesozoic coal-bearing deposits overlap on the deeply erosional surface of the Archean formations. The Mesozoic deposits thickness includes three lower suites. The upper suites, developed in more deep cavities are not found here.

The Mesozoic deposits section is characterized by terrigene rock composition, being an ordinary full set of rocks from conglomerates and gravelstones to argillites and coals, alternating in less regular sequence of rhythms. Sandstones (50-70%) represent the basic rock background.

The Mesozoic deposits section is also characterized by multiordinal cyclic process, typical for basins with frequently changing environmental conditions of sedimentation. The sedimentation cycle begins with coarse-grained and ends with fine-grained rocks. Taking into account the cyclic structure of the sedimentation mass and structural communication of drilling prospecting wells in the Ytymdzhinskaya cavity, sediments of the Jurassic period were identified in the Mesozoic deposits composition. Their total exposed thickness is nearly 1100m. Three suites represent the Mesozoic deposits thickness: the Yukhtinskaya, Duraiskaya and partly, Kabaktinskaya. Upper suites are not found.

The Yukhtinskaya suite (J_1 *jh*). The Yukhtinskaya suite deposits spread in a wide stripe on the north of the Ytimdzhinskaya cavity and on the south, they are developed in a narrow stripe. Rock junction with more ancient formations is tectonic.

In the northern part of the cavity in the Yukhtinskaya suite deposits there are seams and lenses of the middle-late Jurassic and early – Cretaceous tabular and subtabular rarely crosscutting intrusive mass including diorite-granodiorite and monzonite-syenitic formations.

The section of the Yukhtinskaya suite deposits begins with gray and light gray medium and coarse-grained arcose and arcose-quartz sandstones with interlayers and lenses of gravelstones, medium and fine-pebbled conglomerates with

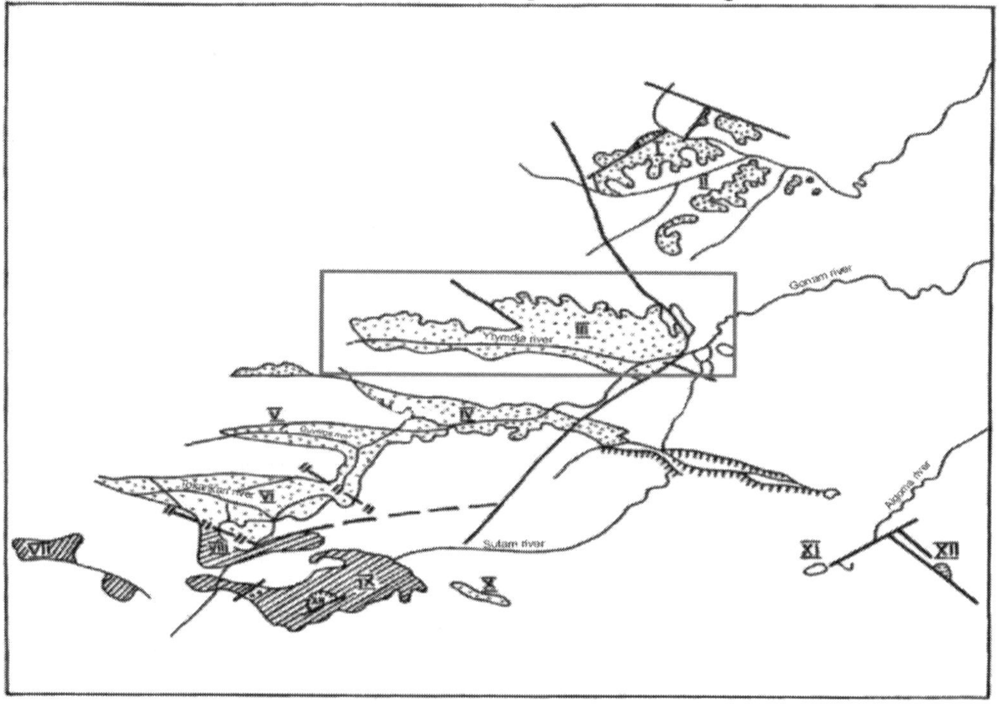

Fig. 1. The arrangement of depressions in the Gonam coal-bearing region

I-Kuranakh-Gynim; II-Gynimo-Semdzhinskaya; **III-Ytymdzhinskaya**; IV - Gyuskangra-Narulak; V-Guivilgra; VI-Tokarikano-Konerkit; VII-Upper Timpton; VIII-Upper-Gonam; IX-Upper Sutama; X-Upper-Daurkan; XI-Chekchoy, XII-Awengur

thickness to 3-10cm. Above the section, the number of coarse-grained material in the sandstones decreases. Thin interlayers of coarse and fine-grained dark-gray siltstones emerge. The Yukhtinskaya suite thickness varies from 230 to 240m.

The Duraiskaya suite (J_2 *dr*). The deposits of the productive Duraiskaya suite conformably overlie everywhere on the Yukhtinskaya suite rocks, especially in the central and northern parts of the cavity [4]. At the base of the suite there are series of coarse and medium-grained sandstones with thickness of up to 10-15 meters including thin interlayers and lenses (to 0,5 – 1,0 m) of fine-pebbled conglomerates and gravelstones. The middle part of the suite is composed mainly of gray and dark-gray series of fine-grained sandstones, alternating with fine and coarse – grained siltstones and medium-grained sandstones. The character of stratification is mostly low wave and horizontal, confined to aleurolitic series, sandstones are characterized by cross-wave and wave bedding. The upper part of the Duraiskaya suite section is composed mainly of fine-grained sandstones with light-gray colour, alternating with interlayers of dark-gray siltstones. Ten coal seams and coal shows were discovered in the formation section. Seven of the coal beds have thickness of 0,7m and more. The suite's thickness is estimated at 330 meters.

The Kabaktinskaya suite (J_3 *κb*) conformably overlies on the rocks of the Duraiskaya formation and is exposed on the whole territory of the cavity. The formation section begins with alternation of the coarse-grained material (medium and coarse- grained sandstones) [7]. Overall, the exposed part of the formation section is presented by thick series (20-30m) of coarse-grained sandstones with light-gray colour, weakly fractured and unstratified with interlayers of fine-grained light-gray sandstones and fine and coarse-grained dark-gray siltstones, weakly fractured with irregular horizontal stratification. More than 25 seams and coal shows were identified in the exposed formation section. 23 of the seams have thickness from 0,7m and more in separate seam intersections. The exposed thickness (partial) of the Kabaktinskaya formation is estimated at 550m. Thus, the total exposed thickness of the Mesozoic deposits in the Ytimdzhinskaya cavity is about 1120m. The quaternary deposits (Q) are also developed everywhere as on the area of the Guvilgrinskaya cavity and are presented by alluvial, diluvial-solifluction and lacustrine - boggy formations.

Tectonics. The Ytymdzhinskaya cavity, bordered from the south by regional South-Ytimdzhinsk fault is a large asymmetric synclinal with low north limb and steep southern limb, cut by the fault. The cavity is oriented in the East-West direction; its sizes are 130x 30 km. The most sunk part of the cavity's foundation is near the South – Ytymdzhinsk regional fault and according to the local geophysic investigations is 0,5 -0,7 km [5].

From the north, the Ytymdzhinskaya cavity is bounded by areal limits of the Proterozoic and Mesozoic sedimentary rocks. Weakly dislocated rocks zone is situated in the Northern and middle parts of the cavity. Here the thickness has simple monoclonal structure frequently complicated by folds, small sized folds of east-west trending [1]. It is possible to observe small interformational movements in the seams, sometimes causing redistribution of the coal material. In this case, in the soil and bed top, well-marked polish faults can be identified. Flexure bends in contrast to linear monoclonal bedding indicate block structure of the basement. They come amid parallel rock fracturing. The bends confirm the hypothesis about long-lasting existence of faults in the basement. Throughout most part of the Ytymdzhinskaya cavity area, low Mesozoic rocks at an angle of $4\text{-}10^0$ (to 20^0) are inclined to the south, to the cavity's center [3].

The zone of rocks with complex dislocation – the southern limb of the synclinal is situated on the right bank of the Ytymdzha. It is narrower than the northern limb, is steep and located in downdip block, to the south of which ridge fault is emerging. The Southern limb width is 1,5 – 3,0 km, it has complex dislocation and is complicated by smaller folds. Dip angles of the second order folds limbs reach 50 and more degrees. Disjunctive faults of the rocks are presented by throws, faults, break-thrusts with steep dip angles. Rather a unique feature of the cavity is intensive show of the Mesozoic magmatism, as a result of which, the Jurassic deposits are broken by numerous subalcalic intrusive sheets, which alternate with sedimentary rocks [9].

3 Results and discussion

Coal-bearing capacity. In the coal-bearing thickness section, studied by the wells, seams and interlayers of coal are almost everywhere (the Duraiskaya and Kabaktinskaya formations) but their distribution both in the section and on the area are uneven [2]. Totally, there are 37 coal seams including more than 25 seams with working thickness (0,7m and more) in the exposed coal-bearing thickness. Total thickness of all coal seams and interlayers of profile 1-1 is more than 40 m, and of working seams – 25m.

In lithologic and stratigraphic section, as was already mentioned, letter and cipher symbols were assigned. The letter symbol represents formation's name and the cipher symbol – counting number of the seam in the formation's section (downwards). Characteristics of coal-bearing capacity of the formations is given below [8]:

The Yukhtinskaya formation is almost coal-free.

The Duraiskaya formation. In the formation section more than 12 coal shows with thickness from 0,20 m to 6,5 m (well 16, seam D15) were discovered. Four of them have working thickness (0,7 m and more), the seams structure is complex and presented mainly by one interlayer. According to their thickness, the seams mainly belong to the category of thin and median. From the five indexed seams of the Duraiskaya formation, the seam D15 is more continuous both in its section and in thickness.

The Kabaktinskaya formation has more than 25 coal shows and eight of them have working values of thickness from 0,7 and more. The seams structure is complex and is presented by one or two interlayers. Coal seams K_{13}, K_8, K_6, K_4 represent the most practical interest as they are more continuous in section and have working values of thickness. K_8 is the most continuous seam. The seams belong to the category of thin and median in their thickness, average thickness is 1,3m.

4 Conclusion

The description of the indexed seams of the formations is given in Table 1.
As a result, it is obvious that the most coal replete is the middle part of the section. Extraction of the coal seams of the section's middle part will demand additional geological exploration.

Table 1. Basic parameters of the indexed coal seams of the Ytymdzhinskaya cavity

Suite	Seam index	Basic parameters				Limits of mining plies thickness variation From – to Average thickness (a number of intersectings)	Bed structure (a number of separation seams)
		Total number of wells penetration	Number of penetrations with thickness to 0,7m/ zero data	Number of penetrations with thickness to 0,7m and more	Seam continuity coefficient		
1	2	3	4	5	6	7	8
Duraiskaya	D_{19}	12	5/6	1	0,1	0-0,86 0,20(12)	Complicated 1
	D_{15}	12	0/0	12	1,0	0,84-6,0 1,80(12)	Complicated 1
	D_{14}	9	1/7	1	0,1	0-0,86 0,13(9)	Complicated 1
	D_{13}	9	1/5	3	0,33	0-1,44 0,44(9)	Complicated 1
	D_{11}	8	1/5	2	0,25	0-0,78 0,23(8)	Simple
	D_7	7	1/4	2	0,29	0-1,20 0,39(7)	Complicated 1-3
Kabaktinskaya	K_{14}	8	5/2	1	0,12	0-1,30 0,45(8)	Complicated 2
	K_{13}	11	1/0	10	0,91	0,36-3,03 1,79(11)	Complicated 1-2
	K_{10}	13	12/0	1	0,1	0,30-1,02 0,50(13)	Complicated 1
	$K_8^{в}$	13	1/5	7	0,54	0-1,20 0,55(13)	Complicated 1-3
	K_8	13	0/0	13	1,0	0,92-1,68 1,27(13)	Complicated 1
	K_6	14	3/1	10	0,71	0-2,97 0,91(14)	Complicated 1
	K_5	15	7/5	3	0,20	0-1,82 0,50(15)	Complicated 1
	$K_4^{в}$	15	0/11	4	0,27	0-1,50 0,28(15)	Complicated 1
	K_4	16	2/5	9	0,56	0-4,78 1,26(16)	Complicated 1-3
	K_3	16	2/9	5	0,31	0-1,32 0,43(16)	Complicated 1-2
	K_1	15	4/10	1	0,1	0-1,20 0,21(15)	Simple

References

1 Verkhovtsev A N Lukyanova Zh K Carrying out specialized works on tracking coal seams with the help of deciphering the AFS in the Gonam coal-bearing area of the South Yakut basin. The report for 1992-1996. Yakutsk. (1996).

2 Zaliznyak M I Report on prospecting works in the southern part of the Gonam area for 1988-1996. Funds of the State Humanitarian University "Yuzhyakutgeologiya". (1997).

3 Minakov V S Verevkin N I Report on the group geological survey and geological study of the m-ba 1: 50000 on the territory of sheets 0-51-132-B, G; 133-AB; 134-AB; 0-51-121,122 (Guvilgra gold-bearing area on the works of the Melemken party for 1981-87). Funds of "Yuzhyakutgeologia".

4 Utrobin D V Geological structure and minerals of the Upper Sutama gold-bearing area. Funds "Yuzhyakutgeologia". (1989).

5 Shiryaev V P Borzykh S D Report on the topic: Structural-tectonic map of the South Yakutian basin 1:500000 with maps-inserts 1:50000 for prospective plots. Funds of "Yuzhyakutgeologia". (1981).

6 Cherepovsky V F Coal base of Russia. Coal basins and fields/hl. Geoinformark, **T.5. Book 2.** 638 p. (1999).

7 Hvorostina A A Potentsial of a subsoil of a frame of the Southern Yakut coal field Neryungri Publishing house Tekhn. in-that SVFU. 238 p. (2012).

8 Mineral resources of the coal industry of Russia. In 2 volumes. **2** (regions and pools). Moskow publishing house. state. mountains. un-that. 448 p. (1999).

9 Cherepovsky V F Coal basins of the region of the Baikal-Amur Mainline. Moskow, Nedra. 157 p. (1984).

E3S Web of Conferences **56**, 04025 (2018) https://doi.org/10.1051/e3sconf/20185604025
VII International Scientific Conference "Problems of Complex Development of Georesources"

Problems of modern technological audit of mining enterprises

Sergei Golovyrin[1], *Mikhail* Kovalev[1], *Mikhail* Lebzin[1] and *Vladimir* Makarov[1*],

[1]FGBOU VO Ural State Mining University, Ekaterinburg, Russia.

Abstract. Modern economic conditions are characterized by spontaneous changes in market conjuncture, by accelerating the pace of scientific and technological progress in industrial production, which tightened the requirements for production activities, economic and organizational management of domestic enterprises. Currently, the most competitive products in the world market are the products of extractive industries, which determines the increased attention to the problems of the functioning of these industries and the clarification of the conditions of their most sustainable development. The unique mineral resource complex of Russia can provide further economic development of the state only on condition of effective activity of the largest mining enterprises. It is important to adapt enterprises to the changing environment, which imposes new planning requirements, on the basis of which not only forecast indicators are calculated, but also the possibility of improving the competitiveness of the organization.

1. Introduction

The need for an effective state industrial policy aimed at the modernization of mining production requires the development of such methods of determining and implementing the main directions of development of engineering and technology as technological audit.

Systematic changes in the volumes of production work in the space and time, variability of the characteristics of the volumes of the development and phasing of obtaining geological information about these characteristics, change in time (due to the duration of mine development) of applied technical means and economic conditions of mining operations determine the complexity of their structure, specific combinations of elements for mining technology, which are diverse in terms of purpose and functioning.

At the same time, along with the positive aspects (dynamism, flexibility, adaptability), the inertia of geotechnology as a system requires a long time to optimize it.

In this case, the project examination should be considered as the first stage of the technological audit of the enterprise, assessing the overall implementation of the project in the Annex to the specific conditions and the ability of the mining enterprise (ME) to its introduction, the efficiency of subsoil use and ensure the required level of industrial safety.

Periodic technology audit allows to solve problems such as determining the optimal

[*] Corresponding author: supermak@yandex.ru

transition point between stages of mining development (including prior adjustments of implemented project solutions) and improving the efficiency of existing technological scheme at an ME and parameters of mining operations by considering the influence of factors of organizational structure of the production processes management (including mainstreaming these issues while designing an ME).

At the same time, in the conditions of market relations, the practical task of the miners is to find such a ratio of the size of the field and the production capacity of the ME, and the associated choice of the opening method, the development system, equipment and technology of mining operations on it, in which the costs necessary to obtain commercial products would be justified in the shortest possible time.

2. Theory

Developing a practical focus on potential mining sciences, in turn, puts the task of developing mechanisms and methods of evaluation the functioning of the ME from the perspective of use in its work the latest achievements of science and practice as beginning with an analysis of the design parameters of the quarry (for example, the study of the possibility of a steeper pit edge construction), and in part of the most urgent environmental issues on the agenda:

1. Finding ways to maximize the possible introduction of non-waste technology of mining (where as a criterion describing the social utility can be used an indicator of the share of productive use of the extracted rock mass, including reduced to the total value of the products obtained from it).

2. Ensuring acceptable environmental safety of their processes (for example, the use of the method of determining and accounting for the energy factor and the amount of environmental payments in the justification of the choice of type and means of open-cut transport).

Along with the definition of promising tasks of technological development in assessing the effectiveness of the ME, it is necessary to pay due attention to the nature of the current processes of the enterprise vital activity. The main event that can provide at the current level of financial capacity of the enterprise its stable operation is the mobilization of its internal reserves (saving of material and energy resources, improving the use of equipment, ordering of wages and other), the search for which is one of the main directions of technological audit. We note the coincidence of this statement of the issue with the opinion of the experts of the international program TACIS (Technical Assistance for the Commonwealth of Independent States) that the very existence of the ME depends on how well it will be able to resist the cost of its production. In this case, an important role in the implementation of this approach to the solution of the problem plays the definition of marketable and technologically possible (in the current mining, geological, technical and economic conditions) productivity of the ME at the present moment. Today, the methods of reliability theory are widely used in the calculation of mining and transport equipment, process lines. In assessing the reliability of the quarry, it is usually considered in the form of a set of overburden and mining ledges with various reserves between them and technological processes, damped reserves of smoked and blasted rock mass. In this case, the calculation of reliability uses its statistical and physical theory, the methods of which can not be directly transferred to calculate and ensure the reliability of a quarry, which is a complex socio-economic system.

One of the main aspects of the performance evaluation of the ME, directly related to the state of its technological systems is the assessment of staff productivity. In modern conditions of functioning of the Russian mining enterprises, productivity increase becomes a means of achievement of balance of interests of owners of the capital and labour, expressed in increase in business value and the price of labour, and also in creation of preconditions for economic growth.

At the same time, it should be noted that the principles of the organization of production in most MEs remain the same - corresponding to the conditions when the growth of labor productivity was considered only as a means of increasing production. This discrepancy has led to the fact that the organizational system of the enterprise has often become a factor hindering its development as a business unit. (As a rule, the measures for equipment reconstruction at the mining enterprise are associated with an increase in production volumes, and in rare cases, with a reduction in production costs).

Since the processes of open-pit mining have been described to a satisfactory degree by now, in order to find reserves for further improvement of their efficiency within the boundaries of the technological schemes used at the ME, in modern economic conditions, it is competent to transfer the focus of the main directions of further research of the mechanisms of functioning of the ME to assess its efficiency, study of the interaction and relationships of the processes of mining operations from the point of view of the functioning of the unified logistics system, in particular, taking into account the impact on its performance and effectiveness of organizational structures as mining management, and the overall organization of the functioning of the ME. Thus, increasing the overall efficiency of mining production and, consequently, improving the competitiveness of the mining enterprise in the market, is the main task of research on the analysis and evaluation of technological systems of the MEs with their mandatory emphasis on the organizational aspects of the ME vital activity.

As an example of a list of such studies of a similar nature and urgent, from the point of view of mining practice, problems of implementation of the stated task can be given:

In the field of technological solutions:

1. Minimization (in the absence of reliable long-term forecasts on the prices of the domestic market for mineral raw materials) of rationing the amount of advance of overburden works, as well as opened and ready to excavate mineral reserves with a shift to the focus on the regulation of methods and parameters of the formation of the working area of mining.
2. The transition in the planning of mining operations from the definition of calendar plans of their development to the construction of conjuncture conditioned stages of formation of the working area of excavation with the mandatory accounting changes in the quality of extracted minerals in determining the mode of staged mining operations, including the temporary preservation of opportunistic low-gain (on a lower level of provision in the period considered simple reproduction) reserves. In this regard, it is advisable and necessary to make legal adjustments to the provisions on their selective testing.
3. Ensuring the management decision-making on the formation of stages of mining operations in the development or correction of the project of the ME by mechanisms of adequate assessment of production performance of the enterprise from the point of view of rationality in the period in question, used in the last technological schemes and parameters of mining processes, and the very reliability for such an assessment and the adequacy of these indicators to justify the decisions (including in terms of compliance with the organizational structure of its technological potential).

In the field of state control over mining operations:

1. Determination of the role and status of technological audit in relation to state supervision (for example, the system of expertise), the principles of division of their competence and powers with the simultaneous integration of areas of their work to ensure the efficiency of the ME, the methodology of accounting for the specific features of the latter by analyzing the need and sufficiency to ensure industrial safety of the

requirements put forward to them, including the use of the apparatus for accounting for possible risks of their activities.

2. Definition and formalization of mechanisms for certification of mining processes (including in accordance with the requirements of the "Technical regulation" Act), aimed primarily at improving the efficiency of economic activity of the mining enterprise by reducing possible damage from injuries of personnel and accidents at the facilities.

In the field of environmental safety:

1. Development of evaluation criteria for the expansion of environmental impact effect and environmental audit of existing MEs by analyzing the performance of enterprises of their environmental programs and the focus (sufficiency) of their technological processes for the full extraction and use of the extracted minerals and rock mass as a whole.

2. Analysis of the technologies used by MEs in terms of identification of their aspects causing negative impact on the environment (including ranking of technologies according to the degree of their environmental danger) and socio-economic justification of measures to neutralize them, also taking into account the long - term prospects of development (as an example-the feasibility of the biological phase of remediation of sludge storage as potential man-made deposits).

3. Determination of mechanisms and conditions for tax and other benefits for the ME (including determination of their composition and size) to stimulate the implementation of environmental programs, including those aimed at obtaining a positive effect in the long term.

World practice shows that the impact of negative factors of the production activity in case of occurrence of various emergency situations depends not only on the level of technology, but also on the quality of the associated management systems, which are aimed primarily at managing the planning process and the achievement of relevant technology processes objectives in the area of production indicators, while ensuring reliability and safety of the processes themselves.

Naturally, any system (and even more so - technological) can be considered effectively functioning only if its continuous improvement is ensured, which, in turn, is achieved through constant analysis and evaluation of its quality. The audit is directed to obtain information for further analysis and correction of the system.

3. Results

Thus, ***technological audit*** is the diagnosis (through the analysis of the technological system) of the enterprise operation in order to identify its increases in production costs. At the same time, a distinctive feature of the process of technological audit is its dynamic nature, unlike control (supervision) or expertise. At the same time, depending on the tasks assigned to the technological audit, the control (examination) and examination procedures can undoubtedly be included in the process of technological audit as one of its stages. It should be emphasized that the technology audit is aimed at assessing the quality and effectiveness of the system as a whole (including its organizational structure), and not to search for specific violations. Any violation in this regard should be considered not as a separate fact, but as a key to finding a discrepancy either in the structure of the system itself or in the process of its implementation. The task of the audit is to find the causes of inconsistencies, the elimination of which will minimize deviations from the project regulations.

Therefore, the process of technological audit is a broader concept in comparison with the traditionally used methods of assessing the activities of enterprises, as not only formulates ("records") the state of their production systems, but also (through the analysis of the enterprise's capabilities) allows to develop (offer) recommendations for improving the efficiency of the enterprise. In addition, the process of technological audit allows to investigate and evaluate the criteria of evaluation of the enterprise, including the requirements imposed on them by the external environment for their justification and completeness in the specific conditions of the enterprise (to determine their "applicability").

Methodically complex technological audit of the enterprise functioning, including along with economic indicators of its activity the criteria characterizing the state of mutual influence of the external environment and technological system of the enterprise, obviously, must meet the following requirements:

1. Provide an assessment of not only the level of material production and opportunities for the implementation of its development prospects, but also closely related issues in the field of industrial safety, environment, organizational management structures, personnel development and, in general, the community of the enterprise location.
2. Ensure the comparability of the results of the assessment of the state of domestic enterprises, both in the national and international market.

Based on the above conditions and objectives of the technological audit, as the main methodological principles of its construction as a process can be identified:

1. Consistency as periodicity and schedule obligations and certainty on the procedures of the study.
2. Objectivity as independence in evaluation and validity according to its criteria, including social positions.
3. Adaptivity as adequacy of the applied research methods to the conditions of construction and organization of technological systems, "sensitivity" to their changes and options, including in relation to the conditions of the external environment of the enterprise.
4. Sufficiency as validity of the depth of research and the scale recommended in its conclusion activities, the establishment of the range of requirements (analyzed) in the audit of the enterprise and their inventory on mandatory and optional ones.

Thus, the main tasks in the field of technological audit, taking into account international practice can be attributed to:

1. Obtaining reliable information about the activities of business entities in the field of functioning of their production systems, increasing their competitiveness in the world market by improving these systems, improving environmental and industrial safety of the latter.
2. Assistance to business entities in the independent regulation of their industrial policy, the formation of priorities for the implementation of preventive measures aimed at maintaining and improving production systems in a changing environment.
3. Reduction of financial risk of business entities, insurance and financial-credit organizations.
4. Creation of a tool for the implementation of the main directions of production process regulation.
5. Integration of activities in the field of technological systems improvement with other areas of activity of enterprises, the introduction of efficiency evaluation of their production systems.
6. Taking into account and applying universally recognized international principles, methods and procedures.
7. Training and involvement of competent specialists in technological design and audit.

Hence, the results of technological audit of production systems should become an objective economic and legal tool to stimulate the economic activity of the enterprise, as it is known that no "straight-line" tightening of requirements for the enterprise can provide its "ideal" work, it will only lead to the inhibition of its processes.

Compulsory technological audit should be performed:

1. In the implementation of international obligations in the field of technical regulation of production processes.
2. On behalf of the state bodies, including justification of possible sanctions against the enterprise.
3. To take into account the technological factor in the privatization of state and municipal enterprises.
4. In the implementation of the bankruptcy proceedings of economic entities.
5. When carrying out mandatory risk insurance.
6. In the preparation and justification of investment projects and programs (when it is provided by the terms of investment), etc.

Mining enterprises as objects of design, construction, operation and management (including auditing, as one of the management tools) belong to the highest category of complexity and are characterized by large amounts of information of various kinds:

1. Geometric, including elements of the deposit and mining technology, considered in the three-dimensional space of the mining branch.
2. Economic, containing indicators and estimates of resource costs, indicators of product value, intensity and efficiency of work, etc.
3. Technological and technical, including parameters and quality characteristics of elements of mining, mining and transport equipment and technological processes.
4. Environmental, reflecting the interaction of technological processes of mining and natural factors, such as subsoil, earth's surface, the surrounding atmosphere, water resources.
5. Social, including the performance and characteristics of industrial danger of technological processes of mining production.

4. Outcomes

1. Improving the overall efficiency of mining production and, consequently, improving the competitiveness of the mining enterprise in the market, is the main task of research on the analysis and evaluation of technological systems of the mining enterprise with their mandatory emphasis on the organizational aspects of the vital activity of the mining enterprise.
2. A technological system can be considered to be effective only if it is continuously improved, which in turn is achieved through continuous analysis and evaluation of its quality.
3. The results of the technological audit of productive systems should become an objective economic and legal tool to stimulate economic activity of the enterprise.
4. The proposed approach to the mining audit methodology allows, on the one hand, to clearly comply with the requirements of state supervisory bodies, on the other hand, it allows to respond flexibly to the emergence of new technologies and new equipment. In fact, on this basis, it is possible to implement the state industrial policy aimed at the modernization of the existing mining industry.

Reference

1. Gusev A.I., Kovalev M.N. Improvement of the industrial safety system. Industrial safety. **12**. P. 2-6. (2003).
2. Gusev A.I., Kovalev M.N. Mining information and analytical bulletin. Published by MGGU. **3**. p.22. (2004).
3. Mining audit and development of the concept of development of mining and processing production (MPP) of JSC "Kombinat Magnezit". Phase 1. Mining audit of technical and economic solutions in the system of MPP: NIR report. / OAO «Kombinat Mgnezit, OOO NVTs «Nedra» - Chelyabinsk, Satka. 101 p. (2002).
4. Pevzner M.E. Mining audit. Textbook for higher school. 3-d edition. Published by Moscow State Mining University. 215 p. (2004).
5. Technological audit act on the Kvaisinskiy inderground mine, developing Volkhokhskiy plot of the Kvaisinsk lead-zink ore deposit, Tskhinval. Ekaterinburg. 74 p. (2012).
6. Anikin A.V., Vorobiev A.E. Mining and technological audit of the gold mining company JSC " Vasilevsky mine». Published by MISIS. (2013).
7. Vorobiev A.E., Danilova E.V. Vestnik RUDN, Engineering research series. **4**. (2013).
8. Sobolev A.O., RF subsoil use expert. Geological audit according to international standards and problems of mining and geological industry in Russia. Gold mining. **178.** (2013).
9. GOST R 57194.3-2016 Technology transfer. Technological audit.
10. Chetverik N.P. Technological audit and construction control: educational manual. Moscow: ANO DPO «APKIRK». 354 p. (2017).

AUTHOR INDEX

AUTHOR INDEX

AUTHOR INDEX